INTERNATIONAL SERIES IN
NATURAL PHILOSOPHY
GENERAL EDITOR: D. TER HAAR

VOLUME 81

THEORY OF COMPLEX NUCLEI

THEORY OF COMPLEX NUCLEI

V. G. SOLOVIEV

Joint Institute for Nuclear Research,
Dubna, U.S.S.R.

Translated by

P. VOGEL

California Institute of Technology,
Pasadena, California, U.S.A.

PERGAMON PRESS

OXFORD · NEW YORK · TORONTO

SYDNEY · PARIS · FRANKFURT

U. K. Pergamon Press Ltd., Headington Hill Hall, Oxford, OX3 0BW, England

U. S. A. Pergamon Press Inc., Maxwell House, Fairview Park, Elmsford, New York 10523, U.S.A.

CANADA Pergamon of Canada Ltd., P.O. Box 9600, Don Mills M3C 2T9, Ontario, Canada

AUSTRALIA Pergamon Press (Aust.) Pty. Ltd., 19a Boundary Street, Rushcutters Bay, N.S.W. 2011, Australia

FRANCE Pergamon Press SARL, 24 rue des Ecoles, 75240 Paris, Cedex 05, France

WEST GERMANY Pergamon Press GmbH, 6242 Kronberg/Taunus, Pferdstrasse 1, Frankfurt-am-Main, West Germany

---

First edition 1976

Library of Congress Cataloging in Publication Data

Solov'ev, Vadim Georgievich.
Theory of complex nuclei.

(International series of natural philosophy; v. 81)
Translation of Teoriia slozhnykh iader.
Bibliography: p.
1. Nuclear physics. I. Title.
QC776.S6413 1975 539.7 75-12885
ISBN 0-08-018053-1

Printed in Hungary

CONTENTS

ABSTRACT

THIS monograph is an introduction to the contemporary theory of complex nuclei. The experimental data characterizing the ground and low-lying excited states of medium and heavy nuclei are systemized and compared with the results of calculations. The achievements and the possibilities of the new methods used for description of the nuclear structure are demonstrated.

The necessary introduction to the theory of complex nuclei is given in Chapters 1 and 2. The choice of the nuclear model Hamiltonian and the Hartree–Fock–Bogolyubov variational principle are discussed in Chapter 3. The independent quasiparticle model is described in Chapters 4–7. Its application to the calculations of the nuclear spectra, nuclear transitions, nuclear equilibrium forms, and nuclear moments of inertia is also described in Chapters 4–7. Chapters 8–10 contain the exposition of the semimicroscopic nuclear vibration theory and the application of the theory to the deformed and spherical nuclei.

FOREWORD

THIS monograph was written with the aim of providing a course in nuclear theory. Main emphasis is given to the description of properties of ground and low-lying states of medium and heavy nuclei.

The scope of nuclear theory is so extensive that it cannot be fully described in a single book. Such parts of the nuclear theory as the formal scattering theory, problems of the two-nucleon interaction, the theory of the interaction with electromagnetic radiation, the theory of α- and β-decays, and several others are sufficiently explained in many nuclear physics textbooks. Special monographs are devoted to the explanation of specific parts of nuclear theory, such as the theory of nuclear matter, the fission theory, the application of group theory in nuclear physics, etc. Naturally, it is unnecessary to repeat here the content of those parts of the nuclear theory which are either sufficiently fully described in textbooks and monographs or which are not directly connected with the main content of this book. If the title of this monograph is to correspond as closely as possible to its content, it should be called "Introduction to certain aspects of the theory of medium and heavy nuclei".

Investigation of the atomic nuclei has two goals: the investigation of the nuclear structure *per se* and the investigation of the elementary interactions manifested in the properties of atomic nuclei. This monograph treats the nuclear structure as such; only the decisive interactions are included in the calculations. Described problems of the α- and β-decay theory and of the nuclear reaction theory are directly related to the nuclear structure, i.e. these processes are used as sources of information about the structure of complex nuclei.

Nuclear theory has reached such a level that the intrinsic unity of our ideas about nuclear structure is sufficiently evident. This allows the use of the deductive method of explanation in the book. It is necessary to note that the monograph describes basic methods of the nuclear many-body problem, which cannot yet form a closed theory. However, these nuclear theory methods form a basis, which makes it possible to perform scientific investigations and to analyze original nuclear physics works.

The general mathematical methods of the nuclear theory are treated in the monograph separately (Chapter 3, § 2; Chapter 4, § 1; and the whole Chapters 5 and 8). Readers not interested in the mathematical basis of the described methods may skip § 2 of Chapter 3, the whole of Chapter 5, and §§ 1 and 3 of Chapter 8. This will not cause any interruption in the continuity of the explanation.

Particular emphasis is given to the systemization of experimental material and to its comparison with the results of calculations. The majority of the theoretical results in tables and figures was recalculated using a unified basis.

The monograph does not contain a full bibliography of the discussed nuclear physics problems. References are given to the more important papers and books and to those works, the results of which are used in the monograph.

This book was writen in 1966–9. Part of the material is included in lectures, given since 1961 in the Dubna branch of the Physics Department of Moscow State University. Some parts were described in lecture courses, given in the IAEA summer school on nuclear theory in Czechoslovakia, 1962; in the JINR school on the structure of complex nuclei in Telavi, Georgia, 1965; in the all-union summer schools in Obninsk, 1966, and Khumsan, 1967, and in lectures given in Hungary, GDR, Italy, Romania, and Japan.

The monograph is intended for theoretical physicists, for the experimentalists working in nuclear physics, and for graduate and upper class students. It is assumed that the reader is familiar with the basic methods of quantum mechanics, statistical and nuclear physics.

I would like to express my deep gratitude to academician N. N. Bogolyubov for his constant interest, which has helped me in writing the monograph.

I would like to acknowledge many helpful discussions with my colleagues in the nuclear theory division of the Joint Institute for Nuclear Research. I am particularly obliged to D. A. Arsenev, V. B. Belyaev, R. V. Jolos, R. A. Eramzhyan, S. I. Fedotov, F. A. Gareev, S. P. Ivanova, I. Khristov, V. K. Lukyanov, L. A. Malov, I. N. Mikhailov, N. I. Pyatov, V. Rybarska, I. Sh. Vashakidze, and A. I. Vdovin, for their help in the explanation of particular problems. I am thankful to my wife, G. M. Solovieva, for her help in preparing the monograph for printing.

INTRODUCTION

NUCLEAR physics is one of the youngest disciplines of science. Up to the second half of the nineteenth century, the atom was believed to be the smallest, nondivisible part of matter. The history of nuclear physics begins with the discovery of the Mendeleyev periodic law. The Mendeleyev table reflects the laws of atomic structure and the existence of nuclear mass number A and nuclear charge Z. The Mendeleyev periodic law played a principal role in the development of nuclear physics.

The development of nuclear physics can be divided into three periods. The first one (1896–1932) is the period in which the most general facts related to the atomic nucleus were discovered. H. Bequerel found radioactivity of uranium in 1896. P. Curie and M. Sklodowska-Curie found new radioactive elements—radium and polonium. Afterwards, three types of radioactive radiation were found: α-, β-, and γ-rays.

In 1904, J. J. Thomson suggested an atomic model, according to which the atom is a positively charged sphere, with electrons moving inside it. The Thomson model was disproved by Rutherford's experiments on α-particle scattering on thin foils. These experiments led to the discovery of atomic nuclei. The atomic planetary model was introduced by E. Rutherford in 1911. According to this model, an atom consists of a positively charged nucleus, with a radius of the order 10^{-12} cm and of electrons distributed around the nucleus with radii 10^{-8} cm. Almost all atomic mass is concentrated in the nucleus. Later, using quantum theory, N. Bohr justified and developed further Rutherford's atomic model. The nuclear transmutations of stable nuclei were discovered by E. Rutherford in 1919. At the same time, F. Aston found stable isotopes and established the basis for the development of mass spectroscopy.

The first quarter of the twentieth century is characterized by great development in physics. The most important achievements were the formulation of the relativity theory and quantum mechanics. These theories radically changed the ideas prevailing at the turn of the twentieth century about the basic laws of nature. They had important revolutionizing impact not only in physics but also in other natural sciences. The establishment of the relativity theory and of quantum mechanics formed the basis for the development of nuclear physics.

The second period in the history of nuclear physics (1932–49) can be described as the prehistory of modern nuclear physics. In 1932, J. Chadwick discovered the neutron, and J. Cockroft and E. Walton made the first nuclear transmutations using artificially accelerated particles. The discovery of the neutron led to the formulation of the proton–neutron model of an atomic nucleus by W. Heisenberg and D. D. Ivanenko. I. Curie and F. Joliot-Curie discovered artificial radioactivity and positron β-decay. These discoveries led to the synthesis of new elements. I. V. Kurchatov discovered nuclear isomerism, and L. Alvarez found nuclear transmutation caused by the capture of the orbital electrons.

The discovery of the fission of uranium nuclei bombarded by neutrons, made in 1938 by Hahn and Strassmann, was very important. G. N. Flerov and K. A. Petrzhak found spontaneous fission of uranium somewhat later. The first nuclear reactor was built and operated in the United States under the guidance of E. Fermi in December 1942. The intensive studies of nuclear fission and neutron interaction with matter formed the scientific basis of nuclear energy production.

The muons were found in cosmic rays in 1938, π-mesons in 1947; K-mesons and hyperons were found somewhat later. The technique of particle acceleration was further developed, and several low-energy accelerators were built. The proton 330 MeV accelerator was built in Berkeley (USA) in 1947, and a 440 MeV proton accelerator was built in Dubna (USSR) in 1949. The nucleon interactions at these energies were studied, and π-meson production caused by the proton–nucleus interaction was discovered. The construction of high-energy particle accelerators and the discovery of many new elementary particles led to the separation of a new discipline from nuclear physics—the physics of elementary particles.

Recent systematic study of nuclear structure and nuclear reaction mechanism can be considered as the third era in the development of nuclear physics. New technical development allowed physical studies on a large scale. New transuranium elements and a large number of new isotopes throughout the periodic system were produced. Experimental facts were accumulated in large numbers, and that brought about the determination of many nuclear properties and quantum characteristics of the ground and excited states of light, medium, and heavy nuclei. The various mechanisms of the nuclear reactions were studied. The development of the α-, β-, and γ-spectroscopy was also significant. The importance of nuclear reactions in the nuclear structure studies is ever-growing. The developed theoretical concepts helped to understand the basic nuclear processes and properties of ground and excited nuclear states.

The contemporary period of the development of nuclear physics is a period of intensive accumulation of experimental facts and their analysis. It is necessary to note, however, that quantitative experimental nuclear structure information is still rather limited.

Figure I.1 shows the nuclear neutron–proton diagram. Nuclei in the region between the lines $B_n = 0$ and $B_p = 0$ should exist in nature (with lifetimes considerably larger than the characteristic nuclear time interval). These nuclei have positive neutron and proton separation energies. It is seen that the experimentally known nuclei form less than a quarter of all the possible nuclei. Physicists are trying to solve the problem of superheavy nuclei. In particular, it is possible that relatively long-lived nuclei exist in the region of $Z = 114$ or 126 and $N = 184$.

The low-lying excited states have been experimentally studied in only about 10% of the existing nuclei. Experimental information about the intermediate region of excitation energies is rather poor.

The methods of neutron spectroscopy give information about the average characteristics of compound states close to the neutron binding energy. Explanation of the nature of such states (How different is the structure of levels with the same spin and parity? What is the

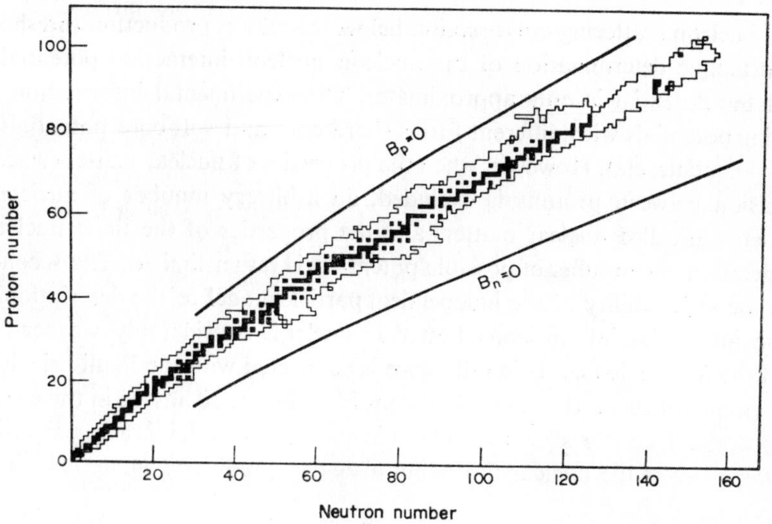

FIG. I.1. Neutron–proton diagram of the atomic nuclei. Dark areas: stable and long-lived nuclei. Light areas: known radioactive nuclei.

manifestation of shell effects? What is the role of the few quasiparticle components of the wave functions?, etc.) is very important. The most useful results are obtained from studies of the $n\gamma$ and $n\alpha$ reactions on resonance states. The study of the analog states and of different giant resonance states provides interesting information about the quasicontinuous part of the spectrum. It is not necessary to stress that the study of the analog states and giant resonance fine structure is only beginning.

The principal importance of elementary particle studies is obvious. Many scientists believe that the fundamental problems of physics will be solved in this way. We shall try to answer the following question here: Is the study of atomic nucleus scientifically equally important? We think that the answer is yes for two reasons. First, the atomic nucleus is the basic and determining part of nature. Almost all the mass of matter is concentrated in nuclei. The nuclear mass and charge determine the structure of the electron cloud and thus the basic physical and chemical properties of atoms. The special features of nuclear structure and nuclear forces are very important in astrophysics because nuclear transmutations play the determining role in stars. The study of the nuclear structure must be of principal scientific importance when nuclei are so important in nature. It should not be forgotten that elementary particle physics was established in connection with the study of nuclear forces and the formulation of nuclear theory. The great variety of nuclear properties, which form an almost inexhaustible source of practical application, should also be mentioned. Thus the first argument is quite general. The second argument is the following one: the study of the elementary interaction process of two particles gives insufficient information about the particles themselves; additional information can be obtained from a study of the systems of interacting particles. The many-body problem gives complementary information about the involved forces. That part of the information, obtained from the solution of the many-body problem, cannot be obtained directly from the two-body problem.

Let us explain the second argument with two examples: the study of the deuteron and of nucleon–nucleon scattering cross-section below the meson production threshold did not lead to the unique determination of the nucleon–nucleon interaction potential (the very concept of the potential is only approximate). The experimental information can be explained using potentials with different forms (hard-core and soft-core potentials, velocity-dependent potentials, etc). However, when the properties of nuclear matter (when the Coulomb repulsion between protons is excluded, an arbitrary number of nucleons forms a stable system—so-called nuclear matter) and the properties of the finite nuclei are taken into consideration, the number of possible potentials is rather limited. The second example deals with the applicability of the independent particle model to the description of atomic nuclei. The interaction of nucleons bound in nuclei is considerably weaker than the interacton of the free nucleons. This difference is connected with the Pauli principle, i.e., we encounter peculiarities of the many-body problem in nuclei absent in the elementary interaction process. When the structure of atomic nuclei is studied, it is necessary to study not only the properties of the nuclear forces but also the conditions under which the forces are acting.

The development of nuclear physics in recent years did not lead to great discoveries; nevertheless, there has been important progress. There are two tendencies in nuclear structure studies. The first tendency is related to the full and detailed description of the ground-state properties of individual nuclei and to the expansion of such a description toward higher excitation energy. The second tendency is related to the description of larger and larger sets of nuclei, moving toward the superheavy nuclei and to nuclei further from the beta stability region.

Our study cannot be limited to one or several nuclei if the nuclear structure is to be understood. Many characteristic features of even–even nuclei are different from those of the neighboring odd–even or odd–odd nuclei. The structure of deformed nuclei is quite different from the structure of spherical nuclei, etc. There are important differences among the deformed nuclei. For example, actinide nuclei undergo fission, while deformed nuclei in the $150<A<190$ region do not. It is evident that fission cannot be studied using the properties of rare-earth isotopes. While nuclei in the $150<A<190$ and $228<A<254$ regions have a prolate ellipsodial form, nuclei in the $50<Z, N<82$ and $28<Z<50, 50<N<82$ have, possibly, oblate form, etc. We could mention other properties which are different for different nuclei. However, it is already clear that the structure of the different nuclei is different even if the forces are always the same. Thus it is necessary to study wide regions of nuclei.

Nuclear structure information, obtained in experiments with intermediate and high-energy particles interacting with nuclei, is ever-increasing. Thus the scattering of fast electrons shows that nuclear charge distribution may deviate from the Fermi distribution. Nuclear K-meson absorption suggests an existence of the neutron skin.

Important nuclear structure information is contained in the muon capture studies. The interaction of π-mesons with nuclei gives information about the short-range correlations of the nucleons in nuclei. The direct knock-out of certain light nuclei during interaction of fast protons with nuclei gives information about the nuclear cluster structure. The mechanisms of high-energy particle interaction with nuclei are very interesting.

Study of hypernuclei should be very useful in future nuclear structure research. In such nuclei, one or several nucleons are replaced by hyperons; the interaction between hyperons and between hyperons and nucleons can act without restrictions. Thus new features of the nuclear structure problem should be seen.

The contemporary period of the accumulation of experimental information is unavoidable. It is possible that it will lead to important discoveries. From the point of view of fundamental physical laws, the atomic nucleus problem is far from being exhausted.

Theoretical nuclear physics began more than 30 years ago after the discovery of the neutron and the realization that the nuclei are built from protons and neutrons. There are two basic difficulties in the formulation of nuclear theory. First: nuclear forces are very complicated and insufficiently known. Second, it is difficult to make a theory of systems composed of a large but finite number of particles when their interaction cannot be characterized by a small parameter (even for simple forces). Therefore the development of nuclear theory followed the line of a search for simple models. In the initial period of nuclear physics, the nucleus was compared to the charged drop or to the degenerate Fermi gas. Later, the word "model" got a wider meaning. Any set of the simplifying assumptions, both physical and mathematical, which allows calculation of nuclear properties with certain accuracy, is called a model. Thus the nuclear problem is transformed into the formulation of a model, which describes the real systems with the highest possible accuracy, on the one hand, and which is mathematically soluble, on the other. Each model stresses one particular aspect of the whole problem. If the model is used for the explanation of experimental facts, only those facts where this particular aspect is important are chosen.

Two basic types of nuclear models were developed. In models with strong interaction the nucleus is treated as an ensemble of tightly bound particles. Models of independent particles form the second type. Nucleons are moving approximately independently in the models of this type.

Two experimental facts were established in the initial period of the study of nuclear structure. The binding energy per particle is approximately constant for all nuclei (with the exception of the very light ones), and nuclear volume is proportional to the number of the nucleons. From these facts it follows that the proton and neutron density is constant in all nuclei. On the other hand, it was found that the interaction between nucleons is of short range and is strong. Therefore, a model with strong interaction, in which the nucleon mean free path is shorter than the nuclear dimensions, was formulated. N. Bohr and Ya. J. Frenkel suggested the charged liquid-drop model. It gave correct results when the stability against deformation was studied; it also established the limit of stability against fission.

N. Bohr has introduced the concept of the compound nucleus in the description of the interaction of incoming nucleons with nuclei. The model is based on the assumption that all nucleons are responsible in the same degree for the properties of a given nuclear state. According to this concept, the incoming nucleon interacts with one or two target nucleons and transfers to them (and through them to the whole nucleus) a large part of its energy. This happens before the particle penetrates the whole nucleus. The incoming particle, having lost most of its energy, is captured by the nucleus. The lifetime of the compound nucleus is long when compared with the time of flight through the nucleus. The compound nucleus has an energy excess—the energy brought in by the incoming particle. After a while, the energy excess (or its considerable part) can concentrate in one nucleon, which can leave the nucleus.

The success of the compound nucleus theory and the discovery and interpretation of the fission phenomenon suggested that the liquid-drop model describes well the real nuclei. However, it turned out that the liquid-drop model is inapplicable to the description of nuclear excited states. The dynamics of nuclear motion responsible for the properties of excited states is much more complicated than the motion of a liquid drop. Besides, the assumption that the mean free path is small was disproved. The mean free path is considerably larger than the distances between nucleons; it is comparable with nuclear dimensions. Nevertheless, the liquid-drop model played an important role in the history of nuclear theory.

The successful application of the Hartree–Fock method in atomic theory could explain interest in nuclear independent particle models. It is obvious that the Hartree–Fock method should give less accurate results in the nucleus than in the atom because the common source of force is missing and nuclear interaction is strong and has a short range. However, the nuclear shell model, formulated by Mayer and Haxel *et al.*,[(1)] was unexpectedly successful. The shell model assumed that a number of nuclear properties could be explained if individual nucleons move independently in an average field. The average field is formed by all other nucleons. The shell model explained not only the enhanced stability of the magic nuclei, but it explained many other experimental features of ground and excited nuclear states and many characteristics of their decay as well.

It should be noted that the average field of atomic electrons and the nuclear average field are substantially different. The nuclear field is induced by the nucleons only; therefore it must be less stable with respect to the deformations and surface vibrations.

Several experimental facts, unexplainable in the framework of the nuclear shell model, were known even in its early period. For example, in a large number of nuclei the intensity of the electrical $E2$ transition is up to one hundred times larger than the intensity of the one-proton transitions. Weakly excited states in even–even nuclei show even more clearly the existence of the collective effects. All these facts were observed in nuclei which have neutron and proton numbers very different from the magic numbers.

The pecularities of nuclei with many nucleons in the unfilled shells were explained by the unified nuclear model developed by Bohr and applied by Bohr and Mottelson.[(2)] The basic assumption of the unified nuclear model is the assumption of the ellipsoidal shape of these nuclei. The lowest excited states in this case are rotational states of the nucleus as a whole. The unified nuclear model takes explicitly into account the degrees of freedom related to the motion of one or several weakly bound nucleons. At the same time, the collective vibrations, related to the changes in the nuclear shape and orientation, were taken over from the liquid-drop model. The unified nuclear model is therefore an intermediate model between the shell and liquid-drop models. However, its basic physical assumptions are much closer to the shell model. All nucleons participate in the collective vibrations to a certain degree. The main role, however, is given to weakly bound nucleons, i.e. nucleons close to the Fermi level.

The unified nuclear model explained a large number of experimental facts and predicted a number of properties of deformed nuclei. The ideas of nuclear collective degrees of freedom were further developed by Davydov and coworkers,[(3)] who have discussed a more general case of nonaxially symmetric nuclei.

The scattering of particles on nuclei is sufficiently described by the nuclear optical model. This model is very similar to the independent particle model. The behavior of the incoming particle is described by its motion in the average nuclear field which has an imaginary (absorptive) part. If during the first collision of the incoming particle with the target nucleon one particle leaves the nucleus, we are dealing with a direct nuclear reaction. On the other hand, if the particles collide again and again, the chances that one particle will leave the nucleus are decreased, and the compound nucleus is formed. Thus, the independent nuclear model includes the two basic forms of nuclear reaction: direct interactions and formation of the compound nucleus.

Models using often contradicting assumptions were used in early nuclear theory. In recent years, however, the models used are complementary rather than contradictory. The properties of ground and low-lying excited states (up to 2 MeV) in medium and heavy nuclei are, at the present time, explained in the framework of the concept of average field plus residual interaction between nucleons.

The interaction between nucleons in the nuclei could be conditionally divided into two parts: the average, or selfconsistent nuclear field and the residual interaction. The average field is the nuclear potential which is formed by all nucleons. The residual interaction is that part of the interaction, which is not included in the average field. Note that some parts of the nucleon interaction cannot contribute to the average field in principle. The residual interactions play an important role in nuclei; they are not weak and cannot be treated by the perturbation method. They change smoothly and slowly when going from one nucleus to its neighbor. The average field determines certain nuclear properties directly. Besides, it governs the residual interactions, i.e., it defines the conditions for the materialization of their effects. The average field is responsible for a number of concrete features of individual nuclei and for the differences between them.

The contemporary state of the nuclear theory is characterized by wide application of mathematical methods and physical ideas of the quantum field theory and of statistical physics. Works on superfluidity,(4) superconductivity,(5, 6) and Fermi-liquid theory(7) were particularly important for the development of nuclear theory. The mathematical methods of the superfluidity and superconductivity theory are very general. They allow one to solve the problem of residual fermion interaction, leading to pair correlations, in a rather general form. Bogolyubov(8) has suggested the possibility of the superfluidity of nuclear matter; later, Bohr *et al.*(9) discussed the existence of the superfluid states in atomic nuclei. The theory of pair correlations of the superconducting type in atomic nuclei was formulated independently by Belyaev(10) and Soloviev.(11, 12) The theory of pairing correlations explained many nuclear properties which were not understood before. Moreover, it gave a basis for broad studies of nuclear structure, based on the microscopical approach. From the large number of papers in this field we would like to mention the work of Migdal(13, 14) based on the Fermi-liquid theory.

Contemporary nuclear theory is not a theory of nuclear models if the term "model" is understood in its usual meaning. Physicists use the word "model" to characterize an approximate method used for the description of a certain restricted class of properties of a large number of nuclei. Only part of the nuclear forces, responsible for the discussed nuclear properties or processes, is taken into account. The remaining forces are either neglected completely or included in a crude way. Thus, the word "nuclear model" means that only part of the nuclear force, particularly important for a certain set of nuclear properties, is taken into account. Note that this part of nuclear force (which is taken into consideration) is changed when other nuclear properties or other nuclei are considered. Therefore, the approximate methods for the description of the main nuclear characteristics change as well, when different characteristics are considered, or when light or heavy nuclei are considered. This important peculiarity of the atomic nucleus is a consequence of its complexity and of the diversity of its properties.

Therefore, "model" in contemporary nuclear theory means an approximate method of description of nuclear properties, which takes into account the most important, i.e. determining for the given properties part of the nuclear force. This monograph is devoted to the description of nuclear models in this sense.

The problems of nuclear theory are so vast that they cannot be fully explained in one book, as Blatt and Weisskopf[15] did some 20 years ago. The experimental material about medium and heavy nuclei is quite rich and diverse. For example, a full description of all experimental facts about nuclei with the mass number $A = 182$, and their analysis[16] forms a book of appreciable volume. Thus a full description of nuclear theory needs several volumes. This problem has been solved, seemingly, by Bohr and Mottelson in their monograph; its first volume has already appeared in print.[17]

A number of the aspects of nuclear theory are fully described in textbooks (see refs. 18–20); some are explained in good monographs (see refs. 21, 22). For example, problems of nuclear matter and of the applications of methods, developed by Brueckner and others, to the finite nuclei, are described in detail in refs. 21–23. Obviously, it is not necessary to include them here.

The description of the theory of ground and excited (up to 2–3 MeV) states of medium and heavy nuclei forms the main content of this monograph. The semimicroscopic approach, based on the selection of interaction (which reflects the most important part of nuclear forces), is used. The nuclear many-body problem can be reduced to the problem of several degrees of freedom. This fact explains the success of the semimicroscopic approach, which is a natural extension of the phenomenological method.

The deductive method is used in this monograph. The interaction Hamiltonian, describing nuclear rotation, the average nuclear field, and residual interaction between nucleons is constructed. Approximate mathematical methods of solution of the nuclear many-body problem (with different parts of the forces, contained in the interaction Hamiltonian) are explained. The structure of ground and excited states of complex nuclei is studied. The interaction Hamiltonian has been chosen in its simplest possible form. We want to test how the large ensemble of the experimental facts can be explained by using a relatively simple interaction, and to see when and what kind of the effective forces must be added.

Large emphasis is given to the systematization of the experimental material and to its comparison with theoretical calculations. This has been done, however, in a rather general way. In this aspect, the book differs from, for example, ref. 24 where all available experimental material about odd-A deformed nuclei is carefully analyzed, systemized, and compared with theory (see also ref. 25).

In conclusion, let us make a few comments about terminology.

The mass number is denoted by A, the number of protons by Z, number of neutrons by N, $A = N+Z$. We use the terms: even–even nuclei for nuclei with an even number of both protons and neutrons; odd-A nuclei for nuclei with either an odd number of neutrons (odd-N nuclei) or an odd number of protons (odd-Z nuclei); and odd–odd nuclei for the nuclei with an odd number of both protons and neutrons. The word complex nuclei is used for nuclei with $A > 40$. The nuclei with a spherical equilibrium shape are called spherical nuclei, those with ellipsoidal shape are called deformed or nonspherical nuclei. When the deformation energy is large, i.e., the nucleus is outside the transitional region, it is called a strongly deformed nucleus.

The system of units $\hbar = c = 1$ is used throughout. The phases of the wave functions have been chosen according to Condon and Shortley[26] (except in special cases). The Clebsch–Gordan coefficients are denoted by $(I_1M_1I_2M_2 \mid I_3M_3)$, the Wigner–Eckart theorem is used in the form given in ref. 27. The matrix elements are denoted by $(\Psi_f \mathfrak{M}(\lambda)\Psi_i)$, where Ψ_i, Ψ_f are wave functions of the initial and final states, respectively, and $\mathfrak{M}(\lambda)$ is an operator of multipolarity λ.

The set of quantum numbers, which characterize a single-particle state, is denoted by f and g. Sometimes the quantum number $\sigma = \pm 1$, which denotes the states related by the time reversal operation, is separated. In such a case, designations $(q\sigma)$, $(s\sigma)$, and $(r\sigma)$ are used. The q denotes both proton and neutron states, s (neutron states only), and r (proton states only). The states, separated according to some particular feature, are denoted by $(\varrho\sigma)$ and $(\nu\sigma)$. The single-particle states in spherical nuclei are denoted by j (the total angular momentum) and m (its projection). The quantum numbers N (main oscillator number) and l (orbital momentum) are not written explicitly. The notation j^n, m^n is used for the neutron system, j^p, m^p for the proton system.

CHAPTER 1

INDEPENDENT PARTICLE MODEL

§ 1. The Hartree–Fock Method

1. Let us make a few general comments first. The atomic nucleus is a system of interacting neutrons and protons, and its structure is determined by the nuclear and Coulomb forces. Generally, there are three types of the interactions between elementary particles: strong, electromagnetic, and weak. All these interactions play an important role in nuclear physics. Nuclear forces belong to the category of the strong interactions. A description of the strongly interacting systems is very complicated (see ref. 28) because it is necessary to deal with a variable particle number and because the system has an infinite number of degrees of freedom. Lack of a fundamental theory of strong interactions makes a consistent description of the two-nucleon interaction impossible. The difficulties are further enhanced when the systems of three, four, or more strongly interacting particles are considered.

The nuclei are relatively weakly bound systems and, therefore, the difficulties of the quantum field theory can be avoided in nuclear theory. The binding energy of a nucleon in the nucleus (i.e., the separation energy of the outer nucleons) is about 8 MeV, and the average kinetic energy of a nucleon is 30 MeV. These energies are much smaller than the nucleon mass ($m \approx 10^3$ MeV), π-meson mass ($m_\pi \approx 140$ MeV), or the mass of the heavier K-, η-, τ-, ϱ-, and ω-mesons. Hence it is possible to treat the nuclei as the systems with a definite number of neutrons and protons and restrict the description to the nonrelativistic treatment.

Introduction of the phenomenological potentials eliminates the difficulties of the quantum field theory in the description of nucleon–nucleon interaction. The parameters containing the central, spin, tensor, and spin–orbital terms are determined from the experimental nucleon–nucleon scattering data. It should be noted, however, that despite the large number of experimental data on the elastic nucleon–nucleon scattering, the data are still insufficient for the unique determination of the nucleon–nucleon potential.

The Coulomb forces in nuclei are extremely important. Many basic nuclear properties are determined by the competition between the nuclear and Coulomb forces.

The weak interactions in nuclei are manifested in the nuclear β-decay and partially in the muon capture. Due to the presence of the weak interactions a large number of nuclei with a given value of A is unstable. By a series of subsequent β-decays these nuclei are transformed into nuclei with a minimal total energy. The number and the abundances of the stable isotopes and the lifetimes of not too heavy radioactive nuclei are, to a certain degree, determined by the characteristics of β-decay.

The nuclear interaction Hamiltonian has the transformation properties of the strong and electromagnetic interactions; it is invariant against time conjugation and space coordinates inversion. The part of the Hamiltonian corresponding to the strong interactions is isotopic (or isobaric) invariant. The total Hamiltonian is not isotope invariant, i.e., charge independent, but nevertheless the corresponding approximate quantum number—isobaric spin T—is very important in nuclear theory.

2. The description of nuclear structure is done in a framework of the many-body problem. The Hartree–Fock method, developed in atomic physics, is one of the basic methods used. The method reduces the problem of many interacting particles to the problem of one particle moving in the outside potential. The method could be used, however, only if the interaction of two nucleons contains the sum of the two-particle potentials and does not contain the three-, four-, or more, particle forces. The wave function of the system is expressed as the antisymmetric product of the single particle wave functions in the Hartree–Fock method. The single-particle wave functions are found from the minimum condition of the total energy of the interacting particles.

The Hartree–Fock method could be successfully applied in nuclear physics only if the nucleon–nucleon interaction potential has no singularities (i.e., when its matrix element, using the independent particles wave functions, is finite). When a strong singularity at $r \to 0$ is present in the nucleon–nucleon potential, the force acting on each of the two particles at short distances will be considerably different from the average field. The particles might then leave their respective orbits in the average field. This means that the average field will describe nuclear properties only if the nucleon–nucleon potential is sufficiently smooth.

The existence of strong repulsion at distances 0.4×10^{-13} cm was established in the high-energy nucleon–nucleon scattering. The corresponding part of the potential is called "hard core." As we mentioned above, the Hartree–Fock method is not directly applicable to singular potentials, i.e., to the potentials with hard core. However, it is possible to use mathematical devices which compensate for the repulsive hard core by a part of the attractive interaction; the Hartree–Fock method could then be used. Usually, the two-particle forces with hard core are divided into the short- and long-range parts. The division is done in such a way that the short-range attractive potential completely compensates for the hard-core effect. With such a definition the short-range part will not cause the scattering of two colliding nucleons. There are no fundamental problems in the application of the Hartree–Fock method to the long-range part of the potential.

The neutrons and protons are spin $\frac{1}{2}$ particles; they are fermions. Therefore the wave function of N-particles must be antisymmetric against permutation of two particles. Such a wave function for N noninteracting particles can be expressed in the form

$$\Psi(\xi_1, \xi_2, \ldots, \xi_N) = \frac{1}{\sqrt{N!}} \det \psi_i(\xi_{i'}), \tag{1.1}$$

where $\xi_{i'}$ denotes variables, characterizing the state of a particle; for example the radius vector $\boldsymbol{r}_{i'}$ and spin $\sigma_{i'}$. Index i denotes the single-particle states; both i and i' are varied from 1 to N. The functions $\psi_i(\xi_{i'})$ are eigenfunctions of the single-particle Hamiltonian:

they form a complete orthonormal system. The wave function of the system of interacting particles is a linear combination of the wave functions (1.1).

The method of second quantization considerably simplifies description of the systems containing interacting particles. Let us describe the method briefly.

Consider system of N noninteracting particles; they can occupy states with wave functions $\psi_i(\xi_{i'})$. Instead of using the total wave function, we can characterize the system by the numbers of particles in different states ψ_i. In this way we have introduced a new representation—second quantizations. The new variables are now the numbers n_i, i.e., occupation numbers of the states i.

The operator a_i describes annihilation of a nucleon in the state i. The Hermitian conjugate operator a_i^+ describes creation of a nucleon in the state i. The operators a_i and a_i^+ satisfy the following anticommutation relations:

$$\left.\begin{aligned} \{a_i, a_{i'}^+\} &= a_i a_{i'}^+ + a_{i'}^+ a_i = \delta_{ii'}, \\ \{a_i, a_{i'}\} &= \{a_i^+, a_{i'}^+\} = 0. \end{aligned}\right\} \tag{1.2}$$

Let us use the vacuum (i.e., the state without particles) as the initial state and apply N different creation operators. The wave function (1.1) is then obtained; it has now the form

$$\Psi(1, 2, \ldots, N) = a_1^+ a_2^+ \ldots a_N^+ \Psi_{00}, \tag{1.3}$$

where Ψ_{00}, the vacuum wave function, is determined by

$$a_i \Psi_{00} = 0. \tag{1.3'}$$

The operators a_i and a_i^+ act on the wave function $\Psi(1, 2, \ldots, N)$ in the following way:

$$a_i \Psi(1, 2, \ldots, N) = \begin{cases} 0, & \text{if} \quad i \neq 1, 2, \ldots, N, \\ \Psi(2, 3, \ldots, N), & \text{if} \quad a_i = a_1; \end{cases}$$

$$a_i^+ \Psi(1, 2, \ldots, N) = \begin{cases} 0 & \text{if} \quad i = 1, 2, \ldots, N, \\ \Psi(1, 2, \ldots, N, i), & \text{if} \quad i \neq 1, 2, \ldots, N. \end{cases}$$

The operator N_i, the particle number operator for the level i, is defined by

$$N_i = a_i^+ a_i. \tag{1.4}$$

It is not difficult to see that $N_i^2 = N_i$; this means that the particle number operator has eigenvalues zero or one. When the level i is occupied, the operator N_i has matrix element equal to one; when the level i is empty, the corresponding matrix element is equal to zero. The operator $a_i a_i^+ = 1 - N_i$ has opposite properties; its matrix element equals zero when the level i is occupied and equals one when the level i is empty. The operator of the total particle number is

$$N = \sum_i N_i = \sum_i a_i^+ a_i. \tag{1.4'}$$

Let us consider one-particle operators. Their matrix elements are completely determined by a matrix in the space of a single particle. The $\langle i | \mathfrak{f} | i' \rangle$ denotes the matrix elements of the single-particle operator, i.e.,

$$\langle i | \mathfrak{f} | i' \rangle \equiv \int dx \psi_i^*(x)\, \mathfrak{f}(x)\, \psi_{i'}(x).$$

The operator $\mathfrak{f}$ can be expressed in the many-particle space in the form

$$\mathfrak{f} = \sum_{ii'} \langle i|\mathfrak{f}|i'\rangle a_i^+ a_{i'} . \tag{1.5}$$

The two-particle operator is determined by a two-particle matrix. It has the form

$$V = \sum_{i_1, i_2, i_3, i_4} \langle i_1, i_2|V|i_3, i_4\rangle a_{i_1}^+ a_{i_2}^+ a_{i_3} a_{i_4} , \tag{1.6}$$

where

$$\langle i_1, i_2|V|i_3, i_4\rangle = \int dx_1\, dx_2 \overset{*}{\psi}_{i_1}(x_1)\, \overset{*}{\psi}_{i_2}(x_2)\, V(x_1, x_2)\, \psi_{i_3}(x_2)\, \psi_{i_4}(x_1).$$

The operator

$$\psi(\boldsymbol{r}, \sigma) = \sum_i \psi_i(\boldsymbol{r}, \sigma) a_i \tag{1.7}$$

annihilates a particle in the point $\boldsymbol{r}$, with spin projection on the z-axis equal to σ. The operators $\psi^+(\boldsymbol{r}, \sigma)$, $\psi(\boldsymbol{r}', \sigma')$ obey the following anticommutation relations:

$$\left.\begin{aligned} \{\psi^+(\boldsymbol{r}, \sigma), \psi(\boldsymbol{r}', \sigma')\} &= \sum_{ii'} \psi_i^*(\boldsymbol{r}, \sigma)\, \psi_{i'}(\boldsymbol{r}', \sigma')\, \{a_i^+, a_{i'}\} \\ &= \sum_i \psi_i^*(\boldsymbol{r}, \sigma)\, \psi_i(\boldsymbol{r}', \sigma') = \delta(\boldsymbol{r}-\boldsymbol{r}')\delta_{\sigma, \sigma'} , \\ \{\psi(\boldsymbol{r}, \sigma), \psi(\boldsymbol{r}', \sigma')\} &= \{\psi^+(\boldsymbol{r}, \sigma), \psi^+(\boldsymbol{r}', \sigma')\} = 0. \end{aligned}\right\} \tag{1.8}$$

Free particles with mass m have the wave functions

$$\psi_{p, \sigma}(\boldsymbol{r}, \sigma') = \frac{1}{\sqrt{\mathfrak{B}}} e^{i\boldsymbol{p}\boldsymbol{r}} \delta_{\sigma, \sigma'} , \tag{1.7'}$$

where $\mathfrak{B}$ is the normalizing volume. The kinetic energy operator is

$$T = \sum_{p, \sigma} \frac{p^2}{2m} a_{p\sigma}^+ a_{p\sigma} , \tag{1.9}$$

and the Coulomb interaction of charged particle could be expressed as

$$\left.\begin{aligned} V_c &= \frac{1}{2} \sum_{\sigma_1, \sigma_2} \int (d\boldsymbol{r}_1)(d\boldsymbol{r}_2)\, \psi^+(\boldsymbol{r}_2, \sigma_2)\psi^+(\boldsymbol{r}_1, \sigma_1) \frac{e^2}{|\boldsymbol{r}_1-\boldsymbol{r}_2|} \psi(\boldsymbol{r}_1, \sigma_1)\, \psi(\boldsymbol{r}_2, \sigma_2) \\ &= \frac{1}{\mathfrak{B}} \sum_{\sigma_1, \sigma_2} \sum_{p_1, p_2, p} \frac{2\pi e^2}{p^2} \overset{+}{a}_{p_1+p, \sigma_1} \overset{+}{a}_{p_2-p, \sigma_2} a_{p_2, \sigma_2} a_{p_1, \sigma_1} . \end{aligned}\right\} \tag{1.9'}$$

3. Let us apply the Hartree–Fock method and find the corresponding selfconsistent field. We shall use the most general Hamiltonian describing the system of interacting particles

$$H = \sum_{g, g'} T(g, g') a_g^+ a_{g'} - \tfrac{1}{4} \sum_{g_1, g_2, g_2', g_1'} G(g_1, g_2; g_2', g_1') \overset{+}{a}_{g_1} \overset{+}{a}_{g_2} a_{g_2'} a_{g_1} . \tag{1.10}$$

The functions $T(g, g')$ and $G(g_1, g_2; g_2', g_1')$ satisfy the following relations:

$$T(g, g') = T^*(g', g), \tag{1.11}$$

$$G(g_1, g_2; g_2', g_1') = -G(g_1, g_2; g_1', g_2') = -G(g_2, g_1; g_2', g_1') = G^*(g_1', g_2'; g_2, g_1). \tag{1.11'}$$

The relations (1.11) and (1.11′) are consequence of the hermiticity of H and of the antisymmetry, expressed by (1.2).

Let g denote the set of quantum numbers which characterize occupied states. Apply a linear canonical transformation of the Fermi amplitudes

$$a_g = \sum_f v(gf)a_f^+, \quad a_g^+ = \sum_f v^*(gf)a_f, \tag{1.12}$$

where f denotes the set of quantum numbers, characterizing both the occupied and empty states.

The new operators a_f and a_f^+ will satisfy relations (1.2) if

$$\eta(g, g') = \sum_f v^*(gf)\, v(g'f) - \delta_{gg'} = 0 \tag{1.13}$$

and

$$\sum_g v^*(gf)\, v(gf') = \delta_{ff'} \tag{1.13′}$$

are valid. The new operators are expressed by

$$a_f = \sum_g v(gf)a_g^+, \quad a_f^+ = \sum_g v^*(gf)a_g. \tag{1.12′}$$

The ground state of the system is determined as the vacuum of the operators a_f, i.e.,

$$a_f \,|\rangle_g = \sum_g v(gf)a_g^+ \,|\rangle_g = 0. \tag{1.14}$$

From (1.14) it follows that

$$|\rangle_g = \prod_g a_g^+ \Psi_{00}, \tag{1.15}$$

because (1.14) shows that all states g are occupied.

The average value of the particle number N in the state (1.15) is equal to

$$N = \sum_g \langle g\,|N|\,g\rangle = \sum_{gf} v^*(gf)\, v(gf). \tag{1.4″}$$

Let us introduce a density operator as a generalization of the particle number N_i,

$$\varrho(f, f') = a_f^+ a_{f'}. \tag{1.16}$$

The average value of $\varrho(g, g')$ in the state (1.15) is

$$\varrho(g, g') = {}_g\langle|\varrho(g, g')|\rangle_g = \sum_f v^*(gf)\, v(g'f). \tag{1.16′}$$

The expectation value of the Hamiltonian operator H in the ground state (1.15) is

$${}_g\langle|H|\rangle_g = \sum_{g, g'} T(g, g')\, \varrho(g, g') - \tfrac{1}{2} \sum_{g_1, g_2, g_2', g_1'} G(g_1, g_2; g_2', g_1')\, \varrho(g_1, g_1')\, \varrho(g_2, g_2'). \tag{1.17}$$

Such expression for the average energy was first derived by Fock.[29] To determine $v(gf)$ and $v^*(gf)$ we shall use the minimum condition of (1.17). Such condition can be written in the form

$$\delta\Big\{{}_g\langle|H|\rangle_g + \sum_{g, g'} \varrho(g, g')\, \lambda(g, g')\Big\} = 0, \tag{1.18}$$

where $\lambda(g, g')$ is the Lagrange multiplier and variations $\delta v(gf)$ and $\delta v^*(gf)$ are independent. The resulting equation is

$$\sum_{g'} \left\{ T(g, g') - \sum_{g_2' g_2} G(g, g_2; g_2', g') \varrho(g_2, g_2') + \lambda(g, g') \right\} v(g'g) = 0. \qquad (1.19)$$

Equation (1.19) is the Schrödinger equation, describing a particle moving in a potential field.

The selfconsistent Hamiltonian is defined by

$$H^{\text{s.c.}} = \sum_{f, f'} \left\{ T(f, f') - \sum_{g_2, g_2'} G(f, g_2; g_2', f') \varrho(g_2, g_2') \right\} a_f^+ a_{f'}. \qquad (1.20)$$

It is possible to use a new representation, where $H^{\text{s.c.}}$ is diagonal. New occupied orbitals are expressed as combinations of the old occupied orbitals, and the new empty orbitals are expressed as combinations of the old empty orbitals in the diagonal representation. The Hamiltonian $H^{\text{s.c.}}$ is invariant with respect to such transformation, because the sum over g_2 in (1.20) is equal to the trace in the subspace of the occupied orbitals. In the diagonal representation

$$T(f, f') - \sum_{g, g'} G(f, g; g', f') \varrho(g, g') = E(f) \delta_{ff'}. \qquad (1.21)$$

The $E(f)$ are the eigenvalues of the Hamiltonian $H^{\text{s.c.}}$. The Hamiltonian $H^{\text{s.c.}}$ in its diagonal representation is denoted H_{av} and treated as the average field Hamiltonian. The eigenvalues $E(f)$ are interpreted as approximate energies of a particle on the level f. The ground-state energy is equal to

$$\mathscr{E}_0 = {}_g\langle |H| \rangle_g = \tfrac{1}{2} \sum_g (T(g) + E(g)) \varrho(g, g). \qquad (1.17')$$

Thus, by using a representation with diagonal density $\varrho(g, g')$, we have introduced the model of independent particles. Each nucleon moves in the average field formed by all nucleons in the nucleus.

When the Hartree–Fock method is used in nuclear physics, it is assumed that

$$G(g, g_2; g_2', g') = \langle g, g_2 | V | g_2', g' \rangle,$$

i.e., that the quantity G is the two-nucleon interaction matrix element.

The ground-state energy, corresponding to the Hamiltonian (1.20), can be brought to the form

$$\mathscr{E}_0 = \sum_{g, g'} \{ T(g, g') - \tfrac{1}{2} \langle g | U | g' \rangle \} \varrho(g, g'), \qquad (1.17'')$$

where

$$\langle g | U | g' \rangle = \sum_{g_2, g_2'} G(g, g_2; g_2', g') \varrho(g_2, g_2').$$

The basic equation (1.19) can be simplified. It is sufficient to use (1.13) only in the diagonal form

$$\eta(g, g) = \sum_f v^*(gf) v(gf) - 1 = 0. \qquad (1.13'')$$

The Lagrange multiplier $\lambda(g, g')$ is then diagonal. Denoting $\lambda(g, g) = -E_g$,

$$\sum_{g'} \{T(g, g') - \langle g | U | g' \rangle\} v(gg') = E_g v(gg), \tag{1.19'}$$

i.e., the Hartree–Fock equation in the usual form.

The Hartree–Fock equations in spherical nuclei are solved by expansion of the wave functions in a series of oscillator wave functions and by subsequent iterations. The total angular momentum and its projection are good quantum numbers in the spherically symmetric systems. Therefore, in expansion of the wave functions, summation is restricted to the main oscillator quantum numbers; the series converges relatively fast. The most difficult part is the calculation of the nucleon–nucleon potential matrix elements $\langle g, g_2 | V | g_2', g' \rangle$. The nonlocal two-particle Tabakin potential is used in many calculations. The binding energies, charge distributions, and single-particle energies of ^{16}O, ^{40}Ca, and of other nuclei were calculated by this method.

When the Hartree–Fock method is applied to the light nuclei with deformed ground or excited states, the approximation of deformed orbits is used. It is assumed that most of the core nucleons do not contribute to the nuclear deformation. The trial wave function contains several creation operators of the particles and holes, which act on the spherically symmetric solution of the Hartree–Fock equations. The wave function is restricted by conditions of parity conservation, axial symmetry, etc. The selfconsistent solutions are obtained by the application of the variational principle to such trial functions. The spectra of ^{20}Ne, ^{24}Mg, and other nuclei were calculated by the deformed orbits method.

The application of the Hartree–Fock method to the complex nonmagic nuclei encounters many serious computational obstacles.

§ 2. The Single-particle Shell Model

1. The Hartree–Fock method is a basis of the nuclear shell model. This model describes noninteracting particles, moving in the common potential well, which is formed by all particles of the nucleus. The energy orbits in the potential well form groups, i.e., shells, divided by considerable energy intervals. The single-particle shell model (i.e., the model of independent particles) is too crude to describe the nuclear structure accurately. It gives, however, a basis for the treatment of nuclear correlations caused by the residual interaction.

The solutions of the Hartree–Fock equations, based on the nucleon–nucleon interaction, are available only for a few light and magic nuclei. Therefore, the average field potential is usually chosen empirically. It is assumed that the behavior of the average field potential (as a function of the radius) and the nuclear density distribution are correlated. Further, the potential should correctly reproduce the magic numbers. Finally, details of the average field potential are determined from the large amount of experimental data.

Let us explain briefly why the independent particle model is applicable to the description of the nuclear properties. The nuclear forces have a short range and they are very strong and basically attractive. The nucleon–nucleon potential can be approximately divided into the very strong, very repulsive part (the hard core, with radius 0.4×10^{-13} cm) and the weaker, longer-range attractive part. The volume, corresponding to the hard cores, occupies only about one-hundredth of the total nuclear volume, when observed nuclear densities are used. That does not mean that the hard core is unimportant—it contributes to the saturation of the nuclear forces and prevents compression of the nuclei. The attractive part of the nuclear forces contributes predominantly to the average field. The repulsion radius, i.e., the relation between the short-range repulsion and attraction, the form of the attractive part, and the Pauli principle, together create conditions necessary for the applicability of the independent particle model as a basis for development of the nuclear theory.

The long-range part of the nucleon–nucleon interaction is considerably reduced by the Pauli principle effect. At the same time the probability of large momentum transfer during collision of two nucleons is small. Let us explain this fact on a simple example. Let the nucleons, with momenta $\mathbf{p}_1$ and $\mathbf{p}_2$, interact via Gaussian interaction

$$V(\boldsymbol{r}_1-\boldsymbol{r}_2) = -V_0 e^{-(|\boldsymbol{r}_1-\boldsymbol{r}_2|^2)/\mu^2}.$$

The probability, that the particles will have momenta $\boldsymbol{p}_1'$ and $\boldsymbol{p}_2'$ in the final state, is equal to the squared matrix element

$$\int e^{-i(\boldsymbol{p}_1\boldsymbol{r}_1+\boldsymbol{p}_2'\boldsymbol{r}_2)} V(\boldsymbol{r}_1-\boldsymbol{r}_2)\, e^{i(\boldsymbol{p}_1\boldsymbol{r}_1+\boldsymbol{p}_2\boldsymbol{r}_2)} (d\boldsymbol{r}_1)\,(d\boldsymbol{r}_2).$$

The integration could be performed in relative variables $\boldsymbol{r} = \boldsymbol{r}_1-\boldsymbol{r}_2$, $\boldsymbol{p} = \boldsymbol{p}_1-\boldsymbol{p}_1'$. The matrix element is equal to

$$-V_0 \int (d\boldsymbol{r}) e^{-i\boldsymbol{p}\cdot\boldsymbol{r}} e^{-r^2/\mu^2} = -V_0 \int (d\boldsymbol{r}) e^{-[\boldsymbol{r}+(i/2)\boldsymbol{p}\mu^2]^2/\mu^2} e^{-p^2\mu^2/4} = -\pi^{3/2}\mu^3 V_0 e^{-p^2\mu^2/4}.$$

It is evident now that the probability of the large momentum transfer is quite small. This result is valid for other forms of the short-range potential too.

At the same time, interactions with small momentum transfer are impossible for all but the nucleons on the highest energy orbits. The neighborhood levels are occupied by other nucleons, and the Pauli principle thus prevents scattering into these levels, i.e., scattering with low momentum transfer.

The wave function of a pair of particles in nuclear matter is very similar to the wave function of free particles (see refs. 21–23). The differences become important only when the distance between the two particles becomes smaller than the average interparticle distance in the nucleus ($d \approx 1.7\times10^{-13}$ cm). Figure 1.1 shows the wave function of relative motion for a pair of particles in the S-state. The curves show the wave functions for non-interacting particles and for a pair participating in the Fermi momentum distribution and interacting via nuclear interaction.

Obviously, the wave function of interacting particles is close to the free particle wave function at large distances. This fact can be understood in the following way: when one particle of the original pair is close to a third particle, the interparticle distance in the original pair is of the order d. But at such a distance the wave function of the original pair is similar to the wave function of noninteracting particles. Thus, most of the collisions with a third particle happen in conditions when the interaction with all remaining particles is unimportant.

As we have just explained, despite the strong nuclear interaction, the wave functions of the nucleons in nuclei are similar to the wave functions of noninteracting particles except

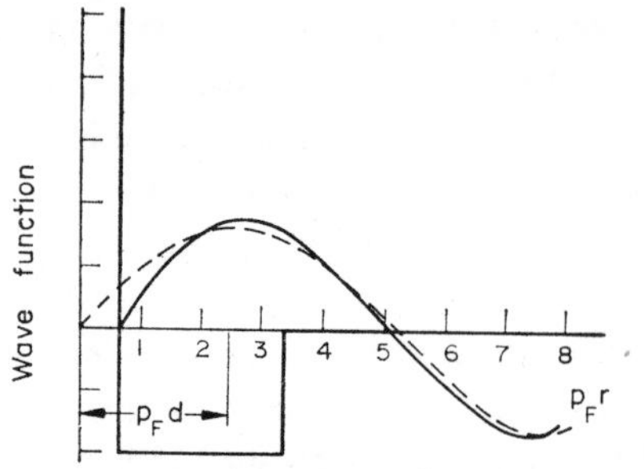

FIG. 1.1. Wave functions of the relative motion for two particles in the S-state. The dashed curve corresponds to the calculated nuclear interaction[30]. Both functions are calculated for relative momentum $p = 0.6p_F$ in the nuclear matter with $p_F = 1.48 \times 10^{13}\ \text{cm}^{-1}$. The continuous line shows the r-dependence of the nuclear potential; d is the average distance to the third particle.

at distances considerably smaller than the average interparticle distance d. The main reason for such behavior of the wave functions is the Pauli principle. It prevents the wave function from responding to the interaction because neighboring Fourier components belong to already occupied states.

2. The square well and harmonic oscillator potentials were originally used as the nuclear average field. The experimental evidence from nucleon scattering on nuclei suggests that the harmonic oscillator potential is more suitable for light nuclei, while the square well potential is more suitable for the heavier nuclei. The real nuclear potential must be finite, must have finite surface thickness (similar to the nuclear density), and the radial dependence should be intermediate between the square well and harmonic oscillator.

Let us consider the sequence of levels in the infinite, spherical harmonic oscillator well. Such potential has the form

$$V(r) = \tfrac{1}{2} m \mathring{\omega}_0^2 r^2, \tag{1.22}$$

where m is the nucleon mass and $\mathring{\omega}_0$ is the classical oscillator frequency. The Schrödinger equation

$$\left(-\frac{1}{2m}\Delta + V(r) - E\right)\psi = 0 \tag{1.23}$$

has the solutions

$$\psi_{nlm} = \frac{u_{nl}(r)}{r} Y_{lm}(\theta, \varphi). \tag{1.24}$$

The spherical functions $Y_{lm}(\theta, \varphi)$ are eigenfunctions of the l^2 operator (l is the orbital angular momentum) and of l_z, the projection of l on the z-axis (with eigenvalue m). The radial part of the wave function $u_{nl}(r)$ satisfies the equation

$$\left\{-\frac{1}{2m}\frac{d^2}{dr^2} + V(r) + \frac{1}{2m}\frac{l(l+1)}{r^2} - E\right\}(u_{nl}) = 0.$$

The corresponding eigenvalues equal (see refs. 31–33)

$$E_N = (N + \tfrac{3}{2})\mathring{\omega}_0, \tag{1.25}$$

where $N = 0, 1, 2, \ldots$ Each of the eigenvalues E_N is degenerate, corresponding to several

l values. If N is an even integer, l has values $0, 2, 4, \ldots, N$; if N is an odd integer, $l = 1, 3, \ldots, N$. The maximum number of particles in the degenerate N state equals

$$n_N = \sum_l 2(2l+1) = (N+1)(N+2).$$

The total number of particles, which fill states from $N = 0$ up to $N = N_0$, equals

$$\sum_N n_N = \tfrac{1}{3}(N_0+1)(N_0+2)(N_0+3).$$

The harmonic oscillator levels are usually denoted by a pair of integer numbers (n, l). The n means that the corresponding l value appears nth time in the level sequence. The l is often denoted by a letter instead of a number, namely

$$\begin{matrix} l = 0, & 1, & 2, & 3, & 4, & 5, & 6, & 7, & 8, \\ s, & p, & d, & f, & g, & h, & i, & j, & k. \end{matrix}$$

So, for example, the sequence begins with 1s, 1p; 2s, 1d; 2p, 1f, etc.

Table 1.1 gives the single-particle energies, maximal particle numbers n_N in each of the degenerate levels, and the total particle numbers $\sum_N n_N$ for the harmonic-oscillator well. The table shows that the shells are filled for the neutron (proton) numbers equal 2, 8, 20, 40, 70, 112, 168, etc.

TABLE 1.1. *Single-particle States in the Harmonic Oscillator Well*

N	$\dfrac{E_N}{\mathring{\omega}_0}$	State (n, l)	n_N	$\sum_n n_N$
0	$\frac{3}{2}$	1s	2	2
1	$\frac{5}{2}$	1p	6	8
2	$\frac{7}{2}$	2s 1d	12	20
3	$\frac{9}{2}$	2p 1f	20	40
4	$\frac{11}{2}$	3s 2d 1g	30	70
5	$\frac{13}{2}$	3p 2f 1h	42	112
6	$\frac{15}{2}$	4s 3d 2g 1i	56	168

These numbers (except the first three) differ from the so-called magic numbers found from the empirical evidence. The magic numbers are 2, 8, 20, 28, 50, 82, and 126; the 126 belongs, strictly speaking, only to the neutron system. These numbers should correspond to the filled shells. Note that the numbers corresponding to the filled shells in the square well differ also from the empirical magic numbers. Modification of the average field potential is therefore necessary.

Let us estimate the frequency $\mathring{\omega}_0$ and relate it to the nuclear radius. The average kinetic and potential energies are equal for all states in the harmonic oscillator. Using (1.22), it is possible to express the sum of the single-particle energies through mean square radius $\langle r^2 \rangle$ (symmetric $N = Z$ case is considered here):

$$\sum_{i=1}^{Z} E_i + \sum_{i=1}^{N} E_i = m\mathring{\omega}_0^2 A \langle r^2 \rangle, \tag{1.26}$$

while, experimentally, $\langle r^2\rangle \approx \frac{3}{5}R^2$, $R = r_0 A^{1/3}$, $r_0 = 1.24\times 10^{-13}$ cm. The sum in (1.26) can be estimated for the nuclear ground state, where all $N = Z \leqslant N_0$ states are filled with neutrons and protons. From (1.25) it follows that

$$2\sum_{i=1}^{A/2} E_i = \sum_{N=0}^{N_0} 2(N+1)(N+2)(N+\tfrac{3}{2})\mathring{\omega}_0 \approx \tfrac{1}{2}(N_0+2)^4\,\mathring{\omega}_0,$$

and further

$$A = 2\sum_{N=0}^{N_0} (N+1)(N+2) \approx \tfrac{2}{3}(N_0+2)^3.$$

The N_0 can be replaced by A now. Equation (1.26) is then

$$\tfrac{1}{2}(\tfrac{3}{2})^{4/3}A^{4/3}\mathring{\omega}_0 = m\mathring{\omega}_0^2 A\,\tfrac{3}{5}R^2$$

and

$$\mathring{\omega}_0 = 41A^{-1/3}\ \text{MeV}. \tag{1.27}$$

Equation (1.27) thus defines the energy difference between the equidistantly spaced levels of the harmonic oscillator potential.

3. Even when more realistic radial dependence of the average field potential is used, it is still impossible to reproduce correctly the numbers corresponding to the filled shells. Therefore it is necessary to introduce a new interaction which breaks the degeneracy of the harmonic oscillator. When the nuclear shell model was developed, it was suggested that sufficiently strong spin–orbital interaction exists and might fulfill such a purpose. The corresponding spin–orbit potential has the form

$$V_{ls} = -V_{ls}(r)\boldsymbol{l}\cdot\boldsymbol{s}, \tag{1.28}$$

where $\boldsymbol{l} = \boldsymbol{r}\times\boldsymbol{p}$, and $\boldsymbol{s}$ is the nuclear spin, $V_{ls}(r) \sim \dfrac{1}{r}\dfrac{dV(r)}{dr}$.

The spin–orbit potential breaks the degeneracy of the single-particle levels with respect to the total angular momentum j. Using the relation

$$j^2 = (\boldsymbol{l}+\boldsymbol{s})^2 = \boldsymbol{l}^2+\boldsymbol{s}^2+2(\boldsymbol{l}\cdot\boldsymbol{s}),$$

we obtain

$$\boldsymbol{l}\cdot\boldsymbol{s} = \tfrac{1}{2}\{j(j+1)-l(l+1)-s(s+1)\} = \begin{cases} \tfrac{1}{2}l & \text{for } j = l+\tfrac{1}{2}, \\ -\tfrac{1}{2}(l+1) & \text{for } j = l-\tfrac{1}{2}. \end{cases}$$

The spin–orbit forces do not cause large changes of the radial wave functions. Therefore, the main effect is the following: The level $j = l+\frac{1}{2}$ is lowered by

$$\tfrac{1}{2}l\langle V_{ls}(r)\rangle_{nl},$$

while the level $j = l-\frac{1}{2}$ is raised by

$$\tfrac{1}{2}(l+1)\langle V_{ls}(r)\rangle_{nl}.$$

The splitting of the two levels is therefore equal to

$$\tfrac{1}{2}(2l+1)\langle V_{ls}(r)\rangle_{nl}. \tag{1.29}$$

The splitting increases when l increases, while $\langle V_{ls}(r)\rangle_{nl}$, i.e., the average value of the

$V_{ls}(r)$ in the state (nl), depends weakly on l. Experimentally observed splittings of the states with $j = l \pm \frac{1}{2}$ give approximately

$$\Delta\mathcal{E}_{ls} \approx -20\boldsymbol{l}\cdot\boldsymbol{s}A^{-2/3}\ \text{MeV}. \tag{1.30}$$

The introduction of the spin–orbit potential is not sufficiently theoretically based. However, many experimental facts confirm the existence of the relatively strong spin–orbit part in the average field potential. Among such facts belongs the splitting of the $j = l \pm \frac{1}{2}$ levels, particularly in nuclei with closed shells plus (or minus) one nucleon. Other evidence comes from the observed polarization effects in the interaction between nucleons and nuclei. These polarization effects confirm the accepted sign and strength of the spin–orbit coupling.

Figure 1.2 shows changes in the energy spectrum caused by the spin–orbit coupling. The numbers of particles, corresponding to the filled shells, coincide now with the magic numbers 2, 8, 20, 28, 50, 82, 126, and 184.

The term "shell" will be further used for the set of states between two magic numbers, the term "subshell" will be used for the degenerate states characterized by the quantum numbers n, l, j. For example, the fourth shell, between neutron numbers (proton numbers) 50 and 82, consists of five subshells: $1g_{7/2}$, $2d_{5/2}$, $2d_{3/2}$, $3s_{1/2}$, $1h_{11/2}$.

The shells are uniquely determined in the harmonic oscillator potential with the spin–orbit part. The ordering of subshells inside the shell is, however, undetermined; it depends on the strength of the spin–orbit coupling.

4. As mentioned earlier, the realistic average field potential should be similar to the nuclear matter distribution. Parameters of such a potential are well determined from the real part of the optical potential, which, in turn, is determined from the whole body of data on nucleon scattering on nuclei. The analytical form of the average field potential is usually chosen as the Woods–Saxon potential.

The Woods–Saxon potential is a spherically symmetric, finite depth potential. The equipotential surface $r = R_0$ corresponds to half of the potential at the nuclear center. It consists of two parts; the central part

$$V(r) = -\frac{V_0^{N,Z}}{1+\exp[(1/a)(r-R_0)]} \tag{1.31}$$

and the spin–orbit coupling

$$V_{ls}(r) = -\zeta\frac{1}{r}\frac{dV(r)}{dr}(\boldsymbol{ls}). \tag{1.32}$$

The usual choice of its parameters(34) is:

$$\left.\begin{aligned} V_0^N &= V_0\left[1-0.63\frac{N-Z}{A}\right], \\ V_0^Z &= V_0\left[1+0.63\frac{N-Z}{A}\right], \end{aligned}\right\} \tag{1.31'}$$

$$V_0 = 53\ \text{MeV}, \quad R_0 = r_0A^{1/3}, \quad r_0 = 1.24\times10^{-13}\ \text{cm},$$

the surface thickness $a = 0.63\times10^{-13}$ cm, spin-orbit coupling strength $\zeta = 0.263$

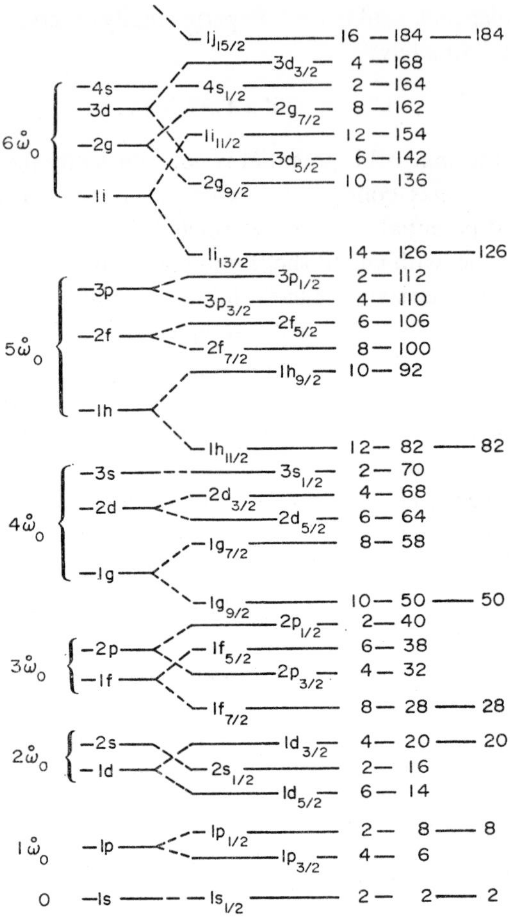

FIG. 1.2. The single-particle level scheme in the potential well, intermediate between the harmonic oscillator and square well. The spin–orbit coupling is also included. The level scheme without the spin–orbit part is shown on the left. Shifts caused by the spin–orbit coupling are shown in the middle. The numbers on the righthand side of the figure denote the number of neutrons (protons) in different subshells, the total number of particles from the well's bottom, and the numbers corresponding to the filled shells.

$\{1+2[(N-Z)/A]\}$ $(10^{-13}$ cm$)^2$. The parameters are sufficiently stable for spherical nuclei within wide range of the atomic numbers A.

The Coulomb potential must be added to the potential (1.31) and (1.32) when the proton levels are calculated. Neglecting the effect of the surface,

$$V_c(r) = \frac{(Z-1)e^2}{r}\begin{cases} \dfrac{3r}{2R_0} - \dfrac{1}{2}(r/R_0)^3, & r \leq R_0, \\ 1, & r > R_0. \end{cases} \tag{1.33}$$

The Woods–Saxon potential and oscillator potentials are compared in Fig. 1.3. The former potential has a flatter bottom and corresponds to the intermediate case between the oscillator and square well. Details of the surface region are particularly important in nuclear reactions. When the nuclear radius parameter is increased in the Woods–Saxon

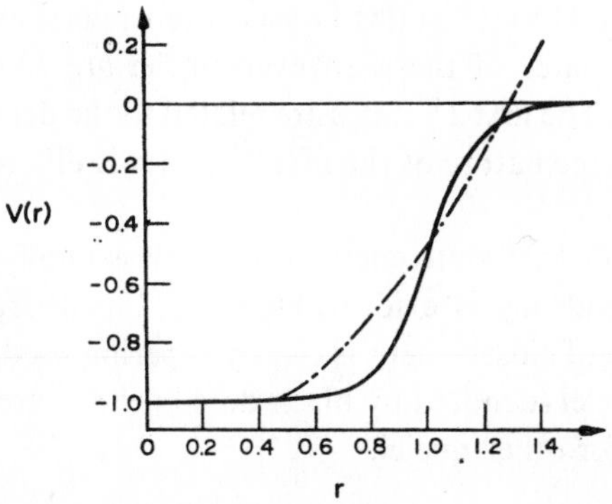

FIG. 1.3. The Woods–Saxon potential with $a = 0.64 \times 10^{-13}$ cm (full curve) and harmonic oscillator potential (dotted and dashed curve). The radius is measured in units of R_0, potential in units of V_0.

potential, the levels with larger l-values are lowered faster than the levels with smaller l-values.

The shells in the Woods–Saxon potential are not changed in comparison to the harmonic oscillator. The position of the subshells depends on the particular choice of the parameters, particularly on the strength of the spin–orbit coupling.

5. A large number of experimental facts exhibit clear-cut shell effects. Nuclei with magic numbers of protons or neutrons have peculiar properties related to the filled shells. (Usually such nuclei are called magic, or doubly-magic, as e.g., $^{208}_{82}\mathrm{Pb}_{126}$).

The shell effects are recognizable in the value of nuclear mass. The nuclear mass is usually expressed in the approximate form

$$\mathscr{E}(N, Z) = Nm_n + Zm_p - B(N, Z), \tag{1.34}$$

and the binding energy in the form

$$B(N, Z) = b_V A - b_S A^{2/3} - \frac{1}{2} b_{\mathrm{sym}} \frac{(N-Z)^2}{A} - \frac{3}{5} \frac{Z^2 e^2}{R_0} - b_{\mathrm{pair}} \frac{\delta}{A}. \tag{1.35}$$

Parameter δ equals 1 for odd–odd nuclei, $\delta = 0$ for odd-A nuclei, and $\delta = -1$ for even–even nuclei. Other parameters in (1.35) have approximate values: $b_V = 16$ MeV, $b_S = 20$ MeV, $b_{\mathrm{sym}} = 25$ MeV, $b_{\mathrm{pair}} = 27$ MeV.

The shell effects cause singularities in the dependence of the binding energy on N or Z. When a nucleon is added to the magic nucleus, its binding energy drops by about 2 MeV, This behavior is particularly apparent in the doubly magic nucleus ^{208}Pb. Similar evidence can be seen in the energies of α- and β-decays. Occasionally, small anomalies exist when the subshells are filled.

The existence of shells affects also the abundance of elements and relative abundancies of the different isotopes for a given element. Such anomalies are related to the sudden decrease of the binding energies for nuclei with numbers N or Z just above magic numbers. The number of the stable and longlived isotopes with $N = 20$, 28, 50, 82, and 126 is larger than for the neighboring N-values. More than one stable odd-A isotone exists for $N = 20$, 50, and 82, while there is only one odd-A stable isotone for other N-values.

The energies of the first $I^\pi = 2^+$ states in even–even nuclei are strongly related to the closed shells. All ground states of the even–even nuclei are $I^\pi = 0^+$ states and the first excited states are $I^\pi = 2^+$. The first 2^+ states are related to the degrees of freedom which are easiest to excite; therefore the nature of the first 2^+ states is different in different even–even nuclei.

The dependence of the first 2^+ state energy on the mass number A is shown in Fig. 1.4. The general decreasing tendency is evident. However, this dependence is not monotonic; the very pronounced effect of closed shells is clearly superimposed on it. The energies of the first 2^+ states in magic nuclei (denoted by black dots in the figure) are considerably higher than the same quantities for other nuclei.

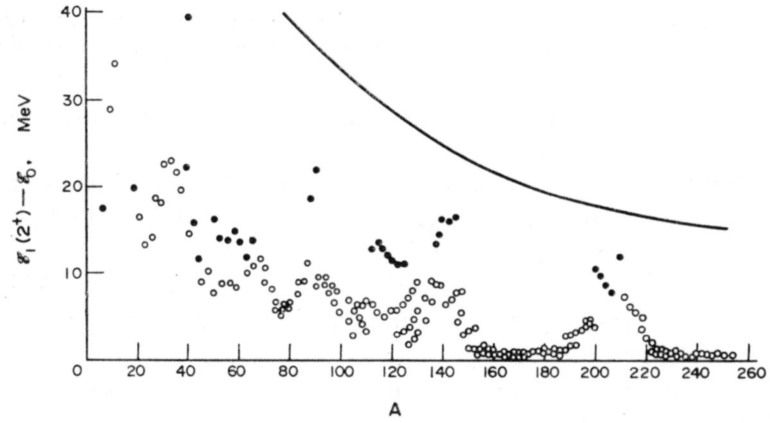

FIG. 1.4. The $\mathscr{E}_1(2^+) - \mathscr{E}_0$ energies of the first excited 2^+ states in the even–even nuclei. The nuclei with magic numbers of neutrons or protons are denoted by black dots, other nuclei by open circles. The curve shows energy values according to the hydrodynamic model.

The magic nuclei have anomalously small level densities up to the neutron binding energies. Therefore, the neutron capture cross-sections are small for both the thermal neutrons and neutrons with energies up to 1 MeV. This is another demonstration of the importance of the shell effects.

The residual interactions are very important in nuclei; therefore it would be senseless to compare the predictions of the independent particle model with the properties of the ground or excited states of real nuclei. Nevertheless, the spins and parities of the odd-A nuclei close to the magic nuclei can be compared with the corresponding predictions.

It will be shown later that the even numbers of neutrons or protons in the nuclear ground states are correlated in pairs, giving the total spin equal to zero and positive parity. Hence, the spin and parity of the odd-A nucleus are determined by the spin and parity of the corresponding average field level occupied by the unpaired particle.

The wave function of the nucleon above closed shells is

$$\psi_{nljm} = \frac{1}{r} u_{nlj}(r) \sum_{m_l, m_s} (l m_l \tfrac{1}{2} m_s \mid jm)\, i^l Y_{lm_l}(\theta, \varphi) \chi_{m_s}, \tag{1.36}$$

where $u_{nlj}(r)$ is the radial part, χ_{m_s} the spin part, $(lm_l\frac{1}{2}m_s|jm)$ the Clebsch–Gordan coefficients. The state, where one particle with momentum j is missing to close the shell, is described as motion of a hole. The formation of the hole state with quantum numbers $nljm$ is equivalent to the annihilation of a particle in the $nlj{-}m$ state.

The single-particle model correctly predicts the ground-state spins and parities of the odd-A nuclei. The only exception is the case when a subshell with a large l-value immediately follows a subshell with a small l-value. The odd nucleon remains on the smaller l-level in such case as a consequence of the residual interaction.

Another success of the single-particle shell model is the explanation of the nuclear isomerism. Isomeric states are nuclear excited states with relatively long lifetimes. The long lifetimes could be related either to the small energy of the de-exciting radiation or to its high multipolarity or to both causes combined. The single-particle shell model predicts existence of the isomeric states in nuclei where l_1 subshell is almost full and the nearest subshell has l_2 value which differs from l_1 by at least three units. A large number of isomers were found experimentally. They are connected with the following subshells: (1) $p_{1/2}-g_{9/2}$; (2) $d_{3/2}-s_{1/2}-h_{11/2}$; (3) $p_{1/2}-f_{5/2}-p_{3/2}-i_{13/2}$.

The single-particle shell model predicts that the isomeric states will be concentrated in nuclei immediately below magic numbers—such prediction is confirmed by the experimental evidence.

The success in the explanation of the spins and parities of the ground and isomeric states in a large number of odd-A spherical nuclei shows that the harmonic oscillator (or Woods–Saxon) potentials with spin–orbit coupling describe the average nuclear field well. The residual interaction does not change the predicted ground state spins of the odd-A spherical nuclei considerably.

The many-particle shell model is the next developmental stage of the single-particle model. This model, which takes the residual interaction into account, is described in a number of textbooks and monographs, for example, in refs. 19, 35, and 36.

§ 3. Isobaric Symmetry

1. The hypothesis of the charge independence of strong interactions and its consequence —the isobaric or isotopic symmetry, are often used in nuclear physics. According to the hypothesis, the nuclear forces acting between the nucleons are independent of their electric charge, i.e., when the (n, p), (n, n), or (p, p) pairs are in the same state, the forces between the nucleons in all pairs are the same. If the hypothesis is true, the neutron and proton could be described as two states of a single particle—the nucleon.

It should be remembered that only strong interactions satisfy isobaric symmetry, while the electromagnetic and weak interactions do not. Therefore, despite the close resemblance between the neutron and proton, there are some features (besides the charge) which make them distinct; for example, they have somewhat different masses and different magnetic moments. Thus, the isobaric symmetry is valid only approximately, and the isobaric spin is not an "exact" quantum number.

To treat the neutron and proton as two states of a nucleon, we shall introduce a new coordinate (in addition to the spatial and spin coordinates), which describes the nucleon charge. The new coordinate has two possible values and is therefore similar to the spin variable—hence its name—isobaric (or isotopic) spin. Because the ordinary and isobaric spins are so similar, the same mathematical apparatus can be used for their description. However, there is no isobaric analog of the orbital angular momentum, and all particles are polarized in isobaric space.

Introducing matrices $\boldsymbol{\tau}$

$$\tau_x = \begin{pmatrix} 0 & 1 \\ 1 & 0 \end{pmatrix}, \quad \tau_y = \begin{pmatrix} 0 & -i \\ i & 0 \end{pmatrix}, \quad \tau_z = \begin{pmatrix} 1 & 0 \\ 0 & -1 \end{pmatrix}$$

the isobaric spin operator is defined as

$$\boldsymbol{t} = \tfrac{1}{2}\boldsymbol{\tau}.$$

The isobaric spin of a nucleon is $\frac{1}{2}$. Its projections are defined as: neutron $t_z = \frac{1}{2}$, proton $t_z = -\frac{1}{2}$. The wave functions of neutron and proton now has one of the following forms:

$$\left.\begin{aligned} \psi_n(\boldsymbol{r}, \sigma_z) &= \psi(\boldsymbol{r}, \sigma_z, t_z = \tfrac{1}{2}) = \psi(\boldsymbol{r}, \sigma_z)\begin{pmatrix} 1 \\ 0 \end{pmatrix} = \psi(\boldsymbol{r}, \sigma_z)\,\chi(t_z = \tfrac{1}{2}), \\ \psi_p(\boldsymbol{r}, \sigma_z) &= \psi(\boldsymbol{r}, \sigma_z, t_z = -\tfrac{1}{2}) = \psi(\boldsymbol{r}, \sigma_z)\begin{pmatrix} 0 \\ 1 \end{pmatrix} = \psi(\boldsymbol{r}, \sigma_z)\,\chi(t_z = -\tfrac{1}{2}). \end{aligned}\right\} \tag{1.37}$$

The wave function of the system of nucleons must be completely antisymmetric against the permutation of the spatial, spin, and isobaric coordinates.

Let $\tau^{\pm}$ denote the matrices

$$\tau^{\pm} = \tfrac{1}{2}(\tau_x \pm i\tau_y).$$

The matrix $\tau^{+} = \begin{pmatrix} 0 & 1 \\ 0 & 0 \end{pmatrix}$ transforms the proton state into the neutron state and gives zero when applied on the neutron state. The matrix $\tau^{-} = \begin{pmatrix} 0 & 0 \\ 0 & 1 \end{pmatrix}$ has opposite properties; it transforms the neutron state into the proton state and annihilates the proton state.

The charge-exchange operator is

$$P^{\tau}_{ii'} = \tfrac{1}{2}(1 + \boldsymbol{\tau}^{(i)} \cdot \boldsymbol{\tau}^{(i')}).$$

The total isobaric spin vector of a system of nucleons is defined by

$$\boldsymbol{T} = \sum_{i=1}^{A} \boldsymbol{t}^{(i)}, \tag{1.38}$$

with the z-component

$$T_z = \sum_{i=1}^{A} t_z^{(i)} = \tfrac{1}{2}(N - Z). \tag{1.39}$$

The eigenvalues of $(\boldsymbol{T})^2$ equal $T(T+1)$. For systems consisting of an even number of nucleons, T is a non-negative integer, i.e., $T = 0, 1, 2, \ldots$; for systems consisting of an odd number of nucleons, $T = \frac{1}{2}, \frac{3}{2}, \ldots$.

The operator $(\boldsymbol{T})^2$ can be expressed as

$$(\boldsymbol{T})^2 = \frac{1}{4}\left(\sum_{i=1}^{A} \boldsymbol{\tau}^{(i)}\right)^2 = A - \frac{A^2}{4} + \sum_{i<i'} P^{\tau}_{ii'}. \tag{1.40}$$

If the isobaric spin part of the wave function is totally symmetric, i.e., $P^{\tau}_{ii'} = 1$, then

$$\sum_{i<i'} P^{\tau}_{ii'} = \frac{A(A-1)}{2} \quad \text{and} \quad (\boldsymbol{T})^2 = \frac{A}{2}\left(\frac{A}{2}+1\right), \quad T = \frac{A}{2}.$$

The isobaric spin has, in this case, its maximal possible value $T_{\max} = A/2$, which corresponds to a system of particles of one type only.

For two particles, P_{12} equals either 1 (symmetric state) or -1 (antisymmetric state). In the former case $T^2 = 2$, $T = 1$, in latter case $T^2 = T = 0$.

Denote the isobaric spin part of the nuclear wave function as $|T, T_z\rangle$. It is then

$$T^{\pm}|T, T_z\rangle = \sqrt{(T \mp T_z)(T \pm T_z + 1)}\,|T, T_z \pm 1\rangle, \tag{1.41}$$

where $T^{\pm} = \frac{1}{2}(T_x + iT_y)$. The operator T^+ increases T_z by one unit, T^- decreases it by one unit.

Two neutrons and two protons on the same average field level have symmetric wave function in spin and spatial coordinate space; therefore they must have antisymmetric wave function in the isobaric spin space. Thus, for the nuclear ground states

$$T = T_z = \tfrac{1}{2}(N - Z), \tag{1.42}$$

i.e., only the neutron excess gives a contribution. Such a function, with $T = T_z$, has the following properties:

$$T^+|T, T\rangle = 0, \quad T^-|T, T\rangle = \sqrt{2T}\,|T, T-1\rangle.$$

2. The isobaric spin T is a good quantum number for systems with the nuclear forces as the only interaction. The presence of the Coulomb force in nuclei breaks the degeneracy of the isobaric multiplets (i.e., the independence of the nuclear energy on the T_z value). Table 1.2 shows two examples of such symmetry breaking for the $T = 1$ multiplets in light nuclei. The purely nuclear binding energies are indeed very close to each other. The similarity of the states is not restricted to the $I^{\pi} = 0^+$ states shown in the table; the excited states with $T = 1$ are similar as well.

TABLE 1.2. *The splitting of the isobaric triplet* ($T = 1$, $I^{\pi} = 0^+$)

Nucleus	Excitation energy (MeV)	T_z	Total binding energy (MeV)	Coulomb energy (MeV)	Nuclear binding energy (MeV)
$^{10}_{4}Be_6$	0	+1	−65.0	3.3	−68.3
$^{10}_{5}B_5$	1.74	0	−63.0	5.5	−68.5
$^{10}_{6}C_4$	0	−1	−60.0	8.3	−68.3
$^{14}_{6}C_8$	0	+1	−105.3	7.4	−112.7
$^{14}_{7}N_7$	2.31	0	−102.4	10.4	−112.8
$^{14}_{8}O_6$	0	−1	−98.7	13.9	−112.6

The Coulomb energy in heavy nuclei is large; therefore a majority of the components of the multiplet, with given A- and T-values, are in the continuous spectrum.

Let us estimate the excitation energy corresponding to the T-value larger than the ground state value $T = T_0 = \frac{1}{2}(N-Z)$. Applying (1.34) and (1.35),

$$\mathcal{E}(A, T, T_z) = \mathcal{E}(A) + \frac{1}{2} b_{\text{sym}} \frac{(N-Z)^2}{A} + \frac{3}{5} \frac{e^2 Z^2}{R_0}. \tag{1.43}$$

For the ground state $T_0 = T_z$, $(N-Z)^2 = 4T_0(T_0+1)$ and $Z = (A/2) - T_z$, thus

$$\mathcal{E}(A, T_0, T_z) = \mathcal{E}(A) + 2b_{\text{sym}} \frac{T(T+1)}{A} + \frac{3}{5} \frac{e^2}{R_0} \left(\frac{A}{2} - T_z\right)^2. \tag{1.43'}$$

The difference between the T_0+1 and T_0 states is then

$$\mathcal{E}(A, T_0+1, T_0) - \mathcal{E}(A, T_0, T_0) = 4b_{\text{sym}} \frac{T_0+1}{A}, \tag{1.44}$$

i.e., approximately $(T+1) \times 10^2/A$ MeV. This quantity is equal, for example, to 7.8 MeV in $^{116}_{50}\text{Sn}$ and 9 MeV in $^{190}_{78}\text{Pt}$, etc.

The Coulomb interaction in nuclei has a twofold effect. The Coulomb term in (1.43) commutes with T but causes the T_z dependence of the nuclear energy. On the other hand, the Coulomb interaction leads to the effects disturbing the T-invariance. The proton radial functions are changed, and the isobaric symmetry violated.

This effect can be understood by using as an example the system consisting of one neutron and one proton moving in the nuclear average field. The wave function of this system is

$$\psi(\xi_1, \xi_2) = \frac{1}{\sqrt{2}} \{\psi_p(\xi_1)\, \chi_{\xi_1}(t_z = -\tfrac{1}{2})\, \psi_n(\xi_2)\, \chi_{\xi_2}(t_z = \tfrac{1}{2})$$
$$- \psi_p(\xi_2)\, \chi_{\xi_2}(t_z = -\tfrac{1}{2})\, \psi_n(\xi_1) \chi_{\xi_1}(t_z = \tfrac{1}{2})\}.$$

Alternatively,

$$\psi(\xi_1, \xi_2) = \tfrac{1}{2}[\psi_p(\xi_1)\, \psi_n(\xi_2) + \psi_p(\xi_2)\, \psi_n(\xi_1)]\, \chi\, (T = 0, T_z = 0)$$
$$+ \tfrac{1}{2}[\psi_p(\xi_1)\, \psi_n(\xi_2) - \psi_p(\xi_2)\, \psi_n(\xi_1)\, \chi\, (T = 1, T_z = 0).$$

Thus, if the $\psi_p(\xi)$ and $\psi_n(\xi)$ wave functions are different, the $T = 1$, $T_z = 0$ state will admix into the $T = 0$, $T_z = 0$ state.

The Coulomb interaction of the protons

$$\sum_{i<i'}^{A} e^2 \frac{(1-\tau_z^{(i)})(1-\tau_z^{(i')})}{4r_{ii'}}$$

can be expressed as

$$\sum_{i<i'}^{A} \frac{e^2}{4r_{ii'}} \left\{ \left(1 + \frac{1}{3} \boldsymbol{\tau}^{(i)} \cdot \boldsymbol{\tau}^{(i')}\right) - (\tau_z^{(i)} + \tau_z^{(i')}) + \left(\tau_z^{(i)} \cdot \tau_z^{(i')} - \frac{1}{3} \boldsymbol{\tau}^{(i)} \boldsymbol{\tau}^{(i')}\right) \right\}. \tag{1.45}$$

The first term (in round brackets) is an isobaric scalar; it does not disturb the T-invariance. The second term has selection rules $\Delta T = 0, \pm 1$ (except $0 \to 0$), and the last term has selection rules $\Delta T = 0, \pm 1, \pm 2$ (except $0 \to 0$, $0 \to 1$). Thus the second (vector) and third (tensor) terms in (1.45) lead to the T-noninvariance.

Another source of the isobaric admixtures is the neutron–proton mass difference $m_n - m_p = 782$ keV. The part of the total Hamiltonian, containing nucleon masses and kinetic energies, is

$$T'_{\text{tot}} = \sum_{i=1}^{A} \frac{1}{2}\{m_n(1+\tau_z^{(i)})+m_p(1-\tau_z^{(i)})\}+\sum_{i=1}^{A}\frac{1}{4}\left\{\frac{1}{m_n}(1+\tau_z^{(i)})+\frac{1}{m_p}(1-\tau_z^{(i)})\right\}\Delta_i,$$

where Δ_i is the Laplace operator of the ith particle. The T'_{tot} can be transformed to

$$T'_{\text{tot}} = A\frac{m_p+m_n}{2}+(m_n-m_p)T_z-\frac{m_n+m_p}{2m_nm_p}\sum_{i=1}^{A}\Delta_i+\frac{m_n-m_p}{4m_nm_p}\sum_{i=1}^{A}\tau_z^{(i)}\Delta_i. \quad (1.46)$$

The first three terms of (1.46) commute with T despite the T_z dependence of the second term (causing the T_z splitting). The last term causes violation of the T-symmetry. It is small and usually neglected.

The effects which break the isobaric symmetry are small in the very light nuclei; they can be treated by the perturbation method. The Coulomb forces in heavy nuclei are not negligible. Their main part, however, does not violate isobaric symmetry. Consequently, effects leading to the admixture of the T_0+1 state into the $T_0 = T_z$ state are basically small. Besides, they are weakened by the factor

$$(T_0, T_0, 1, 0\,|\,T_0+1, T_0) = \frac{1}{T_0+1}, \quad (1.47)$$

which enters the corresponding matrix element (see ref. 37). The problems of isobaric symmetry are treated in a number of review articles, e.g., in ref. 38.

3. The quantum numbers T—isobaric spin and its projection T_z—are widely used for the classification of nuclear states in the nuclear shell model. The concept is particularly fruitful in the description of the light nuclei.

The ground state of nucleus with filled neutron and proton shells has $T = T_z = T_0 = \frac{1}{2}(N-Z)$. The states, having only one particle or one hole above the closed shells, have $T = T_0+\frac{1}{2}$ for a neutron particle or proton hole and $T = T_0-\frac{1}{2}$ for a proton particle or neutron hole. Naturally, a hole and a particle have opposite isobaric spins.

The advanced shell model is more complicated than the extreme single-particle model. The filled shells are treated as an inert core, while the interaction of the remaining particles is taken into account. To characterize the state of the nucleons in an unfilled shell, it is insufficient to use only the total angular momentum I, parity π, and isobaric spin T. New quantum numbers, the seniority $\mathfrak{s}$ and the reduced isospin t, are therefore introduced. Properties of the new quantum numbers are intimately related to the symmetry of the system. According to the Pauli principle, only two nucleons with different isospin projection can occupy the (N, j, l, m) state. A nucleon pair can be in either space–symmetric, isospin–antisymmetric $T = 0$ state, or in a space–antisymmetric, isospin–symmetric $T = 1$ state. Some of the $T = 1$ pairs might have $I^\pi = 0^+$. The seniority $\mathfrak{s}$ is equal to the number of particles which do not participate in the $I^\pi = 0^+$, $T = 1$ pairs. The reduced isospin t is the isospin of the unpaired particles.

The minimal seniority for even nuclei is zero; it corresponds to the ground state of the even–even nuclei. The minimal seniority for the odd-A nuclei is 1; all nucleons, except one, participate then in pairs. In such a case, the nuclear spin and parity are determined by the odd nucleon.

The many-particle shell model often uses the quantum numbers $\mathfrak{S}$ and t. The wave functions are expressed as a mixture of different configurations. Such a model is usually applied to the light nuclei and close to the magic nuclei (see, for example, ref. 19).

§ 4. Electromagnetic Transitions in the Single-particle Model

The properties of electromagnetic radiation are well known; therefore the electromagnetic radiation is often used as a source of nuclear structure information. In nuclear reactions and in α- and β-decays, the nuclei are usually left in excited states. If the excitation energy is less than the nucleon separation energy, the nuclear de-excitation is caused by the electromagnetic interaction (through γ-radiation, internal conversion electrons, or electron–positron pair formation). If the excitation energy is larger than the nucleon separation energy, the nucleon emission and electromagnetic transitions compete. The γ-emission prevails when the nucleon emission is inhibited. The nuclear photoreactions are an important tool for the nuclear structure studies; they are not, however, treated in this monograph.

The theory of nuclear electromagnetic transitions is explained in detail in a number of textbooks, e.g. in refs. 18 and 39. We shall, therefore, give only the basic equations here.

The probability of the electromagnetic transition from the initial state i to the final state f is

$$W_{fi} = 2\pi \sum \int |\langle f| - \int (d\boldsymbol{r}) \boldsymbol{A}\cdot\boldsymbol{j} | i\rangle|^2 \varrho(E_f), \tag{1.48}$$

where $\varrho(E_f) = (p^2/8\pi^3)\, d\Omega$ is the density of final states, $|\boldsymbol{p}| \equiv p = E_i - E_f$ is the wave vector length, $\boldsymbol{j}$ is the current density operator, and $\boldsymbol{A}$ is the vector potential of the electromagnetic field. The gauge is determined by the condition div $\boldsymbol{A} = 0$. The summation in the righthand side of (1.48) includes all polarization states of the photon and the magnetic quantum numbers of the final state; the integration is over all directions of radiation.

The vector potential $\boldsymbol{A}$ contains two parts: one corresponding to the emission and another to the absorption of γ-radiation. If the emission part is decomposed into corresponding multipoles, it is expressed as

$$\boldsymbol{A}_\delta(\boldsymbol{r}) = \sqrt{\frac{2\pi^3}{p}} \sum_{\lambda=1}^{\infty} \sum_{\mu=-\lambda}^{\lambda} i^\lambda (2\lambda+1)\, D^\lambda_{\mu\delta} \{\boldsymbol{A}_M(\lambda\mu) + i\delta \boldsymbol{A}_E(\lambda\mu)\}, \tag{1.49}$$

where $\delta = 1$ and -1 correspond, respectively, to the left and right circular polarization. The rotational $D^\lambda_{\mu\delta}$ functions are described, for example, in refs. 18, 19, and 27. The quantity $\boldsymbol{A}_M(\lambda\mu)$ is the vector potential of the magnetic radiation of multipolarity λ; the $\boldsymbol{A}_E(\lambda\mu)$ is the vector potential of the electric radiation of multipolarity λ. It is

$$\boldsymbol{A}_M(\lambda\mu) = \sqrt{\frac{2}{\pi}}\, j_\lambda(pr) \frac{-i\boldsymbol{r}\times\nabla}{\sqrt{\lambda(\lambda+1)}} Y_{\lambda\mu}, \tag{1.50}$$

$$\boldsymbol{A}_E(\lambda\mu) = \frac{1}{p}\sqrt{\frac{2}{\pi}} \frac{\text{rot}\,(-i\boldsymbol{r}\times\nabla)}{\sqrt{\lambda(\lambda+1)}} j_\lambda(pr) Y_{\lambda\mu}, \tag{1.50'}$$

where $j_\lambda(pr)$ are the spherical Bessel functions. The photons of multipolarity λ, corresponding to the electric and magnetic radiation, have different parity selection rules.

Let the initial and final states be states with definite angular momentum and parity. Let us find the selection rules for the electromagnetic transitions connecting them. The conservation of the total angular momentum gives a simple rule:

$$|I_i - I_f| \leq \lambda \leq I_i + I_f. \tag{1.51}$$

To obtain the parity selection rules we have to remember that the radiation is described by the vector field, which changes sign under the inversion of coordinates. The photons of multipolarity λ are, as was already mentioned, of two types:

(a) magnetic radiation with parity $\pi_M = (-1)^{\lambda+1}$; $l = \lambda$;
(b) electric radiation with parity $\pi_E = (-1)^\lambda$; $l = \lambda \pm 1$.

From the parity conservation, it follows that $\pi_i \pi_f \pi_\gamma = 1$; thus, the parity selection rules are

$$\pi_i \pi_f = (-1)^\lambda \quad \text{for the electric radiation,} \tag{1.52}$$

$$\pi_i \pi_f = (-1)^{\lambda+1} \quad \text{for the magnetic radiation.} \tag{1.52'}$$

The electric multipole transitions, corresponding to the dipole, quadrupole, octupole, etc., are denoted as $E1$, $E2$, $E3$, etc.; the magnetic transitions are denoted as $M1$, $M2$, $M3$, etc.

2. The probability of the electric $E\lambda$-transition, integrated over all directions of the radiation, is

$$W_{fi}(E\lambda) = 4\pi^2 e^2 p \sum_\mu \left| \langle f | \frac{1}{e} \int (d\boldsymbol{r}) \boldsymbol{A}_E(\lambda\mu) \cdot \boldsymbol{j} \, | i \rangle \right|^2. \tag{1.53}$$

The current density operator depends on the nuclear model used. Assume that each nucleon gives an independent contribution to the nuclear current density and to the charge density operators. The charge density operator is then equal to

$$\varrho = \frac{e}{2} \sum_i (1 - \tau_z^{(i)}) \, \delta(\boldsymbol{r} - \boldsymbol{r}_i). \tag{1.54}$$

The current density operator consists of two parts. The first part corresponds to the motion of the charges in the nucleus; the second part corresponds to the magnetic moments of nucleons. It is

$$\boldsymbol{j} = \left\{ \frac{e}{2} \sum_i (1 - \tau_z^{(i)}) \, \delta(\boldsymbol{r} - \boldsymbol{r}_i) \frac{\boldsymbol{p}_i}{m} \right\} + \left\{ \frac{e}{4} (\mu_n + \mu_p) \frac{1}{m} \sum_i \delta(\boldsymbol{r} - \boldsymbol{r}_i) \boldsymbol{\sigma}_i \times \nabla \right.$$
$$\left. - \frac{e}{4} \frac{1}{m} (\mu_p - \mu_n) \sum_i \delta(\boldsymbol{r} - \boldsymbol{r}_i) \tau_z^{(i)} \boldsymbol{\sigma}_i \times \nabla \right\}. \tag{1.55}$$

Here μ_p and μ_n are proton and neutron magnetic moments in nuclear magnetons ($\mu_0 = e/2m$), $\boldsymbol{\sigma}$ is the usual Pauli spin vector, and τ_z is twice the z-component of isospin. The charge and current operators contain parts without isospin operators (isoscalars) and parts linear in τ_z (isovectors). The isobaric spin selection rules for electromagnetic transitions are, therefore,

$$\Delta T = 0, \pm 1. \tag{1.56}$$

The argument of the spherical Bessel functions in (1.50) are always smaller than (or equal to) pR, while

$$(pR)^2 \approx \frac{E^2\,(\text{MeV})}{2\times 10^4} A^{2/3}.$$

So, when the transition energy is smaller than 10 MeV, the long wavelength approximation

$$(pR)^2 \ll 1 \tag{1.57}$$

is valid. In such a case, the spherical Bessel functions can be replaced by powers of pr. Then, using the continuity equation, it is possible to obtain simple expressions for the probability of the electric or magnetic radiation.

The probability of the electric $E\lambda$ radiation in the long wavelength limit is equal to

$$W_{fi}(E\lambda) = 8\pi \frac{\lambda+1}{\lambda[(2\lambda+1)!!]^2} p^{2\lambda+1} B(E\lambda), \tag{1.58}$$

with the reduced probability of the electric transition defined as

$$B(E\lambda) = \sum_{\mu, m_f} \left| \langle f | \int (dr)\, \varrho r^\lambda Y_{\lambda\mu}(\theta, \varphi) \, | i \rangle \right|^2. \tag{1.59}$$

The quantity $B(E\lambda)$ is independent of the transition energy. From (1.59) it follows that the reduced probability of electric radiation depends, in the long wavelength approximation, only on the electric charge density. Therefore, $B(E\lambda)$ can be expressed through the static electric multipole operator

$$Q_{\lambda\mu} = \frac{4}{e} \sqrt{\frac{\pi}{2\lambda+1}} \int (dr)\, \varrho(r)\, r^\lambda Y_{\lambda\mu}(\theta, \varphi) \tag{1.60}$$

as

$$B(E\lambda) = e^2 \sum_{\mu, m_f} \left| \langle f | \sqrt{\frac{2\lambda+1}{16\pi}}\, Q_{\lambda\mu} \, | i \rangle \right|^2. \tag{1.59'}$$

Thus, the probability of the nuclear electric multipole transition is proportional to the squared nondiagonal matrix element of the electric multipole moment operator. The energy dependence of the transition probability is given by the $p^{2\lambda+1}$ factor in (1.58). This means that the transition probability is proportional to the $(2\lambda+1)$th power of the transition energy. The diagonal matrix elements of the operator (1.60) determine the static nuclear electric moments.

The probability of the magnetic radiation in the long wavelength limit is equal to

$$W_{fi}(M\lambda) = 8\pi \frac{\lambda+1}{\lambda[(2\lambda+1)!!]^2} p^{2\lambda+1} B(M\lambda), \tag{1.61}$$

with the reduced probability of the magnetic radiation defined as

$$B(M\lambda) = \sum_{\mu, m_f} \left| \langle f | \frac{-i}{(\lambda+1)} \int r^\lambda \mathbf{r} \times \nabla Y_{\lambda\mu} \mathbf{j}(dr) \, | i \rangle \right|^2. \tag{1.62}$$

The probability of electric radiation $E\lambda$ is about c^2/v^2 larger than the probability of the magnetic $M\lambda$ radiation.

The reduced electric and magnetic transition probabilities depend sensitively on the nuclear wave functions. Therefore, when experimentally determined, they give very valuable nuclear structure information.

3. Consider the electric transitions corresponding to a change of the state of individual nucleons in nuclei. The reduced transition probability is then equal to

$$B(E\lambda) = \sum_{\mu, m_f} \left| \langle f | \sum_{i'=1}^{A} e \frac{1-\tau_z^{(i')}}{2} r_{i'}^{\lambda} Y_{\lambda\mu}(\theta_{i'}, \varphi_{i'}) | i \rangle \right|^2, \tag{1.59''}$$

where i' denotes all nuclear nucleons and r_i' is the distance from the nuclear center of mass. Assume that the nucleus consists of the even–even core, which does not participate in the transition, and one unpaired nucleon. The wave function of the odd nucleon is, according to § 2,

$$\Psi_{nljm} = \frac{1}{r} u_{nl_j}(r) \sum_{m_l, m_s} \left(lm_l \frac{1}{2} m_s | jm \right) i^l Y_{lm_l}(\theta, \varphi) \chi_{m_s}.$$

To account for the relative motion of the nucleon (mass m, charge $(1-\tau_z)/2$, radius vector $\boldsymbol{r}_1$), and of the core (mass $m(A-1)$, charge $Z-(1-\tau_z)/2$, radius vector $\boldsymbol{r}_0$), we shall introduce the concept of the effective charge. Remember that the center of mass radius vector is $\boldsymbol{R}_0 = (1/A)[\boldsymbol{r}_1 + \boldsymbol{r}_0(A-1)]$. Then

$$\sum_{i'} \left(\frac{1-\tau_z^{(i')}}{2} \right) \boldsymbol{r}_{i'}^{\lambda} = \frac{1-\tau_z}{2} (\boldsymbol{r}_1 - \boldsymbol{R}_0)^{\lambda} + \left(Z - \frac{1-\tau_z}{2} \right) (\boldsymbol{r}_0 - \boldsymbol{R}_0)^{\lambda} = e_{\text{eff}}^{(\lambda)} \boldsymbol{r}^{\lambda}, \tag{1.63}$$

where $\boldsymbol{r} = \boldsymbol{r}_1 - \boldsymbol{r}_0$ is the vector determining the position of the odd nucleon with respect to the core. The effective charge is then defined by

$$e_{\text{eff}}^{(\lambda)} = \frac{1-\tau_z}{2} \left(\frac{A-1}{A} \right)^{\lambda} + \left(Z - \frac{1-\tau_z}{2} \right) \left(\frac{-1}{A} \right)^{\lambda}. \tag{1.64}$$

So, the effective charge for the $E1$ transitions is $e_{\text{eff}}^{(1)} = N/A$ for protons and $e_{\text{eff}}^{(1)} = -Z/A$ for neutrons; for the $E2$ transitions, $e_{\text{eff}}^{(2)} = 1-(A+N)/A^2$ for protons, and $e_{\text{eff}}^{(2)} = Z/A^2$ for neutrons. For transitions of higher multipolarity ($\lambda > 2$) the recoil effect is small and often neglected; thus $e_{\text{eff}}^{(\lambda)} \approx 1$ for protons and $e_{\text{eff}}^{(\lambda)} \approx 0$ for neutrons. Another source of the effective charge, the core polarization, is often more important. This effect will be discussed in the following chapters.

The $E\lambda$ transition operator contains isobaric spin operators and has, therefore, definite T-selection rules. In the particular case of the $E1$ transitions, the corresponding operator is proportional to

$$\sum_{i'} \frac{1-\tau_z^{(i')}}{2} \boldsymbol{r}_{i'} = \frac{1}{2} \left(\frac{N-Z}{A} \boldsymbol{r} - \tau_z \boldsymbol{r} \right). \tag{1.65}$$

It is obvious that the first term has nonvanishing matrix elements when $T_i = T_f$, the second term, when $T_f = T_i + 1$. The $T_f = T_i$ transitions are $((N-Z)/A)^2$ times inhibited when compared with the $T_f \pm T_i+1$ transitions. The selection rules are, therefore,

$$\Delta T = 0, \pm 1, \quad \text{if} \quad T_z \neq 0;$$
$$\Delta T = \pm 1, \quad \text{if} \quad T_z = 0 \quad \text{(i.e. if } N = Z).$$

It is often useful to have an estimate of the $B(E\lambda)$ value for a single-particle transition. To obtain it, we shall use (1.59″), wave functions (1.36) as the initial and final states, and the effective charge $e_{\text{eff}}^{(\lambda)}$. The angular variables are excluded by using the usual vector addition formulae. The result is

$$B(E\lambda); j_i \to j_f) = (e_{\text{eff}}^{(\lambda)})^2 \frac{e^2}{4\pi}(2\lambda+1)(j_i \tfrac{1}{2} \lambda 0 \,|\, j_f \tfrac{1}{2})^2 \left| \int dr\, u_{n_f l_f j_f}(r)\, r^\lambda u_{n_i l_i j_i}(r) \right|^2. \quad (1.66)$$

The radial matrix element can be estimated, according to Weisskopf, if we assume that the wave function is constant inside the nuclear volume and equal to zero outside. The radial matrix element is then equal to $3R_0^\lambda(\lambda+3)^{-1}$, where R_0 is the nuclear radius. The Clebsch–Gordan coefficient in (1.66) is close to unity and can be left out. So, finally,

$$B(E\lambda)_{\text{s.p.}} = \frac{2\lambda+1}{4\pi}\left(\frac{3}{3+\lambda}\right)^2 R_0^{2\lambda} e^2 \quad (1.67)$$

is the single-particle unit of the reduced $E\lambda$ transition probability.[15, 39] For $\lambda = 2$ and 3

$$B(E2)_{\text{s.p.}} = 3e^2 A^{4/3}\, 10^{-53}\ \text{cm}^4, \quad (1.67')$$

$$B(E3)_{\text{s.p.}} = 4.2 e^2 A^2\, 10^{-79}\ \text{cm}^6. \quad (1.67'')$$

The notation $B(E\lambda)_{\text{s.p.u.}}$ will be used below when the reduced transition probabilities are expressed in the single-particle units.

The probability of the $E\lambda$ transition in the single-particle units is

$$W(E\lambda) = (e_{\text{eff}}^{(\lambda)})^2 e^2 18 \frac{(\lambda+1)(2\lambda+1)}{\lambda[(2\lambda+1)!!]^2} (j_i \tfrac{1}{2}\lambda 0 \,|\, j_f \tfrac{1}{2})^2 p^{2\lambda+1} \frac{R_0^{2\lambda}}{(\lambda+3)^2}. \quad (1.68)$$

This expression differs from the Weisskopf unit

$$W_W(E\lambda) = 18 e^2 \frac{\lambda+1}{\lambda[(2\lambda+1)!!]^2} p^{2\lambda+1} \frac{R_0^{2\lambda}}{(\lambda+3)^2} \quad (1.68')$$

only by a multiplicative factor.

4. Before discussing magnetic transitions, let us derive an equation for the magnetic moment of an odd-A nucleus in the single-particle shell model.

The nuclear magnetic moment operator has two parts: one corresponds to the current, caused by the charge motion, and the other is caused by the intrinsic magnetic moment of the nucleons:

$$\mu = \sum_{i=1}^{A} (g_i^{(l)} l + g_i^{(s)} s_i)\mu_0. \quad (1.69)$$

The $g_i^{(l)}$ is the gyromagnetic ratio of the orbital motion, and $g_i^{(s)}$ is the spin gyromagnetic ratio. The g factors have the following values for free nucleons $(l): g^{(l)} = 1$, (0) for protons (neutrons), $g^{(s)} = 5.5856$ (-3.8263) for protons (neutrons).

The nuclear magnetic moment is determined as the average value of the z-component of μ in the $j_z = m = j$ state. The magnetic moment is expressed in nuclear magnetons μ_0:

$$\mu = gj = \int (dr)\psi^*_{j,\,m=j}\mu_z\psi_{j,\,m=j}. \tag{1.70}$$

The magnetic moment in the extreme single-particle model is determined by the odd nucleon; the nucleons in pairs obviously do not contribute. To obtain an explicit formula for μ, we shall use a formula valid for any vector $\boldsymbol{V}$:

$$\langle j, m\,|\boldsymbol{V}|\,j, m\rangle = \langle j, m\,|\boldsymbol{j}|\,j, m\rangle \frac{\langle j, m\,|\boldsymbol{j}\cdot\boldsymbol{V}|\,j, m\rangle}{\langle j, m\,|\boldsymbol{j}^2|\,j, m\rangle}.$$

In the $m = j$ case the z-component of $\boldsymbol{V}$ equals

$$\langle j, m\,|V_z|\,j, m\rangle = (j+1)^{-1}\langle j, m=j\,|\boldsymbol{j}\cdot\boldsymbol{V}|\,j, m=j\rangle.$$

Using μ instead of $\boldsymbol{V}$

$$\mu = \langle j, m=j\,|\mu_z|\,j, m=j\rangle = (j+1)^{-1}\langle j, m=j\,|\boldsymbol{\mu}\cdot\boldsymbol{j}|\,j, m=j\rangle.$$

and the relations

$$2\boldsymbol{j}\cdot\boldsymbol{l} = \boldsymbol{j}^2+\boldsymbol{l}^2-\boldsymbol{s}^2, \quad 2\boldsymbol{j}\cdot\boldsymbol{s} = \boldsymbol{j}^2+\boldsymbol{s}^2-\boldsymbol{l}^2$$

a final formula for μ is obtained:

$$\begin{aligned} \mu &= (j-\tfrac{1}{2})g^{(l)}+\tfrac{1}{2}g^{(s)} && \text{for} \quad j = l+\tfrac{1}{2}, \\ \mu &= \frac{j}{j+1}\left\{\left(j+\frac{3}{2}\right)g^{(l)}-\frac{1}{2}g^{(s)}\right\} && \text{for} \quad j = l-\tfrac{1}{2}. \end{aligned} \tag{1.71}$$

The above expressions can be further simplified. For the odd-N nucleus $\mu_n = -1.91\mu_0$, and

$$\begin{aligned} \mu &= \mu_n && \text{for} \quad j = l+\tfrac{1}{2}, \\ \mu &= -\frac{j}{j+1}\mu_n && \text{for} \quad j = l-\tfrac{1}{2}, \end{aligned} \tag{1.72}$$

for the odd-Z nucleus $\mu_p = 2.79\mu_0$ and

$$\begin{aligned} \mu &= (j-\tfrac{1}{2})+\mu_p && \text{for} \quad j = l+\tfrac{1}{2}, \\ \mu &= \frac{j}{j+1}\left\{\left(j+\frac{3}{2}\right)-\mu_p\right\} && \text{for} \quad j = l-\tfrac{1}{2}, \end{aligned} \tag{1.73}$$

The values of the magnetic moments (1.72) and (1.73) are usually called Schmidt values; curves in the μ versus j plot are called Schmidt lines. The experimental values of the magnetic moments do not agree with (1.72) or (1.73). They are, however, inside the Schmidt lines (the few exceptions are several light nuclei). The residual interaction has a profound effect on the magnetic moments; the extreme single-particle shell model gives, therefore, only a very crude estimate of their value.

The amplitude of the $M1$ transition is determined by the nondiagonal matrix element of the μ-operator. The operator μ is rewritten in the form

$$\mu = \sum_{i=1}^{A}\left\{\frac{1-\tau_z^{(i)}}{2}(g_p s_i + l_i) + \frac{1+\tau_z^{(i)}}{2} g_n s_i\right\} = \frac{1}{2}I + 0.38\sum_{i=1}^{A} s_i - \sum_{i=1}^{A}\tau_z^{(i)}(4.7 s_i + 0.5 l_i).$$

The isospin selection rules $\Delta T = 0, \pm 1$ are easily recognizable in the above expression. The $\Delta T = \pm 1$ transitions are enhanced by a factor of ten when compared with the $\Delta T = 0$ transitions.

The single-particle unit of the reduced transition probability of the magnetic single-proton transitions, according to the Weisskopf approximation, is determined by

$$B(M\lambda)_{\text{s.p.}} = \frac{10e^2}{4\pi m^2}\left(\frac{3}{\lambda+3}\right)^2 R_0^{2\lambda-2}, \tag{1.74}$$

and the corresponding probability of the magnetic multipole radiation is

$$W_{\text{W}}(M\lambda) = 18\frac{10e^2}{m^2}\frac{\lambda+1}{\lambda[(2\lambda+1)!!]^2}p^{2\lambda+1}\frac{R_0^{2\lambda-2}}{(\lambda+3)^2}. \tag{1.75}$$

The Moszkowski units[(40)] $W_{\text{M}}(M\lambda)$ are sometimes used for magnetic transitions; they are related to the Weisskopf units by

$$W_{\text{W}}(M\lambda) = 10\left(\frac{\lambda+2}{\lambda+3}\right)^2\left(\mu_p\lambda - \frac{\lambda}{\lambda+1}\right)^{-2} W_{\text{M}}(M\lambda). \tag{1.76}$$

5. The probability of the electric and magnetic radiation decreases rapidly when the multipolarity increases. This fact follows from the estimates (1.68) and (1.75) and from the long wavelength approximation (1.57). Therefore, only one or two multipoles participate usually in the nuclear electromagnetic transition; namely, the multipoles with the minimal values λ (i.e., $\lambda = |I_f - I_i|$ or $\lambda = |I_f - I_i| + 1$), which satisfy the angular momentum and parity selection rules. The parity selection rules would allow combinations $M\lambda + E(\lambda+1)$ or $E\lambda + M(\lambda+1)$. If the radiation $\lambda = |I_f - I_i|$ is electric, the admixture of the magnetic radiation is usually small and rarely observed. If, however, the $\lambda = |I_f - I_i|$ radiation is magnetic, the admixture of the corresponding $\lambda+1$ electric transition can be quite large. The estimates discussed above give the following ratios of the $M\lambda$ and $E(\lambda+1)$ radiation probabilities:

$$\frac{W_{\text{W}}(M\lambda)}{W_{\text{W}}(E(\lambda+1))} = \frac{10}{m^2p^2R_0^4}\frac{(\lambda+1)^2(\lambda+4)^2(2\lambda+3)^2}{\lambda(\lambda+2)(\lambda+3)^2}. \tag{1.76'}$$

This ratio is, for heavy nuclei and transition energies, 1 MeV in the range 10–100. In a number of cases, however, the measured mixed $M1$ and $E2$ transitions have comparable intensities of both components.

To give some idea of the order of magnitude of the quantities related to the electromagnetic transitions, Table 1.3 gives half-lives of the electric and magnetic γ-transitions for different energies and multipolarities. The numbers in the table were calculated from (1.68′) and (1.75); they correspond to the nucleus $A = 165$, $Z = 67$. The half-lives for a nucleus

$A = 125$ are longer: the change is approximately 20% for the $E1$ transitions, 50% for the $E2$ transitions, and 200% for the $E3$ transitions. On the other hand, the half-lives of the $A = 235$ nucleus are shorter in approximately same ratios.

TABLE 1.3. *Half-lives (in seconds) of the electric and magnetic transitions with different energy and multipolarity*

$E_\gamma = p$ (keV)	Electric transitions					Magnetic transitions			
	$E1$	$E2$	$E3$	$E4$	$E5$	$M1$	$M2$	$M3$	$M4$
10	2×10^{-10}	10^{-1}	10^{7}	10^{15}	10^{25}	2×10^{-8}	3	10^{8}	10^{16}
100	2×10^{-13}	10^{-6}	7	8×10^{7}	10^{15}	2×10^{-11}	3×10^{-5}	90	5×10^{0}
300	8×10^{-15}	4×10^{-9}	3×10^{-3}	5×10^{3}	4×10^{9}	9×10^{-13}	10^{-7}	4×10^{-2}	2×10^{3}
500	2×10^{-15}	3×10^{-10}	10^{-4}	50	2×10^{7}	2×10^{-13}	8×10^{-9}	10^{-3}	3×10^{2}
750	5×10^{-16}	4×10^{-11}	6×10^{-6}	1	2×10^{5}	6×10^{-14}	10^{-9}	7×10^{-5}	7
1000	2×10^{-16}	10^{-11}	7×10^{-7}	8×10^{-2}	10^{4}	2×10^{-14}	3×10^{-10}	9×10^{-6}	5×10^{-1}
1500	6×10^{-17}	10^{-12}	5×10^{-8}	2×10^{-3}	10^{2}	7×10^{-15}	4×10^{-11}	6×10^{-7}	2×10^{-2}
2000	3×10^{-17}	3×10^{-13}	6×10^{-9}	2×10^{-4}	6	3×10^{-15}	9×10^{-12}	8×10^{-8}	10^{-3}

The internal conversion becomes sizable for transitions with energy of 200 keV or less. The corresponding corrections considerably decrease the half-lives of γ-transitions of high multipolarities.

The estimates in Table 1.3 or nomographs in ref. 39 are useful tools for the analysis of experimental data. The real γ-transitions are often faster or slower than the estimates based on the Weisskopf or Moszkowski units. Such deviations are related to particular features of the nuclear states connected by the transitions.

§ 5. Single-particle Model of the Deformed Nucleus

1. The nucleus in the shell model was assumed to have a spherical shape. Therefore, in § 2, particles moved in a spherically symmetric potential. There are, however, convincing arguments (discussed in Chapter 2) that nuclei with the neutron and proton numbers sufficiently far from the magic numbers have nonspherical, axially symmetric ellipsoidal shapes.

The classification of the single-particle states depends on the symmetry of the average potential. The single-particle states in spherical nuclei are characterized by their energy, parity, total angular momentum j, and its projection m; states with different m-values are degenerate, i.e., have the same energy. In deformed, axially symmetric ellipsoidal nuclei, the single-particle states are characterized by their energy and parity and by the projection K of the full angular momentum on the nuclear symmetry axis; the total angular momentum j is not a valid quantum number. In nuclei without axial symmetry both j and K lose their meaning as good quantum numbers.

Suppose that a nucleus has a form of the ellipsoid of revolution in a certain time. If its shape is not changed very fast, the nucleons will move in orbits corresponding to the nonspherical potential. Such a particle motion will act as a feedback and will help to stabilize the nuclear shape. The correlated particle motion could result in a slow change of the orientation of the ellipsoid in space, without changing its form: i.e., in the nuclear rotation. The rotating nucleus is characterized by its kinetic energy and angular momentum. If the rotational frequencies are small when compared to the characteristic frequencies of the internal motion, the two modes of motion can be treated as approximately independent.

In reality, the rotational frequencies of the deformed nuclei are often considerably smaller than the frequencies of the surface vibrations and of the internal motion. Consequently, the adiabatic approximation, which neglects the coupling of the rotational motion to the internal and vibrational motion, can be used. The problem of the single-particle motion is then reduced to a motion in an axially symmetric, quadrupole deformed average field.

Each spherical subshell j is split in the deformed potential: i.e., the $(2j+1)/2$-fold degeneracy over $|m|$ is broken. Each level is, however, still doubly degenerate; states with $\pm m$ have the same energy. This degeneracy is a quite general property of all Hamiltonians invariant against time conjugation.

The time conjugation is an operation which transforms the wave function ψ into $\tilde{\psi}$. Any experiment performed at time t on the ψ gives results exactly identical to those of the same experiment performed at time $-t$ on $\tilde{\psi}$. The time conjugation operator $\mathfrak{T}$ is not a unitary operator. It can be expressed as a product of a unitary operator and either the operator which interchanges the ordering of the operators (Schrödinger's definition) or the operator of the complex conjugation (Schwinger's definition).

The nuclear Hamiltonian is invariant under time conjugation, i.e.,

$$H\mathfrak{T}-\mathfrak{T}H = 0. \tag{1.77}$$

Another important property of the $\mathfrak{T}$-operator is its anticommutation relation with the total angular momentum operator I,

$$\mathfrak{T}\boldsymbol{I}+\boldsymbol{I}\mathfrak{T} = 0. \tag{1.77'}$$

If Ψ^I_M is an eigenstate of H, the $\mathfrak{T}\Psi^I_M=\zeta\Psi^I_{-M}$ is also an eigenstate, corresponding to the same energy. Both states have identical quantum numbers except the sign of M; the ζ is a phase factor. Note that the degeneracy with respect to the sign of M is valid in the nonaxial fields, too.

2. The motion of a particle in an axially symmetric potential (with additional symmetry plane, perpendicular to the symmetry axis) was described by Nilsson.[41]

The potential has the form of the anisotropic harmonic oscillator; it also contains the spin–orbit coupling and another term proportional to l^2, which should flatten the bottom part of the potential, i.e., bring it closer to the square well.

The Hamiltonian is

$$H_{\text{av}} = H_0^{\text{av}}+C_N\boldsymbol{ls}+D_N\boldsymbol{l}^2, \tag{1.78}$$

with

$$H_0^{\text{av}} = -\frac{1}{2m}\Delta'+\frac{m}{2}(\omega_{x'}^2x'^2+\omega_{y'}^2y'^2+\omega_{z'}^2z'^2). \tag{1.78'}$$

The x', y', z' are particle coordinates in the body-fixed coordinate system. The eigenvalues of H_{av} must give, in the special case of the spherical symmetry, the correct sequence of the shell model states. This condition can be fulfilled by an appropriate choice of the parameters C_N and D_N in (1.78).

The deformation parameter δ can be conveniently introduced in the case of the axial symmetry by the relations

$$\left.\begin{aligned} \omega_{x'}^2 &= \omega_{y'}^2 = \omega_0^2(\delta)\left(1+\tfrac{2}{3}\delta\right), \\ \omega_{z'}^2 &= \omega_0^2(\delta)\left(1-\tfrac{4}{3}\delta\right). \end{aligned}\right\} \tag{1.79}$$

The equipotential surfaces have a constant volume if

$$\omega_{x'}\omega_{y'}\omega_{z'} = \text{const}, \tag{1.79'}$$

and, therefore,

$$\omega_0(\delta) = \mathring{\omega}_0\left(1-\tfrac{4}{3}\delta^2-\tfrac{16}{27}\delta^3\right)^{-1/6}. \tag{1.79''}$$

The $\mathring{\omega}_0$ is the $\omega_0(\delta)$ value at $\delta = 0$; according to (1.27) $\mathring{\omega}_0 = 41A^{-1/3}$ MeV. The deformation parameter δ is related to the other used deformation parameter β by

$$\delta \approx 0.95\beta.$$

Let us use dimensionless coordinates, defined by

$$x = \sqrt{m\omega_0}\,x', \quad y = \sqrt{m\omega_0}\,y', \quad z = \sqrt{m\omega_0}\,z',$$

and divide H_0^{av} into the spherical part $\mathring{H}_0$ and H_δ—the part proportional to the deformation:

$$H_0^{av} = \mathring{H}_0 + H_\delta, \tag{1.80}$$

$$\mathring{H}_0 = \frac{\omega_0}{2}(-\Delta + r^2), \tag{1.80'}$$

$$H_\delta = \omega_0\delta\frac{4}{3}\sqrt{\frac{\pi}{5}}\,r^2 Y_{20}. \tag{1.80''}$$

It is advantageous to use a representation, where $\mathring{H}_0$, l^2, l_z, and s_z are diagonal; the corresponding quantum numbers are N (total number of the oscillator quanta), l, Λ, and Σ. The operator $j_z = l_z + s_z$ commutes with the Hamiltonian H_{av}; the corresponding quantum number is $K = \Lambda + \Sigma$. The parity is, naturally, also conserved. Thus each eigenstate of H_{av} is characterized by the quantum numbers K and π. Obviously,

$$\mathring{H}_0\,|Nl\Lambda\Sigma\rangle = \left(N+\tfrac{3}{2}\right)\omega_0\,|Nl\Lambda\Sigma\rangle.$$

The $\boldsymbol{l}\cdot\boldsymbol{s}$ operator has matrix elements $\langle N'l'\Lambda'\Sigma'\,|\boldsymbol{l}\cdot\boldsymbol{s}|\,Nl\Lambda\Sigma\rangle$ with the following selection rules:

$$N = N', \quad l = l', \quad \Lambda = \begin{cases} \Lambda' \\ \Lambda' \pm 1, \end{cases} \quad \Sigma = \begin{cases} \Sigma' \\ \Sigma' \mp 1. \end{cases}$$

The operator $r^2 Y_{20}$ has these selection rules:

$$\Lambda = \Lambda', \quad \Sigma = \Sigma', \quad l = \begin{cases} l' \\ l' \pm 2, \end{cases} \quad N = \begin{cases} N' \\ N' \pm 2. \end{cases}$$

The coupling between the N and $(N\pm 2)$ oscillator shells was neglected in ref. 41; the corresponding states are usually divided by a large energy interval $2\mathring{\omega}_0$. In a number of cases, however, the coupling between the N and $N\pm 2$ shells must be included.

3. Another representation is also often used (see ref. 41). It is based on a different deformation parameter, denoted ε, and its related frequency $\omega(\varepsilon)$. It is

$$\left.\begin{aligned} \omega_{x'} &= \omega_{y'} = \omega_0(\varepsilon)\left(1+\tfrac{1}{3}\varepsilon\right) \\ \omega_{z'} &= \omega_0(\varepsilon)\left(1-\tfrac{2}{3}\varepsilon\right). \end{aligned}\right\} \tag{1.81}$$

The parameters ε and δ are connected by the relations

$$\left.\begin{aligned} \varepsilon &= \delta+\tfrac{1}{6}\delta^2+O(\delta^3), \\ \omega_0(\varepsilon) &= \mathring{\omega}_0[1+\tfrac{1}{9}\varepsilon^2+O(\varepsilon^3)]. \end{aligned}\right\} \tag{1.81'}$$

The transformation to the dimensionless coordinates transforms the original ellipsoid into a sphere in the new coordinate system. The new coordinates are defined by

$$\xi = x'\sqrt{m\omega_{x'}}, \quad \eta = y'\sqrt{m\omega_{y'}}, \quad \zeta = z'\sqrt{m\omega_{z'}}. \tag{1.82}$$

The Hamiltonian is expressed in the new system as

$$H_0^{\text{av}} = H_\xi+H_\eta+H_\zeta, \quad \text{with} \quad H_\xi = \frac{\omega_{x'}}{2}\left(-\frac{\partial^2}{\partial\xi^2}+\xi^2\right),$$

etc. The Hamiltonian H_0^{av} is diagonal in the $|n_\xi\rangle\,|n_\eta\rangle\,|n_\zeta\rangle$ representation, i.e.,

$$H_\xi\,|n_\xi\rangle = (n_\xi+\tfrac{1}{2})\omega_{x'}\,|n_\xi\rangle.$$

The H_0^{av} can be, alternatively, divided into "spherical" and deformed parts as

$$H_0^{\text{av}} = \mathring{H}_0+H_\varepsilon, \tag{1.83}$$

where

$$\mathring{H}_0 = \omega_0(\varepsilon)\,\tfrac{1}{2}(-\Delta+\xi^2+\eta^2+\zeta^2), \tag{1.83'}$$

$$H_\varepsilon = \frac{\varepsilon}{6}\,\omega_0(\varepsilon)\left\{\left(-\frac{\partial^2}{\partial\xi^2}+\xi^2\right)+\left(-\frac{\partial^2}{\partial\eta^2}+\eta^2\right)-2\left(-\frac{\partial^2}{\partial\zeta^2}+\zeta^2\right)\right\}. \tag{1.83''}$$

The angular momentum operators can be defined in the new representation. For example, the operator $\boldsymbol{l}_t$ has components

$$(l_t)_{x'} = -i\left(\eta\frac{\partial}{\partial\zeta}-\zeta\frac{\partial}{\partial\eta}\right), \quad \text{etc.}$$

It is natural to use a basis, where $\mathring{H}_0$, $(\boldsymbol{l}_t)^2$, $(l_t)_z$, $(\boldsymbol{s})^2$, and s_z are diagonal. The base vectors are $|N_t l_t \Lambda_t \Sigma\rangle$, where $l_t(l_t+1)$ and Λ_t are eigenvalues of the operators $\boldsymbol{l}_t^2$ and $(l_t)_z$ respectively. It is then

$$\mathring{H}_0\,|N_t t_t \Lambda_t \Sigma\rangle = \frac{\omega_0}{2}(-\Delta+\xi^2+\eta^2+\zeta^2)\,|N_t t_t \Lambda_t \Sigma\rangle = \left(N_t+\frac{3}{2}\right)\omega_0\,|N_t l_t \Lambda_t \Sigma\rangle.$$

Equation (1.78) now has the form:

$$H_{av} = H_t + H'_t. \tag{1.84}$$

where

$$H_t = \mathring{H}_0 + H_\varepsilon + C_N l_t \cdot s + D_N l_t^2, \tag{1.84'}$$

$$H'_t = C_N (l - l_t) \cdot s + D_N (l^2 - l_t^2). \tag{1.84''}$$

The operator H_t has vanishing matrix elements between states with the different N-values. This means that N is a good quantum number for the operator H_t. The matrix elements of H_t in the $|N_t l_t \Lambda_t \Sigma\rangle$ representation are identical with the matrix elements of H_{av} in the $|Nl\Lambda\Sigma\rangle$ basis; it is necessary, of course, to use ε, $\omega_0(\varepsilon)$ instead of δ, $\omega_0(\delta)$. The H'_t can be treated as a perturbation and expanded into a power series in ε; this term was completely neglected in the early calculations.

Thus, as we have just shown, the interaction between the N and $N+2$ shells can be approximately included when the δ, $\omega_0(\delta)$ parameters are replaced by the ε, $\omega_0(\varepsilon)$.

4. Let us describe some details of the calculations. Instead of the parameters C_N and D_N, the dimensionless quantities μ and $\varkappa$ are used:

$$\varkappa = -\frac{1}{2}\frac{C_N}{\omega_0}, \qquad \mu = \frac{2D_N}{C_N}, \tag{1.85}$$

and a new deformation parameter η, which depends on $\varkappa$, is introduced by

$$\eta = \frac{\delta}{\varkappa}\frac{\omega_0(\delta)}{\mathring{\omega}_0} = \frac{\delta}{\varkappa}\left(1 - \frac{4}{3}\delta^2 - \frac{16}{27}\delta^3\right)^{-1/6} \tag{1.85'}$$

The deformation dependent part of the H_{av} is

$$H_\delta = \delta\omega_0\left(-\frac{4}{3}\right)\sqrt{\frac{\pi}{5}}\, r^2 Y_{20} = \varkappa\mathring{\omega}_0\eta\left(-\frac{4}{3}\sqrt{\frac{\pi}{5}}\right) r^2 Y_{20},$$

and

$$H_{av} - \mathring{H}_0 = \varkappa\mathring{\omega}_0 \mathfrak{U}, \tag{1.86}$$

where

$$\mathfrak{U} = \eta\left(-\frac{4}{3}\sqrt{\frac{\pi}{5}}\right) r^2 Y_{20} - 2 l \cdot s - \mu l^2. \tag{1.86'}$$

The dimensionless matrix $\mathfrak{U}$ is then diagonalized in the selected representation for many values of the deformation parameter η.

The quantum numbers K and π do not characterize the single-particle states fully; the asymptotic quantum numbers can complement them. The quantum numbers l and j are relevant for the shapes with small deformation. The degeneracies of the $\delta = 0$ situation are broken by the deformation effects; for small $\delta > 0$ the states with larger K-values have larger energies. At intermediate deformations the spherical quantum numbers l, j lose their meaning; a state K is a superposition of the spherical states with different l-values. The situation is simplified again when the deformation is large and the $l \cdot s$ and l^2 parts of the

Hamiltonian can be treated as a perturbation. The unperturbed potential is then the anisotropic harmonic oscillator, and the single-particle states are characterized by the quantum numbers N, n_z (the number of the oscillator quanta along z'-axis) and Λ.

Accordingly, each single-particle state is characterized by a set of quantum numbers $K^\pi[Nn_z\Lambda]$. For a given N-shell and for $\delta > 0$, the lowest $K = \frac{1}{2}$ state has $n_z = N$, next $K = \frac{1}{2}$ state has $n_z = N-1$, etc. This rule directly follows from the expression for the energy levels of anisotropic oscillator (neglecting the $\boldsymbol{l}\cdot\boldsymbol{s}$ and $\boldsymbol{l}^2$ terms):

$$E_0 = \omega_0(\varepsilon)\left[\left(N+\frac{3}{2}\right)+\varepsilon\,\frac{n_\xi+n_\eta-2n_\zeta}{3}\right]. \tag{1.87}$$

When the n_z value of a level is found, it is easy to determine the $\Lambda = K\pm\frac{1}{2}$ value. The Λ is even (odd) when $N-n_z$ is even (odd). The K, N, n_z quantum numbers therefore determine the Λ and the Σ numbers completely. Note that the lowest $K = K_1$ state of the N-shell has $n_z = N-K_1+\frac{1}{2}$.

The notation $K^\pi[Nn_z\Lambda]$ is rather cumbersome. Consequently, simplified notations $Nn_z\Lambda\uparrow$ and $Nn_z\Lambda+$ for the $K = \Lambda+\frac{1}{2}$ case, or $Nn_z\Lambda\downarrow$ and $Nn_z\Lambda-$ for the $K = \Lambda-\frac{1}{2}$ are often used.

The quantum numbers n_z and Λ are unique characteristics of the single particle states. When the parameters $\varkappa$ and μ are varied, the spherical ($\varepsilon = 0$) basis is changed, too. If such a modification is accompanied by a change in the sequence of the K^π states for $\varepsilon \neq 0$, the corresponding n_z and Λ are also interchanged. Hence the asymptotic quantum numbers n_z and Λ and the corresponding spherical subshell are uniqely related.

The energy eigenvalues of the total Hamiltonian H_{av} are

$$E(Nn_z\Lambda\pm) = (N+\tfrac{2}{3})\,\omega_0(\delta)+\varkappa\mathring{\omega}_0 r(Nn_z\Lambda\pm), \tag{1.87}$$

where $r(Nn_z\Lambda\pm)$ are eigenvalues of the matrix $\mathfrak{U}$ obtained in its diagonalization. The single-particle eigenfunction is equal to

$$\varphi_K(Nn_z\Lambda\pm) = \sum_{l\Lambda} a_{l\Lambda}\,|Nl\Lambda\Sigma\rangle = \sum_l \left(a_{l,K-\frac{1}{2}}\,|Nl(K-\tfrac{1}{2})+\rangle + a_{l,K+\frac{1}{2}}\,|Nl(K+\tfrac{1}{2})-\rangle\right). \tag{1.88}$$

It is often useful to express the wave functions in the jK representation, i.e., in the basis characterized by the quantum numbers N, l, j, and K. It is then

$$\varphi_K(Nn_z\Lambda\Sigma) = \sum_j d_j\,|NljK\rangle. \tag{1.88'}$$

The coefficients d_j and $a_{l\Lambda}$ are related by

$$d_j = \sum_{\Lambda\Sigma} a_{l\Lambda}\langle Nl\Lambda\Sigma|NljK\rangle. \tag{1.88''}$$

Two constraints are used for the selection of parameters $\varkappa$ and μ. The eigenvalues at $\delta = 0$ should correctly reproduce the shell model level sequence, and, for the nonzero equilibrium δ values, the ground states of the odd-A deformed nuclei should be correctly predicted. To fulfill these conditions it is necessary to allow variation of μ from one shell to another and to introduce separate shifts of the individual spherical subshells.

Let us see how the functions φ_K are transformed under the time conjugation. The phase is chosen according to ref. 26. The set of quantum numbers is denoted by $(q\sigma)$; $\sigma = \pm 1$ denotes the states, conjugated under time reversal. Equation (1.88′) for $\sigma = +1$ now reads

$$\varphi_K(q) = \sum_j d_j^q |NljK\rangle \equiv |q+\rangle. \qquad (1.88''')$$

Applying the time conjugation operator,

$$\mathfrak{T}|q+\rangle = |q-\rangle = (-1)^{K+\frac{1}{2}}\varphi_{-K}(q), \qquad (1.89)$$

where

$$\varphi_{-K}(q) = \sum_j d_j^q(-1)^{l+j-\frac{1}{2}}|Nlj-K\rangle. \qquad (1.89')$$

Application of $\mathfrak{T}$ on the $|q-\rangle$ function gives

$$\mathfrak{T}|q-\rangle = (-1)^{-K+\frac{1}{2}}(-1)^{K+\frac{1}{2}}\varphi_K(q) = (-1)\varphi_K(q) = (-1)|q+\rangle. \qquad (1.89'')$$

The single-particle wave function is not, therefore, an eigenfunction of $\mathfrak{T}$. Operator $\mathfrak{T}$ transforms $|q+\rangle$ into $|q-\rangle$ and $|q-\rangle$ into $-|q+\rangle$; operator $\mathfrak{T}^2$ multiplies the wave function by a factor of -1.

To demonstrate how the structure of a wave function depends on the deformation, the neutron state $\frac{7}{2}^-$ [714] is shown below at different δ-values. For $\delta = 0$, i.e., in the spherical case, the wave function corresponds to the $f_{7/2}$ $(m = \frac{7}{2})$ state.

For $\delta = 0.1$:

$$\varphi_{7/2}(714\downarrow) = -0.567|733+\rangle + 0.781|754-\rangle - 0.121|774-\rangle - 0.223|753+\rangle + 0.058|773+\rangle.$$

For $\delta = 0.2$:

$$\varphi_{7/2}(714\downarrow) = -0.348|733+\rangle + 0.882|754-\rangle - 0.216|774-\rangle - 0.216|753+\rangle + 0.092|773+\rangle.$$

For $\delta = 0.5$:

$$\varphi_{7/2}(714\downarrow) = -0.205|733+\rangle + 0.904|754-\rangle - 0.332|774-\rangle - 0.140|753+\rangle + 0.100|773+\rangle.$$

The structure of the state becomes more complex for the small deformations, but later, when the asymptotic region is reached, it is simplified again. At larger deformations, considerable simplification is achieved if the quantum number Λ is used.

A modification of the Nilsson potential was suggested in ref. 42. A supplementary term, proportional to the average value of the operator $\boldsymbol{l}^2$, $\langle \boldsymbol{l}^2\rangle = N(N+3)/2$ (averaging over all states within the N shell), was introduced there. The Hamiltonian H_{av} is now

$$H_{av} = H_0^{av} + C_N \boldsymbol{l}\cdot\boldsymbol{s} + D_N\{\boldsymbol{l}^2 - \langle \boldsymbol{l}^2\rangle\}. \qquad (1.90)$$

The level scheme for the actinide nuclei was constructed with only one pair of the $\varkappa$- and μ-values, and the number of shifts and parameters was reduced in the $150 \leqslant A \leqslant 190$ region, when the Hamiltonian (1.90) was applied.

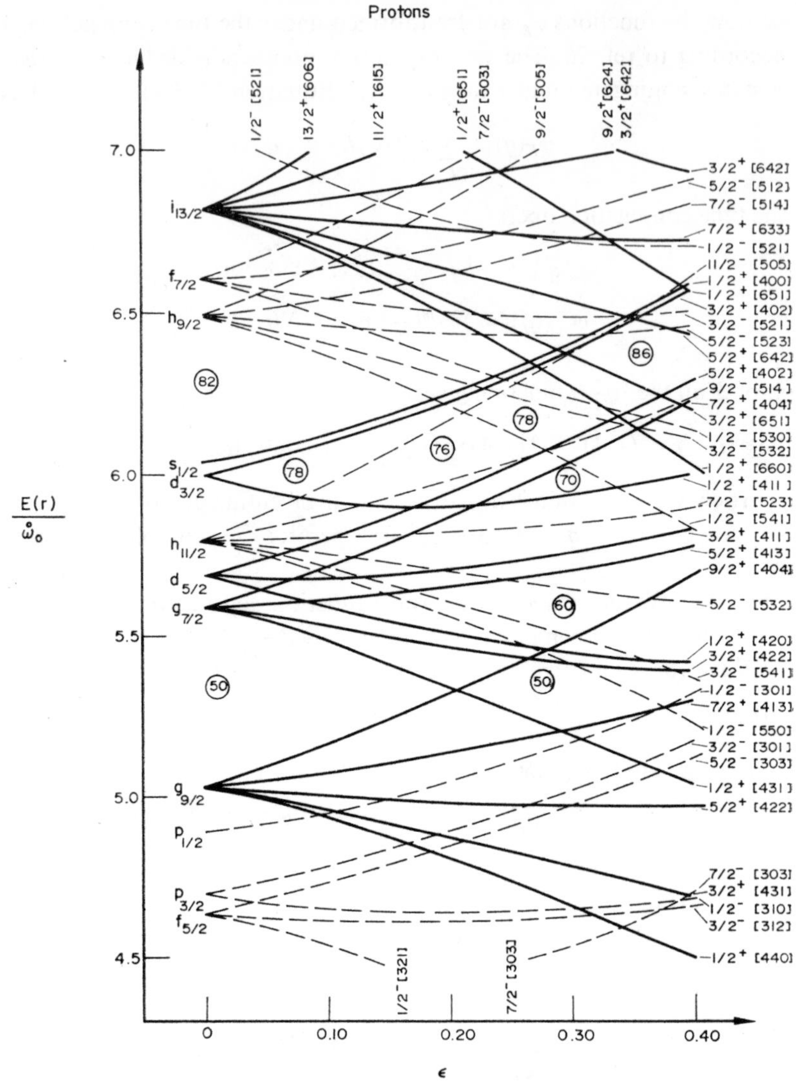

FIG. 1.5. Single-proton levels of the Nilsson potential for $44 < Z < 88$.

Refs. 41 and 43 show single-particle level schemes of the Nilsson potential and give tables of corresponding $a_{l\Lambda}$ values. Modified results, obtained with adjusted parameters and with additional shifts, are listed in refs. 44–47.

Figures 1.5–1.8 show the single-particle level schemes. The parameters, given in Table 1.4, were chosen in such a way that the experimentally found ground state $K^{\pi}[Nn_z\Lambda]$ characteristics of the odd-A deformed nuclei are reproduced correctly. Each state is characterized by the value K of the angular momentum projection on the nuclear symmetry axis, by parity π, and by the asymptotic quantum numbers $[Nn_z\Lambda]$. The encircled numbers of the neutrons or protons show gaps in the level spacings. The energy is in units of $\mathring{\omega}_0 = 41/A^{1/3}$ MeV.

Several spherical subshells are not shown in the figures. Their positions (at $\varepsilon = 0$) are:

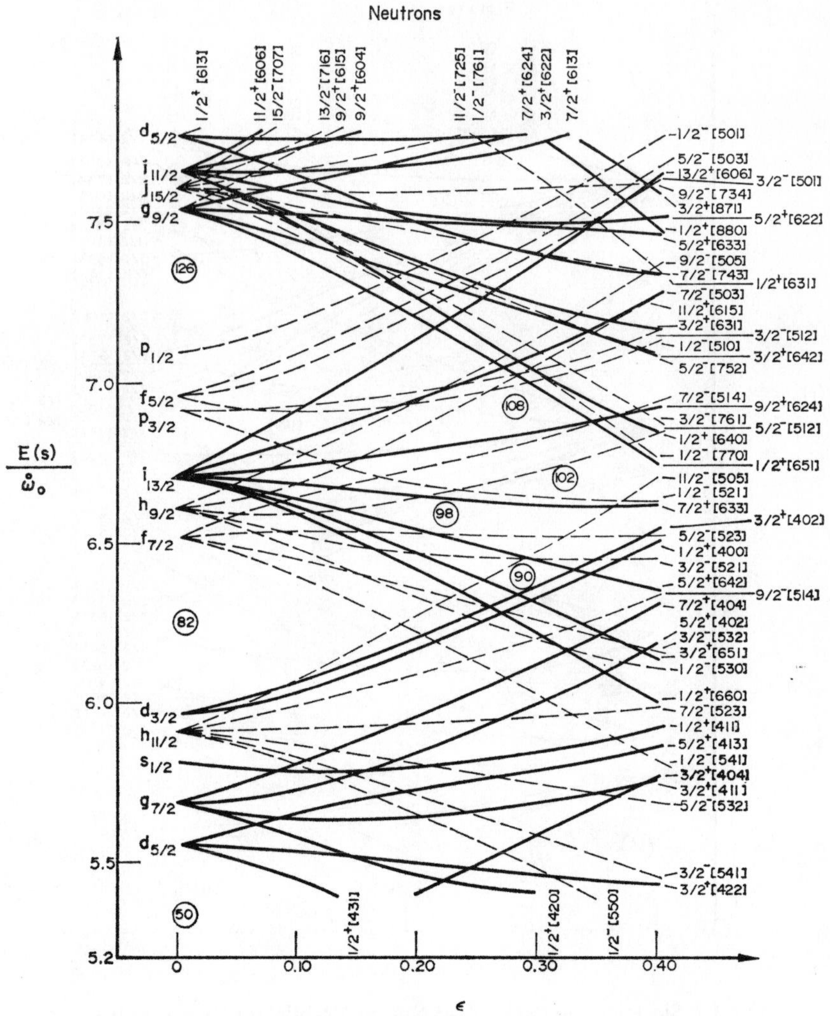

FIG. 1.6. Single-neutron levels of the Nilsson potential for $58 < N < 136$.

in Fig. 1.5, proton states with $Z < 28$: $f_{5/2}$, $-4.2\mathring{\omega}_0$; $d_{3/2}$, $-3.7\mathring{\omega}_0$; $s_{1/2}$, $-3.5\mathring{\omega}_0$; and $d_{5/2}$, $3.37\mathring{\omega}_0$. In Fig. 1.6, neutron states with $N < 50$: $g_{9/2}$, $-5.1\mathring{\omega}_0$; $p_{1/2}$, $-4.78\mathring{\omega}_0$; $f_{5/2}$, $-4.7\mathring{\omega}_0$; $p_{3/2}$, $-4.6\mathring{\omega}_0$; $f_{7/2}$, $-4.24\mathring{\omega}_0$.

5. The reduced transition probabilities $B(E\lambda)$ and $B(M\lambda)$ of the electric and magnetic transitions between single-particle states were calculated in ref. 41 using the wave functions (1.88).

The reduced transition probabilities of the electromagnetic transitions are defined by (1.59) and (1.62). They are determined by the equation

$$B(\lambda, I_i \to I_f) = \frac{1}{2I_i+1} \sum_{\mu, M_i, M_f} |\langle I_f M_f | \mathfrak{M}(\lambda\mu) | I_i M_i \rangle|^2. \tag{1.91}$$

The multipole operator $\mathfrak{M}(\lambda\mu)$ must be transformed into the body-fixed coordinate system

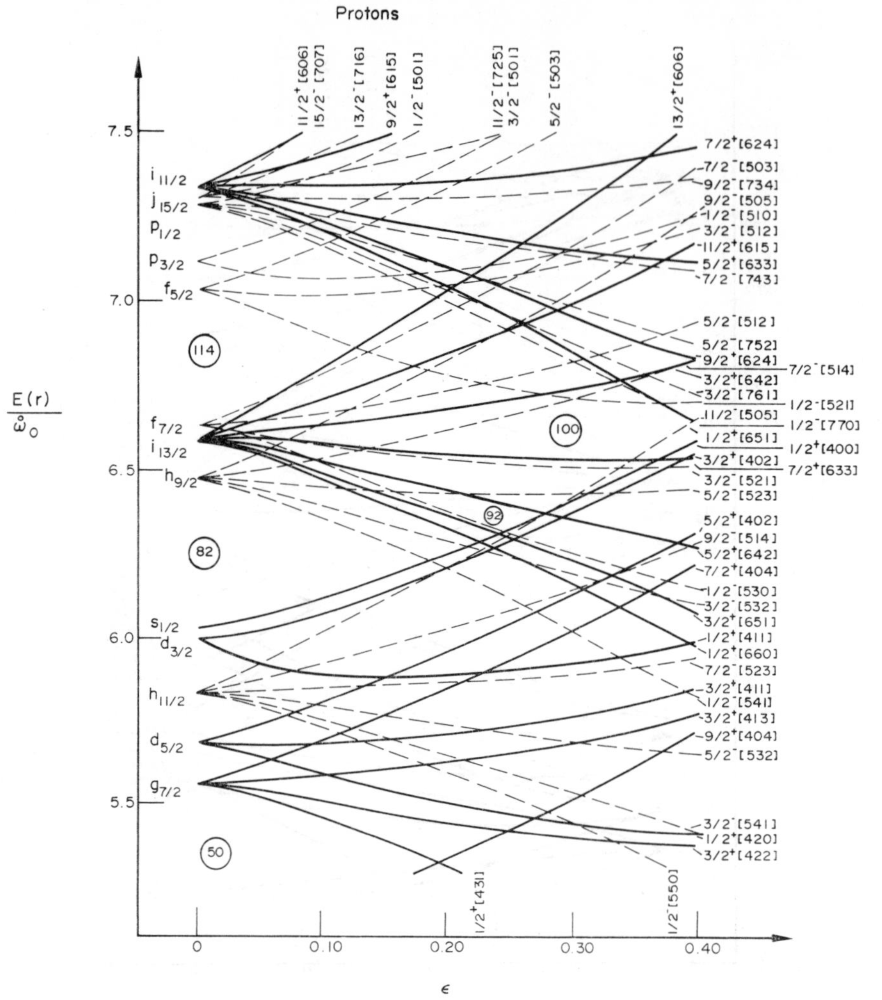

FIG. 1.7. Single-proton levels of the Nilsson potential for $90 < Z < 102$.

(x', y', z'). This is accomplished by

$$\mathfrak{M}(\lambda\mu) = \sum_{\mu'} D^{\lambda}_{\mu\mu'}(\theta_i)\, \mathfrak{M}'(\lambda\mu'), \tag{1.91'}$$

where $D^{\lambda}_{\mu\mu'}(\theta_i)$ are the rotational D-functions.

The following expressions are obtained for the reduced transition probabilities of the transitions between the single particle $I_i K_i^{\pi_i}[N_i n_{zi} \Lambda_i]$ and $I_f K_f^{\pi_f}[N_f n_{zf} \Lambda_f]$ states:

$$B(E\lambda) = \frac{2\lambda+1}{4\pi} \times \frac{e^2}{(m\omega_0)^{\lambda}} \left[1+(-1)^{\lambda}\frac{Z}{A^{\lambda}}\right]^2 |(I_i K_i \lambda K_f - K_i \,|\, I_f K_f)$$
$$+ b_{E\lambda}(-1)^{I_f+K}\ (I_i K_i \lambda - K_f - K_i \,|\, I_f - K_f)|^2\, G^2_{E\lambda}, \tag{1.92}$$

$$B(M\lambda) = \frac{2\lambda+1}{16\pi m^2} \times \frac{e^2}{(m\omega_0)^{\lambda-1}} |(I_i K_i \lambda - K_f - K_i \,|\, I_f K_f)$$
$$+ b_{M\lambda}(-1)^{I_f+K_f}\, (I_i K_i \lambda - K_f - K_i \,|\, I_f - K_f)\,|^2\, G^2_{M\lambda}. \tag{1.93}$$

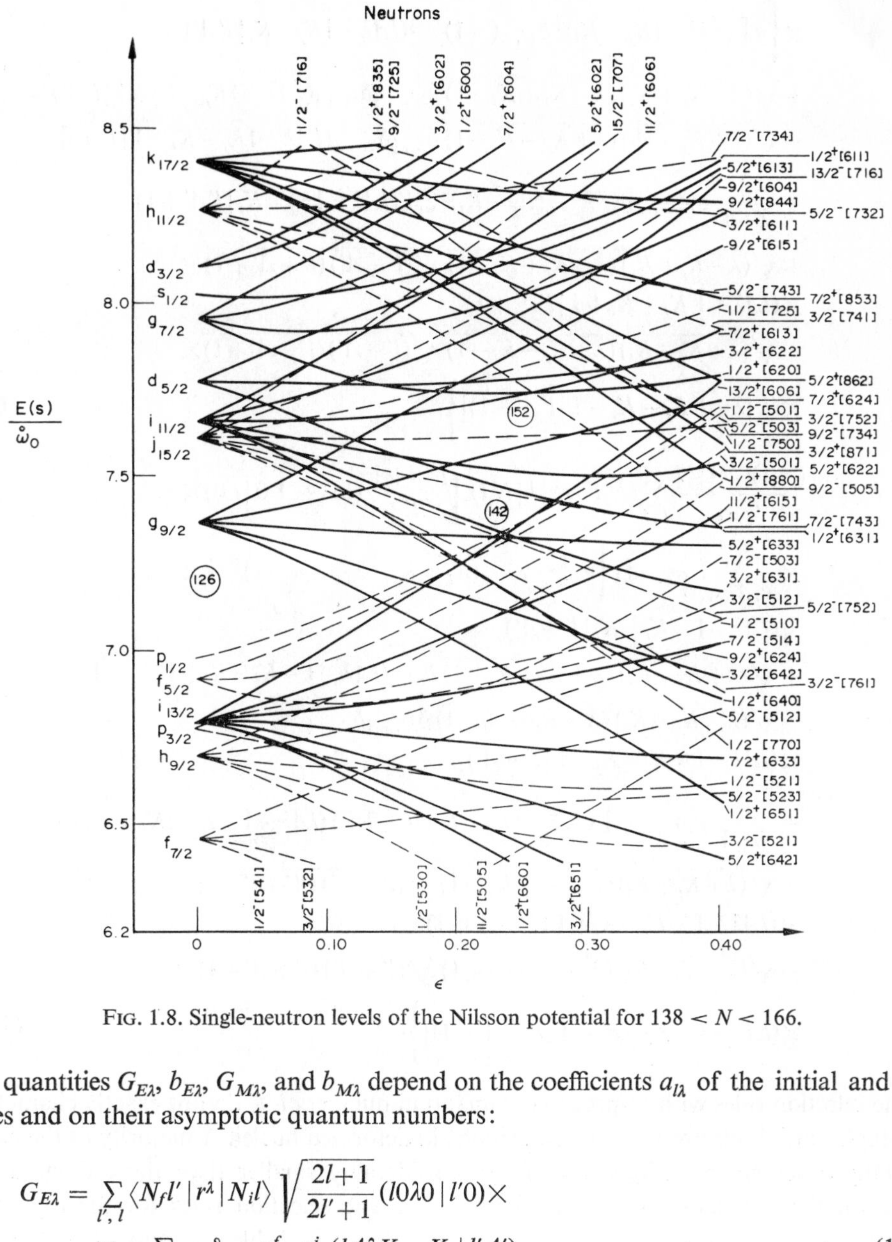

FIG. 1.8. Single-neutron levels of the Nilsson potential for $138 < N < 166$.

The quantities $G_{E\lambda}$, $b_{E\lambda}$, $G_{M\lambda}$, and $b_{M\lambda}$ depend on the coefficients $a_{l\Lambda}$ of the initial and final states and on their asymptotic quantum numbers:

$$G_{E\lambda} = \sum_{l',\, l} \langle N_f l' \,|\, r^\lambda \,|\, N_i l\rangle \sqrt{\frac{2l+1}{2l'+1}}\,(l0\lambda 0\,|\,l'0)\times$$
$$\times \sum_{\Lambda',\, \Lambda,\, \Sigma',\, \Sigma} \delta_{\Sigma',\,\Sigma}\, a^f_{l'\Lambda'} a^i_{l\Lambda} (l\Lambda\lambda K_f - K_i \,|\, l'\Lambda'), \tag{1.92'}$$

$$b_{E\lambda} = \frac{(-1)^{K_f+N_f+\frac{1}{2}}}{G_{E\lambda}} \sum_{l'l} \langle N_f l' \,|\, r^\lambda N_i l\rangle \sqrt{\frac{2l+1}{2l'+1}}\,(l0\lambda 0\,|\,l'0)\times$$
$$\times \sum_{\Lambda',\, \Lambda,\, \Sigma',\, \Sigma} \delta_{\Sigma,\, -\Sigma'}\, a^f_{l'\Lambda'} a^i_{l\Lambda} (l\Lambda\lambda - K_f - K_i \,|\, l' - \Lambda'), \tag{1.92''}$$

$$G_{M\lambda} = \sum_{l'l} \langle N_f l' \,|\, r^{\lambda-1} \,|\, N_i l\rangle (l0\lambda-1\,0\,|\,l'0) \sqrt{\frac{2l+1}{2l'+1}} \sum_{\Lambda',\, \Lambda,\, \Sigma',\, \Sigma} a^f_{l'\Lambda'} a^i_{l\Lambda}\times$$

$$\times\Big\{g_s[\sqrt{\lambda^2-(K_f-K_i)^2}\,\delta_{\Sigma',\Sigma}(-1)^{\Sigma-\frac{1}{2}}(l\Lambda\lambda-1K_f-K_i\,|\,l'\Lambda')$$
$$+\sqrt{(\lambda-K_f+K_i)(\lambda-K_f+K_i-1)}\,\delta_{\Sigma',-\frac{1}{2}}\delta_{\Sigma,+\frac{1}{2}}(l\Lambda\lambda-1K_f-K_i+1\,|\,l'\Lambda')$$
$$-\sqrt{(\lambda+K_f-K_i)(\lambda+K_f-K_i-1)}\,\delta_{\Sigma',\frac{1}{2}}\delta_{\Sigma,-\frac{1}{2}}(l\Lambda\lambda-1K_f-K_i-1\,|\,l'\Lambda')]$$
$$+\frac{2}{\lambda+1}g_l\delta_{\Sigma',\Sigma}[\sqrt{\lambda^2-(K_f-K_i)^2}\,2\Lambda'(l\Lambda\lambda-1K_f-K_i\,|\,l'\Lambda'+1)$$
$$+\sqrt{(\lambda-K_f+K_i)(\lambda-K_f+K_i-1)}\sqrt{(l'-\Lambda')(l'+\Lambda'+1)}\times$$
$$\times(l\Lambda\lambda-1\,K_f+K_i+1\,|\,l'\Lambda'+1)$$
$$-\sqrt{(\lambda+K_f-K_i)(\lambda+K_f-K_i-1)}\sqrt{(l'+\Lambda')(l'-\Lambda'+1)}\times$$
$$\times(l\Lambda\lambda-1\,K_f-K_i-1\,|\,l'\Lambda'-1)]\Big\}, \tag{1.93'}$$

$$b_{M\lambda}=\frac{(-1)^{K'+\frac{1}{2}+N_f}}{G_{M\lambda}}\sum_{l',l}\langle N_fl'\,|\,r^{\lambda-1}\,|\,N_il\rangle\sqrt{\frac{2l+1}{2l'+1}}(l0\lambda-1\,0\,|\,l'0)\times$$
$$\times\sum_{\Lambda',\Lambda,\Sigma',\Sigma}a^f_{l'\Lambda'}a^i_{l\Lambda}\Big\{g_s[\sqrt{\lambda^2-(K_f+K_i)^2}\,\delta_{\Sigma,-\Sigma'}(-1)^{\Sigma+\frac{1}{2}}\times$$
$$\times(l\Lambda\lambda-1-K_f-K_i\,|\,l'-\Lambda')$$
$$+\sqrt{(\lambda+K_f+K_i)(\lambda+K_f+K_i-1)}\,\delta_{\Sigma',\frac{1}{2}}(l\Lambda\lambda-1-K_f-K_i+1\,|\,l'-\Lambda')$$
$$-\sqrt{(\lambda-K_f-K_i)(\lambda-K_f-K_i-1)}\,\delta_{\Sigma',-\frac{1}{2}}\delta_{\Sigma,-\frac{1}{2}}\times$$
$$\times(l\Lambda\lambda-1-K_f-K_i-1\,|\,l'-\Lambda')]$$
$$+\frac{2}{\lambda+1}g_l\delta_{\Sigma,-\Sigma'}[\sqrt{\lambda^2-(K_f+K_i)^2}(-2\Lambda')(l\Lambda\lambda-1-K_f-K_i\,|\,l-\Lambda')$$
$$+\sqrt{(\lambda+K_f+K_i)(\lambda+K_f+K_i-1)}\sqrt{(l'+\Lambda')(l'-\Lambda'+1)}\times$$
$$\times(l\Lambda\lambda-1-K_f-K_i+1\,|\,l'-\Lambda'+1)$$
$$-\sqrt{(\lambda-K_f-K_i)(\lambda-K_f-K_i-1)}\sqrt{(l'-\Lambda')(l'+\Lambda'+1)}\times$$
$$\times(l\Lambda\lambda-1-K_f-K_i-1\,|\,l'-\Lambda'-1)]\Big\}. \tag{1.93''}$$

The selection rules with respect to the quantum numbers $I^{\pi}K$ do not exactly characterize the single-particle electromagnetic transitions in deformed nuclei. A majority of the experimentally observed transitions have rates considerably smaller than the estimated rates based on the Weisskopf–Moszkowski formulae. The selection rules for the asymptotic quantum numbers, formulated in ref. 43 and shown in Tables 1.5 and 1.6, help one to understand the situation.

The asymptotic quantum numbers are not exact; therefore, the corresponding selection rules are valid only approximately. Their violation does not mean that the transition is completely forbidden; its probability is, however, much smaller. For each unit of forbiddenness the transition probability is decreased on the average a hundred times. For transitions which are allowed according to the asymptotic selection rules, eqns. (1.92) and (1.93) give values close to the Weisskopf–Moszkowski estimates.

TABLE 1.4. *Parameters of the Nilsson potential*

Neutrons					Protons			
N	$\varkappa$	μ	Subshell shifts in $\mathring{\omega}_0$	Region of nuclei	N	$\varkappa$	μ	Region of nuclei
3	0.0637	0.35			2	0.0637	0	
4	0.0637	0.35	$d_{3/2}+0.10$		3	0.0637	0.60	
5	0.0637	0.42		$58 \leqslant N \leqslant 136$	4	0.0637	0.60	$44 \leqslant Z \leqslant 88$
6[(a)]	0.0660	0.30	$g_{9/2}+0.18$		5	0.0637	0.60	
7	0.0660	0.30			6	0.0637	0.30	
4	0.0660	0.30			3	0.0577	0.65	
5	0.0660	0.30			4	0.0577	0.65	
6	0.0660	0.30		$138 \leqslant N \leqslant 156$	5	0.0577	0.65	$90 \leqslant Z \leqslant 102$
7	0.0660	0.30			6	0.0577	0.65	
8	0.0660	0.30			7	0.0577	0.65	

[(a)] $\varkappa = 0.0637$ and $\mu = 0.42$ used for the $i_{13/2}$ subshell.

TABLE 1.5. *Selection rules for asymptotic quantum numbers; electric transitions in deformed nuclei*

Multipole	ΔK	ΔN	Δn_z	$\Delta\Lambda$
$E1$	1	± 1	0	0
	0	± 1	± 1	0
	2	$0, \pm 2$	0	2
$E2$	1	$0, \pm 2$	± 1	1
	0	$0, \pm 2$	$0, \pm 2$	0
	3	$\pm 1, \pm 3$	0	3
	2	$\pm 1, \pm 3$	± 1	2
$E3$	1	$\pm 1, \pm 3$	$0, \pm 2$	1
	0	$\pm 1, \pm 3$	$\pm 1, \pm 3$	0

TABLE 1.6. *Selection rules for asymptotic quantum numbers: magnetic transitions in deformed nuclei*

Multipole	ΔK	ΔN	Δn_z	$\Delta\Lambda$
$M1$	1	$0, \pm 2$	$0, \pm 1$	0, 1
	0	0	0	0
	2	$\pm 1, \pm 3$	$0, \pm 1$	1, 2
$M2$	1	$\pm 1, \pm 3$	$0, \pm 1, \pm 2$	0, 1
	0	$\pm 1, \pm 3$	$0, \pm 1$	$0, \pm 1$
	3	$0, \pm 2, \pm 4$	$0, \pm 1$	2, 3
$M3$	2	$0, \pm 2$	$0, \pm 1, \pm 2$	1, 2
	1	$0, \pm 2, \pm 4$	$0, \pm 1, \pm 2, \pm 3$	0, 1, 2
	0	$0, \pm 2, \pm 4$	$0, \pm 1, \pm 2$	$0, \pm 1$

6. The nonspherical nuclei have the shape of an ellipsoid of revolution. It is, however, possible that some transitional nuclei (transitional between the spherical nuclei and deformed nuclei) have shapes of a triaxial ellipsoid. It is also possible that the shapes of excited states differ from the ground-state shapes and that some excited states have triaxial ellipsoidal forms. It is, therefore, interesting to follow(ref. 48) and find the eigenvalues and eigenfunctions of the potential [(1.78) and (1.78')] for $\omega_{x'} \neq \omega_{y'} \neq \omega_{z'}$.

The nonaxial shape is characterized by two parameters: β' and γ. For $\beta > 0$ the $\gamma = 0°$ situation corresponds to the axially symmetric prolate ellipsoid, the $\gamma = 60°$ corresponds to the oblate ellipsoid. When $\gamma \neq 0°$ and $\gamma \neq 60°$ the ellipsoid has no axial symmetry; the projection K on any axis is not a conserved quantity. The frequencies $\omega_{x'}, \omega_{y'}, \omega_{z'}$ are connected with the deformation parameters β' and γ by

$$\left.\begin{aligned} \omega_{x'} &= \frac{\omega_0(\beta')}{1-\sqrt{5/4\pi}\;\beta'\cos(\gamma+\pi/3)}, \\ \omega_{y'} &= \frac{\omega_0(\beta')}{1-\sqrt{5/4\pi}\;\beta'\cos(\gamma-\pi/3)}, \\ \omega^{z'} &= \frac{\omega_0(\beta')}{1+\sqrt{5/4\pi}\,\beta'\cos\lambda}, \end{aligned}\right\} \tag{1.94}$$

where

$$\omega_0(\beta') = \mathring{\omega}_0[(1-\tfrac{1}{2}\sqrt{5/4\pi}\beta')^2(1+\sqrt{5/4\pi}\beta')]^{1/3} \tag{1.94'}$$

and

$$\omega_{x'}\omega_{y'}\omega_{z'} = \mathring{\omega}_0^3.$$

Newton(48) made the transformation (1.82), chose the appropriate basis, and performed the corresponding diagonalization. His potential at $\gamma = 0°$ coincides with the Nilsson potential. Eigenvalues and eigenfunctions of the Hamiltonian (1.90), using the parameters of Table 1.4, were tabulated in ref. 49 for γ between 0° and 60° and β' between 0 and 0.3. At $\gamma = 0°$ the corresponding level schemes, naturally, coincide with those in Fig. 1.5–1.8.

How do the energy levels and wave functions depend on the parameter γ? To answer this question, we shall fix one value of $\beta = \beta' > 0$ and change γ. Suppose that the Oz' axis is the symmetry axis at $\gamma = 0°$ and that then $K = K_1$. When γ is changed between 0° and 30°, the corresponding single-particle state will contain, besides the prevailing $K = K_1$ component, components corresponding to the other K-values. If the γ is changed further, the oblate shape with Oy', ($Oy' \perp Oz'$) as the symmetry axis is reached at $\gamma = 60°$. The K is again a good quantum number; here, however, it has a different value K_2, which is prevailing in the whole 30–60° interval of the parameter γ. Such behavior is demonstrated in Table 1.7, which shows the K-components of the lowest state originating in $i_{13/2}$ subshell for $\beta = 0.20$ and different γ values. The corresponding state at $\gamma = 0°$ is $\frac{1}{2}^+[660]$; the symmetry axis is Oz'. At $\gamma = 60°$ the corresponding state is $\frac{13}{2}^+[606]$ and the symmetry axis is Oy'.

Figure 1.9 gives the proton level scheme for nuclei with $150 < A < 190$, calculated for $\beta' = 0.30$, $\varkappa = 0.0637$, and $\mu = 0.60$. The figure shows how the corresponding energies depend on the deformation parameter γ. The states corresponding to $\gamma = 0°$ are shown to the left, the states corresponding to $\gamma = 60°$ are shown to the right.

The levels of the same parity and with the same N-values do not cross—they can approach each other and diverge again. The point of closest approach is called the quasi-

TABLE 1.7. *Contributions of the components with different K-values to the wave function of the lowest state originating in the $i_{13/2}$ subshell. (Calculations performed at $\beta = 0.2$.)*

γ in degrees	*K*-value $\frac{1}{2}$	$\frac{3}{2}$	$\frac{5}{2}$	$\frac{7}{2}$	$\frac{9}{2}$	$\frac{11}{2}$	$\frac{13}{2}$	Relative to the axis
0	1.0	0	0	0	0	0	0	
10	0.733	0.223	0.040	0.003	2×10^{-4}	4×10^{-6}	4×10^{-8}	
20	0.600	0.300	0.085	0.014	0.001	6×10^{-5}	10^{-6}	*Oz'*
30	0.532	0.320	0.118	0.027	0.003	3×10^{-4}	8×10^{-6}	
30	10^{-4}	8×10^{-6}	2×10^{-3}	7×10^{-7}	0.056	2×10^{-7}	0.942	
40	10^{-5}	5×10^{-8}	5×10^{-4}	8×10^{-9}	0.025	8×10^{-10}	0.975	*Oy'*
50	10^{-7}	6×10^{-10}	3×10^{-5}	4×10^{-12}	0.006	4×10^{-12}	0.994	
60	0	0	0	0	0	0	1.0	

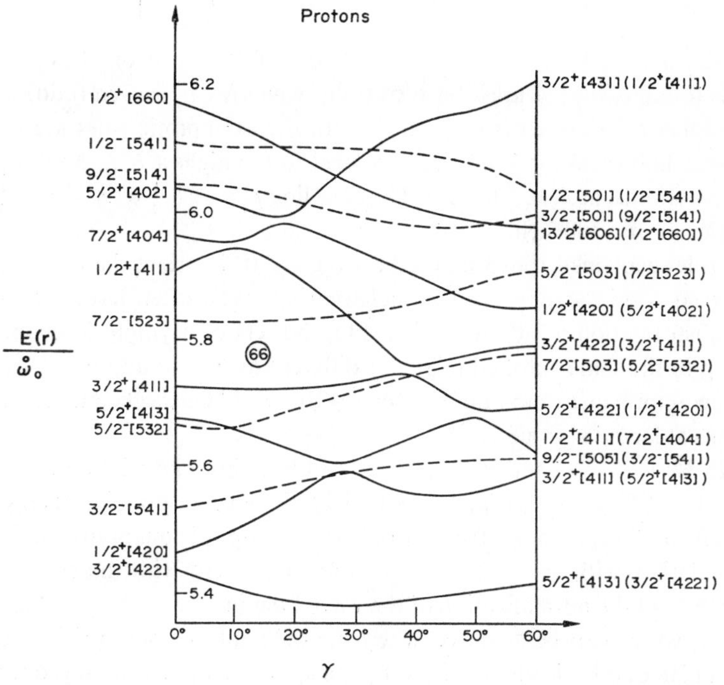

FIG. 1.9. Single proton states of the Nilsson–Newton potential for $150 < A < 190$. The dependence on the γ-parameter (in degrees) for $\beta = 0.3$ is shown. The asymptotic quantum numbers $K^\pi[Nn_z\Lambda]$ corresponding to $\gamma = 0°$ and $\gamma = 60°$ are shown to the left and right, respectively. The quantum characteristics in round brackets to the right give the corresponding level at $\gamma = 0°$ as explained in the text.

crossing point. In a narrow region around the quasicrossing point, the two states are strongly mixed; further away from it, the levels behave as if they actually crossed each other, and the mixing is minimal. So, when moving along a continuous line from $\gamma = 0°$ to $\gamma = 60°$ the different sections have different structures. Let us, as an example, follow the sections of different curves corresponding to the $\frac{7}{2}^+[404]$ state. The first quasicrossing point is at

$\gamma = 10°$. The state resembling $\frac{7}{2}^+[404]$ follows the lower curve, while the continuous curve after $\gamma = 10°$ resembles the $\frac{1}{2}^+[411]$ state. The next quasicrossing point (with the curve starting as $\frac{3}{2}^+[411]$ at $\gamma = 0°$) is at $\gamma = 40°$. Finally, the last quasi-crossing point is at $\gamma = 50°$, this time with the curve starting as $\frac{5}{2}^+[422]$ at $\gamma = 60°$. Thus, the state which starts as $\frac{7}{2}^+[404]$ at $\gamma = 0°$ goes through three quasicrossing points and is changed into the $\frac{1}{2}^+[411]$ state at $\gamma = 60°$. To realize these changes more clearly, the corresponding starting states are shown in parentheses to the right in Fig. 1.9. Figure 1.9 shows that the energies of many states depend only weakly on the γ-parameter. There are, however, states which change by more than 1 MeV when the γ is changed between 0° and 60°.

It is somewhat more difficult to find the asymptotic quantum numbers for the $\varepsilon < 0$ case, than it is in the $\varepsilon > 0$ case. Each subshell is split in such a way that the state with $K = \frac{1}{2}$ is highest, followed by the $K = \frac{3}{2}$ state, etc. When the $\varepsilon = -|\varepsilon|$ is large in the absolute value, the $\boldsymbol{l}\cdot\boldsymbol{s}$ and $\boldsymbol{l}^2$ terms can be neglected, and the asymptotic relation

$$E(q) = \omega_0(\varepsilon)\left[(N+\tfrac{3}{2})-|\varepsilon|\frac{n_\xi+n_\eta-2n_\zeta}{3}\right]$$

is valid. Obviously, the lowest $K = \frac{1}{2}$ state has $n_z = 0$, the next one has $n_z = 1$, etc. The $\Lambda = K\pm\frac{1}{2}$ is found easily; Λ must be even (odd) when $N-n_z$ is even (odd).

When the spherical subshell is split and $\varepsilon > 0$, the asymptotic rules are simple: Λ is increased by one unit from the lowest $K = \frac{1}{2}$ level to the highest K_{max} level, n_z is decreased from its maximal value at $K = \frac{1}{2}$ to zero. The coupling $K = \Lambda+\frac{1}{2}$ or $K = \Lambda-\frac{1}{2}$ is then conserved in all levels of a subshell.

When the rules explained above are used for the $\varepsilon < 0$ situation, the $K = \Lambda\pm\frac{1}{2}$ coupling is not conserved within a subshell. Two neighboring levels often have the same Λ-value. When the transformation $\gamma = 0° \rightarrow 60°$ is made, the sets of asymptotic quantum numbers corresponding to the same subshell might be different at $\gamma = 0°$ and $\gamma = 60°$. The correctness of the asymptotic quantum numbers for $\varepsilon < 0$ should be checked by calculation of the wave functions at large $|\varepsilon|$ values.

Nuclei with $34 < Z < 82$ and $50 < N < 82$ can have both positive and negative equilibrium deformations. Therefore, Fig. 1.10 and 1.11 show corresponding level diagrams calculated for deformations ε from -0.4 to $+0.4$. The asymptotic quantum numbers were tested by a calculation with $\varepsilon = -0.8$, where the component with the asymptotic Λ-value gives at least 90% of the normalization of the wave function.

7. Accuracy, which can be achieved when the different characteristics of the deformed nuclei are calculated, is basically restricted by the approximations employed for the description of the single-particle states, i.e., by the application of the Nilsson potential. This problem will be discussed in the following chapters, where the residual interaction is included.

Hopefully, the accuracy can be improved if finite depth, finite surface thickness potentials are used. A similar problem was encountered for spherical nuclei and was solved by the application of the Woods–Saxon potential. The first calculations of its nonspherical modification were done in ref. 50. Numerical integration of the corresponding differential equations was used in ref. 51; the single-particle states in the deformed Woods–Saxon potential were calculated also in refs. 52–57. The $150 < A < 190$ interval was divided into four sections and the $226 < A < 258$ interval into three sections; the corresponding single-particle

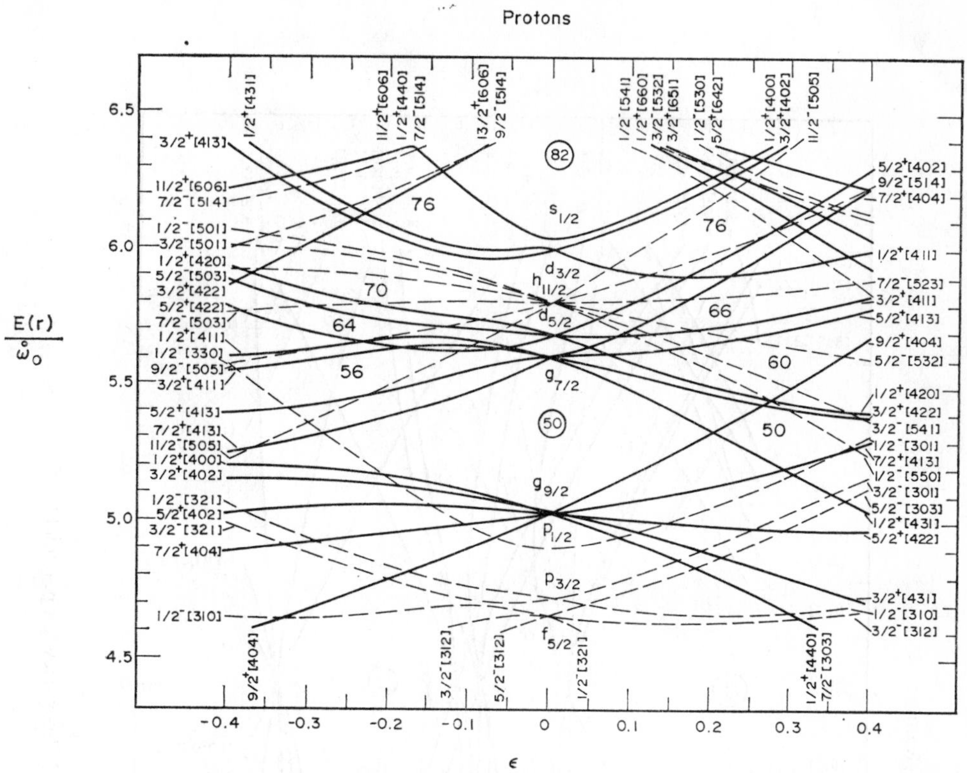

FIG. 1.10. Single-proton states of the Nilsson–Newton potential for $34 < Z < 82$.

energies and wave functions were calculated in refs. 55 and 56 using the adjusted values of the parameters.

The nonspherical Woods–Saxon potential has two terms

$$V(r, \beta, \theta) = \frac{-V_0^{N,Z}}{1+\exp\{(1/a)[r-R_0(1+\beta Y_{20}(\theta))]\}}, \tag{1.95}$$

$$V_{ls}(r, \beta, \theta) = 2\zeta(\boldsymbol{p}\times\boldsymbol{s})\,\mathrm{grad}\, V(r, \beta, \theta), \tag{1.96}$$

where $\boldsymbol{p}$ is the nucleon momentum, β is the deformation parameter, and the other symbols have the same meaning as in (1.31) and (1.32). The spin–orbit coupling usually includes the nuclear nonsphericity and finite surface thickness.

The Coulomb potential

$$V_c(r, \beta, \theta) = \frac{3(Z-1)e^2}{4\pi R_0^3}\int \frac{(d\boldsymbol{r}')}{|\boldsymbol{r}-\boldsymbol{r}'|}\left\{1+\exp\left[\frac{r'-R_0(1+\beta Y_{20}(\theta'))}{a}\right]\right\}^{-1} \tag{1.97}$$

is added to (1.95) and to (1.96) when the single-proton states are calculated.

The solution of the Schrödinger equation with potentials (1.95), (1.96), and (1.97) may be expressed as superpositions

$$\Phi = \sum_{n,l,j,K} a_{nlj}^K \varphi_{nlj}^K, \tag{1.98}$$

where φ_{nlj}^K are the eigenfunctions of the spherical Woods–Saxon potential.

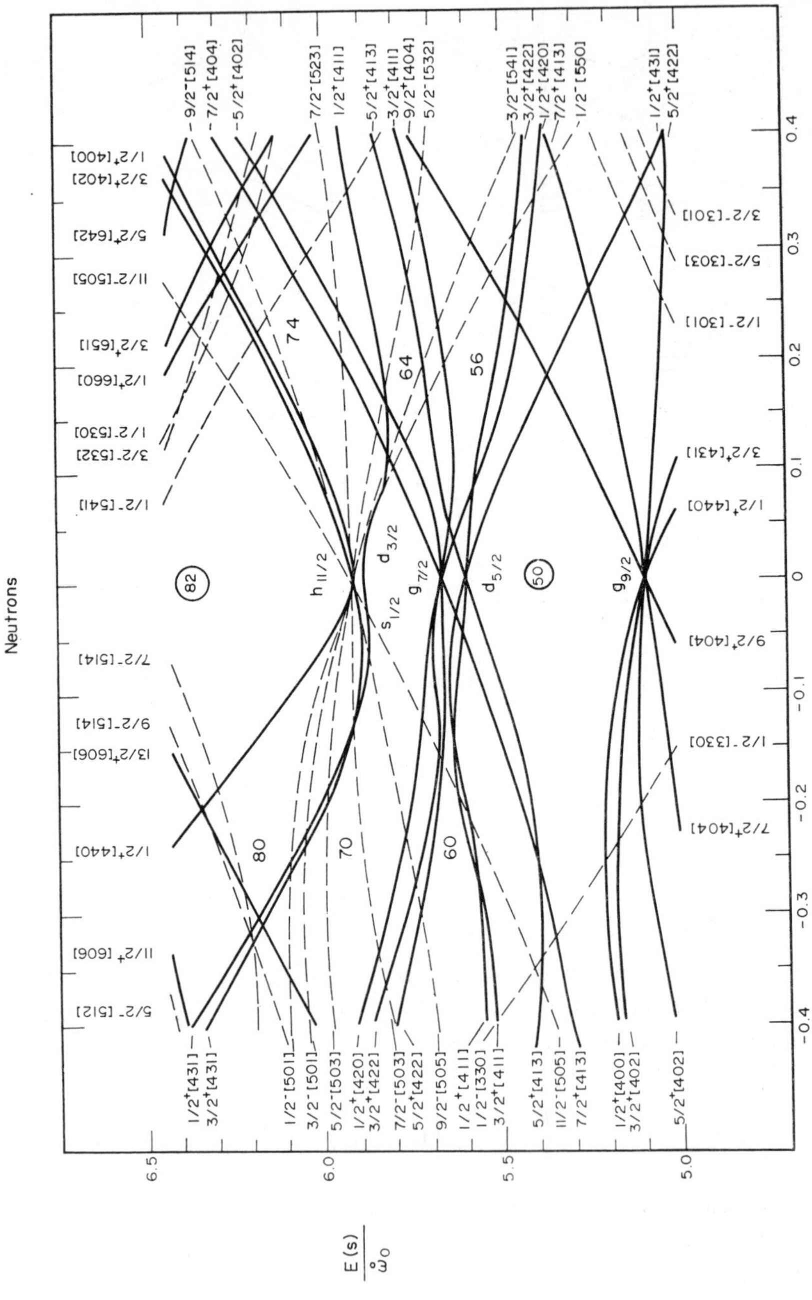

FIG. 1.11. Single-neutron states of the Nilsson–Newton potential for $50 < N < 82$.

Parameters (1.31′) are often used; but they are sometimes slightly modified in order to achieve a correct sequence of the single-particle levels in deformed nuclei. For example, ref. 55 uses $r_0 = 1.24\times10^{-13}$ cm and $\zeta = (0.356\text{–}0.365)\ (10^{-13}\ \text{cm})^2$ for protons and $r_0 = 1.26\times\times10^{-13}$ cm and $\zeta = (0.360\text{–}0.370)\ (10^{-13}\ \text{cm})^2$ for neutrons. The single-particle level diagrams for $A = 155$ and $A = 181$, shown in Figs. 1.12 and 1.13 illustrate the results.

What are the differences in behavior of single-particle states in the Nilsson and Woods–Saxon potentials?

The asymptotic quantum numbers and the corresponding selection rules are very useful

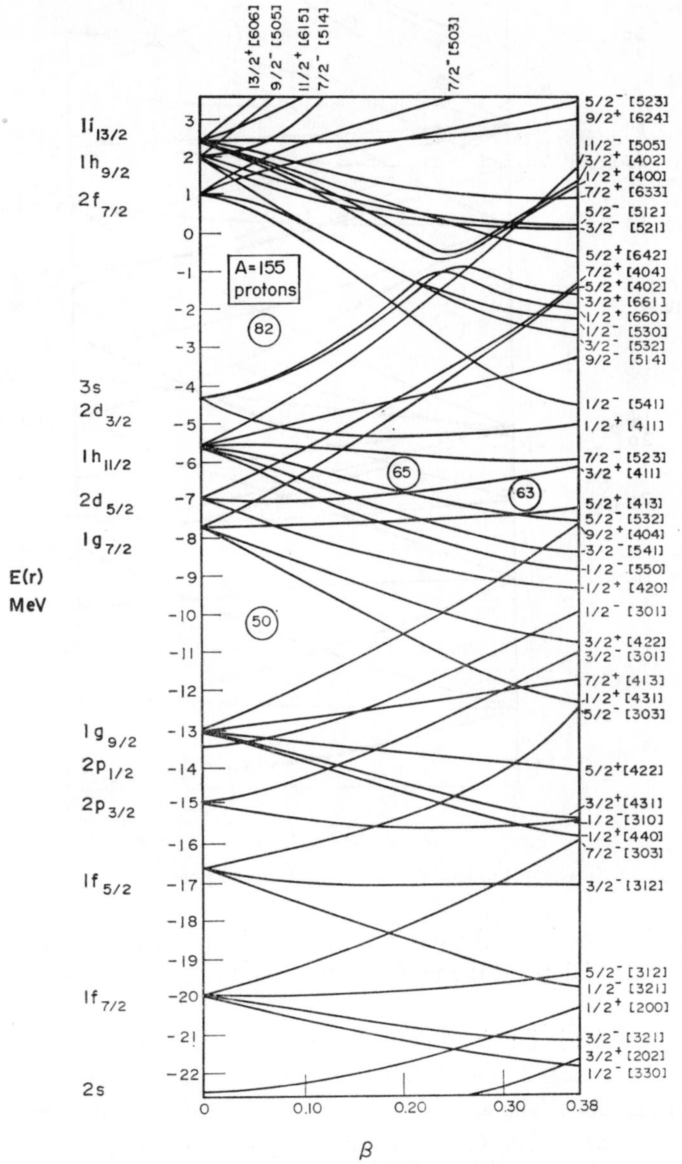

FIG. 1.12. Single-proton levels of the Woods–Saxon potential for $A = 155$.

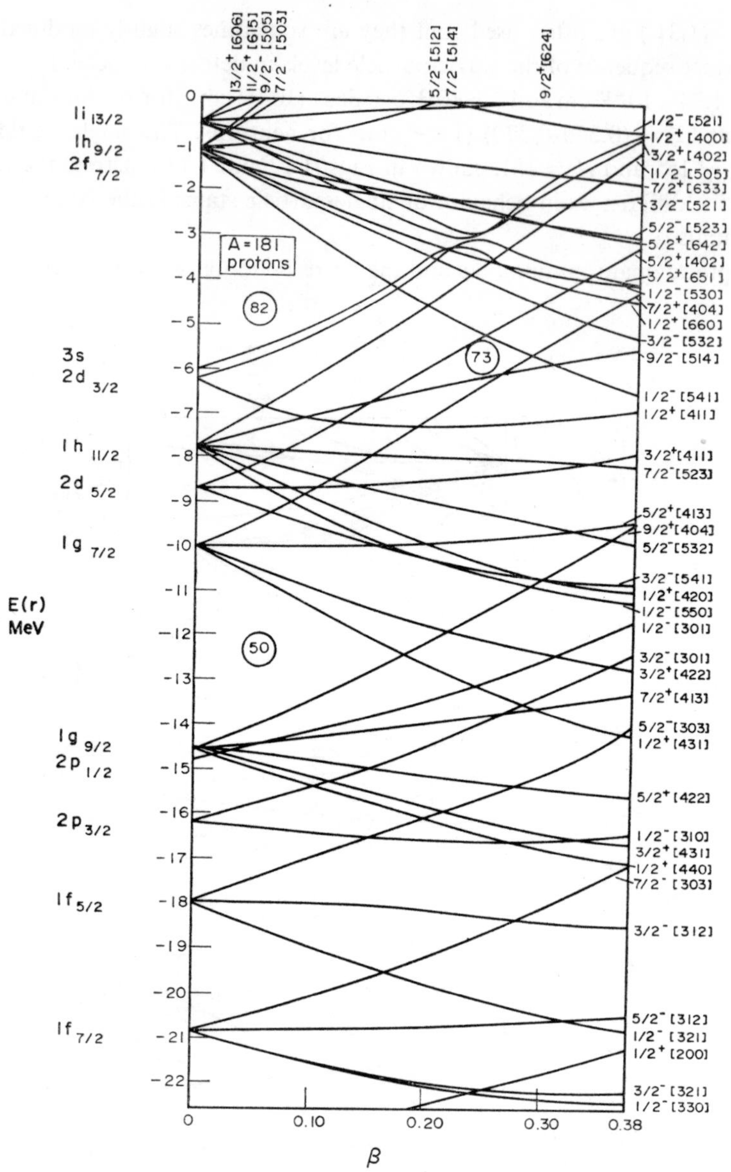

FIG. 1.13. Single-proton levels of the Woods–Saxon potential for $A = 181$.

when analyzing the experimental data. The asymptotic quantum numbers have the same validity in both potentials; and the useful features and commonly employed designations are identical.

The Nilsson potential is infinitely deep, hence it cannot be used for description of the processes where the nuclear surface is important, e.g. for calculations of the cross-sections in direct nuclear reactions. The wave functions of the Woods–Saxon potential have correct behavior on the nuclear surface. Consequently, many matrix elements calculated with these wave functions differ considerably from those calculated with the Nilsson wave functions.

Another difference between the two potentials is in their A-dependence. The Nilsson potential depends on A through the energy unit $\mathring{\omega}_0 = 41/A^{1/3}$ MeV. On the other hand, the sequence of the single-particle levels can change with A in the Woods–Saxon potential because different states depend on the nuclear radius in different ways. Table 1.8 shows how various states change when A is varied from 165 to 155 or 181; the Nilsson levels are also shown. It is evident that the level scheme of the Woods–Saxon potential is essentially A-dependent, e.g. the ordering of 642 ↑ and 523 ↓ states is different in $A = 155$ and in $A = 165$. Figures 1.12 and 1.13 demonstrate the same effect for all proton states, e.g. at the same deformation $\beta = 0.28$ the levels $\frac{9}{2}^-[514]$, $\frac{7}{2}^+[404]$ and $\frac{5}{2}^+[402]$ have rather different positions in the level diagrams for $A = 155$ and for $A = 181$. On the other hand, the eigenfunctions of the Woods–Saxon potential depend on A only weakly.

TABLE 1.8. *The A-dependence of single-neutron energies. Calculations performed at β = 0.25 for the Nilsson* (N) *and Woods–Saxon* (WS) *potentials*

State	$E(s)-E(521\downarrow)$ (MeV)				
	$A = 156$	$A = 155$		$A = 181$	
		N	WS	N	WS
624 ↑	1.45	1.47	1.75	1.41	1.80
512 ↑	0.65	0.66	0.80	0.63	0.75
633 ↑	0.10	0.10	0.40	0.10	0.50
521 ↓	0	0	0	0	0
523 ↓	−0.80	−0.82	−0.95	−0.78	−1.17
642 ↑	−1.05	−1.07	−0.60	−1.02	−0.42
521 ↑	−1.20	−1.22	−1.25	−1.16	−1.22

The coupling of the N and $N\pm 2$ shells was mostly neglected in published calculations with the Nilsson potential. The Woods–Saxon potential calculations, e.g. in refs. 51 and 54–57, include this coupling, and the levels with the same K-values never cross. At deformations, where levels with identical K come very close, the corresponding states are strongly mixed. The mixing, however, is confined to a very narrow interval of β, $\Delta\beta = 0.002$–0.01. At deformations more remote from the quasicrossing points, the wave functions behave as if they were really crossing each other. For example, the upper level, starting from the $2d_{3/2}$ subshell in Fig. 1.13 ($\frac{3}{2}^+[402]$) and the $K = \frac{3}{2}$ level from $1i_{13/2}$ subshell ($\frac{3}{2}^+[651]$) approach each other at $\beta = 0.25$. The curves in the figure never cross; but the $\frac{3}{2}^+[402]$ characteristics are used for the upwards-going level which is a "continuation" of the curve starting from the $2d_{3/2}$ subshell.

The generalization of the Woods–Saxon potential in the triaxial case is quite evident:

$$V(\mathbf{r}, \beta, \gamma) = -\frac{V_0}{1+\exp\{(1/a)\,[r-R(\beta, \gamma, \theta, \varphi)]\}}, \tag{1.99}$$

where

$$R(\beta, \gamma, \theta, \varphi) = R_0[1+\beta\cos\gamma P_2(\theta)+\sqrt{2}\beta\sin\gamma P_2^2(\theta)\cos 2\varphi].$$

The energies and wave functions calculated with the potential consisting of three terms (1.99), (1.96), and (1.97) were calculated in ref. 54.

Various deformation parameters β, β', δ, and ε are used in calculations of the characteristics of nonspherical nuclei. Parameters δ, ε, and β' are defined in (1.79''), (1.81'), and (1.94'), respectively. They are related to the frequency ω through the volume conservation condition of the equipotential surfaces. Ratio ω_z/ω_y can be calculated in different representations and the following relations between different deformation parameters deduced:

$$\left.\begin{aligned} \beta' &= 2\sqrt{4\pi/5}\,\frac{\varepsilon}{3-\varepsilon}, & \varepsilon &= \frac{\frac{3}{2}\sqrt{5/4\pi}\,\beta'}{1+\frac{1}{2}\sqrt{5/4\pi}\beta'}; \\ \delta &= \frac{\varepsilon-\frac{1}{6}\varepsilon^2}{1+\frac{2}{9}\varepsilon^2}, & \varepsilon &= \frac{1-\sqrt{1-\frac{2}{3}\delta-\frac{8}{9}\delta^2}}{\frac{1}{3}+\frac{4}{9}\delta}. \end{aligned}\right\} \tag{1.100}$$

Table 1.9 gives parameters ε, β', and δ calculated for the prolate and oblate shapes and $\gamma = 0$. The quadrupole moment is equal to

$$Q_0 = \frac{3}{\sqrt{5\pi}} ZR_0^2 \frac{\omega_0^2}{\omega_0^2(\beta')} \beta'(1+0.16\beta'), \tag{1.101}$$

TABLE 1.9. *Deformation parameters ε, β and δ defined in (1.100)*

ε	β'	δ	ε	β'	δ	ε	β'	δ
0.02	0.021	0.020	0.30	0.352	0.279	−0.16	−0.160	−0.163
0.04	0.043	0.040	0.32	0.378	0.296	−0.18	−0.179	−0.184
0.06	0.065	0.059	0.34	0.405	0.313	−0.20	−0.198	−0.205
0.08	0.087	0.079	0.36	0.432	0.330	−0.22	−0.217	−0.225
0.10	0.109	0.098	0.38	0.460	0.345	−0.24	−0.235	−0.246
0.12	0.132	0.117	0.40	0.487	0.360	−0.26	−0.253	−0.267
0.14	0.155	0.136	0.50	0.634	0.434	−0.28	−0.271	−0.288
0.16	0.179	0.155	−0.02	−0.021	−0.020	−0.30	−0.288	−0.309
0.18	0.202	0.173	−0.04	−0.041	−0.040	−0.32	−0.305	−0.330
0.20	0.226	0.192	−0.06	−0.062	−0.060	−0.34	−0.322	−0.350
0.22	0.251	0.210	−0.08	−0.082	−0.081	−0.36	−0.340	−0.371
0.24	0.276	0.227	−0.10	−0.102	−0.101	−0.38	−0.356	−0.392
0.26	0.301	0.245	−0.12	−0.122	−0.122	−0.40	−0.373	−0.412
0.28	0.326	0.262	−0.14	−0.141	−0.143	−0.50	−0.453	−0.513

where $\omega_0^2(\beta')$ is given by (1.94').

Another deformation parameter β is defined by

$$R = R_0(1+\beta Y_{20}(\theta)). \tag{1.102}$$

The quadrupole moment expressed through β is of the form

$$Q_0 = \frac{3}{\sqrt{5\pi}} ZR_0^2\beta[1+0.36\beta+\ldots]. \tag{1.103}$$

By equating (1.101) and (1.103), we find a formula connecting β and β'. For small β,

$$\beta = \beta'(1-0.2\beta'). \tag{1.104}$$

The deformation dependence of the radius vector in the Woods–Saxon potential is given by (1.102). However, a body with a sharp surface and a body with a finite surface thickness have different quadrupole moments. To account for this difference, we can use the approximate relation

$$\beta_{WS} \approx 1.1\beta. \tag{1.105}$$

The independent particle model, discussed in this chapter, forms a basis for the inclusion of residual interactions. Successful choice of the single-particle energies and wave functions is an important step toward further development of the theory and correct description of the ground and excited nuclear states.

CHAPTER 2

UNIFIED NUCLEAR MODEL

§ 1. Rotational Wave Functions and Spectra

1. Magic nuclei are spherical at equilibrium. Nuclei with only a few particles outside closed shells have also spherical shapes in their ground states. The lowest 2^+ states in even–even nuclei are related to quadrupole vibrations of the nuclear surface; they represent the degrees of freedom, which are the easiest to excite. The features just described characterize the "vibrational" nuclei—nuclei with only a few particles or holes in unfilled shells and with a spherical equilibrium form.

The spherical nuclear shape becomes less and less stable when the number of particles or holes in unfilled shells is increased. The outside nucleons interact by the residual interaction; the interaction results in a correlated motion of the particles which, in turn, leads to a nuclear nonsphericity. The likelihood of a stable deformed nuclear shape is a fast increasing function of the number of particles in unfilled shells. Consequently, nuclei with many neutrons and protons in unfilled shells have nonspherical, ellipsoidal shapes. The first 2^+ states of such even–even nuclei have very small energy; the sequence of the 2^+, 4^+, 6^+, etc., levels can be interpreted as a rotational band corresponding to the rotation of the whole nucleus. Nuclei with these properties are called "rotational" nuclei.

The correlated motion of the nucleons causes not only a static nuclear deformation, but leads to a number of other collective properties as well. The deformed nuclei have large quadrupole moments as a consequence of this ordered motion of many particles. The reduced $E2$ transition probabilities for the transitions from the first 2^+ states to the ground states in even–even nuclei increase as the number of particles in unfilled shells inreases. When the last neutron and proton shells are approximately half-filled, the corresponding reduced $E2$ transition probabilities exceed the single-particle values by more than a hundred times. Figure 2.1 shows the experimental reduced transition probabilities (in single-particle units) for excitation of the first 2^+ states in even–even nuclei. The large $B(E2)_{\text{s.p.u.}}$ values for $150 < A < 190$ and $A > 226$ are apparent. Remember that just for these nuclei the first 2^+ state has very low energy, as shown in Fig. 1.4. The nuclear spectra in these regions have a clearly recognizable rotational structure: this is another proof that they have a nonspherical equilibrium form.

Many properties of the rotational and vibrational spectra follow from quite general physical assumptions and from corresponding symmetry relations. They can be understood without a detailed understanding of the interaction between the nucleons. These facts are

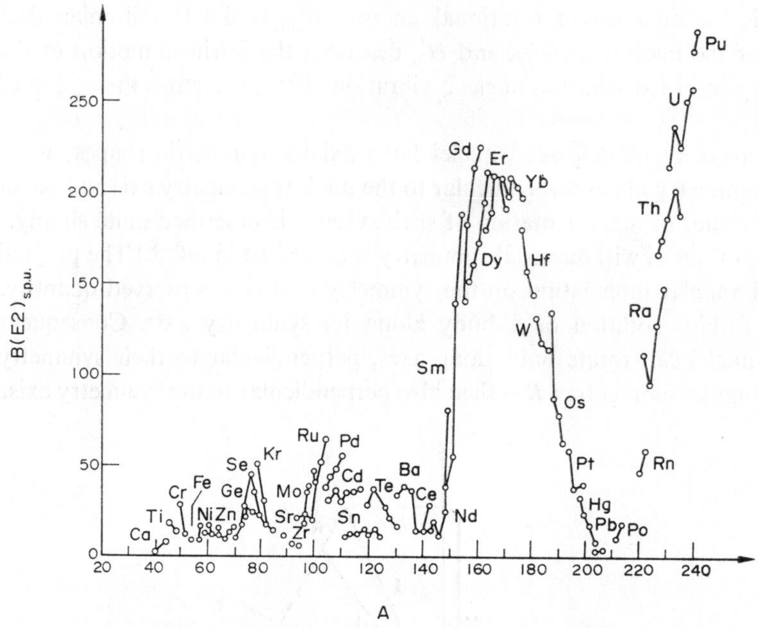

FIG. 2.1. Reduced transition probabilities $B(E2)_{\text{s.p.u.}}$ for excitation of first $I^\pi = 2^+$ states in even–even nuclei (according to ref. 60).

used in a phenomenological description of low-lying excited nuclear states in the unified nuclear model. The phenomenological approach is described in detail, e.g. in refs. 17 and 58–60.

2. The unified nuclear model is based on two assumptions: the first one claims that nuclei with many particles in unfilled shells have the form of an axially symmetric ellipsoid. The orientation of the ellipsoid in space is described by specially introduced collective variables. The second assumption is related to the adiabaticity of the collective motion. It is assumed that the nuclei rotate so slowly that individual nucleons can follow such a motion adiabatically. The condition of adiabaticity is expressed as

$$\omega_{\text{rot}} \ll \omega_{\text{vib}} \ll \omega_{\text{in}}, \tag{2.1}$$

i.e., the rotational frequencies are considerably smaller than the vibrational frequencies, which, in turn, are much smaller than the frequencies of intrinsic motion. The nuclear motions can be then divided into three approximately independent modes: the intrinsic motion, vibrational motion, and the rotation of the whole nucleus. Correspondingly, the nuclear wave function is a product of the intrinsic function $\varphi_K(q)$, vibrational function φ_{vib}, and rotational function $D(\theta_e)$, i.e.,

$$\Psi \sim D(\theta_e)\,\varphi_K(q)\varphi_{\text{vib}}, \tag{2.2}$$

where θ_e are the Euler angles describing the nuclear orientation. The nuclear Hamiltonian is then approximately separated into

$$H = H'_{\text{in}} + H_{\text{rot}} + H_{\text{vib}}. \tag{2.3}$$

Here H_{rot} is the operator of rotational energy, H_{vib} is the Hamiltonian describing the vibrations of the nuclear surface, and H'_{in} describes the intrinsic motion of the nucleons. We are not concerned with the nuclear vibrations here; therefore the φ_{vib} in (2.2) will be omitted.

A large majority of deformed nuclei have axially symmetric shapes, which have an additional symmetry plane perpendicular to the nuclear symmetry axis and passing through the nuclear center of mass. Rotation of such systems is described quite simply. (The more general case of nuclei without axial symmetry is described in ref. 3.) The projection $I_3 = K$ of the total angular momentum on the symmetry axis is a conserved quantity. Quantum mechanics forbids rotation of a body along its symmetry axis. Consequently, axially symmetric nuclei can rotate only along axes, perpendicular to their symmetry axis. The rotational angular momentum $\boldsymbol{R}$ is then also perpendicular to the symmetry axis. Figure 2.2

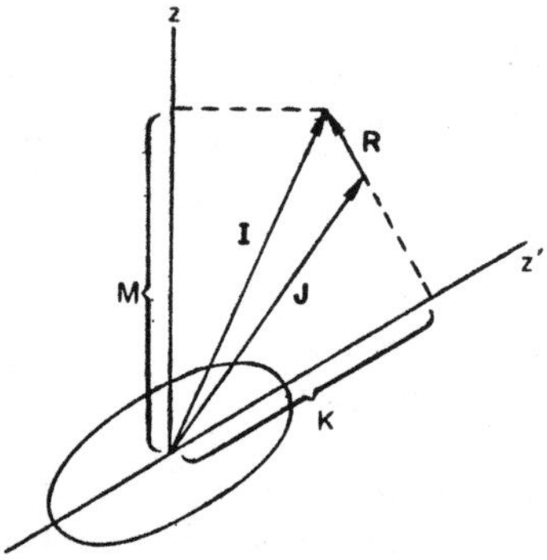

FIG. 2.2. Coupling of the angular momenta in an axially symmetric nonspherical nucleus.

shows all relevant vectors and their couplings. The coordinate system x, y, z is fixed in space (laboratory system), while the coordinate system x', y', z' is coupled to the nucleus (body-fixed system). The z'-axis is the nuclear symmetry axis. The total angular momentum $\boldsymbol{I} = \boldsymbol{J}+\boldsymbol{R}$ has a projection M on the z-axis in the laboratory system and a projection K on the nuclear symmetry axis. $\boldsymbol{J}$ denotes the angular momentum of the intrinsic motion.

The rotational Hamiltonian H_{rot} of a rigid rotation is

$$H_{rot} = \frac{1}{2\mathcal{J}}(R_{x'}^2+R_{y'}^2)+\frac{1}{2\mathcal{J}_{z'}}R_{z'}^2,$$

where, due to the axial symmetry, $R_{z'} = 0$ and two moments of inertia are equal, i.e., $\mathcal{J}_{x'} = \mathcal{J}_{y'} = \mathcal{J}$. Using the equality $\boldsymbol{R} = \boldsymbol{I}-\boldsymbol{J}$,

$$H_{rot} = \frac{1}{2\mathcal{J}}(I^2+J^2-2\boldsymbol{I}\cdot\boldsymbol{J}) \quad \text{and} \quad 2\boldsymbol{I}\cdot\boldsymbol{J} = (I_+J_-+I_-J_+)+2K^2,$$

where

$$I_\pm = I_{x'}\pm iI_{y'}, \quad J_\pm = J_{x'}\pm iJ_{y'}.$$

Therefore,

$$H_{rot} = \frac{I(I+1)}{2\mathcal{J}}-\frac{1}{2\mathcal{J}}(I_+J_-+I_-J_+)+\frac{1}{2\mathcal{J}}\{J(J+1)-2K^2\}.$$

The term $(1/2\mathcal{J})\{J(J+1)-2K^2\}$ depends only on the intrinsic motion; hence, it can be included in H'_{in}. The total Hamiltonian (2.3) is now

$$H = H_{\text{in}}+T_{\text{rot}}+H_{\text{cor}}+H_{\text{vib}}, \tag{2.4}$$

with

$$T_{\text{rot}} = \frac{I(I+1)}{2\mathcal{J}}, \tag{2.4'}$$

$$H_{\text{cor}} = -\frac{1}{2\mathcal{J}}(I_+J_-+I_-J_+). \tag{2.5}$$

The operators $I_\pm$, $J_\pm$ have nonvanishing matrix elements only for states with the same I; they raise or lower the K-value by one unit. The term H_{cor} describes the coupling of intrinsic and rotational motions. It is usually referred to as the "Coriolis coupling" because it has the same origin as the potential energy of classical Coriolis forces. In situations where H_{cor} is important, the quantum number K is not conserved.

3. Let us construct the rotational wave function (the function $D(\theta_e)$ in (2.2)) for an axially symmetric nucleus, with additional symmetry plane perpendicular to the nuclear symmetry axis. As a first step, we shall assume that the H_{cor} interaction can be neglected. If $K = 0$, the rotating system is spinless, and the rotational function $D(\theta_e)$ is proportional to the spherical function $Y_{IM}(\vartheta, \varphi)$.

In the general case, the known transformation properties of the functions Ψ^I_M are used; the functions Ψ^I_M in the fixed laboratory system are related to the same functions in the rotating coordinate system by the relation

$$\Psi^I_M = \sum_K D^I_{MK}(\theta_e)\Psi'^I_K,$$

where Euler angles θ_e couple the axes x, y, z with the axes x', y', z'. The properties of the rotational functions D^I_{MK} are well known; they are described, for example, in refs. 17–19 or 27.

The operator $\boldsymbol{I}$ acts only on the function $D^I_{MK}(\theta_e)$, while $\boldsymbol{J}$ acts only on the intrinsic function $\varphi_K(q)$. The following relations are then valid:

$$\begin{aligned} I^2D^I_{MK}(\theta_e) &= I(I+1)\,D^I_{MK}(\theta_e), \\ I_zD^I_{MK}(\theta_e) &= MD^I_{MK}(\theta_e), \\ I_{z'}D^I_{MK}(\theta_e) &= KD^I_{MK}(\theta_e). \end{aligned}$$

The states with $+K$ and $-K$ must have the same energy as a consequence of the time-reversal invariance. Hence, the wave function sought should be of the form

$$\Psi^I_{MK} = \chi_1 D^I_{MK}(\theta_e)\,\varphi_K(q)+\chi_2 D^I_{M,-K}\varphi_{-K}(q). \tag{2.2'}$$

The parameters χ_1 and χ_2 can be found from the normalization condition, time-reversal invariance, and parity conservation. Obviously, they can differ from each other only by a phase factor. The wave function must be, on the other hand, symmetric under rotation by 180° around an axis perpendicular to the symmetry axis (a consequence of mirror symmetry). This means that the normalized wave function is

$$\Psi^I_{MK} = \frac{1}{\sqrt{2}}(1+R^{x'})\sqrt{\frac{2I+1}{8\pi^2}}\,D^I_{MK}(\theta_e)\,\varphi_K(q), \tag{2.2''}$$

where $R^{x'}$ denotes rotation by 180° around the x'-axis.

The operator $R^{x'}$ is a product of two factors R_e^{-1} and R_i. The operator R_e rotates the nucleus together with the body-fixed coordinate system; it acts on the rotational part of the wave function only. The operator R_i rotates the nucleus relative to the body-fixed coordinate system; it acts on the intrinsic part of the wave function only. Consequently,

$$R_e D^I_{MK}(\theta_e) = (-1)^{I-K}\,D^I_{M,\,-K}(\theta_e),$$
$$R_e^{-1} D^I_{MK}(\theta_e) = (-1)^{I+K}\,D^I_{M,\,-K}(\theta_e),$$
$$R_i\varphi_K(q) = \varphi_{-K}(q),$$

where, according to (1.88″) and (1.89′),

$$\varphi_K(q) = \sum_J d^q_J\,|NlJK\rangle, \qquad \varphi_{-K}(q) = \sum_J d^q_J(-1)^{l+J-\frac{1}{2}}\,|NlJ-K\rangle.$$

Evidently, $R^{x'}$ is equal to

$$R^{x'} = R_e^{-1}R_i. \tag{2.6}$$

After some simple manipulations, the wave function fulfilling all the mentioned symmetry relations is determined by the relation

$$\Psi^I_{MK} = \sqrt{\frac{2I+1}{16\pi^2}}\,\{D^I_{MK}(\theta_e)\,\varphi_K(q)+(-1)^{I+K}\,D^I_{M,-K}(\theta_e)\,\varphi_{-K}(q)\}. \tag{2.7}$$

The particular case $K = 0$ needs somewhat special handling. The operator R_i acting on $\varphi_K = 0$ gives $R_i\varphi_{K=0}(q) = \gamma\varphi_{K=0}(q)$. The wave function (2.7) is nonvanishing if

$$\gamma(-1)^I = 1. \tag{2.8}$$

The angular momenta in rotational bands, therefore, have values

$$I = 0, 2, 4, 6, \ldots, \quad \text{for} \quad \gamma = 1,$$
$$I = 1, 3, 5, 7, \ldots, \quad \text{for} \quad \gamma = -1.$$

Note that the quantum numbers γ and π (parity) are not identical. The properly normalized $K = 0$ wave function is

$$\Psi^I_{M0} = \sqrt{\frac{2I+1}{8\pi^2}}\,D^I_{M0}(\theta_e)\,\varphi_0(q). \tag{2.7'}$$

4. How do the rotational spectra of nonspherical nuclei look? The energy of a rotational state is determined, according to (2.4′), by the equation

$$E_I = \frac{1}{2\mathcal{J}}I(I+1). \tag{2.9}$$

The ground states of the even–even nuclei have $K^{\pi} = 0^{+}$ and $\gamma = +1$; the I has, consequently, only even values. Figure 2.3 shows the ground-state rotational band of ^{170}Hf (this is one of the longer known rotational bands). The $1/2\mathcal{J}$ can be determined from the energy of the 2^{+} state, remaining energies can be calculated by (2.9). These calculated energies are shown in the second column in Fig. 2.3. The $I(I+1)$ rule is approximately valid for $I < 10$; for a correct description of the higher rotational states it is necessary to

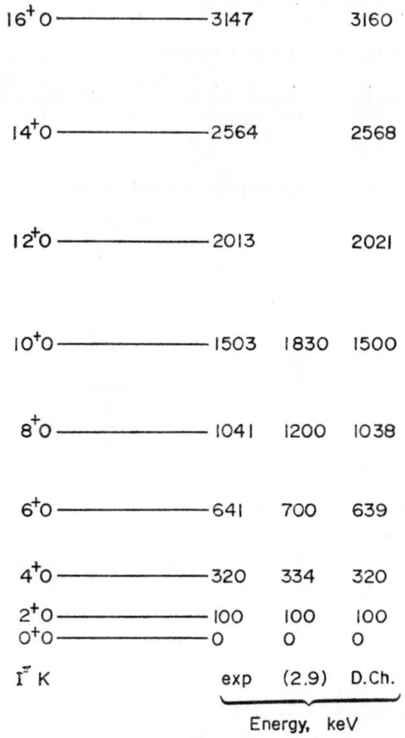

FIG. 2.3. Ground-state rotational band of ^{170}Hf. The experimental data are from ref. 6; the second column gives energies calculated according to (2.9); the last column gives energies calculated in the Davydov–Chaban model.[62]

include Coriolis interaction, coupling of rotations with vibrations, and centrifugal stretching.[62, 63]

Figure 2.4 demonstrates the degree of validity of the $I(I+1)$ rule (2.9) in even–even nonspherical nuclei. It shows the experimentally determined quantities E_I/E_2, together with the corresponding theoretical values (denoted by horizontal lines). The $I(I+1)$ rule is evidently satisfied rather well for the first few states of strongly deformed nuclei; when the angular momentum I increases, the deviations from the $I(I+1)$ rule also increase. Equation (2.9) is violated already for $I = 6$ states in transitional nuclei; the violation for larger I is quite substantial there.

The quantity γ is equal to -1 for the excited states $K^{\pi} = 0^{-}$ states of even–even nuclei; spin I has thus only odd values. For the excited bands with $K \neq 0$ I has values K, $K+1$, $K+2$, etc. Figure 2.5 demonstrates all three types of rotational bands; it shows the

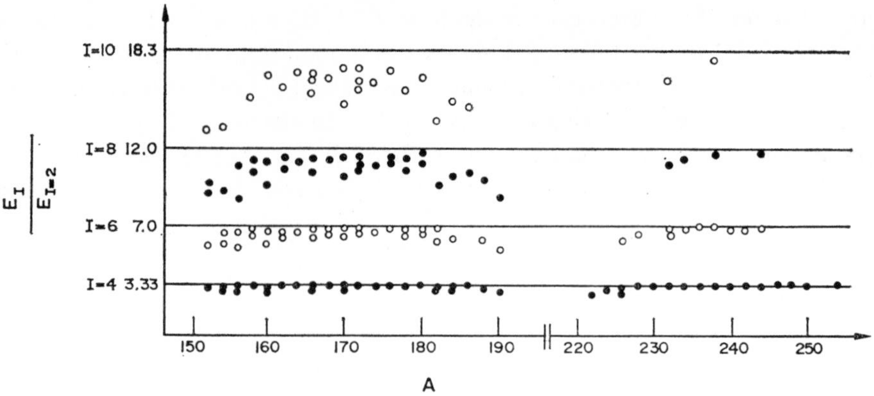

FIG. 2.4. Ratios E_I/E_2 for the ground-state rotational bands of even–even nonspherical nuclei (data from ref. 60).

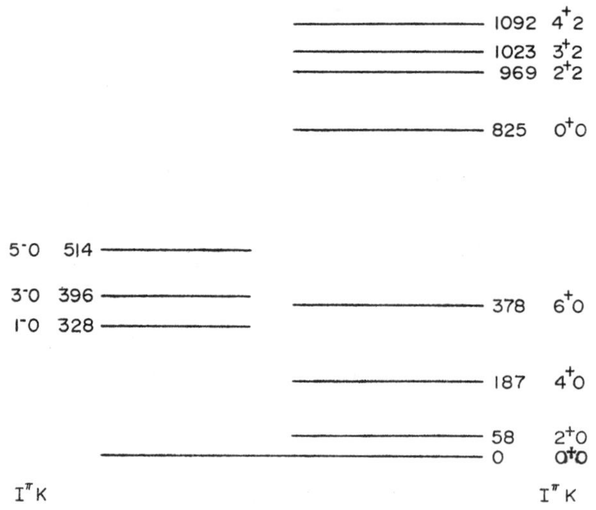

FIG. 2.5. Rotational bands in ^{228}Th.

ground state band and the $K^\pi = 0^-$ and $K^\pi = 2^+$ bands of ^{228}Th. Note that some noncollective $K^\pi = 0^-$ bands have $\gamma = +1$; they have, therefore, even values of I, i.e., $I = 0, 2, 4, \ldots$

The $K = 0$ bands in the odd–odd nuclei are somewhat more complicated. The parameter γ can have both values $\gamma = \pm 1$ there, and the intrinsic wave function $\varphi_{K=0}(qq')$ is

$$\varphi_{K=0}(qq') = \varphi_K^p(q)\,\varphi_{-K}^n(q') \pm \varphi_{-K}^p(q)\,\varphi_K^n(q'). \tag{2.10}$$

There are, therefore, two $K = 0$ bands in odd–odd nuclei; one corresponds to $\gamma = 1$, another to $\gamma = -1$. The two bands are shifted relative to each other. Such a situation is demonstrated in Fig. 2.6, which shows the $K^\pi = 0^-$ bands in ^{166}Ho.

The Coriolis coupling must be included when rotational bands in odd-A deformed nuclei are analyzed. The operator $I_\pm$ has the following matrix elements in the basis of the wave

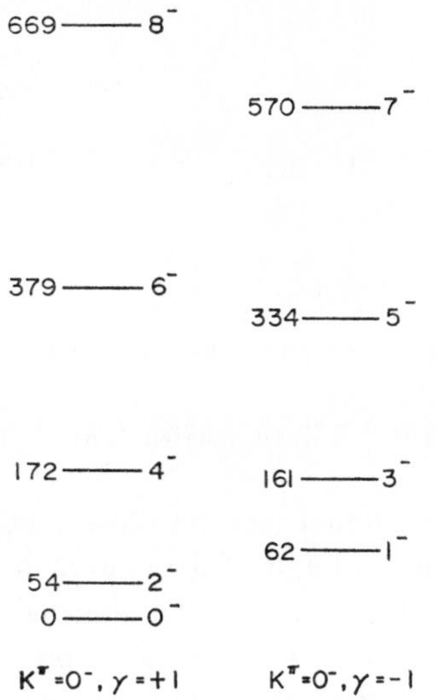

FIG. 2.6. Rotational bands in ^{166}Ho.

functions (2.7):

$$(\Psi^{I*}_{MK} I_{\pm} \Psi^{I}_{MK\pm 1}) = \sqrt{(I \mp K)(I \pm K+1)}.$$

Obviously, the operator $I_{\pm}$ has diagonal matrix elements for $K = \frac{1}{2}$ states only. (The wave function contains then $K = \frac{1}{2}$ and $K = -\frac{1}{2}$ components, for $K = -\frac{1}{2}$ the quantity $K+1$ is $\frac{1}{2}$.) Using (2.7), it is easy to realize that for $K = \frac{1}{2}$

$$(\Psi^{I*}_{M\frac{1}{2}} H_{\mathrm{cor}} \Psi^{I}_{M,\,-\frac{1}{2}}) = \frac{1}{2\mathcal{J}}\, a(-1)^{I+\frac{1}{2}}\,(I+\tfrac{1}{2}), \qquad (2.11)$$

where a, the so-called "decoupling parameter", is determined by

$$a = -(\varphi_{K=\frac{1}{2}} J_{+}\, \varphi_{K-=\frac{1}{2}}). \qquad (2.12)$$

The decoupling parameter a is expressed through components of the Nilsson wave function (1.88) as

$$a = (-1)^{N} \sum_{l} (a_{l0}^{2} + 2\sqrt{l(l+1)}\, a_{l0} a_{l1}). \qquad (2.12')$$

If wave functions, calculated with the Woods–Saxon potential are used, the decoupling parameter depends on the mass number A; in some cases such A-dependence is substantial.

We have shown that the Coriolis interaction contributes to the energy of $K = \frac{1}{2}$ bands already in the first-order perturbation theory. The energy of rotational states is, consequently, determined by the modified formula

$$E_{I} = \frac{1}{2\mathcal{J}} \{I(I+1) + \delta_{K,\,\frac{1}{2}}\, a(-1)^{I+\frac{1}{2}}(I+\tfrac{1}{2})\}. \qquad (2.13)$$

Figure 2.7 shows two $K = \frac{1}{2}$ and one $K = \frac{7}{2}$ rotational bands in ^{171}Lu. The distortion of the $K = \frac{1}{2}$ bands is clearly visible.

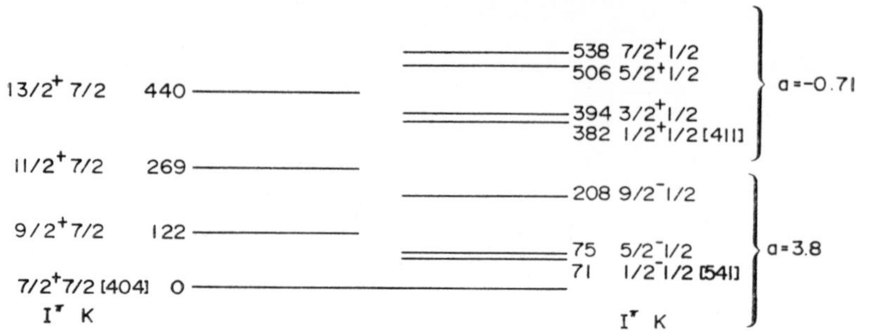

FIG. 2.7. Rotational bands in ^{171}Lu.

The experimental values of rotational energies allow a sufficiently accurate determination of the moments of inertia of ground and of many excited states in deformed nuclei. The quantity $\mathcal{J}_{\text{rig}}$, i.e., the moment of inertia of a rotating rigid body, is obtained when independent particle model with an anisotropic oscillator potential is used in the calculation. The experimental moments of inertia are substantially smaller $\mathcal{J}_{\text{exp}} \approx (0.2\text{–}0.6)\, \mathcal{J}_{\text{rig}}$. This discrepancy is related to the fact that the residual interactions are neglected in the independent particle model; improved calculations will be described later in Chapter 7.

§ 2. Probabilities of Electromagnetic Transitions in Rotational Model

1. To represent the nuclear state as a product of the intrinsic and rotational wave functions is, in most cases, a fairly good approximation. Using such a product wave function it is possible to deduce a number of simple relations between the electromagnetic transition probabilities of transitions within a rotational band or between different rotational bands. These rules also connect the electromagnetic transition probabilities with the static quadrupole moments and gyromagnetic ratios.

We shall derive here such relations using the wave functions (2.7) and neglecting the Coriolis interaction. The rules, first formulated in ref. 64, are known as "Alaga rules."

As a first step, the matrix element of the multipole operator $\mathfrak{M}(\lambda\mu)$ must be calculated. Let $\mathfrak{M}(\lambda\mu)$ denote the operator in the laboratory system, and $\mathfrak{M}'(\lambda\mu)$ denotes the same operator in the body-fixed coordinate system. The two operators are related by

$$\mathfrak{M}(\lambda\mu) = \sum_{\mu'} D^{\lambda}_{\mu\mu'}(\theta_e)\, \mathfrak{M}'(\lambda\mu')$$

provided the $\mathfrak{M}'(\lambda\mu')$ and $D^{\lambda}_{\mu\mu'}$ commute. The matrix element of $\mathfrak{M}(\lambda\mu)$ using the wave functions (2.7) is then

$$(\Psi^{I_f*}_{M_fK_f}\, \mathfrak{M}(\lambda\mu) \Psi^{I_i}_{M_iK_i}) = \frac{(I_iM_i\lambda\mu \mid I_fM_f)}{\sqrt{2I_f+1}} \langle I_fK_f \mid \mathfrak{M}(\lambda\mu) \mid I_iK_i\rangle, \tag{2.14}$$

where

$$\langle I_fK_f \mid \mathfrak{M}(\lambda\mu) \mid I_iK_i\rangle = (2I_i+1)^{1/2} \{(I_iK_i\lambda K_f-K_i \mid I_fK_f) \langle K_f \mid \mathfrak{M}'(\lambda, K_f-K_i) \mid K_i\rangle$$
$$+(-1)^{I_i+K_i}(I_i-K_i\lambda K_f+K_i \mid I_fK_f) \langle K_f \mid \mathfrak{M}'(\lambda, K_f+K_i) \mid -K_i\rangle\} \tag{2.14'}$$

The last term in (2.14′) vanishes for $\lambda < K_i+K_f$. Formula (2.14) shows the matrix elements factored into terms depending on the intrinsic wave functions φ_K and the geometrical coefficients depending on I_i, I_f.

2. Let us discuss in more detail the electric $E2$ transitions within a rotational band. The simplest, yet important, case involves $K = 0$ bands in even–even nuclei. The reduced transition probability is then

$$B(E2;\, I_i \to I_f) = \frac{1}{2I_i+1} \sum_{\mu,\, M_i,\, M_f} \left|(\Psi^{I_f*}_{M_f0}\mathfrak{M}_E(2\mu)\Psi^{I_i}_{M_i0})\right|^2 \\ = (I_i020\,|\,I_f0)^2\,|(\varphi_0^*\mathfrak{M}_E(20)\varphi_0)|^2. \tag{2.15}$$

The diagonal matrix element of the $\mathfrak{M}_E(20)$ operator is proportional to the intrinsic nuclear quadrupole moment Q_0:

$$(\varphi_0^*\mathfrak{M}_E(20)\varphi_0) = \sqrt{\frac{5}{16\pi}}\,eQ_0.$$

The reduced $E2$ transition probability is, therefore, expressed through Q_0 by the relation

$$B(E2;\, I_i \to I_f) = \frac{5}{16\pi}\,e^2Q_0^2(I_i020\,|\,I0)^2. \tag{2.15′}$$

It is now evident that nuclei with large intrinsic quadrupole moments have large $E2$ transition probabilities and vice versa.

The nuclear electric quadrupole moment is defined as the expectation value of the quadrupole moment operator

$$Q_0 = \frac{4}{e}\sqrt{\frac{\pi}{5}}\int (d\mathbf{r})\,\varrho(r)\,r^2Y_{20}(\theta, \varphi)$$

in the state (2.7) with the $M = K$, i.e.,

$$eQ = \sqrt{\frac{16\pi}{5}}\left(\Psi^{I*}_{KK}\int d\mathbf{r}\varrho(r)\,r^2Y_{20}(\theta, \varphi)\Psi^{I}_{KK}\right) = \frac{3K^2-I(I+1)}{(I+1)(2I+3)}\,Q_0, \tag{2.16}$$

where $\varrho(r)$ is the nuclear charge density. The quadrupole moment Q is, obviously, equal to zero if $I = 0$ or $I = \frac{1}{2}$.

The Q_0 and the nuclear deformation parameter β are related. Assume that the nuclear surface is determined by

$$R = R_0(1+\beta Y_{20}(\theta, \varphi))$$

and that the charge density is constant inside the nucleus. The intrinsic quadrupole moment is then

$$Q_0 = \frac{3}{\sqrt{5\pi}}\,ZR_0^2\beta(1+0.36\beta+\ \ldots). \tag{2.17}$$

Equation (2.17) can be used to determine the equilibrium deformation parameter β_0 [see (1.102) and (1.103)] if the experimental value of Q_0 is known. Remember, however, that such a β_0 is model dependent; we have assumed that Q_0 is a static moment and that the nucleus has a sharp surface and a constant charge density.

The $E2$ transition probability for transitions within a $K \neq 0$ rotational band is equal to

$$B(E2;\, I_i \to I_f) = \frac{5}{16\pi}\, e^2 Q_0^2 (I_i K 2 0 \mid I_f K)^2 . \tag{2.18}$$

The intensities of transitions between states with different I_i and I_f values depend on the same value of Q_0. Consequently, ratios of the transition probabilities depend only on the geometrical factors; e.g., the branching ratio for the transitions from the level I to $I-1$ and $I-2$, respectively, is

$$\frac{B(E2;\, I\to I-1)}{B(E2;\, I\to I-2)} = \frac{(IK20\,|\,I-1K)^2}{(IK20\,|\,I-2K)^2} = \frac{2K^2(2I-I)}{(I+1)(I-1+K)(I-1-K)} . \tag{2.19}$$

This and analogous rules (Alaga rules) agree usually well with the experimental data; they are very useful when rotational bands are constructed from the set of experimentally determined levels.

3. Let us discuss the magnetic dipole moments and transitions. The magnetic moment operator is defined in (1.96) as

$$\boldsymbol{\mu} = \sum_{i=1}^{A} (g_i^{(l)} \boldsymbol{l}_i + g_i^{(s)} \boldsymbol{s}_i).$$

The above expression can be divided into a term explicitly related to the rotational motion and another term, related to the intrinsic motion, i.e.,

$$\mu = g_R \boldsymbol{R} + \sum_{i}^{A} (g_i^{(l)} \boldsymbol{l}_i' + g_i^{(s)} \boldsymbol{s}_i'),$$

where $\boldsymbol{l}_i'$ and $\boldsymbol{s}_i'$ are operators of the intrinsic motion and $\boldsymbol{J} = \sum_i (\boldsymbol{l}_i' + \boldsymbol{s}_i')$. The quantity g_R is the gyromagnetic ratio of the rotational motion. Defining

$$\mu' = \sum_i (g_i^{(l)} \boldsymbol{l}_i' + g_i^{(s)} \boldsymbol{s}_i'),$$

gives

$$\mu = g_R \boldsymbol{I} + (\mu' - g_R \boldsymbol{J}). \tag{2.20}$$

The vector operator $\mu' - g_R \boldsymbol{J}$ is transformed into the body-fixed coordinate system according to the formula

$$(\mu_x' - g_R J_x) = \sum_{x'} D^1_{xx'}(\theta_e)\,(\mu_{x'}' - g_R J_{x'}).$$

The x-component of μ is then

$$\mu_x = g_R I_x + \sum_{x',\, i} D^1_{xx'}(\theta_e) \{(g_i^{(l)} - g_R)(l_i')_{x'} + (g_i^{(s)} - g_R)(s_i')_{x'}\}. \tag{2.20'}$$

The nuclear magnetic moment is defined as the average value of μ_z in the $M = K$ state. After simple manipulation the following formula is obtained:

$$\mu = g_R I + (g_K - g_R) \frac{K^2}{I+1} \{1 + (2I+1)(-1)^{I+\frac{1}{2}} b_0 \delta_{K,\frac{1}{2}}\}. \tag{2.21}$$

Here

$$Kg_K = Kg^{(l)} + (g^{(s)} - g^{(l)})(\varphi_K^* s_{z'} \varphi_K) \tag{2.22}$$

and

$$(g_K - g_R) b_0 = (g^{(l)} - g_R)(\varphi_{K=\frac{1}{2}}^* J_+ \varphi_{K=-\frac{1}{2}}) + (g^{(s)} - g_R)(\varphi_{K=\frac{1}{2}}^* s_+ \varphi_{K=-\frac{1}{2}}). \tag{2.23}$$

The expressions (2.22) and (2.23) can be rewritten using the Nilsson expansion coefficients $a_{l\Lambda}$:

$$K \frac{g_K - g^{(l)}}{g^{(s)} - g^{(l)}} = \sum_l \frac{1}{2} (a_{l,\,K-\frac{1}{2}}^2 - a_{l,\,K+\frac{1}{2}}^2), \tag{2.22'}$$

$$(g_K - g_R) b_0 = -(-1)^N \left((g^{(s)} - g_R) \sum_l a_{l0}^2 + 2(g^{(l)} - g_R) \sum_l \sqrt{l(l+1)}\, a_{l0} a_{l1} \right) \tag{2.23'}$$

Equation (2.21) shows that $K = \frac{1}{2}$ states have an additional term b_0, analogous to the decoupling parameter a in the rotational energy equation (2.13).

The gyromagnetic ratio g_R can be found in simple situations. Assume that the core has no spin polarization; g_R is then

$$g_R = \frac{\mathcal{J}_p}{\mathcal{J}_p + \mathcal{J}_n}, \tag{2.24}$$

i.e., it is equal to the ratio of the proton moment of inertia to the total moment of inertia. Assuming further that the moments of inertia are proportional to the number of participating particles, one obtains that

$$g_R = Z/A. \tag{2.24'}$$

The reduced probability of $M1$ transitions within a rotational band is determined by the formula

$$B(M1; I_i \to I_f) = \frac{3}{16\pi} \times \frac{e^2}{m^2} K^2 (g_K - g_R)^2 (I_i K 1 0 \mid I_f K)^2 (1 + b_0 \delta_{K,\frac{1}{2}} (-1)^{I_> + \frac{1}{2}})^2, \tag{2.25}$$

where $I_>$ is the larger of I_i and I_f values. The quantity $(g_K - g_R)$ can be found from the experimentally determined $B(M1)$ values in $K \neq \frac{1}{2}$ bands. This serves as a test of fulfillment of the adiabaticity condition; the $g_K - g_R$ values determined from different $I+1 \to I$ transitions must be equal.

It is simple to find the ratio of the $M1$ and $E2$ probabilities of transitions within a rotational band. Using (1.58), (1.61), (2.18), and (2.25),

$$\frac{W(M1; I \to I-1)}{W(E2; I \to I-1)} = \frac{20}{3} \left(\frac{1}{mpQ_0} \right)^2 (I+1)(I-1)(g_K - g_R)^2 (1 + \delta_{K,\frac{1}{2}} b_0 (-1)^{I+\frac{1}{2}})^2. \tag{2.26}$$

It was already mentioned that deformed nuclei have large Q_0 values. Consequently, the admixtures of $E2$ multipole into the $M1$ transitions are enhanced by at least an order of magnitude when compared to typical single-particle transitions in spherical nuclei. The quantity

$$\frac{g_K - g_R}{Q_0} (1 + \delta_{K,\frac{1}{2}} (-1)^{I+\frac{1}{2}} b_0)$$

can be determined in correlation experiments by measuring the angular correlations of two subsequently emitted γ-rays or by measuring the angular distribution of γ-rays emitted in Coulomb excitation.

Thus, by measuring the magnetic moment and at least three γ-transition probabilities for transitions within a rotational band, four parameters: Q_0, g_R, g_K, and b_0, can be determined.

4. We have discussed transitions within a rotational band. The wave function (2.7) leads to simple rules for transitions involving different rotational bands, too. They can be used as tests of the validity of our basic assumptions (particularly, the adiabaticity of the rotational motion); they can be also employed as a tool for determination of unknown quantum numbers of the involved states.

Assume that $\lambda < K_i + K_f$ and use eqn. (2.14′). The matrix element of the $\mathfrak{M}(\lambda\mu)$ operator is then determined by the formula

$$\langle I_f K_f \,|\, \mathfrak{M}(\lambda\mu) \,|\, I_i K_i \rangle = (2I_i+1)^{1/2} \, (I_i K_i \lambda K_f - K_i \,|\, I_f K_f) \, (\varphi^*_{K_f} \mathfrak{M}'(\lambda, K_f - K_i) \varphi_{K_i})$$

and the reduced transition probability is

$$B(\lambda;\, I_i \to I_f) = \frac{1}{2I_i+1} \, |\langle I_f K_f \,|\, \mathfrak{M}(\lambda\mu) \,|\, I_i K_i \rangle|^2. \tag{2.27}$$

Note that the same matrix element $(\varphi^*_{K_f} \mathfrak{M}'(\lambda, K_f - K_i)\varphi_{K_i})$ enters the expressions for the transition probabilities from all I_i states of the band K_i, to all I_f states of the band K_f. The ratio of two reduced transition probabilities depends, therefore, only on the geometrical factors, i.e.,

$$\frac{B(\lambda;\, I_i K_i \to I_f K_f)}{B(\lambda;\, I_i K_i \to I'_f K_f)} = \frac{(I_i K_i \lambda K_f - K_i) \,|\, K_f I_f)^2}{(I_i K_i \lambda K_f - K_i \,|\, I'_f K_f)^2}\,. \tag{2.28}$$

Measurement of the relative intensities of the γ-transitions is a commonly used method of the K_i determination. For example, the ratios of $E1$ transition intensities leading to the ground state band of even-even nuclei are:

$$\frac{B(E1;\, 1 \to 0)}{B(E1;\, 1 \to 2)} = \begin{cases} \dfrac{(1010\,|\,00)^2}{(1010\,|\,20)^2} = 0.5, & \text{when} \quad K_i = 0, \\[2ex] \dfrac{(111-1\,|\,00)^2}{(111-1\,|\,20)^2} = 2, & \text{when} \quad K_i = 1. \end{cases}$$

It is obvious that the quantity K_i can be simply determined from the measured relative intensities. Note that analogous relations are valid for the β-decay intensities.

The Clebsch–Gordan coefficients in (2.27) vanish if the relation

$$|K_i - K_f| \leq \lambda \tag{2.29}$$

is not valid. Transitions for which (2.29) is invalid are called "K-forbidden" transitions. They could be observed because our basic assumptions are valid only approximately. The "real" nuclear wave functions contain small correction terms, corresponding to the change of nuclear shape during rotation, nonadiabaticity, admixtures of different configurations, etc.; these small components make the K-forbidden transitions possible but slow. According to the experimental evidence, the K-forbidden transitions are almost a hundred times slowed down for each unit of the forbiddenness $|K_i - K_f| - \lambda$.

Figure 2.8 shows two K-forbidden transitions, in ^{177}Lu and in ^{180}Hf. The 116 keV $E3$ transition in ^{177}Lu goes from the $I^\pi K = \frac{23}{2}^- \frac{23}{2}$ level with an energy 970 keV, to the $I^\pi K = \frac{17}{2}^+ \frac{7}{2}$ level with an energy 854 keV. The transition is five times K-forbidden $(\frac{23}{2} - \frac{7}{2} - 3 = 5)$; it is 10^8 times slower than the corresponding single-particle value. The 57 keV $E1$ transition in ^{180}Hf goes from the $I^\pi K = 8^- 8$ state at 1142 keV to the $I^\pi K = 8^+ 0$ state at 1085 keV. It is seven times K-forbidden and 10^{15} times slower than the corresponding single-particle value. Similar K-forbiddeness can be observed in nuclear β-decays.

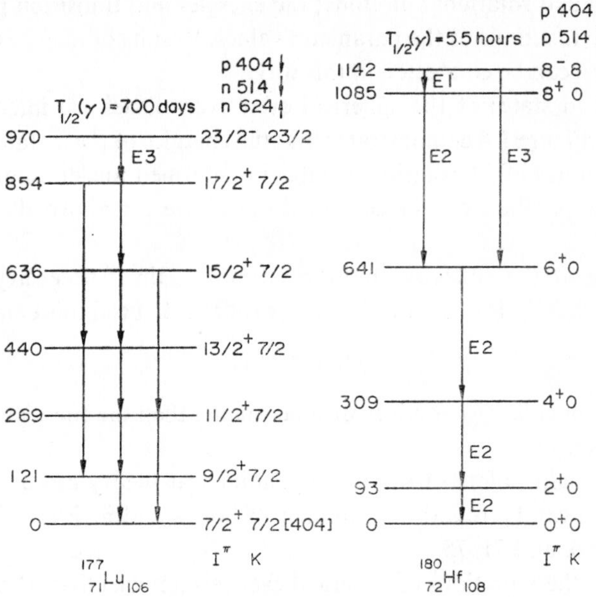

FIG. 2.8. K-forbidden γ-transitions in ^{177}Lu and ^{180}Hf.

The considerable hindrance of K-forbidden transitions demonstrates once again that the adiabatic rotational model describes a wide class of the nuclear excited states quite accurately.

The electromagnetic transitions between rotational states belonging to different rotational bands are discussed in ref. 65. The experimental data on energies and lifetimes of the rotational states and the corresponding quadrupole moments and gyromagnetic ratios are collected and systemized in refs. 66–70.

§ 3. Vibrations of the Nuclear Surface

1. The rotation of the whole nucleus and the nuclear surface vibrations are two examples of nuclear collective motions. The correct description of the rotational states in nonspherical nuclei is an important success of the unified nuclear model. In this section we shall discuss

nuclear vibrations, i.e., the vibrations of the nuclear surface which conserve orientation of the nucleus in space. The shape of deformed nuclei deviates from the spherical form statically. If such a static deviations exist, the dynamic deviations from the equilibrium form might exist as well.

The phenomenological description of nuclear vibrations is less accurate than the description of nuclear rotations. It gives, however, a useful classification of the corresponding states and allows deduction of rules, which connect the energies and transition probabilities of different states. The phenomenological method describes, besides, the coupling of the vibrations and rotations, the change of the nuclear form, and inertia during rotation, etc.

Each nucleus is characterized by a set of parameters in the phenomenological model of nuclear vibrational and rotational motions; the energies and transition probabilities of the collective states are functions of the parameter values. Results of many experiments, related to the same nucleus, can be connected in this way.

The lowest excited states of the spherical even–even nuclei are interpreted as nuclear surface vibrations. Figure 1.4 demonstrates that the energies of these states are considerably larger than the energies of 2^+ rotational states in deformed nuclei. Similarly, the energies of states described as vibrations in deformed nuclei are far above the lowest rotational levels.

The adiabaticity of the collective motion is once again a necessary condition of the phenomenological description of vibrations. For spherical nuclei this condition states

$$\omega_{\text{vib}} \ll \omega_{\text{in}}, \tag{2.30}$$

that is, the energies of collective motion must be smaller than the energies related to changes of intrinsic structure.

As before, only main points of the phenomenological theory are quoted here; a more thorough discussion can be found in a number of books and review articles, for example, in refs. 17, 19, 58–60, and 71–75.

2. Let us discuss the vibrations of spherical even–even nuclei first. Deviations of the nuclear mass distribution from the spherical form are determined by a set of parameters $\alpha_{\lambda\mu}$. In the idealized case of uniform density the sharp nuclear surface is defined by

$$R(\theta, \varphi) = R_0\left\{1+\sum_{\lambda=2}^{\infty}\sum_{\mu=-\lambda}^{\lambda}\alpha_{\lambda\mu}Y_{\lambda\mu}(\theta, \varphi)+O(\alpha^2)\right\}, \tag{2.31}$$

where $\alpha_{\lambda\mu} = (-1)^{\mu}\alpha^*_{\lambda-\mu}$ (because R must be real); R_0 is the average nuclear radius. The $\lambda = 0$ and $\lambda = 1$ multipoles are omitted; $\lambda = 0$ corresponds to isotropic compression and expansion of the nucleus and $\lambda = 1$ corresponds to displacement of nuclear center of mass or to the motion of neutron center of mass relative to the proton center of mass.

Vibrations with sufficiently small amplitudes are always harmonic; their Hamiltonian is

$$H_{\text{vib}} = \sum_{\lambda\mu}\left\{\tfrac{1}{2}B_\lambda\,|\dot{\alpha}_{\lambda\mu}|^2+\tfrac{1}{2}C_\lambda\,|\alpha_{\lambda\mu}|^2\right\}, \tag{2.32}$$

and the corresponding Lagrangian is

$$\mathcal{L}_{\text{vib}} = \sum_{\lambda\mu}\left\{\tfrac{1}{2}B_\lambda\,|\dot{\alpha}_{\lambda\mu}|^2-\tfrac{1}{2}C_\lambda\,|\alpha_{\lambda\mu}|^2\right\}.$$

The generalized momentum is defined in the usual way as

$$\pi_{\lambda\mu} = \frac{\partial}{\partial \dot{\alpha}_{\lambda\mu}} \mathcal{L}_{\text{vib}} = B_\lambda \dot{\alpha}^+_{\lambda\mu} = (-1)^\mu B_\lambda \dot{\alpha}_{\lambda, -\mu}.$$

We shall introduce a quantized theory now. The $\alpha_{\lambda\mu}$ and $\pi_{\lambda\mu}$ are then operators with the commutation relations

$$[\alpha_{\lambda\mu}, \pi_{\lambda\mu}] = i, \tag{2.33}$$

which means that

$$\pi_{\lambda\mu} = -i \frac{\partial}{\partial \alpha_{\lambda\mu}} \left(\text{or } \alpha_{\lambda\mu} = i \frac{\partial}{\partial \pi_{\lambda\mu}} \right).$$

The Hamiltonian (2.32) is then rewritten as

$$H_{\text{vib}} = \sum_{\lambda\mu} \left\{ \frac{1}{2B_\lambda} |\pi_{\lambda\mu}|^2 + \tfrac{1}{2} C_\lambda |\alpha_{\lambda\mu}|^2 \right\} \tag{2.32'}$$

and the equations of motion are

$$\dot{\pi}_{\lambda\mu} = -\frac{\partial H_{\text{vib}}}{\partial \alpha_{\lambda\mu}} = -C_\lambda \alpha^+_{\lambda\mu} \quad \text{or} \quad \ddot{\alpha}_{\lambda\mu} + \frac{C_\lambda}{B_\lambda} \alpha_{\lambda\mu} = 0. \tag{2.34}$$

The harmonic vibration frequencies are

$$\omega_\lambda = \sqrt{C_\lambda / B_\lambda}, \tag{2.35}$$

and the energy spectrum is determined by

$$E^{(n)}_\lambda = (n_\lambda + \tfrac{1}{2}) \omega_\lambda. \tag{2.36}$$

The vibrational quanta, so called "phonons," are characterized by the angular momentum λ, parity $(-1)^\lambda$, mass parameter B_λ, and restoring force constant C_λ.

The parameters B_λ and C_λ can be calculated if nuclei are treated as drops of an ideal, incompressible liquid with irrotational flow. The mass parameter is then

$$(B_\lambda)_h = \frac{1}{\lambda} \frac{3}{4\pi} A m R_0^2. \tag{2.37}$$

The restoring force parameter has two components: one corresponding to the surface tension and another to the Coulomb energy:

$$(C_\lambda)_h = (\lambda - 1)(\lambda + 2) R_0^2 \sigma - \frac{\lambda - 1}{2\lambda + 1} \frac{3}{2\pi} \frac{Z^2 e^2}{R_0}, \tag{2.38}$$

σ is the surface tension coefficient; from the binding energy formula (1.34) we know that $4\pi R_0^2 \sigma \approx 20 A^{2/3}$ MeV. The hydrodynamical parameters do not give a correct description of nuclear vibrations. The phenomenological method uses B_λ and C_λ as free parameters; they are determined from the experimental data. Each even–even nucleus has its own parameter values.

The nuclear surface vibrations are described as phonons. The corresponding operators are defined by

$$\left.\begin{aligned} Q^+_{\lambda\mu} &= \sqrt{\frac{\omega_\lambda B_\lambda}{2}}\left(\alpha_{\lambda\mu} - i\frac{(-1)^\mu}{\omega_\lambda B_\lambda}\pi_{\lambda,\,-\mu}\right), \\ Q_{\lambda\mu} &= \sqrt{\frac{\omega_\lambda B_\lambda}{2}}\left((-1)^\mu \alpha_{\lambda,\,-\mu} + i\frac{1}{\omega_\lambda B_\lambda}\pi_{\lambda\mu}\right). \end{aligned}\right\} \tag{2.39}$$

The $\alpha_{\lambda\mu}$ and $\pi_{\lambda\mu}$ are expressed through phonon operators by

$$\left.\begin{aligned} \alpha_{\lambda\mu} &= (2\omega_\lambda B_\lambda)^{-1/2}(Q^+_{\lambda\mu} + (-1)^\mu Q_{\lambda,\,-\mu}), \\ \pi_{\lambda\mu} &= i\sqrt{\frac{\omega_\lambda B_\lambda}{2}}((-1)^\mu Q^+_{\lambda,\,-\mu} - Q_{\lambda\mu}). \end{aligned}\right\} \tag{2.39$'$}$$

The phonons are bosons; their commutation relations are

$$[Q_{\lambda\mu}, Q^+_{\lambda'\mu'}] = \delta_{\lambda\lambda'}\delta_{\mu\mu'}. \tag{2.40}$$

The ground state of an even–even nucleus is the phonon vacuum, i.e.,

$$Q_{\lambda\mu}\Psi_0 = 0. \tag{2.41}$$

The excited states are:

One-phonon state: $$Q^+_{\lambda\mu}\Psi_0 \equiv |n_\lambda = 1\rangle; \tag{2.42}$$

Two-phonon state: $$\frac{1}{\sqrt{2}}Q^+_{\lambda\mu}Q^+_{\lambda\mu}\Psi_0 \equiv |n_\lambda = 2\rangle; \tag{2.42$'$}$$

n_λ-phonon state: $$\frac{1}{\sqrt{n_\lambda!}}(Q^+_{\lambda\mu})^{n_\lambda}\Psi_0 \equiv |n_\lambda\rangle, \tag{2.42$''$}$$

while

$$Q^+_{\lambda\mu}|n_\lambda\rangle = \sqrt{n_\lambda+1}\,|n_\lambda+1\rangle, \tag{2.43}$$
$$Q_{\lambda\mu}|n_\lambda\rangle = \sqrt{n_\lambda}\,|n_\lambda-1\rangle. \tag{2.43$'$}$$

The vibrational Hamiltonian H_{vib} is now [using (2.39′) and (2.35)] rewritten as

$$H_{\text{vib}} = \sum_{\lambda\mu}\omega_\lambda\{Q^+_{\lambda\mu}Q_{\lambda\mu} + \tfrac{1}{2}\}. \tag{2.44}$$

Vibrational states are characterized by the phonon number n_λ, angular momentum I, and parity π. Figure 2.9 shows schematically the harmonic spectra of quadrupole ($\lambda = 2$) and octupole ($\lambda = 3$) vibrations. Note that the phonon energy, generally, increases with λ.

The first excited states of even–even spherical nuclei are the 2^+ states; they correspond to quadrupole vibrations. The $n_{\lambda=2} = 1$ state has $I^\pi = 2^+$ and energy ω_2, the $n_{\lambda=2} = 2$ state is a degenerate triplet $I^\pi = 0^+, 2^+, 4^+$ with energy $2\omega_2$, the $n_{\lambda=2} = 3$ is a quintet, etc. The states with $n_{\lambda=2} = 4$ are not determined by I completely (some I-values enter the multiplet several times); additional quantum numbers must be introduced in such a case.

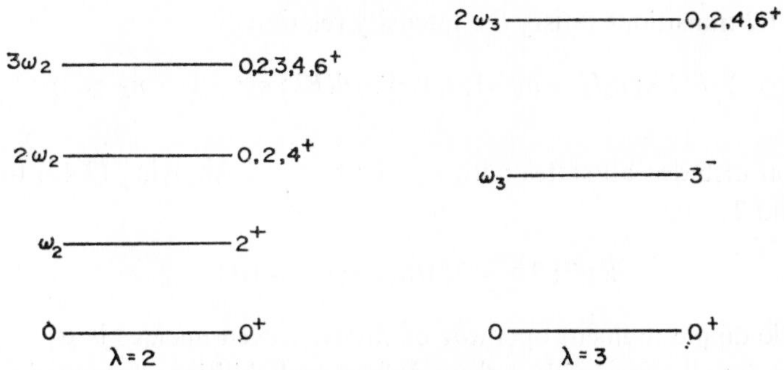

FIG. 2.9. Energy spectrum of harmonic vibrations.

The one-phonon octupole ($\lambda = 3$) state has energy ω_3 and $I^\pi = 3^-$, the $n_{\lambda=3} = 2$ states have energy $2\omega_3$ and $I^\pi = 0^+, 2^+, 4^+, 6^+$, etc. Phonons with different multipolarity can combine. For example, one quadrupole and one octupole phonon can give five states $I^\pi = 1^-, 2^-, 3^-, 4^-, 5^-$.

3. Let us derive expressions for electromagnetic transitions in the harmonic approximation of the vibrational model. The collective variables are connected with the electric multipole moments $\mathfrak{M}_E(\lambda\mu)$. The electric multipole moment of the nonspherical liquid drop is equal to

$$\mathfrak{M}_E(\lambda\mu) = \frac{3}{4\pi} ZeR_0^\lambda \alpha_{\lambda\mu} + O(\alpha^2). \tag{2.45}$$

When $Q_{\lambda\mu}$ and $Q^+_{\lambda\mu}$ operators are used instead of $\alpha_{\lambda\mu}$, we have

$$\mathfrak{M}_E(\lambda\mu) = \frac{3}{4\pi} ZeR_0^\lambda \frac{1}{\sqrt{2\omega_\lambda B_\lambda}} (Q^+_{\lambda\mu} + (-1)^\mu Q_{\lambda,-\mu}). \tag{2.45'}$$

Operator (2.45′) changes the number of oscillator quanta by one unit; this means that the selection rules for the electric transitions between vibrational states are

$$\Delta n_\lambda = \pm 1. \tag{2.46}$$

The reduced $E\lambda$ transition probability for the vibrational states is then

$$B(E\lambda; n_\lambda = 1 \to n_\lambda = 0) = \left(\frac{3}{4\pi} ZeR_0^\lambda\right)^2 \frac{1}{2\sqrt{B_\lambda C_\lambda}}. \tag{2.47}$$

In the particular case of $E2$ transitions the reduced probability can be expressed in the single-particle units ($B(E2)_{\text{s.p.}} = \frac{9}{20}\pi e^2 R_0^4$):

$$B(E2; n_2 = 1 \to n_2 = 0) = \frac{9}{16\pi^2} Z^2 e^2 R_0^4 \frac{1}{2\sqrt{B_2 C_2}}. \tag{2.47'}$$

$$B(E2; n_2 = 1 \to n_2 = 0) = \frac{25}{4\pi} \frac{Z^2}{2\sqrt{B_2 C_2}} B(E2)_{\text{s.p.}}. \tag{2.48}$$

Figure 2.1 showed that the $E2$ transitions from the first vibrational 2^+ states in even–even nuclei are enhanced 10–50 times when compared with the single-particle values.

The electric $E2$ transitions satisfy the intensity relation

$$\sum_f B(E\lambda; n_\lambda I_i \to n_\lambda - 1, I_f) = n_\lambda B(E\lambda; n_\lambda = 1 \to n_\lambda = 0). \tag{2.49}$$

The summation includes all states with $n_\lambda - 1$ phonons. Applying (2.49) to the transition from the second 2_2^+ state,

$$B(E2; 2_2^+ \to 2_1^+)/B(E2; 2_1^+ \to 0^+) = 2. \tag{2.49'}$$

The magnetic dipole moment operator of an even–even nucleus is $\mu = g_R \boldsymbol{I}$; the g_R has the same value for all states with a given λ. The probabilities of magnetic $M1$ transitions between vibrational states vanish in the harmonic approximation. The different aspects of the $M1$ transitions in the vibrational model deserve detailed analysis[76].

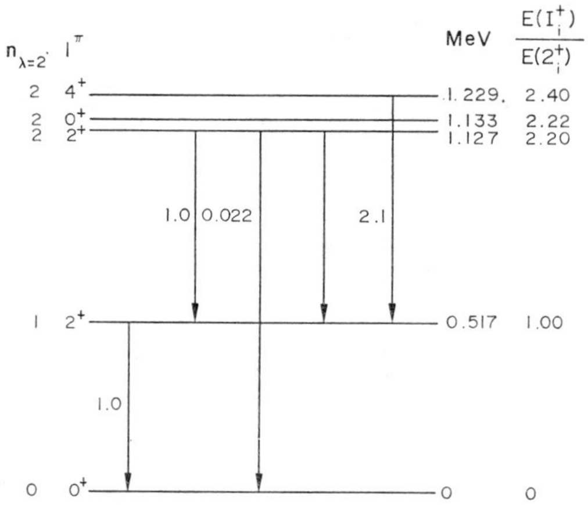

FIG. 2.10. Low-lying states in $^{106}_{46}Pd_{60}$. The relative $B(E2)$ values in units of $B(E2; n_2 = 1 \to n_2 = 0)$ are given next to the transitions.

The relations for energies and relative intensities can be used to test how correctly the vibrational model describes the excited states of even–even nuclei. Figure 2.10 shows a typical example, the spectrum of ^{106}Pd and the corresponding γ-transition intensities in units of $B(E2; n_2 = 1 \to n_2 = 0)$.

The whole body of experimental data confirms the vibrational character of the first 2^+ states in spherical even–even nuclei. The two-phonon states are somewhat more complicated. The $I^\pi = 0^+, 2^+, 4^+$ triplet should be degenerate (at energy $2\omega_2$) in the harmonic approximation. The example in Fig. 2.10 shows that all three states are actually present, but the triplet is split. Similarly, the $E2$ transitions from the second 2_2^+ state to the ground state are observed, but are very slow, and the forbidden $M1$ transitions connecting 2_2^+ with 2_1^+ are also strongly hindered. The relative $E2$ transition probabilities agree qualitatively with the predictions of the harmonic model; but deviations of 30–50% are often encountered.

Thus, the harmonic vibrational model gives basically a correct description of the lowest excited states in even–even spherical nuclei. It is even sometimes possible to explain excited states in odd-A nuclei, related to the coupling of the odd nucleon to the vibrations of even–even core.

The effect of the anharmonic corrections was studied, for example, in refs. 77 and 78. It was, however, impossible to explain fully the existing discrepancies between the predictions of the model and experimentally determined energies and transition probabilities.

4. Let us discuss the surface vibrations of even–even nonspherical nuclei. The intrinsic angular momentum is not conserved in these nuclei; the excitations are, therefore, characterized by the parity $(-1)^{\lambda}$ and angular momentum projection μ. It is convenient to use the coordinate system with the main axes of the ellipsoid as the coordinate axes.

The relevant variables for the quadrupole vibrations are α_{20} and $\alpha_{22} = \alpha_{2,-2}$. They can be related to the deformation parameters β and γ by

$$\left.\begin{aligned} \alpha_{20} &= \beta \cos\gamma, \\ \alpha_{22} = \alpha_{2,-2} &= \frac{1}{\sqrt{2}} \beta \sin\gamma. \end{aligned}\right\} \tag{2.50}$$

We shall assume that the equilibrium shape is axially symmetric (characterized by the deformation parameter β_0) and that the quadrupole vibrations have small amplitudes and are, therefore, harmonic. The collective Hamiltonian is then

$$H_{\text{vib}} + T_{\text{rot}} = T_{\text{rot}} + H^{\beta}_{\text{vib}} + H^{\gamma}_{\text{vib}}, \tag{2.51}$$

where

$$T_{\text{rot}} = \frac{1}{6B_2\beta_0^2}(I^2 - I^2_{z'}), \tag{2.52}$$

$$H^{\beta}_{\text{vib}} = -\frac{1}{2B_2}\frac{1}{\beta^4}\frac{\partial}{\partial\beta}\left(\beta^4\frac{\partial}{\partial\beta}\right) + \frac{1}{2}C_{\beta}(\beta-\beta_0)^2, \tag{2.53}$$

$$H^{\gamma}_{\text{vib}} = -\frac{1}{2B_2\beta_0^2}\left\{\frac{1}{\gamma}\frac{\partial}{\partial\gamma}\left(\gamma\frac{\partial}{\partial\gamma}\right) - \frac{I^2_{z'}}{4\gamma^2}\right\} + \frac{1}{2}C_{\gamma}\beta_0^2\gamma^2 \tag{2.53'}$$

(see, for example, ref. 79). The variables are (approximately) separable and the solutions of the Schrödinger equation are expressed as products of three functions: one depends on the variable β, the other depends on γ, and the third one depends on the Euler angles.

The vibrations, related to the α_{20} variable, preserve the axial symmetry. They are called "β-vibrations" and have quantum numbers $K^{\pi} = 0^+$. The vibrations, related to the α_{22} variable, are called "γ-vibrations"; they correspond to $K^{\pi} = 2^+$ and the axial symmetry is not preserved during the vibration. The energy of the collective states, corresponding to the Hamiltonian (2.51), consists of three terms

$$E_{\text{v.r.}} = E_I + E^{\beta}_{\text{vib}} + E^{\gamma}_{\text{vib}}. \tag{2.54}$$

The rotational energy is equal to

$$E_I = \frac{I(I+1) - K^2}{6B_2\beta_0^2}, \tag{2.55}$$

the energy of the $K^{\pi} = 0^+$ β-vibrations is

$$E^{\beta}_{\text{vib}} = \omega_{\beta}(n_{\beta} + \tfrac{1}{2}), \tag{2.56}$$

where $\omega_\beta = \sqrt{C_\beta/B_2}$, $n_\beta = 0, 1, 2, \ldots$, the energy of the $K^\pi = 2^+$ γ-vibrations is

$$E^\gamma_{\text{vib}} = \omega_\gamma(n_\gamma + 1), \tag{2.57}$$

where $\omega_\gamma = \sqrt{C_\gamma/B_2}$, $n_\gamma = 0, 1, 2, \ldots$.

The reduced transition probability for the electric excitation of a one-phonon state is determined by

$$B(E\lambda;\, K_i I_i n_{\lambda\mu} = 0 \rightarrow K_f I_f n_{\lambda\mu} = 1)$$
$$= \left(\frac{3}{4\pi} ZeR_0^\lambda\right)^2 \frac{\omega^{\lambda\mu}}{2C_{\lambda\mu}} (I_i K_i \lambda \pm \mu \mid I_f K_f \pm \mu)^2 \,[1 + \delta_{K_i 0}(1 - \delta_{\mu 0})], \tag{2.58}$$

where $\omega^{20} = \omega_\beta$, $\omega^{22} = \omega_\gamma$, $C_{20} = C_\beta$, $C_{22} = C_\gamma$. Note that (2.58) connects all states in the ground-state rotational band with all states in the rotational band of the vibrational state.

The experimental reduced $E2$ transition probabilities for excitation of the quadrupole vibrational states are typically

$$B(E2)_{\text{s.p.u.}} = 2 - 10. \tag{2.59}$$

Vibrational states with negative parity and $K^\pi = 0^-, 1^-, 2^-, 3^-$ (octupole vibrations) should exist in deformed nuclei according to the phenomenological theory. Many such states were actually observed, particularly the $K^\pi = 0^-$ states in actinide nuclei. The experimental $E3$ probabilities are, typically,

$$B(E3)_{\text{s.p.u.}} = 2 - 20. \tag{2.60}$$

Figure 2.5 shows three types of the collective vibrational states in ^{228}Th and the corresponding rotational bands. The octupole state $K^\pi = 0^-$ is the lowest vibrational state in this nucleus; it is followed by the β-vibrational state at 825 keV and γ-vibrational state at 969 keV. The very low energy of the octupole state in ^{228}Th is unusual; as a rule the octupole vibrational levels are somewhat higher than the quadrupole states. The $K^\pi = 0^-$ state in ^{228}Th, and similar states in other light actinide isotopes, do not follow the general pattern of phenomenological description of the nuclear surface vibrations.

§ 4. Coupling of the Collective and Intrinsic Motion

1. As we have shown, the predictions of the adiabatic rotational model are in rather good agreement with the experimentally determined energies and probabilities of electromagnetic transitions. The experimental data are, however, accurate enough that systematic deviations from the theoretical predictions can be found. Theoretically, we would expect correction terms in equations for energies and transition probabilities, which depend on the total angular momentum I. The corrections are related to the Coriolis forces, nuclear centrifugal stretching, and to the coupling of rotations and vibrations.

The nuclear rotations are slow in the adiabatic rotational model. Hence, the correction terms can be found if the corresponding matrix elements are expanded in a power series of angular momentum. The usual adiabatic model is identical with the lowest order terms; if the model is applicable, the series converges rapidly. The coefficients in the correction terms are often not calculated, but determined from experimental data.

The wave function $\varphi_K(q)$, describing the intrinsic motion, is not an eigenfunction of the total angular momentum operator. To avoid violation of the angular momentum conservation law it is necessary to project out a component of $\varphi_K(q)$ with the definite value of I. In a crude way such a projection is achieved by multiplication with the rotational function $D^I_{MK}(\theta_e)$. All equations contain additional expressions when the projection of the angular momentum is performed more accurately; for example, the formula for the rotational energy [eqn. (2.9) with the $I(I+1)$ dependence] will contain terms with a more complicated I-dependence.

The coupling of the collective and intrinsic motion is described, for example, in refs. 17, 60–63, and 79–87.

Let us see how the Coriolis interaction affects energies in a rotational band. To do that, we shall calculate the energy corrections in the second-order perturbation theory. They are:

$$\begin{aligned}\Delta E_I(q) &= \sum_{K',q'} \frac{|(\Psi^{I*}_{MK'}(q')\,H_{\rm cor}\Psi^I_{MK}(q))|^2}{E(K,q)-E(K',q')} \\ &= -\frac{1}{4\mathcal{J}^2}\sum_{q'}\left\{\frac{|\langle K+1,q'\,|J_+|\,Kq\rangle|^2}{E(K+1,q')-E(K,q)}[I(I+1)-K(K-1)]\right. \\ &\quad \left.+\frac{|\langle K-1,q'\,|J_-|\,K,q\rangle|^2}{E(K-1,q')-E(K,q)}[I(I+1)-K(K+1)]\right\},\end{aligned}$$

where q denotes all additional quantum number sexcept $I^\pi K$. The I-dependent part of the above formula is then

$$\Delta E_I(q) = -\frac{I(I+1)}{\mathcal{J}^2}\sum_{q'}\frac{|\langle K\pm1,q'\,|J_\pm|\,Kq\rangle|^2}{E(K\pm1,q')-E(K,q)}+\dots\,. \tag{2.61}$$

It is evident that the second-order corrections in $H_{\rm cor}$ are proportional to $I(I+1)$, i.e., they lead only to a renormalization of the moment of inertia. To obtain "real" correction terms, higher-order perturbation theory must be used. Note that the perturbation theory is applicable, obviously, if and only if $E(K\pm1,q')-E(K,q) \gg 1/2\mathcal{J}$.

The wave function of a state with a certain K-value will contain admixtures with $K\pm1$, $K\pm2$, etc., when the Coriolis interaction is taken into account. In some cases, the energy denominators $E(K'q')-E(Kq)$ may be small and the matrix elements $\langle K'q'\,|J_\pm|\,Kq\rangle$ large at the same time; the K cannot be used as a quantum number then, because the admixtures with other K-values are large. One of such cases, ^{183}W, was studied in refs. 85 and 88. The calculations were exact, i.e., the perturbation theory was not utilized. The mentioned studies have proven that the phenomenological theory with Coriolis coupling explains the observed energy and transition probabilities.

The Coriolis coupling in odd-A nuclei changes values of the gyromagnetic ratios g_K and g_R, and for $K = \frac{1}{2}$ states, values of the quantity $(g_K-g_R)b_0$. Let us give, for completeness, the expressions for these correction terms:

$$\Delta g_R = -\frac{1}{2\mathcal{J}}\sum_{q'}\left\{\frac{\langle Kq|\,J_+\,|K-1,q\rangle\langle K-1,q'|\,(g^{(l)}-g_R)l_-+(g^{(s)}-g_R)s_-\,|Kq\rangle}{E(K,q)-E(K-1,q')}\right.$$
$$\left.+\frac{\langle Kq|\,(g^{(l)}-g_R)l_-+(g^{(s)}-g_R)s_-\,|K+1,q'\rangle\langle K+1,q'|\,J_+\,|Kq\rangle}{E(K,q)-E(K+1,q')}\right\}, \tag{2.62}$$

$$\Delta g_K = \frac{1}{2\mathcal{J}}\sum_{q'}\left\{\frac{\langle Kq|\,(g^{(l)}-g_R)l_-+(g^{(s)}-g_R)s_-\,|K+1,q'\rangle\langle K+1,q'|\,J_+\,|Kq\rangle}{E(K,q)-E(K+1,q')}\right.$$
$$\left.-\frac{\langle Kq|\,J_+\,|K-1,q'\rangle\langle K-1,q'|\,(g^{(l)}-g_R)l_-+(g^{(s)}-g_R)s_-\,|Kq\rangle}{E(K,q)-E(K-1,q')}\right\}, \tag{2.63}$$

$$\Delta((g_K-g_R)b_0)$$
$$= -\frac{g^{(s)}-g^{(l)}}{\mathcal{J}}\sum_{q'}\frac{\langle K=\tfrac{1}{2},q|\,s_{z'}\,|K=\tfrac{1}{2},q'\rangle\langle K=\tfrac{1}{2},q'|\,J_+\,|K=-\tfrac{1}{2},q\rangle}{E(K=\tfrac{1}{2},q)-E(K=\tfrac{1}{2},q')}. \tag{2.64}$$

2. The Coriolis coupling included in higher orders of the perturbation theory changes eq. (2.9) for the ground-state rotational bands of even–even deformed nuclei into

$$E_I = \mathfrak{A}I(I+1)+\mathfrak{B}I^2(I+1)^2+\mathfrak{C}I^3(I+1)^3, \tag{2.65}$$

with $\mathfrak{A} = 1/2\mathcal{J}$. In well-deformed nuclei the empirical values of $\mathfrak{B}/\mathfrak{A}$ are about 10^{-3} and of $\mathfrak{C}/\mathfrak{A}$ about 10^{-5}–10^{-6}. The values of $\mathfrak{B}$ for nuclei with $150 < A < 190$ are shown in Fig. 2.11. The sudden increase of $\mathfrak{B}$ in the transitional region is apparent; $\mathfrak{C}$ behaves in a similar way.

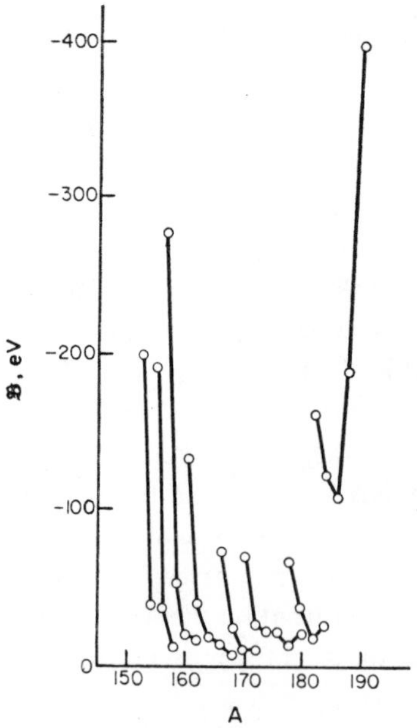

FIG. 2.11. Experimental values of the $\mathfrak{B}$-parameter in (2.65). The 2^+, 4^+, and 6^+ states were used; the experimental energies are from ref. 60.

Formula (2.65) must be modified for the $K = 1$ and $K = 2$ states. The additional term for $K = 1$ is

$$\Delta E_I = (-1)^{I+1}(I+1)\{\mathfrak{A}_2+\mathfrak{B}_2 I(I+1)+\ldots\}, \tag{2.66}$$

and for $K = 2$

$$\Delta E_I = (-1)^I(I-1)(I+1)(I+2)\{\mathfrak{B}_4+\mathfrak{C}_4 I(I+1)+\ldots\}. \tag{2.66'}$$

Let us return to the ground-state rotational band. For the angular momenta $I > 8$, eqns. (2.9) or (2.65) fail; the power expansion in $I(I+1)$ is, apparently, inapplicable. Rotational spectrum of ^{170}Hf was shown in Fig. 2.3; note that at $I = 10$ deviation from (2.9) is already 327 keV.

Rotational bands can be characterized by the I dependence of the ratio of the effective moments of inertia, i.e.,

$$\frac{\mathcal{J}_I^{\text{eff}}}{\mathcal{J}_{I+2}^{\text{eff}}} = \frac{E_{I+2}-E_I}{4I+6}\bigg/\frac{E_I-E_{I-2}}{4I-2}.$$

The above discussed band in ^{170}Hf is analyzed again in Fig. 2.12. The $\mathcal{J}_I^{\text{eff}}/\mathcal{J}_{I+2}^{\text{eff}}$ according to (2.65) with and without the $\mathfrak{C}$ term are shown together with the predictions of the Davydov–Chaban model and with the corresponding experimental values. It is obvious that (2.65) poorly describes the high angular momentum states.

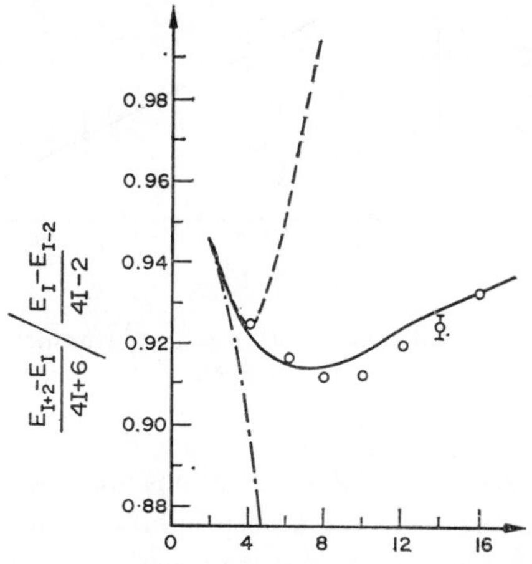

FIG. 2.12. Ratios of the effective moments of inertia $\mathcal{J}_I^{\text{eff}}/\mathcal{J}_{I+2}^{\text{eff}}$ for the ground-state rotational band in ^{170}Hf. The dots denote the experimental values, the dashed curve was calculated according to (2.65) with $\mathfrak{B}/\mathfrak{A} = -2.7+10^{-3}$ and $\mathfrak{C} = 0$. The dot-and-dashed curve was calculated according to (2.65) with $\mathfrak{B}/\mathfrak{A} = -3.1\times10^{-3}$ and $\mathfrak{C}/\mathfrak{A} = 1.8\times10^{-5}$. The full curve is calculated in Davydov–Chaban model with $\gamma = 0$ and $\mu = 0.29$. The figure is reproduced from ref. 61.

Davydov and Chaban[62] formulated a special model for the description of the rotational bands in even–even deformed nuclei. The model contains a new nonadiabaticity parameter μ, the variable γ is replaced by its effective value, and the variable β remains free. The μ^2 value is equal to twice the square of ratio of the zero-point β-vibration amplitude and the β^2 value. The $\mu = 0$ corresponds to the adiabatic approximation.

The values of γ_{eff} and μ are found in tables published in ref. 89, when experimental ratios of energies of three collective levels are known in the studied nucleus. The same tables then give energies of the remaining collective states. Figures 2.3 and 2.12 demonstrate that the Davydov–Chaban model works quite well; note, however, that in other cases (e.g. in ^{172}Hf, ^{166}Yb, or ^{164}Yb) the agreement between the Davydov–Chaban model and experiment is not as spectacular as in ^{170}Hf.

The Coriolis interaction has an analogous effect in odd-A nuclei as in the even–even nuclei, particularly, the $K=\frac{1}{2}$ and $K=\frac{3}{2}$ bands have additional sign-changing terms

$$\begin{aligned} E_I = {} & \mathfrak{A}I(I+1)+\mathfrak{B}I^2(I+1)^2+\mathfrak{C}I^3(I+1)^3 \\ & +\delta_{K,\frac{1}{2}}(-1)^{I+\frac{1}{2}}\left(I+\tfrac{1}{2}\right)\{\mathfrak{A}_1+\mathfrak{B}_1I(I+1)+\mathfrak{C}_1I^2(I+1)^2\} \\ & +\delta_{K,\frac{3}{2}}(-1)^{I+\frac{3}{2}}\left(I-\tfrac{1}{2}\right)\left(I+\tfrac{1}{2}\right)\left(I+\tfrac{3}{2}\right)\{\mathfrak{B}_3+\mathfrak{C}_3I(I+1)\}. \end{aligned} \tag{2.67}$$

The $\mathfrak{B}_1$ and $\mathfrak{B}_3$ are of similar magnitude as $\mathfrak{B}$, the $\mathfrak{C}_1$ and $\mathfrak{C}_3$ are of similar magnitude as $\mathfrak{C}$; $\mathfrak{B}$ is about hundred times larger than $\mathfrak{C}$.

Figure 2.13 shows the $K=\frac{3}{2}$ rotational band in ^{159}Tb; it shows how the quantity

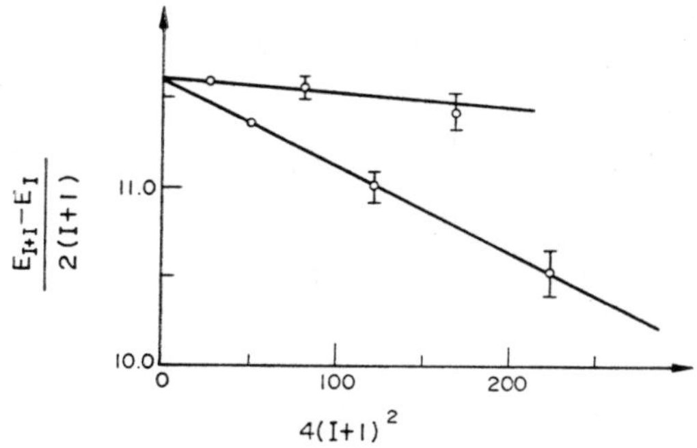

FIG. 2.13. Analysis of the rotational band of the $\frac{3}{2}+[411]$ ground state in ^{159}Tb. The figure is reproduced from ref. 90.

$(E_{I+1}-E_I)/2(I+1)$ depends on $4(I+1)^2$. Calculations are performed according to (2.67); the two straight lines are related to the term proportional to $\mathfrak{B}_3$, the upper line corresponds to positive $(-1)^{I+\frac{3}{2}}$, the lower line to the negative values. Equation (2.67) evidently agrees well with the experimental data.

3. The admixtures in the wave function, which correspond to different K-values, affect the intensities of electromagnetic transitions in deformed nuclei too.

Equations (2.15′) and (2.18) describe the transitions inside a rotational band quite well. The coupling of the rotational and intrinsic degrees of freedom will add corrections to the expression for the transition matrix elements. Their effect is, however, small, because the main term has a collective character and is, consequently, dominant.

The electromagnetic transitions between different rotational bands are considerably more sensitive to admixtures. While in allowed transitions the admixtures are usually unimportant, they can play a decisive role in forbidden transitions.

Let us consider a transition between two rotational bands with $|K_i - K_f| = \lambda + \nu$; $\nu \geqslant 0$ is an integer (the transition is ν times K-forbidden). The corresponding reduced transition probability is equal to

$$B(\lambda;\, I_iK_i \to I_fK_f) = (I_iK_i \pm \nu\lambda K_f - K_i \mp \nu \,|\, I_fK_f)^2 \times |\, M_0 + \{I_f(I_f+1) - I_i(I_i+1)\}M_1 + \delta_{K_i,\frac{1}{2}}(-1)^{I_i+\frac{1}{2}}(I_i+\tfrac{1}{2})M_2 + \delta_{K_f,\frac{1}{2}}(-1)^{I_f+\frac{1}{2}}(I_f+\tfrac{1}{2})M_3 \,|^2, \quad (2.68)$$

where the parameters M_1, M_2, M_3 are related to the admixtures in the wave function; they were calculated in ref. 80. The upper sign in (2.68) is used when $K_i < K_f$, the lower sign when $K_i > K_f$. The formula is also applicable for K-allowed transitions, i.e., when $|K_i - K_f| = \lambda$ and $\nu = 0$. The expression for $B(\lambda)$ in the case, when $|K_i - K_f| < \lambda$, is quite complicated; it is also quoted in ref. 80.

The formula (2.68) for the $E1$ transitions is much simpler, i.e.,

$$B(E1;\, I_iK_i \to I_f,\, K_i+1) = (I_iK_i11 \,|\, I_fK_i+1)^2 \,|\, M_0 + M_1\{I_f(I_f+1) - I_i(I_i+1)\} \,|^2. \quad (2.69)$$

The $E1$ transitions in deformed nuclei are often hindered and, therefore, quite sensitive to admixtures. Nevertheless, by including the first-order corrections in I into the matrix element it is possible to explain the relative intensities of many $E1$ transitions in odd-A deformed nuclei[91].

Formula (2.68), applied to the $E2$ transitions from the γ-vibrational band to the ground-state band in even–even nuclei, is of the form

$$B(E2;\, I_iK_i = 2 \to I_f,\, K_f = 0) = (I_i22-2 \,|\, I_f0)^2 \,|\, M_0 + M_1\{I_f(I_f+1) - I_i(I_i+1)\} \,|^2. \quad (2.70)$$

Parameters M_0 and M_1 are related to the parameter z_2 introduced in ref. 92;

$$z_2 = \frac{2M_1}{M_0 - 4M_1}. \quad (2.71)$$

A large number of electromagnetic transitions in deformed nuclei have now been analyzed and the corresponding parameters were found. These analyses show that the parameters, calculated from various relative transition intensities, generally agree with each other. This means that formula (2.68) basically describes correctly electromagnetic transitions between two rotational bands. At the same time, in many K-allowed cases the intensity rules (2.28) do not agree with the experimental data. Relative intensities of many K-forbidden transitions were also satisfactorily explained in ref. 81.

The just described explanation of the intensities of electromagnetic transitions, in which the corresponding matrix element is expanded into a power series of total momentum I, contains no assumptions about the intrinsic structure of the involved states. The success of this approach shows the fruitfulness of the phenomenological method, which, however, gives no information about the intrinsic structure of the participating nuclear states.

CHAPTER 3

SEMIMICROSCOPIC METHOD IN NUCLEAR THEORY

§ 1. Introduction

1. The phenomenological method uses collective coordinates $\alpha_{\lambda\mu}$ for the description of complex nuclei. The excited nuclear states are interpreted as rotations of the whole nucleus or as vibrations of the nuclear surface. Each nucleus is characterized by a certain number of parameters. The observable quantities, for instance, the excitation energies, multipole moments, and electromagnetic transition probabilities, depend on the values of the parameters, which are determined from experimental data for each nucleus independently. The phenomenological method and its successful features were briefly described in the preceding chapter.

However, such a description has also serious limitations and shortcomings. The adiabaticity condition (2.1) must be, first of all, fulfilled. But strictly speaking, this condition is not valid in odd-A and odd–odd nuclei. The condition $\omega_{\text{vib}} \ll \omega_{\text{in}}$ is not valid in even–even deformed nuclei, and in even–even spherical nuclei for all states, except the first 2^+ states. On the other hand, the phenomenological theory can be formulated in such a way that the condition $\omega_{\text{rot}} \ll \omega_{\text{vib}}$ is not necessary.

The phenomenological description cannot predict or even explain the variations of nuclear characteristics when the number of particles in the nucleus is changed.

It is natural to assume that the properties of quadrupole and octupole states should change slowly when the number of nucleons is changed, and, particularly, that the sequence of collective states should be the same in all nuclei. The experimental data, however, contradict this assumption. For example, the $K^\pi = 0^-$ states are below the β- and γ-vibrational states in even–even isotopes of thorium, uranium, and plutonium, and above them in other deformed nuclei. Similarly the γ-vibrations have lower energies than the β-vibrations in even–even dysprosium and erbium isotopes, etc. These strong variations of energies of quadrupole and particularly of octupole states cannot be explained by the phenomenological theory.

There are other facts unexplainable by phenomenological methods. The excitation spectrum in many even–even spherical nuclei is very different from the harmonic quadrupole spectrum. For instance, the first excited state in ^{90}Zr has spin zero, not two; in other nuclei (^{70}Ge, ^{140}Ce) the first 0^+ states are very close to the first 2^+ states. It is impossible to account for the splitting of the two-phonon quadrupole triplet and the corresponding electromagnetic transition probabilities.

Also, the use of collective coordinates excludes the interpretation of collective vibrational states as a result of coherent motion of many nucleons. This is, perhaps, the most serious shortcoming of the phenomenological description, because it renders useless the whole complex of experimental information about the structure of vibrational states.

Not only the previously mentioned limitations of the phenomenological approach but also the desire to find an explanation of nuclear states, starting from the interactions between particles, has led to the formulation of a new approach in nuclear theory.

2. Studies of nuclear matter and calculations of main nuclear characteristics in the Hartree–Fock method are among the most important microscopic developments in nuclear theory. In them, the information from the nucleon–nucleon scattering is used, and basic nuclear properties (e.g., density of nuclear matter or binding energy per nucleon) are calculated.

The experimental data from the nucleon–nucleon scattering are insufficient for a unique determination of the nucleon–nucleon potential, which consists of central, spin–orbit, and tensor terms, and contains up to 30 parameters. Several empirical potentials, including the Yale,[93] Hamada–Johnston,[94] the nonlocal Tabakin potential,[95] and others, explain the experimental data equally well. All these potentials have a strong short-range repulsion ("hard" or "soft" core) and a weaker long-range attraction.

Often used are the potentials[96,97] based on the exchange of π-mesons and heavier ϱ-, ω-, η-, and other mesons. The short-range repulsion and the short-range spin–orbit interaction can be explained by the exchange of heavy vector mesons.

The nucleon–nucleon potential is not well determined at short distances between the interacting particles. Hopefully, the properties of nuclei are insensitive to this part. Moreover, the whole concept of describing the nucleon–nucleon interaction by a potential is only an approximation, particularly at short distances.

Only two-particle forces are generally considered in nuclear theory. Consideration of two-particle forces rests on an assumption; there is no proof that three-, four-, or more particle forces do not exist. Neither is there any direct experimental evidence that such interactions are necessary. Moreover, the observed correlations of three, four, or more particles may be caused by two-particle interactions.

It is important to remember that the nucleon–nucleon scattering provides information about the "on-mass shell" behavior of the nucleon–nucleon amplitude (i.e., when the momenta of initial and final states obey the energy and momentum conservation conditions). It is necessary to know this amplitude off-mass shell, when nuclear characteristics are calculated. The corresponding experimental information can be obtained, for example, by studying the *bremsstrahlung* in p–p scattering.

Let us turn our attention now to nuclear matter. The energy per particle in nuclear matter, B, is equal to the coefficient of the volume energy in the semiempirical mass formula (1.35); i.e., B is about 16 MeV. The nucleon density inside a nucleus is about one-sixth nucleon per $(10^{-13}\text{ cm})^3$. This corresponds to the radius $r_0 \approx 1.1\times10^{-13}$ cm and Fermi momentum $p_F \approx 1.36\times10^{13}\text{ cm}^{-1}$. Nuclear matter theory tries to explain the empirical values of B and p_F by means of the experimental nucleon–nucleon scattering data.

To achieve this, it was necessary to develop a corresponding mathematical apparatus (see ref. 23). The theory uses, in reality, a perturbation approach; the small parameter is the ratio between the hard-core radius and the nucleon radius. Major difficulties are related to the presence of the infinite repulsive core in the nucleon–nucleon potential; they can be overcome in the framework of the Brueckner–Goldstone formalisms.[23, 98–100]

Despite the progress described above in the formal solution of many-body problems, the attempts to obtain correct values of nuclear density and of energy per nucleon at the same time, were unsuccessful. The values of B equal 8–13 MeV (instead of 16 MeV) were obtained when nucleon–nucleon potentials and correct nuclear density were used; many results of B and r_0 calculations are collected in ref. 101. The three-body correlations, the more complicated diagrams, and tensor forces, etc., did not remove this discrepancy[102]. The mentioned results are not, however, final. Systematic studies of potentials with soft-core might be more successful.

The calculations of the basic characteristics of nuclear matter have excluded several potentials which satisfactorily describe nucleon–nucleon scattering. Among them, for example, are the hard-core potentials.

Calculations of the nuclear binding energies, of charge and mass distribution, and of the single-particle energy levels in the Hartree–Fock method, are also very significant. The methods employed in these calculations are described in refs. 103–105. Again, experimental information obtained in nucleon–nucleon scattering is used. Main complications are related to the calculation of matrix elements; it is, therefore, advantageous to use a nonlocal, two-particle potential, as in ref. 95, or corresponding effective potentials as in refs. 106 and 107. Brueckner, Eden, Ulehla, Kerman, Baranger, Mamasakhlisov, and others[108–114] obtained interesting results. It turns out again that it is quite difficult to get correct binding energies and nuclear radii at the same time.

The Hartree–Fock calculations are more and more common; the number of studied nuclei is ever growing. Application of these methods to nonmagic nuclei is, however, rather complicated. Another promising approach to the calculation of basic nuclear characteristics is the K-harmonic method[115].

3. Other directions in nuclear theory can be characterized as semimicroscopic; they involve the concept of effective nuclear interaction. Such an effective interaction is determined by various methods. Nucleon–nucleon potentials are used for calculation of effective interaction in ref. 116. In other papers, the average field is isolated first, and the effective interaction is calculated or postulated later. Naturally, there is considerable arbitrariness in the choice of effective interaction; the criterion of correctness must be in agreement with the experimental data. Some limitations on the choice of effective interaction are imposed by the general symmetry relations. They are, in most cases, trivial; the only exceptions are the limitations caused by gauge invariance of the theory.

One variant of the semimicroscopic approach is the so-called "theory of finite Fermi-systems"[13, 14]. It uses an empirical "effective" interaction of the δ-function type. The interaction strengths depend on the local nuclear density, i.e., they are different inside and outside the nuclear surface. It is possible to calculate various nuclear characteristics by using such an effective interaction and a proper nuclear average field.

Most popular semimicroscopic theories use nuclear model Hamiltonians. For each class of nuclear effects one finds the corresponding, most important component of nuclear forces; only this component is then used in the calculations. Such a semimicroscopic approach is a natural generalization of the phenomenological theories. Its success is related to its limitation; it attempts to explain properties of only a few low-lying nuclear states. The nuclear many-body problem is thus reduced to a problem with a limited number of degrees of freedom.

The parameters of nuclear average field are chosen by analyzing a large collection of experimental data supplied by nuclear reactions and other nuclear structure studies. Determination of the effective residual nucleon–nucleon interaction is a somewhat more complicated problem. It is solved, in this particular variant of semimicroscopic theory, by a heavy reliance on experimental data. Schematically, the following procedure is used: nuclear spectra are qualitatively analyzed and a trial form of the effective interaction is chosen. Properties of a whole group of nuclei are then calculated and compared with the available experimental data, and suggestions for supplementary experiments are made. New experimental data make possible the adjustment of the effective interaction, followed by additional calculations; then the whole procedure is repeated.

The before-mentioned semimicroscopic approach was developed after the theory of pairing correlations of the superfluid type was formulated. The nuclear model Hamiltonian was divided into the nuclear average field, interactions causing pairing correlations, and the residual interaction. The residual interaction was usually represented by multipole–multipole interaction, surface δ-interaction, etc.

The theory described above is often called the "superfluid nuclear model." The rotational states are treated phenomenologically, and a number of parameters obtained from analysis of the experimental data are used for their handling. The vibrational states, on the other hand, are described microscopically; they are interpreted as a coherent effect of many particles. All excited states, including vibrational states, involving change in the intrinsic structure, are explained in a unified way. The adiabaticity condition $\omega_{vib} \ll \omega_{in}$ is never used.

The superfluid nuclear model describes low-lying excited states (2–4 MeV) in medium and heavy nuclei. One cannot expect that the model will accurately predict the nuclear binding energies; instead, it is designed to describe nuclear spectra, i.e., differences of the energies of nuclear ground and of excited states. Similarly, the model describes relative α-, β-, and γ-transition probabilities better than the corresponding absolute rates.

4. Before explaining the nuclear superfluid model, let us enumerate the main experimental facts not explained in the independent particle model:

(1) Existence of energy gap in spectra of even–even nuclei (only intrinsic excitations are counted, i.e., rotational states are excluded). Such a gap is not observed in spectra of odd-A or odd–odd nuclei. Figure 3.1 shows spectra of the nonrotational states in several deformed nuclei. Note that the first excited states in odd-A and odd–odd nuclei have energies typically 10–100 keV, while in even–even nuclei they are usually above 1 MeV. Analogous situations are observed in all nuclei, including the spherical nuclei.

(2) Moments of inertia of nonspherical nuclei. Experimental values of the moments of inertia in many nuclei were deduced from measured energies of rotational states. Moments of inertia two to three times larger than the experimental values were calculated in the independent particle model. Another unexplained fact is the variation of moments of inertia between the odd-A nuclei and even–even nuclei. The moments of inertia of odd-A nuclei are considerably larger; the difference far exceeds the contribution of a single nucleon.

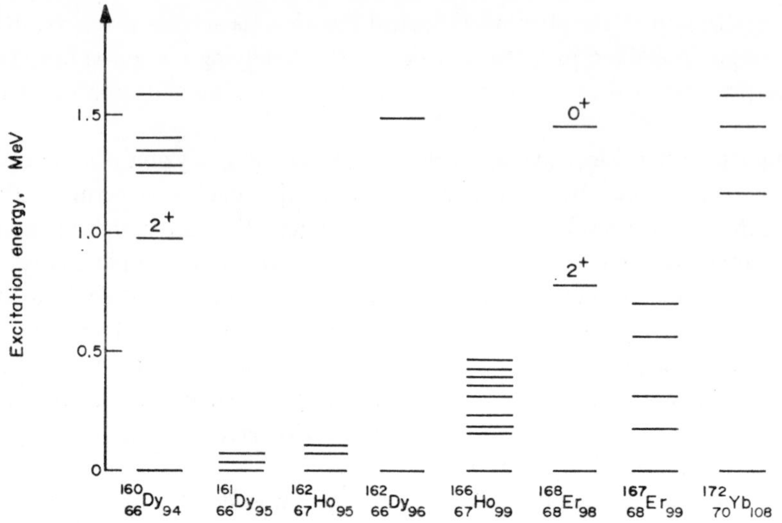

FIG. 3.1. Excitation spectra of several deformed nuclei (only states corresponding to excitation of intrinsic degrees of freedom are shown). 2^+ and 0^+ denote the γ- and β-vibrational states, respectively.

(3) It is difficult to explain nuclear equilibrium form by means of the independent particle model. Actually, this model predicts that all nonmagic nuclei are deformed, while in reality the transition point from the spherical to the ellipsoidal form is reached only when about a quarter of the states in the last unfilled shell are occupied.

(4) The experimentally determined density of the single-particle states in odd-A deformed nuclei is twice as large as the density calculated from the Nilsson potential. The density of the single-particle states in the independent particle model is connected with the nuclear matter density; consequently, the mentioned discrepancy is unchanged when different average field potentials are used.

(5) Many β-transitions, strongly forbidden in the independent particle model (because they correspond to a change of state of several particles at once), have unhindered rates.

(6) In using the independent particle model one encounters a number of difficulties in describing α-decay. Nuclear radii, calculated from the α-decay lifetimes, are larger than the radii determined in nuclear reactions. The calculated reduced α-decay probabilities for ground-state–ground-state transitions in even–even nuclei strongly fluctuate, while experimentally only small fluctuations are observed. The decrease in reduced α-decay probabilities for unfavored transitions relative to favored ones is not explained.

All these shortcomings of the independent particle model have stimulated development of the nuclear superfluid model.

This chapter lays the foundation for our attempted deductive exposition of the nuclear theory. The variational principle is used in § 2 for isolation both of the nuclear average field and the pairing interactions which lead to superfluid correlations. The nuclear model Hamiltonian is discussed in § 3.

§ 2. The Hartree–Fock–Bogolyubov Variational Principle†

1. As we have mentioned earlier, the Hartree–Fock variational method is among the basic methods of solution of the many-body problem. Its application in atomic and molecular physics was particularly fruitful; it is also used often in nuclear matter studies. It must be remembered, however, that the corresponding trial wave functions describe independent particles; hence pairing and more complicated correlations are excluded.

A new variational principle, which is a natural generalization of the Hartree–Fock method, was suggested by Bogolyubov.(117) The minimum is sought on a broader class of wave functions in the Bogolyubov method; the trial function contains not only wave functions of the individual particles, but also wave functions of pairs of particles. The whole method is usually called the Hartree–Fock–Bogolyubov Variational Principle. Let us describe the corresponding equations.

The Hamiltonian of interacting nucleons is

$$H = \sum_{f,f'} T'(f,f')a_f^+ a_{f'} - \tfrac{1}{4}\sum_{f_1,f_2,f_2',f_1'} G(f_1,f_2;f_2',f_1')a_{f_1}^+ a_{f_2}^+ a_{f_2'} a_{f_1'}, \tag{3.1}$$

where f is the set of quantum numbers which describe the state of a nucleon. The creation and annihilation operators fulfill the usual anticommutation relations

$$a_{f_1}^+ a_{f_2} + a_{f_2} a_{f_1}^+ = \delta_{f_1,f_2}, \quad a_{f_1} a_{f_2} + a_{f_2} a_{f_1} = 0. \tag{3.2}$$

Further,

$$T'(f, f') = T(f, f') - \lambda\delta_{f,f'}, \tag{3.3}$$

and λ denotes the chemical potential.

The real functions $T(f, f')$ and $G(f_1, f_2; f_2', f_1')$ have the following symmetry properties:

$$\left.\begin{aligned} T(f, f') &= T(f', f), \\ G(f_1, f_2, f_2', f_1') &= -G(f_1, f_2; f_1', f_2') = G(f_1', f_2'; f_2, f_1). \end{aligned}\right\} \tag{3.4}$$

Functions $T(f, f')$ and $G(f_1, f_2; f_2', f_1')$ may be complex in the most general cases; the symmetry relations (3.4) are then replaced by

$$\left.\begin{aligned} T(f, f') &= T^*(f', f), \\ G(f_1, f_2; f_2', f_1',) &= -G(f_1, f_2; f_1', f_2') = G^*(f_1', f_2'; f_2, f_1). \end{aligned}\right\} \tag{3.4'}$$

Let us apply a linear canonical transformation of the Fermi operators:

$$a_f = \sum_g \{u(fg)\alpha_g + v(fg)\alpha_g^+\}. \tag{3.5}$$

† This section may be omitted during the first reading of the book.

The commutation relations (3.2) are not violated if functions $u(fg)$ and $v(fg)$ satisfy the following set of relations:

$$\left.\begin{aligned} \zeta(f,f') &\equiv \sum_g \{u(fg)\,u^*(f'g)+v(fg)\,v^*(f'g)\}-\delta_{f,f'} = 0,\\ \eta(f,f') &\equiv \sum_g \{u(fg)\,v(f'g)+v(fg)\,u(f'g)\} = 0,\\ \eta^*(f,f') &\equiv \sum_g \{u^*(fg)\,v^*(f'g)+v^*(fg)\,u^*(f'g)\} = 0, \end{aligned}\right\} \tag{3.6}$$

where

$$\zeta(f_1, f_2) = \zeta^*(f_2, f_1); \qquad \eta(f_1, f_2) = \eta(f_2, f_1).$$

Using (3.6) it is easy to find the inverse transformation

$$\alpha_g = \sum_f \{u^*(fg)a_f + v(fg)a_f^+\}. \tag{3.7}$$

Now we can find another set of relations, which functions $u(fg)$ and $v(fg)$ must satisfy:

$$\left.\begin{aligned} &\sum_f \{u^*(fg_1)u(fg_2)+v^*(fg_1)\,v(fg_2)\} = \delta_{g_1,g_2},\\ &\sum_f \{v^*(fg_1)u(fg_2)+v^*(fg_2)u(fg_1)\} = 0,\\ &\sum_f \{v(fg_1)u^*(fg_2)+v(fg_2)u^*(fg_1)\} = 0. \end{aligned}\right\} \tag{3.6'}$$

We shall define the ground state of a system containing an even number of particles as the vacuum of the α_g operators, i.e.,

$$\alpha_g|\ \rangle_0 = 0, \quad {}_0\langle\ |\alpha_g^+ = 0. \tag{3.8}$$

Let us show that the explicit form of the wave function $|\ \rangle_0$ is

$$|\ \rangle_0 = n \exp\left\{-\tfrac{1}{2}\sum_{f,f'} \Omega(f,f')a_f^+ a_{f'}^+\right\}\Psi_{00}, \tag{3.8'}$$

where $\Omega(f,f') = \sum_g (u^*(fg))^{-1}\,v(f'g)$, n is the normalizing factor, and Ψ_{00} is the vacuum of the a_f particles, i.e., $a_f\Psi_{00} = 0$. Obviously, the function $\Omega(f,f')$ must be antisymmetric. From (3.6′) it follows that

$$v(fg) = -\sum_{g_2 f_2} u^*(fg_2)^{-1}\,v(f_2g_2)u^*(f_2g).$$

If this $v(fg)$ is substituted in formula for $\Omega(f,f')$,

$$\begin{aligned} \Omega(f',f) = \sum_g v(fg)\,u^*(f'g)^{-1} &= -\sum_{g,g_2,f_2} u^*(fg_2)^{-1}\,v(f_2g_2)\,u^*(f_2g)u^*(f'g)^{-1}\\ &= -\sum_{g_2} u^*(fg_2)^{-1}\,v(f'g_2) = -\Omega(f,f'). \end{aligned}$$

Equation (3.8) can be transformed using (3.7) to give

$$\left\{a_{f_2}+\sum_{g,f} u^*(f_2g)^{-1}\,v(fg)a_f^+\right\}|\ \rangle_0 = \left\{a_{f_2}+\sum_f \Omega(f_2,f)a_f^+\right\}|\ \rangle_0 = 0.$$

Now it can be multiplied by M^{-1} from left

$$M^{-1}\Big\{a_{f_2}+\sum_{f}\Omega(f_2, f)a_f^+\Big\}MM^{-1}|\,\rangle_0 = 0,$$

where

$$M = \exp\Big\{-\tfrac{1}{2}\sum_{f,f'}\Omega(f, f')a_f^+a_{f'}^+\Big\}.$$

Further,

$$M^{-1}a_f^+M = a_f^+,$$
$$M^{-1}a_{f_2}M = a_{f_2}-\sum_{f_2'}\Omega(f_2, f_2')a_{f_2'}^+.$$

Hence

$$a_{f_2}M^{-1}|\;\rangle_0 = 0;$$

and, therefore,

$$M^{-1}|\;\rangle_0 = \Psi_{00},$$

This means, that, apart from normalization,

$$|\;\rangle_0 = M\Psi_{00} = \exp\Big\{-\tfrac{1}{2}\sum_{f,f'}\Omega(f, f')a_f^+a_{f'}^+\Big\}\Psi_{00}$$

(for details, see ref. 118).

We need to know now the expectation value of the Hamiltonian H in the state (3.8). To find it, we replace operators a_f, a_f^+ in (3.1) by α_g, α_g^+. After using the anticommutation relations (3.2) and the definition (3.8),

$$\langle|H|\rangle_0 = \sum_{f,f'}T'(f, f')\sum_g v^*(fg)\,v(f'g)-\tfrac{1}{2}\sum_{f_1, f_2, f_2', f_1'}G(f_1, f_2; f_2', f_1')\times$$
$$\times\Big\{\sum_g v^*(f_1g)\,v(f_1'g)\sum_{g'}v^*(f_2g')\,v(f_2'g')$$
$$+\tfrac{1}{2}\sum_g u^*(f_1'g)\,v^*(f_2g)\sum_{g'}u(f_1'g')\,v(f_2'g')\Big\}. \quad (3.9)$$

Functions $u(fg)$ and $v(fg)$ are determined from the variational principle, i.e., from the condition that $\langle|H|\rangle_0$ is minimal. The condition is, symbolically,

$$\delta\Big\{\langle|H|\rangle_0+\sum_{f_1,f_2}[\lambda(f_2, f_1)\,\zeta(f_1, f_2)+\mu^*(f_2, f_1)\,\eta(f_1, f_2)+\mu(f_2, f_1)\,\eta^*(f_1, f_2)]\Big\} = 0, \quad (3.10)$$

where we have introduced Lagrange multipliers $\lambda(f_1, f_2)$ and $\mu(f_1, f_2)$; the variations $\delta u(fg)$, $\delta u^*(fg)$, $\delta v(fg)$, and $\delta v^*(fg)$ are then independent. The multipliers assure validity of the auxiliary conditions (3.6) and (3.6′). The chemical potential λ is another Lagrange multiplier; it assures that the average number of particles is conserved, i.e., that

$$N = \sum_f\langle|a_f^+a_f|\rangle_0 = \sum_{f,g}v^*(fg)\,v(fg). \quad (3.11)$$

Thus, we have formulated the new variational principle: the functions $u(fg)$ and $v(fg)$ are solutions of the stationarity (extremum) equations; $|\ \rangle_0$ is the ground state wave function and $\langle|H|\rangle_0$ is the nuclear ground state energy.

Let us find the equations determining the functions $u(fg)$ and $v(fg)$. The variations in (3.10) are

$$\frac{\delta}{\delta u^*(fg)}\langle|H|\rangle_0+\sum_{f''}[\lambda(f,f'')u(f''g)+\mu(f,f'')v^*(f''g)+\mu(f'',f)v^*(f''g)]=0,$$

$$\frac{\delta}{\delta v^*(fg)}\langle|H|\rangle_0+\sum_{f''}[\lambda(f,f'')v(f''g)+\mu(f,f'')u^*(f''g)+\mu(f'',f)u^*(f''g)]=0$$

(plus the corresponding complex-conjugate equations). Let us define

$$\mathfrak{C}_1(f,f')\equiv\sum_g\left\{v(f'g)\frac{\delta}{\delta u^*(fg)}\langle|H|\rangle_0+u(f'g)\frac{\delta}{\delta v^*(fg)}\langle|H|\rangle_0\right\}+\mu(f,f')+\mu(f',f)=0,$$

$$\mathfrak{C}_2(f,f')\equiv\sum_g\left\{u^*(f'g)\frac{\delta}{\delta u^*(fg)}\langle|H|\rangle_0+v^*(f'g)\frac{\delta}{\delta v^*(fg)}\langle|H|\rangle_0\right\}+\lambda(f,f')=0,$$

and, excluding the Lagrange multipliers, we obtain the basic equations:

$$\left.\begin{aligned}\mathfrak{A}_0(f,f')&=\mathfrak{C}_1(f,f')-\mathfrak{C}_1(f',f)=0,\\ \mathfrak{B}_0(f,f')&=\mathfrak{C}_2(f,f')-\mathfrak{C}_2^*(f',f)=0.\end{aligned}\right\}\tag{3.12}$$

The functions $\mathfrak{A}_0(f,f')$ and $\mathfrak{B}_0(f,f')$ are not independent; they are related by the equation

$$\sum_{f,f'}\{u(fg)v(f'g')\mathfrak{A}_0^*(f,f')+u^*(f'g')v^*(fg)\mathfrak{A}_0(f,f')$$
$$+[u(fg)u^*(f'g')-v^*(f'g)v(fg')]\mathfrak{B}_0(f,f')\}=0.\tag{3.13}$$

Hence, if $\mathfrak{A}_0(f,f')=0$ then $\mathfrak{B}_0(f,f')=0$ automatically. Consequently, it is sufficient to solve only the first equation (3.12).

2. Let us define new functions

$$\Phi(f_1,f_2)=\langle|a_{f_1}a_{f_2}|\rangle,\quad \varrho(f_1,f_2)=\langle|a_{f_1}^+a_{f_2}|\rangle,\tag{3.14}$$

which evidently satisfy the symmetry relations

$$\Phi(f_1,f_2)=-\Phi(f_2,f_1),\quad \varrho^*(f_2,f_1)=\varrho(f_1,f_2).\tag{3.14'}$$

The definitions are general; $|\ \rangle$ is an arbitrary state. If, particularly, the averaging in (3.14) is over the vacuum (3.8) of the new particles (quasiparticles), the functions (3.14) are

$$\left.\begin{aligned}\Phi_0(f_1,f_2)&\equiv\langle|a_{f_1}a_{f_2}|\rangle_0=\sum_g u(f_1g)v(f_2g),\\ \Phi_0^*(f_1,f_2)&\equiv\langle|a_{f_1}^+a_{f_2}^+|\rangle_0=\sum_g u^*(f_1g)v^*(f_2g),\\ \varrho(f_1,f_2)&\equiv\langle|a_{f_1}^+a_{f_2}|\rangle_0=\sum_g v^*(f_1g)v(f_2g).\end{aligned}\right\}\tag{3.14''}$$

The ground-state energy $\langle|H|\rangle_0$ is now rewritten as

$$\langle|H|\rangle_0 = \sum_{f,f'} T(f,f')\,\varrho_0(f,f') - \tfrac{1}{2}\sum_{f_1,f_2,f_2',f_1'} G(f_1,f_2;f_2',f_1')\times \times\{\varrho_0(f_1,f_1')\,\varrho_0(f_2,f_2') + \tfrac{1}{2}\Phi_0^*(f_1,f_2)\,\Phi_0(f_1',f_2')\} \tag{3.9'}$$

and the functions $\mathfrak{A}_0(f,f')$ and $\mathfrak{B}_0(f,f')$ are

$$\mathfrak{A}_0(f_1,f_2) = \sum_f \{\xi(f_1,f)\Phi_0(f,f_2) + \Phi_0(f_1,f)\,\xi(f_2,f)\} + \tfrac{1}{2}\sum_{f_1',f_2',f}\Phi_0(f_1',f_2')\{G(f_1,f;f_2',f_1')\,\varrho_0(f,f_2) + G(f,f_2;f_2',f_1')\,\varrho_0(f,f_1)\} - \tfrac{1}{2}\sum_{f_1',f_2'}\Phi_0(f_1',f_2')\,G(f_1,f_2;f_2',f_1'), \tag{3.15}$$

$$\mathfrak{B}_0(f_1,f_2) = \sum_f \{\xi(f_2,f)\,\varrho_0(f_1,f) - \xi(f,f_1)\,\varrho_0(f,f_2)\} - \tfrac{1}{2}\sum_{f_1',f_2',f}\{G(f_2,f;f_2',f_1')\,\Phi_0(f_1',f_2')\,\Phi_0^*(f_1,f) - G(f_1,f;f_2',f_1')\,\Phi_0(f_2,f)\,\Phi_0^*(f_1',f_2')\}, \tag{3.16}$$

$$\xi(f,f') = T'(f,f') - \sum_{f_1,f_2} G(f,f_1;f_2,f')\,\varrho_0(f_1,f_2). \tag{3.17}$$

We shall introduce another new notation, the correlation function

$$C_{f_1,f_2} = \tfrac{1}{2}\sum_{f_1',f_2'} G(f_1,f_2;f_2',f_1')\,\Phi(f_2',f_1'). \tag{3.18}$$

Now, using an arbitrary state $|\ \rangle$ we find functions $\mathfrak{A}(f_1,f_2)$ and $\mathfrak{B}(f_1,f_2)$:

$$\mathfrak{A}(f_1,f_2) = \sum_f \{\xi(f_1,f)\,\Phi(f,f_2) + \Phi(f_1,f)\,\xi(f_2,f)\} - \sum_f \{C_{f_1,f}\,\varrho(f,f_2) + C_{f,f_2}\,\varrho(f,f_1)\} + C_{f_1,f_2}, \tag{3.15'}$$

$$\mathfrak{B}(f_1,f_2) = \sum_f \{\xi(f_2,f)\,\varrho(f_1,f) - \xi(f,f_1)\,\varrho(f,f_2)\} + \sum_f \{C_{f_2,f}\Phi^*(f_1,f) - C_{f_1,f}^*\Phi(f_2,f)\}, \tag{3.16}$$

where

$$C_{f_1,f_2}^* = \tfrac{1}{2}\sum_{f_1',f_2'} G(f_1,f_2;f_2',f_1')\,\Phi^*(f_2',f_1').$$

The equations

$$\mathfrak{A}(f_1,f_2) = 0, \quad \mathfrak{B}(f_1,f_2) = 0, \tag{3.12'}$$

are more general than their special form (3.12). Corresponding state, involved in the averaging, is not defined; it can be more complicated than the state (3.8).

The functions $\Phi_0(f_1,f_2)$ and $\varrho_0(f_1,f_2)$ are defined by equations

$$\mathfrak{A}_0(f_1,f_2) = 0 \tag{3.12''}$$

and

$$N = \sum_f \varrho_0(f,f). \tag{3.11'}$$

We can solve these equations directly for $\Phi_0(f_1, f_2)$ and $\varrho_0(f_1, f_2)$, i.e., we can omit the functons $u(fg)$ and $v(fg)$. We must, however, remember that $\Phi_0(f_1, f_2)$ and $\varrho_0(f_1, f_2)$ are not independent; they are connected by relations which will be derived now.

It is advantageous to use combined indexes $\omega=(f, \mu)$, $\mu=0, 1$; $\mu=(g, \tau)$, $\tau=0, 1$ and denote

$$\varphi_{g0}(f, 0) = v^*(fg), \quad \varphi_{g0}(f, 1) = u(fg),$$
$$\varphi_{g1}(f, 0) = u^*(fg), \quad \varphi_{g1}(f, 1) = v(fg).$$

Equations (3.6) have a simple form now:

$$\sum_{\eta} \varphi_{\eta}^*(\omega)\, \varphi_{\eta}(\omega') = \delta_{\omega, \omega'}. \tag{3.6''}$$

All relevant quantities are contained in the matrix

$$F(\omega, \omega') = \sum_{\eta} \varphi_{\eta}^*(\omega)\, \varphi_{\eta}(\omega') \bar{n}_{\eta}, \tag{3.19}$$

in which $\bar{n}_{g,1} = 1$, $\bar{n}_{g,0} = 0$. From (3.6'') it follows that

$$F(\omega, \omega') = \begin{vmatrix} \varrho_0(f', f) & -\Phi_0(f, f') \\ \Phi_0^*(f, f') & \delta_{f,f'} - \varrho_0(f, f') \end{vmatrix}. \tag{3.19'}$$

The operator F is defined in such a way that $\varphi_{\eta}(\omega)$ are its eigenvectors and n_{η} its eigenvalues. Because the eigenvalues are equal to zero or unity, the operator F is a projection operator, i.e., it obeys the relation

$$F^2 = F. \tag{3.19''}$$

This equation then gives the relations between $\Phi_0(f, f')$ and $\varrho_0(f, f')$ which we were looking for:

$$\left.\begin{aligned} \varrho_0(f_1, f_2) &= \sum_{f'} \{\varrho_0(f_1, f')\, \varrho_0(f', f_2) + \Phi_0^*(f', f_1)\, \Phi_0(f', f_2)\}, \\ &\sum_{f'} \{\varrho_0(f_1, f')\, \Phi_0(f', f_2) + \varrho_0(f_2, f')\, \Phi_0(f', f_1)\} = 0. \end{aligned}\right\} \tag{3.20}$$

The above equations were first derived in refs. 119 and 120.

3. The solution of equations (3.10) corresponds to a minimum of the energy if the second variation is positive. Such a condition can be expressed as

$$\sum_{f, g} \{\mathfrak{C}_1\, \delta v^*(fg)\, \delta v(fg) + \mathfrak{C}_2\, \delta u^*(fg)\, \delta u(fg)\} > 0. \tag{3.21}$$

The condition (3.21) is satisfied if $\mathfrak{C}_1$ and $\mathfrak{C}_2$ have positive eigenvalues; the quantities $\mathfrak{C}_1$, $\mathfrak{C}_2$, $\delta u(fg)$, and $\delta v(fg)$ are determined from the eigenvalue equation.

The Hartree–Fock–Bogolyubov variational principle is a generalization of the Hartree–Fock method; consequently, Hartree–Fock solutions are always among the solutions of the more general method. When do the Hartree–Fock functions not give energy minimum? Evidently, if

$$\sum_{f, g} \{\mathfrak{C}_1\, \delta v^*(fg)\, \delta v(fg) + \mathfrak{C}_2\, \delta u^*(fg)\, \delta u(fg)\} < 0, \tag{3.21'}$$

where $\mathfrak{C}_1$, $\mathfrak{C}_2$, $\delta u(fg)$, and $\delta v(fg)$ are determined in the Hartree–Fock eigenvalue problem. If (3.21′) is valid, we have to seek the minimum, using a wider class of trial functions or, in other words, the validity of (3.21′) is the criterion for the existence of pairing correlations (see ref. 121).

Let us discuss two special cases. First, we shall consider an interaction between the fermions of the form

$$G(f_1, f_2; f_2', f_1') = \tfrac{1}{2}V(f_1, f_2)\,(\delta_{f_1 f_1'}\delta_{f_2 f_2'} - \delta_{f_1 f_2'}\delta_{f_2 f_1'}), \tag{3.22}$$

where $f = (\boldsymbol{r}, \sigma)$, $\boldsymbol{r}$ is the radius vector, and σ is spin. Let us assume that

$$V(f_1, f_2) = V(|\boldsymbol{r}_1 - \boldsymbol{r}_2|, \sigma_1, \sigma_2), \quad T(f_1, f_2) = T(r)\delta_{f_1, f_2}.$$

The expectation value of the Hamiltonian equals

$$\begin{aligned}\langle |H| \rangle = \sum_f T(f)\,\varrho(f, f) + \tfrac{1}{2}\sum_{f_1, f_2} V(f_1, f_2)\,\Phi_0^*(f_1, f_2)\,\Phi_0(f_1, f_2)\\ + \tfrac{1}{2}\sum_{f_1, f_2} V(f_1, f_2)\,\{\varrho_0(f_1, f_1)\,\varrho_0(f_2, f_2) - \varrho_0(f_1, f_2)\,\varrho_0^*(f_1, f_2)\}.\end{aligned} \tag{3.23}$$

This expression differs from the analogous formula in the Hartree–Fock method by

$$+\tfrac{1}{2}\sum_{f_1, f_2} V(f_1, f_2)\,\Phi_0^*(f_1, f_2)\,\Phi_0(f_1, f_2),$$

i.e., by the term describing pairing correlations.

Equation (3.15) is then

$$\begin{aligned}\Big\{T(f_1) + T(f_2) + V(f_1, f_2) + \sum_f \varrho_0(f, f_1)\,[V(f_1, f) + V(f_2, f)]\Big\}\,\Phi_0(f_1, f_2)\\ - \sum_f \{V(f_1, f)\,\Phi_0(f_1, f)\,\varrho_0(f, f_2) - V(f_2, f)\,\Phi_0(f_2, f)\,\varrho_0(f, f_1)\}\\ - \sum_f \{\Phi_0(f_1, f)\,V(f_2, f)\,\varrho_0(f, f_2) - \Phi_0(f_2, f)\,V(f_1, f)\,\varrho_0(f, f_1)\} = 0.\end{aligned} \tag{3.24}$$

Equation (3.16) is satisfied automatically. To give more meaning to individual terms in (3.24), let us rewrite the equation symbolically as

$$\{T_1 + T_2 + U_1 + U_2 + V(1, 2)\}\Phi_{12} + \sum \{\mathfrak{C}_{1i}\Phi_{2i} - \mathfrak{C}_{2i}\Phi_{1i}\} - \sum_i \{Z_{1i}\varrho_{i2} - Z_{2i}\varrho_{i1}\} = 0. \tag{3.24′}$$

Here $U_i = \sum_{r, \sigma} V(|\boldsymbol{r} - \boldsymbol{r}_i|\,\sigma, \sigma_i)\,\varrho_0(\boldsymbol{r}, \sigma; \boldsymbol{r}, \sigma)$ is the potential at $\boldsymbol{r}_i$ caused by all particles;

$$\mathfrak{C}_{ii'} = V(|\boldsymbol{r}_i - \boldsymbol{r}_{i'}|, \sigma_i, \sigma_{i'})\,\varrho_0(\boldsymbol{r}_i, \sigma_i; \boldsymbol{r}_{i'}, \sigma_{i'});$$

and $Z_{2i} = V(|\boldsymbol{r}_2 - \boldsymbol{r}_i|, \sigma_2, \sigma_i)\,\Phi_0(\boldsymbol{r}_2, \sigma_2; \boldsymbol{r}_i\,\sigma_i)$ describes the interaction of two particles. The last term in (3.24′) describes the effect of the environment on the wave function of a pair.

We shall derive the basic equation of superconductivity theory as the second example.

The index f denotes now $(\boldsymbol{p}, \sigma)$, where $\boldsymbol{p}$ is momentum and σ is spin. The kinetic energy is $T(f, f') = T(p)\delta_{f, f'}$. Let us choose

$$v(f\omega) = v_f \delta_{f, \omega}, \quad u(f\omega) = u_f \delta_{f, -\omega}, \tag{3.25}$$

where

$$u_f = u_{-f} = u_p, \quad v_f = -v_{-f} = v_p \delta_{\sigma, \frac{1}{2}}, \quad u_p^2 + v_p^2 = 1.$$

Then

$$\varrho(f_1, f_2) = v_{f_1}^2 \delta_{f_1, f_2}, \quad \Phi(f_1, f_2) = u_{f_1} v_{f_2} \delta_{f_1, -f_2}.$$

Equation (3.15) for $f = -f'$ is of the form

$$2\xi(p) u_p v_p - (u_p^2 - v_p^2) \sum_{p''} G(p, -p; \; -p'', p'') u_{p''} v_{p''} = 0, \tag{3.26}$$

where
$$\xi(p) = T(p) + \sum_{p''} \{G(p, p''; \; p'', p) + G(p, -p''; \; -p'', p)\} v_{p''}^2.$$

This is an equation equivalent to the compensation of "dangerous" graphs in superconductivity theory, suggested in ref. 6.

4. The functions $\varrho(f_1, f_2)$ and $\Phi(f_1, f_2)$ can be brought to a simpler form. Let us apply a unitary transformation

$$a_f = \sum_{f'} a'_{f'} U_{f'f}, \quad a_f^+ = \sum_{f'} U_{ff'}^+ a_{f'}'^+. \tag{3.27}$$

Such a transformation does not change commutation properties (3.2). The functions ϱ and Φ are transformed in the following way:

$$\varrho(f_1, f_2) = \sum_{f_1', f_2'} \varrho'(f_1', f_2') U_{f_1 f_1'}^+ U_{f_2' f_2}, \tag{3.28}$$

$$\Phi(f_1, f_2) = \sum_{f_1', f_2'} \Phi'(f_1', f_2') U_{f_1', f_1}, U_{f_2', f_2}, \tag{3.29}$$

or, using the matrix notation,

$$\varrho = U^+ \varrho' U, \tag{3.28'}$$

$$\Phi = \tilde{U} \Phi' U, \tag{3.29'}$$

where

$$\tilde{U}_{f, f'} = U_{f', f}.$$

Thus, ϱ is transformed as an Hermitian operator and Φ as a second-order antisymmetric tensor.

Let us find the canonical form of Φ using an Hermitian operator

$$\Theta = \Phi^+ \Phi = \Phi \Phi^+ \tag{3.30}$$

with transformation properties

$$\Theta = U^+ \Phi'^+ U^* \cdot \tilde{U} \Phi' U = U^+ \Theta' U \tag{3.30'}$$

(because $U^+ U = 1$ and, therefore, $U^* \tilde{U} = 1$). Matrix Θ will be diagonalized now; $\Theta(i)$ denotes its eigenvalues (they can be degenerate), $\mathcal{L}^{(i)}$ denotes the corresponding subspace of eigenvectors. From

$$\Phi \Theta = \Theta \Phi \tag{3.31}$$

it follows that for diagonal Θ

$$\Phi(i_1, i_2) \{\Theta(i_1) - \Theta(i_2)\} = 0. \tag{3.31'}$$

Consequently, Φ has nonvanishing components corresponding to the same eigenvalue

$\Theta(i)$ only, i.e., it has nonvanishing matrix elements between states belonging to the same subspace $\mathcal{L}^{(i)}$.

Let us transform Φ to the canonical form in each subspace $\mathcal{L}^{(i)}$. Let $\boldsymbol{l}$ be an eigenvector, i.e.,

$$\Phi \boldsymbol{l} = L\boldsymbol{l}, \tag{3.32}$$

and, therefore,

$$\Theta(i) = |L|^2. \tag{3.33}$$

If $\boldsymbol{l} \neq \boldsymbol{l}'$, the corresponding eigenfunctions are orthogonal, $\boldsymbol{l}^* \boldsymbol{l}' = 0$. Besides, if $L \neq 0$, then $-L$ is an eigenvalue with eigenfunction $\boldsymbol{l}^*$.

Let us define two real vectors $\boldsymbol{b}_1$ and $\boldsymbol{b}_2$ as

$$\boldsymbol{l} = \frac{1}{\sqrt{2}}(\boldsymbol{b}_1 + i\boldsymbol{b}_2), \quad \boldsymbol{l}^* = \frac{1}{\sqrt{2}}(\boldsymbol{b}_1 - i\boldsymbol{b}_2),$$

then $\boldsymbol{b}_1^2 = \boldsymbol{b}_2^2 = 1$ and $\boldsymbol{b}_1\boldsymbol{b}_2 = 0$. We shall, further, use the real orthogonal transformation U_0. Evidently

$$\tilde{U}_0 = U_0^+ = U^{-1}$$

and Φ transforms as

$$\Phi = U_0^{-1}\Phi' U_0.$$

The operator Φ in the new axes acts as

$$\Phi \boldsymbol{b}_1 = iL\boldsymbol{b}_2, \quad \Phi \boldsymbol{b}_2 = -iL\boldsymbol{b}_1. \tag{3.34}$$

Denoting $-iL = \mathfrak{L}$, we have Φ in the canonical form

$$\begin{pmatrix} 0 & & & & & \\ & 0 & \mathfrak{L}_1 & & & \\ & -\mathfrak{L}_1 & 0 & & & \\ & & & 0 & \mathfrak{L}_2 & \\ & & & -\mathfrak{L}_2 & 0 & \\ & & & & & \ddots \end{pmatrix} \tag{3.35}$$

All matrix elements outside squares vanish. Reference 122 has proved that the numbers $\mathfrak{L}$ are real for a real antisymmetric tensor Φ.

The set of quantum numbers f can be divided into $f = q\sigma$; states with $\sigma = \pm 1$ are transformed into each other under time conjugation. For example, σ might denote the sign of the angular momentum projection on nuclear symmetry axis. The function Φ is, according to (3.35), equal to

$$\Phi(f, f') = \Phi(f)\delta_{f, -f'} = \Phi(q)\delta_{q, q'}\delta_{\sigma, -\sigma'}. \tag{3.36}$$

From (3.20) (see ref. 123) it follows that

$$\varrho(f, f') = \varrho(f)\delta_{f, f'} = \varrho(q)\delta_{q, q'}\delta_{\sigma, \sigma'}, \tag{3.37}$$

$$\varrho(q) = \varrho^2(q) + \Phi^*(q)\Phi(q), \tag{3.38}$$

and from $\varrho^*(f_2, f_1) = \varrho(f_1, f_2)$ it follows that

$$\varrho(q) = \varrho^*(q).$$

Particularly, if $\Phi(q) = \Phi^*(q)$, then

$$\varrho(q) = \varrho^2(q),$$

and, therefore, $\varrho = 1$ or 0.

Thus, we have shown that it is possible to diagonalize $\varrho(f_1, f_2)$ and bring $\Phi(f_1, f_2)$ to the canonical form at the same time.

5. When the matrices ϱ_0 and Φ_0 are diagonal, the basic equations (3.12) are considerably simplified. The average value of the Hamiltonian in state (3.8) is then equal to

$$\langle | H | \rangle_0 = \sum_f \left\{ T'(f) - \tfrac{1}{2} \sum_{f'} G(f, f'; f', f)\, \varrho_0(f') \right\} \varrho_0(f) - \sum_{q,\, q'} G(q+, q-; q'-, q'+)\, \Phi_0^*(q)\, \Phi_0(q'). \tag{3.39}$$

The basic equations are

$$2\xi(q)\, \Phi_0(q) - (1 - 2\varrho_0(q)) \sum_{q'} G(q+, q-; q'-, q'+)\, \Phi_0(q') = 0, \tag{3.40}$$

$$N = 2 \sum_q \varrho_0(q), \tag{3.41}$$

where

$$\xi(q) = T'(q) - \sum_{q',\, \sigma'} G(q+, q'\sigma'; q'\sigma', q+)\, \varrho_0(q') \equiv E(q) - \lambda.$$

The quantities $E(q)$ are the single-particle energies.

We have seen how the selfconsistent, average nuclear field and the pairing interaction, involving pairs of particles in time-reversal conjugated states, can be isolated from the most general two-particle interaction. This procedure is, in fact, an important step in the foundation of the independent particle model. The fact that the average field can be isolated is connected neither with the magnitude nor with the form of the interaction potential.

Two points, however, should be remembered. The basic equations (3.15) and (3.16) are only approximately valid; the correlation function $\langle a_{f_1}^+ a_{f_2}^+ a_{f_2'} a_{f_1'} \rangle$ is expressed through ϱ and Φ by the canonical transformation in only an approximate way. The results are, therefore, not exact. Convergence of this method depends on the interaction; it was not studied in depth. The only exception is the model Hamiltonian of superconductivity theory, for which Bogolyubov(124) found the asymptotic expressions. The calculations discussed above are valid for all nonsingular interactions. Their applicability to nucleon–nucleon potentials was never, as in the Hartree–Fock method, strictly proven.

The following assumption is used in nuclear theory as an axiom: The average nuclear field corresponds to a representation, in which the density matrix $\varrho_0(f, f')$ is diagonal for the ground states of several even–even nuclei. All residual interactions in such a case are reduced to the pairing interactions of the superfluid type. No other residual interactions are necessary.

The Bogolyubov canonical transformation (3.5) gives us an apparatus applicable to any correlations of pairs of nucleons. When the transformation is applied, the relevant objects are quasiparticles and the ground state is the corresponding quasiparticle vacuum. The ground state can be treated without interaction between quasiparticles if both ϱ and Φ

are diagonal. In such a representation, moreover, ϱ and Φ are diagonal even for many excited states. On the other hand, there is a whole class of excited states in which the functions ϱ and Φ are not diagonal; the wave functions of these states are superpositions of states corresponding to the diagonal density matrix ϱ.

The general Bogolyubov canonical transformation can be expressed as a product of three simpler transformations:[122]

$$B = U_2 B_s U_1, \tag{3.42}$$

the unitary transformation U_1 brings the functions ϱ and Φ to the canonical form (the operators a_f are expressed as a linear combination of the same operators; the transformation does not include a_f^+); the second transformation B_s is the special Bogolyubov transformation

$$a_{q\sigma} = u_q \alpha_{q,\,-\sigma} + \sigma v_q \alpha_{q\sigma}^+, \tag{3.43}$$

and, finally, U_2 is again a unitary transformation which does not mix α_f with α_f^+.

Let us transform (3.40) and (3.41) to a more common form. Denote

$$\Phi(q) = u_q v_q = \Phi^*(q), \tag{3.36'}$$

$$\varrho(q) = v_q^2, \tag{3.37'}$$

where u_q and v_q are real functions. The condition (3.38) now reads

$$u_q^2 + v_q^2 = 1 \tag{3.44}$$

and (3.40) is of the form

$$2\xi(q) u_q v_q - (u_q^2 - v_q^2) \sum_{q'} G(q+, q-; q'-, q'+) u_{q'} v_{q'} = 0, \tag{3.40'}$$

with $\xi(q) = E(q) - \lambda$.

The correlation function $C_{f,\,-f}$ is usually denoted C_q if the function $\Phi_0(f, f')$ is of the form (3.36). The C_q equals

$$C_q = \sum_{q'} G(q+, q-; q'-, q'+) u_{q'} v_{q'}. \tag{3.45}$$

Functions u_q^2 and v_q^2 can be expressed as

$$u_q^2 = \frac{1}{2}\left\{1 + \frac{\xi(q)}{\varepsilon(q)}\right\}, \quad v_q^2 = \frac{1}{2}\left\{1 - \frac{\xi(q)}{\varepsilon(q)}\right\} \tag{3.46}$$

and, using (3.40'),

$$u_q v_q = \frac{1}{2} \frac{C_q}{\varepsilon(q)}. \tag{3.46'}$$

Now, the square of (3.46') can be compared with the product constructed according to (3.46). From such an equation it follows that

$$\varepsilon(q) = \sqrt{C_q^2 + \xi(q)^2}. \tag{3.46''}$$

Formula (3.46') can be substituted in (3.40') and (3.46) in (3.41), and this system of fundamental equations derived:

$$C_q = \frac{1}{2} \sum_{q'} G(q+, q-; q'-, q'+) \frac{C_{q'}}{\sqrt{C_{q'}^2 + \xi(q')^2}}, \tag{3.47}$$

$$N = \sum_q \left\{1 - \frac{\xi(q)}{\varepsilon(q)}\right\}. \tag{3.48}$$

Equation (3.47) generally has two solutions. The trivial solution $C_q = 0$ corresponds to the normal state; the $C_q \neq 0$ solution corresponds to the superfluid state. If $C_q = 0$, then

$$u_q = 1 - \theta_F(q), \quad v_q = \theta_F(q), \tag{3.49}$$

where $\theta_F(q)$ is the usual step function; it is equal to one if the energy $E(q)$ of the state q is less than the Fermi energy E_F, and it is equal to zero, otherwise. The real system is described, naturally, by the solution corresponding to the lowest energy of the system.

One limiting case is interesting. If the interaction matrix element $G(q+, q-; q'-, q'+)$ converges to zero and the level density increases, the solution of (3.47) is[11]

$$C_q = \text{const}\, \frac{G(q+, q-; q_0-, q_0+)}{G(q_0+, q_0-; q_0-, q_0+)} \exp\left\{\frac{\text{const}}{G(q_0+, q_0-; q_0-, q_0+)}\right\}, \tag{3.50}$$

where $\xi(q_0) = E_F$. The same expression for the correlation function was obtained in the superfluidity theory.[6] From (3.50) it follows that the system of particles interacting via $G(q+, q-, q'-, q'+)$ cannot be solved by the perturbation methods, i.e., expanded into a power series of the interaction strength.

§ 3. The General Form of the Nuclear Model Hamiltonian

1. We want to explain nuclear theory deductively. The nuclear theory based on "first principles" does not exist; thus, fully consistent deductive description is impossible. It is, however, possible to explain many properties of the nuclear ground and excited states in the framework of the superfluid nuclear model. The model, as we have already noted, is based on a model Hamiltonian. We shall construct such a Hamiltonian in this section.

Nuclear rotations are described phenomenologically. The nuclear Hamiltonian must, consequently, contain terms describing the rotational kinetic energy and the Coriolis coupling, i.e., coupling of the intrinsic motion to the rotation. These two terms are [see (2.4′) and (2.5)]:

$$T_{\text{rot}} + H_{\text{cor}} = \frac{I(I+1)}{2\mathcal{J}} - \frac{1}{2\mathcal{J}}(I_+J_- + I_-J_+) \tag{3.51}$$

Note that several attempts to describe nuclear rotations microscopically were recently made in refs. 125–127.

Another part of our nuclear Hamiltonian will contain terms describing the nuclear average field and the interactions leading to the pairing superfluid correlations. These terms have been discussed in the preceding section. In this instance the Hamiltonian contains

$$H_1 = H_{\text{av}} + H_{\text{pair}} + T_{\text{rot}} + H_{\text{cor}}. \tag{3.52}$$

We have discussed the problems related to the average field in §§ 2 and 5 of Chapter 1. Let us repeat very briefly: the H_{av} consists of two parts—neutron and proton average fields. The form of the corresponding potentials is chosen so as to reproduce the whole body of experimental nuclear properties.

The employed potentials and, therefore, the energy eigenvalues and eigenfunctions, describe the actual average nuclear field better in the case of deformed nuclei than in the case of spherical nuclei. The potential in deformed nuclei is chosen in such a way that the density matrix is almost diagonal, i.e., the average field includes part of the residual interaction. This circumstance explains the success of the Nilsson potential in description of the levels of deformed nuclei. It explains also the high degree of single-particle character of the ground states in odd-A deformed nuclei.

The potential of the average field in spherical nuclei is based only on the interactions of nucleons in completely filled shells; the interaction of the particles in unfilled shells is not included. The average field is not fully selfconsistent because it does not contain the effects related to the change in the occupation of unfilled shells. It is, therefore, sometimes impossible to describe several odd-A isotopes of the same element using a single level scheme. For the same reason even the low-lying states in spherical nuclei contain considerable admixtures of the more complicated configurations.

The interaction, which causes pairing superfluid correlations, has the form

$$H_{\text{pair}} = -\sum_{q,\, q'} G(q+, q-; q'-, q'+) a^{+}_{q+} a^{+}_{q-} a_{q'-} a_{q'+} . \tag{3.53}$$

Such an interaction scatters the pairs of particles from the time-reversal conjugated states $(q'+,\, q'-)$ into states $(q+,\, q-)$. In deformed nuclei the interaction (3.53) transfers two particles from one average field level to another one, i.e., the particles undergo virtual transitions pairwise.

The interaction Hamiltonian (3.53) contains that specific part of the residual interaction, which depends only on the quantum numbers of the two states q and q'. Strong correlation of two nucleons is achieved only if the nucleons are in the $S = 0$ and isospin $T = 1$ states, and if the two particles have degenerate energies and identical quantum numbers (except $\sigma = \pm 1$); this fact explains the special role of the interaction (3.53). The short-range forces act more effectively in states with zero total angular momentum than in other states. The contributions of different particle pairs are added coherently in the former case. Various components of the short-range nuclear forces and the specific importance of the interaction (3.53) were analyzed in refs. 17 and 128.

The pairing–correlation–causing forces are short-range forces. They can be, therefore, approximated by the $\delta(\boldsymbol{r}-\boldsymbol{r}')$ force. This means that they are constant in the momentum representation and their matrix elements for different single-particle states are approximately identical. Consequently, we can assume that $G(q+,\, q-,\, q'-,\, q'+)$ is independent of q and q', i.e.,

$$G(q+, q-; q'-, q'+) = G. \tag{3.54}$$

While a general theoretical analysis often uses the Hamiltonian (3.53), the actual calculations of nuclear characteristics usually use the approximation (3.54).

The superfluid pairing correlations between protons and neutrons are missing in the medium and heavy nuclei. The neutron and proton systems can be, therefore, treated separately[(12)].

Let us summarize. The short-range part of the residual nucleon–nucleon interaction, leading to the superfluid pairing correlations in the medium and heavy nuclei, can be expressed as

$$H_{\text{pair}} = -G_N \sum_{s,s'} a^+_{s+}a^+_{s-}a_{s'-}a_{s'+} - G_Z \sum_{r,r'} a^+_{r+}a^+_{r-}a_{r'-}a_{r'+}. \tag{3.55}$$

Here G_N, G_Z are neutron, respectively proton, pairing interaction constants; $a^+_{s\sigma}$, $a_{s\sigma}$ are the nucleon creation and annihilation operators. The quantum numbers $(s\sigma)$ denote the neutron states, quantum numbers $(r\sigma)$ denote the proton states.

As the next step in the development of our theory we shall make a transformation from the particles to the quasiparticles. The quasiparticle operator is defined as

$$\alpha_{q\sigma} = u_q a_{q,\,-\sigma} + \sigma v_q a^+_{q\sigma}. \tag{3.56}$$

The quasiparticle operator coincides with the particle (hole) operator for the states far above (below) the Fermi level. On the other hand, the quasiparticle is a superposition of the particle and hole for the states near the Fermi level.

2. Let us discuss the form of the quasiparticle interaction potentials. The quasiparticle interaction is not identical with the free nucleon interaction because the average field and pairing interaction were already separated by the quasiparticle transformation. Thus, it is quite difficult to determine the quasiparticle interaction.

The full nuclear model Hamiltonian is equal to

$$H = H_{\text{av}} + H_{\text{pair}} + H' + T_{\text{rot}} + H_{\text{cor}}, \tag{3.57}$$

where H' describes the basic part of the quasiparticle interaction. Such a division of the Hamiltonian is somewhat artificial. It is usually assumed that H_{av} describes the nuclear average field completely including the corresponding parts of H_{pair} and H'. The separation of the average field is, in reality, done not in H, but in the expectation value of H.

The general central residual interaction is equal to

$$V(|\boldsymbol{r}_1 - \boldsymbol{r}_2|) + V_\sigma(|\boldsymbol{r}_1 - \boldsymbol{r}_2|)(\boldsymbol{\sigma}^{(1)} . \boldsymbol{\sigma}^{(2)}). \tag{3.58}$$

Each of the potentials can be expanded into a series of spherical harmonics:

$$V(|\boldsymbol{r}_1 - \boldsymbol{r}_2|) = \sum_{\lambda=0}^{\infty} R_\lambda(r_1, r_2) \frac{4\pi}{2\lambda+1} \sum_{\mu=-\lambda}^{\lambda} (-1)^\mu Y_{\lambda\mu}(\theta_1, \varphi_1) Y_{\lambda,\,-\mu}(\theta_2, \varphi_2), \tag{3.59}$$

$$V_\sigma(|\boldsymbol{r}_1 - \boldsymbol{r}_2|)(\boldsymbol{\sigma}^{(1)} . \boldsymbol{\sigma}^{(2)}) = \sum_{n=0}^{\infty} R^\sigma_n(r_1, r_2) \frac{4\pi}{2n+1} \sum_{\lambda=n,\,n\pm1} (-1)^{n+1-\lambda} \times$$

$$\times \sum_{\mu=-\lambda}^{\lambda} (-1)^\mu \{\boldsymbol{\sigma}^{(1)} . \boldsymbol{Y}_n(\theta_1, \varphi_1)\}_{\lambda\mu} \{\boldsymbol{\sigma}^{(2)} . \boldsymbol{Y}_n(\theta_2, \varphi_2)\}_{\lambda,\,-\mu}, \tag{3.60}$$

where

$$\{\boldsymbol{\sigma} . \boldsymbol{Y}_n(\theta_1, \varphi_1)\}_{\lambda\mu} = \sum_{\varrho=0,\,\pm1} \sum_{p=-n}^{n} (1\varrho np \mid \lambda\mu)\, \sigma_\varrho Y_{n\varrho}(\theta_1, \varphi_1). \tag{3.61}$$

The position and spin of a particle are determined by r_i, θ_i, φ_i, $\sigma^{(i)}$. The functions $R_\lambda(r_1, r_2)$ and $R^\sigma_n(r_1, r_2)$ describe the radial dependence. The most general central interaction is thus expressed as a series of the multipoles and spin-multipoles.

The existence of a static quadrupole deformation in the rare-earth and actinide elements suggests that the quadrupole–quadrupole part of interaction is particularly important. The quasiparticle interaction is equal to the portion of the general interaction which remains after the average field separation. The long-range part of the residual nucleon–nucleon interaction can be, therefore, approximated by the few terms in series (3.59) and (3.60).

The effective nuclear forces, based on the nucleon–nucleon scattering data, were studied in ref. 116. It was shown that the quadrupole–quadrupole component of the effective interaction is particularly large.

The multipole–multipole interaction is described by the Hamiltonian

$$H_Q = -\tfrac{1}{2}\sum_{\lambda,\,\mu\geqslant 0}\{\varkappa_n^{(\lambda)}Q_{\lambda\mu}^{+}(n)\,Q_{\lambda\mu}(n)+\varkappa_p^{(\lambda)}Q_{\lambda\mu}^{+}(p)\,Q_{\lambda\mu}(p)+\varkappa_{np}^{(\lambda)}(Q_{\lambda\mu}^{+}(n)\,Q_{\lambda\mu}(p)+Q_{\lambda\mu}^{+}(p)\,Q_{\lambda\mu}(n))\}, \tag{3.62}$$

which contains, besides the neutron–neutron and proton–proton interactions, also the neutron–proton interaction. The quadrupole–quadrupole and octupole–octupole interactions are the most important ones. The $\lambda = 0$ terms contribute mostly to the average field because the nuclear density vibrations are small. The dipole–dipole $\lambda = 1$ term describes relative motion of the protons against the neutrons. Such a term determines the properties of the giant dipole resonance, but does not affect properties of the low-lying nuclear states. This term must be, however, included in the calculation of the $E1$ transition probabilities. Finally, the expression (3.62) is not the most general equation. It does not include the corresponding exchange terms.

In the standard nuclear structure calculations the H_Q contains quadrupole–quadrupole and octupole–octupole interactions; the $\lambda = 4$ hexadecapole terms were also included in several recent studies.

The quadrupole interaction in the deformed nuclei includes $\mu = 0$ and 2 projections; the $\lambda = 3$ octupole interaction includes $\mu = 0$, 1, 2, and 3. The multipole moment operator is equal to (see ref. 129)

$$Q_{\lambda\mu}(n) = \sum_{q,\,q',\,\sigma,\,\sigma'}\langle q\sigma|\mathfrak{f}^{\lambda\mu}|q'\sigma'\rangle a_{q\sigma}^{+}a_{q'\sigma'} = \sum_{q,\,q',\,\sigma}\{\mathfrak{f}^{\lambda\mu}(q,q')a_{q\sigma}^{+}a_{q'\sigma}+\sigma\bar{\mathfrak{f}}^{\lambda\mu}(q,q')a_{q\sigma}^{+}a_{q',-\sigma}\}, \tag{3.63}$$

where

$$\mathfrak{f}^{\lambda\mu=0} = r^{\lambda}Y_{\lambda 0},\quad \mathfrak{f}^{\lambda\mu\neq 0} = \frac{1}{\sqrt{2}}r^{\lambda}(Y_{\lambda\mu}+(-1)^{\mu}Y_{\lambda,-\mu}), \tag{3.64}$$

$$\mathfrak{f}^{\lambda\mu}(q,q') = \langle q+|\mathfrak{f}^{\lambda\mu}|q'+\rangle = \langle q-|\mathfrak{f}^{\lambda\mu}|q'-\rangle = \langle q'+|\mathfrak{f}^{\lambda\mu}|q+\rangle, \tag{3.65}$$

$$\bar{\mathfrak{f}}^{\lambda\mu}(q,q') = \langle q+|\mathfrak{f}^{\lambda\mu}|q'-\rangle = -\langle q'+|\mathfrak{f}^{\lambda\mu}|q-\rangle = \langle q'-|\mathfrak{f}^{\lambda\mu}|q+\rangle, \tag{3.65'}$$

$$\mathfrak{f}^{\lambda\mu}(q,q') = \mathfrak{f}^{\lambda\mu}(q',q),\quad \bar{\mathfrak{f}}^{\lambda\mu}(q,q') = -\bar{\mathfrak{f}}^{\lambda\mu}(q',q). \tag{3.65''}$$

The matrix elements $\mathfrak{f}^{\lambda\mu}(q,q')$ and $\bar{\mathfrak{f}}^{\lambda\mu}(q,q')$ fulfill the corresponding selection rules for each value of λ and μ. The $\mathfrak{f}^{\lambda\mu}(q_1,q_2)$ are nonvanishing if $K_1 \pm \mu = K_2$. When $\mu \neq 0$, the matrix elements $\bar{\mathfrak{f}}^{\lambda\mu}(q_1,q_2)$ must be included; the selection rules for them are $K_1+K_2 = \pm\mu$. (The K_1 and K_2 are the correponding projections on the nuclear symmetry axis.)

In spherical nuclei the quadrupole excitations have the quantum numbers $I^{\pi} = 2^{+}$ and $\lambda = 2$. The octupole states correpond to $I^{\pi} = 3^{-}$ and $\lambda = 3$. The multipole moment operator for spherical nuclei is equal to

$$Q_{\lambda\mu}(n) = \sum_{j, j'; m, m'} \langle j'm' | i^\lambda Y_{\lambda, -\mu} w_\lambda | jm \rangle a^+_{j'm'} a_{jm}$$

$$= \frac{1}{\sqrt{2\lambda+1}} \sum_{j,j'; m, m'} (-1)^{j-m} (j'm'j-m | \lambda -\mu) \mathfrak{f}^\lambda(j', j) a^+_{j'm'} a_{jm}, \tag{3.66}$$

where

$$\mathfrak{f}^\lambda(j', j) = \langle j' | i^\lambda Y_\lambda w_\lambda | j \rangle \equiv \langle N'l'j' | i^\lambda Y_\lambda w_\lambda | Nlj \rangle, \tag{3.67}$$

$w_\lambda = (m\mathring{\omega}_0)^{\lambda/2} r^\lambda$, and $\mathring{\omega}_0$ is the oscillator frequency.

The Hamiltonian H_Q can be brought to the isotopically covariant form

$$H_Q = -\tfrac{1}{2} \sum_{\lambda, \mu \geq 0} \sum_{\substack{q_1, q_1', \\ \sigma_1, \sigma_1'}} \sum_{\substack{q_2, q_2', \\ \sigma_2, \sigma_2'}} \{\chi^*_{q_1}(t_z^{(1)}) \chi^*_{q_2}(t_z^{(2)}) (\varkappa_0^{(\lambda)} + \varkappa_1^{(\lambda)} \boldsymbol{\tau}^{(1)} \boldsymbol{\tau}^{(2)}) \times$$

$$\times \chi_{q_2'}(t_z^{(2')}) \chi_{q_1'}(t_z^{(1')})\} \langle q_1\sigma_1 | \mathfrak{f}^{\lambda\mu} | q_1'\sigma_1' \rangle \times$$

$$\times \langle q_2\sigma_2 | \mathfrak{f}^{\lambda\mu} | q_2'\sigma_2' \rangle a^+_{q_1\sigma_1}(t_z^{(1)}) a_{q_1'\sigma_1'}(t_z^{(1')}) a^+_{q_2\sigma_2}(t_z^{(2)}) a_{q_2'\sigma_2'}(t_z^{(2')}). \tag{3.68}$$

The $\chi_q(t_z)$ are the isospin wave functions defined in § 3 of Chapter 1. The multipole interaction constants $\varkappa_p^{(\lambda)}$ (proton–proton), $\varkappa_n^{(\lambda)}$ (neutron–neutron), and $\varkappa_{np}^{(\lambda)}$ (neutron–proton) are related to the constants $\varkappa_0^{(\lambda)}$ and $\varkappa_1^{(\lambda)}$ by

$$\varkappa_n^{(\lambda)} = \varkappa_p^{(\lambda)} = \varkappa_0^{(\lambda)} + \varkappa_1^{(\lambda)}, \quad \varkappa_{np}^{(\lambda)} = \varkappa_0^{(\lambda)} - \varkappa_1^{(\lambda)}. \tag{3.69}$$

The assumption $\varkappa_1^{(\lambda)} = 0$ is often used. In such a case,

$$\varkappa_n^{(\lambda)} = \varkappa_p^{(\lambda)} = \varkappa_{np}^{(\lambda)} \equiv \varkappa^{(\lambda)}. \tag{3.70}$$

The spin–multipole interaction is also simplified if $\varkappa_\sigma^{(\lambda)}(n) = \varkappa_\sigma^{(\lambda)}(p) = \varkappa_\sigma^{(\lambda)}(np) \equiv \varkappa_\sigma^{(\lambda)}$; it has then the following form:

$$H_\sigma = -\tfrac{1}{2} \sum_{\lambda=1, 2, \ldots} \sum_{\mu=-\lambda}^{\lambda} \sum_{n=\lambda, \lambda\pm 1} \varkappa_\sigma^{(\lambda)} T_\sigma^+(\lambda\mu n) T_\sigma(\lambda\mu n), \tag{3.71}$$

with

$$T_\sigma(\lambda\mu n) = \sum_{\substack{q, q', \\ \sigma, \sigma}} \langle q\sigma | r^n [\{\boldsymbol{\sigma} \boldsymbol{Y}_n\}_{\lambda\mu} + (-1)^\mu \{\boldsymbol{\sigma} \boldsymbol{Y}_n\}_{\lambda, -\mu}] | q'\sigma' \rangle a^+_{q\sigma} a_{q'\sigma'}, \tag{3.72}$$

the $\{\boldsymbol{\sigma}\boldsymbol{Y}_n\}_{\lambda\mu}$ are defined in (3.61). In the particular case of the spin–quadrupole interaction $\lambda = 2$, $n = 2$ and

$$\left.\begin{aligned} (\boldsymbol{\sigma}\boldsymbol{Y}_2)_{20} &= \frac{1}{\sqrt{2}} (\sigma_+ Y_{2,-1} - \sigma_- Y_{2,1}), \\ (\boldsymbol{\sigma}\boldsymbol{Y}_2)_{22} &= -\sqrt{\tfrac{2}{3}} \sigma_z Y_{22} + \tfrac{1}{5} \sigma_+ Y_{21}, \end{aligned}\right\} \tag{3.61'}$$

$$T_\sigma(2\mu 2) = \sum_{\substack{q, q', \\ \sigma, \sigma}} \langle q\sigma | r^2 [\{\boldsymbol{\sigma}\boldsymbol{Y}_2\}_{2\mu} + (-1)^\mu \{\boldsymbol{\sigma}\boldsymbol{Y}_2\}_{2, -\mu}] q'\sigma' \rangle a^+_{q\sigma} a_{q'\sigma'}$$

$$= \sum_{q, q', \sigma} \{t^{2\mu}(q, q') a^+_{q\sigma} a_{q'\sigma} + \sigma \bar{t}^{2\mu}(q, q') a^+_{q\sigma} a_{q', -\sigma}\}, \tag{3.73}$$

where

$$t^{2\mu}(q, q') \equiv \langle q+| r^2[\{\sigma Y_2\}_{2\mu}+(-1)^{\mu}\{\sigma Y_2\}_{2,-\mu}]|q'+\rangle$$
$$= \langle q-| r^2[\{\sigma Y_2\}_{2\mu}+(-1)^{\mu}\{\sigma Y_2\}_{2,-\mu}]|q'-\rangle, \qquad (3.74)$$
$$\bar{t}^{2\mu}(q, q') \equiv \langle q+| r^2[\{\sigma Y_2\}_{2\mu}+(-1)^{\mu}\{\sigma Y_2\}_{2,-\mu}]|q'-\rangle$$
$$= -\langle q-| r^2[\{\sigma Y_2\}_{2\mu}+(-1)^{\mu}\{\sigma Y_2\}_{2,-\mu}]|q'+\rangle, \qquad (3.74')$$
$$t^{2\mu}(q, q') = -t^{2\mu}(q', q), \quad \bar{t}^{2\mu}(q, q') = \bar{t}^{2\mu}(q', q). \qquad (3.74'')$$

The spin quadrupole interaction was studied in ref. 130 for the spherical nuclei and in ref. 131 for the deformed nuclei.

It is important to stress that the approximate second quantization method discussed later in this book is not related to the concrete form of $Q_{\lambda\mu}$ or $T_{\sigma}(\lambda\mu n)$; only the factorization of the interactions H_Q or H_{σ} is important for its applicability. While the angular dependence of the interaction is undoubtedly correct, the radial dependence of H_Q and H_{σ} is inherited from the phenomenological description of the nuclear surface vibrations and thus based on rather weak arguments.

3. Another type of the residual interaction was suggested in ref. 132. The interaction is known as the surface δ interaction:

$$V(|\mathbf{r}_1-\mathbf{r}_2|) = -2F(u_0R_0)^{-4}(1-\eta-\eta P_{\sigma})\delta(r_1-R_0)\,\delta(r_2-R_0)\,\delta(\cos\theta_{12}-1). \qquad (3.75)$$

In ref. 132 it was assumed that two quasiparticles can interact only if they occupy the same position on the nuclear surface. The parameter F is the singlet interaction constant ($P = -1$ in the singlet state). The interaction strength in the triplet state ($P = +1$) is characterized by an additional parameter $F(1-2\eta)$. The u_0 is the amplitude of the radial function on the nuclear surface of radius R_0. Such a quantity is practically state-independent. The radial integral multiplied by $(u_0R_0)^{-4}$ is usually assumed to be equal to unity.

The singlet interaction is predominant in the low-lying states of heavy nuclei. The surface δ-interaction contains then only one constant F. The monopole part in the particle–particle interaction channel of (3.75) is the pairing interaction causing superfluid correlations. It fulfills the approximation (3.54) automatically.

The angular part of the interaction (3.75) can be expanded into series of spherical harmonics. The surface δ-interaction in the $P = -1$ case has then the form

$$V(|\mathbf{r}_1-\mathbf{r}_2|) = -4\pi F(R_0u_0)^{-4}\,\delta(r_1-R_0)\,\delta(r_2-R_0)\sum_{\lambda,\mu} Y^*_{\lambda\mu}(\theta, \varphi)\,Y_{\lambda\mu}(\theta_2, \varphi_2). \qquad (3.76)$$

The surface δ-interaction was employed in calculations of the single closed shell nuclei,[134] of the *sd* shell nuclei, and of the deformed nuclei.[135, 136]

The finite Fermi system theory, developed by Migdal,[13, 14, 137] does not use the quasiparticle interaction Hamiltonian explicitly. The theory uses instead an "effective" interaction, and the amplitudes in the particle–particle and particle–hole channels are calculated. We shall follow the exposition used in this book and find the Hamiltonian H_M characteristic for the effective interaction in the particle–hole channel. We must remember, however, that the integral equations derived from the interaction Hamiltonian should be identical (with the exception of exchange terms) with the basic equations of the finite Fermi system theory.

The Hamiltonian of the effective quasiparticle interaction can be expressed as

$$H_M = \sum_{f_1, f_2, f_2', f_1'} V_M(f_1, f_2; f_2', f_1') a_{f_1}^+ a_{f_2}^+ a_{f_2'} a_{f_1'}, \tag{3.77}$$

$$V_M(f_1, f_2; f', f_1') = \int (dr_1)(dr_2)\, \varphi_{f_1}^*(r_1)\, \varphi_{f_2}(r_2)\, V_M(|r_1 - r_2|)\, \varphi_{f_2'}(r_2)\, \varphi_{f_1'}(r_1), \tag{3.78}$$

where

$$V_M(|\boldsymbol{r}_1 - \boldsymbol{r}_2|) = \mathcal{F}^{\omega}(r_1)\, \delta(\boldsymbol{r}_1 - \boldsymbol{r}_2), \tag{3.79}$$

$$\mathcal{F}^{\omega}(r_1) = \frac{\pi^2}{m^* p_F}\{f + g\sigma^{(1)}.\sigma^{(2)} + (f' + g'\sigma^{(1)}.\sigma^{(2)})\boldsymbol{\tau}^{(1)}.\boldsymbol{\tau}^{(2)}\}, \tag{3.80}$$

f, f', g, g' are certain functions of angle between the initial and final momenta of the interacting quasiparticles, m^* is the effective nucleon mass, and p_F is the Fermi momentum. The expression (3.80) is expanded into series of Legendre polynomials

$$\mathcal{F}^{\omega} = \sum_l \mathcal{F}_l^{\omega} P_l(\cos\theta_{12}), \tag{3.81}$$

and only the first nonvanishing term is included in most applications. It is further assumed that the function $\mathcal{F}^{\omega}(r)$ depends on the radial coordinate in the following way:

$$\mathcal{F}^{\omega}(r) = \mathcal{F}^{\omega} \frac{\mathcal{F}_{ex}^{\omega} - \mathcal{F}^{\omega}}{1 + e^{-1/a(r - R_0)}}, \tag{3.82}$$

i.e., the function has the value $\mathcal{F}^{\omega}$ inside the nucleus and $\mathcal{F}_{ex}^{\omega}$ outside the nucleus. The quantity $1/a$ is close to the surface thickness value used in the Woods–Saxon potential (1.31). The zero-order harmonics in (3.81) are usually chosen as:[14]

$$\left.\begin{aligned} f_0 &= 0.5 \pm 0.3, \quad (f_0)_{ex} = -2 \pm 0.5;\\ f_0' &= 0.7 \pm 0.2, \quad (f_0')_{ex} = 1 \pm 0{,}5;\\ g_0 &= 0.5, \quad |g_0 - g_0'| \lesssim 0.1. \end{aligned}\right\} \tag{3.83}$$

In the framework of the Migdal theory, the interaction (3.79) is universal. This means that the parameters (3.83) should be used for all nuclei. The theory of the finite Fermi systems was able to reproduce many results obtained previously by other methods, and many new results were also obtained.

The interaction constants of the effective interaction (3.80) may be calculated. The employed method[138] uses the expressions for the energy of nuclear matter and finite nuclei.

Another residual interaction,

$$V = \frac{V_0}{A}(\sigma^{(1)}.\sigma^{(2)}) + \frac{V_1}{A}(\boldsymbol{\tau}^{(1)}.\boldsymbol{\tau}^{(2)})(\sigma^{(1)}.\sigma^{(2)}), \tag{3.84}$$

was introduced in ref. 139 for the description of the core spin polarization. It contains two parameters V_0 and V_1, and it is usually treated in perturbation theory.

It is very difficult to decide which additional quasiparticle interactions may be important. For example, attempts to describe the spin splitting of the two-, three-, and many-particle states were up till now unsuccessful. The problem of a more complete form of the residual interaction presumably will be solved when the higher excited states will be analyzed. The model Hamiltonian described above is not complete. It will be supplemented and modified as our knowledge of nuclear structure grows.

CHAPTER 4

THE MODEL OF INDEPENDENT QUASIPARTICLES

§ 1. Superfluid Pairing Correlations of Nucleons

1. Let us explain the mathematical apparatus employed for the description of superfluid pairing correlations in atomic nuclei. The theory was originally developed by Bogolyubov and by Bardeen, Cooper, and Schrieffer. The basic equations of the superfluid nuclear model can be derived in a number of possible ways; we shall follow a method based on the variational principle. The theory has been explained, for example, in refs. 12, 17, 123, 128, and 140.

The Hamiltonian describing the interaction of the nucleons will be, for our purposes,

$$H_0 = H_{\text{av}} + H_{\text{pair}}. \tag{4.1}$$

The oscillator potential with spin–orbit coupling or the Woods–Saxon potential (1.31) and (1.32) will be utilized as the average field for the spherical nuclei. For deformed nuclei we shall use either Nilsson potential (1.78) and (1.90) or the deformed Woods–Saxon potential (1.95) and (1.96). The methods of handling the pairing correlations are very general. They do not depend on the symmetry properties or on the concrete form of the average field. We shall, therefore, first derive the basic equations in a general form and then modify these equations in order to obtain a particular form convenient for the spherical or deformed nuclei. In studies of pairing correlations it is necessary to separate the quantum number with eigenvalues $\sigma = \pm 1$ from the full set of quantum numbers. States which differ only by the sign of σ are conjugated under the time-reversal transformation. The quantum number σ can represent, for example, the sign of the angular momentum projection on the nuclear symmetry axis. We shall use $(q\sigma)$ to denote the single-particle levels of the average field. $E(q)$ will denote the corresponding energies. The neutron states will be denoted by $(s\sigma)$, and the proton states by $(r\sigma)$.

The superfluid neutron–proton correlations are missing in the medium and heavy nuclei.[12] The average field potentials are also constructed separately, and independent Schrödinger equations are solved for neutrons and protons. Hence, the neutron and proton systems are treated separately in the model of independent quasiparticles. The Hamiltonian (4.1) can be thus divided into the neutron and proton parts:

$$H_0 = H_0(n) + H_0(p). \tag{4.1'}$$

We have shown in § 2 of Chapter 3 that the problem can be solved generally for the pairing interaction $G(q+, q-; q'-, q'+)$. However, we shall use the approximation (3.54), i.e., we shall replace the functions $G(q+, q-; q'-, q'+)$ by a single constant from the very beginning. The pairing interaction is then characterized by two parameters. The quantity G_N describes the neutron system and G_Z describes the proton system. Using our notation we rewrite the Hamiltonian (4.1′) as

$$\left.\begin{aligned} H_0(n) &= \sum_{s\sigma} \{E_0(s)-\lambda_n\}a^+_{s\sigma}a_{s\sigma} - G_N \sum_{s,\, s'} a^+_{s+}a^+_{s-}a_{s'-}a_{s'+}, \\ H_0(p) &= \sum_{r\sigma} \{E_0(r)-\lambda_p\}a^+_{r\sigma}a_{r\sigma} - G_Z \sum_{r,\, r'} a^+_{r+}a^+_{r-}a_{r'-}a_{r'+}, \end{aligned}\right\} \tag{4.2}$$

where the $E_0(s)$ and $E_0(r)$ are the unrenormalized single-particle energies.

The mathematical approximations used in the description of pairing correlations lead to the nonconservation of the particle number. To counteract this effect, we shall demand that the particle number is conserved in the average, i.e., that the conditions

$$N = \sum_{s\sigma} \langle| a^+_{s\sigma}a_{s\sigma} |\rangle; \qquad Z = \sum_{r\sigma} \langle| a^+_{r\sigma}a_{r\sigma} |\rangle, \tag{4.3}$$

are valid. The symbol $\langle| \ldots |\rangle$ denotes the averaging over the studied state. The Lagrange multipliers are introduced in order to ensure the fulfillment of the relations (4.3). These Lagrange multipliers λ_n and λ_p are usually called "chemical potentials." It is convenient to add the terms $-\lambda_n N$ and $-\lambda_p Z$ to the Hamiltonian from the start. This means that the single-particle energies are counted not from the zero energy but from the λ_n and λ_p energy values, i.e., from the values close to the Fermi-level energy in the corresponding neutron or proton systems.

2. Let us, for instance, consider the neutron system. The Hamiltonian (4.2) is then

$$H_0(n) = \sum_{s\sigma} \{E_0(s)-\lambda_n\}a^+_{s\sigma}a_{s\sigma} - G_N \sum_{s,\, s'} a^+_{s+}a^+_{s-}a_{s'-}a_{s'+}.$$

The creation and annihilation operators obey the usual anticommutation relations

$$a^+_{s\sigma}a_{s'\sigma'} + a_{s'\sigma'}a^+_{s\sigma} = \delta_{ss'}\delta_{\sigma\sigma'}, \tag{4.4}$$

$$a_{s\sigma}a_{s'\sigma'} + a_{s'\sigma'}a_{s\sigma} = 0, \tag{4.4′}$$

$$a^+_{s'\sigma'}a^+_{s\sigma} + a^+_{s\sigma}a^+_{s'\sigma'} = 0. \tag{4.4″}$$

The linear canonical transformation of the $a^+_{s\sigma}$ and $a_{s\sigma}$ operators is used in order to replace the particle operators by the quasiparticle operators. Such a canonical transformation is defined by the expressions

$$a_{s\sigma} = u_s\alpha_{s,\,-\sigma} + \sigma v_s\alpha^+_{s\sigma}, \qquad a^+_{s\sigma} = u_s\alpha^+_{s,\,-\sigma} + \sigma v_s\alpha_{s\sigma}. \tag{4.5}$$

The new operators $\alpha^+_{s\sigma}$ and $\alpha_{s'\sigma'}$ will satisfy the relations (4.4), (4.4′), and (4.4″), i.e., these operators will describe fermions, if the equation

$$\eta_s = u_s^2 + v_s^2 - 1 = 0 \tag{4.6}$$

is valid for all real functions u_s and v_s.

Equation (4.6) can be used for finding the inverse transformation of (4.5), i.e., of the relation

$$\alpha_{s\sigma} = u_s a_{s,\,-\sigma} + \sigma v_s a^+_{s\sigma}. \tag{4.7}$$

The ground state of a system consisting of an even number of neutrons is, by definition, the quasiparticle vacuum. The corresponding wave function Ψ_0 can be determined from the obvious equations

$$\alpha_{s\sigma}\Psi_0 = 0, \quad \Psi_0^* \alpha^+_{s\sigma} = 0, \tag{4.8}$$

valid for all $(s\sigma)$.

Let us find the expectation value of $H_0(n)$ in the state Ψ_0. The $\langle | a^+_{s+} a^+_{s-} a_{s'-} a_{s'+} | \rangle$ and $\langle | a^+_{s\sigma} a_{s\sigma} | \rangle$ are easily found when equations (4.8) and the commutation relations (4.4), (4.4′), and (4.4″) are used. Thus,

$$\langle | a^+_{s\sigma} a_{s\sigma} | \rangle_0 = \langle | (u_s \alpha^+_{s,\,-\sigma} + \sigma v_s \alpha_{s\sigma})(u_s \alpha_{s,\,-\sigma} + \sigma v_s \alpha^+_{s\sigma}) | \rangle_0 = v_s^2 \langle | \alpha_{s\sigma} \alpha^+_{s\sigma} | \rangle_0 = v_s^2,$$

$$\begin{aligned}\langle | a^+_{s+} a^+_{s-} a_{s'-} a_{s'+} | \rangle_0 &= \langle | (u_s \alpha^+_{s-} + v_s \alpha_{s+})(u_s \alpha^+_{s+} - v_s \alpha_{s-})(u_{s'} \alpha_{s'+} - v_{s'} \alpha^+_{s'-})(u_{s'} \alpha_{s'-} + v_{s'} \alpha^+_{s'+}) | \rangle_0 \\ &= v_s v_{s'} \langle | (u_s \alpha_{s+} \alpha^+_{s+} - v_s \alpha_{s+} \alpha_{s-})(u_{s'} \alpha_{s'+} \alpha^+_{s'+} - v_{s'} \alpha^+_{s'-} \alpha^+_{s'+}) | \rangle_0 \\ &= u_s v_s u_{s'} v_{s'} - v_s^2 v_{s'}^2 \langle | \alpha_{s+} \alpha_{s-} \alpha^+_{s'-} \alpha^+_{s'+} | \rangle_0 = u_s v_s u_{s'} v_{s'} - \delta_{ss'} v_s^4,\end{aligned}$$

and the expectation value of $H_0(n)$ is equal to

$$\langle | H_0(n) | \rangle_0 = 2 \sum_s \{E_0(s) - \lambda_n\} v_s^2 - G_N \Big(\sum_s u_s v_s\Big)^2 - G_N \sum_s v_s^4.$$

Let us remember that the average field potential was found empirically, and, consequently, contains the contributions of different terms of the nuclear Hamiltonian. Particularly, the pairing interactions also contribute to the average field. Hence, the term $-G_N \sum_s v_s^2 v_{s'}^2$, which has the form of the second term in (3.39), should be included in the average nuclear Hamiltonian. Therefore, the single-particle energies have to be renormalized:

$$E(s) = E_0(s) - \frac{G_N}{2} v_s^2. \tag{4.9}$$

The term $-G_N \sum_s v_s^4$ describes the coupling of the nuclear average field with the characteristics of the pairing correlations. For example, when G_N is changed, the single-particle levels should be also changed as shown in (4.9). If the renormalization (4.9) is utilized, the pairing interaction has, in certain approximation, no effect on the single-particle levels of the average field.

We shall rewrite the expression for the average value of $H_0(n)$ over Ψ_0 using (4.9) as

$$\langle | H_0(n) | \rangle_0 = 2 \sum_s \{E(s) - \lambda_n\} v_s^2 - G_N \Big(\sum_s u_s v_s\Big)^2. \tag{4.10}$$

The functions u_s and v_s will be determined from the minimum condition of (4.10). The additional Lagrange multipliers μ_s ensure a validity of the supplementary condition (4.6). The variations δu_s and δv_s then become formally independent. The energy has an extremum if

$$\delta\Big\{\langle | H_0(n) | \rangle_0 + \sum_s \mu_s \eta_s\Big\} = 0. \tag{4.11}$$

Variation of (4.11) leads to the two equations

$$4\{E(s)-\lambda_n\}v_s-2G_N u_s\sum_{s'} u_{s'}v_{s'}-2\mu_s v_s = 0,$$
$$-2G_N v_s\sum_{s'} u_{s'}v_{s'}-2\mu_s u_s = 0.$$

The quantity μ_s will be excluded. Therefore, the first equation is multiplied by u_s, the second equation is multiplied by v_s, and the resulting equations subtracted. The result is

$$2\{E(s)-\lambda_n\}u_s v_s-(u_s^2-v_s^2)G_N\sum_{s'} u_{s'}v_{s'} = 0. \tag{4.11'}$$

This equation should be supplemented by the equation

$$N = 2\sum_s v_s^2, \tag{4.12}$$

which follows from (4.3). The quantity $2v_s^2$ is the particle density on the level s; $2u_s^2 = 2(1-v_s^2)$ is the corresponding hole density.

Equation (4.11) has two solutions. The first solution, $u_s v_s = 0$, is trivial and corresponds to the independent particles. The functions v_s and u_s are then equal to the step function, i.e.,

$$u_s = 1-\theta_F(s), \quad v_s = \theta_F(s), \tag{4.13}$$

where $\theta_F(s) = 1$ if $E(s) < \lambda_n$, and $\theta_F(s) = 0$ if $E(s) > \lambda_n$. The corresponding wave function is

$$\Psi_0^0 = \prod_{s<F} a_{s+}^+ a_{s-}^+ \Psi_{00}, \tag{4.14}$$

where

$$a_{s\sigma}\Psi_{00} = 0. \tag{4.14'}$$

Symbol $s < F$ means that $E(s) < \lambda_n$. All levels up to the Fermi level are, evidently, filled up. The remaining states are empty.

The other solution of (4.11) is nontrivial. It is characterized by the correlation function†

$$C_n = G_N\sum_s u_s v_s. \tag{4.15}$$

The solutions are sought in the form

$$u_s^2 = \frac{1}{2}\left\{1+\frac{E(s)-\lambda_n}{\varepsilon(s)}\right\}, \quad v_s^2 = \frac{1}{2}\left\{1-\frac{E(s)-\lambda_n}{\varepsilon(s)}\right\}. \tag{4.16}$$

By substituting (4.15) and (4.16) in (4.11),

$$u_s v_s = \frac{1}{2}\frac{C_n}{\varepsilon(s)}. \tag{4.17}$$

† The correlation function is often denoted by Δ in the literature.

On the other hand, from (4.16) it follows that

$$u_s^2 v_s^2 = \frac{1}{4}\,\frac{\varepsilon(s)^2 - \{E(s)-\lambda_n\}^2}{\varepsilon(s)^2}.$$

This expression can be compared with the square of (4.17) to obtain

$$\varepsilon(s) = \sqrt{C_n^2 + \{E(s)-\lambda_n\}^2}. \tag{4.18}$$

The expression (4.17) is substituted in (4.15) and the resulting equation is divided by C_n. Keeping (4.12) in mind, we obtain, finally, the system of equations which determine the quantities C_n and λ_n:

$$1 = \frac{G_N}{2}\sum_s \frac{1}{\sqrt{C_n^2 + \{E(s)-\lambda_n\}^2}}, \tag{4.19}$$

$$N = \sum_s \left\{1 - \frac{E(s)-\lambda_n}{\sqrt{C_n^2+\{E(s)-\lambda_n\}^2}}\right\}. \tag{4.20}$$

The ground-state energy can be found from (4.10) and (4.15) as

$$\mathscr{E}_0^n = \sum_s 2E(s)v_s^2 - C_n^2/G_N. \tag{4.21}$$

The quantity $\mathscr{E}_0^n$ has no absolute meaning. It can be, however, used for calculations of the relative quantities; for example, for calculations of the excitation energies, for studies of the dependence of ground-state energy on G_N or deformation, etc.

Let us find the wave function of the ground state. Assumingly, Ψ_0 has the form

$$\Psi_0 = \prod_{s'} (u'_{s'} + v'_{s'} a^+_{s'+} a^+_{s'-}) \Psi_{00}$$

and (4.8) can be used to find the unknown quantities u'_s and v'_s. Namely,

$$\begin{aligned}\alpha_{s+}\Psi_0 &= (u_s a_{s-} + v_s a^+_{s+}) \prod_{s'} (u'_{s'} + v'_{s'} a^+_{s'+} a^+_{s'-})\Psi_{00}\\ &= \prod_{s'\neq s} (u'_{s'} + v'_{s'} a^+_{s'+} a^+_{s'-})(-u_s v'_s a^+_{s+} + v_s u'_s a^+_{s+})\Psi_{00}\\ &= (u'_s v_s - u_s v'_s) a^+_{s+} \prod_{s'\neq s} (u'_{s'} + v'_{s'} a^+_{s'+} a^+_{s'-})\Psi_{00} = 0.\end{aligned}$$

From this relation it follows that $u_s = u'_s$ and $v_s = v'_s$, and that the ground-state wave function is, therefore,

$$\Psi_0 = \prod_s (u_s + v_s a^+_{s+} a^+_{s-})\Psi_{00}. \tag{4.22}$$

The same expression for Ψ_0 can be derived from (3.8′) directly.

The pairing interaction of nucleons is attractive and the value of G_N is usually sufficiently large. The nuclear ground state is, therefore, the superfluid state, i.e., it has $C \neq 0$. This means that the energy of the nontrivial, superfluid solution of (4.11) is substantially lower than the energy corresponding to the trivial, independent particle solution.

The proton system is characterized by analogous quantities, i.e., by the correlation function C_p, chemical potential λ_p, and ground-state energy $\mathcal{E}_0^p$. The basic equations have, naturally, the same form as (4.19)–(4.22).

Figure 4.1 shows the particle pair density v_s^2 for the superfluid and normal solutions. It is evident that the pairs of particles in the interacting system do not stay all the time below the Fermi level; they spend some time also on the levels above the Fermi level. Figure 4.2 shows, schematically, the distribution of particles on the single-particle levels for the situations corresponding to the state Ψ_{00} (the model of independent particles), and

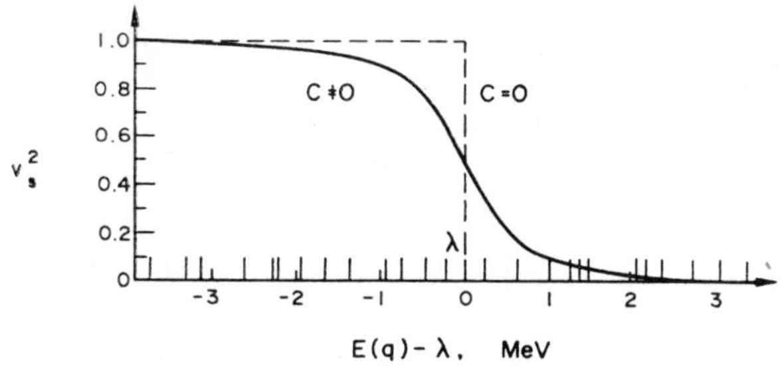

FIG. 4.1. Distribution of the particle pair density among the single-particle levels. The continuous curve corresponds to the superfluid state. The dashed curve corresponds to the normal state. The abscissa shows the energy relative to the chemical potential λ. The short vertical lines show the positions of the single-particle levels.

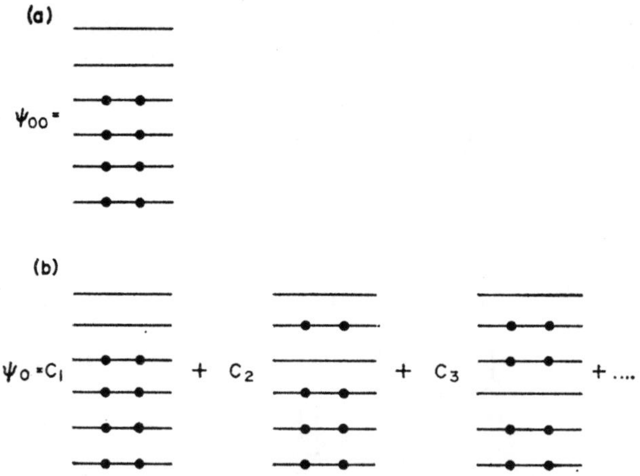

FIG. 4.2. Distribution of the particles among the single-particle levels. (a) Independent particle model, and (b) independent quasiparticle model (projection).

to the state Ψ_0 (the model of independent quasiparticles). The figure actually shows the projection of the wave function onto the subspace with the definite particle number. The ground-state energy of the superfluid state is smaller than the energy of Ψ_{00} state. This means that the loss of kinetic energy (particles spend some time above the Fermi level, i.e., in states with a higher energy) is smaller than the gain C^2/G in potential energy.

3. The nucleon interaction can cause pairing superfluid correlations if the interaction is attractive and if the interaction constants G_N (or G_Z) are not too small. The nontrivial

solutions of (4.19) exists if the inequality

$$\frac{G_N}{2}\sum_s \frac{1}{|E(s)-\lambda_n|} > 1 \tag{4.23}$$

is valid. The condition (4.23) is fulfilled in most nonmagic medium and heavy nuclei. This means that the corresponding nuclear ground states are superfluid.

As we have mentioned already, there are no superfluid neutron–proton pairing correlations in medium and heavy nuclei. Such correlations may be present if

$$|\lambda_n-\lambda_p| < 2C, \tag{4.24}$$

i.e., if the difference of the neutron and proton chemical potentials is smaller than $2C$.[123] The condition (4.24) is not satisfied in medium and heavy nuclei.

The problem of pairing correlations in light nuclei is quite involved. It is clear, however, that the pairing correlations play a less important role there than in the heavier nuclei. The whole complex of problems was studied in a number of papers, for example in refs. 141–145.

The study[145] showed that the ground-state energy of a light nucleus is minimal if the pairing superfluid neutron–neutron and proton–proton correlations are present and the neutron–proton correlations are absent. Thus, it was shown that there are no superfluid pairing neutron–proton correlations in light nuclei in the framework of the independent quasiparticle model.

The approximate mathematical methods employed in the superfluid nuclear model lead to nonconservation of the particle number. Let us estimate the fluctuations of the neutron particle number in the ground state. In order to do that, we shall calculate the mean square fluctuations of the particle number in the state Ψ_0, i.e.,

$$(\Delta N)^2 = \left\langle \left| \left(\sum_{s\sigma} a^+_{s\sigma}a_{s\sigma}\right)^2 \right| \right\rangle_0 - \left| \left\langle \left| \sum_{s\sigma} a^+_{s\sigma}a_{s\sigma} \right| \right\rangle_0 \right|^2 = \sum_s 4u_s^2v_s^2 = \sum_s \frac{C_n^2}{C_n^2+\{E(s)-\lambda_n\}^2}. \tag{4.25}$$

Evidently, the particle number fluctuations are not small. The $(\Delta N)^2$ converges, however, to zero if C_n approaches zero.

The condition of the particle number conservation is replaced by the (4.3), i.e., the average particle number has the correct value. The chemical potential λ_n is introduced for such a purpose. Obviously,

$$\lambda_n = \frac{\delta\langle|H_0(n)|\rangle_0}{\delta N}, \tag{4.26}$$

i.e., the λ_n is equal to the change in the ground state energy when a neutron is added.

4. It is instructive to discuss a particular model in which (4.19) and (4.20) can be solved analytically. Suppose that the single-particle level density is constant and equals $\bar{\varrho}$ in the energy interval between E_1 and E_2. Such a case resembles in some respects the single-particle spectrum of a deformed nucleus. The E_1 and E_2 have only a conditional meaning. Evidently, λ should not be close to either of them. If there are no single-particle levels below E_1 and above E_2, we have a model of an isolated shell. The total number of levels is $\Omega = \bar{\varrho}(E_2-E_1)$ and the maximal number of particles is $N_{\max} = 2\Omega$.

The summation in (4.19) and (4.20) can be replaced by integration, and the new equations now have the form

$$1 = \frac{G\bar{\varrho}}{2} \int_{E_1}^{E_2} \frac{dE}{\sqrt{C^2+(E-\lambda)^2}}, \tag{4.27}$$

$$N = \bar{\varrho} \int_{E_1}^{E_2} \left\{1 - \frac{E-\lambda}{\sqrt{C^2+(E-\lambda)^2}}\right\} dE. \tag{4.27'}$$

After the integration is performed, the result is

$$C = \frac{1}{\sinh \frac{2}{G\bar{\varrho}}} \sqrt{(E_1-\lambda)^2+(E_2-\lambda)^2-2(E_1-\lambda)(E_2-\lambda)\cosh \frac{2}{G\bar{\varrho}}},$$

$$\frac{N-\Omega}{\bar{\varrho}} = \sqrt{C^2+(E_1-\lambda)^2} - \sqrt{C^2+(E_2-\lambda)^2}.$$

Substituting for the C in the second expression and using the properties of hyperbolic functions,

$$\frac{E_1+E_2}{2} - \lambda = \frac{N-\Omega}{2\bar{\varrho}} \frac{1}{\tanh \frac{1}{G\bar{\varrho}}} = \frac{N-\Omega}{2\bar{\varrho}} \frac{e^{2/G\bar{\varrho}}+1}{e^{2/G\bar{\varrho}}-1}. \tag{4.28}$$

Finally, the expression (4.28) is utilized in the expression for C

$$C = \frac{\sqrt{N(2\Omega-N)}}{\bar{\varrho}(e^{1/G\bar{\varrho}}-e^{-1/G\bar{\varrho}})}. \tag{4.29}$$

It is evident now that there are no pairing correlations if $N = 0$ or $N = 2\Omega$. The $N = 2\Omega$ situation corresponds to the full shell and it illustrates that the pairing correlations are absent not only if (4.23) is not valid, but also if there are no close-lying single-particle levels above the Fermi level.

The correlation function is small if $N = 2$ or if $N = 2(\Omega-1)$. It increases with the increasing number of particles or holes. The maximal value is reached when the shell is half-filled, i.e., if $N = \Omega$. Equations (4.28) and (4.29) then result in

$$C = \frac{\Omega}{\bar{\varrho}} \frac{1}{e^{1/G\bar{\varrho}}-e^{-1/G\bar{\varrho}}}, \quad \lambda = \frac{E_1+E_2}{2},$$

i.e., the chemical potential is in the middle of the shell. Our model is similar to the situation in deformed nuclei if N is not close to 2Ω. (The quantities E_1 and E_2 are the cut-off parameters, and they have more or less arbitrary values.) In the actual situation $\bar{\varrho}G < 1$ and, approximately,

$$C \approx \frac{1}{\bar{\varrho}} \sqrt{N(2\Omega-N)}e^{-1/G\bar{\varrho}}, \tag{4.29'}$$

$$\frac{E_1+E_2}{2} - \lambda \approx \frac{N-\Omega}{2\bar{\varrho}}. \tag{4.28'}$$

The above equations prove again that the superfluid solution corresponds to a lower energy than the normal solution. The particle number density distribution

$$v^2(E) = \frac{1}{2}\left\{1-\frac{E-\lambda}{\sqrt{C^2+(E-\lambda)^2}}\right\}$$

resembles Fig. 4.1 if $G\bar{\varrho} < 1$. On the other hand, if $G\bar{\varrho} > 1$, the particles are practically isotropically distributed among all single-particle levels in the available interval.

5. Let us turn our attention to the spectrum of excited states of a neutron superfluid system. A number of simple relations will be useful later on. The commutation relations between the particle and quasiparticle operators are

$$\left.\begin{aligned} [a_{q\sigma}, \overset{+}{\alpha}_{q_2\sigma_2}] &= [\alpha_{q_2\sigma_2}, \overset{+}{a}_{q\sigma}] = u_q\delta_{qq_2}\delta_{\sigma,\,-\sigma_2}, \\ [\overset{+}{a}_{q\sigma}, \overset{+}{\alpha}_{q_2\sigma_2}] &= [\alpha_{q_2\sigma_2}, a_{q\sigma}] = \sigma v_q\delta_{qq_2}\delta_{\sigma\sigma_2}. \end{aligned}\right\} \tag{4.30}$$

The commutator of $H_0(n)$ and $\overset{+}{\alpha}_{s_2\sigma_2}$ is then

$$[H_0(n), \overset{+}{\alpha}_{s_2\sigma_2}] = \{E(s_2)-\lambda_n\}\,(u_{s_2}\overset{+}{a}_{s_2,\,-\sigma_2}-\sigma v_{s_2}a_{s_2\sigma_2})+G_N v_{s_2}(\delta_{\sigma_2-}\overset{+}{a}_{s_2+}+\delta_{\sigma_2+}\overset{+}{a}_{s_2-})\sum_{s'} a_{s'-}a_{s'+}+$$

$$+G_N u_{s_2}(\delta_{\sigma_2+}a_{s_2+}-\delta_{\sigma_2-}a_{s_2-})\sum_{s'} \overset{+}{a}_{s'+}\overset{+}{a}_{s'-}\,. \tag{4.30'}$$

First, it is necessary to calculate the lowest excited states of a system consisting of an even number of neutrons. We have shown above that the ground state is the quasiparticle vacuum, i.e., it contains no quasiparticles. It is easy to see that the states with one quasiparticle correspond to the system with an odd number of particles. The lowest excited states, therefore, are the states with one broken pair, i.e., the states with two quasiparticles.

The wave function of a two-quasiparticle state is

$$\Psi_0(s_1, s_2) = \overset{+}{\alpha}_{s_1\sigma_1}\overset{+}{\alpha}_{s_2\sigma_2}\Psi_0\,. \tag{4.31}$$

The excitation energy, i.e., the energy difference between the two-quasiparticle state and the ground state, equals (if $s_1 \neq s_2$)

$$\begin{aligned} &\langle|\alpha_{s_2\sigma_2}\alpha_{s_1\sigma_1}H_0(n)\,\overset{+}{\alpha}_{s_1\sigma_1}\overset{+}{\alpha}_{s_2\sigma_2}|\rangle_0-\langle|\,H_0(n)\,|\rangle_0 \\ &\quad = \langle|\alpha_{s_2\sigma_2}[H_0(n), \overset{+}{\alpha}_{s_2\sigma_2}]\,|\rangle_0+\langle|\alpha_{s_1\sigma_1}[H_0(n), \overset{+}{\alpha}_{s_1\sigma_1}]\,|\rangle_0 \\ &\quad + \langle|\alpha_{s_2\sigma_2}\alpha_{s_1\sigma_1}[[H_0(n), \overset{+}{\alpha}_{s_1\sigma_1}], \overset{+}{\alpha}_{s_2\sigma_2}]\,|\rangle_0 \approx \varepsilon(s_1)+\varepsilon(s_2) > 2C_n\,. \end{aligned} \tag{4.32}$$

We can see that the excited states in a system consisting of an even number of neutrons are separated from the ground state by a gap of magnitude $2C_n$. Naturally, a similar gap exists for an even number of protons and its magnitude is $2C_p$.

The existence of the energy gap separating the excited noncollective states of even–even nuclei from the corresponding ground states is one of the main features of nuclear spectra. A generalization of the above conclusions is straightforward. We can expect four-, six-, and more quasiparticle states among the higher excited states of an even system.

Another situation is encountered in systems consisting of an odd number of neutrons. The ground state of such a system is the one-quasiparticle state with the quasiparticle on the Fermi level s_F. The state s_F is defined in such a way that all single-particle states below

s_F are full and all single-particle states above s_F are empty in the independent particle model.

The ground state and a number of excited states in odd-A nuclei are one-quasiparticle states. If the single quasiparticle occupies a level above s_F, the state is called the particle state; and if the quasiparticle state is below s_F, the state is called the hole state. The one-quasiparticle state is described by the wave function

$$\Psi_0(s_2) = \alpha^+_{s_2\sigma_2}\Psi_0 = a^+_{s_2,-\sigma_2}\prod_{s\neq s_2}(u_s+v_s a^+_{s+}a^+_{s-})\Psi_{00}, \tag{4.33}$$

and its energy is equal to

$$\langle|\alpha_{s_2\sigma_2}H_0(n)\alpha^+_{s_2\sigma_2}|\rangle_0 = \langle|H_0(n)|\rangle_0 + \langle|\alpha_{s_2\sigma_2}[H_0(n), \alpha^+_{s_2\sigma_2}]|\rangle_0 \approx \langle|H_0(n)|\rangle_0 + \varepsilon(s_2). \tag{4.34}$$

The excitation energies can be found from (4.34). Evidently

$$\langle|\alpha_{s\sigma}H_0(n)\alpha^+_{s\sigma}|\rangle_0 - \langle|\alpha_{s_F\sigma'}H_0(n)\alpha^+_{s_F\sigma'}|\rangle_0 \approx \varepsilon(s)-\varepsilon(s_F), \tag{4.35}$$

i.e., there is no gap in the spectrum of odd-A nuclei, and the excitation energies can be arbitrarily small.

The higher excited states are again more complicated, and we expect three-, five-, and more quasiparticle states among them.

Thus, the superfluid pairing correlations cause a qualitative difference between the spectra of even–even nuclei and the spectra of odd-A nuclei. These results do not depend on the form of the average field and on the involved quantum numbers. The results are valid in both spherical and deformed nuclei, and even in the axially asymmetric nuclei.

6. The basic equations (4.19) and (4.20) for the excited quasiparticle states should be modified to achieve a conservation of the average particle number without violating the Pauli principle. In order to do this, we have to calculate the average values of $H_0(n)$ and $H_0(p)$ and solve the variational problem for each state separately.

The effect of unpaired particles on the superfluid properties of a system is known as the "blocking effect." The superfluid state of a nucleus is caused by the pairing Hamiltonian (4.2). If any doubly degenerate level is occupied by a single nucleon, the corresponding level is excluded from all further considerations because, according to (4.2), the nucleons fill all levels pairwise. Thus the blocking effect means that all single-particle levels which contain the quasiparticles should not participate in the calculations of the superfluid characteristics. This means further that the chemical potentials and the correlation functions for the states with a different number of quasiparticles or with quasiparticles in different states are not identical. The blocking effect theory was developed and described in refs. 12, 45, 123, and 146–148.

Let us start our discussion with the system of an odd number of neutrons. The expectation value of $H_0(n)$ in the one-quasiparticle state $\alpha^+_{s_2\sigma_2}\Psi_0$ is equal to

$$\langle|\alpha_{s_2\sigma_2}H_0(n)\alpha^+_{s_2\sigma_2}|\rangle_0 = E(s_2)-\lambda_n+\frac{G_N}{2}v^2_{s_2}+2\sum_{s\neq s_2}\{E(s)-\lambda_n\}v_s^2-G_N\Big(\sum_{s\neq s_2}u_s v_s\Big)^2. \tag{4.34'}$$

The term $(G_N/2)v^2_{s_2}$ is related to the renormalization of the single-particle level s_2; however, it is small and can be neglected.

The average energy (4.34′) has a minimum if

$$2\{E(s)-\lambda_n\}u_s v_s - G_N(u_s^2 - v_s^2) \sum_{s' \neq s_2} u_{s'} v_{s'} = 0.$$

The correlation function is defined by

$$C_n(s_2) = G_N \sum_{s \neq s_2} u_s v_s, \tag{4.36}$$

and the basic equations are

$$1 = \frac{G_N}{2} \sum_{s \neq s_2} \frac{1}{\sqrt{C_n(s_2)^2 + \{E(s) - \lambda_n(s_2)\}^2}}, \tag{4.37}$$

$$N = 1 + \sum_{s \neq s_2} \left\{1 - \frac{E(s) - \lambda_n(s_2)}{\sqrt{C_n(s_2)^2 + \{E(s) - \lambda_n(s_2)\}^2}}\right\}. \tag{4.38}$$

The wave function and the energy of the one-quasiparticle state (with the quasiparticle on the level s_2) are given by

$$\Psi_0(s_2) = \overset{+}{a}_{s_2\sigma_2} \prod_{s \neq s_2} (u_s(s_2) + v_s(s_2)\, \overset{+}{a}_{s+} \overset{+}{a}_{s-}) \Psi_{00}, \tag{4.33'}$$

$$\mathscr{E}_0(s_2) = E(s_2) + \sum_{s \neq s_2} 2E(s)\, v_s^2(s_2) - \frac{C_n^2(s_2)}{G_N}. \tag{4.39}$$

Thus, each one-quasiparticle state is characterized by its own quantities $C_n(s_2)$ and $\lambda_n(s_2)$. The excitation energy is determined by

$$\mathscr{E}_0(s_2) - \mathscr{E}_0(s_F). \tag{4.35'}$$

The wave functions (4.33′) corresponding to various s_2 values are all orthogonal. The quantities $u_s(s_2)$ and $v_s(s_2)$ are calculated according to (4.16) using the proper values of $C_n(s_2)$ and $\lambda_n(s_2)$.

For completeness we shall give here also the basic formulae for the three-quasiparticle states with quasiparticles on levels s_1, s_2, and s_3. The correlation function and the basic equations are then:

$$C_n(s_1, s_2, s) = G_N \sum_{s \neq s_1, s_2, s_3} u_s(s_1, s_2, s_3)\, v_s(s_1, s_2, s_3), \tag{4.40}$$

$$1 = \frac{G_N}{2} \sum_{s \neq s_1, s_2, s_3} \frac{1}{\varepsilon(s \mid s_1, s_2, s_3)}, \tag{4.41}$$

$$N = 3 + \sum_{s \neq s_1, s_2, s_3} 2v_s^2(s_1, s_2, s_3), \tag{4.42}$$

where

$$\varepsilon(s \mid s_1, s_2, s_3) = \sqrt{C_n^2(s_1, s_2, s_3) + \{E(s) - \lambda_n(s_1, s_2, s_3)\}^2}.$$

The wave function and the energy are given by

$$\Psi_0(s_2, s_1, s_3) = \overset{+}{a}_{s_3\sigma_3} \overset{+}{a}_{s_2\sigma_2} \overset{+}{a}_{s_1\sigma_1} \prod_{s \neq s_1, s_2, s_3} (u_s(s_1, s_2, s_3) + v_s(s_1, s_2, s_3) \overset{+}{a}_{s+} \overset{+}{a}_{s-}) \Psi_{00}, \tag{4.43}$$

$$\mathscr{E}_0(s_1, s_2, s_3) = E(s_1) + E(s_2) + E(s_3) + \sum_{s \neq s_1, s_2, s_3} 2E(s)\, v_s^2(s_1, s_2, s_3) - \frac{C_n^2(s_1, s_2, s_3)}{G_N}. \tag{4.44}$$

As before, equations (4.16) with corresponding $C_n(s_1, s_2, s_3)$ and $\lambda_n(s_1, s_2, s_3)$ are used for the calculation of the quantities $u_s(s_1, s_2, s_3)$ and $v_s(s_1, s_2, s_3)$.

Let us repeat the same procedure for the two-quasiparticle states (4.31) with $s_1 \neq s_2$. The expectation value of $H_0(n)$ in the state (4.31) is equal to

$$\langle|\alpha_{s_2\sigma_2}\alpha_{s_1\sigma_1}H_0(n)\overset{+}{\alpha}_{s_1\sigma_1}\overset{+}{\alpha}_{s_2\sigma_2}|\rangle_0 = E(s_1)-\lambda_n+E(s_2)-\lambda_n+2\sum_{s\neq s_1,\, s_2}\{E(s)-\lambda_n\}v_s^2-G\Big(\sum_{s\neq s_1\, s_2}u_s v_s\Big)^2$$

and the variational principle leads to the equation

$$2\{E(s)-\lambda_n\}u_s v_s-G_N(u_s^2-v_s^2)\sum_{s'\neq s_1,\, s_2}u_{s'}v_{s'} = 0.$$

The correlation function is defined by

$$C_n(s, s_2) = G_N\sum_{s\neq s_1,\, s_2}u_s(s_1, s_2)\, v_s(s_1, s_2) \tag{4.45}$$

and the basic equations, energy, and wave function have the familiar forms:

$$1 = \frac{G_N}{2}\sum_{s\neq s_1,\, s_2}\frac{1}{\sqrt{C_n^2(s_1, s_2)+\{E(s)-\lambda_n(s_1, s_2)\}^2}}, \tag{4.46}$$

$$N = 2+\sum_{s\neq s_1,\, s_2}\left[1-\frac{E(s)-\lambda_n(s_1, s_2)}{\sqrt{C_n^2(s_1, s_2)+\{E(s)-\lambda_n(s_1, s_2)\}^2}}\right], \tag{4.47}$$

$$\mathscr{E}_0(s_1, s_2) = E(s_1)+E(s_2)+2\sum_{s\neq s_1,\, s_2}E(s)\, v_s^2(s_1, s_2)-\frac{C_n^2(s_1, s_2)}{G_N}, \tag{4.48}$$

$$\Psi_0(s_1, s_2) = \overset{+}{a}_{s_1\sigma_1}\overset{+}{a}_{s_2\sigma_2}\prod_{s\neq s_1,\, s_2}\{u_s(s_1, s_2)+v_s(s_1, s_2)\overset{+}{a}_{s+}\overset{+}{a}_{s-}\}\Psi_{00}. \tag{4.31'}$$

The excitation energy of the two-quasiparticle state is determined by

$$\mathscr{E}_0(s_1, s_2)-\mathscr{E}_0, \tag{4.32'}$$

The quantities $\varepsilon(s\,|\,s_1, s_2)$, $u_s(s_1, s_2)$, and $v_s(s_1, s_2)$ depend on the quantum numbers s_1 and s_2 of the occupied quasiparticle states. The wave functions (4.31′) are mutually orthogonal and they are orthogonal to the ground state (4.22) if $s_1 \neq s_2$.

The two-quasiparticle states with $K^\pi = 0^+$ and both quasiparticles on the same level are somewhat special. They are not orthogonal to each other and to the ground state. There is one superfluous state among all 0^+ states, usually called "spurious state." The methodological difficulties related to the particle number nonconservation are particularly concentrated in the 0^+ states. The spurious state can be excluded and the orthogonality of the 0^+ states established if a part of the quasiparticle interaction is taken into account.

7. The equations of the pairing correlation theory given above are particularly suitable for the description of deformed nuclei. In spherical nuclei it is more convenient to modify the equations and include the $(2j+1)$ degeneracy explicitly. Let us give the corresponding formulae here.

The single-particle wave functions of the spherical nuclei are characterized by the quantum numbers $Nljm$. These wave functions are the eigenfunctions of the total angular momentum operator. The quantum number σ can be identified with the sign of m, and the quantum numbers Nl will be dropped for simplicity. The eigenfunctions of angular momentum operators include an arbitrary phase factor. The phase factor in ref. 149 is selected in such a way that the functions $\alpha_{jm}^{+}\Psi_0$ and $\alpha_{j,-m}^{+}\Psi_0$ have the same sign. In other papers, e.g. in 150, 151, the phase is chosen according to Condon and Shortley,(26) where $\alpha_{jm}^{+}\Psi_0$ corresponds to $(-1)^{j-m}\alpha_{j,-m}^{+}\Psi_0$.

Using the phase of ref. 149, the $H_0(n)$, the canonical transformation, and the wave function of the quasiparticle vacuum are:

$$H_0(n) = 2\sum_{j,\,m>0}\{E_0(j)-\lambda_n\}a_{jm}^{+}a_{jm} - G_N\sum_{jj'}\sum_{m,\,m'>0} a_{jm}^{+}a_{j,-m}^{+}a_{j',-m'}a_{j'm'}, \tag{4.2'}$$

$$a_{j\sigma m} = u_j\alpha_{j,-\sigma m} + \sigma v_j\alpha_{j\sigma m}^{+}, \tag{4.5'}$$

$$\Psi_0 = \prod_{j,\,m>0}(u_j + v_j a_{jm}^{+}a_{j,-m}^{+})\Psi_{00}. \tag{4.22'}$$

If the phase is chosen according to ref. 26, then

$$H_0(n) = \sum_{j,\,m}\{E_0(j)-\lambda_n\}a_{jm}^{+}a_{jm} - \frac{G_N}{4}\sum_{j,j'}\sum_{m,\,m'}(-1)^{j-m}(-1)^{-j'+m'}a_{jm}^{+}a_{j,-m}^{+}a_{j',-m'}a_{j'm'}, \tag{4.2''}$$

$$a_{jm} = u_j\alpha_{jm} + (-1)^{j-m}v_j\alpha_{j,-m}^{+}, \tag{4.5''}$$

$$\Psi_0 = \prod_{j,\,m>0}(u_j + (-1)^{j-m}v_j a_{jm}^{+}a_{j,-m}^{+})\Psi_{00}. \tag{4.22''}$$

In the following discussion, unless specifically stated, the phase convention of Condon and Shortley(26) is used.

Let us give the basic expressions (naturally independent of the phase) for the ground state of a system consisting of an even number of neutrons. The correlation function is given by

$$C_n = G_N\sum_j (j+\tfrac{1}{2})u_j v_j, \tag{4.15'}$$

and the quantities u_j, v_j, and $\varepsilon(j)$ do not depend explicitly on m, i.e.,

$$v_j^2 = \frac{1}{2}\left\{1-\frac{E(j)-\lambda_n}{\varepsilon(j)}\right\}, \quad u_j^2 = 1-v_j^2, \tag{4.16'}$$

$$\varepsilon(j) = \sqrt{C_n^2+\{E(j)-\lambda_n\}^2}. \tag{4.18'}$$

The basic equations and the ground state energy are:

$$\frac{G_N}{2}\sum_j \frac{j+\frac{1}{2}}{\sqrt{C_n^2+\{E(j)-\lambda_n\}^2}} = 1, \tag{4.19'}$$

$$N = \sum_j\left(j+\frac{1}{2}\right)\left\{1-\frac{E(j)-\lambda_n}{\sqrt{C_n^2+\{E(j)-\lambda_n\}^2}}\right\}, \tag{4.20'}$$

$$\mathcal{E}_0 = \sum_j (2j+1)E(j)v_j^2 - \frac{C_n^2}{G_N}. \tag{4.21'}$$

The one-quasiparticle states are described by $\alpha^+_{j_2m_2}\Psi_0$. The expectation value of $H_0(n)$ is then equal to

$$\langle \alpha_{j_2m_2}H_0(n)\alpha^+_{j_2m_2}\rangle_0 = E_0(j_2)-\lambda+(2j_2-1)\left\{E_0(j_2)-\lambda_n-\frac{G_N}{2}v^2_{j_2}\right\}v^2_{j_2}$$
$$+\sum_{j\neq j_2}(2j+1)\left\{E_0(j)-\lambda_n-\frac{G_N}{2}v^2_j\right\}v^2_j$$
$$-G_N\left\{\sum_{j\neq j_2}\left(j+\frac{1}{2}\right)u_jv_j+\left(j_2-\frac{1}{2}\right)u_{j_2}v_{j_2}\right\}^2. \quad (4.34'')$$

After renormalizing the single-particle energies the minimum condition of (4.34'') results in

$$(2j+1)\{E(j)-\lambda_n\}u_jv_j-G_N(j+\tfrac{1}{2})(u^2_j-v^2_j)\left\{\sum_{j'\neq j_2}(j'+\tfrac{1}{2})u_{j'}v_{j'}+(j_2-\tfrac{1}{2})u_{j_2}v_{j_2}\right\}=0.$$

The correlation function is defined by

$$C_n(j_2) = G_N\left\{\sum_{j\neq j_2}(j+\tfrac{1}{2})u_jv_j+(j-\tfrac{1}{2})u_{j_2}v_{j_2}\right\}. \quad (4.36')$$

The basic equations, energy, and wave function are determined by the following expressions:

$$1 = \frac{G_N}{2}\left\{\frac{j_2-\frac{1}{2}}{\varepsilon(j_2\,|\,j_2)}+\sum_{j\neq j_2}\frac{j+\frac{1}{2}}{\varepsilon(j\,|\,j_2)}\right\}, \quad (4.37')$$

$$N = 1+(2j_2-1)\,v^2_{j_2}(j_2)+\sum_{j\neq j_2}(2j+1)\,v^2_j(j_2), \quad (4.38')$$

$$\mathscr{E}^n_0(j_2) = E(j_2)+(2j_2-1)\,E(j_2)\,v^2_{j_2}(j_2)+\sum_{j\neq j_2}(2j+1)\,E(j)\,v^2_j(j_2)-\frac{C^2_n(j_2)}{G_N}, \quad (4.39')$$

$$\Psi_0(j_2) = a^+_{j_2m_2}\prod_{\substack{m'\neq m_2\\ m'>0}}(u_{j_2}(j_2)+(-1)^{j_2-m'}v_{j_2}(j_2)a^+_{j_2m'}a^+_{j_2,-m'})\times$$
$$\times\prod_{\substack{j\neq j_2\\ m>0}}(u_j(j_2)+(-1)^{j-m}v_j(j_2)a^+_{jm}a^+_{j,-m})\Psi_{00}, \quad (4.33'')$$

where

$$\varepsilon(j\,|\,j_2) = \sqrt{C^2_n(j_2)+\{E(j)-\lambda_n(j_2)\}^2},\quad v^2_j(j_2) = \frac{1}{2}\left\{1-\frac{E(j)-\lambda_n(j_2)}{\varepsilon(j\,|\,j_2)}\right\},$$
$$u^2_j(j_2) = 1-v^2_j(j_2).$$

Let us give here also the main formulae for the two-quasiparticle states. In the $j_1\neq j_2$ case the wave function of the two-quasiparticle state is

$$\Psi_0(JM\,|\,j_1,j_2) = \sum_{m_1,\,m_2}(j_1m_1j_2m_2\,|\,JM)\,\Psi_0(j_1,j_2), \quad (4.31'')$$

where

$$\Psi_0(j_1, j_2) = \alpha^+_{j_1m_1}\alpha^+_{j_2m_2}\Psi_0 = a^+_{j_1m_1}a^+_{j_2m_2} \prod_{\substack{m'\neq m_1\\ m'>0}} (u_{j_1}(j_1, j_2)+(-1)^{j_1-m'} v_{j_1}(j, j_2)a^+_{j_1m'}a^+_{j_1, -m'})\times$$
$$\times \prod_{\substack{m''\neq m_2\\ m''>0}} (u_{j_2}(j_1, j_2)+(-1)^{j_2-m''} v_{j_2}(j_1, j_2)a^+_{j_2m''}a^+_{j_2, -m''})\times$$
$$\times \prod_{\substack{j\neq j_1, j_2\\ m>0}} (u_j(j_1, j_2)+(-1)^{j-m} v_j(j_1, j_2)a^+_{jm}a^+_{j, -m})\Psi_{00}.$$

The correlation function is equal to

$$C_n(j_1, j_2) = G_N\Big\{(j_1-\tfrac{1}{2})\, u_{j_1}(j_1, j_2)\, v_{j_1}(j_1, j_2)$$
$$+(j_2-\tfrac{1}{2})\, u_{j_2}(j_1, j_2)\, v_{j_2}(j_1, j_2)+\sum_{j\neq j_1, j_2} (j+\tfrac{1}{2})\, u_j(j_1, j_2)\, v_j(j_1, j_2)\Big\}, \quad (4.45')$$

and the basic equations are:

$$\frac{2}{G_N} = \frac{j_1-\frac{1}{2}}{\varepsilon(j_1\,|\,j_1, j_2)}+\frac{j_2-\frac{1}{2}}{\varepsilon(j_2\,|\,j_1, j_2)}+\sum_{j\neq j_1, j_2}\frac{j+\frac{1}{2}}{\varepsilon(j\,|\,j_1, j_2)}, \quad (4.46')$$
$$N = 2+(2j_1-1)\, v^2_{j_1}(j_1, j_2)+(2j_2-1)\, v^2_{j_2}(j_1, j_2)+\sum_{j\neq j_1, j_2} (2j+1)\, v^2_j(j_1, j_2), \quad (4.47')$$

where

$$\varepsilon(j\,|\,j_1, j_2) = \sqrt{C^2_n(j_1, j_2)+\{E(j)-\lambda_n(j_1, j_2)\}^2}.$$

The energy of the two-quasiparticle state is equal to

$$\mathscr{E}^n_0(j_1, j_2) = E(j_1)+E(j_2)+(2j_1-1)\, E(j_1)\, v^2_{j_1}(j_1, j_2)$$
$$+(2j_2-1)\, E(j_2)\, v^2_{j_2}(j_1, j_2)+\sum_{j\neq j_1, j_2} (2j+1)\, E(j)\, v^2_j(j_1, j_2)-\frac{C^2_n(j_1, j_2)}{G_N}. \quad (4.48')$$

The corresponding expressions are somewhat modified if $j_1 = j_2$, but $m_1 \neq m_2$:

$$\Psi_0(JM\,|\,j_1, j_1) = \sum_{m_1, m_2} (j_1m_1j_1m_2\,|\,JM)\, \Psi_0(j_1, j_1), \quad (4.31''')$$

$$\Psi_0(j_1, j_1) = a^+_{j_1m_1}a^+_{j_1m_2} \prod_{\substack{m'\neq m_1, m_2\\ m'>0}} (u_{j_1}(j_1, j_1)+(-1)^{j_1-m'} v_{j_1}(j_1, j_1)a^+_{j_1m'}a^+_{j_1, -m'})\times$$
$$\times \prod_{\substack{j\neq j_1\\ m>0}} (u_j(j_1, j_1)+(-1)^{j-m} v_j(j_1, j_1)a^+_{jm}a^+_{j, -m})\Psi_{00},$$

$$C(j_1, j_1) = G_N\Big\{(j_1-\tfrac{3}{2})\, u_{j_1}(j_1, j_1)\, v_{j_1}(j_1, j_1)+\sum_{j\neq j_1} (j+\tfrac{1}{2})\, u_j(j_1, j_1)\, v_j(j_1, j_1)\Big\}, \quad (4.45'')$$

$$\frac{2}{G_N} = \frac{j_1-\frac{3}{2}}{\varepsilon(j_1,|\,j_1, j_1)}+\sum_{j\neq j_1}\frac{j+\frac{1}{2}}{\varepsilon(j\,|\,j_1, j_1)}, \quad (4.46'')$$

$$N = 2+(2j_1-3)\, v^2_{j_1}(j_1, j_1)+\sum_{j\neq j_1} (2j+1)\, v^2_j(j_1, j_1), \quad (4.47'')$$

$$\mathscr{E}^n_0(j_1, j_1) = 2E(j_1)+(2j_1-3)\, E(j_1)\, v^2_{j_1}(j_1, j_1)$$
$$+\sum_{j\neq j_1} (2j+1)\, E(j)\, v^2_j(j_1, j_1)-\frac{C^2_n(j_1, j_1)}{G_N}. \quad (4.48'')$$

Similar formulae are valid for the odd–odd nuclei. The wave function of a state with angular momentum J, projection M, and with the neutron quasiparticle in the subshell j_1^n and the proton quasiparticle in the subshell j_2^p can be expressed as:

$$\Psi_0(JM\,|\,j_1^n, j_2^p) = \sum_{m_1^n,\, m_2^p} (j_1^n m_1^n j_2^p m_2^p\,|\,JM)\,\Psi_0(j_1^n, j_2^p), \tag{4.49}$$

where $$\Psi_0(j_1^n, j_2^p) = a^+_{j_2^p m_2^p} a^+_{j_1^n m_1^n} \prod_{\substack{j^p \\ m^p>0}}{}' \left[u_{j^p}(j_2^p) + (-1)^{j^p - m^p} v_{j^p}(j_2^p) a^+_{j^p m^p} a^+_{j^p,\,-m^p}\right] \times$$

$$\times \prod_{\substack{j^n \\ m^n>0}} \left[u_{j^n}(j_1^n) + (-1)^{j^n - m^n} v_{j^n}(j_1^n) a^+_{j^n m^n} a^+_{j^n,\,-m^n}\right] \Psi_{00}.$$

The symbol $'$ means that the terms with $j^n = j_1^n$, $m^n = \pm m_1^n$ and $j^p = j_2^p$, $m^p = \pm m_2^p$ are omitted in the corresponding products.

The mathematical apparatus of the superfluid nuclear model was developed and described in many papers, among others in refs. 10–12, 44, 45, 123, 146–157, and 158–160.

§ 2. Pairing Energies and Correlation Functions

1. Let us describe the process of fitting the values of interaction constants G_N and G_Z and discuss the main characteristics of the superfluid states.

We are restricting our attention to the medium and heavy nuclei; hence, we shall consider only the nuclei in $100 < A < 256$ mass region. The numerical calculations discussed below have used the following single-particle level schemes: the Nilsson potential with the $\langle l^2 \rangle$ term, described in Chapter 1, § 5, and the Woods–Saxon potential of ref. 55 (part of the corresponding level schemes was shown in Chapter 1, § 5).

The pairing interaction constants of the nuclear superfluid model are found by comparing the calculated and experimental nuclear mass differences (pairing energies). The pairing neutron and proton energies are defined by the relations

$$P_N(Z, N) = \tfrac{1}{4}\{3\mathcal{E}_0(Z, N-1) + \mathcal{E}_0(Z, N+1) - 3\mathcal{E}_0(Z, N) - \mathcal{E}_0(Z, N-2)\}, \tag{4.50}$$

$$P_Z(Z, N) = \tfrac{1}{4}\{3\mathcal{E}_0(Z-1, N) + \mathcal{E}_0(Z+1, N) - 3\mathcal{E}_0(Z, N) - \mathcal{E}_0(Z-2, N)\}; \tag{4.50$'$}$$

where N and Z are even numbers of neutrons and protons, respectively. The nuclear masses in (4.50) and (4.50′) are denoted by $\mathcal{E}_0(Z-i,\, N-i')$; in the calculations the nuclear masses are replaced by $\mathcal{E}_0^n(N-i')$ from (4.50) and by $\mathcal{E}_0^p(N-i)$ from (4.50′). The theoretical masses are calculated according to the expressions (4.21) and (4.39) if the number of particles is even or, respectively, odd.

Sometimes, the experimental data are insufficient for the application of (4.50) and (4.50′). In such a case, a less precise formula for the determination of the pairing energy must be used:

$$P_N = \tfrac{1}{2}\{2\mathcal{E}_0(Z, N-1) - \mathcal{E}_0(Z, N) - \mathcal{E}_0(Z, N-2)\}. \tag{4.50$''$}$$

The formula (4.50″) is also advisable in the transitional region where we should try to exclude from P_N or P_Z the effects related to the different equilibrium deformations of involved nuclei. For similar reasons, the less precise formula (4.50″) is also used for nuclei close to the magic numbers.

The fitting procedure of the interaction constants G_N and G_Z is straightforward. As a first step, the pairing energies P_N or P_Z are found from the experimental data (the whole complex of problems related to the experimental nuclear masses, binding energies and pairing energies is explained in monograph[(161)]). Then the theoretical pairing energies are calculated using (4.21) and (4.29) and varying the values of G_N or G_Z. Finally, by comparing the two sets of P_N and P_Z, the correct values of G_N and G_Z are found.

What is the dependence of G on the nuclear mass number A? We can make a general statement: the single-particle matrix elements of the short-range forces are inversely proportional to the volume of the system, i.e., they are proportional to A^{-1}. Such a statement can be easily proved for the δ-force assuming that the single-particle wave functions are constant inside the nucleus and vanish outside. From the above statement it follows that

$$G = \text{const}/A. \tag{4.51}$$

The theoretical pairing energies are compared with the values calculated from the experimental nuclear masses in Figs. 4.3 and 4.4. The experimental masses are from ref. 162 and the experimental errors were used to assign the corresponding error bars to the pairing energies. The single-particle level schemes shown in Figs. 1.5–1.8 were used in the calculations of the theoretical pairing energies. The following values of the pairing interaction constants were found:

$$G_N = \frac{23}{A} \text{ Mev}, \qquad G_Z = \frac{27}{A} \text{ MeV} \tag{4.52}$$

for $150 \leqslant A < 256$;

$$G_N = \frac{20}{A} \text{ MeV}, \qquad G_Z = \frac{23}{A} \text{ MeV} \tag{4.53}$$

for $100 < A < 150$.

The agreement between the theory and experiment can be further improved if small variations of the pairing interaction constants are allowed. Here are the corresponding intervals:

$$\text{For } 150 \leqslant A < 256: \quad G_N = \frac{22.5 \pm 0.5}{A} \text{ MeV}, \qquad G_Z = \frac{26.5 \pm 0.5}{A} \text{ MeV}. \tag{4.52'}$$

$$\text{For } 100 < A < 150: \quad G_N = \frac{19.5 \pm 0.5}{A} \text{ MeV}, \qquad G_Z = \frac{22.5 \pm 0.5}{A} \text{ MeV}. \tag{4.53'}$$

The interval $100 < A < 256$ contains both spherical and deformed nuclei, and the deformed nuclei there have different equilibrium forms. The whole interval was, therefore, divided into several smaller sections, each containing nuclei of approximately identical forms. In the transition regions, where the nuclear deformations change abruptly, the corresponding deformation energy contributes partially to the pairing energy (for a discussion of this point, see ref. 163).

As seen in Figs. 4.3 and 4.4, the theoretical and experimental pairing energies are in a reasonable agreement. The mass number dependence (4.51) of the constant G is thus

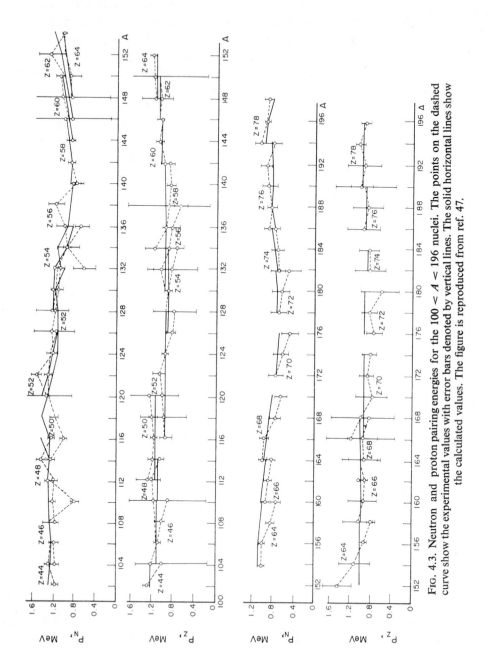

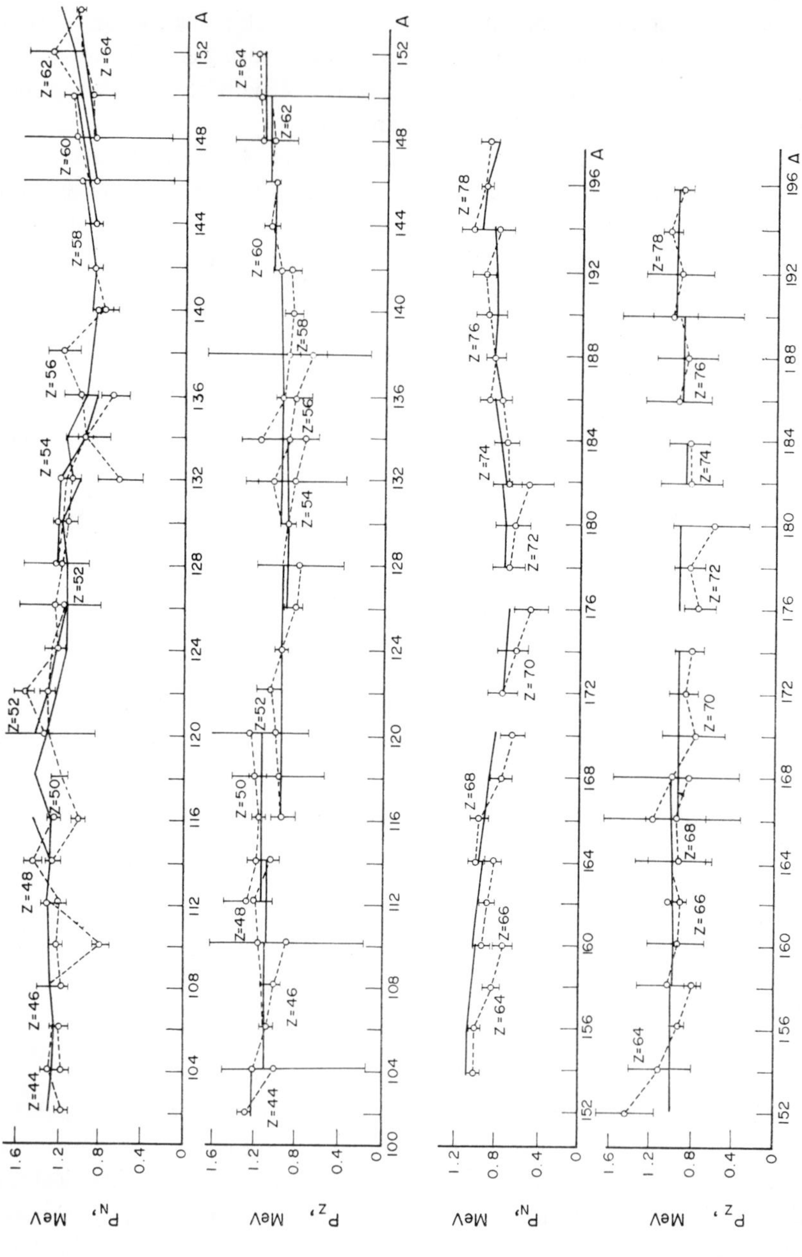

FIG. 4.3. Neutron and proton pairing energies for the $100 < A < 196$ nuclei. The points on the dashed curve show the experimental values with error bars denoted by vertical lines. The solid horizontal lines show the calculated values. The figure is reproduced from ref. 47.

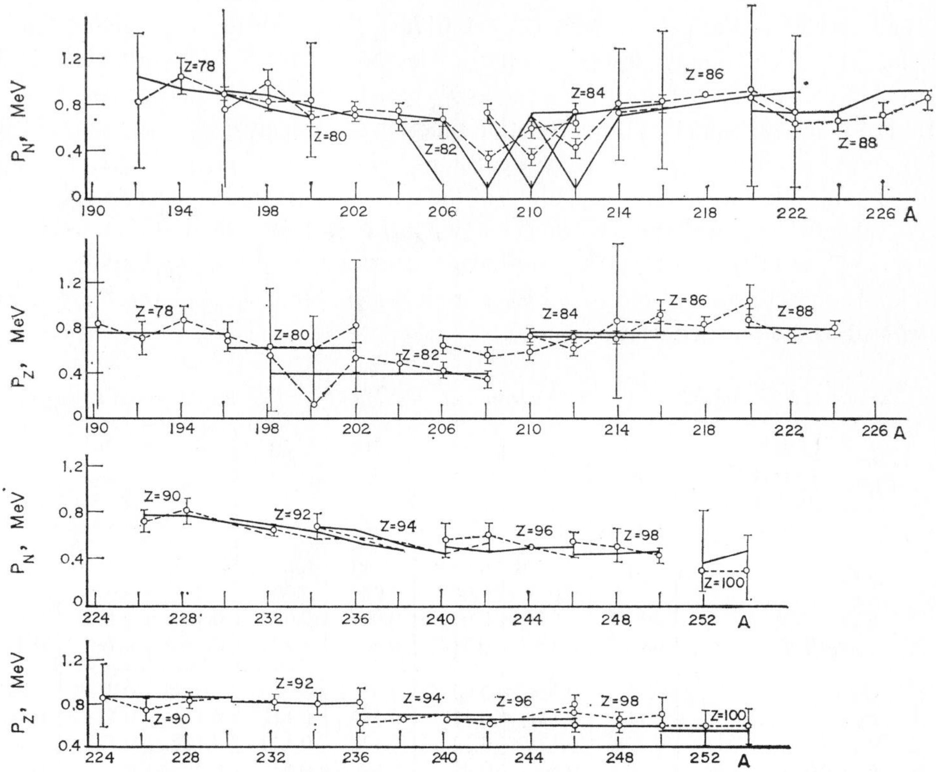

FIG. 4.4. Neutron and proton pairing energies for the $190 < A < 254$ nuclei. The notation is the same as in Fig. 4.3. The figure is reproduced from ref. 47.

supported by the fact that the same values of the parameters G_NA and G_ZA can be used in the whole wide interval $150 \leqslant A < 256$. The found values of G_N and G_Z are quite firmly established, and can be thus applied with a sufficient confidence to the transitional nuclei inside the $150 \leqslant A < 256$ interval. However, for $A < 150$ the G_NA and G_ZA values are considerably smaller than in heavier nuclei and the G_N and G_Z are also closer to each other.

The G_N and G_Z values given in (4.52) and (4.53) were calculated in ref. 47. The summation in (4.19) and (4.20) involved 42 single-particle levels. Calculations reported in ref. 12 include 36 single-particle levels. The values $G_NA = 26$–27 and $G_ZA = 28$–29 were obtained for the deformed nuclei with $152 \leqslant A \leqslant 188$ and $225 \leqslant A \leqslant 255$. We can thus conclude that the values of G_NA and G_ZA depend on the cut-off, i.e., on the number of single-particle levels in (4.19), (4.20), and other equations.

The contribution of distant states in (4.19) decreases only as $(E(q)-\lambda)^{-1}$ when the approximation $G = \text{const}$ is used. Such an approximation overemphasizes the contributions of distant states. The "real" matrix element $G(q+, q-; q'-, q'+)$ depends on q and q' and decreases when q and q' are divided by a large energy interval. It is, therefore, necessary to study how strongly the correlation functions depend on the corresponding cut-off in (4.19), (4.20), (4.37), (4.38), etc. In order to do that, the correlation functions C_p and $C_p(r_F)$ for the proton $Z = 94$ and 93 systems were calculated using different sets of the single-particle levels. The results are shown in Table 4.1 for two variants of the interaction constants G_Z.

The first variant was calculated with $G_Z^0 = 0.0176\mathring{\omega}_0$ (i.e., with the value giving the best agreement for calculations with 42 single-particle levels), and the G_Z in the other variant was adjusted in such a way that the pairing energy P_N remained approximately constant.

Table 4.1 clearly shows that the correlation function C_p strongly depends on the cut-off in (4.19) and (4.20) if the G_Z value is kept constant. The value of the pairing energy P_Z is changed in a similar way. The calculations of the nuclear superfluid model relate, however, the G_Z value to the experimental pairing energy P_Z. It is evident that, if G_Z is fitted in such a way, the C_p and the ratio $C_p(r_F)/C_p$ are independent of the cut-off, provided that at least 8–10 levels on both sides of the Fermi level are included. Similar calculations in spherical nuclei lead to the same conclusions.

TABLE 4.1. *Effect of the cut-off on the correlation function of $Z = 94$ and 93 proton systems*

Number of levels	total	18	18	30	30	30	36	36	42
	above F	14	8	8	20	14	14	20	20
	below F	3	9	21	9	15	21	15	21
G_Z[a]		0.0176	0.0176	0.0176	0.0176	0.0176	0.0176	0.0176	0.0176
P_Z		0.021	0.039	0.069	0.065	0.091	0.091	0.091	0.110
C_p		0.037	0.053	0.079	0.077	0.085	0.100	0.101	0.119
$C_p(r_F)/C_p$		0.005	0.34	0.71	0.66	0.72	0.78	0.76	0.81
G_Z		0.0302	0.0248	0.0206	0.0208	0.0200	0.0188	0.0188	0.0176
P_Z		0.111	0.111	0.112	0.112	0.111	0.112	0.112	0.110
C_p		0.128	0.121	0.120	0.122	0.120	0.121	0.121	0.119
$C_p(r_F)/C_p$		0.70	0.80	0.82	0.80	0.82	0.81	0.81	0.81

[a] The G_Z, P_Z, and C_p values are in units of $\mathring{\omega}_0$.

Thus, to summarize: the superfluid characteristics are independent of the cut-off in (4.19), (4.20), etc., if all single-particle levels in energy interval at least 4–5 MeV on both sides of the Fermi level are included. It is not necessary to introduce any additional cut-off parameters. The pairing constants are automatically renormalized if the number of included levels is changed.

It would be interesting to study the parameters G_N and G_Z in a broad region of nuclei in calculations based on the Woods–Saxon potential. To minimize the effect of cut-off, the calculations[164] included all single-particle levels from the bottom of the potential well up to 5–10 MeV in the quasicontinuous spectrum. The results showed that the $1/A$ dependence is valid in the whole $100 < A < 250$ region with a reasonable accuracy. The fitted values were $G_N A = 17$–20, $G_Z A = 19$–25. The pairing constants of deformed nuclei are 10–20% larger than the corresponding constants for spherical nuclei. This result suggests that the parameters GA increase with an increasing nuclear surface. The same conclusion was reached in ref. 165, where the authors assumed that GA is proportional to the nuclear surface area.

2. Let us discuss the values of correlation functions C_n, $C_n(s_F)$, C_p, and $C_p(r_F)$, and of chemical potentials λ_n, $\lambda_n(s_F)$, λ_p, and $\lambda_p(r_F)$ for the even–even and odd-A nuclei. All these quantities were calculated in ref. 47. The calculations used the single-particle level schemes from Figs. 1.5–1.8, and the G_N and G_Z values from (4.52) and (4.53). The results are collected

in Tables 4.2 and 4.3. The studied nuclei in the $100 < A < 256$ interval are, as before, divided into groups with approximately equal deformations. The ground-state correlation functions of the odd-A nuclei are smaller than the same quantities in even–even nuclei (i.e., in the states without quasiparticles). Such a decrease is caused by the blocking effect. The reduction is about 10–15% for spherical, nonmagic nuclei, and 15–30% for deformed nuclei.

It is interesting to find the energy differences between the normal and superfluid states of the systems with an even number of nucleons. Such a quantity is equal to

$$\mathscr{E}_0(G=0)-\mathscr{E}_0(G^0) = 2\left\{\sum_{q\leq q_F} E(q)u_q^2 - \sum_{q>q_F} E(q)v_q^2\right\} + \frac{C^2}{G}. \tag{4.54}$$

TABLE 4.2. *Correlation functions and chemical potentials for the neutron systems*

Even system $N+1$				Odd system N		Even system $N+1$				Odd system N	
N		C_n (MeV)	λ_n ($\mathring{\omega}_0$)	C_n (s_F) (MeV)	λ_n (s_F) ($\mathring{\omega}_0$)	N		C_n (MeV)	λ_n ($\mathring{\omega}_0$)	C_n (s_F) (MeV)	λ_n (s_F) $\mathring{\omega}_0$
57	$\beta_0 = 0$	1.31	5.627	1.06	5.599	107	$\beta_0 = 0.20$	0.85	6.834	0.62	6.810
59		1.37	5 670	1.15	5.648	109		0.87	6.886	0.64	6.868
61		1.34	5.715	1.13	5.698	111		0.90	6.933	0.72	6.901
63		1.38	5.757	1.19	5.747	113		0.89	6.977	0.73	6.953
65		1.39	5.800	1.18	5.776	115		0.84	7.024	0.70	7.003
67		1.31	5.836	1.14	5.809	117		0.77	7.085	0.55	7.072
69		1.34	5.873	1.18	5.850	119		0.84	7.149	0.48	7.106
71		1.33	5.908	1.19	5.888						
73		1.28	5.943	1.17	5.924	113	$\beta_0 = 0$	1.09	6.903	1.00	6.893
75		1.20	5.977	1.10	5.960	115		1.03	6.933	0.94	6.914
77		1.05	6.013	0.97	5.997	117		0.96	6.965	0.87	6.948
79		0.81	6.052	0.75	6.028	119		0.85	6.999	0.79	6.975
81		0	6.098	0	6.071	121		0.70	7.035	0.63	7.013
83		0.79	6.433	0	6.416	123		0.46	7.084	0.35	7.069
85		1.01	6.468	0.72	6.445	125		0	7.148	0	7.068
87		1.16	6.502	0.94	6.491	127		0.58	7.470	0	7.456
89		1.22	6.536	1.05	6.525	129		0.76	7.500	0.54	7.488
						131		0.88	7.529	0.71	7.518
89	$\beta_0 = 0.27$	1.15	6.400	1.00	6.383	133		0.95	7.557	0.82	7.548
91		1.12	6.443	0.96	6.419	135		1.00	7.584	0.89	7.577
93		1.07	6.488	0.92	6.463						
95		1.01	6.537	0.84	6.516	135	$\beta_0 = 0.24$	0.86	7.286	0.76	7.270
97		0.95	6.591	0.73	6.569	137		0.83	7.320	0.73	7.306
99		0.88	6.647	0.65	6.608	139		0.77	7.357	0.66	7.334
101		0.86	6.704	0.64	6.679	141		0,69	7.400	0.57	7.382
103		0.82	6.764	0.55	6.738	143		0.62	7.453	0.40	7.430
105		0.81	6.827	0.53	6.788	145		0.60	7.511	0.32	7.477
107		0.82	6.894	0.48	6.875	147		0.62	7.564	0.40	7.534
109		0.94	6.952	0.63	6.908	149		0.57	7.617	0.34	7.599
111		1.03	6.997	0.81	6.975	151		0.63	7.668	0.37	7.642
						153		0.69	7.711	0.50	7.685
						155		0.73	7.747	0.59	7.731
						157		0.76	7.782	0.63	7.783
						159		0.75	7.816	0.62	7.799

Figures 4.5 and 4.6 show this energy gain caused by the pairing correlations. The correlation functions are also shown for comparison. The figures clearly demonstrate that the correlation functions C_n and C_p decrease slowly when new particles are added. An exception from this rule are the magic nuclei and the nuclei immediately adjacent to them. The correlation functions vanish in magic nuclei and are very small in their close neighborhood. Small fluctuations of C elsewhere are caused by fluctuations of the single-particle level density near the Fermi level.

TABLE 4.3. *Correlation functions and chemical potentials for the proton systems*

		Even system $Z+1$		Odd system Z	
Z		C_p (MeV)	λ_p ($\mathring{\omega}_0$)	$C_p(r_F)$ (MeV)	$\lambda_p(r_F)$ ($\mathring{\omega}_0$)
43	$\beta_0 = 0$	1.38	5.004	1.13	4.972
45		1.30	5.058	1.12	5.029
47		1.07	5.116	0.94	5.087
49		0	5.200	0	5.158
51		0.90	5.489	0	5.468
53		1.13	5.530	0.82	5.514
55		1.17	5.576	0.95	5.563
57		1.16	5.617	0.98	5.607
59		1.18	5.652	1.04	5.633
61		1.22	5.687	1.10	5.671
63		1.28	5.722	1.14	5.709
59	$\beta_0 = 0.27$	0.97	5.588	0.69	5.565
61		1.03	5.655	0.73	5.610
63		1.04	5.720	0.76	5.690
65		1.03	5.788	0.72	5.761
67		1.09	5.857	0.69	5.821
69		0.98	5.925	0.65	5.892
71		0.99	5.985	0.71	5.955
73		1.14	6.040	0.91	6.007

		Even system $Z+1$		Odd system Z	
Z		C_p (MeV)	λ_p ($\mathring{\omega}_0$)	$C_p(r_F)$ (MeV)	$\lambda_p(r_F)$ ($\mathring{\omega}_0$)
69	$\beta_0 = 0.20$	1.17	5.878	0.88	5.856
71		1.01	5.938	0.79	5.916
73		0.96	6.002	0.67	5.966
75		1.00	6.062	0.71	6.033
77		1.07	6.118	0.83	6.085
75	$\beta_0 = 0$	0.99	5.950	0.87	5.938
77		0.84	6.004	0.70	5.961
79		0.67	6.060	0.55	6.030
81		0	6.124	0	6.076
83		0.69	6.385	0	6.363
85		0.86	6.425	0.62	6.407
87		0.96	6.464	0.78	6.449
89		1.01	6.503	0.85	6.491
87	$\beta_0 = 0.24$	0.95	6.292	0.84	6.277
89		0.92	6.332	0.79	6.312
91		0.88	6.374	0.74	6.357
93		0.81	6.421	0.67	6.393
95		0.77	6.473	0.60	6.452
97		0.73	6.529	0.54	6.491
99		0.69	6.590	0.46	6.572
101		0.72	6.652	0.42	6.615
103		0.76	6.706	0.55	6.678

The energy differences $\mathcal{E}_0(G = 0) - \mathcal{E}_0(G^0)$ have quite different behavior. The differences are very close to the C-value in deformed nuclei, but they exceed the correlation functions considerably in the transitional nuclei. Finally, in spherical nuclei the energy differences (4.54) strongly fluctuate within the 3–5 MeV interval.

Let us note that the total additional pairing binding energy of an even–even nucleus can be calculated, evidently, by adding the corresponding neutron and proton contributions.

3. Let us now discuss the characteristic features of the superfluid states. As mentioned before, the quantity $2v_q^2$ represents the density of particles on the level q. Figure 4.7 shows the v_s^2 and $1/\varepsilon(s)$ values [see (4.18)] for $N = 106$. It is evident that the particle density is considerably larger than zero, up to energies 3–4 MeV above the Fermi level. It is also clear that

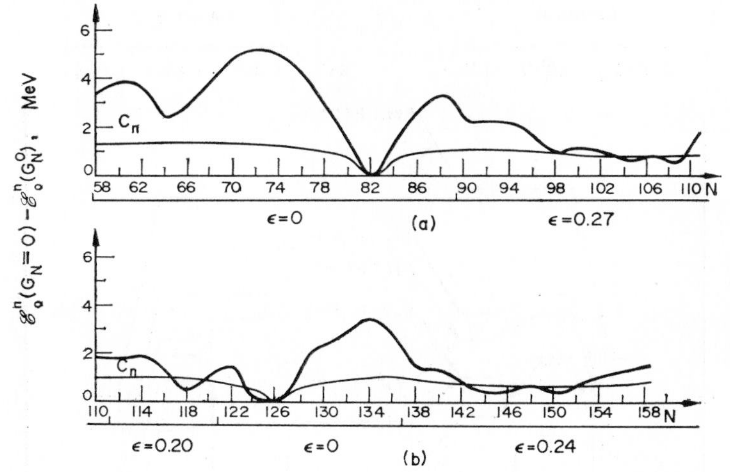

FIG. 4.5. The energy differences $\mathscr{E}_0^n(G_N = 0) - \mathscr{E}_0^n(G_N^0)$ and the correlation functions C_n for the ground states of systems containing an even number of neutrons. The equilibrium deformations are shown below the axes.

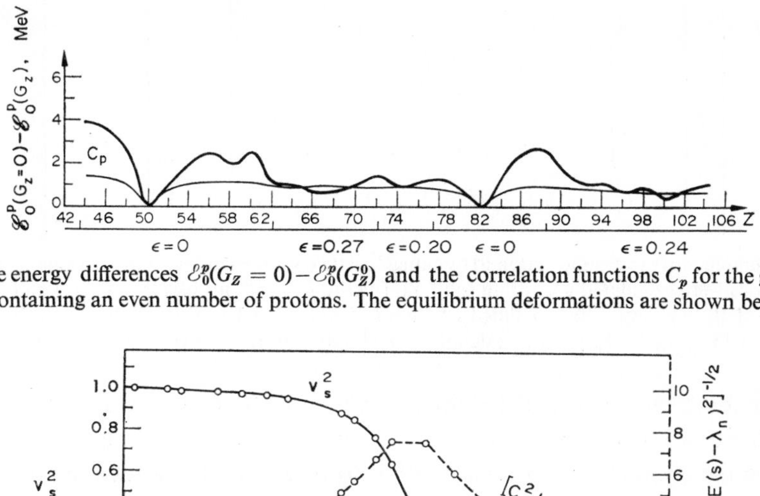

FIG. 4.6. The energy differences $\mathscr{E}_0^p(G_Z = 0) - \mathscr{E}_0^p(G_Z^0)$ and the correlation functions C_p for the ground states of systems containing an even number of protons. The equilibrium deformations are shown below the axes.

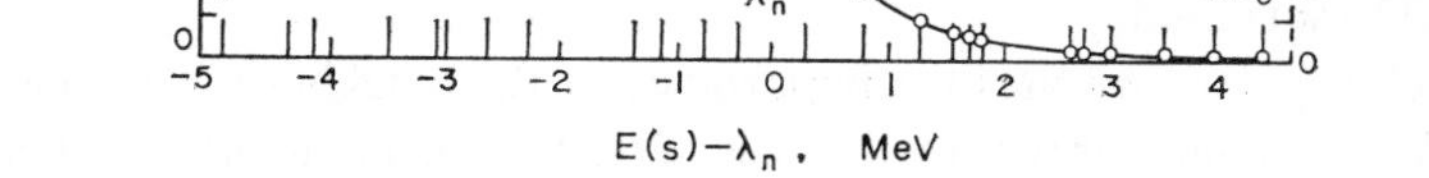

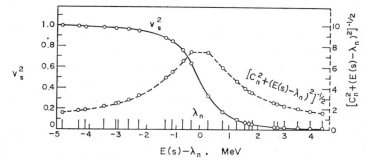

FIG. 4.7. The v_s^2 functions (the scale is to the left) and $[C_n^2+(E(s)-\lambda_n)^2]^{-1/2}$ (the scale is to the right) and their dependence on the single-particle energy for the system of 106 neutrons and deformation $\beta_0 = 0.26$. The energy is counted from the chemical potential λ. Short vertical lines show the positions of the single-particle levels.

the states at distance 4 MeV or more from the Fermi level give nonnegligible contributions to the righthand side of (4.19).

The energy dependence of the ground state and of a number of two-quasiparticle states on the G-value is rather interesting. Figure 4.8 gives an example of such a dependence. It shows

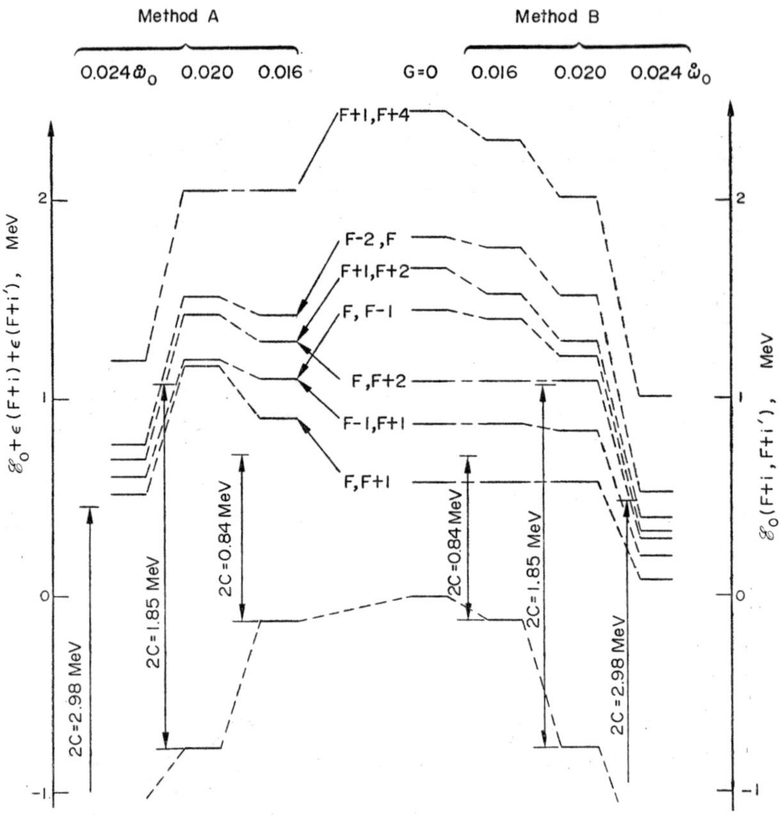

FIG. 4.8. Dependence of the ground state and two-quasiparticle state energies on the G-value. The two-quasiparticle energies were calculated according to (4.32) (method A without blocking effect) and according to (4.32′) (method B, blocking effect included). Letter F denotes the Fermi level; $F+1$, $F+2$, ..., denotes particle levels, $F-1$, $F-2$, ..., denotes hole levels, and C is the correlation function.

the corresponding energies calculated for four values of $G: G = 0$, 0.016, 0.020, and 0.024 $\mathring{\omega}_0$. The value $G = 0.020$ represents a realistic value for a deformed nucleus $A = 180$. The ground-state energy was calculated according to (4.21), while the two-quasiparticle state energies were calculated in two different ways. Method A on the left side uses (4.32), i.e., it does not include the blocking effect. In method B the blocking effect was taken into account and (4.32′) was used.

Figure 4.8 clearly demonstrates the importance of the blocking effect. The two variants differ particularly strongly for the small values of G. The calculations by method A give then unphysically large energies for a number of excited states. The $(F, F+1)$ two-quasiparticle states (one quasiparticle is on the Fermi level, another immediately above it) are especially strongly affected by the blocking effect, and their energy becomes smaller than the energy gap $2C$.

The correlation functions of two-quasiparticle states are smaller than the ground state C-values—this reduction is caused by the blocking effect. The tendency is demonstrated in Table 4.4, where the ratios $C_n(F\pm i, F\pm i')/C_n$ are shown for the $N = 94$ and 96 neutron systems. The superfluidity of two-quasiparticle states is, evidently, considerably smaller than

TABLE 4.4. *Ratio of the correlation functions of various two-quasiparticle states to the ground-state correlation function*

Calculated for a deformed nucleus with $\beta_0 = 0.3$

N	94	96
$C_n(F, F+1)/C_n$	0.65	0.56
$C_n(F, F+2)/C_n$	0.67	0.70
$C_n(F-1, F)/C_n$	0.66	0.62
$C_n(F-2, F)/C_n$	0.70	0.62
$C_n(F-1, F+1)/C_n$	0.65	0.58
$C_n(F-1, F+2)/C_n$	0.67	0.70
$C_n(F-2, F-1)/C_n$	0.70	0.63
$C_n(F+1, F+2)/C_n$	0.68	0.71
$C_n(F-2, F-3)/C_n$	0.76	0.68
$C_n(F+2, F+3)/C_n$	0.78	0.82

the ground-state superfluidity. It should be noted, however, that (4.46) and (4.47) are not exact. They tend to underestimate the correlation function.

Another important effect is the deformation dependence of the correlation function C. A typical example is shown in Fig. 4.9, which gives the value of C for 104 neutrons and for deformation interval $-0.5 \leq \beta \leq 0.5$. Figure 4.9 shows that the correlation function has the minimum at the equilibrium deformation $\beta_0 = 0.3$. The maximum of C at $\beta = 0$ is related to a higher single-particle level density when the chemical potential λ_n is close to the position of $h_{9/2}$ and $i_{13/2}$ subshells. There are some indications, mentioned, for example, in ref. 166, that the correlation functions at large deformations are actually larger than those in Fig. 4.9. It seems that such an increase of C with deformation may be explained, assuming that G is proportional to the nuclear surface area.[165]

Let us turn our attention to the one-quasiparticle states of odd-A deformed nuclei. The ground states of such nuclei correspond to the states with a single quasiparticle on the Fermi level F. The hole states contain a quasiparticle on levels $F-1$, $F-2$, etc., below the Fermi level, while the particle states have a quasiparticle on levels $F+1$, $F+2$, etc., above F.

The quasiparticle excitation energies depend on the pairing constant G. A typical such

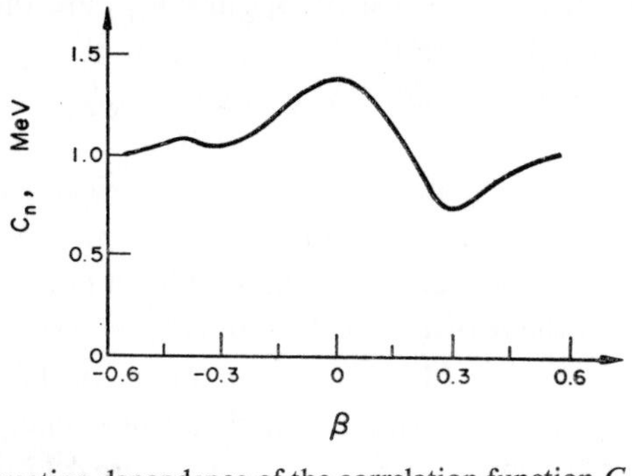

FIG. 4.9. Deformation dependence of the correlation function C_n for $N = 104$.

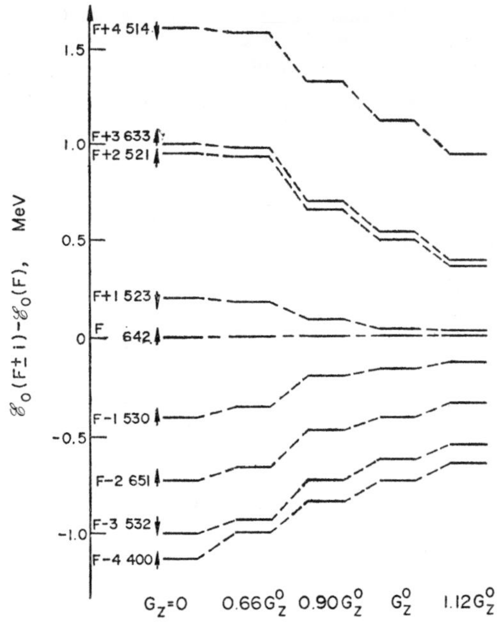

FIG. 4.10. Dependence of the one-quasiparticle state energy on the pairing interaction constant G. Calculated for 93 protons. The ground state energy is defined as zero, the particle (hole) states are plotted above (below). The $G_Z = G_Z^0$ corresponds to the real nuclear forces.

dependence, for 93 protons, is shown in Fig. 4.10. The levels at $G_Z = 0$ are determined by the average field potential only. The energy of ground state is defined as zero, the hole states are plotted below the ground state, the particle states above it. The figure demonstrates how the pairing correlations increase the level density. The low-lying states are affected most; their excitation energy for $G = G_Z^0$ (G_Z^0 is the realistic value) is about two times smaller than in the $G_Z = 0$ case.

4. Further discussion will deal with the superfluid states in spherical nuclei. The quantity $(2j+1)v_j^2$ determines the number of particles in subshell j $(0 \leq v_j^2 \leq 1)$. The chemical potential λ is generally below the single-particle energy $E(j)$ if the subshell j is empty. As the subshell is filled, the λ moves upwards and, finally, has a larger value than $E(j)$ when the subshell is full. Figure 4.11 shows the quantities λ_n and the function v_j^2 for the neutron numbers $N = 66, 68, 70, 74, 78$; the subshells $s_{1/2}$ and $h_{11/2}$ are then filled. At $N = 68$ the value of λ is closer to the $E(s_{1/2})$ than to the $E(h_{11/2})$ even if the $h_{11/2}$ subshell is being filled. Similarly, at $N = 78$ the λ_n value is 0.5 MeV above the $E(d_{3/2})$ subshell, even if $d_{3/2}$ would be empty without pairing.

The just-described behavior of chemical potential in spherical nuclei may cause somewhat unusual properties of several states. Hence, in some cases, the $(F, F+1)$ two-quasiparticle states may not be the lowest two-quasiparticle states. Or, in odd-A nuclei, the ground state I^π values may be different for the realistic G values and for $G = 0$.

In the case of spherical nuclei, (4.19) uses the quantity $(j+\frac{1}{2})/\varepsilon(j)$ instead of the quantity $1/\varepsilon(j)$. The $1/\varepsilon(j)$ depends on the single-particle energy $E(j)$ similarly to the analogous quantity for the deformed nuclei (Fig. 4.7). Figure 4.12 demonstrates the impor-

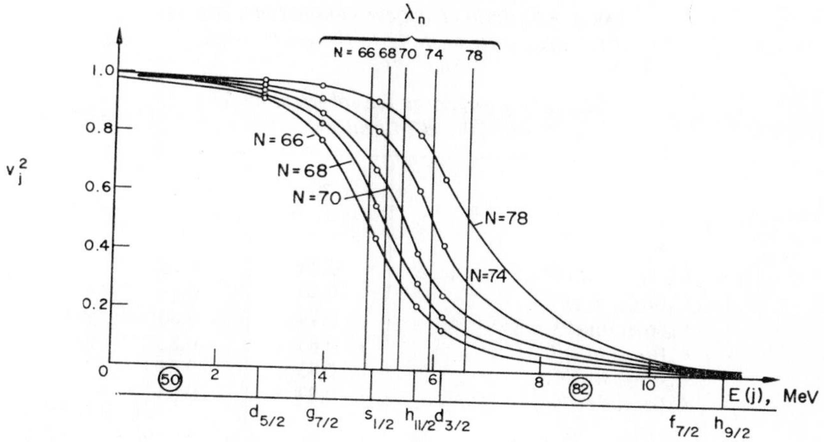

FIG. 4.11. The v_j^2 function and the chemical potential λ_n for the spherical nuclei with $N = 66, 68, 70, 74, 78$. The energy in MeV is plotted on abscissa, with the single-particle levels shown below. High vertical lines denote λ_n.

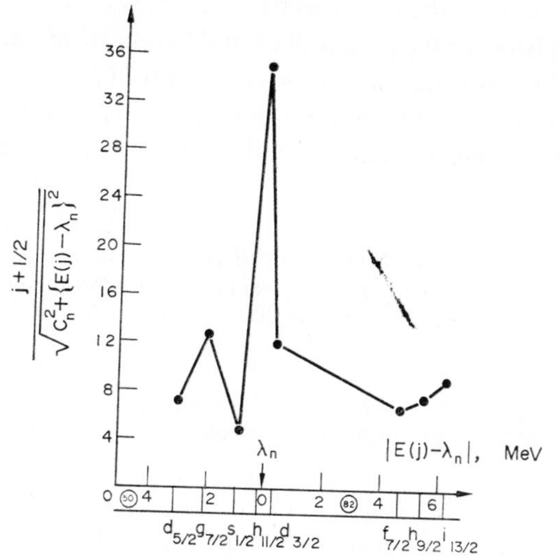

FIG. 4.12. Dependence of the quantity $(j+\frac{1}{2})/\sqrt{C_n^2+\{E(j)-\lambda_n\}^2}$ on the single-particle energy for the spherical nucleus with $N = 74$. The energy is counted from λ_n.

tance of the weighing factor $(j+\frac{1}{2})$ in the case of spherical nuclei. It shows the quantity $(j+\frac{1}{2})\big/\sqrt{C_n^2+\{E(j)-\lambda_n\}^2}$ for the quasiparticle vacuum and $N = 74$; the influence of the weighing factor is evident. For example, the $s_{1/2}$ subshell contributes less than the $g_{7/2}$ subshell in (4.19′) even though $s_{1/2}$ is closer to the Fermi level. It is also clear that the distant states, particularly those with large degeneracy $j+\frac{1}{2}$, give large contributions in the summations (4.19′) and (4.20′).

As in the case of deformed nuclei, the blocking effect makes the correlation functions $C(j_1, j_2)$ of two-quasiparticle states smaller than the ground state value of C. Table 4.5 shows the ratios $C_n(j_1, j_2)/C_n$ for the spherical nuclei with 70 and 74 neutrons. Again, the

TABLE 4.5. *Ratio of the correlation functions of two-quasiparticle states to the ground-state value of C*

Calculations performed for spherical nuclei with 70 and 74 neutrons

N	70	74
$C_n(h_{11/2}, h_{11/2})/C_n$	0.78	0.78
$C_n(h_{11/2}, d_{3/2})/C_n$	0.80	0.79
$C_n(s_{1/2}, h_{11/2})/C_n$	0.79	0.80
$C_n(g_{7/2}, h_{11/2})/C_n$	0.83	0.82
$C_n(g_{7/2}, s_{1/2})/C_n$	0.85	0.84
$C_n(g_{7/2}, d_{3/2})/C_n$	0.84	0.83

reduction of C-value is considerable. The blocking effect, however, has less pronounced influence in spherical nuclei.

The correlation functions of the one-quasiparticle states of odd-A spherical nuclei have the values intermediate between the ground state and two-particle state C-values. The quantities $C(j)$ depend weakly on j; they slowly increase with $E(j)$ moving away from E_F. The chemical potentials $\lambda(j)$ of the hole states are shifted above E_F, while for the particle states the chemical potentials are shifted below E_F. All these findings are demonstrated in Table 4.6.

TABLE 4.6. *Correlation functions, chemical potentials, and excitation energies calculated in the independent particle model $E(j) - E_F$, and in the independent quasiparticle model $\mathscr{E}_0(j) - \mathscr{E}_0^{min}(j')$, for the spherical nucleus $N = 69$*

l_j	$E(j)$ (MeV)	$C_n(j)$ (MeV)	$\lambda_n(j)$ (MeV)	$E(j)-E_F$ (MeV)	$\mathscr{E}_0(j)-\mathscr{E}_0^{min}(j')$ (MeV)	Energy (MeV)	
						^{119}Sn	^{121}Te
$g_{9/2}$	0	1.39	6.41	6.8	–	–	–
$d_{5/2}$	3.8	1.34	6.40	3.0	1.5	0.92	–
$g_{7/2}$	4.9	1.32	6.38	1.9	0.65	0.84	–
$s_{1/2}$	6.0	1.26	6.33	0.8	0	0	0
$h_{11/2}$	6.8	1.25	6.22	0	0.03	0.09	0.29
$d_{3/2}$	7.4	1.27	6.18	0.6	0.28	0.02	0.21
$f_{7/2}$	12	1.36	6.12	5.2	–	–	–

The spin and parity of the one-quasiparticle states are determined by the I^π quantum numbers of the corresponding subshell. However, unlike in the case of deformed nuclei, the lowest one-quasiparticle state in spherical nuclei does not necessarily correspond to the state expected from the independent particle model.

Such an effect can be observed if the single-particle level scheme has a subshell with small degeneracy, i.e., with small $2j'+1$, immediately below (above) a subshell with large degene-

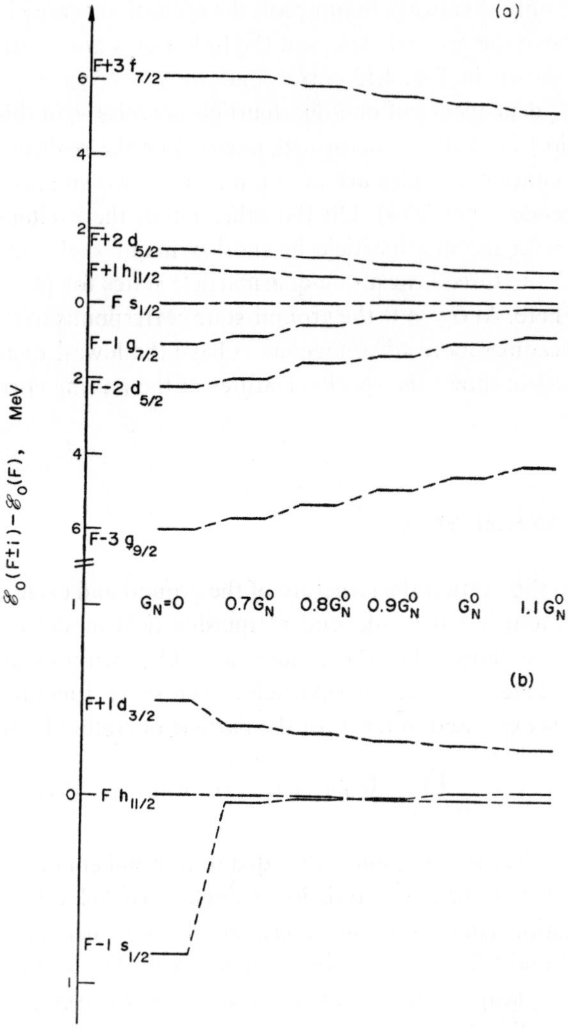

FIG. 4.13. Dependence of the one-quasiparticle state energies on the pairing interaction constant G. Calculations performed for spherical nuclei with 65 (a) and 69 (b) neutrons. The particle states are shown above the ground state, the hole states are shown below the ground state. The $G_N = G_N^0$ corresponds to the real nuclear forces.

racy $2j+1$. When the large j subshell is almost empty (full), the lowest one-quasiparticle states have the quasiparticle in subshell with small degeneracy $2j'+1$. Table 4.6 shows this effect very clearly. According to the independent particle model, the $h_{11/2}$ subshell begins to be filled starting from $N = 67$. However, the $N = 69$ system has the lowest energy if the single quasiparticle is in $s_{1/2}$ subshell. Thus, we have a simple explanation for the I^π values of the ground states in ^{119}Sn and ^{121}Te; the corresponding experimental data are also shown in Table 4.6.

Figure 4.13 shows how the energy of the one-quasiparticle state depends on the value of pairing interaction strength G. The level positions at $G_N = 0$ are determined by the single-particle level scheme only. As always in our plot, the excitation energy is shown; the particle states are plotted above the ground state and the hole states are plotted below the ground state. The situation shown in Fig. 4.13 corresponds to $N = 65$; at $G_Z = 0$ the odd particle is in $s_{1/2}$ state. The G_N dependence of one-quasiparticle energies is, in this case, similar to the dependence shown in Fig. 4.10 for deformed nuclei. For the realistic value $G_N = G_N^0$ the one-quasiparticle excitation energies are about factor of two smaller than in the $G_N = 0$ case (i.e., for independent particles). On the other hand, the excitation energies of one-quasiparticle states with the quasiparticle in another major shell are changed very little. Figure 4.13(b) shows the three lowest one-quasiparticle states for $N = 69$. The situation is somewhat different there. At $G_N = 0$ the ground state corresponds to the $h_{11/2}$ quasiparticle. As G_N increases it becomes more advantageous to have the lowest quasiparticle on the $s_{1/2}$ subshell. Thus, the figure shows the specific features of the pairing correlations in spherical nuclei.

§ 3. Quasiparticle Excited States

We shall discuss in this section the structure of the ground and excited states of deformed and spherical nuclei and use the independent quasiparticle model as a framework. The preceding sections have shown that these states are either without quasiparticles or that they have one, two, three, or more quasiparticles. The wave function $\alpha^+_{q_1\sigma_1}\alpha^+_{q_2\sigma_2}\ldots\Psi_0$ of a quasiparticle state is expressed in terms of the particle operators by the formula

$$a^+_{q_1,-\sigma_1}a^+_{q_2,-\sigma_2}\ldots\prod_{q\neq q_1,q_2,\ldots}(u_q(q_1,q_2,\ldots)+v_q(q_1,q_2,\ldots)a^+_{q+}a^+_{q-})\Psi_{00}. \quad (4.55)$$

This means that the spins, parities, and other quantum numbers of the quasiparticle state are determined by the quantum characteristics of single-particle levels occupied by unpaired nucleons. The excitation energies of the quasiparticle state depend, for fixed values of pairing constants G_N^0 and G_Z^0, on the density of single-particle states in the vicinity of Fermi level; the ordering of quasiparticle states is determined by the average field.

It is important to realize that the simple relation between the quantum characteristics of the nuclear states and quantum characteristics of the single-particle levels in an average nuclear field is intimately connected with the existence of pairing correlations. Without the pairing superfluid correlations, even a weak residual interaction would have destroyed the simple one-particle structure of the low-lying states in odd-A nuclei.

The mutual interaction of quasiparticles and the couplings between the quasiparticle states and the rotational and vibrational states are ignored in the model of independent quasiparticles. The nuclear wave function is a product of the corresponding neutron and proton wave functions.

As a consequence of the quasiparticle structure, simple selection rules for change in the quasiparticle number and in the quasiparticle states govern the α-, β-, γ-transitions and the direct nuclear reactions. Transitions, in which the change in the number of quasiparticles, or the change in the distribution of quasiparticles between the initial and final states do not exceed the number compatible with the corresponding transition operator, are allowed according to the quasiparticle selection rules. The nuclear transitions which do not fulfill the quasiparticle selection rules are called "F-forbidden".(12, 167) The quasiparticle selection rules, discussed later, are very useful in analyzing the experimental data.

2. Let us consider the quasiparticle states in odd-A deformed nuclei.

The ground state and a number of excited states have one-quasiparticle structure in these nuclei. The states are characterized by the quantum numbers $K^{\pi}[Nn_z\Lambda]$ of the single-particle level occupied by the quasiparticle.

The wave function of a one-quasiparticle state in odd-N nucleus is a product of two factors:

$$\Psi_0(s_2)\Psi_0, \tag{4.56}$$

i.e., of the one-quasiparticle neutron wave function (quasiparticle is on the s_2 orbit) described in (4.33′), and of the proton quasiparticle vacuum (4.2). Similarly, in odd-Z nucleus, the one-quasiparticle state, with the quasiparticle on the r_2 orbit, is determined by

$$\Psi_0\Psi_0(r_2). \tag{4.56′}$$

The excitation energies of, e.g., odd-N nucleus, are determined by the differences (4.35′) in the energies of one-quasiparticle states. The calculated spectra of all isotones with the same number of neutrons are identical if the Nilsson model single-particle energies are used. On the other hand, if the Woods–Saxon potential represents the average field, the single-particle levels will change with A, and the calculated one-quasiparticle excitation energies in different isotones will not be equal.

By analyzing the existing experimental material on noncollective states in odd-A deformed nuclei, the predominantly one-quasiparticle structure of the ground states and of a number of excited states was clearly brought forth. The experimental excitation energies of the most clearly established one-quasiparticle states of the odd-A deformed nuclei are collected in Tables 4.7–4.10. The quantum numbers of the one-quasiparticle states are shown in the first line; the element and its mass number A are given in the first column. All known energies of one-quasiparticle states are shown (the ground states are denoted by 0); the tables do not contain rotational states and the states in which the one-quasiparticle component is smaller than 80%. The experimental data are from the review articles and monographs,(43, 168–174) and from many original papers(175–192).

The data in Tables 4.7–4.10 demonstrate that the independent quasiparticle model correctly describes the quantum characteristics of the ground states of odd-A nuclei. There are however, several exceptions, e.g., for $N = 91$ and 95. Such accidents can happen if two single particle levels of the average field cross in the vicinity of the Fermi level.

It is also evident that the independent quasiparticle model correctly describes the general features of one-quasiparticle spectra in odd-A deformed nuclei. The one-to-one correspondence between the sequence of single-particle levels of the average field and the sequence of one-quasiparticle states is clearly established. Each orbit enters first as a particle state. Its energy decreases with increasing $N(Z)$ and at some value of $N(Z)$ the orbit corresponds to the Fermi level, i.e., it forms the ground state. Finally, the orbit becomes a hole state and its energy increases when $N(Z)$ increases.

The calculated energies of one-quasiparticle states agree more closely with the experi-

TABLE 4.7. *Energies of one-quasiparticle states of odd-N deformed nuclei in the interval* $151 \leqslant A \leqslant 189$ (*in* keV)†

Element	A	Quantum numbers of one-particle states														N
		$\frac{3}{2}+$ [402]	$\frac{11}{2}-$ [505]	$\frac{3}{2}+$ [651]	$\frac{3}{2}-$ [521]	$\frac{5}{2}+$ [642]	$\frac{5}{2}-$ [523]	$\frac{7}{2}+$ [633]	$\frac{1}{2}-$ [521]	$\frac{5}{2}-$ [512]	$\frac{7}{2}-$ [514]	$\frac{9}{2}+$ [624]	$\frac{1}{2}-$ [510]	$\frac{3}{2}-$ [512]	$\frac{7}{2}-$ [503]	
Nd	151				162		0									
Sm	153		65	0	125	91	322		698							91
Gd	155	267	119	105	0	85	321		556							
Sm	155				0	25	338		824							
Gd	157	477	426		0	64	437		704							93
Dy	159		356		0	178	310		534							
Er	161		396		0		172									
Gd	159	743	681		0	68	146		506	875						
Dy	161			551	75	0	26		365							95
Er	163	463	444		104	22	0		346	609						
Gd	161				313		0		356							
Dy	163		495		421	251	0	286	351	720						
Er	165		591	1031	243	47	0	117	297	608						97
Yb	167				188	30	0		212							
Dy	165						535	0	108	184						
Er	167		1052		753		585	0	208	348	1049					99
Yb	169				657	584	570	0	24	191	960					
Er	169		1394				850	245	0	92	822					
Yb	171							95	0	122	835					101
Hf	173							165	0	107						
Er	171								194	0	531	398	706	906		
Yb	173							351	399	0	399		1031	1340		103
Hf	175								125	0	348					
Yb	175							995	913	633	0	260	511	809		
Hf	177							746	560	509	0	324	567	911	1060	105
W	179							447	222	430	0	309	705			
Yb	177									619	109	0	333	709	1226	
Hf	179									518	215	0	375	721		107
W	181								746	366	409	0	385	726	807	
Os	183											0	171			
Hf	181											68	0	255	670	109
W	183												0	209	453	
Os	185												0	128		
W	185											716[a]	24	0	237	111
Os	187										1080[c]	557[b]	0	9.8	100	
W	187												145	0	351	
Os	189												36	0		113

[a] R. F. CASTEN, P. KLEINHEINZ, P. DALY, and B. ELBEK, *Mat. Fys. Medd. Dan. Vid. Selsk.* **38** (13) (1972).
[b] K. AHLGREN and P. DALY, *Nucl. Phys.* **A**, **189**, 368 (1972).
[c] R. THOMPSON and R. K. SHELINE, *Phys. Rev.* C, **7**, 1247 (1973).
† Remaining data are reproduced from ref. 24.

TABLE 4.8. *Energies of one-quasiparticle states of odd-Z deformed nuclei in the* 151 ⩽ *A* ⩽ 189 *interval* (*in* keV)

Element	*A*	Quantum numbers of one-particle states													*Z*
		$\frac{3}{2}-$ [541]	$\frac{5}{2}-$ [532]	$\frac{5}{2}+$ [413]	$\frac{3}{2}+$ [411]	$\frac{7}{2}-$ [523]	$\frac{1}{2}+$ [411]	$\frac{7}{2}+$ [404]	$\frac{9}{2}-$ [514]	$\frac{5}{2}+$ [402]	$\frac{3}{2}+$ [402]	$\frac{1}{2}-$ [541]	$\frac{11}{2}-$ [505[	$\frac{1}{2}+$ [400]	
Eu	153		98	0	103										
	155	1095	105	0	246	614									63
	157			0											
Tb	155		227	271	0	250									
	157		327	328[a]	0	572[a]	923[b]			837[b]					65
	159		364	348	0		974[b]	777[b]		1020[b]					
	161		482	315	0	418	990[b]	944[b]							
	163				0										
Ho	159			650		0	206								
	161	1458[e]	826	760[c]	299[c]	0	211	252[d]		447[d]		526[d]			
	163			876		0	298	440							
	165		1056[e]	995	360	0	429	715							67
	167					0									
	169					0									
Tm	161						0								
	163						0								
	165					161[f]	0	81[f]				182[f]			
	167					293	0	179							69
	169					379	0	316							
	171				676[g]	425	0	636		913					
	173					510	0								
Lu	169					493	97[h]	0	439[h]	170[h]		30			
	171					662	208	0	470	296		70			
	173						425	0		358		128			71
	175						505	0	395	343		350			
	177						570	0	150	458					
Ta	177						488[i]	0	74	70		217			73
	179						520	0	30	239		750			
	181						615	0	6	482					
	183							0	73	459					
Re	181						826		262	0		432			
	183						1102	851	496	0	1034	702	1309		75
	185						880	872	387	0	835	1045	1303		
	187						626		206	0	773				
	189									0					
Ir	185										0				
	187										0				
	189										0		372	94	77
	191										0		171	82	

[a] G. WINTER, L. FUNKE, K. H. KAUN, P. KEMNITZ, and H. SODAN, *Nucl. Phys.* A, **176**, 609 (1971).
[b] J. S. BOYNO and J. R. HUIZENGA, *Phys. Rev.* C, **6**, 1411 (1972).
[c] B. HARMATZ and T. H. HANDLEY, *Nucl. Phys.* A, **191**, 497 (1972).
[d] J. L. WOOD and D. S. BRENNER, *Nucl. Phys.* A, **185**, 58 (1972).
[e] G. MAURON, J. KERN, and O. HUBER, *Nucl. Phys.* A, **181**, 489 (1972).
[f] J. GIZON, A. GIZON, S. A. HJORTH, D. BARNEOUD, S. ANDRE, and J. TRECHERNE, *Nucl. Phys.* A, **193** 193 (1972).
[g] R. L. GRAHAM, J. S. GEIGER, and M. W. JOHNS, *Can. J. Phys.* **50**, 513 (1972).
[h] C. FOIN, D. BARNEOUD, S. A. HJORTH, and R. BETHOUX, *Nucl. Phys.* A, **199**, 129 (1973).
[i] B. L. ADER and N. N. PERRIN, *Nucl Phys.* A, **197**, 593 (1972).

TABLE 4.9. *Energies of one-quasiparticle states of odd-N deformed nuclei in the* $229 \leqslant A \leqslant 257$ *interval* (*in* keV)†

Element	A	Quantum numbers of one-particle states													N
		$\frac{1}{2}-$ [501]	$\frac{3}{2}+$ [631]	$\frac{5}{2}-$ [752]	$\frac{5}{2}+$ [633]	$\frac{7}{2}-$ [743]	$\frac{1}{2}+$ [631]	$\frac{5}{2}+$ [622]	$\frac{7}{2}+$ [624]	$\frac{9}{2}-$ [734]	$\frac{1}{2}+$ [620]	$\frac{7}{2}+$ [613]	$\frac{3}{2}+$ [622]	$\frac{9}{2}+$ [615]	
Th	229	536[a]	0.1[a]		0	20[a]	262[a]								139
U	231			0											
Th	231	551[b]	228[e]	183[b]	0	386[e]	(273)[e]								141
U	233	572[a]	312		0		399								
Th	233	538[d]	336[d]		(261)[d]		0[b, d]	6[b, d]	278[d]		(1097)[b]				143
U	235	659[e) f]	393	633	333	0	0.08	129	492	822	(1273)		(1418)	(1436)	
Pu	237		(370)[a]	655[a]	439[a]	0	145	280[a, g]	474[a]						
U	237	865[g]	668[g]	(1000)[a]	468[a]	274	0	158	428[a]		1080[a]	1125[a]	1160[a]		145
Pu	239	(470)[i]				392	0	286	512[h,]						
Cm	241						0								
U	239	816[a]	853[a]				133	0	173		693[g]	680[a]	728[a]		147
Pu	241	967[j]	(809)[i]			320[a]	163	0	172		(752)[k]				
Cm	243	729[m]				280[a]	87	0	(133)[a, m]						
Pu	243						397[k]	288[l]	0	402[l]	(629)[k]	(570)[k]	(813)[k]		149
Cm	245	913[a]	950[a]			644[n]	358	251	0	388[n]	741[m]	722[o]			
Cm	247	(958)[m]				(506)[m]	227[m]	(287)[m]	0	404[m]	400[a]				151
Cf	249						145	(379)[r]	0	417					
Fm	251								0[s]	541[s]	385[s]	191[s]			
Cm	249									0	50[p]	208[m]			153
Cf	251						544[q]		434[q]	0	106[q]	178			
Fm	253									0					
Cf	253										0		242		155
Fm	255										0				
Fm	257												0		157

† The parentheses denote tentative data. For additional references, see Table 4.10.

TABLE 4.10. *Energies of one-quasiparticle states of odd-Z deformed nuclei in the* $231 \leqslant A \leqslant 253$ *interval (in* keV)†

Element	*A*	Quantum numbers of one-particle states								*Z*
		$\frac{1}{2}+$ [400]	$\frac{3}{2}+$ [651]	$\frac{1}{2}-$ [530]	$\frac{5}{2}+$ [642]	$\frac{5}{2}-$ [523]	$\frac{3}{2}-$ [521]	$\frac{7}{2}+$ [633]	$\frac{7}{2}-$ [514]	
Pa	231		84[t]	0	183[t]	174[e]				
	233	169[u]	86[u]	0	238[u]	257[u]				91
	235			0						
Np	235				0	49	756[v]			
	237	332		268[w]	0	60	514[w]			93
	239				0	75				
Am	239				187[g]	0	(558)[g]			
	241				206	0	474[g]			
	243				84	0	267	466		95
	245				0	28		327		
Bk	245						0			
	247					295[g]	0[g]			97
	249	378[x]		569[x]	389	(496)[y]	9	0		
Es	251						0[r]		(490)	
	253							0		99

† The parentheses denote tentative data.

[a] J. R. ERSKINE, private communication, 1972.

[b] T. GROTDAL, J. LIMSTRAND, K. NYBØ, K. SKÅR, and T. F. THORSTEINSEN, *Nucl. Phys.* A, **189** (1972) 592.

[c] A. ARTNA-COHEN, *Nucl. Data Sheets* B **6**, 225 (1971).

[d] T. VON EGIDY, O. W. B. SCHULT, D. RABENSTEIN, J. R. ERSKINEB, O. A. WASSON, R. E. CHRIEN, D. BREITING, B. P. SHARMA, H. A. BAADER, and H. R. KOCH, *Phys. Rev.* C, **6**, 266 (1972).

[e] F. A. RICKEY, E. T. JURNEY, and H. C. BRITT, *Phys. Rev.* C, **5**, 2072 (1972).

[f] T. H. BRAID, R. R. CHASMAN, J. R. ERSKINE, and A. M. FRIEDMAN, *Phys. Rev.* C, **1**, 275 (1970).

[g] Y. A. ELLIS and M. R. SCHMORAK, *Nucl. Data Sheets* B **8**, 345 (1972).

[h] K. S. KRANE, C. E. OLSEN, and W. A. STEYERT, *Phys. Rev.* C, **5**, 1671 (1972).

[i] F. T. PORTER, I. AHMAD, M. S. FREEDMAN, R. F. BARNES, R. K. SJOBLOM, F. WAGNER, Jr., and P. K. FIELDS, *Phys. Rev.* C, **5**, 1738 (1972).

[j] TH. W. ELZE and J. R. HUIZENGA, *Phys. Rev.* C, **3**, 234 (1971).

[k] T. H. BRAID, R. R. CHASMAN, J. R. ERSKINE, and A. M. FRIEDMAN, *Phys. Rev.* C, **6**, 1374 (1972).

[l] P. R. FIELDS, I. AHMAD, A. M. FRIEDMAN, J. LERNER, and D. N. METTA, *Nucl. Phys.* A, **160**, 460 (1971).

[m] T. H. BRAID, R. R. CHASMAN, J. R. ERSKINE, and A. M. FRIEDMAN, *Phys. Rev.* C, **4**, 247 (1971).

[n] W.-D. SCHMIDT-OTT, R. W. FINK, P. VENUGOPALA RAO, *Z. Physik* **245**, 191 (1971).

[o] S. A. BARANOV, V. M. SHATINSKIJ, V. M. KULAKOV, and YU. F. RODIONOV, *JETP* **64**, 1970 (1973).

[p] C. E. BEMIS, Jr. ,and J. HALPERIN, *Nucl. Phys.* A, **121**, 433 (1968).

[q] I. AHMAD, F. T. PORTER, M. S. FREEDMAN, R. F. BARNES, R. K. SJOBLOM, F. WAGNER, Jr., J. MILSTED, and P. R. FIELDS, *Phys. Rev.* C, **3**, 390 (1971).

[r] I. AHMAD, R. K. SJOBLOM, R. F. BARNES, E. P. HORWITZ, and P. R. FIELDS, *Nucl. Phys.* A, **140**, 141 (1970).

[s] C. E. BEMIS, Jr., P. F. DITTNER, D. C. HENSLEY, C. D. GOODMAN, and R. J. SILVA, *Proceedings of the International Conference on Heavy Ion Physics, Dubna*, 1971, p. 175.

[t] E. BROWNE and F. ASARO, *Phys. Rev.* C, **7**, 2545 (1973).

(u) C. Sebille, G. Bastin, Chin Fan Leang, and R. Piepenbring, *C. R.* **270**, 354 (1970).
(v) U. Jäger, H. Münzel, and G. Pfennig, *Phys. Rev.* C, **7,** 1627 (1973).
(w) Th. W. Elze and J. R. Huizenga, *Phys. Rev.* C, **1,** 328 (1970).
(x) R. W. Hoff, J. E. Evans, L. G. Mann, J. F. Wild, and R. W. Longheed, *Bul. Am. Phys. Soc.* **16,** 494 (1971)
(y) S. A. Baranov, V. M. Shatinskij, V. M. Kulakov, and Yu. F. Rodionov, *JETP* **63**, 373 (1972); translation in *Soviet Physics JETP* **36,** 199 (1973).

---

mental data than the energies directly obtained from the single-particle level schemes using the Nilsson or Woods–Saxon potentials. However, the quantitative description of the excitation energies in the framework of the independent quasiparticle model, is rather inaccurate.[44, 152]

The density of one-quasiparticle states of odd-A deformed nuclei obtained from the analysis of experimental data[193] is about two times larger than the density of single-particle levels of the Nilsson potential. The modification of the Nilsson potential (by adding the term $\langle l^2\rangle$) or its replacement by the Woods–Saxon potential do not change the average density of single-particle states. As we have mentioned before, the pairing correlations increase the density of one-quasiparticle states; the density of states with excitation energy below 0.5 MeV is increased approximately twice.[12, 44] The whole body of experimental data in the rare-earth region was analyzed again in refs. 169 and 170. It was shown that the density of one-quasiparticle states is increased by a factor of 2 for neutrons and by a factor of 1.9 for protons when compared with the level diagrams of the Nilsson potential. Thus, the pairing superfluid correlations give a natural explanation of the experimental density of one-quasiparticle states in deformed odd-A nuclei.

The three-, five-, and many-quasiparticle states should exist among the higher excited states of odd-A nuclei. There are two types of the three-quasiparticle states. All three quasiparticles may be neutrons or protons (we shall denote such states as ($3n$) or ($3p$)), or two quasiparticles can be neutrons and one a proton, and vice versa (we shall denote such states as ($2n$, p) or ($2p$, n)). The wave functions, correlation functions, and basic equations of the first type were discussed in § 1 of this chapter [eqns. (4.40)–(4.44)]. The wave function and the excitation energy of the second type of three-quasiparticle state, for example, of ($2n$, p), are:

$$\Psi_0(s_1, s_2)\,\Psi_0(r_1), \tag{4.57}$$

$$\mathcal{E}_0(s_1, s_2)+\mathcal{E}_0(r_1)-\mathcal{E}_0^n-\mathcal{E}_0^p(r_F). \tag{4.58}$$

The neutron quasiparticles are in neutron orbits s_1 and s_2, the proton quasiparticle is in proton orbit r_1. [All quantities are defined in eqns (4.21), (4.31′), (4.33′), (4.39), and (4.48)]. Equation (4.58) indicates that the three-quasiparticle states will have the excitation energies approximately $2C$ or higher.

Direct experimental evidence is available only for the ($2n$, p) and ($2p$, n) three-quasiparticle states. The three-quasiparticle states, or, more exactly, the fact that the observed state has primarily three-quasiparticle structure, can be immediately recognized in two cases: if the state has a large K-value or if the state has a certain $\log ft$ value.

Examples of the states of the first type are collected in Table 4.11. The table contains the three-quasiparticle states which have the K-values larger than any single-particle K-value

TABLE 4.11. *High spin three-quasiparticle states in odd-A deformed nuclei*

Nucleus	K^π	Configuration	Energy (MeV) Experimental	Energy (MeV) Calculated
$^{247}_{97}Bk_{150}$	$\frac{23}{2}-$	$p633\uparrow\ n734\uparrow\ n624\downarrow$		1.1
$^{245}_{97}Bk_{148}$	$\frac{19}{2}+$	$p633\uparrow\ n622\uparrow\ n624\downarrow$		1.1
$^{183}_{75}Re_{108}$	$\frac{21}{2}-$	$p402\uparrow\ n624\uparrow\ n514\downarrow$		2.0
	$\frac{25}{2}+$	$p514\uparrow\ n624\uparrow\ n514\downarrow$	1.907	2.3
$^{181}_{75}Re_{106}$	$\frac{21}{2}-$	$p402\uparrow\ n624\uparrow\ n514\downarrow$		1.4
	$\frac{25}{2}+$	$p514\uparrow\ n624\uparrow\ n514\downarrow$		2.0
$^{179}_{73}Ta_{106}$	$\frac{23}{2}-$	$p404\downarrow\ n624\uparrow\ n514\downarrow$		1.4
	$\frac{25}{2}+$	$p514\uparrow\ n624\uparrow\ n514\downarrow$		1.4
$^{179}_{72}Hf_{107}$	$\frac{25}{2}-$	$n624\uparrow\ p404\downarrow\ p514\uparrow$	1.106	1.4
$^{177}_{72}Hf_{105}$	$\frac{23}{2}+$	$n514\downarrow\ p404\downarrow\ p514\uparrow$	1.315	1.4
$^{175}_{72}Hf_{103}$	$\frac{21}{2}+$	$n512\uparrow\ p404\downarrow\ p514\uparrow$		1.4
$^{177}_{71}Lu_{106}$	$\frac{23}{2}-$	$p404\downarrow\ n624\uparrow\ n514\downarrow$	0.9696	~1.0
	$\frac{15}{2}+$	$p404\downarrow\ n514\downarrow\ n510\uparrow$	1.3565	1.7
$^{175}_{71}Lu_{104}$	$\frac{21}{2}+$	$p404\downarrow\ n512\uparrow\ n514\downarrow$		~1.5
$^{165}_{67}Ho_{98}$	$\frac{19}{2}+$	$p523\uparrow\ n523\downarrow\ n633\uparrow$		~1.7

in the possible interval of the corresponding single-particle level scheme. The energies of the three states observed in refs. 194 and 195 are in a reasonable agreement with the estimates which do not include the spin splitting of three-quasiparticle states.

The second possibility of recognizing the three-quasiparticle states was suggested in ref. 167. It involves the three-quasiparticle states populated in allowed unhindered (*au*) β-decays. The fast β-decays to one-quasiparticle states are absent in these cases. The three-quasiparticle states observed in fast β-transitions are described in Table 4.12; the experimental data are from the refs. 181 and 196–200. Such a method of identification of the three-quasiparticle states is very efficient.

As we have mentioned before, the only experimentally observed three-quasiparticle-states were the states of the second type; there is no experimental evidence about three-quasiparticle states with all the particles of the same kind. It is, however, very difficult to excite such states because they are not populated in β-transitions going from one-quasiparticle state, nor are they excited in single-nucleon transfer reactions. Presumably, the easiest way to observe such states is to utilize two-nucleon transfer reactions on odd-*A* target nuclei.

3. Now we shall discuss the quasiparticle states in even-*A* deformed nuclei. The ground states of even–even nuclei, according to the model of independent quasiparticles, are states without quasiparticles. The excited states can have two, four, or more, quasiparticles. The ground state, and many excited states, of odd–odd nuclei have a two-quasiparticle structure.

There are two types of two-quasiparticle states in even–even nuclei: the neutron states (denoted by $2n$) and the proton states (denoted by $2p$). The wave functions are the simple

TABLE 4.12. *Three-quasiparticle states in odd-A deformed nuclei populated by the fast (au) β-transitions*

Nucleus	K^π	Configuration	Energy (MeV) Experimental	Energy (MeV) Calculated	Parent nucleus	log *ft* experiment
^{179}W	$\frac{3}{2}+$	$n514\downarrow, p514\uparrow, p402\uparrow$	0.720[a]	1.4	^{179}Re $p402\uparrow$	5.2[a]
	$\frac{7}{2}+$		1.680[a]			5.1[a]
^{177}Lu	$\frac{7}{2}+$	$p514\uparrow, n514\downarrow, n624\uparrow$	1.148	~1.2	^{177}Yb $n624\uparrow$	5.4
	$\frac{11}{2}+$		1.230			4.2
	$\frac{7}{2}+$		1.241			4.4
^{175}Hf	$\frac{9}{2}+$, $\frac{5}{2}+$	$n514\downarrow, p514\uparrow, p404\downarrow$	–	~1.4	^{175}Ta $p404\downarrow$	–
^{175}Yb	$\frac{3}{2}+$	$n514\downarrow, p514\uparrow, p411\downarrow$	1.497	1.6	^{175}Tm $p411\downarrow$	5.0
	$\frac{1}{2}+$		1.891			5.0
	$\frac{3}{2}+$	$n523\downarrow, p523\uparrow, p411\downarrow$	1.792	~3	^{175}Tm $p411\downarrow$	5.7
	$\frac{1}{2}+$		2.039			5.3
^{173}Hf	$\frac{9}{2}+$	$n514\downarrow, p514\uparrow, p404\downarrow$	–	1.7–2.2	^{173}Ta $p404\downarrow$	–
^{167}Yb	$\frac{9}{2}+$	$n523\downarrow, p523\uparrow, p404\downarrow$	–	~2	^{167}Lu $p404\downarrow$	–
^{165}Er	$\frac{3}{2}+$	$n523\downarrow, p523\uparrow, p411\downarrow$	1.427	1.4	^{165}Tm $p411\downarrow$	5.1
^{163}Er	$\frac{3}{2}+$	$n523\downarrow, p523\uparrow, p411\downarrow$	1.540	1.4	^{163}Tm $p411\downarrow$	5.4
	$\frac{1}{2}+$		1.803			5.2
^{163}Dy	$\frac{1}{2}+$	$n523\downarrow, p523\uparrow, n411\uparrow$	0.884	~1.2	^{163}Tb $p411\uparrow$	4.9
^{161}Ho	$\frac{1}{2}-$, $\frac{5}{2}-$	$p523\uparrow, n521\uparrow, n523\downarrow$	–	~1.8	^{161}Er $n521\uparrow$	–

[a] R. ARLT *et al.*, Izv. AN USSR. phys. ser. **37**, 929 (1973).

products, i.e.,

$$\Psi_0(s_1, s_2)\Psi_0, \tag{4.59}$$

$$\Psi_0\Psi_0(r_1, r_2), \tag{4.59'}$$

All of the relevant quantities were discussed earlier in § 1 of this chapter.

The states with different angular momentum orientation, i.e., the state $K = K_1+K_2$ and the state $K' = |K_1-K_2|$ are degenerate in the model of independent quasiparticles. The two mentioned states correspond to the parallel $\Sigma = \Sigma_1+\Sigma_2$ and antiparallel $\Sigma' = \Sigma_1-\Sigma_2$ orientation of the spin. Their degeneracy is broken by the quasiparticle interaction. According to the Gallagher's rule, the $\Sigma = 0$ states (i.e. states with antiparallel spins) have lower energies in even–even nuclei, while the $\Sigma = 1$ states (parallel spins) have higher energies. The spin splitting energy was calculated in refs. 201 and 202 by the perturbation theory using the Nilsson wave functions and the following quasiparticle interaction potential:

$$V_{12} = -4\pi g_{12}\delta(\mathbf{r}_1-\mathbf{r}_2)(1+\eta_{12}\boldsymbol{\sigma}^{(1)}\boldsymbol{\sigma}^{(2)}). \tag{4.60}$$

The calculations have proven the applicability of Gallagher's rules in all cases, except in the $K = 0$ state. It is, however, impossible to explain all splittings of two-quasiparticle states using the perturbation approach and only one set of parameters g_{12} and η_{12}.

The two-quasiparticle states in even–even deformed nuclei were analyzed in refs. 203–206 and in other studies. The energies of all two-quasiparticle states up to 2 MeV for practically all even–even nuclei with $150 \leqslant A \leqslant 188$ and $228 \leqslant A \leqslant 254$ were calculated in refs. 46, 203, 207, and 208.The spectra of several even-deformed nuclei were analyzed in lectures.(25) Calculations of the two-quasiparticle states based on the Woods–Saxon potential are also available.

It is important to stress that no experimental evidence on levels of even–even deformed nuclei contradicts the conclusions based on the model of independent quasiparticles. Moreover, new experimental data on two-quasiparticle states agree surprisingly well with the calculations performed in 1960–1. Naturally, we are talking about the low-lying states with energies 1.5–2.5 MeV, where the interaction H_Q is not very important.

Table 4.13 shows a number of experimentally observed two-quasiparticle states. The $\Sigma = 0$ states are mainly there, only for ^{168}Er both the $\Sigma = 0$ and $\Sigma = 1$ states are shown. The energies calculated without the spin splitting effect agree reasonably well with the experimental data. The majority of tabulated states are the $(F, F+1)$ states; it is easy to see (using the $2C$ values in Tables 4.2 and 4.3) that these states have energies inside the gap $2C$. Thus this fact demonstrates the importance of the blocking effect.

The four-quasiparticle states should exist among the higher excited states of even–even nuclei. Again, they can be of two types: $(4n)$ or $(4p)$ states on the one hand, and $(2n, 2p)$ states on the other hand. The corresponding equations for the four-quasiparticle states can be easily derived using the methods developed in § 1 of this chapter. The four-quasiparticle states $(2n, 2p)$ with quasiparticles on levels s_1, s_2, r_1, r_2 have the wave functions and the energies

$$\Psi_0(s_1, s_2)\,\Psi_0(r_1, r_2), \tag{4.61}$$

$$\mathscr{E}_0(s_1, s_2)+\mathscr{E}_0(r_1, r_2)-\mathscr{E}_0^n+\mathscr{E}_0^p. \tag{4.62}$$

The corresponding symbols are defined in (4.21), (4.31′), and (4.48).

The high spin four-quasiparticle states of the $(2n, 2p)$ type are the easiest to be recognized experimentally. Table 4.14 shows some states of this type; their energy was calculated without the spin-splitting effect. The isomer state in ^{178}Hf, observed in ref. 209, agrees surprisingly well with the theoretical prediction. Some experimental indications(210) suggest that the $K^\pi = 3^+$ state at 2135 keV in ^{166}Er is of a predominantly four-quasiparticle structure.

The ground state and a number of excited states of odd–odd nuclei are the two-quasiparticle states in which one quasiparticle is the neutron quasiparticle and the second is the proton quasiparticle. The wave function can be expressed as

$$\Psi_0(s_1)\,\Psi_0(r_1), \tag{4.63}$$

and the energy is equal to

$$\mathscr{E}_0(s_1)+\mathscr{E}_0(r_1)-\mathscr{E}_0(s_F)-\mathscr{E}_0(r_F); \tag{4.64}$$

the corresponding quantities are defined in (4.33′) and (4.39).

TABLE 4.13. *Two-quasiparticle states of even–even deformed nuclei*

Nucleus	K^π	Configuration	Energy (MeV)	
			Calculated	Experimental
^{252}Cf	5−	*pp*633↑ 521↑	0.9	1.070
^{246}Cm	8−	*nn*734↑ 624↓	1.1	1.181
^{244}Cm	6+	*nn*622↑ 624↓	0.9	1.042
^{234}U	6−	*nn*633↓ 743↑	1.2	1.417
	3+	*nn*633↓ 631↓	1.4	1.495
^{184}Pt	8−	*nn*514↓ 624↑	1.7	1.838
^{182}Os	8−	*nn*514↓ 624↑	1.7	1.833
^{184}W	7−	*pp*514↑ 402↑	1.4	1.503
^{182}W	4−	*nn*624↑ 510↑	1.5	1.554
^{180}W	8−	*nn*514↓ 624↑	1.5	1.531
^{178}Hf	8−	*pp*514↑ 404↓	1.2	1.148
	8−	*nn*514↓ 624↑	1.4	1.480
^{176}Hf	6+	*nn*512↑ 514↓	1.4	1.335
	8−	*pp*514↑ 404↓	1.4	1.562
	3+	*nn*514↓ 521↓	1.6	1.647
^{176}Yb	8−	*nn*514↓ 624↑	1.1	1.038
^{174}Yb	6+	*nn*514↓ 512↑	1.6	1.530
	5−	*pp*514↑ 411↓	1.8	1.900
^{172}Yb	3+	*nn*521↓ 512↑	1.3	1.174
	3+	*pp*411↓ 404↓	1.8	1.664
^{168}Er	4− 3−	*nn*633↑ 521↓	1.1	1.095 1.543
^{166}Er	6−	*nn*523↓ 633↑	1.7	1.785
^{164}Dy	6−	*nn*523↓ 633↑	1.7	1.680

(contd.)

Continuation of Table 4.13.

Nucleus	K^π	Configuration	Energy (MeV)	
			Calculated	Experimental
^{162}Dy	5−	$nn642\uparrow$ $523\downarrow$	1.7	1.485
	4+	$nn521\uparrow$ $523\downarrow$	1.9	1.536
^{160}Dy	4+	$nn521\uparrow$ $523\downarrow$	1.8	1.694
^{158}Dy	4+	$nn521\uparrow$ $523\downarrow$	2.0	1.895
^{156}Gd	4+	$pp413\downarrow$ $411\uparrow$	1.6	1.511
	3−	$nn642\uparrow$ $523\downarrow$	2.0	1.935

TABLE 4.14. *Four-quasiparticle states of* (2*n*, 2*p*) *type in even–even deformed nuclei*

Nucleus	K^π	Configuration	Energy (MeV)	
			Experi-mental	Calculated
^{254}Cf	13−	$p633\uparrow$ $p521\uparrow$ $n613\uparrow$ $n615\downarrow$		~2.5
^{248}Cf	13+	$p633\uparrow$ $p521\uparrow$ $n734\uparrow$ $n624\downarrow$		~2
^{178}Hf	16+	$p514\uparrow$ $p404\downarrow$ $n514\downarrow$ $n624\uparrow$	≈2.5	2.4
^{176}Hf	14−	$p514\uparrow$ $p404\downarrow$ $n512\uparrow$ $n514\downarrow$		2.9
^{176}Yb	16+	$p514\uparrow$ $p404\downarrow$ $n514\downarrow$ $n624\uparrow$		~3
^{174}Yb	11−	$p514\uparrow$ $p411\downarrow$ $n514\downarrow$ $n512\uparrow$		~3.5

The two possible states $K = K_1^n + K_1^p$ and $K' = |K_1^n - K_1^p|$ are degenerate in the model of the independent quasiparticles. The state with parallel spins ($\Sigma = 1$) has, according to the Gallagher–Moszkowski rules,[211] a lower energy than the state with antiparallel spins ($\Sigma = 0$). The calculations by Pyatov[202] have supported the Gallagher–Moszkowski rules in all cases, except for the $K^\pi = 0^-$ states. It was, however, impossible to explain the spin-splitting energies with only one set of parameters in the interaction (4.60).

The experimental energies, and the predicted configurations for some two-quasiparticle states of odd–odd nuclei, are collected in Table 4.15. It is evident that the lowest states really have a two-quasiparticle structure.

The spectra of odd–odd nuclei are quite complex. Their two-quasiparticle structure is disturbed by the Coriolis interaction, and the spin–spin interaction is also important. Nevertheless, the independent quasiparticle model is reliable and can be used as a basis for inclusion of the additional interactions.

4. This paragraph will discuss the quasiparticle states of spherical nuclei. Let us start with the one-quasiparticle states in odd-A nuclei. The wave function of such a state, with

TABLE 4.15. *Two-quasiparticle states of odd–odd deformed nuclei*

Nucleus	K^π	Energy (MeV) (exp.)	$F \pm i$; $Nn_z\Lambda\Sigma$ Neutron	Proton
^{254}Es	7^+	0	F 613↑	F 633↑
	0^+			
	2^-		F−1 620↑	F 633↑
	1^-			
^{250}Bk	2^-	0	F 620↑	F 521↑
	1^-			
	4^+	35.5	F 620↑	F+1 633↑
	3^+			
	7^+	85.2	F+1 613↑	F+1 633↑
	0^+			
	5^-	98.5	F+1 613↑	F 521↑
	2^-			
^{244}Am	6^-	0	F 624↓	F 523↓
	1^-	69		
^{242}Am	0^-	0	F 622↑	F 523↓
	5^-	48.6		
^{238}Np	2^+	0	F 631↓	F 642↑
	3^+	87		
	3^-	135	F 631↓	F+1 523↓
	2^-			
	6^-	294	F+2 624↓	F+1 523↓
	1^-			
	5^-	337	F+1 622↑	F+1 523↓
	0^-			
^{234}Pa	4^+	0	F 743↑	F 530↑
	3^+			
	0^-	70	F+1 631↓	F 530↑
	1^-	162		
^{188}Re	1^-	0	F−1 512↓	F 402↑
	4^-	272		
	3^-	170	F−2 510↑	F 402↑
	2^-			
	6^-	172	F 503↑	F 402↑
	1^-	290		
^{186}Re	1^-	0	F 512↑	F 402↑
	4^-			
	3^-	203	F−1 510↑	F 402↑
	2^-			
^{184}Re	3^-	0	F 510↑	F 402↑
	2^-			
	8^+	188	F+2 503↑	F−1 514↑
	1^+			

Nucleus	K^π	Energy (MeV)	$F \pm i$; $Nn_z\Lambda\Sigma$ Neutron	Proton
^{176}Ta	1^-	0	F 512↑	F−1 404↓
	6^-			
	0^+	100	F−2 633↑	F−1 404↓
	7^+			
^{176}Lu	7^-	0	F 514↓	F 404↓
	0^-	127		
	3^-	662	F+2 520↑	F 404↓
	4^-	791		
^{172}Lu	4^-	0	F 512↓	F 404↓
	3^-			
	1^-			
	6^-	41.4	F+1 512↑	F 404↓
	0^+			
	7^+	65.9	F−1 633↑	F 404↓
^{172}Tm	2^-	0	F 512↑	F 411↓
	3^-			
	1^-	410	F−1 521↑	F 411↓
	0^-	475		
^{170}Tm	1^-	0	F 521↓	F 411↓
	0^-	114		
	3^+	183	F−1 633↓	F 411↓
	4^+			
	2^-	204	F+1 512↑	F 411↓
	3^-	447		
^{166}Ho	7^-	5	F 633↑	F 523↑
	0^-	0		
	3^+	191	F+1 521↓	F 523↑
	4^+	372		
	6^+	294	F+2 512↑	F 523↑
	1^+			
	1^+	425	F−1 523↓	F 523↑
	6^+			
^{158}Tb	3^-	0	F 521↑	F 411↑
	0^-			
	7^-	340	F−2 505↑	F 411↑
	4^-			
^{156}Eu	0^+	0	F+1 642↑	F 413↓
	5^+			
	1^-		F 521↑	F 413↓
	4^-			
	1^+		F+2 523↑	F+2 523↑
	6^+			

a quasiparticle in subshell j_2, is given by the formula

$$\Psi_0(j_2)\Psi_0, \tag{4.65}$$

and its energy is equal to

$$\mathscr{E}_0(j_2)-\mathscr{E}_0^{\min}(j). \tag{4.66}$$

The corresponding quantities are defined in (4.22″), (4.33″), and (4.39′). The $\mathscr{E}_0^{\min}(j)$ denotes the ground-state energy of the spherical odd-A nucleus. The quantum numbers of the one-quasiparticle state are identical with the quantum numbers of the corresponding subshell occupied by the quasiparticle.

Tables 4.16 and 4.17 show a part of the experimental information available on the one-quasiparticle states in spherical odd-A nuclei with $89 \leqslant A \leqslant 149$; the experimental data were taken from the refs. 172, 212, and 213. The most valuable information is extracted from the single-nucleon transfer reactions (see, for example, ref. 214). By analyzing these experiments we can obtain not only the energy, angular momentum I, and parity π of the excited states, but also the contribution of any particular one-quasiparticle component to the complex nuclear state.

Tables 4.16 and 4.17 show, for each odd number $N(Z)$ of neutrons (protons), the I^π values of the ground state predicted by the independent particle ($G = 0$) and independent quasiparticle ($G = G_0$) models. The corresponding single-particle level schemes, i.e., the subshell positions in the independent particle model, are shown in Figs. 1.5 and 1.6 for $\varepsilon = 0$. The subshell positions agree in a qualitative way with the observed sequence of the one-quasiparticle states. There are, however, many exceptions to this correlation.

The pairing correlations affect the one-quasiparticle states substantially. The particular case $N = 69$, analyzed in Table 4.6, and a comparison of the two predictions ($G = 0$ and $G = G^0$) of the ground state I^π values with the experimental data in Table 4.16 and 4.17 clearly shows that the pairing correlations improve the agreement between theory and experiment. As we have already mentioned, the ground state $I^\pi lj$ characteristics in odd-A nuclei are changed by the pairing correlations if a subshell with large degeneracy $2j+1$ is almost empty (full) and if another subshell, with small degeneracy $2j'+1$, is immediately below (above) it. Many experimental data can be thus understood without inclusion of any additional interaction.[215]

The energies of one-quasiparticle states calculated in the independent quasiparticle model generally agree with the experimental data better than the results of calculations based on the independent particle model.[214] There is, however, much unexplained data. In some cases it is even necessary to shift the single-particle subshells in order to explain the sequence of the states in different odd-N isotopes of the same element.

It is difficult to determine experimentally the admixtures to the one-quasiparticle component of low-lying states in odd-A spherical nuclei. The existing experimental information suggests that the low-lying states have a predominantly one-quasiparticle structure; the admixtures are, however, more important in odd-A spherical nuclei than in deformed nuclei discussed previously. In several cases the nuclei with magic numbers of nucleons do not even have the "completely closed" shells.[214]

Despite the overall success of the independent quasiparticle model, many observed states cannot be fully understood within the framework of such a model. This failure is related

TABLE 4.16. *One-quasiparticle states of odd-N nuclei with* $101 \leqslant A \leqslant 149$ *(energies in* keV)

Element	A	$I^\pi lj$ of single-particle states								I^π of ground states		
		$\frac{5}{2}+$ $d_{5/2}$	$\frac{7}{2}+$ $g_{7/2}$	$\frac{1}{2}+$ $s_{1/2}$	$\frac{11}{2}-$ $h_{11/2}$	$\frac{3}{2}+$ $d_{3/2}$	$\frac{7}{2}-$ $f_{7/2}$	$\frac{3}{2}-$ $p_{3/2}$	$\frac{5}{2}-$ $f_{5/2}$	$G_N = 0$	$G_N = G_N^0$	N
Mo	101			0								
Pd	105	0	306	344[b]	486	280				$\frac{7}{2}+$	$\frac{7}{2}+$	59
Cd	107	0	205[a]		933[a]	366[a]						
Pd	107	0	307	115	210	390[a]						
Cd	109	0	205	60	500[a]	289[a]				$\frac{7}{2}+$	$\frac{7}{2}+$	61
Pd	109	0		110	188	300						
Cd	111	247	419	0	396	340				$\frac{7}{2}+$	$\frac{7}{2}+$	63
Sn	113	405[c]	79	0	740[c]	490[c]						
Pd	111			0								
Cd	113	310[a]	500[a]	0	210	300						
Sn	115	990[c]	619	0	730[c]	495[c]				$\frac{1}{2}+$	$\frac{1}{2}+$	65
Te	117	274[d]		0								
Cd	115	350[c]		0	180	200[a]						
Sn	117	1020[c]	730	0	317	158				$\frac{11}{2}-$	$\frac{1}{2}+$	67
Te	119			0	300							
Cd	117	430[a]		0	133							
Sn	119	1090[e]		0	89	24				$\frac{11}{2}-$	$\frac{1}{2}+$	69
Te	121			0	294	212						
Sn	121	1120[f]	923[f]	60[f]		0						
Te	123			0	248	159				$\frac{11}{2}-$	$\frac{11}{2}-$	71
Sn	123			150[g]	0	25[g]						
Te	125	462	636[h]	0	150	35				$\frac{11}{2}-$	$\frac{11}{2}-$	73
Sn	125	1261[h]	1364[g]	217[g]	0	28[g]						
Te	127	473[h]	686[h]	61[h]	88[h]	0				$\frac{11}{2}-$	$\frac{3}{2}+$	75
Xe	129			0	236	40						
Ba	131	317[i]		0		108[i]						

Te	129	544[h]	813[h]	170	106	0				$\frac{11}{2}-$	$\frac{3}{2}+$	77
Xe	131	364	637[j]	80	164	0						
Ba	133			0	288	12						
Te	131	642[h]	944[h]	297	182	0						
Xe	133	530[k]		263[j]		0						
Ba	135	481[l]		218	268	0				$\frac{3}{2}+$	$\frac{3}{2}+$	79
Xe	135	1260[n]	1131[n]	200[n]	527	0						
Ba	137	1290[m]	2230[m]	281	662	0						
Ce	139	1330[m]	2550[m]	250[m]	740	0				$\frac{3}{2}+$	$\frac{3}{2}+$	81
Ba	139						0	620[o]	1420[o]			
Ce	141						0	670[p]	1530[p]			
Nd	143						0	737[q]				
Sm	145						0	894[r]	1659[r]	$\frac{7}{2}-$	$\frac{7}{2}-$	83
Nd	145						0	67				
Sm	147						0	189	122	$\frac{7}{2}-$	$\frac{7}{2}-$	85
Gd	149						0	352[s]	164[s]			
Nd	147								0	$\frac{7}{2}-$	$\frac{7}{2}-$	87
Sm	149						0		22			

[a] J. Rivier and R. Mout, *Nucl. Phys.* A, **177,** 379 (1971).
[b] M. Behar and Z. W. Grabowski, *Nucl. Phys* A, **196,** 412 (1973).
[c] D. G. Fleming *et al.*, *Nucl. Phys.* A, **163,** 401 (1971).
[d] H. F. Brinckmann *et al.*, *Nucl. Phys.* A, **193,** 257 (1973).
[e] P. H. Stelson *et al.*, *Nucl. Phys.* A, **190,** 197 (1972).
[f] R. F. Casten *et al.*, *Nucl. Phys.* A, **180,** 49 (1972).
[g] P. L. Carson and L. C. McIntyre, *Nucl. Phys.* A, **198,** 289 (1972).
[h] J. Blachot *et al.*, *Phys. Rev.* C, **4,** 214 (1971).
[i] B. Harmatz and T. H. Handley, *Nucl. Phys.* A, **191,** 497 (1972).
[j] A. Backlin *et al.*, *Nucl. Phys.* A, **181,** 76 (1972).
[k] R. N. Saxena and H. D. Sharma, *Nucl. Phys* A, **171,** 593 (1971).
[l] R. B. Begzhanov *et al.*, Izv. AN USSR, phys. ser. **35,** 2099 (1971).
[m] A. Chaumeaux *et al.*, *Nucl. Phys.* A, **164,** 176 (1970).
[n] E. Achterberg *et al.*, *Phys. Rev.* C, **5,** 1759 (1972).
[o] H. Clement *et al.*, *Phys. Rev. Lett.* **27,** 526 (1971).
[p] S. Gales *et al.*, *Nucl. Phys.* A, **202,** 534 (1973).
[q] P. Neuman *et al.*, *Phys. Rev.* C, **7,** 290 (1973).
[r] J. Adam *et al.*, preprint JINR, Dubna, R-2581, 1966 (unpublished).
[s] K. Vilskij *et al.*, preprint JINR, Dubna, R6-3128, 1967 (unpublished).

TABLE 4.17. *One-quasiparticle states of odd-Z nuclei with* $89 \leqslant A \leqslant 149$ (*energies in* keV)

Element	A	$I^{\pi}lj$ of single-particle states								I^{π} of ground states		Z
		$\frac{3}{2}-$ $p_{3/2}$	$\frac{1}{2}-$ $p_{1/2}$	$\frac{9}{2}+$ $g_{9/2}$	$\frac{7}{2}+$ $g_{7/2}$	$\frac{5}{2}+$ $d_{5/2}$	$\frac{11}{2}-$ $h_{11/2}$	$\frac{3}{2}+$ $d_{3/2}$	$\frac{1}{2}+$ $s_{1/2}$	$G_z = 0$	$G_z = G_z^0$	
Y	89		0	910						$\frac{1}{2}-$	$\frac{1}{2}-$	39
	91		0	551								
Nb	91			0						$\frac{9}{2}+$	$\frac{9}{2}+$	41
	93	808[a]	30	0								
	95	770[b]	235[b]	0								
Tc	93		390[c]	0						$\frac{9}{2}+$	$\frac{9}{2}+$	43
	99	500[d]	143	0								
Rh	101	3053[e]	0	157[e]						$\frac{9}{2}+$	$\frac{9}{2}+$	45
	103	298	0	93[e]								
Ag	105		0	53[f]						$\frac{9}{2}+$	$\frac{9}{2}+$	47
	107	325	0	126								
	109	312	0	133[f]								
	111	290[f]	0	130								
In	111			0						$\frac{9}{2}+$	$\frac{9}{2}+$	49
	113		393[g]	0								
	115	600[g]	336	0								
	117	587	315	0								
Sb	119				270	0	1.37	700[h]	644[h]	$\frac{7}{2}+$	$\frac{7}{2}+$	51
	121				37	0		507	573			
	123				0	160		543[i]	713[i]			
	125				0	332						

J	123					0		178	149			
	125					0						
	127				57	0		203	375	$\frac{7}{2}+$	$\frac{7}{2}+$	53
	129				0	27						
	131				0	150						
Cs	131				79	0		220[k]	120[k]			
	133				0	81	605[j]	384	437	$\frac{7}{2}+$	$\frac{7}{2}+$	55
	135				0	250[k]		408[k]				
	137				0	450[k]		820[k]				
La	137				0	10	1004[l]		708[l]	$\frac{7}{2}+$	$\frac{7}{2}+$	57
	139				0	166	1430[l]	1559[l]	1206[l]			
Pr	141				145	0				$\frac{5}{2}+$	$\frac{5}{2}+$	59
	143				0	57						
Pm	147				0	91				$\frac{5}{2}+$	$\frac{5}{2}+$	61
	149				0	114	240					

(a) MATTHEWS *et al.*, *Phys. Rev.* C, **4,** 1876 (1971).
(b) R. J. RILEY *et al.*, *Phys. Rev.* C, **4,** 1864 (1971).
(c) M. GRECESKU *et al.*, *Annual Report 1970*, *RIfPh Stockholm*, p. 12.
(d) P. D. BOND *et al.*, *Nucl. Phys.* A, **179,** 389 (1972).
(e) B. NYMAN *et al.*, *Physica scripta* **5,** 13 (1972).
(f) G. BERZINS *et al.*, *Nucl. Phys.* A, **126,** 273 (1969).
(g) C. V. WEIFFENBACH and R. TICKLE, *Phys. Rev.* C, **3,** 1668 (1971).
(h) R. L. KERNELL *et al.*, *Nucl. Phys.* A, **176,** 449 (1971).
(i) E. BARNARD *et al.*, *Nucl. Phys* A, **172,** 215 (1971).
(j) S. KIKUCHI, *Nucl. Phys.* A, **171,** 480 (1971).
(k) E. ACHTERBERG *et al.*, *Phys. Rev.* C, **5,** 1759 (1972).
(l) F. T. BAKER and R. TICKLE, *Phys. Rev.* C, **5,** 182 (1972).

to the incomplete selfconsistency of the average field; the average field potential does not include contributions from the mutual interaction of nucleons in unfilled shells.

The accuracy of interpreting the low-lying excited states of spherical nuclei with one closed shell as the one-quasiparticle states was tested in ref. 216. The tests used the inverse-gap method. The study has proven that the pairing correlations of the superfluid type are responsible for many observed features of low-lying states.

Experimental information about the three-quasiparticle states of spherical nuclei is very limited. The existing data were analyzed in ref. 217.

Let us now turn our attention to the two-quasiparticle states of spherical nuclei. The wave function of a state with quasiparticles in subshells j_1, j_2 is determined by the formula

$$\Psi_0(JM \mid j_1, j_2)\Psi_0, \tag{4.67}$$

the excitation energy is equal to

$$\mathscr{E}_0(j_1, j_2) - \mathscr{E}_0. \tag{4.68}$$

All corresponding quantities were defined in (4.21′), (4.22″), (4.31″), and (4.48′). Both quasiparticles are of the same kind—neutrons or protons.

Several two-quasiparticle states are shown in Table 4.18; the experimental data are taken from ref. 172 and the calculations were performed in ref. 215. The energies of several $(F-1, F)$ two-quasiparticle states are smaller than the energies of $(F+1, F)$ states; such a situation was never encountered in deformed nuclei. Table 4.18 shows as an example the 5^- state in ^{118}Sn. The two-quasiparticle states were analyzed, for example, in ref. 217; the calculated energies agree satisfactorily in many cases with the experimental data.

TABLE 4.18. *Two-quasiparticle states of even–even spherical nuclei*

Nucleus	I^π	Configuration				Energy (MeV) Experimental	Energy (MeV) Calculated
^{138}Ce	7^-	$nn\ F-1$	$h_{11/2}$,	F	$d_{3/2}$	2.12	2.2
^{136}Ba	7^-	$nn\ F-1$	$h_{11/2}$,	F	$d_{3/2}$	2.04	2.2
^{120}Sn	7^-	$nn\ F$	$h_{11/2}$,	$F+1$	$d_{3/2}$	2.50	2.8
^{118}Sn	7^-	$nn\ F$	$h_{11/2}$,	$F+1$	$d_{3/2}$	2.57	3.1
	5^-	$nn\ F-1$	$s_{1/2}$,	F	$h_{11/2}$	2.32	2.6
^{116}Sn	5^-	$nn\ F$	$s_{1/2}$,	$F+1$	$h_{11/2}$	2.35	2.7
	6^-					2.77	
^{94}Mo	6^+	$nn\ F$	$d_{5/2}$,	$F+1$	$g_{7/2}$	2.42	2.8
^{90}Zr	5^-	$pp\ F$	$p_{1/2}$,	$F+1$	$g_{9/2}$	2.32	2.8
	4^-					2.74	

The ground state and many excited states of the odd–odd spherical nuclei are the twoquasiparticle states. Such an interpretation explains the existence of many isomeric states.[218]

The quasiparticle interaction which causes the spin splittings of nuclear states is more important in spherical nuclei than it was in deformed nuclei. The situation is particularly complicated in case of many-quasiparticle states because it is then necessary to diagonalize large matrices. The group-theoretical methods are often used in classification of such states (see refs. 19, 35, 36, and 219).

CHAPTER 5

MATHEMATICAL DESCRIPTION OF THE PAIRING CORRELATIONS†

§ 1. Compensation of the Dangerous Graphs

1. The so-called "method of compensation of dangerous graphs" can be used in the derivation of the basic equations in the theory of superconductivity. When the canonical transformation (3.43) is applied and the Hamiltonian expressed in terms of new operators $\alpha_{g\sigma}$, it contains the terms corresponding to the creation and annihilation from vacuum of two quasiparticles with a zero total momentum. N. N. Bogolyubov has formulated the method of compensation of dangerous graphs which excludes such terms from the Hamiltonian and from the corresponding expressions in higher orders of perturbation theory. The stated method is equivalent to the Hartree–Fock–Bogolyubov variational principle. The method is often used in the description of superfluid pairing correlations in atomic nuclei. We shall, therefore, describe its main features in this section.

The principle of compensation of dangerous graphs is expressed by the equation

$$\langle| \alpha_{g_1}\alpha_{g_2}, H |\rangle_0 = 0; \tag{5.1}$$

the Hamiltonian H was defined in (3.1) and the operators α_g in (3.5).

The state $| \rangle_0$ is the vacuum of the new amplitudes α_g, i.e.,

$$\alpha_g |\rangle_0 = 0, \quad {}_0\langle| \alpha_g^+ = 0.$$

The whole method was formulated in ref. 120.

Equation (5.1) can be expressed in another form, namely as

$$\left.\begin{aligned} \mathfrak{A}_0(f_1, f_2) &\equiv \langle| [a_{f_1}a_{f_2}, H] |\rangle_0 = 0, \\ \mathfrak{B}_0(f_1, f_2) &\equiv \langle| [a_{f_1}^+ a_{f_2}, H] |\rangle_0 = 0. \end{aligned}\right\} \tag{5.2}$$

† This chapter can be omitted when the book is read for the first time.

Let us show that (5.2) follows from (5.1). Really,

$$\begin{aligned}\mathfrak{A}_0(f_1, f_2) &\equiv \sum_{g_1, g_2} \langle|[(u(f_1g_1)\alpha_{g_1}+v(f_1g_1)\overset{+}{\alpha}_{g_1})\,(u(f_2g_2)\alpha_{g_2}+v(f_2g_2)\overset{+}{\alpha}_{g_2}),\, H]|\rangle_0 \\ &= \sum_{g_1, g_2} \{u(f_1g_1)\,u(f_2g_2)\langle|\alpha_{g_1}\alpha_{g_2}H-H\alpha_{g_1}\alpha_{g_2}|\rangle_0 \\ &\quad +u(f_1g_1)\,v(f_2g_2)\langle|\alpha_{g_1}\overset{+}{\alpha}_{g_2}H-H\alpha_{g_1}\overset{+}{\alpha}_{g_2}|\rangle_0 \\ &\quad +v(f_1g_1)\,u(f_2g_2)\langle|\overset{+}{\alpha}_{g_1}\alpha_{g_2}H-H\overset{+}{\alpha}_{g_1}\alpha_{g_2}|\rangle_0 \\ &\quad +v(f_1g_1)\,v(f_2g_2)\langle|\overset{+}{\alpha}_{g_1}\overset{+}{\alpha}_{g_2}H-H\overset{+}{\alpha}_{g_1}\overset{+}{\alpha}_{g_2}|\rangle_0\}.\end{aligned}$$

But from (3.8) it follows that

$$\langle|H\alpha_{g_1}\alpha_{g_2}|\rangle_0 = \langle|H\overset{+}{\alpha}_{g_1}\alpha_{g_2}|\rangle_0 = \langle|\overset{+}{\alpha}_{g_1}\alpha_{g_2}H|\rangle_0 = \langle|\overset{+}{\alpha}_{g_1}\overset{+}{\alpha}_{g_2}H|\rangle_0 = 0$$

and, besides,

$$\langle|\alpha_{g_1}\overset{+}{\alpha}_{g_2}H-H\alpha_{g_1}\overset{+}{\alpha}_{g_2}|\rangle_0 = -\langle|\overset{+}{\alpha}_{g_2}\alpha_{g_1}H-H\overset{+}{\alpha}_{g_2}\alpha_{g_1}|\rangle_0 = 0.$$

Hence,

$$\mathfrak{A}_0(f_1, f_2) = \sum_{g_1, g_2} \{u(f_1g_1)\,u(f_2g_2)\langle|\alpha_{g_1}\alpha_{g_2}H|\rangle_0 - v(f_1g_1)\,v(f_2g_2)\langle|H\overset{+}{\alpha}_{g_1}\overset{+}{\alpha}_{g_2}|\rangle_0 = 0.$$

The second equation (5.2) can be proven similarly.

Let us generalize (5.2) and assume that the implicit averaging is over an arbitrary state $|\ \rangle$. If the functions $\Phi(f_1, f_2)$ and $\varrho(f_1, f_2)$ are time independent, the resulting equations

$$\left.\begin{aligned}\mathfrak{A}(f_1, f_2) &\equiv \langle|[a_{f_1}a_{f_2}, H]|\rangle = 0, \\ \mathfrak{B}(f_1, f_2) &\equiv \langle|[a^+_{f_1}a_{f_2}, H]|\rangle = 0,\end{aligned}\right\} \tag{5.3}$$

are exact. It is easy to see that

$$\begin{aligned}\mathfrak{A}(f_1, f_2) &= \sum_f \{T'(f_1, f)\langle|a_fa_{f_2}|\rangle+T'(f_2, f)\langle|a_{f_1}a_f|\rangle\} \\ &\quad -\tfrac{1}{2}\sum_{f_1', f_2', f} \{G(f_1, f; f_2', f_1')\langle|a^+_f a_{f_2}a_{f_2'}a_{f_1'}|\rangle+G(f, f_2; f_2', f_1')\langle|a^+_f a_{f_1}a_{f_2'}a_{f_1'}|\rangle \\ &\quad -\tfrac{1}{2}\sum_{f_1', f_2'} G(f_1, f_2; f_2', f_1')\langle|a_{f_1'}a_{f_2'}|\rangle,\end{aligned} \tag{5.4}$$

$$\begin{aligned}\mathfrak{B}(f_1,f_2) &= \sum_f \{T'(f_2, f)\langle|a^+_{f_1}a_f|\rangle-T(f, f_1)\langle|a^+_f a_{f_2}|\rangle\} \\ &\quad +\tfrac{1}{2}\sum_{f_1', f_2', f} \{G(f_1, f; f_2, f_1')\langle|a^+_{f_1'}a^+_{f_2'}a_fa_{f_2}|\rangle+G(f_2, f; f_2', f_1')\langle|a^+_{f_1}a^+_f a_{f_2'}a_{f_1'}|\rangle\}.\end{aligned} \tag{5.5}$$

In order to find the explicit formulae for the $\mathfrak{A}(f_1, f_2)$ and $\mathfrak{B}(f_1, f_2)$ we have to use some approximate method and express the average values $\langle|a^+_{f_1'}a^+_{f_2'}a_fa_{f_2}|\rangle$ and $\langle|a^+_f a_{f_2}a_{f_2'}a_{f_1'}|\rangle$ through the functions $\varrho(f_1, f_2)$ and $\Phi(f_1, f_2)$ introduced in § 2 of Chapter 3.

The Bogolyubov general canonical transformation (3.5) gives a prescription how to express the average values over the quasiparticle vacuum through the functions ϱ_0 and Φ_0.

The following equations are valid:

$$\langle|a_{f'}^{+}a^{+}a_{f}a_{f_2}|\rangle_0 = \varrho_0(f_1', f_2)\,\varrho_0(f_2', f) - \varrho_0(f_1', f)\,\varrho_0(f_2', f_2) + \Phi_0^*(f_2', f_1')\,\Phi_0(f, f_2), \quad (5.6)$$

$$\langle|a_f^{+}a_{f_2}a_{f_2'}a_{f_1'}|\rangle_0 = \varrho_0(f_1, f_2)\,\Phi_0(f_2', f_1) - \varrho_0(f, f_2')\,\Phi_0(f_2, f_1') + \varrho_0(f_1, f_1')\,\Phi_0(f_2, f_2'). \quad (5.6')$$

Let us assume that (5.6) and (5.6′) are valid for averaging over an arbitrary state $|\,\rangle$. In such a case we obtain for $\mathfrak{A}(f_1, f_2)$ and $\mathfrak{B}(f_1, f_2)$ the explicit expressions, identical with (3.15′) and (3.16′).

The equations

$$\mathfrak{A}_0(f_1, f_2) = 0 \quad \text{and} \quad \mathfrak{B}_0(f_1, f_2) = 0$$

are identical with (3.12) derived in § 2 of Chapter 3 by using the variational principle.

Equations (5.3) are more general than (3.12). We shall show in Chapter 8 that (5.3) describe also collective states interpreted as the nuclear surface vibrations.

2. Let us apply the method of compensation of dangerous graphs to the case of a special Bogolyubov transformation.

We shall assume that the Hamiltonian of our system is equal to

$$H = \sum_f \{T(f) - \lambda\} a_f^{+} a_f - \tfrac{1}{4} \sum_{f_1, f_2, f_2', f_1'} G(f_1, f_2; f_2', f_1') a_{f_1}^{+} a_{f_2}^{+} a_{f_2'} a_{f_1'}, \quad (5.7)$$

because the kinetic energy $T(f, f')$ can be also diagonalized.

The quantum numbers f are divided into q and σ, with $\sigma = \pm 1$. Both notations f and $q\sigma$ will be used.

We shall apply the special canonical transformation of the Fermi amplitudes:

$$a_{q\sigma} = u_q \alpha_{q, -\sigma} + \sigma v_q \alpha_{q\sigma}^{+} \quad (5.8)$$

with

$$u_q^2 + v_q^2 = 1.$$

The Hamiltonian (5.7) is expressed in the α-representation as

$$\begin{aligned} H = \langle|H|\rangle_0 &+ \sum_{f, f'} \{h_{11}(f, f')\alpha_f^{+} \alpha_{f'} + h_{20}(f, f')\alpha_f^{+}\alpha_{f'}^{+} \\ &+ h_{02}(f, f')\alpha_f \alpha_{f'}\} + \sum_{f_1, f_2, f_2', f_1'} \{h_{22}(f_1, f_2; f_2', f_1')\alpha_{f_1}^{+}\alpha_{f_2}^{+}\alpha_{f_2'}\alpha_{f_1'} \\ &+ h_{31}(f_1, f_2; f_2', f_1')\alpha_{f_1}^{+}\alpha_{f_2}^{+}\alpha_{f_2'}^{+}\alpha_{f_1'} + h_{13}(f_1, f_2; f_2', f_1')\alpha_{f_1}^{+}\alpha_{f_2}\alpha_{f_2'}\alpha_{f_1'} \\ &+ h_{40}(f_1, f_2; f_2', f_1')\alpha_{f_1}^{+}\alpha_{f_2}^{+}\alpha_{f_2'}^{+}\alpha_{f_1'}^{+} + h_{04}(f_1, f_2; f_2', f_1')\alpha_{f_1}\alpha_{f_2}\alpha_{f_2'}\alpha_{f_1'}\}. \end{aligned} \quad (5.7')$$

The Hamiltonian is a hermitian operator, hence

$$h_{20} = h_{02}, \quad h_{31} = h_{13}, \quad h_{40} = h_{04} \quad (5.9)$$

and

$$\begin{aligned} \langle|H|\rangle_0 &= \sum_f \Big\{T(f) - \lambda - \tfrac{1}{2} \sum_{f'} G(f, f'; f', f) v_{q'}^2 \Big\} v_q^2 \\ &- \sum_{q, q'} G(q+, q-; q'-, q'+) u_q v_q u_{q'} v_{q'}. \end{aligned} \quad (5.10)$$

Thus we can see that the Bogolyubov canonical transformation (5.8) separates the nucleon interaction in time-reversal conjugated states from the general Hamiltonian.

The function $h_{20}(f, f')$ is equal to

$$h_{20}(f, f') = \sigma' u_q v_{q'} \Big\{ (T(q)-\lambda)\delta_{qq'}\delta_{\sigma, -\sigma'} - \sum_{f_2} G(q, -\sigma, f_2; f_2, f') v_{q_2}^2 \Big\} + \tfrac{1}{2}(u_q u_{q'} + \sigma\sigma' v_q v_{q'}) C_{f, f'}, \tag{5.11}$$

where

$$C_{f, f'} = \tfrac{1}{2} \sum_{q_2, \sigma_2} G(f, f'; q_2, -\sigma_2, q_2\sigma_2)\sigma_2 u_{q_2} v_{q_2} = \sum_{q_2} G(f, f'; q_2-, q_2+) u_{q_2} v_{q_2}. \tag{5.12}$$

The principle of compensation of dangerous diagrams demands that

$$h_{20}(q+, q-) = 0. \tag{5.13}$$

From the above expressions it follows that

$$\xi(q) u_q v_q - \tfrac{1}{2}(u_q^2 - v_q^2) C_q = 0, \tag{5.14}$$

where

$$\left.\begin{aligned} C_q &= \sum_{q_2} G(q+, q-; q_2-, q_2+) u_{q_2} v_{q_2}, \\ \xi(q) &= T(q) - \lambda - \sum_{f_2} G(q+, f_2; f_2, q+) v_{q_2}^2. \end{aligned}\right\} \tag{5.12'}$$

Assuming

$$u_q^2 = \frac{1}{2}\left(1 + \frac{\xi(q)}{\varepsilon(q)}\right), \quad v_q^2 = \frac{1}{2}\left(1 - \frac{\xi(q)}{\varepsilon(q)}\right),$$

we can find from (5.14) that

$$u_q v_q = \frac{C_q}{2\varepsilon(q)}$$

and, substituting in (5.12'),

$$C_q = \frac{1}{2} \sum_{q_2} G(q+, q-; q_2-, q_2+) \frac{C_{q_2}}{\varepsilon(q_2)}. \tag{5.15}$$

Equations (5.15) are identical with (3.47) obtained from the variational principle.

The function $h_{11}(f, f')$ is equal to

$$h_{11}(f, f') = \Big\{ (T(f)-\lambda)\delta_{ff'} - \sum_{f_2} G(f, f_2; f_2, f') v_{q_2}^2 \Big\} \times \\ \times (u_q u_{q'} - \sigma\sigma' v_q v_{q'}) + C_{f, q', -\sigma'}(u_q v_{q'} + v_q u_{q'}).$$

Particularly,

$$h_{11}(q+, q+) = \xi(q)(u_q^2 - v_q^2) + 2C_q u_q v_q. \tag{5.16}$$

By using the expressions derived above, we write

$$h_{11}(q+, q+) = \frac{\xi^2(q) + C_q^2}{\varepsilon(q)} = \varepsilon(q). \tag{5.17}$$

Finally, we shall state here the expressions for h_{22}, h_{31}, and h_{40} in (5.7′), namely:

$$h_{40}(f_1, f_2; f_2', f_1') = -\tfrac{1}{4}G(f_1, f_2; q_2'-\sigma_2', q_1'-\sigma_1')\sigma_2'\sigma_1' u_{q_1}u_{q_2}v_{q_2'}v_{q_1}, \tag{5.18}$$

$$h_{31}(f_1, f_2; f_2', f_1') = -\tfrac{1}{2}G(f_1, f_2; q_1'-\sigma_2', f_1')\times$$
$$\times(\sigma_2' u_{q_1}u_{q_2}v_{q_2'}u_q + \sigma_1\sigma_2\sigma_1' v_{q_1}v_{q_2}u_{q_2}v_{q_1}). \tag{5.19}$$

$$\begin{aligned} h_{22}(f_1, f_2; f_2', f_1') = -\tfrac{1}{4}\{&G(f_1, f_2; f_2', f_1')(u_{q_1}u_{q_2}u_{q_2'}u_{q_1'} + \sigma_1\sigma_2\sigma_2'\sigma_1' v_{q_1}v_{q_2}v_{q_2'}v_{q_1'}) \\ &+G(f_1, q_2'-\sigma_2'; f_1', q_2-\sigma_2)(\sigma_2\sigma_2' u_{q_1}v_{q_2}v_q u_q + \sigma_1\sigma_1' v_{q_1}u_{q_2}u_{q'}v_{q_1}) \\ &-G(f_2, q_2'-\sigma_2'; f_1', q_1-\sigma_1)(\sigma_1\sigma_2' v_{q_1}u_{q_2}v_{q_2'}u_{q_1} + \sigma_2\sigma_1' u_{q_1}v_{q_2}u_{q_2'}v_{q_1'})\} \end{aligned} \tag{5.20}$$

The expressions for h_{40} and h_{31}, which satisfy all the necessary symmetry relations, were given in ref. 220.

3. Let us discuss the pairing interaction of, e.g., neutrons. The Hamiltonian (4.2) is equal to

$$H_0(n) = \sum_{s\sigma}\{E_0(s)-\lambda_n\}a_{s\sigma}^+a_{s\sigma} - G_N\sum_{s,\,s'} a_{s+}^+a_{s-}^+a_{s'-}a_{s'+}\,.$$

Let us rewrite $H_0(n)$ in the α-representation and use the principle of compensation of dangerous diagrams to bring $H_0(n)$ to a more compact form.

After application of the canonical transformation (5.8) our Hamiltonian is transformed into

$$H_0(n) = H_0^0(n) + H_0^\beta(n) + H_0'(n) + H_0''(n), \tag{5.21}$$

where

$$\begin{aligned} H_0^0 &= \sum_s\{2(E_0(s)-\lambda_n) - G_N v_s^2\}v_s^2 - G_N\sum_{s,\,s'} u_s v_s u_{s'} v_{s'} \\ &+\sum_s\left\{(E_0(s)-\lambda_n - G_N v_s^2)(u_s^2 - v_s^2) + 2G_N u_s v_s \sum_{s'} u_{s'}v_{s'}\right\}\sum_\sigma \alpha_{s\sigma}^+\alpha_{s\sigma}^+ \\ &+\sum_s\left\{2(E_0(s)-\lambda_n - G_N v_s^2)u_s v_s - G_N\sum_{s'} u_{s'}v_{s'}(u_s^2 - v_s^2)\right\}(\alpha_{s-}^+\alpha_{s+}^+ + \alpha_{s+}\alpha_{s-}). \end{aligned} \tag{5.22}$$

The other terms are given below.

The term $G_N\sum_s v_s^4$ describes the contribution of the nucleon interaction $G_N\sum_s a_{s+}^+a_{s-}^+a_{s'-}a_{s'+}$ to the nuclear average field. To exclude this term we have used a renormalization

$$E(s) = E_0(s) - \frac{G_N}{2}\, v_s^2$$

in § 1 of Chapter 4. Such a renormalization was performed in the expression for the average energy of the quasiparticle vacuum. On the other hand, it is also possible to perform the variation over δv_s and δu_s and thus obtain the corresponding equations first. The mentioned renormalization will be in this case different, namely,

$$E(s) = E_0(s) - G_N v_s^2\,. \tag{5.23}$$

This form of renormalization must be used in order to obtain the basic equations of the method of compensation of dangerous graphs.

The expression in front of the term $(\alpha_{s-}^{+}\alpha_{s+}^{+}+\alpha_{s+}\alpha_{s-})$ must vanish. From this condition it follows that

$$2\{E(s)-\lambda_n\}u_s v_s-(u_s^2-v_s^2)G_N\sum_{s'}u_{s'}v_{s'}=0, \tag{5.24}$$

identically with (4.11′). Using (4.16), (4.17), and (5.24),

$$\{E(s)-\lambda_n\}(u_s^2-v_s^2)+2u_s v_s G_N\sum_{s'}u_{s'}v_{s'}=\frac{\{E(s)-\lambda_n\}^2}{\varepsilon(s)}+\frac{C^2}{\varepsilon(s)}=\varepsilon(s).$$

In order to simplify (5.21) we shall employ the following notation:

$$B(s,s')=\sum_{\sigma}\alpha_{s\sigma}^{+}\alpha_{s'\sigma}, \tag{5.25}$$

$$A(s,s')=\frac{1}{\sqrt{2}}\sum_{\sigma}\sigma\alpha_{s'\sigma}\alpha_{s,-\sigma}=A(s',s), \tag{5.26}$$

where

$$A(s,s)=\sqrt{2}\alpha_{s+}\alpha_{s-},\qquad A^{+}(s,s)=\sqrt{2}\alpha_{s-}^{+}\alpha_{s+}^{+}. \tag{5.26′}$$

It is somewhat more convenient to rewrite $H_0(n)$ in such a form that the operators are not normally ordered (all α^+ operators are to the left of α operators in the normal product). The expressions $H_0^0(n)$, $H_0^\beta(n)$, $H_0'(n)$, and $H_0''(n)$ have then the following explicit form:

$$H_0^0(n)=\sum_s 2\{E(s)-\lambda_n\}v_s^2-\frac{C_n^2}{G_N}+\sum_s\varepsilon(s)B(s,s), \tag{5.27}$$

$$H_0^\beta(n)=-\frac{G_N}{2}\sum_{s,s'}[u_s^2A^+(s,s)-v_s^2A(s,s)][u_{s'}^2A(s',s')-v_{s'}^2A^+(s',s')], \tag{5.28}$$

$$H_0'(n)=-G_N\Big[\sum_s u_s v_s B(ss)\Big]^2, \tag{5.29}$$

$$H_0''=-\frac{G_N}{\sqrt{2}}\sum_{s,s'}(u_s^2-v_s^2)u_{s'}v_{s'}[A^+(s,s)B(s',s')+B(s',s')A(s,s)]. \tag{5.30}$$

§ 2. The Selfconsistent Field Method and Alternative Ways of the Treatment of Pairing Correlations

1. Let us generalize the method of the preceding section and include explicitly the time-dependent processes. We shall use the Heisenberg representation where the amplitudes $a_f(t)$ depend explicitly on the time variable. The new time-dependent functions

$$\left.\begin{aligned}\varrho_t(f_1,f_2)&=\langle|a_{f_1}^{+}(t)\,a_{f_2}(t)|\rangle,\\ \Phi_t(f_1,f_2)&=\langle|a_{f_1}(t)\,a_{f_2}(t)|\rangle,\end{aligned}\right\} \tag{5.31}$$

are introduced; the averaging involves the time-independent ground state. As usual

$$a_f(t)=e^{iHt}a_f e^{-iHt},$$

The following exact equations are equivalent to the equation of motion:

$$i\frac{\partial \varrho_t(f_1,f_2)}{\partial t} = \langle|[a^+_{f_1}(t)\, a_{f_2}(t),\, H]|\rangle, \tag{5.32}$$

$$i\frac{\partial \Phi_t(f_1,f_2)}{\partial t} = \langle|[a_{f_1}(t)\, a_{f_2}(t),\, H]|\rangle. \tag{5.33}$$

Let us rewrite the above written equations for systems described by the Hamiltonian (3.1). After simple manipulations we find that

$$\begin{aligned} i\frac{\partial \varrho_t(f_1,f_2)}{\partial t} &= \sum_f \{T'(f_2,f)\,\varrho_t(f_1,f) - T'(f,f_1)\,\varrho_t(f,f_2)\} \\ &+\tfrac{1}{2}\sum_{f,f_2',f_1'} \{G(f_1',f_2';f,f_1)\,\langle|a^+_{f_1'}(t)\,a^+_{f_2'}(t)\,a_f(t)\,a_{f_2}(t)|\rangle \\ &-G(f_2,f;f_2',f_1')\,\langle|a^+_{f_1}(t)\,a^+_{f}(t)\,a_{f_2'}(t)\,a_{f_1'}(t)|\rangle\}, \end{aligned} \tag{5.32'}$$

$$\begin{aligned} i\frac{\partial \Phi_t(f_1,f_2)}{\partial t} &= \sum_f \{T'(f_1,f)\,\Phi_t(f,f_2) + T'(f_2,f)\,\Phi_t(f_1,f)\} \\ &+\tfrac{1}{2}\sum_{f_1',f_2'} G(f_1,f_2;f_2',f_1')\,\Phi_t(f_1',f_2') \\ &-\tfrac{1}{2}\sum_{f,f_2',f_1'} \{G(f_1,f;f_2',f_1')\,\langle|a^+_{f}(t)\,a_{f_2}(t)\,a_{f_2'}(t)\,a_{f_1'}(t)|\rangle \\ &+G(f,f_2;f_2',f_1')\,\langle|a^+_{f}(t)\,a_{f_1}(t)\,a_{f_2'}(t)\,a_{f_1'}(t)|\rangle\}. \end{aligned} \tag{5.33'}$$

Now we need, naturally, the equations for $\langle a^+_{f_1}(t)\,a^+_{f_2}(t)\,a_{f_2'}(t)\,a_{f_1'}(t)\rangle$ and $\langle a^+_{f_1}(t)\,a_{f_2}(t)\,a_{f_2'}(t)\,a_{f_1'}(t)\rangle$. It is easy to see that they contain correlation functions of even a higher order.

A closed system of approximate equations can be realized if, at a certain stage, the higher correlation functions are approximated by the lower functions. In the selfconsistent field method, (5.32′) and (5.33′) are used, and the functions $\langle a^+_{f_1'}(t)\,a^+_{f_2'}(t)\,a_{f_2'}(t)\,a_{f_1'}(t)\rangle$ and $\langle a^+_{f_1}(t)\,a_{f_2}(t)\,a_{f_2'}(t)\,a_{f_1'}(t)\rangle$ are approximately replaced by $\varrho_t(f_1, f_2)$ and $\Phi_t(f_1, f_2)$.

Let us assume that the wave function involved in the averaging in (5.31) and in higher correlation functions is an eigenfunction of the operator

$$n_g = \alpha^+_g(t)\,\alpha_g(t), \tag{5.34}$$

and that

$$\bar{n}_g = \langle|\alpha^+_g(t)\,\alpha_g(t)|\rangle. \tag{5.34'}$$

Such an assumption, strictly speaking, can be made only for one fixed value of t, because $\alpha^+_g(t)$ and $\alpha_g(t)$ depend on time. However, if the main part of the Hamiltonian has the form

$$\sum_g \nu(g)\,\alpha^+_g(t)\,\alpha_g(t),$$

then the equations of motion have the approximate form

$$i\frac{\partial \alpha_g(t)}{\partial t} = \nu(g)\,\alpha_g(t), \qquad -i\frac{\partial \alpha^+_g(t)}{\partial t} = \nu(g)\,\alpha^+_g(t).$$

This means that

$$i\frac{\partial}{\partial t}(\alpha_g^+(t)\alpha_g(t)) = -\alpha_g^+(t)\alpha_g(t)\,\nu(g)+\alpha_g^+(t)\alpha_g(t)\,\nu(g) = 0$$

and, consequently,

$$\alpha_g^+(t)\alpha_g(t) = \text{const.}$$

In our case, the time dependence of a_f^+ and a_f is compensated for by the time dependence of u_q and v_q.

In the above-mentioned approximation, the following equations hold:

$$\varrho_t(f_1, f_2) = \sum_g \{v^*(f_1g)\,v(f_2g)\,(1-\bar{n}_g)+u^*(f_1g)\,u(f_2g)\bar{n}_g\}, \tag{5.35}$$

$$\Phi_t(f_1, f_2) = \sum_g \{u(f_1g)\,v(f_2g)\,(1-\bar{n}_g)+v(f_1g)\,u(f_2g)\bar{n}_g\}, \tag{5.36}$$

and the higher correlation functions $\langle|a_{f_1}^+(t)\,a_{f_2}(t)\,a_f(t)\,a_{f_1'}(t)|\rangle$ and $\langle|a_{f_1}^+(t)\,a_{f_2}^+(t)\,a_{f_2'}(t)\,a_{f_1'}(t)|\rangle$ are expressed through $\varrho_t(f_1, f_2)$ and $\Phi_t(f_1, f_2)$ using the formulae analogous to (5.6) and (5.6′). By substituting these formulae in (5.32′) and (5.33′) we obtain the following self-consistent field equations:

$$i\frac{\partial\varrho_t(f_1, f_2)}{\partial t} = \mathfrak{B}(f_1, f_2), \quad i\frac{\partial\Phi_t(f_1, f_2)}{\partial t} = \mathfrak{A}(f_1, f_2), \tag{5.37}$$

where the functions $\mathfrak{A}(f_1, f_2)$ and $\mathfrak{B}(f_1, f_2)$ are defined in (3.15′) and (3.16′).

Reference 120 proved that the equation $(d/dt)\bar{n}_g = 0$ is true for all the solutions of the system (5.37). The relations (3.13) and (3.20) are essential for the proof. This means that an arbitrary set $\ldots n_g \ldots$ is conserved, and, particularly, the system $\bar{n}_g = 0$ is also conserved. Such a system corresponds to the ground state of an even system discussed in the preceding section.

2. We shall discuss some general properties of the Hamiltonian

$$H = \sum_{q\sigma} T(q)a_{q\sigma}^+a_{q\sigma} - \tfrac{1}{4}\sum_{\substack{q,\,q'\\ \sigma,\,\sigma'}} G(q\sigma;\, q'\sigma')a_{q\sigma}^+a_{q,\,-\sigma}^+a_{q',\,-\sigma'}a_{q'\sigma'} \tag{5.38}$$

in this section. The interaction matrix elements are $G(q\sigma; q'\sigma') = G(q\sigma, q, -\sigma; q', -\sigma', q'\sigma')$. Let us define the occupation numbers

$$N_{q\sigma} = a_{q\sigma}^+a_{q\sigma}$$

and show that the difference $N_{q+}-N_{q-}$ is an integral of motion. Actually, it is easy to see that

$$a_{q-}a_{q+}(N_{q+}-N_{q-})-(N_{q+}-N_{q-})a_{q-}a_{q+} = 0,$$

$$a_{q+}^+a_{q-}^+(N_{q+}-N_{q-})-(N_{q+}-N_{q-})a_{q+}^+a_{q-}^+ = 0,$$

and hence

$$H(N_{q+}-N_{q-})-(N_{q+}-N_{q-})H = 0.$$

Consequently,

$$\frac{d}{dt}(N_{q+}(t)-N_{q-}(t)) = 0. \tag{5.39}$$

Let us prove that if Ψ_H corresponds to the lowest eigenvalue of the Hamiltonian (5.38), the following equation is valid for all values of q:[124]

$$(N_{q+}-N_{q-})\Psi_H = 0. \tag{5.40}$$

It is well known (Kramer's theorem) that all eigenstates of the time-reversal invariant Hamiltonian are doubly degenerate.

To prove (5.40) we shall assume that its opposite is true. Let Ψ_H be an eigenfunction of $N_{q+}-N_{q-}$ which corresponds to the eigenvalues $+1$, 0, -1. Let P_0, P_+, P_- denote the set of states q for which the following relations are true:

$$(N_{q+}-N_{q-})\Psi_H = 0, \quad q \in P_0,$$
$$(N_{q+}-N_{q-}-1)\Psi_H = 0, \quad q \in P_+,$$
$$(N_{q+}-N_{q-}+1)\Psi_H = 0, \quad q \in P_-.$$

If $q \in P_+$ then $N_{q+} = 1$ and $N_{q-} = 0$. On the other hand, if $q \in P_-$ then $N_{q+} = 0$ and $N_{q-} = 1$. The Ψ_H can, therefore, be represented in the form $\Psi_H = \Psi_{P_0}\Psi_{P_+}\Psi_{P_-}$, where

$$\Psi_{P_+} = \prod_{q\in P_+} \delta(N_{q+}-1)\,\delta(N_{q-}); \qquad \Psi_{P_-} = \prod_{q\in P_-} \delta(N_{q+})\,\delta(N_{q-}-1).$$

The function Ψ_{P_0} is an eigenfunction of $N_{q\sigma}$ for which $q \in P_0$. From the relations

$$a_{q-}a_{q+}\delta(N_{q+}-1)\,\delta(N_{q-}) = 0, \quad a_{q-}a_{q+}\delta(N_{q+})\,\delta(N_{q-}-1) = 0,$$
$$a^+_{q+}a^+_{q-}\delta(N_{q+}-1)\,\delta(N_{q-}) = 0, \quad a^+_{q+}a^+_{q-}\delta(N_{q+})\,\delta(N_{q-}-1) = 0,$$

it follows that

$$a_{q-}a_{q+}\Psi_{P_+}\Psi_{P_-} = 0, \quad a^+_{q+}a^+_{q-}\Psi_{P_+}\Psi_{P_-} = 0$$

for all $q \in P_+$ or $q \in P_-$. Consequently,

$$H\Psi_H = \Big\{\sum_{q\in P_+} T(q) + \sum_{q\in P_-} T(q) + \sum_{q\in P_0,\,\sigma} T(q)N_{q\sigma} - \tfrac{1}{4}\sum_{q\in P_0,\,\sigma}\ \sum_{q'\in P_0,\,\sigma'} G(q\sigma;\, q'\sigma')a^+_{q\sigma}a^+_{q,\,-\sigma}a_{q',\,-\sigma'}a_{q'\sigma'}\Big\}\Psi_H,$$

and

$$(\Psi^*_H H\Psi_H) = \sum_{q\in P_+} T(q) + \sum_{q\in P_-} T(q) + \Big(\Psi^*_{P_0}\Big\{\sum_{q\in P_0,\,\sigma} T(q)N_{q\sigma} - \tfrac{1}{4}\sum_{q\in P_0,\,\sigma}\ \sum_{q'\in P_0,\,\sigma'} G(q\sigma,\, q'\sigma')a^+_{q\sigma}a^+_{q,\,-\sigma}a_{q',\,-\sigma'}a_{q'\sigma'}\Big\}\Psi_{P_0}\Big). \tag{5.41}$$

The set $P_+ + P_-$ can be divided into two subsets Q_+ and Q_- such that $P_+ + P_- = Q_+ + Q_-$. The division is done in such a way that Q_+ contains those q from $P_+ + P_-$ for which $T(q) = E(q)-\lambda \geq 0$. Naturally, Q_- contains the states with $T(q) < 0$. Hence,

$$(\Psi^*_H H\Psi_H) = \sum_{q\in Q_+} |T(q)| - \sum_{q\in Q_-} |T(q)| + \Big[\Psi^*_{P_0}\Big\{\sum_{q\in P_0,\,\sigma} T(q)N_{q\sigma} - \tfrac{1}{4}\sum_{q\in P_0,\,\sigma}\ \sum_{q'\in P_0,\,\sigma'} G(q\sigma,\, q'\sigma')a^+_{q\sigma}a^+_{q,\,-\sigma}a_{q',\,-\sigma'}a_{q'\sigma'}\Big\}\Psi_{P_0}\Big]. \tag{5.41'}$$

After all these transformations we can prove our theorem. Our assumption can be reduced to the claim that the sets P_+ and P_- are not empty and that

$$(\Psi_H^* H \Psi_H) \leqslant (\psi^* H \psi) \tag{5.42}$$

for an arbitrary function ψ. We shall demand that ψ fulfills an additional condition

$$(N_{q+} - N_{q-})\psi = 0.$$

The function ψ can be then represented as

$$\psi = \Psi_{P_0} \Psi_{Q_+} \Psi_{Q_-}$$

and

$$\Psi_{Q_+} = \prod_{q \in Q_+, \sigma} \delta(N_{q\sigma})\, \delta(N_{q,\,-\sigma}), \quad \Psi_{Q_-} = \prod_{q \in Q_-, \sigma} \delta(N_{q\sigma} - 1)\, \delta(N_{q,\,-\sigma} - 1)$$

(the $q+$ and $q-$ belong to the Q_+ or Q_- at the same time in such a case). After calculating the average value of H in the state ψ,

$$(\psi^* H \psi) = -2 \sum_{q \in Q_-} |T(q)| - \tfrac{1}{4} \sum_{q \in Q_-, \sigma} G(q\sigma, q\sigma) + \Bigg[\Psi_{P_0}^* \Bigg\{ \sum_{q \in P_0, \sigma} T(q) N_{q\sigma}$$

$$- \tfrac{1}{4} \sum_{q \in P_0, \sigma} \sum_{q' \in P_0, \sigma'} G(q\sigma; q'\sigma') a_{q\sigma}^+ a_{q,\,-\sigma}^+ a_{q',\,-\sigma'} a_{q'\sigma'} \Bigg\} \Psi_{P_0} \Bigg]. \tag{5.43}$$

From (5.41′) and (5.43) it follows that

$$(\Psi_H^* H \Psi_H) > (\psi^* H \psi),$$

i.e., we have found a contradiction with our assumption (5.42). Thus we have shown that the wave function which corresponds to the lowest eigenvalue of the Hamiltonian H obeys condition (5.40). The proof was done for a system containing an even number of particles. From (5.40) it follows that the ground state of a system with the Hamiltonian (5.38) is the state without quasiparticles. We have also demonstrated that the single-particle states with a quasiparticle (they belong to the sets P_+ and P_-) are excluded from further considerations; the interaction between particles is not effective for such states. This fact is evident in (5.41); its consequences are known as the blocking effect in the nuclear physics application.

3. Another mathematical method, suggested originally by Bogolyubov *et al.*,[221] is very useful in the treatment of the pairing correlations. Let us, once more, have a system of interacting particles with the Hamiltonian (5.38). Let us define an auxiliary Hamiltonian

$$H' = \sum_{q\sigma} T(q) a_{q\sigma}^+ a_{q\sigma} - \tfrac{1}{4} \sum_{\substack{q,\, q' \\ \sigma,\, \sigma'}} G(q\sigma; q'\sigma') \times$$

$$\times \{ \Phi_\sigma^*(q) a_{q'-\sigma'} a_{q'\sigma'} + \Phi_{\sigma'}(q') a_{q\sigma}^+ a_{q,\,-\sigma}^+ - \Phi_\sigma^*(q)\, \Phi_{\sigma'}(q') \}. \tag{5.44}$$

The operators with identical values of q enter into H' additively. Hence, its eigenfunctions Ψ_0 can be expressed in a product form $\Psi_0 = \prod_q \Psi_0(q)$.

The functions $\Phi_\sigma(q)$ are determined in such a way that the average value of H over Ψ_0 is equal to the minimal eigenvalue of H', i.e.,

$$(\Psi_0^* H \Psi_0) = (\Psi_0^* H' \Psi_0).$$

12

Using the fact that Ψ_0 is a simple product, we write

$$(\Psi_0^* H \Psi_0) \equiv \langle | H | \rangle = \sum_{q\sigma} T(q) \langle a_{q\sigma}^+ a_{q\sigma} \rangle - \frac{1}{4} \sum_{\substack{q,\,q' \\ \sigma,\,\sigma'}} G(q\sigma;\, q'\sigma') \langle | a_{q\sigma}^+ a_{q,\,-\sigma}^+ | \rangle \langle a_{q',\,-\sigma'} a_{q'\sigma'} | \rangle.$$

By comparing with $\langle | H' | \rangle$ we now find that

$$\Phi_\sigma(q) = \langle | a_{q,\,-\sigma} a_{q\sigma} | \rangle. \tag{5.45}$$

The Hamiltonian H is obviously equal to $H = H' + H''$, where

$$H'' = \frac{1}{4} \sum_{\substack{q,\,q' \\ \sigma,\,\sigma'}} G(q\sigma;\, q'\sigma') \{a_{q\sigma}^+ a_{q,\,-\sigma}^+ - \Phi_0^*(q)\} \{a_{q',\,-\sigma'} a_{q'\sigma'} - \Phi_{\sigma'}(q')\}.$$

It can be shown[221] that $\langle | H'' | \rangle$ converges to zero if the volume of the system increases to infinity and the density stays constant.

Thus, if $\Phi_\sigma(q)$ is determined from the minimum condition of the eigenvalue of H' and if H'' is neglected, then we again obtain the well-known approximation discussed in § 2 of Chapter 3 and in § 1 of this chapter.

The studies of the pairing superfluid correlations in nuclei utilize different modifications of the Bardeen–Cooper–Schrieffer method. Sometimes it is possible to apply the exact methods to the solution of a simplified model problem.

An interesting approach to the solution of the systems governed by the Hamiltonian (5.38) was suggested in ref. 222. Another method proposed in ref. 223 was applied to the lead isotopes,[224] and the Hamiltonian (4.2) was used in that problem.

Often the approximation $G(q+,\, q-,\, q'-,\, q'+) = G$, i.e., the assumption of a constant matrix element, is not used. In ref. 225 the correlation functions C_q and chemical potentials λ were calculated for the deformed actinide nuclei. The interaction

$$G(q\sigma, q'\sigma') = G_0 \langle q+, q- | \delta(\mathbf{r}) | q'-, q'+ \rangle$$

was used together with the single-particle Nilsson wave functions. The quantity $G(q\sigma,\, q'\sigma')$ depends relatively strongly on q and q'. Nevertheless, the correlation functions C_q depend primarily on the density of single-particle states. The average values of C_q do not differ strongly from the C-values calculated with the constant G in ref. 208. The finite-range forces were used for calculations of pairing properties in refs. 216 and 226.

§ 3. The Quasiaverages and the Projection Methods

1. Nuclear physics often uses methods violating certain conservation laws. Such a violation is a result of attempts to separate the simplest modes of motion from the complex motions of the many-body system. The conservation laws simplify the solutions of the quantum mechanical two- or three-body problems because they decrease the number of independent variables. In the many-body problem we are dealing with too many variables, and the conservation laws play a completely different role. They complicate the solution of the problem. Such a complication is often inconvenient and therefore one or another

conservation laws are excluded. In each case, when a conservation law is violated, a modified Hamiltonian is defined and its corresponding wave functions are used for the calculation of properties of the real systems.

Let us explain the whole point using the example of the center-of-mass motion in the nuclear shell model. The Hamiltonian of the system is equal to

$$H = \sum_i \frac{p_i^2}{2m} + \sum_{i<j} v_{ij}. \tag{5.46}$$

Its eigenfunctions are simultaneously eigenfunctions of the total momentum $\boldsymbol{P} = \sum_i \boldsymbol{p}_i$. They can be represented as

$$\psi_{Pn}(\boldsymbol{x}_i) = e^{iPX}\psi_{0n}(\boldsymbol{x}_i),$$

where $\boldsymbol{X} = \frac{1}{N}\sum_i \boldsymbol{x}_i$— is the center-of-mass coordinate. Each state is characterized by its total momentum $\boldsymbol{P}$ and by the quantum number n characterizing the relative motion of the particles. Obviously $\boldsymbol{P}\psi_{\boldsymbol{P}n} = \boldsymbol{P}\psi_{\boldsymbol{P}n}$, and the corresponding energy equals

$$E_{\boldsymbol{P}n} = E_n + \frac{\boldsymbol{P}^2}{2Nm},$$

where E_n is the intrinsic energy.

The difficulties related to the conservation laws become apparent when the wave functions of the independent particle model are constructed. It is impossible to describe a bound system of particles without localizing such particles in a certain region of space. But wave functions of such a localized system are not eigenfunctions of the total momentum. Thus, when the wave functions of the independent particle model are constructed, the conservation law of the total momentum is violated.

The independent particle model wave functions Ψ^M, localized in space, may be represented as a superposition of states with different values of the total momentum, i.e.,

$$\psi_n^M = \sum_P a_P e^{iPX}\psi_{0n}, \tag{5.47}$$

where a_P are the expansion coefficients, and $\boldsymbol{P}\psi_{0n} = 0$. Hence ψ_{0n} is an eigenfunction of H. The wave function (5.47) is not an eigenfunction of the Hamiltonian (5.46) because it corresponds to the localized system instead of having the plane wave dependence on the total momentum.

Let us modify the Hamiltonian (5.46) in such a way that the wave function (5.47) becomes an exact eigenfunction of the new Hamiltonian. In order to do that we simply subtract the kinetic energy of center-of-mass motion from H and obtain

$$H' = \sum_i \frac{p_i^2}{2m} + \sum_{i>j} v_{ij} - \sum_{ij} \frac{\boldsymbol{p}_i\boldsymbol{p}_j}{2Nm}. \tag{5.48}$$

The eigenvalues of H' for the states ψ_{Pn} are equal to E_n, i.e., they are independent of $\boldsymbol{P}$. Any linear combination of such functions, corresponding to the same value of E_n, is an exact eigenfunction of H'. Thus, particularly, the wave function (5.47) is an exact eigenfunction of H'.[227]

The Hamiltonian H' formally describes a system with $3N$ degrees of freedom, but only the $3N-3$ intrinsic degrees of freedom determine all eigenvalues of H'. The remaining three degrees of freedom describe the center-of-mass motion and cause an unphysical degeneracy of all eigenvalues of H'.

Another example of the violation of conservation laws is connected with deformed nuclei. It is impossible to describe a nonspherical system without fixing its orientation in space. By doing this we violate the conservation law of total angular momentum. This means that the Nilsson model wave function is in reality a superposition of states with different total angular momenta, i.e.,

$$\psi = \sum_I a_I \psi_I . \tag{5.49}$$

The corresponding modified Hamiltonian is obtained when the rotational energy $\boldsymbol{I}^2/2\mathcal{J}$ is subtracted ($\mathcal{J}$ is the moment of inertia). The states with different total angular momentum are degenerate. In this way we obtain rotational bands composed of nuclear states with the same state of intrinsic motion but with different angular momenta.

This case differs in one important aspect from the preceding one. In the former case we have subtracted the center-of-mass motion, which is well known (it depends on the total mass only). In the case of rotation, the subtracted term contains the unknown moment of inertia. In reality we have to subtract from H an unknown function $f(I^2)$ which is often expanded in a power series in I^2 and the corresponding coefficients are calculated from certain additional physical assumptions.

Thus, as we have seen, the wave functions which violate conservation laws are used for a more precise description of nuclear states. A modified Hamiltonian is defined such that the states corresponding to different eigenvalues of the originally conserved quantity are degenerate. Linear combinations of such states form the exact eigenfunctions of the modified Hamiltonian.[227]

2. The mathematical basis for methods employing the violation of certain conservation laws was developed by Bogolyubov. He introduced the concept of "quasiaverages" and found how to work with them. Let us describe briefly the main features of this method.

To understand better the concept of quasiaverages we shall consider another example. We shall discuss an ideal isotropic ferromagnet. The Hamiltonian of the system describing an ideal isotropic ferromagnet is

$$H = -\tfrac{1}{2} \sum_{x_1, x_2} \mathcal{F}(x_1 - x_2)\,(\boldsymbol{S}_{x_1} \boldsymbol{S}_{x_2}). \tag{5.50}$$

Here x_i are the lattice points, $\boldsymbol{S}_{x_i}$ is the spin vector and $\mathcal{F}(x_1-x_2)$ is a positive function. Each component of the total spin $\boldsymbol{S} = \sum_{x_i} \boldsymbol{S}_{x_i}$ is the constant of motion. The total spin components obey the usual commutation relations

$$S_1 S_2 - S_2 S_1 = i S_3 .$$

From this equation it follows that

$$i\,\mathrm{Sp}(S_3 e^{-H/\Theta}) = \mathrm{Sp}\{(S_1 S_2 - S_2 S_1) e^{-H/\Theta}\},$$

where Θ is the temperature and Sp denotes the trace of a matrix. Because S_1 commutes with H, the following relations are valid:

$$\mathrm{Sp}(S_2 S_1 e^{-H/\Theta}) = \mathrm{Sp}(S_2 e^{-H/\Theta} S_1) = \mathrm{Sp}(S_1 S_2 e^{-H/\Theta})$$

and, therefore,

$$\mathrm{Sp}(S_3 e^{-H/\Theta}) = 0.$$

Similarly,

$$\mathrm{Sp}(S_1 e^{-H/\Theta}) = \mathrm{Sp}(S_2 e^{-H/\Theta}) = 0.$$

The magnetization vector per unit volume is equal to $\mathfrak{M} = (\mu/V)\boldsymbol{S}$. From the above written relations it follows that

$$\langle \mathfrak{M} \rangle = \lim_{V \to \infty} \frac{\mathrm{Sp}(\mathfrak{M} e^{-H/\Theta})}{\mathrm{Sp}(e^{-H/\Theta})} = 0.$$

As expected, the average value of the vector $\mathfrak{M}$ vanishes for all temperatures Θ. This is in accordance with the isotropy of our dynamical system with respect to the group of spin rotation.

Let us now consider a temperature Θ below the Curie point. The vector $\mathfrak{M}$ is not equal to zero but its direction is arbitrary, i.e., our system is degenerate. Suppose that an outside magnetic field $\delta \boldsymbol{e}$ is turned on, where $\delta > 0$, $\boldsymbol{e}^2 = 1$. This means that our Hamiltonian (5.50) should be modified to

$$H_{\delta e} = H + \delta(\boldsymbol{e}\mathfrak{M})V. \tag{5.51}$$

The average value $\langle \mathfrak{M} \rangle = \boldsymbol{e} M_\delta$ in this case, and M_δ remains finite even if the intensity of the outside magnetic field approaches zero, i.e., $M = \lim_{\delta \to 0} M_\delta$.

Let us introduce the concept of quasiaverages for this system. For an arbitrary dynamical quantity B, which is a linear combination of the products

$$S_{x_1}^{\alpha_1}(t_1) \ldots S_{x_r}^{\alpha_r}(t_r),$$

the quasiaverage value of B is defined as

$$\langle B \rangle_q = \lim_{\delta \to 0} \langle B \rangle_{\delta e}, \tag{5.52}$$

where $\langle B \rangle_{\delta e}$ is the ordinary average value of B determined in the system with the Hamiltonian (5.51). Hence, the existence of degeneracies causes a dependence of the quasiaverages on an arbitrary direction $\boldsymbol{e}$. Further,

$$\langle B \rangle = \int \langle B \rangle_q \, d\boldsymbol{e}; \tag{5.53}$$

or, evidently, the ordinary average value is the quasiaverage value integrated over all directions.

While the ordinary average values

$$\langle S_{x_1}^{\alpha_1}(t_1) \ldots S_{x_r}^{\alpha_r}(t_r) \rangle$$

are invariant with respect to the group of rotations, the quasiaverages are only covariant, i.e., during rotation the vector $\boldsymbol{e}$ must be also rotated if the expression

$$\langle S_{x_1}^{\alpha_1}(t_1) \ldots S_{x_r}^{\alpha_r}(t_r) \rangle_q$$

should remain constant.

This means that the quasiaverages do not follow the selection rules which for the ordinary average values are caused by the invariance with respect to spin rotations. The direction of the magnetization vector $\boldsymbol{e}$ characterizes the degeneracy of the statistical equilibrium state. The degeneracy is broken if the direction of $\boldsymbol{e}$ is fixed, for example, along the z-axis. All quasiaverages acquire then certain values. In other words, the degeneracy of the statistical equilibrium state with respect to the group of spin rotation can be broken if the system Hamiltonian is modified by an additional noninvariant term $\delta \mathfrak{M}_z V$ with infinitely small strength δ.

The quasiaverages can be defined in a general case. Let us consider a system with the Hamiltonian H, and let us add an infinitely small term to H. The additional term corresponds to outside fields or sources, and it violates the additive conservation laws. As a result of this addition, we obtain a modified Hamiltonian $H_{\delta,\, \delta \to 0}$. Suppose that all average values $\langle A \rangle$ of the operator

$$A = \ldots a^{+}_{q_i \sigma_i}(t_i) \ldots a_{q'_i \sigma'_i}(t'_i) \ldots \tag{5.54}$$

receive infinitely small increments. In such a case the corresponding state is not degenerate. On the other hand, if some average values of (5.54) receive finite increments when H is changed into $H_{\delta,\, \delta \to 0}$, then the states over which the averaging is done are degenerate.

In the case of degeneracy, it is useful to introduce the quasiaverage values

$$\langle A \rangle_q^H = \lim_{\delta \to 0} \langle A \rangle^{H_\delta} \tag{5.55}$$

instead of the ordinary average values. The quasiaverages do not need to obey all selection rules related to the additive conservation laws. The additions to H should be chosen in such a way that the conservation laws, which break the degeneracy, are violated.

In order to apply any type of perturbation theory to the systems with degenerate states, it is necessary to break the degeneracy first. This means that we should not work, for example, with the ordinary Green functions based on ordinary average values. Instead we should work with the Green functions based on the quasiaverages which do not obey all selection rules.

An important situation for our purposes is the case when the degeneracy is related to the conservation of the total number of particles. Let H be the system Hamiltonian containing the term λN; N is the operator of the total number of particles and naturally $[H, N] = 0$. If φ is an arbitrary real number and $U = e^{i(\varphi/2)N}$, then $H = U^{+}HU$. The Hamiltonian is then invariant with respect to the transformation

$$a_{q'\sigma'} \to U^{+} a_{q'\sigma'} U = e^{i(\varphi/2)} a_{q'\sigma'}. \tag{5.56}$$

The relation (5.56) is easy to prove if we realize that

$$a_{q'\sigma'} N = a_{q'\sigma'} \sum_{q\sigma} a^{+}_{q\sigma} a_{q\sigma} = (N+1) a_{q'\sigma'},$$

$$a_{q'\sigma'} e^{i(\varphi/2)N} = e^{i(\varphi/2)(N+1)} a_{q'\sigma'}.$$

Let us consider an average value of the Heisenberg operators over a certain state

$$\langle a^+_{q_1\sigma_1}(t_1) \dots a^+_{q_i\sigma_i}(t_i)\, a_{q'_1\sigma'_1}(t'_1) \dots a_{q'_{i'}\sigma'_{i'}}(t'_{i'})\rangle. \tag{5.57}$$

After applying the transformation (5.56),

$$e^{i(\Delta/2)\varphi}\langle a^+_{q_1\sigma_1}(t_1) \dots a_{q'_1\sigma'_1}(t'_1) \dots\rangle,$$

where Δ is the difference between the number of annihilation and creation operators (Δ must be even; the average value vanishes for the odd Δ). From this it follows that

$$(1-e^{i(\Delta/2)\varphi})\,\langle a^+_{q_1\sigma_1}(t_1) \dots a_{q'_1\sigma'_1}(t'_1) \dots\rangle = 0 \tag{5.57'}$$

because the average value (5.57) vanishes if the number of annihilation and creation operators are not equal. The same selection rules are applicable to the Green functions.

Let us consider a system with the Hamiltonian H of (5.38), and let us define the auxiliary Hamiltonian H' (5.44). This Hamiltonian is invariant with respect to the transformation

$$a_{q\sigma} \to e^{i(\varphi/2)}a_{q\sigma}, \quad a^+_{q\sigma} \to e^{-i(\varphi/2)}a^+_{q\sigma}, \quad \Phi \to e^{i\varphi}\Phi. \tag{5.56}$$

Bogolyubov has shown [228] that the quasiaverages of the Hamiltonian H are equal to the ordinary averages over eigenfunctions of the auxiliary Hamiltonian H', i.e.,

$$\langle \dots a^+_{q_i\sigma_i}(t'_i) \dots a_{q'_i\sigma'_i}(t'_i) \dots\rangle^H_q = \langle \dots a^+_{q_i\sigma_i}(t_i) \dots a_{q'_i\sigma'_i}(t'_i) \dots\rangle^{H'}. \tag{5.58}$$

The ordinary average values are obtained from the quasiaverages after the additional integration over an arbitrary angle:

$$\langle \dots a^+_{q_i\sigma_i}(t_i) \dots a_{q'\sigma'}(t'_i) \dots\rangle^H = \frac{1}{2\pi}\int_0^{2\pi} d\varphi \langle \dots a^+_{q_i\sigma_i}(t_i) \dots a_{q'_i\sigma'_i}(t'_i) \dots\rangle^H_q. \tag{5.58'}$$

Similarly to the previously discussed cases, the quasiaverages are obtained if the Hamiltonian (5.38) is complemented by infinitesimal terms which remove the degeneracy:

$$H_\delta = H - \frac{\delta}{2}\sum_{q\sigma} g(q\sigma)\,(a_{q,\,-\sigma}a_{q\sigma} + a^+_{p\sigma}a^+_{q,\,-\sigma}), \tag{5.58''}$$

where $g(q\sigma)$ is a real regular function (it is often assumed that $G(q\sigma, q'\varrho') = g(q\varrho)\,g(q'\sigma')$).

Now we shall determine the quasiaverages as

$$\begin{aligned}\langle \dots a^+_{q_i\ i}(t_i) \dots a_{q'_i\sigma'_i}(t'_i) \dots\rangle_q &= \lim_{\substack{\delta\to 0\\ \delta>0}} \langle \dots a^+_{q_i\sigma_i}(t_i) \dots a_{q'_i\sigma'_i}(t'_i) \dots\rangle^{H_\delta}\\ &= \langle \dots a^+_{q_i\sigma_i}(t_i) \dots a_{q'_i\sigma'_i}(t'_i) \dots\rangle^{H'} e^{i(\Delta/2)\varphi}.\end{aligned}$$

It is now evident that the quasiaverages depend on an arbitrary phase angle φ. The selection rules related to the conservation of particle number are not valid for the quasiaverages. To get a well-defined value of the quasiaverages, we have to fix the φ-value. Let us choose $\varphi = 0$; this is a convenient choice because all equal-time quasiaverages become real, i.e.,

$$\langle \dots a^+_{q\sigma}(t) \dots a_{q'\sigma'}(t) \dots\rangle \quad \text{is real.} \tag{5.59}$$

The perturbation theory must be modified in the case of degeneracy. This is done in the following way: instead of dividing H into $H = H_0+H_1$ and making expansion in H_1, we must use a different division $H = H_0'+H_1'$, where $H_0' = H_0+\Delta$ and $H_1' = H_1-\Delta$. The additional term Δ is similar to the infinitesimal term in (5.58″). In such a way certain additive conservation laws are not valid for H_0' any more, for example, the conservation of the particle number. And we have anomalous terms, as for example $\langle a_{q,\,-\sigma} a_{q\sigma}\rangle$ already in the zeroth approximation.

3. We have stressed repeatedly that the nuclear theory often uses mathematical methods violating certain conservation laws. The accuracy of such methods may be improved by the application of "projection operators."

Let H be the system Hamiltonian and N some Hermitian operator commuting with H. The common eigenfunction $\Psi_{N,\,\xi}$ is obviously a solution of the equations

$$H\Psi_{N,\,\xi} = E_{N,\,\xi}\Psi_{N,\,\xi}, \quad N\Psi_{N,\,\xi} = N_0\Psi_{N,\,\xi}, \tag{5.60}$$

or, in other words, $\Psi_{N,\,\xi}$ is a solution of the variational problem

$$\delta(\Psi^*H\Psi) = 0, \quad (\Psi^*\Psi) = 1 \tag{5.60'}$$

with the supplementary condition

$$N\Psi = N_0\Psi.$$

Suppose that the conservation law related to the operator N is violated. The function Ψ is replaced in such a case by Ψ_A and Ψ_A is not an eigenfunction of N. The variational problem is also different, namely,

$$\delta(\Psi_A^*H\Psi_A) = 0, \quad (\Psi_A^*\Psi_A) = 1, \quad (\Psi_A^*N\Psi_A) = N_0. \tag{5.61}$$

Thus the conservation law is valid only in the average.

Let us define a projection operator P_N with the following property:

$$P_{N_0}\Psi_{N,\,\xi} = \delta_{N,\,N_0}\Psi_{N,\,\xi}. \tag{5.62}$$

This means that the projection operator separates out of Ψ_A just the part with $N = N_0$, i.e.,

$$P_{N_0}\Psi_A = \Psi_A^{N_0} \neq 0, \quad (1-P_{N_0})\Psi_A \neq 0, \tag{5.63}$$

where

$$\Psi_A = \sum_N d_N\Psi_N, \tag{5.64}$$

i.e., the d_N are the expansion coefficients.

The properly normalized function $\Psi_A^{N_0}$ is a solution of the problem (5.61). At the same time, $\Psi_A^{N_0}$ is closer to the solution Ψ_N of the exact problem (5.60′) than the function Ψ_A. This feature makes the projection operators very attractive for nuclear theory applications(229–233).

Let us construct the projected functions for the nuclear pairing superfluid correlation problem. Let us consider, for example, the neutron system with the Hamiltonian (4.2),

$$H_0(n) = \sum_{s\sigma}\{E(s)-\lambda_n\}a_{s\sigma}^+a_{s\sigma} - G_N\sum_{s,\,s'} a_{s+}^+a_{s-}^+a_{s'-}a_{s'+}.$$

The wave function of the ground state of an even system has the familiar form (4.22), i.e.,

$$\Psi_0 = \prod_s (u_s + v_s a_{s+}^+ a_{s-}^+)\Psi_{00}.$$

The wave function Ψ_0 is not an eigenfunction of the particle number operator.

Let us denote

$$\beta_s = a_{s-}a_{s+}, \quad \beta_s^+ = a_{s+}^+ a_{s-}^+,$$

and, obviously,

$$\beta_s\beta_s = \beta_s^+\beta_s^+ = 0,$$

and, therefore,

$$e^{d_s\beta_s^+} = 1 + d_s\beta_s^+. \tag{5.65}$$

The ground-state wave function can be transformed in the following way:

$$\Psi_0 = \prod_s (u_s + v_s\beta_s^+)\Psi_{00} = \prod_s u_s\left(1 + \frac{v_s}{u_s}\beta_s^+\right)\Psi_{00}$$

$$= \left(\prod_s u_s\right)\prod_s e^{(v_s/u_s)\beta_s^+}\Psi_{00} = \left(\prod_s u_s\right)e^{\sum_s (v_s/u_s)\beta_s^+}\Psi_{00}.$$

The resulting exponential function can be expanded in a power series

$$\Psi_0 = \left(\prod_s u_s\right)\sum_n \frac{1}{n!}\left(\sum_s \frac{v_s}{u_s}\beta_s^+\right)^n \Psi_{00}$$

and, remembering that the system under consideration consists of N_0 (N_0 is an even number) particles, we write

$$\Psi_0 = \left(\prod_s u_s\right)\sum_{(N/2)} \frac{1}{(N/2)!}\left(\sum_s \frac{v_s}{u_s}\beta_s^+\right)^{N/2}\Psi_{00}, \tag{5.66}$$

which means that

$$\Psi_0 = \sum_{N/2} \Psi_0'^N,$$

where

$$\Psi_0'^N = \frac{1}{(N/2)!}\left(\prod_s u_s\right)\left\{\sum_s \frac{v_s}{u_s}\beta_s^+\right\}^{N/2}\Psi_{00}. \tag{5.66'}$$

The Ψ_0 evidently contains besides the N_0 component also components with N_0+2, N_0+4, etc., particles.

The projected function must be, finally, normalized. For $N = N_0$ (denoting $\gamma \equiv N_0/2$)

$$\Psi_0^{N_0} = \left\{\sum_{s_1 < s_2 < \ldots < s_\gamma} \frac{v_{s_1}^2 v_{s_2}^2 \ldots v_{s_\gamma}^2}{u_{s_1}^2 u_{s_2}^2 \ldots u_{s_\gamma}^2}\right\}^{-1/2} \sum_{s_1 < s_2 < \ldots < s_\gamma} \frac{v_{s_1}v_{s_2}\ldots v_{s_\gamma}}{u_{s_1}u_{s_2}\ldots u_{s_\gamma}}\beta_{s_1}^+\beta_{s_2}^+\ldots\beta_{s_\gamma}^+\Psi_{00}; \tag{5.67}$$

by $s_1 < s_2 < \ldots < s$ we mean that $E(s_1) < E(s_2) < \ldots < E(s)$.[234]

The ground-state energy for N_0 particles equals

$$\mathcal{E}_0^{N_0} = \left\{ \sum_{s_1 < s_2 < \dots < s_\gamma} \frac{v_{s_1}^2 v_{s_2}^2 \dots v_{s_\gamma}^2}{u_{s_1}^2 u_{s_2}^2 \dots u_{s_\gamma}^2} \right\}^{-1} \times$$
$$\times \left\{ 2 \sum_{s_1 < s_2 < \dots < s_\gamma} \left([E(s_1)+E(s_2)+ \dots +E(s_\gamma)] \frac{v_{s_1}^2 v_{s_2}^2 \dots v_{s_\gamma}^2}{u_{s_1}^2 u_{s_2}^2 \dots u_{s_\gamma}^2} \right) \right.$$
$$\left. - G_N \sum_{s, s'} \frac{v_s v_{s'}}{u_s u_{s'}} \sum_{s_2 < s_3 < \dots < s_\gamma \neq s, s'} \frac{v_{s_2}^2 v_{s_3}^2 \dots v_{s_\gamma}^2}{u_{s_2}^2 u_{s_3}^2 \dots u_{s_\gamma}^2} \right\}. \tag{5.68}$$

It is not difficult to generalize the above achieved results and find the projected wave functions for excited two-quasiparticle states and for odd-A systems.

The projected wave functions and the corresponding matrix elements are rather cumbersome; it is difficult to work with them. Reference 235 suggested a simpler but less accurate projection technique. Only components with $N \neq N_0$ and $|N-N_0| \leqslant \Delta$ are excluded in this method. If Δ is large enough, the effect of the particle number nonconservation is sufficiently reduced.

4. The calculation of average values using the projected wave functions leads often to complicated analytical expressions. Thus, it is sometimes useful to employ some approximate methods. Let us explain an alternative variational method for derivation of the pairing theory equations, suggested by Bayman.[236]

We assume that the Hamiltonian H and the particle number operator N commute. The wave functions are determined from the variational equations

$$\delta\{(\Psi^* H \Psi) - \mathcal{E}(\Psi^* \Psi)\} = 0, \quad N_0 = (\Psi^* N \Psi), \tag{5.69}$$

where $\mathcal{E}$ is the energy of the system and N_0 is the number of particles.

Let us define a new function

$$\Psi(z) = z^N \Psi, \tag{5.70}$$

where z is a complex variable. It is easy to see that the expression

$$\frac{1}{2\pi i} \oint \frac{dz}{z} \frac{1}{z^{N_0}} \Psi(z) \tag{5.71}$$

is the projection of Ψ onto the subspace with N_0 particles. The integration in (5.71) is over a closed contour which contains the point $z = 0$.

The average value of an arbitrary operator over the state (5.71) is equal to

$$\langle O \rangle = \frac{\dfrac{1}{2\pi i} \oint \dfrac{dz}{z^{N_0+1}} (\Psi^* O \Psi(z))}{\dfrac{1}{2\pi i} \oint \dfrac{dz}{z^{N_0+1}} (\Psi^* \Psi(z))}. \tag{5.72}$$

Let us denote

$$O(z) = \frac{(\Psi^* O \Psi(z))}{(\Psi^* \Psi(z))}, \tag{5.73}$$

$$\Gamma(z) = \ln \frac{(\Psi^* \Psi(z))}{z^{N_0}}, \tag{5.74}$$

and, consequently,

$$\frac{1}{z^{N_0}}(\Psi^* O\Psi(z)) = O(z)\frac{(\Psi^*\Psi(z))}{z^{N_0}} = O(z)e^{\Gamma(z)}.$$

This means that the average value (5.72) can be rewritten as

$$\langle O\rangle = \frac{\oint (dz/z)\,O(z)e^{\Gamma(z)}}{\oint (dz/z)e^{\Gamma(z)}} \tag{5.72'}$$

Now comes the time to introduce approximate methods. As an approximation we may calculate the integrals by means of the saddle-point method. If the function $e^{\Gamma(z)}$ has a sharp maximum in a real saddle point, then

$$\oint e^{\Gamma(z)}\,dz \approx \frac{1}{\sqrt{2\pi\Gamma''(z_0)}}\,e^{\Gamma(z_0)},$$

and the saddle point can be determined from the condition $(d/dz)\Gamma(z)\Big|_{z=z_0} = 0$. Let us assume that $\Gamma(z)$ has, indeed, a sharp maximum and find z_0:

$$\frac{d}{dz}\Gamma(z) = \frac{d}{dz}\{\ln(\Psi^*\Psi(z)) - N_0 \ln z\} = \frac{(d/dz)(\Psi^* z^N\Psi)}{(\Psi^*\Psi(z))} - \frac{N_0}{z} = \frac{1}{z}\left\{\frac{(\Psi^* N\Psi(z))}{(\Psi^*\Psi(z))} - N_0\right\}.$$

from the condition

$$\frac{d}{dz}\Gamma(z)\Big|_{z=z_0} = 0$$

it follows that

$$N_0 = \frac{(\Psi^* N\Psi(z_0))}{(\Psi^*\Psi(z_0))}. \tag{5.75}$$

This means that $z = z_0$ is a saddle point of $\Gamma(z)$ if the average particle number is equal to N_0. Let us find the second derivative:

$$\frac{d^2}{dz^2}\Gamma(z)\Big|_{z=z_0} = \frac{1}{z_0^2}\,\Delta N_{z_0}^2.$$

The function $\Gamma(z)$ can be expanded around z_0:

$$\Gamma(z) = \Gamma(z_0) + \frac{1}{2}\left(\frac{z-z_0}{z_0}\right)^2 \Delta N_{z_0}^2 + \dots \tag{5.76}$$

It is possible to prove that there is only one saddle point, and that the saddle point is indeed on the real axis.

Assume that some matrix element of an operator $O(z)$ depends on z more smoothly than $\Gamma(z)$. The expression (5.72′) is then approximately equal to

$$\begin{aligned}\langle O\rangle &= O(z_0) - \frac{1}{2}\frac{1}{\Delta N_{z_0}^2}\left\{\frac{\partial^2}{\partial\varrho^2}O(z_0 e^{\varrho})\right\}_{\varrho=0} + \dots \\ &= O(z_0) - \frac{1}{2}\frac{1}{\Delta N_{z_0}^2}\{\langle O(N-N_0)^2\rangle_{z_0} - O(z_0)\,\Delta N_{z_0}^2\} + \dots\,.\end{aligned} \tag{5.77}$$

Now we can calculate the average values of $O(z_0)$, $N(z_0)$, and $\Delta N^2_{z_0}$ for the concrete case of a system with the Hamiltonian (4.2). For convenience we shall use instead of Ψ, $\Psi(z)$ another function Ψ^z_0. If the operator O commutes with N, then

$$(\Psi_0^* O z^N \Psi_0) = (\Psi_0^* z^{N/2} O z^{N/2} \Psi_0),$$

and, using (5.66), we find that

$$z^{N/2}\Psi_0 = \left(\prod_q u_q\right) \sum_{N/2} \frac{z^{N/2}}{(N/2)!} \left(\sum_q \frac{v_q}{u_q} \beta_q^+\right)^{N/2} \Psi_{00}$$

$$= \left(\prod_q u_q\right) \exp\left(\sum_q z \frac{v_q}{u_q} \beta_q^+\right) \Psi_{00} = \prod_q (u_q + z v_q \beta_q^+) \Psi_{00} \equiv \Psi_0^z, \tag{5.78}$$

and hence

$$(\Psi_0^* O z^N \Psi_0) = (\Psi_0^{*z} O \Psi_0^z). \tag{5.79}$$

Obviously, in such a case

$$(\Psi_0^{*z} \Psi_0^z) = \prod_q (u_q^2 + z^2 v_q^2) = 1, \tag{5.80}$$

$$(\Psi_0^{*z} N \Psi_0^z) = \left(\Psi_{00}^* \prod_{q'} (u_{q'} + z v_{q'} \beta_{q'}) \sum_{q_2\sigma_2} a^+_{q_2\sigma_2} a_{q_2\sigma_2} \prod_q (u_q + z v_q \beta_q^+) \Psi_{00}\right)$$

$$= 2 \sum_{q_2} z^2 v_{q_2}^2 \prod_{q \neq q_2} (u_q^2 + z^2 v_q^2) = \prod_q (u_q^2 + z^2 v_q^2) 2 \sum_{q'} \frac{z^2 v_{q'}^2}{u_{q'}^2 + z^2 v_{q'}^2}, \tag{5.81}$$

and we obtain for H in the form (4.2)

$$(\Psi_0^{*z} H_0 \Psi_0^z) = \prod_{q'} (u_{q'}^2 + z^2 v_{q'}^2) \left[2 \sum_q \{E(q) - \lambda\} \frac{z^2 v_q^2}{u_q^2 + z^2 v_q^2} - G_N \left(\sum_q \frac{z u_q v_q}{u_q^2 + z_q^2 v_q^2} \right)^2 \right.$$

$$\left. - G_N \sum_q \left(\frac{z^2 v_q^2}{u_q^2 + z^2 v_q^2} \right)^2 \right]. \tag{5.82}$$

By using the normalization condition (5.80) we find from (5.81) that at the saddle point $z = z_0$ the following equation is valid:

$$N_0 = N(z_0) = 2 \sum_q \frac{z_0^2 v_q^2}{u_q^2 + z_0^2 v_q^2}. \tag{5.81'}$$

The point z_0 enters always in combination with v_q^2; therefore we can put $z_0 = 1$. Furthermore, the expressions (5.81′) and (5.82) are not changed if u_q and v_q are multiplied by the same constant factor. Thus, we can write $u_q^2 + v_q^2 = 1$. The average value of H_0 is then equal to

$$H_0(z_0 = 1) = 2 \sum_q \{E(q) - \lambda\} v_q^2 - G_N \left(\sum_q u_q v_q \right)^2 - G_N \sum_q v_q^4, \tag{5.82'}$$

and

$$N_0 = 2 \sum_q v_q^2. \tag{5.81''}$$

After the variation is executed we obtain once more the familiar equations of the pairing correlation theory.

The corrections to $O(z_0)$ in (5.77) have been calculated in ref. 237 for the interaction (4.2). We shall give only the final results here:

$$\langle N \rangle = N(z_0 = 1) - \frac{1}{2} \frac{\Delta N^3_{z_0}}{\Delta N^2_{z_0}};$$

the correction to $H_0(z_0 = 1)$ is equal to

$$-\frac{1}{\Delta N^2_{z_0=1}} \frac{C^2 - W^2}{G},$$

where

$$C = G \sum_q u_q v_q, \quad W = G \sum_q u_q v_q (u_q^2 - v_q^2), \quad \Delta N^2_{z_0=1} = 4 \sum_q u_q^2 v_q^2.$$

If u_q^2 and v_q^2 depend on q sufficiently smoothly, then the following inequalities are valid:

$$\Delta N^3_{z_0} \ll \Delta N^2_{z_0}, \quad W \ll C.$$

In such a case we can write

$$N_0 = N(z_0 = 1) = 2 \sum_s v_s^2,$$

$$\langle H_0 \rangle = H_0(z_0 = 1) - \frac{1}{\Delta N^2_{z_0=1}} \frac{C^2}{G} = 2 \sum_q \{E(q) - \lambda\} v_q^2 - G\left(1 + \frac{1}{\Delta N^2_{z_0=1}}\right) \left(\sum_q u_q v_q\right)^2$$
$$= 2 \sum_q \{E(q) - \lambda\} v_q^2 - G_{\text{eff}} \left(\sum_q u_q v_q\right)^2. \tag{5.83}$$

This means that the original equations can be improved if the interaction constant G is replaced by an effective value

$$G_{\text{eff}} = G\left(1 + \frac{1}{\Delta N^2_{z_0=1}}\right). \tag{5.84}$$

The value of G is, however, found from a comparison of calculated and experimental pairing energies. Consequently, the renormalization (5.84) is unnecessary for the calculations of the ground states. It can be nevertheless quite useful for the calculation of properties of the quasiparticle excitations. The quantity ΔN^2 is smaller for the two-quasiparticle state than for the ground state. Therefore the G_{eff} is somewhat larger for the excited states when compared with the G_{eff} for the ground state. When such an effect is taken into account, the accuracy of the calculation is improved.[(238)]

5. We have discussed in point 3 of this section a method where the projection of the wave functions on the subspace with N_0 particles was made after the solution of the variational problem. A more accurate method was suggested in ref. 239 (the so-called FBCS method).

The variational procedure was applied to the minimization of the average value of the energy operator calculated by means of the projected wave functions. This means that the projection was made first and the variational procedure was performed afterwards. The studies 225, 239 and 240 have shown that such a method is somewhat more accurate than the standard canonical transformation method with blocking effect or than the ordinary projection methods. They have also shown that in the cases where the canonical transformation has only a trivial solution $C = 0$ the FBCS method gives a nontrivial solution ($C \neq 0$). Thus the vanishing correlation functions of some two-quasiparticle states in deformed nuclei, obtained earlier in ref. 44, are a consequence of inaccuracies of the employed mathematical methods.

Another approach which conserves the particle number was suggested in ref. 241. The expression $\langle a_{q+}^{+}a_{q-}^{+}a_{q'-}a_{q'+}\rangle$ is decomposed in the same way as in the canonical transformation method. The wave functions of N_0 and $N_0 \pm 2$ particles are, however, treated separately. The method has been tested on simple models which allow exact solution. Such tests suggest that the method is somewhat more accurate than the Bogolyubov canonical transformation. Another improvement in the solution of the nuclear pairing problem was suggested in ref. 242. The Hamiltonian (4.2) was supplemented by an additional term $\lambda_2 N^2$. The problem of particle number nonconservation in the Bardeen–Cooper–Schrieffer method has been discussed repeatedly, for example, in refs 243–245.

§ 4. Model Studies of the Accuracy of Approximate Methods

1. Exactly solvable models are very important in theoretical physics, particularly in the quantum field theory (see, e.g., refs. 246–250). A number of exactly solvable problems have been employed in nuclear theory for testing of the various mathematical approximate methods. In this section we shall study the accuracy of the approximate methods discussed in the preceding and this chapters.

We shall use our usual pairing Hamiltonian applicable for deformed nuclei

$$H = \sum_{q\sigma} E(q) a_{q\sigma}^{+} a_{q\sigma} - G \sum_{q,\, q'} a_{q+}^{+} a_{q-}^{+} a_{q'-} a_{q'+}\,. \tag{5.85}$$

The seniority operator is defined by

$$\mathfrak{s} = \sum_{q} \mathfrak{s}_q^2, \tag{5.86}$$

where

$$\mathfrak{s}_q = N_{q+} - N_{q-}, \quad N_{q\sigma} = a_{q\sigma}^{+} a_{q\sigma}, \quad \sigma = \pm 1. \tag{5.86'}$$

The eigenvalues of $\mathfrak{s}$ are equal to the number of levels occupied only by one particle, i.e., they are equal to the number of unpaired particles. The operators $\mathfrak{s}$ and $\mathfrak{s}_q$ commute with the Hamiltonian (5.85). This means that the interaction in (5.85) does not change the number and state of unpaired particles. This circumstance is very important for the exact solution of the problem.

The pairing Hamiltonian of spherical nuclei has a somewhat different form:

$$H = \sum_{j,\, m} E(j) a_{jm}^{+} a_{jm} - \tfrac{1}{4} G \sum_{j,\, j'} \sum_{m,\, m'} (-1)^{j-m} (-1)^{j'-m'} a_{jm}^{+} a_{j,\, -m}^{+} a_{j',\, -m'} a_{j'm'}\,. \tag{5.85'}$$

We shall define the quasispin operators by

$$S_{+} = \sum_{j} S_{+}^{j}, \quad S_{+}^{j} = \sum_{m>0} (-1)^{j-m} a_{jm}^{+} a_{j,\, -m}^{+}, \tag{5.87}$$

$$S_{-} = \sum_{j} S_{-}^{j}, \quad S_{-}^{j} = \sum_{m>0} (-1)^{j-m} a_{j,\, -m} a_{jm}, \tag{5.87'}$$

$$S_{z} = \sum_{j} S_{z}^{j}, \quad S_{z}^{j} = \sum_{m>0} \{a_{jm}^{+} a_{jm} - a_{j,\, -m} a_{j,\, -m}^{+}\}, \tag{5.87''}$$

where obviously $(S_{-}^{j})^{+} = S_{+}^{j}$.

The operator S_+^j describes the creation of a pair of particles with total angular momentum zero in subshell j and the operator S_-^j describes annihilation of such a pair. The quasispin operators have the same commutation relations as the ordinary spin operators.[251]

The particle number operator equals

$$N = \sum_j N^j, \quad N^j = 2 \sum_{m>0} a_{jm}^+ a_{jm},$$

and denoting

$$\Omega = \sum_j \Omega^j, \quad \Omega^j = j + \tfrac{1}{2}$$

we can rewrite S_z as

$$S_z = \tfrac{1}{2} \sum_{j,\, m>0} \{a_{jm}^+ a_{jm} + a_{j,\,-m}^+ a_{j,\,-m} - 1\} = \tfrac{1}{2}(N - \Omega). \tag{5.88}$$

By $\bar{S}$, $\bar{S}_z$, and $\bar{N}^j$ we shall denote the eigenvalues of the operators S, S_z, and N^j. From (5.88) it follows that

$$\bar{S}_z^j = \tfrac{1}{2}(\bar{N}^j - \Omega^j), \tag{5.88'}$$

and therefore

$$\bar{S}^j \geqslant |\bar{S}_z^j|,$$

i.e., $|\bar{S}_z^j|$ is maximal (i.e., equal to $\frac{1}{2}\Omega^j$) if the subshell j is either empty ($\bar{N}^j = 0$) or completely full ($\bar{N}^j = 2\Omega^j$). When the operator S_+^j is applied to the state $|\bar{S}^j, \bar{S}_z^j\rangle$ then the $\bar{S}^j$ value is not changed and the $\bar{S}_z^j$ value is increased by one unit.

The seniority operator determining the number of unpaired particles is defined by

$$\mathfrak{s} = \sum \mathfrak{s}^j, \quad \mathfrak{s}^j = \Omega^j - 2\bar{S}^j. \tag{5.89}$$

The Hamiltonian (5.85′) may be expressed in terms of the operator S in the following way:

$$H = \sum_j E(j)\Omega^j + 2\sum_j E(j) S_z^j - GS_+S_-. \tag{5.85''}$$

The Hamiltonian is now diagonalized in the basis

$$|\xi\mathfrak{s}\bar{S}\bar{S}_z\rangle, \tag{5.90}$$

where ξ denotes all the remaining quantum numbers characterizing the state. Using the standard rules of angular momentum theory,

$$\begin{aligned} GS_+S_- |\xi\mathfrak{s}\bar{S}\bar{S}_z\rangle &= GS_+ \sqrt{(\bar{S}-\bar{S}_z+1)(\bar{S}+\bar{S}_z)}\, |\xi\mathfrak{s}\bar{S}(S_z-1)\rangle \\ &= G(\bar{S}+\bar{S}_z)(\bar{S}-\bar{S}_z+1)\, |\xi\mathfrak{s}\bar{S}\bar{S}_z\rangle = G\{\bar{S}(\bar{S}+1) - \bar{S}_z(\bar{S}_z-1)\}\, |\xi\mathfrak{s}\bar{S}\bar{S}_z\rangle. \end{aligned}$$

The expectation value of H in state (5.90) is equal to

$$\langle \bar{S}_z\bar{S}\mathfrak{s}\xi | H | \xi\mathfrak{s}\bar{S}\bar{S}_z\rangle = \sum_j E(j)\Omega^j + 2\sum_j E(j) \langle \bar{S}_z\bar{S}\mathfrak{s}\xi | S_z^j | \xi\mathfrak{s}\bar{S}\bar{S}_z\rangle - G\{\bar{S}(\bar{S}+1) - \bar{S}_z(\bar{S}_z-1)\}. \tag{5.91}$$

The state without quasiparticles is the ground state of a system with an even number of particles. The seniority of such a state is $\mathfrak{s} = 0$, and, therefore, from (5.89) it follows that

$\bar{S}^j = \frac{1}{2}\Omega^j$, $\bar{S} = \frac{1}{2}\Omega$. We can calculate the ground-state energy using (5.88′):

$$\langle S_z\bar{S}\mathfrak{s} = 0\,\xi\,|\,H\,|\,\xi\mathfrak{s} = 0\,\bar{S}S_z\rangle = \sum_j E(j)\Omega^j + 2\sum_j E(j)\,\langle S_z\bar{S}\mathfrak{s} = 0\,\xi\,|\,S_z^j\,|\,\xi\mathfrak{s} = 0\,\bar{S}S_z\rangle$$
$$-\tfrac{1}{4}G\bar{N}(2\Omega - \bar{N} + 2). \tag{5.91′}$$

The ground state and many excited states of an odd system have a single, unpaired particle. Suppose that the unpaired particle is in the j_1 subshell, then $\mathfrak{s} = 1$, $\mathfrak{s}^{j_1} = 1$, $\bar{S}^{j_1} = \frac{1}{2}(\Omega^{j_1} - 1)$. When a particle occupies the $j_1 m_1$ state, the j_1, $\pm m_1$ state is blocked for any pair; this means that this state is in fact missing from the interaction (5.85′) (the so-called blocking effect).

The excited states of an even system have one, two, or more, pairs broken. If one pair is broken, the seniority is equal to two, $\mathfrak{s} = 2$ and $\bar{S} = \frac{1}{2}(\Omega - 2)$. If the unpaired nucleons occupy different subshells j_1 and j_2, then $\bar{S}^{j_1} = \frac{1}{2}(\Omega^{j_1} - 1)$ and $\bar{S}^{j_2} = \frac{1}{2}(\Omega^{j_2} - 1)$. If both unpaired nucleons are in the same subshell j_1, then $\bar{S}^{j_1} = \frac{1}{2}(\Omega^{j_1} - 2)$.

Generally the state with $\mathfrak{s}$ unpaired nucleons has $\bar{S} = \frac{1}{2}(\Omega - \mathfrak{s})$. The average value of H is then equal to

$$\langle S_z\bar{S}\mathfrak{s}\xi\,|\,H\,|\,\xi\mathfrak{s}\bar{S}S_z\rangle = \sum_j E(j)\Omega^j + 2\sum_j E(j)\,\langle S_z\bar{S}\mathfrak{s}\xi\,|\,S_z^j\,|\,\xi\mathfrak{s}\bar{S}S_z\rangle$$
$$-\tfrac{1}{4}G(\bar{N} - \mathfrak{s})\,(2\Omega - \bar{N} - \mathfrak{s} + 2). \tag{5.91″}$$

Our aim in this section is to compare the exact solutions of simple systems with the Hamiltonians (5.85) or (5.85′) with the approximate solutions explained earlier in the last two chapters. The renormalization of the single-particle energies $E(j) = E_0(j) - \frac{1}{2}Gv_j^2$ is not done in the exact solution. This means that we have to be careful to compare with the approximate solutions which do not include such a renormalization. For example, the formula (4.21′) for the ground-state energy of an even system should be replaced by

$$\mathscr{E}_0 = \sum_j (2j+1)\,E(j)v_j^2 - \frac{C^2}{G} - G\sum_j \left(j + \frac{1}{2}\right)v_j^4. \tag{5.92}$$

Similarly, the formula (4.48) for the energy of a two-quasiparticle state in a deformed nucleus is replaced by

$$\mathscr{E}_0(q_1, q_2) = E(q_1) + E(q_2) + 2\sum_{q \neq q_1,\, q_2} E(q)v_q^2 - \frac{C^2(q_1, q_2)}{G} - G\sum_{q \neq q_1,\, q_2} v_q^4. \tag{5.93}$$

We shall compare two approximate methods explained in § 1 of Chapter 4 with the exact solutions. Method A is the standard Bardeen–Cooper–Schrieffer method and the blocking effect is not included. Method B includes the blocking effect. Naturally, both methods predict the same ground-state properties.

2. We shall discuss the following model (suggested by Mottelson in ref. 157): $\bar{N}$ identical particles in the same subshell j with total degeneracy $2\Omega = 2j+1$. This is the limiting case of strong coupling (in the strong coupling case the pairing constant is larger than the average energy separation of single-particle states.) The Hamiltonian (5.85′) in our model is

$$H = E\sum_m a_m^+ a_m - \frac{G}{4}\sum_{m,\, m'} (-1)^{1-m-m'}\, a_m^+ a_{-m}^+ a_{-m'} a_{m'}. \tag{5.94}$$

Its average value over the state (5.90) is equal to

$$\langle \bar{S}_z\bar{S}\mathfrak{s}|\, H\, |\mathfrak{s}\bar{S}\bar{S}_z\rangle = E\bar{N} - G\{\bar{S}(\bar{S}+1) - \bar{S}_z(\bar{S}_z-1)\}. \tag{5.95}$$

The exact energy (denoted by $\mathcal{E}^T$) of the seniority $\mathfrak{s} = 0$ state is

$$\mathcal{E}_0^T(\mathfrak{s}=0) = E\bar{N} - \tfrac{1}{4}G\bar{N}(2\Omega - \bar{N}+2), \tag{5.96}$$

and the excited state with $\mathfrak{s} \neq 0$ has the energy

$$\mathcal{E}^T(\mathfrak{s}) = E\bar{N} - \tfrac{1}{4}G(\bar{N}-\mathfrak{s})(2\Omega - \bar{N} - \mathfrak{s}+2). \tag{5.96'}$$

The excitation energy is the difference of the above two expressions, i.e.,

$$\mathcal{E}^T(\mathfrak{s}) - \mathcal{E}^T(\mathfrak{s}=0) = \tfrac{1}{4}G\mathfrak{s}(2\Omega - \mathfrak{s}+2), \tag{5.97}$$

and, in the particular case $\mathfrak{s} = 2$,

$$\mathcal{E}^T(\mathfrak{s}=2) - \mathcal{E}_0^T(\mathfrak{s}=0) = G\Omega. \tag{5.97'}$$

What are the corresponding energies in the approximate methods A and B? The ground-state energy is equal to

$$\mathcal{E}_0 = 2\Omega E v^2 - \frac{C^2}{G} - G\Omega v^4. \tag{5.92'}$$

Equations (4.19′) and (4.20′) determining the C- and λ-values have a simple form

$$2/G = \Omega/\varepsilon, \quad \bar{N} = 2\Omega v^2;$$

and, consequently,

$$\varepsilon = \sqrt{C^2+(E-\lambda)^2} = G\Omega/2, \quad v^2 = \bar{N}/2\Omega.$$

From the expression

$$2v^2 = \bar{N}/\Omega = 1-(E-\lambda)/\varepsilon$$

we find that

$$\left.\begin{aligned} E-\lambda &= \tfrac{1}{2}G(\Omega-\bar{N}),\\ C^2 &= \varepsilon^2-(E-\lambda)^2 = \tfrac{1}{4}G^2(2\Omega-\bar{N})\bar{N},\\ \mathcal{E}_0 &= E\bar{N} - \tfrac{1}{4}G\bar{N}(2\Omega-\bar{N}+\bar{N}/\Omega). \end{aligned}\right\} \tag{5.92''}$$

The energies of two-quasiparticle states, calculated according to method A, are equal to

$$\mathcal{E}_0^a(\mathfrak{s}=2) = \mathcal{E}_0 + 2\varepsilon = \mathcal{E}_0 + G\Omega. \tag{5.98}$$

The corresponding expressions are somewhat more complicated in method B. The energy of a two-quasiparticle state is given by

$$\mathcal{E}_0^b(\mathfrak{s}=2) = 2E + 2(\Omega-1)Ev^2 - \frac{C^2}{G} - G(\Omega-2)v^4. \tag{5.99}$$

while (4.46″) and (4.47″) can be written as

$$2/G = (\Omega-2)/\varepsilon, \quad \bar{N} = 2+2(\Omega-2)v^2,$$

and hence

$$\varepsilon = \frac{\Omega-2}{2}G, \quad v^2 = \frac{1}{2}\frac{\bar{N}-2}{\Omega-2}, \quad E-\lambda = \tfrac{1}{2}G(\Omega-\bar{N}),$$

$$C^2 = \tfrac{1}{4}G^2(\bar{N}-2)(2\Omega-\bar{N}-2),$$

$$\mathcal{E}_0^b(\mathfrak{s}=2) = EN - \tfrac{1}{4}G(\bar{N}-2)\left(2\Omega-\bar{N}-2+\frac{\bar{N}-2}{\Omega-2}\right). \tag{5.99'}$$

The excitation energy of a two-quasiparticle state is:

Method A

$$\mathcal{E}_0^a(\mathfrak{s}=2)-\mathcal{E}_0 = G\Omega; \tag{5.100}$$

Method B

$$\mathcal{E}_0^b(\mathfrak{s}=2)-\mathcal{E}_0 = G\Omega - G\left(1-\frac{\bar{N}^2}{4\Omega}+\frac{(\bar{N}-2)^2}{\Omega-2}\right). \tag{5.100'}$$

By comparing the exact expression (5.97′) with the above written approximations, we see that method A gives the same excitation energy as the exact solution, while method B gives somewhat worse results.

Another related problem is the problem of the correlation of motion of four particles. At the present time no effective method of describing such correlations in light nuclei exists. It is very difficult to describe four-particle correlations microscopically.[252, 253] It is therefore quite interesting to see if the Hamiltonian (3.53) cannot automatically cause correlations of four particles (two protons and two neutrons). Such a problem was solved in the framework of the discussed model in ref. 118. The following exact expressions for the correlation energy of four particles and for the pairing energy were obtained:

$$P_4(4n) = \tfrac{1}{2}\{2\mathcal{E}_0^T(4n+2)-\mathcal{E}_0^T(4n)-\mathcal{E}_0^T(4n+4)\} = G, \tag{5.101}$$

$$P_2(4n) = \tfrac{1}{2}\{2\mathcal{E}_0^T(4n+1)-\mathcal{E}_0^T(4n)-\mathcal{E}_0^T(4n+2)\} = \tfrac{1}{2}G\Omega. \tag{5.102}$$

The interaction (3.53) evidently leads to some type of four-particle correlations. However, while the pairing correlations grow with increasing degeneracy Ω, the value of the four-particle correlation energy is constant.

3. The accuracy of calculations with nucleons of one kind interacting via the Hamiltonian (5.85′) was tested in ref. 254. The matrix of the interaction (5.85′) was diagonalized numerically for spherical nuclei with one closed shell. The single-particle scheme contained 4–5 subshells of the last unfilled shell.

Figures 5.1 and 5.2 show the excitation energies of seniority two ($\mathfrak{s}=2$) states. Figure 5.1 demonstrates that method A calculates the two-quasiparticle energies with an accuracy of 10–15%. The modified method of ref. 242 is not better, but the method of ref. 245 is quite successful. Figure 5.2 shows that method A can result in certain cases in a wrong sequence of two-quasiparticle states.

The energies of seniority one ($\mathfrak{s}=1$) states in ^{59}Ni calculated in ref. 254 are collected in Table 5.1. This nucleus has three neutrons above the doubly magic core. The table shows that the results obtained by using method B are closer to the exact values than the results of method A. Nevertheless, even method A gives sufficiently accurate results for the $\mathfrak{s}=1$

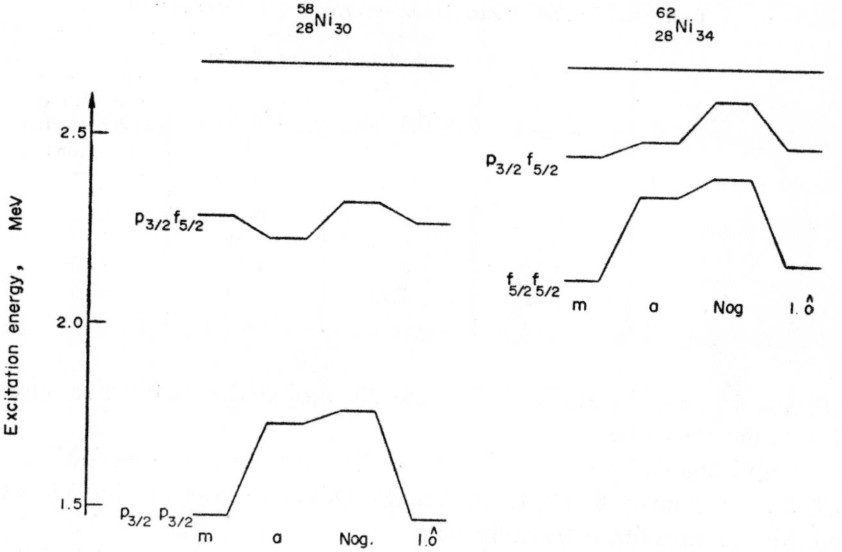

FIG. 5.1. Lowest seniority $\mathfrak{s} = 2$ excited states in the even–even isotopes ^{58}Ni and ^{62}Ni: m denotes the exact results of ref. 254; a denotes the results of method A; (149) Nog denotes the results of Nogami method, (242) and $1.\hat{0}$ the method of ref. 245. The following parameters were used in the calculations: $G = 0.331$ MeV and the single-particle energies $E(p_{3/2}) = 0$, $E(f_{5/2}) = 0.78$ MeV, $E(p_{1/2}) = 1.56$ MeV, $E(g_{9/2}) = 1.52$ MeV.

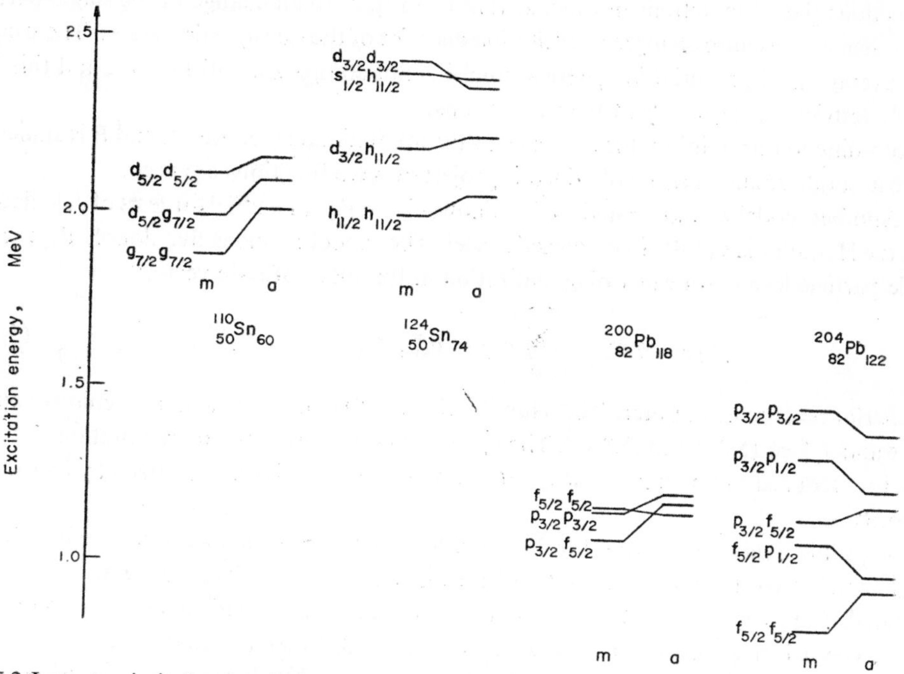

FIG. 5.2. Lowest seniority $\mathfrak{s} = 2$ excited states in ^{110}Sn, ^{124}Sn, ^{200}Pb, and ^{204}Pb calculated exactly (m) in ref. 254 and (a) according to method A in ref. 149. The parameters for tin isotopes are: $G = 0.187$ MeV, $E(d_{5/2}) = 0$, $E(g_{7/2}) = 0.22$ MeV, $E(s_{1/2}) = 1.90$ MeV, $E(d_{3/2}) = 2.20$ MeV, and $E(h_{11/2}) = 2.80$ MeV. The parameters for lead isotopes are: $G = 0.111$ MeV, $E(p_{1/2}) = 0$, $E(f_{5/2}) = -0.57$ MeV, $E(p_{3/2}) = -0.90$ MeV, $E(i_{13/2}) = -1.63$ MeV, and $E(f_{7/2}) = -2.35$ MeV.

TABLE 5.1. *Excitation energies of seniority $\mathfrak{s} = 1$ states in* ^{59}Ni

States	Calculated energy (MeV)			
	Exact	Method A	Method B	Calculation with projected functions
$p_{3/2}$	0	0	0	0
$f_{5/2}$	0.46	0.34	0.41	0.44
$p_{1/2}$	1.26	0.95	1.10	1.15
$g_{9/2}$	4.03	3.76	3.86	4.04

states. It should be noted that calculations with the projected wave functions give energies very close to the exact values.

The overlap integrals between the exact and the projected wave functions are equal to 0.98–0.99 for the seniority $\mathfrak{s} = 0$, 1, and 2 states. This fact shows how very close the exact and approximate wave functions really are.

It should be remembered that the adequacy of calculating the energies of the excited noncollective states in spherical nuclei depends on several factors. One of them is the accuracy of the mathematical solution of the problem with the Hamiltonian (5.85′). Equally important are the effects related to the accuracy of the average field and to neglected residual interactions [i.e., interactions not included in (5.85′)]. A small change in the single-particle level scheme may cause changes in excitation energies of the quasiparticle states. The usually used average field potential in spherical nuclei is not completely satisfactory, and this fact mainly restricts the accuracy of the calculations.

According to our opinion, the accuracy of the approximate methods A and B is sufficient. More accurate results may be obtained if projected wave functions are used.

4. Another model was suggested by Pawlikowski and Rybarska[(255)] for tests of calculations with the Hamiltonian (5.85) in deformed nuclei. The model includes five doubly-degenerate single-particle levels. An exact diagonalization in the space of base vectors

$$|q_1\sigma_1, q_2\sigma_2, \ldots, q_n\sigma_n\rangle = a^+_{q_1\sigma_1}, a^+_{q_2\sigma_2}, \ldots, a^+_{q_n\sigma_n}\Psi_{00} \tag{5.103}$$

was performed on a computer. The Hamiltonian (5.85) commutes with the seniority operators $\mathfrak{s}$ and $\mathfrak{s}_a$ [see (5.86) and (5.86′)]. The system of equations in the representation (5.103) is therefore reduced to uncoupled subsystems. Each subsystem is characterized by its value of seniority $\mathfrak{s}$.

The problem has been solved for 4, 5, and 6 particles distributed among five equidistant single-particle levels; additional shifts of certain single-particle levels were also sometimes included. The energy unit in such a model is the quantity ΔE, the distance of the equidistant levels. The pairing constant was chosen as $G = 0.5$, 0.8, 1.0, and 1.25 of ΔE.

Some results are shown in Fig. 5.3. The figure gives excitation energies of seniority $\mathfrak{s} = 2$ states for six particles. The calculation was performed for $G = \Delta E$. The figure shows the exact energies, the energies calculated by methods A and B and the results obtained from methods A and B combined with the projection procedure (denoted as *ap* and *bp*). The exact

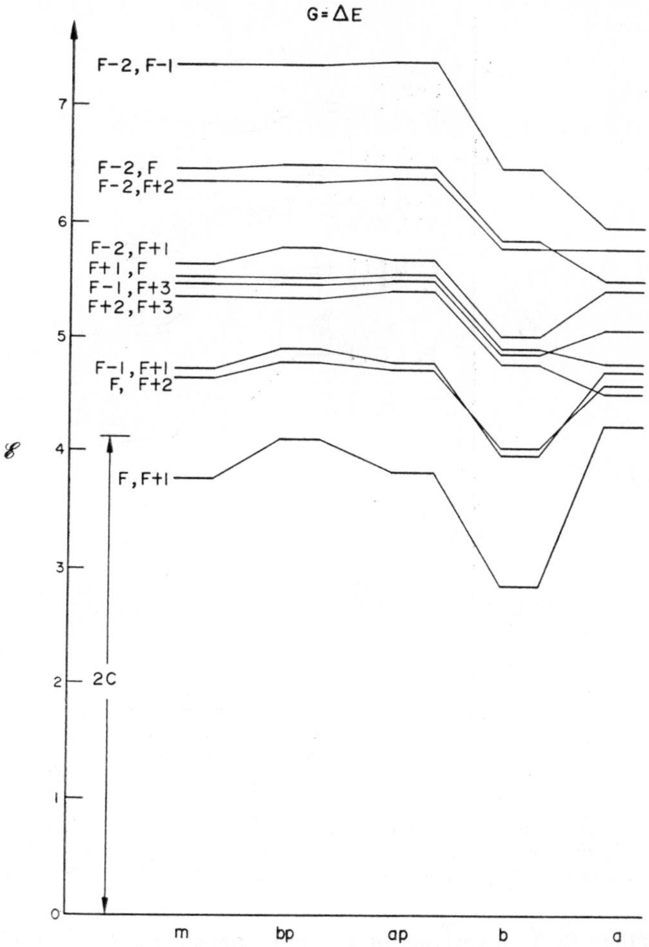

FIG. 5.3. Excitation energies of the seniority $\mathfrak{s} = 2$ states calculated in the model of ref. 255 for $G = \Delta E$ and the equidistant level scheme. The following methods were used: exact (m), methods A and B, and methods A and B combined with projection procedure (denoted ap and bp). Energies in ΔE units. The $2C$ value is shown for comparison.

energy of the $(F, F+1)$ state is less than $2C$; this fact demonstrates the importance of the blocking effect.

The calculations by method B give the correct sequence of the $\mathfrak{s} = 2$ states. The excitation energies are, however, rather far from the exact values. On the other hand, calculations by method A sometimes predict a wrong sequence of the states. As expected, the calculations *ap* and *bp* with projected wave functions are quite close to the exact values, particularly for large G-values.

In actual calculations the G-value is determined from a comparison of the calculated and the experimental pairing energies. It is, therefore, of interest to compare the exact and approximate excitation energies of seniority $\mathfrak{s} = 2$ states with G determined in the just described way. Such a comparison is done in Fig. 5.4 for $\mathfrak{s} = 2$ states, six particles, and equidistant level scheme. The exact calculations were performed for $G = \Delta E$, the calculated

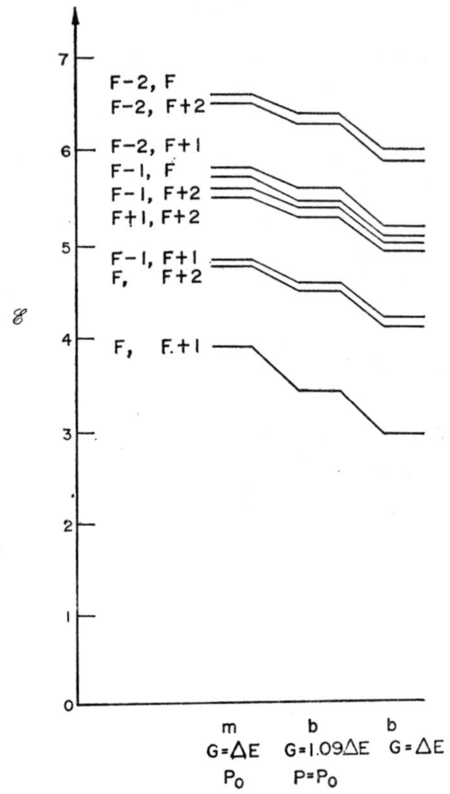

FIG. 5.4. Excitation energies of the seniority $\mathfrak{s} = 2$ states calculated in the model of ref. 255 for the equidistant level scheme. The following methods were used: Exact calculation (m) with $G = \Delta E$ and pairing energy $P = P_0$; method B with $G = \Delta E$; and method B with $G = 1.09\Delta E$ and $P = P_0$. Energies in ΔE units.

pairing energy is then P_0. Two variants of method B are shown: $G = 1.09\Delta E$ (which results in $P = P_0$) and $G = \Delta E$. It is seen that the calculations performed with pairing constant G fixed by the $P = P_0$ condition have about half the error. This means that the calculations based on the experimental pairing energies have improved accuracy.

Let us discuss the systems with an odd number of particles. Figure 5.5 shows the excitation energies of seniority $\mathfrak{s} = 1$ states for five particles and $G = \Delta E$ and $G = 1.25\Delta E$. The calculations were done exactly, and according to methods A and B. Several facts may be deduced from the figure. In the exact calculations the level density increases with increasing G. Calculations by method B are quite successful; they give the correct sequence and correct distances between the levels. On the other hand, method A predicts too large compression of the levels.

To test the accuracy of the wave functions we have compared the pair density distributions in the ground and excited states calculated exactly and by using the approximate methods. Such a comparison is more effective than the calculation of the overlap integrals.

Table 5.2 shows the pair density distribution for the ground state of a system with six particles. The calculations were performed for $G = 1.25\Delta E$ and $G = 0.5\Delta E$. The function v_q^2 describes the pair distribution rather well. Calculations with projected wave functions are not superior to the simple method A.

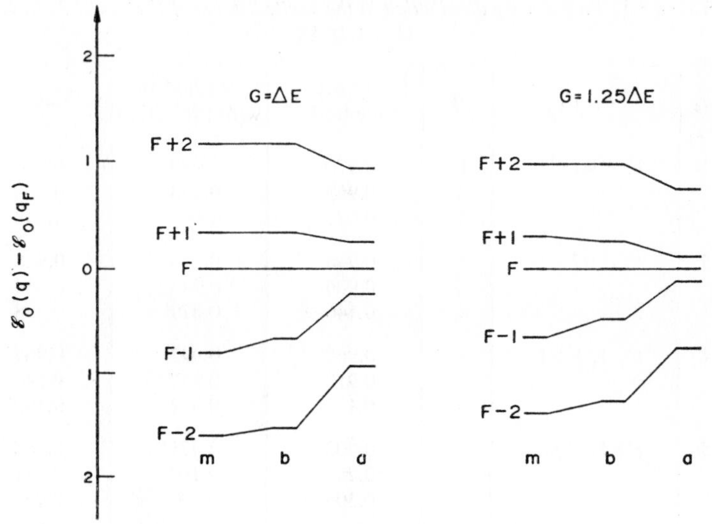

FIG. 5.5. Excitation energies of the seniority $\mathfrak{s} = 1$ states calculated in the model of ref. 255 for five particles. The third level was shifted upwards by $\frac{1}{2}\Delta E$. The following methods were used: exact calculation (*m*), and methods A (*a*) and B (*b*). The particle levels are shown above zero, the hole levels below zero. Energies are in units of ΔE.

TABLE 5.2. *Pair density distribution in the ground state of the model*[255]

q	$F \pm i$	$G = 1.25\Delta E$			$G = 0.5\Delta E$		
		$\bar{N}$ (exact)	$(\bar{N}_q)_p$ (*ap*)	v_q^2 (*a*)	N_q (exact)	$(\bar{N}_q)_p$ (*ap*)	v_q^2 (*a*)
1	$F-2$	0.877	0.889	0.853	0.979	0.995	0.983
2	$F-1$	0.810	0.812	0.767	0.959	0.987	0.958
3	F	0.675	0.661	0.631	0.888	0.932	0.807
4	$F+1$	0.398	0.410	0.455	0.126	0.072	0.208
5	$F+2$	0.240	0.228	0.294	0.048	0.013	0.044

A similar comparison, this time for the seniority $\mathfrak{s} = 2$ states, is done in Table 5.3. It is seen that the calculations performed by the simple method B describe well the pair distribution in the excited states. The pairing correlations in the $(F, F+1)$ state are more inhibited in method B than in the exact calculations. The opposite situation was observed for the other two-quasiparticle states. Calculations with projected wave functions excessively reduce the pairing correlations. The above-mentioned analyses thus show that method B gives results sufficiently close to the exact values.

We should realize that in the model introduced by ref. 255 the calculations with approximate methods are less accurate than the calculations of deformed nuclei. This inaccuracy is caused by a very small number of single-particle levels and relatively large value of pairing constant G. Accuracy tests of the model are described in papers 12, 234, 255, and 256. In ref. 256 the calculations were extended to include eight and nine twice-degenerate levels.

TABLE 5.3. *Pair density distribution in the excited states of the model 255 at* $G = 1.25\Delta E$

q_1, q_2	F_1, F_2	q	Exact method	Method B with projection	Method B
3, 4	$F, F+1$	1	0.970	0.999	0.986
		2	0.953	0.998	0.966
		3	0.077	0.003	0.048
3, 5	$F, F+2$	1	0.948	0.978	0.928
		2	0.906	0.943	0 834
		4	0.146	0.079	0.238
2, 4	$F-1, F+1$	1	0.968	0.992	0.967
		3	0.910	0.957	0.840
		5	0.122	0.051	0.193
4, 5	$F+1, F+2$	1	0.902	0.921	0.868
		2	0.793	0.789	0.711
		3	0.305	0.290	0.421
2, 3	$F-1, F$	1	0.966	0.985	0.965
		4	0.794	0.804	0.685
		5	0.240	0.211	0.350

The quasiparticle excited states in strongly deformed nuclei are calculated with 10–20% accuracy. This accuracy is limited by the crude description of the average field more than by the mathematical approximations (e.g., the blocking effect). If a better accuracy is desirable, it is possible to use G_{eff} of (5.84), as discussed in ref. 237, or to work with the projected wave functions.

CHAPTER 6

NUCLEAR TRANSITIONS IN THE INDEPENDENT QUASIPARTICLE MODEL

§ 1. Effect of Pairing Correlations on α-Decay Probability

1. Alpha-decay is a process in which the parent nucleus emits an α-particle; naturally, the daughter nucleus has two neutrons and two protons less than the parent nucleus. Symbolically,

$$Z^A \rightarrow (Z-2)^{A-4} + {}^4_2\text{He}. \tag{6.1}$$

If the daughter nucleus is left in its ground state, all the available energy Q_f,

$$Q_f = M(A) - M(A-4) - M({}^4\text{He}), \tag{6.2}$$

is distributed between the α-particle and the daughter nucleus. The kinetic energy of each part is inversely proportional to its mass.

The α-particle has zero spin and, therefore, the α-decay selection rules are very simple. The angular momentum of the α-particle is restricted to the interval

$$|I_i - I_f| \leq l \leq I_i + I_f, \tag{6.3}$$

where $I_i(I_f)$ are angular momenta of the initial (final) states. The l is further restricted by the parity conservation. The l-value is even if the parities of the initial and final states are identical, and l is odd if these parities are different.

The transition probability per one second is related to the half-life by the well-known relation

$$w = \frac{\ln 2}{t_{1/2}}. \tag{6.4}$$

The daughter nucleus could be left in an excited state. The total transition probability is then $w = \sum_i w_i$ and the partial half-lives are $t^i_{1/2} = \ln 2/w_i$. The observed half-lives of α-radioactive nuclei vary in a very broad interval. Its limits are not firmly established but depend on the available techniques of the α-particle detection.

From the measured nuclear masses it follows that atomic nuclei with $A > 150$ (and many neutron-deficient nuclei with $A < 150$) are α-radioactive. However, the number of nuclei where α-decay can be used for the nuclear structure studies is quite limited.

An important feature of the α-decay is the exponential dependence of the transition probability on the decay energy. Consequently, α-decay can be used only for studies of the low-lying states. For example, if the structure of the ground and the excited states at 1 MeV are identical, the α-decay to the ground state is 10^6 times more intense than the α-decay to the excited state.

The systematics of α-decay probabilites and energies is explained in many specialized publications, e.g., in refs. 18, 19, and 257–259.

To calculate the α-decay probability one divides the process into two parts: probability of the formation of α-particle from four nucleons in the nucleus and the potential barrier penetration. The α-decay probability is symbolically expressed as

$$w_I = \sum_l \gamma_{Il}^2 P_l, \tag{6.5}$$

where P_l is the barrier penetrability and γ_{Il}^2 is the reduced width. The α-decay theory was developed, among others, in papers 260–266. The characteristic features of the α-decay of deformed nuclei were discussed, e.g., in refs. 267–272.

In order to characterize the nuclear structure of the states connected by the α-decay it is important to find a quantity independent of the energy of the decay (i.e., a quantity similar to the reduced transition probability of the electromagnetic processes or to the ft value of the nuclear β-decay). The hindrance factor F_α is such a quantity. According to its definition, the hindrance factor of α-decay between the ground states of even–even nuclei is equal to unity. The hindrance factor for the α-transition to a certain excited state of an even–even nucleus is equal to the ratio between the reduced width of the ground-state transition and the reduced width leading to the excited state at $l = 0$. This means that

$$F_\alpha = \frac{(\gamma_{00}^2)_g}{(\gamma_{00}^2)_{ex}}. \tag{6.6}$$

The hindrance factor for the α-transitions in odd-A nucleus is defined in a similar way; only the ground-state reduced width is replaced by the average ground-state reduced width of the neighboring even–even nuclei, i.e.,

$$F_\alpha = \frac{\frac{1}{2}\{(\gamma_{00}^2)_g^A + (\gamma_{00}^2)_g^{A+2}\}}{(\gamma_{00}^2)^{A+1}}. \tag{6.6'}$$

Generally, the hindrance factor of any transition characterizes the reduction of the α-decay probability compared to the α-decay between the ground states at equivalent decay energies and at $l = 0$.

In investigations of the nuclear structure the α-decay serves as a tool for the study of the nuclear states involved in the α-transition. In such a case, it is useful to employ the formalism of ref. 273. The study is focused on the matrix elements connecting the wave functions of the parent and daughter nuclei.

In this section we shall study the effect of pairing superfluid correlations on the α-decay probability. This problem has been solved in ref. 274 for spherical nuclei, and in refs. 12, 273, and 275 for deformed nuclei. A number of numerical calculations have been performed in refs. 276–278 and elsewhere.

2. The reduced width of the α-decay of the parent nucleus with wave function $\Psi_i(N, Z)$ into the daughter nucleus with wave function $\Psi_f(N-2, Z-2)$ is determined by

$$\gamma_{II}^2 = \left| \left(\Psi_f^*(N-2, Z-2) \frac{1}{4} \sum_{\substack{r, r', \\ \sigma_1, \sigma_1'}} \sum_{\substack{s, s', \\ \sigma_2, \sigma_2'}} W^{II}_{\sigma_1\sigma_1';\ \sigma_2\sigma_2'}(rr'|ss') a_{r\sigma_1} a_{r'\sigma_1'} a_{s\sigma_2} a_{s'\sigma_2'} \Psi_i(N, Z) \right) \right|^2. \quad (6.7)$$

The operator

$$\frac{1}{4} \sum_{\substack{r, r', \\ \sigma_1, \sigma_1'}} \sum_{\substack{s, s', \\ \sigma_2, \sigma_2'}} W^{II}_{\sigma_1\sigma_1';\ \sigma_2\sigma_2'}(rr'|ss') a_{r\sigma_1} a_{r'\sigma_1'} a_{s\sigma_2} a_{s'\sigma_2'} \quad (6.8)$$

connects the parent and daughter wave functions. The function W describes the probability of formation of an α-particle from two protons in states r and r' and two neutrons in states s and s'. The function W fulfills several symmetry relations, for example,

$$W^{II}_{\sigma_1\sigma_1';\ \sigma_2\sigma_2'}(rr'|ss') = -W^{II}_{\sigma_1'\sigma_1;\ \sigma_2\sigma_2'}(r'r|ss') = -W^{II}_{\sigma_1\sigma_1';\ \sigma_2'\sigma_2}(rr'|s's). \quad (6.8')$$

Let us find the expression for γ_{00} for the α-transitions between the ground states of even-even deformed nuclei. We shall use the independent quasiparticle model. The wave function is represented as

$$\Psi(N, Z) = \Psi_0(N)\Psi_0(Z),$$

where the function Ψ_0 was determined by (4.22). After simple manipulations,

$$\gamma_{00} = \sum_{r, s} W_{+-;\ +-}(rr|ss)\xi_r\xi_s, \quad (6.9)$$

where, for example,

$$\xi_s(N) = u_s(N-2)\, v_s(N) \prod_{s' \neq s} (u_{s'}(N-2)\, u_{s'}(N) + v_s(N-2)\, v_{s'}(N)). \quad (6.10)$$

The $u_s(N-2)$ belongs to the daughter nucleus with $N-2$ neutrons, and $v_s(N)$ belongs to the parent nucleus with N-neutrons. We may often use the approximate expressions for ξ_s and ξ_r:

$$\xi_s(N) = u_s(N-2)\, v_s(N), \quad (6.10')$$

because each factor of the product in (6.10) has a value close to unity.

On the other hand, if γ_{00} is calculated by means of the independent particle model [i.e., with Ψ_0 in the form (4.14)],

$$\gamma_{00} = W_{+-;\ +-}(r_F r_F|s_F s_F), \quad (6.11)$$

where r_F and s_F denote the quantum numbers of the Fermi levels for protons and neutrons. This means that, if the pairing correlations are not present, the α-particle is formed from the two protons and two neutrons in the last full levels of the average field. The probability of forming an α-particle in the nucleus is proportional to the overlap integral of the corresponding wave functions. Hence, the reduced widths γ_{00} in (6.11) should considerably differ for various isotopes because the quantum characteristics of the Fermi level vary. This conclusion is, however, contradicted by the experimental data.

The pairing superfluid correlations change the above-stated conclusion; the α-particle is formed from nucleon pairs occupying many single-particle levels. Figure 6.1 shows the weight factors ξ_r for the proton transitions $Z = 100 \rightarrow Z = 98$ and $Z = 98 \rightarrow Z = 96$ in even–even nuclei. It is evident that even the states far from the Fermi level contribute sub-

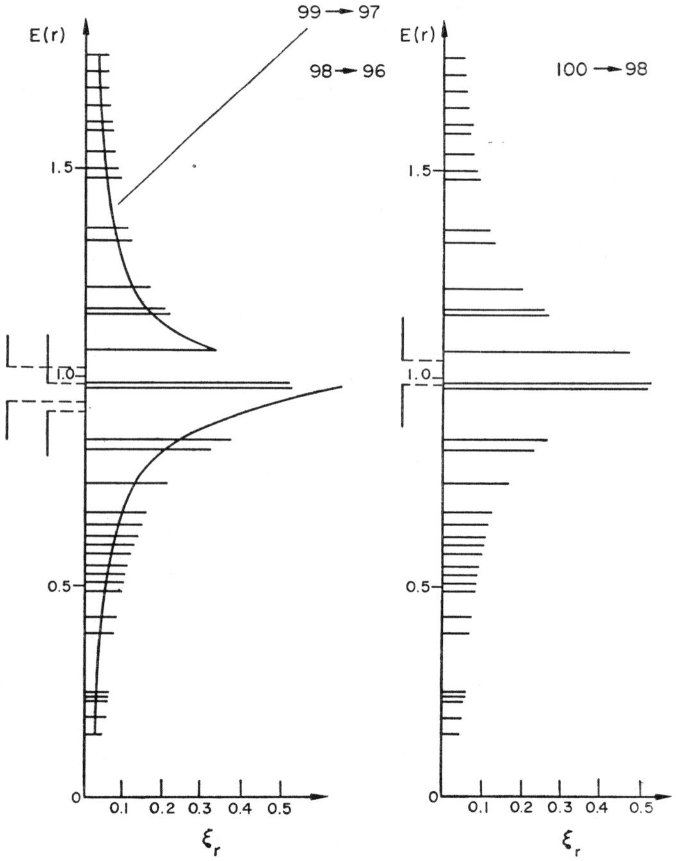

FIG. 6.1. The values of $\xi_r = v_r(Z)\,u_r(Z-2)$ for the α-decay of even–even nuclei with $Z = 100$ and $Z = 98$ and for the favored α-decay of the odd-Z nucleus $Z = 99$. The single-particle energies are plotted on the ordinate in units of $\mathring{\omega}_0$. The length of horizontal lines shows the ξ_r-values for each orbit. The chemical potentials λ for the parent and daughter nuclei are shown by dashed lines to the left (the shorter lines refer to $Z = 100 \rightarrow 98$ and $98 \rightarrow 96$, the longer lines to $Z = 99 \rightarrow 97$). The vertical lines adjacent to the dashed lines show the C_p values (for parent nuclei pointing upward, for daughter nuclei pointing downwards). The full curve on the left diagram shows ξ_r values for the decay $99 \rightarrow 97$.

stantially. The comparison of the ξ_r values for the two transitions in Fig. 6.1 shows that ξ_r are changed negligibly. In other words, ξ_r is a slowly varying function of the chemical potential (i.e., the Fermi level position). Thus, the reduced widths of the ground state to ground state α-decays for different nuclei vary slowly.

The α-decay probability is enhanced by pairing correlations. To estimate this enhancement, we shall use the following approximation: we will assume that the diagonal terms

$W_{+-;\,+-}(rr\,|\,ss)$ are independent of r and s, i.e.,

$$W_{+-;\,+-}(rr\,|\,ss) = W, \tag{6.12}$$

$$W_{+-;\,\sigma_2\sigma_2'}(rr\,|\,ss') = W_{\sigma_2\sigma_2'}(\,|\,ss'). \tag{6.12'}$$

Such an approximation is rather crude, as shown in ref. 275; the different diagonal terms in W can vary by factor of 20 to 50. Nevertheless, it is useful for a rough estimate of the role of pairing correlations.

The reduced width γ_{00} is expressed in approximation (6.12) as

$$\gamma_{00} = W \sum_r \xi_r \sum_s \xi_s. \tag{6.13}$$

The expression $\sum \xi_s$ can be approximated by

$$\sum_s \xi_s \approx \sum_s u_s v_s = \frac{C_n}{G_N}$$

The quantities $\sum_r \xi_r \sum_s \xi_s$ are naturally equal to unity in the independent particle model. If the model of independent quasiparticles is used, and the summation in the Woods–Saxon potential includes all single-particle levels from the bottom of the well to zero binding energy, the quantity $(\gamma_{00}/W)^2$ has values 2000–3000. If the summation is extended to include the single-particle states in quasicontinuum, the $(\gamma_{00}/W)^2$ value is increased by 5–10%.

Thus, the pairing correlations increase by three orders of magnitude the α-decay probability for transitions between ground states of even–even deformed nuclei. When this increase is taken into account, the discrepancy in nuclear radii obtained from α-decays and from nuclear reactions is somewhat reduced.

Similar enhancement of the α-decay probability in spherical nuclei is less pronounced. For example, the α-decay probability of ^{212}Po is increased by pairing correlations only 10 times. Such a small enhancement, found in ref. 274, is undoubtedly related to the restricted set of single-particle levels used there.

3. Let us study the hindrance factors F_α for α-decay leading to the two-quasiparticle states of even–even nuclei and for α-transitions between the one-quasiparticle states in odd-A deformed nuclei. The reduced width for α-decay from the ground state $I^\pi K = 0^+0$ of an even–even deformed nucleus to a two-quasiparticle neutron (s_1, s_2) state (with $s_1 \neq s_2$) is equal to

$$\gamma_{II}(s_1, s_2) = \sum_r W^{II}_{+-,\,\sigma_2\sigma_2'}(rr\,|\,s_1 s_2)\, \xi_r v_{s_1}(N)\, v_{s_2}(N). \tag{6.14}$$

Thus, the α-particle is formed by many proton pairs, as shown in Fig. 6.1, but only from neutrons in the states s_1 and s_2. The α-decay rate is proportional to $v_{s_1}^2$ and $v_{s_2}^2$ and, therefore, decreases if s_1 and s_2 are above the Fermi level.

The hindrance factor is expressed by

$$F_\alpha(s_1, s_2) = \frac{\left(\sum_{r,s} W_{+-;\,+-}(rr\,|\,ss)\xi_r\xi_s\right)^2}{\left(\sum_r W_{+-;\,\sigma_2\sigma_2'}(rr\,|\,s_1 s_2)\,\xi_r v_{s_1}(N)\, v_{s_2}(N)\right)^2}. \tag{6.15}$$

The effect of pairing correlations may be separated in the approximation (6.12), (6.12′). We then obtain

$$F_\alpha(s_1, s_2) = \left(\frac{W}{W_{\sigma_2\sigma_2'}(\quad|s_1s_2)^2}\right)^2 \frac{\left(\sum_s \xi_s\right)^2}{v_{s_1}^2 v_{s_2}^2}. \tag{6.15′}$$

The coefficient $\left(\sum_s \xi_s\right)^2 v_{s_1}^{-2} v_{s_2}^{-2}$ is equal to 200–500 for the α-decay to the two-quasiparticle states with energy below 2 MeV. Thus, the pairing correlations cause a reduction of two orders of magnitude in the α-decay probability for transitions to two-quasiparticle states.

The β- and γ-transitions are classified according to their multipolarity. The centrifugal barrier and its related angular momenta are less important in the case of α-decay. The α-transitions are classified differently. The classification is based on the increased rate of α-decay in case when the α-particle is formed by correlated neutron and proton pairs. Such α-decays are called "favored." The α-transitions between the ground states of even–even nuclei belong to this group. The α-decay rate is considerably reduced if the α-particle is formed from unpaired nucleons. Such an α-decay is called "unfavored." Hence, the classification is based on the collective properties related to the pairing superfluid correlations.

Let us explain the α-decay classification on the example of an odd-A deformed nucleus. The α-decay is favored if the state of the odd unpaired nucleon is not changed by the transition. The α-particle is then formed from many proton and neutron pairs. Figure 6.2 shows the α-decay of ^{249}Cf to levels of ^{245}Cm. The favored transition goes to the 388 keV level. The state of the 734↑ quasiparticle is unchanged in this case. Experimentally, 84% α-particles excite the 388 keV level; only 1.9% decays go to the ground state. The α-decays to the 624↓ and 622↑ states are unfavored. The hindrance factor F_α is equal to 2.0 for the favored transition, but the same quantity equals 9000 for the 624↓ ground state and 270 for the 622↑ state. Similar situation was observed in other odd-A nuclei.

Let us estimate the hindrance factor $F_\alpha(q)$ for favored α-transitions. We shall calculate the reduced width of a favored α-decay from the nucleus with $N+1$ neutrons and single quasiparticle on the s_2 level. Obviously, the wave functions (4.33′) should be used. The result is

$$\gamma_{II}(s_2) = \sum_r \sum_{s \neq s_2} W^{II}_{+-;\,+-}(rr|ss)\, \xi_r u_s(N-1, s_2)\, v_s(N+1, s_2), \tag{6.16}$$

where $v_s(N+1, s_2)$ means that the corresponding quantity is calculated for the parent nucleus. The blocking effect causes a reduction of $\gamma_{II}(s_2)$ when compared to γ_{00} [(eq. (6.9)]. The $\xi_r = u_r(Z-1, r_2)\, v_r(Z+1, r_2)$ values for the α-transition $Z = 99 \to 97$ are shown in Fig. 6.1 by full curve. The reduction is caused by the decrease in the ξ_r values (because $C_p(r_2)$ is smaller than C_p) and by the absence of the level occupied by the quasiparticle from the summation.

The hindrance factor F_α of the favored α-decays can be calculated explicitly in approximation (6.12); as a result

$$F_\alpha(s_2) = \frac{\frac{1}{2}\{\gamma_{00}^2(N)+\gamma_{00}^2(N+2)\}}{\gamma_{00}^2(N+1, s_2)} = \frac{1}{2}\,\frac{\left(\sum_s \xi_s(N)\right)^2 + \left(\sum_s \xi_s(N+2)\right)^2}{\left(\sum_{s\neq s_2} \xi_s(N+1, s_2)\right)^2}. \tag{6.17}$$

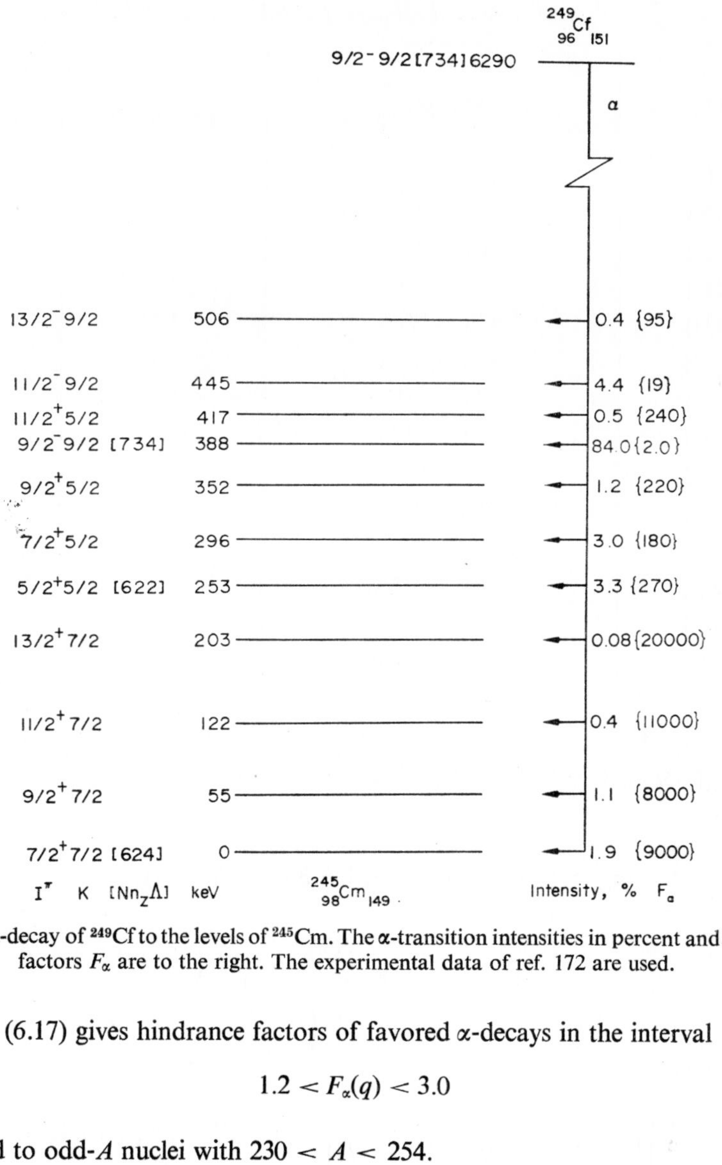

FIG. 6.2. Alpha-decay of ^{249}Cf to the levels of ^{245}Cm. The α-transition intensities in percent and the hindrance factors F_α are to the right. The experimental data of ref. 172 are used.

The formula (6.17) gives hindrance factors of favored α-decays in the interval

$$1.2 < F_\alpha(q) < 3.0 \tag{6.17'}$$

when applied to odd-A nuclei with $230 < A < 254$.

The calculated and experimental $F_\alpha(q)$ values are collected in Table 6.1. The experimental data from refs. 172, 188, 189, and 279 were used. The experimental hindrance factors are really within the interval (6.17′), and their agreement with the calculated estimates is satisfactory.

Let us turn our attention to the unfavored α-decays of odd-A deformed nuclei. When the nucleus with $N+1$ neutrons and the quasiparticle in the state s_2 decays into the $N-1$ nucleus with quasiparticle in the state s_1 ($s_1 \neq s_2$), the reduced width is determined by the formula

$$\gamma_{II}(s_2 \to s_1) = \sum_r W_{+-;\,\sigma_2,\,-\sigma_2'}(rr\,|\,s_1 s_2)\,\xi_r u_{s_2}(N-1, s_1)\,v_{s_1}(N+1, s_2), \tag{6.18}$$

TABLE 6.1. *Favored α-decays of odd-A deformed nuclei*

State	α-Decay	Hindrance factor $F_\alpha(q)$	
		Experiment	Calculations
$\frac{7}{2}^+$ [613]	^{255}Fm → ^{251}Cf	1.2	1.5
$\frac{1}{2}^+$ [620]	^{253}Fm → ^{249}Cf	3.3	1.8
$\frac{1}{2}^+$ [620]	^{251}Cf → ^{247}Cm	1.8	1.7
$\frac{9}{2}^-$ [734]	^{249}Cf → ^{245}Cm	2.0	2.0
$\frac{7}{2}^+$ [624]	^{245}Cm → ^{241}Pu	1.3	1.7
$\frac{5}{2}^+$ [622]	^{243}Cm → ^{239}Pu	1.6	1.7
$\frac{5}{2}^+$ [622]	^{241}Pu → ^{237}U	1.3	1.7
$\frac{1}{2}^+$ [631]	^{241}Cm → ^{237}Pu	2.7	2.0
$\frac{1}{2}^+$ [631]	^{239}Pu → ^{235}U	3.0	2.0
$\frac{7}{2}^+$ [633]	^{253}Es → ^{249}Bk	1.8	1.7
$\frac{7}{2}^+$ [633]	^{249}Bk → ^{245}Am	1.3	1.7
$\frac{3}{2}^-$ [521]	^{245}Bk → ^{241}Am	2.4	1.7
$\frac{5}{2}^-$ [523]	^{243}Am → ^{239}Np	1.3	1.5
$\frac{5}{2}^-$ [523]	^{241}Am → ^{237}Np	1.3	1.5

where $u_{s_2}(N-1, s_1)$ corresponds to the daughter-nucleus. The α-particle in unfavored α-decays of an odd-N nucleus is formed from many proton pairs and from neutrons in states s_1 and s_2.

The hindrance factor F_α of unfavored α-transitions in odd-A nuclei can be calculated in the approximation (6.12). The effect of pairing correlations is then separated. As a result we get

$$F_\alpha = \left(\frac{W}{W_{\sigma_2, -\sigma_2'}(\ |s_1 s_2)}\right)^2 \frac{1}{2} \frac{\left(\sum_s \xi_s(N)\right)^2 + \left(\sum_s \xi_s(N+2)\right)^2}{u_{s_2}^2(N-1, s_1)\, v_{s_1}^2(N+1, s_2)}. \tag{6.19}$$

The pairing factors

$$\frac{1}{2}\left\{\left(\sum_s \xi_s(N)\right)^2 + \left(\sum_s \xi_s(N+2)\right)^2\right\} u_{s_2}^{-2}(N-1, s_1)\, v_{s_1}^{-2}(N+1, s_2)$$

were calculated for odd-A deformed nuclei. The following values (or interval limits) were obtained: 50–130 for the decays to the ground and hole states, 200–800 for the decays to the $F+2$ states, and $> 10^3$ for the decays to the $F+3$ and higher particle states. We see that the unfavored α-transitions to the particle excited states are strongly inhibited in comparison with the transitions to the hole states.

Pairing correlations affect the F_α value strongly. It is, therefore, advisable to systemize the quantity $\left(\dfrac{W}{W_{\sigma_2, -\sigma_2'}(\ |s_1 s_2)}\right)^2$, which is directly related to the structure of the states s_1 and s_2, instead of systemizing the F_α values (as done in ref. 280).

4. We shall now discuss the α-decay of odd–odd nuclei. The α-decays of odd–odd nuclei in the independent quasiparticle model may be divided into three groups: favored–favored (the transitions are favored in both proton and neutron systems), favored–unfavored (the transition in one system is favored and in the other system unfavored), and unfavored-unfavored transitions (the transition is unfavored in both systems).

Let us give here the expressions for reduced widths of α-decays of $(Z+1, N+1)$ odd–odd deformed nuclei:

Favored–favored transition; quasiparticles in states (r_2, s_2):

$$\gamma_{II}(r_2, s_2) = \sum_{r\neq r_2} \sum_{s\neq s_2} W^{II}_{+-;\,+-}(rr\,|\,ss)\, u_r(Z-1, r_2)\, v_r(Z+1, r_2)\, u_s(N-1, s_2)\, v_s(N+1, s_2). \tag{6.20}$$

Unfavored–favored transition; quasiparticles are in states (r_2, s_2) and (r_1, s_2) where $r_1 \neq r_2$:

$$\gamma_{II}(r_2, s_2 \to r_1, s_2) = \sum_{s\neq s_2} W_{\sigma,\,-\sigma';\,+-}(r_1r_2\,|\,ss)\, u_{r_2}(Z-1, r_1)\times$$
$$\times v_{r_1}(Z+1, r_2)\, u_s(N-1, s_2)\, v_s(N+1, s_2). \tag{6.20'}$$

Unfavored–unfavored transition; quasiparticles are in states (r_2, s_2) and (r_1, s_1), where $r_1 \neq r_2$, $s_1 \neq s_2$:

$$\gamma_{II}(r_2, s_2 \to r_1, s_1) = W_{\sigma_1,\,-\sigma_1';\,\sigma_2,\,-\sigma_2'}(r_1r_2\,|\,s_1s_2)\, u_{r_2}(Z-1, r_1)\times$$
$$\times v_{r_1}(Z+1, r_2)\, u_{s_2}(N-1, s_1)\, v_{s_1}(N+1, s_2). \tag{6.20''}$$

The neutron and proton systems are independent in the independent particle model. Hence we may try to use the F_α values of odd-A nuclei to describe the odd–odd nuclei. To do that, we shall assume that $W_{\sigma\sigma;\,\sigma_2\sigma_2'}(rr'\,|\,ss')$ is separable, i.e.,

$$W_{\sigma\sigma';\,\sigma_2\sigma_2'}(rr'\,|\,ss') = W^p_{\sigma\sigma'}(rr')\, W^n_{\sigma_2\sigma_2'}(ss'). \tag{6.21}$$

The hindrance factors for the α-decays of odd-odd $(Z+1,\ N+1)$ deformed nuclei can be expressed in the approximations (6.12), (6.12)′, and (6.21) by the following relations:

Favored–favored transitions; quasiparticles are in states (r_2, s_2):

$$F_\alpha = \frac{1}{4}\, \frac{\left(\sum_r \xi_r(Z)\right)^2 + \left(\sum_r \xi_r(Z+2)\right)^2}{\left(\sum_{r\neq r_2} \xi_r(Z+1, r_2)\right)^2}\; \frac{\left(\sum_s \xi_s(N)\right)^2 + \left(\sum_s \xi_s(N+2)\right)^2}{\left(\sum_{s\neq s_2} \xi_s(N+1, s_2)\right)^2}. \tag{6.22}$$

Unfavored–favored transition; quasiparticles are in states (r_2, s_2) and (r_1, s_2) where $r_1 \neq r_2$;

$$F_\alpha = \frac{1}{4}\left(\frac{W^p}{W^p_{\sigma_1,\,-\sigma_1'}(r_1r_2)}\right)^2 \frac{\left(\sum_r \xi_r(Z)\right)^2 + \left(\sum_r \xi_r(Z+2)\right)^2}{u^2_{r_2}(Z-1, r_1)\, v^2_{r_1}(Z+1, r_2)}\; \frac{\left(\sum_s \xi_s(N)\right)^2 + \left(\sum_s \xi_s(N+2)\right)^2}{\left(\sum_{s\neq s_2} \xi_s(N+1, s_2)\right)^2}. \tag{6.22'}$$

Unfavored–unfavored transition; quasiparticles are in states (r_2, s_2) and (r_1, s_1), where $r_1 \neq r_2$, $s_1 \neq s_2$:

$$F_\alpha = \frac{1}{4}\left(\frac{W^p}{W^p_{\sigma,\,-\sigma'}(r_1 r_2)}\right)^2 \frac{\left(\sum_r \xi_r(Z)\right)^2 + \left(\sum_r \xi_r(Z+2)\right)^2}{u^2_{r_2}(Z-1, r_1)\, v^2_{r_1}(Z+1, r_2)} \times$$
$$\times\left(\frac{W^n}{W^n_{\sigma_2,\,-\sigma_2'}(s_1 s_2)}\right)^2 \frac{\left(\sum_s \xi_s(N)\right)^2 + \left(\sum_s \xi_s(N+2)\right)^2}{u^2_{s_2}(N-1, s_1)\, v^2_{s_1}(N+1, s_2)}. \tag{6.22''}$$

Table 6.2 shows the experimentally determined hindrance factors F_α (see ref. 281) for α-decays of two odd–odd deformed nuclei; the calculated F_α values are also shown. The experimental F_α values of unfavored α-transitions of odd-A nuclei were used in the calculations. Equations (6.22), (6.22′), and (6.22″) were employed. The agreement between the experimentally determined hindrance factors F_α and the theoretical estimates is evidently satisfactory.

TABLE 6.2. *Hindrance factors F_α of the α-decays of odd–odd deformed nuclei*

Parent nucleus			Daughter nucleus			F_α		
Nucleus	K^π	Configuration	Nucleus	K	Configuration	Energy (keV)	Experimental	Calculated
^{254}Es	7^+	$p633\uparrow,\ n613\uparrow$	^{250}Bk	2^-	$p521\uparrow,\ n620\uparrow$	0	$> 1.3\times10^5$	7×10^5
				4^+	$p633\uparrow,\ n620\uparrow$	36	1300	4×10^3
				7^+	$p633\uparrow,\ n613\uparrow$	86	2.8	2.6
				5^-	$p521\uparrow,\ n613\uparrow$	99	140	180
^{242}Am	5^-	$p523\downarrow,\ n622\uparrow$	^{238}Np	2^+	$p642\uparrow,\ n631\downarrow$	0	$> 6\times10^4$	3×10^6
				3^-	$p523\downarrow,\ n631\downarrow$	136	2200	4×10^3
				5^-	$p523\downarrow,\ n622\uparrow$	342	1.6	2.6
				6^+	$p523\downarrow,\ n743\uparrow$	462	200	220

Thus we have seen that pairing superfluid correlations are important for the correct understanding of both the absolute α-decay probabilities and the hindrance factors.

§ 2. Beta-decay

1. Nuclear β-decay is realized as the β^--decay

$$n \to p+e^-+\tilde{\nu}, \tag{6.23}$$

or as the β^+-decay

$$p \to n+e^++\nu, \tag{6.23'}$$

or, finally, as the electron capture

$$p+e^- \to n+\nu. \tag{6.23''}$$

The β-decay theory was originally formulated by Fermi. Further important developments in the theory are related to the discovery of the parity nonconservation in weak interactions and to the formulation of the universal weak interaction theory. The β-decay theory and the basic related experimental facts were described repeatedly, for example, in refs. 18, 19, or 282–287.

Nuclear β-decay is interesting for two reasons. Firstly, it can be used in studies of the weak interactions and for determination of the corresponding interaction parameters. Secondly, it represents an important tool for investigations of the structure of spherical and deformed nuclei. This section will present the main equations of the β-decay theory. The effect of the pairing correlations on β-decay probabilities will be also studied and the main characteristic features of the β-decay of spherical and deformed nuclei will be discussed.

The β-decay is a process of first order in the weak interaction constant. The weak interaction Hamiltonian has its usual form

$$H_\beta = \sum_{i=1}^{5} \{(\bar{\psi}_p O_i \psi_n)(\bar{\psi}_e O_i'(G_i + G_i'\gamma_5)\psi_\nu) + \text{c.c.}\}, \tag{6.24}$$

where $\bar{\psi}_p = \psi_p^* \gamma_4$, ψ_n, ψ_e, ψ_ν are the operator wave functions of a proton, neutron, electron, and neutrino, respectively. The operators O_i, O_i' are transformed as scalars, vectors, second-rank tensors, axial vectors, and pseudoscalars under the Lorentz transformation. The Hamiltonian (6.24) contains thus 20 real constants. However, the number of constants is reduced to five if the invariance of H_β under time reversal is assumed and the two-component theory of neutrino is used. The number of constants is further reduced when the theory of universal weak interactions is applied. According to this theory, which is supported by an extensive experimental evidence, the β-decay Hamiltonian contains only vector and axial vector terms. The Hamiltonian then has the familiar form

$$H_\beta = \frac{G_V}{\sqrt{2}} \sum_{\mu=1}^{4} \{(\bar{\psi}_p \gamma_\mu (1+\lambda\gamma_5)\psi_n)(\bar{\psi}_e \gamma_\mu (1+\gamma_5)\psi_\nu) + \text{c.c.}\}, \tag{6.25}$$

where

$$\lambda = G_A/G_V, \quad \gamma_4 = \beta, \quad \gamma_m = -i\beta\alpha_m, \quad \alpha_m = \begin{pmatrix} 0 & \sigma_m \\ \sigma_m & 0 \end{pmatrix}, \quad m = 1, 2, 3.$$

In the nuclear theory the interaction Hamiltonian H_β is often rewritten as the sum of the Fermi and Gamow–Teller terms, i.e.,

$$H_\beta = \frac{G_V}{\sqrt{2}} \left\{ \sum_i \tau^{+(i)} (\gamma_\mu)_i \, (\bar{\psi}_e(\mathbf{r}) \, \gamma_\mu (1+\gamma_5) \, \psi_\nu(\mathbf{r})) + \text{c.c.} \right\}$$
$$+ \frac{G_A}{\sqrt{2}} \left\{ \sum_i \tau^{+(i)} (\gamma_\mu \gamma_5)_i \, (\bar{\psi}_e(\mathbf{r}) \, \gamma_\mu (1+\gamma_5) \, \psi_\nu(\mathbf{r})) + \text{c.c.} \right\}. \tag{6.26}$$

The quantity $\tau^{+(i)}$ denotes the isospin vector component which transforms the ith neutron into a proton. The G_V and G_A are the vector and axial vector interaction constants. The matrix element describing the β-transitions between the initial and final nuclear states is equal to

$$M_\beta = \langle f | H_\beta | i \rangle = M_V + M_A, \tag{6.27}$$

where M_V and M_A denote the Fermi and Gamow–Teller matrix elements.

Matrix elements of the Dirac operators γ_4 and $\gamma_m\gamma_5$ $(m = 1, 2, 3)$ are of the order of unity. On the other hand, the matrix elements of the operators γ_m and $\gamma_4\gamma_5$ are of the order v/c (v is the nucleon velocity at the Fermi level). Consequently, the expression (6.27) contains both terms independent of v/c and terms proportional to v/c. The matrix elements M_V and M_A are usually expanded in a power series of $|\boldsymbol{p}_e - \boldsymbol{p}_\nu| r_i$. The difference $\boldsymbol{p}_e - \boldsymbol{p}_\nu$ is the difference of electron and neutrino momenta and r_i is approximately equal to the nuclear radius. Both parameters v/c and $|\boldsymbol{p}_e - \boldsymbol{p}_\nu| R$ are of the order of $\frac{1}{10}$. It is hence sufficient to include only the largest terms compatible with the selection rules. Such an expansion in powers of parameters v/c and $|\boldsymbol{p}_e - \boldsymbol{p}_\nu| R$ then leads to the standard classification of the β-transitions. We shall denote the allowed, first forbidden, second forbidden, etc., β-transitions by a, 1, 2, etc.

The selection rules are expressed in terms of the spin difference $\Delta I = |I_i - I_f|$ and of the parity change $\pi_i\pi_f$ (this product equals $+1$ if the parities are identical and -1 if the parities are different). The parity nonconservation of the weak interaction and the parity selection rules in classification of nuclear β-transitions are not in contradiction. When using the parity selection rules we do not consider the parity of the whole system, only the parity of the initial and final nuclear states is taken into account. The nuclear structure is mostly determined by the strong and electromagnetic interactions and, consequently, each nuclear state is characterized by its parity. The admixtures of the parity nonconserving components in nuclear wave functions are extremely small, of order 10^{-5} to 10^{-7} (see ref. 288). Such small effects can be neglected in the nuclear structure studies.

Table 6.3 contains a collection of the selection rules in ΔI and parity change $\pi_i\pi_f$ for the

TABLE 6.3. *Selection rules of the allowed and first-forbidden β-transitions*

Classification	Matrix element	$\pi\,\pi_f$	Selection rules
a	$\langle 1\rangle$	1	$\Delta I = 0$
	$\langle\boldsymbol{\sigma}\rangle$	1	$\Delta I = 0, \pm 1$ except $0 \to 0$
	$\langle\boldsymbol{r}\rangle$	-1	$\Delta I = 0, \pm 1$ except $0 \to 0$
	$\langle\boldsymbol{\sigma}\gamma_5\rangle$	-1	$\Delta I = 0, \pm 1$ except $0 \to 0$
	$\langle\gamma_5\rangle$	-1	$\Delta I = 0$
1	$\langle\boldsymbol{\sigma r}\rangle$	-1	$\Delta I = 0$
	$\langle\boldsymbol{\sigma}\times\boldsymbol{r}\rangle$	-1	$\Delta I = 0, \pm 1$ except $0 \to 0$; $\frac{1}{2} \to \frac{1}{2}$
	B_{ij}	-1	$\Delta I = 0, \pm 1$ except $0 \to 0$; $\frac{1}{2} \to \frac{1}{2}$; $0 \to 1$
1*	B_{ij}	-1	$\Delta I = \pm 2$, except $0 \to 0$; $\frac{1}{2} \to \frac{1}{2}$; $0 \to 1$

allowed and first-forbidden transitions. The table also shows symbolically the corresponding nuclear matrix elements. The symbol B_{ij} denotes $\int (d\boldsymbol{r})\,\{\sigma_i x_j + \sigma_j x_i - \frac{2}{3}\delta_{ij}(\boldsymbol{\sigma}\cdot\boldsymbol{r})\}$. The first forbidden transitions with $\Delta I = \pm 2$ and $\pi_i\pi_f = -1$ are called "unique", and they are denoted by 1^*.

The isobaric spin selection rules are also very important. The β-decay Hamiltonian (6.26) contains the matrices $\tau^{+(i)}$ and $\tau^{-(i)}$ with the selection rules

$$\Delta T = 0, \quad \pm 1; \qquad \Delta T_z = \pm 1. \tag{6.28}$$

The matrix element of the allowed Fermi type transition has the form

$$\langle f|\sum_i (\tau_x^{(i)} \pm i\tau_y^{(i)})\,|i\rangle = \langle f|\,T_\pm\,|i\rangle,$$

and its square equals

$$|\langle f|\,T_\pm\,|i\rangle|^2 \sim T(T+1) - T_z^i T_z^f .$$

The operator $T_\pm$ does not change the total isobaric spin. Hence, the selection rules in this case are

$$\Delta T = 0, \quad \Delta T_z = \pm 1. \tag{6.29}$$

2. If the only measured quantity is the electron (or positron) energy, the nuclear β-decay probability is equal to

$$W_{if}^{(n)}(E)\,dE = \frac{1}{2\pi^3}\,S_{if}^{(n)}(E)\,F(\mp Z, E)\,pE(E_0 - E)^2\,\eta\,dE. \tag{6.30}$$

Here, the electron (positron) energy is between the E and $E+dE$ values; the summations over all electron and neutrino polarization have been already performed. The meaning of individual symbols in (6.30) is the following: n is the degree of forbiddenness, E_0 is the maximum electron energy, and the factor η is related to the summation over all possible final spin states and averaging over all possible initial spin states. The Fermi function $F(\mp Z, E)$ includes the corrections for the Coulomb interaction between the electron (positron) and the nucleus. The function F is tabulated in refs. 289 and 290. The nuclear wave functions enter through

$$(S_{if}^{(n)}(E))^{1/2} = (\Psi^*(f)\,H_\beta\Psi(i)). \tag{6.31}$$

Let us rewrite the Hamiltonian (6.26) in the second quantization formalism, i.e., express it through the nucleon creation and annihilation operators. For the deformed nuclei,

$$H_\beta = \sum_{s,\,\sigma}\sum_{r,\,\sigma'}\{\langle r\sigma'|\,\Gamma_\beta\,|s\sigma\rangle a_{r\sigma'}^{+}a_{s\sigma} + \langle s\sigma|\,\Gamma_\beta\,|r\sigma'\rangle a_{s\sigma}^{+}a_{r\sigma'}\}; \tag{6.32}$$

and for the spherical nuclei,

$$H_\beta^\lambda = \frac{1}{\sqrt{2\lambda+1}}\sum_{j^p,\,m^p}\sum_{j^n,\,m^n}\{(-1)^{j^n-m^n}(j^n m^n j^p - m^p\,|\,\lambda - \mu)\langle j^p|\,\Gamma_\beta\,|j^n\rangle a_{j^p m^p}^{+}a_{j^n m^n}$$
$$+(-1)^{j^p-m^p}(j^p m^p j^n - m^n\,|\,\lambda - \mu)\langle j^n|\,\Gamma_\beta\,|j^p\rangle a_{j^n m^n}^{+}a_{j^p m^p}\} \tag{6.32'}$$

The symbol λ denotes the multipolarity of the β-transition (for example, $\lambda = 0$ for the Fermi transitions and $\lambda=1$ for the Gamow–Teller transitions). The $\langle r\sigma'|\Gamma_\beta|s\sigma\rangle$ and $\langle j^n|\Gamma_\beta|j^p\rangle$ denote the single-particle beta-decay matrix elements and j^p, j^n denote the total angular momenta of the proton, and neutron single-particle states, respectively.

What is the energy dependence of the nuclear matrix elements? The experimentally determined electron spectrum is usually expressed as the Fermi–Kurie plot, i.e., as the function

$$\sqrt{\frac{W_{if}(E)}{F(\mp Z, E)pE}} = \frac{1}{\sqrt{2\pi^3}}(E_0 - E)\,\{S_{if}^{(n)}(E)\}^{1/2}. \tag{6.33}$$

The quantity $S_{if}^{(0)}$ for allowed transitions is energy independent and equal to

$$S_{if}^{(0)} = \zeta\{G_V^2 |\langle i|\tau^{\pm}|f\rangle|^2 + G_A^2 |\langle i|\sigma, \tau^{\pm}|f\rangle|^2\}. \tag{6.34}$$

The ζ characterizes a change of the states in the final nucleus compared to the states of the initial nucleus. It also includes a correction for the purity of the states. The function $S_{if}^{(0,1)}$ is independent of energy and hence the $\sqrt{\dfrac{W_{if}(E)}{F(\mp Z, E)pE}}$ depends for the allowed transitions and for the first-forbidden transitions with $\Delta I = 0, \pm 1$ on the energy linearly. However, the energy dependence (6.33) is not linear for the first-forbidden $\Delta I = \pm 2$ transitions and for the β-transitions of a higher order.

The β-decay probability is obtained when the expression (6.30) is integrated over energy. The partial half-life t is then equal to

$$t^{-1} = \frac{1}{\ln 2}\int_{m_e}^{E_0} W_{if}(E)\,dE = \frac{\eta}{2\pi^3 \ln 2}\int_{m_e}^{E_0} pE(E_0-E)^2 F(\mp Z, E)\, S_{if}^{(n)}(E)\,dE, \tag{6.35}$$

where m_e is the electron mass. The quantity t^{-1} is proportional to E_0^5, when $S_{if}^{(n)}$ is energy independent.

It is advantageous to characterize the nuclear matrix elements by the product ft analogous to the reduced transition probabilities of the electromagnetic transitions. The function f depends on the nuclear charge and on the end-point energy, and it is defined as

$$f \equiv f(Z, E_0) = \int_{m_e}^{E_0} F(\mp Z, E)\,pE(E_0-E)^2\,dE. \tag{6.36}$$

The function f has been repeatedly tabulated; for example, in refs. 289–291. The f calculated for the forbidden transitions includes corrections for deviation of the dependence (6.33) from the linear function. The ft values vary in a broad interval; it is therefore customary to use $\log ft$ and measure t in seconds.

Let us discuss the magnitude of the constants G_N and G_A. Only the Fermi matrix element $\langle\tau^{\pm}\rangle$ contributes to the β-decay between two $I^{\pi}=0^+$ states. The nuclear matrix element of such a transition evidently equals unity and, consequently, the magnitude of the constant G_V can be determined from the measured ft value. The ft values were carefully measured for the transitions between the $I^{\pi} = 0^+$ states in the decays $^{14}O \rightarrow {}^{14}N$, $^{34}Cl \rightarrow {}^{34}S$, $^{42}Sc \rightarrow {}^{42}Ca$, and others. From these measurements the G_V was determined as

$$G_V = (1.40 \pm 0.20)\times 10^{-49}\ \text{erg cm}^3. \tag{6.37}$$

The axial-vector interaction constant G_A can be found from the ratio of the ft values for β-decay between two $I^{\pi} = 0^+$ states and for the neutron β-decay, i.e., from the ratio

$$\left(\frac{G_A}{G_V}\right)^2 = \frac{1}{3}\left\{\frac{2(ft)_{0+\rightarrow 0+}}{(ft)_n} - 1\right\}. \tag{6.38}$$

The experimentally found ratio equals

$$G_A/G_V = -1.18 \pm 0.03. \tag{6.38'}$$

The measured asymmetry of the electrons emitted by polarized nuclei and the measured β–γ circular polarization correlation allow one to determine the contribution of the Fermi matrix element to mixed β-transitions with $\Delta I = 0$.

The isobaric spin is unchanged in pure Fermi-type allowed transitions between two $I^\pi = 0^+$ states. This fact is often utilized for the experimental determination of the isobaric spin impurities in the nuclear wave functions. Table 6.4 contains the $\log ft$ values of the $0^+ \to 0^+$, $\Delta T = 1$ β-transitions; it uses the experimental data of the refs. 172 and 292. Such β-transitions are approximately 10^4 times inhibited when compared with the ordinary allowed transitions. This fact shows the high isobaric spin purity of the low-lying states in the studied nuclei. The admixtures of the $T+1$ components to the basic T-component are of the order 10^{-4}.

TABLE 6.4. *Beta-transitions between the $I^\pi = 0^+$ states with $\Delta T = 1$*

Parent nucleus	T_i	Daughter nucleus	Energy $\mathcal{E}$ MeV	T_f	$\log ft$
$^{66}_{31}Ga$	2	$^{66}_{30}Zn$	0	3	7.9
$^{156}_{63}Eu$	15	$^{156}_{64}Gd$	0	14	9.8
			1.048	14	10.2
$^{170}_{71}Lu$	14	$^{170}_{70}Yb$	0	15	9.7
$^{234}_{93}Np$	24	$^{234}_{92}U$	0	25	9.5

3. Let us study the effect of the pairing superfluid correlations on the β-decay probabilities. The problem was discussed in refs. 12, 153, and 293. Evidently, we have to calculate the matrix elements of the Hamiltonian (6.32) with the wave functions (4.22), (4.31′), and (4.33′). For example, let us take the β-decay of an odd-$N+1$ deformed nucleus with a single quasiparticle in the state $s_1\sigma_1$. The final nucleus has an odd number of protons $(Z+1)$ and a single quasiparticle in the state $r_2\varrho_2$. The corresponding matrix element is of the form

$$\mathcal{M}_\beta(ns_1 \to pr_2) = (\Psi_0^*(N)\Psi_0^*(r_2, Z+1)\, H_\beta \Psi_0(Z)\Psi_0(s_1, N+1)) = \langle r_2|\Gamma_\beta|s_1\rangle R^{1/2}, \quad (6.39)$$

where $R = R_N R_Z$ and

$$R_N = u_{s_1}^2(N) \prod_{s \neq s_1} (u_s(N)\, u_s(N+1) + v_s(N)\, v_s(N+1))^2, \quad (6.40)$$

$$R_Z = u_{r_2}^2(Z) \prod_{r \neq r_2} (u_r(Z)\, u_r(Z+1) + v_r(Z)\, v_r(Z+1))^2. \quad (6.41)$$

The products in (6.40) and (6.41) describe the change in the superfluid characteristics of the initial and final nuclei. These products are usually close to unity and thus may be omitted.

We shall explain here how to calculate the corrections R_N and R_Z for an arbitrary number of quasiparticles. The corrections R_N (or R_Z) are nonvanishing if the number of quasiparticles in the neutron (proton) system changes by one in the transition from the initial state to the final state. The state of the remaining quasiparticles should not change during this transition. The correction R_N is equal to u_s^2 or v_s^2. The state s is a single-particle state in which a quasiparticle either disappeared or was formed. The quantities u_s^2 and v_s^2 belong to the system with a smaller number of quasiparticles. The equality $R_N = u_s^2$ is true when the number of pairs in the neutron system does not change during the β-decay. On the other hand, $R_N = v_s^2$ if the number of pairs has been changed by one pair. The proton system is treated independently of the neutron system. The R_N and R_Z are calculated in the same way; the same rules apply in both the spherical and the deformed nuclei.

Let us illustrate the above-stated rules on several examples. The R_N and R_Z for the β-transitions between one-quasiparticle states will be discussed first. Let us take the transition ${}^{181}_{72}\mathrm{Hf}_{108+1} \rightarrow {}^{181}_{72+1}\mathrm{Ta}_{108}$. The number of pairs is identical in the initial and final states; hence $R_N = u_s^2$, $R_Z = u_r^2$. Another case is the transition ${}^{161}_{66+1}\mathrm{Ho}_{94} \rightarrow {}^{161}_{68}\mathrm{Er}_{92+1}$. The initial state has 94 paired neutrons and 66 paired protons, while the final state has 92 paired neutrons and 68 paired protons. Hence $R_N = v_s^2$ and $R_Z = v_r^2$. Another example is the transition from the two-quasiparticle state with $K^\pi = 6^-$ and configuration $p523\uparrow$, $n642\uparrow$ in ${}^{162}_{66+1}\mathrm{Ho}_{94+1}$. The final state is the two-quasiparticle state with $K^\pi = 5^-$ and configuration $n523\downarrow$, $n642\uparrow$ in ${}^{162}_{66}\mathrm{Dy}_{94+2}$. Evidently in this case $R_N = u^2_{523\downarrow}$ and $R_Z = u^2_{523\uparrow}$; the quasiparticle $n642\uparrow$ has not changed during the β-decay process.

Analogous situation is encountered in the β-transitions involving the three-quasiparticle states. Let us take, for example, the β-decay of ${}^{177}_{70+1}\mathrm{Lu}_{104+2}$. The initial state has $K^\pi = \frac{23}{2}-$ and configuration $n514\downarrow+n624\uparrow+p404\downarrow$. The final state is the $K^\pi = \frac{23}{2}+$ state with configuration $n514\downarrow+p514\uparrow+p404\downarrow$ in ${}^{177}_{70+2}\mathrm{Hf}_{104+1}$. The $642\uparrow$ neutron has changed in the $514\uparrow$ proton during the β-transition; the remaining quasiparticles stay unchanged. The pairing corrections are $R_N = u^2_{624\uparrow}$, $R_Z = u^2_{514\uparrow}$.

The corrections R_N and R_Z have the values 0 or 1 in the independent particle model (i.e., if $G_N = G_Z = 0$). The β-decay is possible if $R_N = 1$; all the neutrons except one must be unchanged. On the other hand, the β-decays with $R_N = 0$ are strictly forbidden in the independent particle model.

These features suggest an additional classification of β-decays. All β-transitions are divided into three groups according to the values of R_N and R_Z:

$$\left.\begin{array}{lll} \text{(I)} & R_N(G_N = 0) = 1, & 0.5 \leqslant R_N(G_N = G_N^0) < 1, \\ \text{(II)} & R_N(G_N = 0) = 0, & 0 < R_N(G_N = G_N^0) \leqslant 0.5, \\ \text{(III)} & R_N(G_N = 0) = 0, & R_N(G_N = G_N^0) = 0. \end{array}\right\} \quad (6.42)$$

Analogous division is performed for R_Z. We shall call the "particle" transitions those transitions in which a quasiparticle disappears or is formed on the level q where $E(q) > \lambda$. Similarly, the "hole" transitions involve quasiparticles with $E(q) < \lambda$.

The first group in (6.42) includes:

(a) β-decays between the nuclear ground states;
(b) particle transitions with unchanged number of pairs;
(c) hole transitions with number of pairs changed by one.

The transitions belonging to the second group in (6.42) are:

(a) hole transitions with unchanged number of pairs;
(b) particle transitions with number of pairs changed by one.

The β-transitions of the second group have nonvanishing transition probabilities in the superfluid nuclear model. They are, however, strictly forbidden in the independent particle model. The pairing corrections R_N and R_Z of the first and second groups are close to each other when the low-lying excited states are involved, but they differ considerably when the excitation energy increases.

More than thirty β-transitions of deformed nuclei belonging to the second group were experimentally found. The observation of such transitions proves once more an existence of the short-range interaction and related pairing correlations.

As we have stated above, the first and second groups contain the transitions with only one particle changed in each system. On the other hand, to the third group belong the transitions where:

(a) the number of quasiparticles in the proton (neutron) systems is changed by more than one;
(b) the number of quasiparticles is changed by one but other quasiparticles change their state at the same time.

Transitions of the third group are called F-forbidden.

Let us assume that the initial and final states in the following example are pure two-quasiparticle states. If the $I^\pi K = 2^- \ 2$ state at 1290 keV in ^{182}W has the configuration $p514\uparrow - p402\uparrow$, then its β-decay to the $I^\pi K = 3^- \ 3$ state in ^{182}Ta (configuration $p404\downarrow - n510\uparrow$) is F-forbidden. The experimental $\log ft$ value for this transition is 8.3.[294] Thus the transition is approximately 100 times slower than a typical allowed β-decay. We shall show later that the octupole–octupole interaction brings the impurities to the $I^\pi K = 2^- \ 2$ state in ^{182}W. The β-decay is realized owing to these small admixtures of other configurations. Thus, if the admixtures are taken into account, the β-decay is not F-forbidden.

The correction R changes the expression for the ft value:

$$ft = \frac{2\pi^2 \ln 2}{|\langle s | \Gamma_\beta | r \rangle|^2} \eta^{-1} R^{-1}. \tag{6.43}$$

The $R < 1$, hence the pairing correlations decrease the β-decay probabilities when compared to the independent particle model.

4. Let us discuss the characteristic features of the β-decay of deformed nuclei. The calculations performed use wave functions (2.7), but $\varphi_{\pm K}(q)$ are replaced by the functions (4.22), (4.31′), and (4.33′).

The β-transition intensities for the population of the levels of the same rotational band satisfy several simple intensity rules as a consequence of the rotational model. If the β-transition is K-allowed, i.e., if

$$|K_i - K_f| \leqslant \lambda \tag{6.44}$$

(λ is the degree of forbiddenness), the ft values are related by

$$\frac{ft(I_i K_i \to I_{1f} K_f)}{ft(I_i K_i \to I_{2f} K_f)} = \frac{(I_i K_i \lambda K_i - K_f | I_{2f} K_f)^2}{(I_i K_i \lambda K_i - K_f | I_{1f} K_f)^2}. \tag{6.45}$$

The experimental ratios are compared with the right-hand side of (6.45) in Table 6.5. The agreement is evidently quite satisfactory.

TABLE 6.5. *Ratios of the ft vales for the β-transitions to the levels of the same rotational band*

Initial nucleus	$I_i^{\pi i} K_i$	Final nucleus	Relative decay intensity to the state (in %)		$\frac{ft(I_i^{\pi i}K_i \to 2^+0)}{ft(I_i^{\pi i}K_i \to 0^+0)} = \frac{(I_iK_i\lambda K_f - K_i \vert 00)^2}{(I_iK_i\lambda K_f - K_i \vert 20)^2}$	
			0^+	2^+	Theory	Experiment
^{164}Ho	1^+ 1	^{164}Er	35	14	2.0	1.8±0.5
^{170}Tm	1^- 1	^{170}Yb	76	24	2.0	1.9±0.2
^{172}Tm	2^- 2	^{172}Yb	23	41	0.7	0.63
^{176}Lu	1^- 0	^{176}Hf	42	58	0.5	0.56±0.16
^{178}Ta	1^+ 1	^{178}Hf	66	30	2.0	2.0±0.2
^{180}Ta	1^+ 1	^{180}Hf	60	27	2.0	2.0±0.3
^{180}Ta	1^+ 1	^{180}W	6.9	3.2	2.0	1.8±0.3
^{186}Re	1^- 1	^{186}W	6	2	2.0	1.8±0.5
^{186}Re	1^- 1	^{186}Os	70	22	2.0	2.1±0.2

A beta-transition is called *K*-forbidden if the condition (6.44) is not valid. The probabilities of the *K*-forbidden β-transitions are nonvanishing because *K* is not an exact quantum number. Small admixtures of the states with other *K*-values are caused by the Coriolis interaction or by the nonadiabaticity of rotational motion. The existing experimental evidence shows that the probabilities of the *K*-forbidden transitions are for each unit of $|K_i - K_f| - \lambda$ approximately a hundred times smaller than the analogous *K*-allowed transitions. It is interesting that the degree of hindrance caused by the *K*-forbiddenness is approximately the same for the β-transitions and for the electromagnetic transitions.

Figure 6.3 shows the *K*-forbidden β-decay of the ^{176}Lu. The initial state is the $I^\pi K = 7^- 7$ state, and the final state is the rotational state $I^\pi K = 6^+ 0$ in ^{176}Hf. The transition is six times forbidden ($K_i - K_f - \lambda = 7-0-1 = 6$), and its $\log ft = 18.7$. Thus the transition is approximately 10^{12} times slowed down. Figure 6.3 also shows for comparison the *K*-allowed β-transitions to the ground state and to the first rotational state. The corresponding $\log ft = 6.8$ and 6.5, respectively.

The concept of the Λ-forbidden β-transitions[203] is useful when β-decays to the two-quasiparticle states in even–even nuclei are analysed. The β-transition between the two-quasiparticle states is possible only if one of the quasiparticles does not change its state. If the relative coupling of the two-quasiparticles has changed in the β-decay, the transition is hindered unless

$$\lambda \geqslant \Lambda_i + \Lambda_f. \tag{6.46}$$

The transitions which do not fulfill the condition (6.46) are Λ-forbidden; the degree of forbiddenness is determined in a similar way to the case of the *K*-forbidden transitions.

Let us consider, for example, the β-decay of the $K^\pi = 3^+$ state in ^{168}Tm. The configuration of this initial state is $n633\uparrow - p411\downarrow$ ($\Lambda_i = 2$). Two final states in ^{168}Er are excited: the $K^\pi = 3^-$ state at 1543 keV has the configuration $n633\uparrow - n521\downarrow$ and $\Lambda_f = 2$; the other state at 1095 keV has $K^\pi = 4^-$ and configuration $n633\uparrow + n521\downarrow$ ($\Lambda_f = 4$). The angular momentum projections are parallel in the $K^\pi = 4^-$ state and antiparallel in the $K^\pi = 3^-$ state and in the $K^\pi = 3^+$ initial state. Because the relative coupling has changed and the condition (6.46) is not satisfied, the β-decay to the $K^\pi = 4^-$ state is Λ-forbidden; its

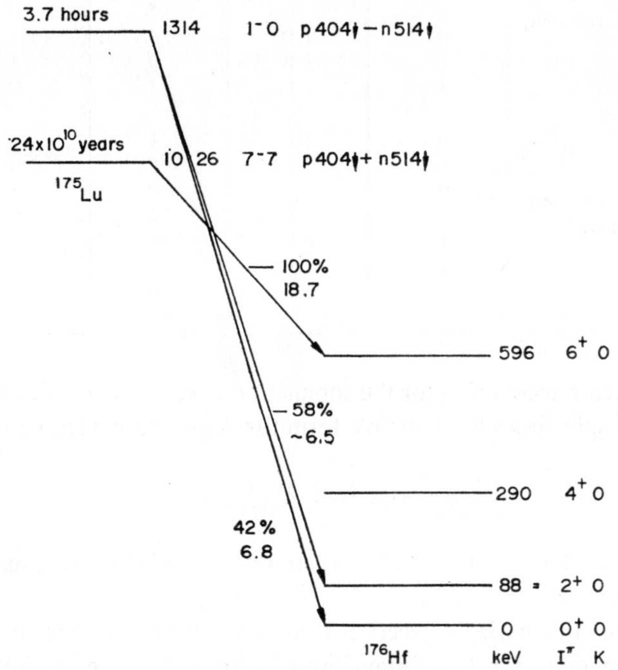

FIG. 6.3. The K-forbidden β-decay of the 7^-7 state and the β-decay of the 1^-0 state in ^{176}Lu. The final states are members of the ground-state rotational band in ^{176}Hf. Experimental data from refs. 203 and 295.

$\log ft = 7.7$. At the same time, $\log ft = 6.2$ for the $K^\pi = 3^-$ state with unchanged coupling. Thus the Λ-forbiddenness slowed down the decay rate 30 times.

Let us generally discuss the β-transitions between the quasiparticle states in deformed nuclei. The selection rules in Table 6.3 are insufficient for the detailed classification of these transitions. Consequently, as in the case of the electromagnetic transitions, we shall formulate the selection rules in terms of the asymptotic quantum numbers. Such selection rules are shown in Table 6.6. We can see that the asymptotic quantum numbers N, n_z, and Λ should not change in the allowed transitions, and in the first-forbidden transition they can change by one unit.

The β-decays between the quasiparticle states which do not violate the asymptotic selection rules are called "unhindered." They are denoted by *au* and 1*u*. The β-decay probabilities are 10–1000 times smaller if the asymptotic selection rules are violated. Such β-transitions are called "hindered" and are denoted by *ah* and 1*h*.

TABLE 6.6. *Asymptotic selection rules for the allowed and first forbidden β-transitions in deformed nuclei*

β-transition	ΔK	ΔN	Δn_z	$\Delta\Lambda$
Allowed, *a*	0	0	0	0
	1	0	0	0
First forbidden, 1	0	1	1	0
	0	−1	−1	0
	0	±1	0	±1
	1	±1	0	±1
	1	±1	±1	0
First forbidden unique, 1*	0	1	1	0
	0	−1	−1	0
	0	±1	0	±1
	1	1	1	0
	1	−1	−1	0
	1	±1	0	1
	2	±1	0	1

The statistical factor accounting for the summation over the final spin states and averaging over the initial spin states has a simple form for deformed nuclei, namely,

$$\eta = (I_i K_i \lambda K_f - K_i \mid I_f K_f)^2, \tag{6.47}$$

with the exception of the $I^\pi = 0^\pm \rightarrow I^\pi = 1^\pm$ transitions, where the expression (6.47) must be multiplied by 2.

Let us now discuss the pairing correction factors R_N and R_Z. Figures 6.4–6.7 summarize the quantities R_N and R_Z for the β-transitions between the states without quasiparticles in even systems and the one-quasiparticle states in odd systems. The transitions $N+1 \rightleftarrows N$, $Z+1 \rightleftarrows Z$, in which the number of pairs is unchanged, are shown in Fig. 6.4(a) and 6.5(a). Figures 6.4(b) and 6.5(b) show, on the other hand, the transitions $N-1 \rightleftarrows N$, $Z-1 \rightleftarrows Z$, in which the number of pairs is changed. The points corresponding to different transitions of the same odd system are connected by a full line. For example, the curve denoted by $N+1 = 91$ in Fig. 6.4(a) corresponds to the transitions between the different one-quasiparticle states in the 91 neutron system, on the one hand, and the state without quasiparticles in the 90 neutron system, on the other hand. Similarly, the curve denoted by $N-1 = 105$ in Fig. 6.4(b) corresponds to the transitions between the state without quasiparticles at $N = 106$ and the following one-quasiparticle states with 105 neutrons: the hole states $F-6$, $F-5$, $F-4$, $F-3$, $F-2$, and $F-1$, the ground state F (denoted by ⊙), and the particle states $F+1$, $F+2$, $F+3$, $F+4$, $F+5$, $F+6$.

The calculations of the R_N and R_Z correction terms were performed with the single-particle level schemes shown in Fig. 1.5–1.8. The following equilibrium deformations were used: $\beta_0 = 0.28$ for $150 < A < 180$, $\beta_0 = 0.23$ for $180 < A < 190$, and $\beta_0 = 0.24$ for $228 < A < 254$. The G_N and G_Z value of (4.51′) were employed.

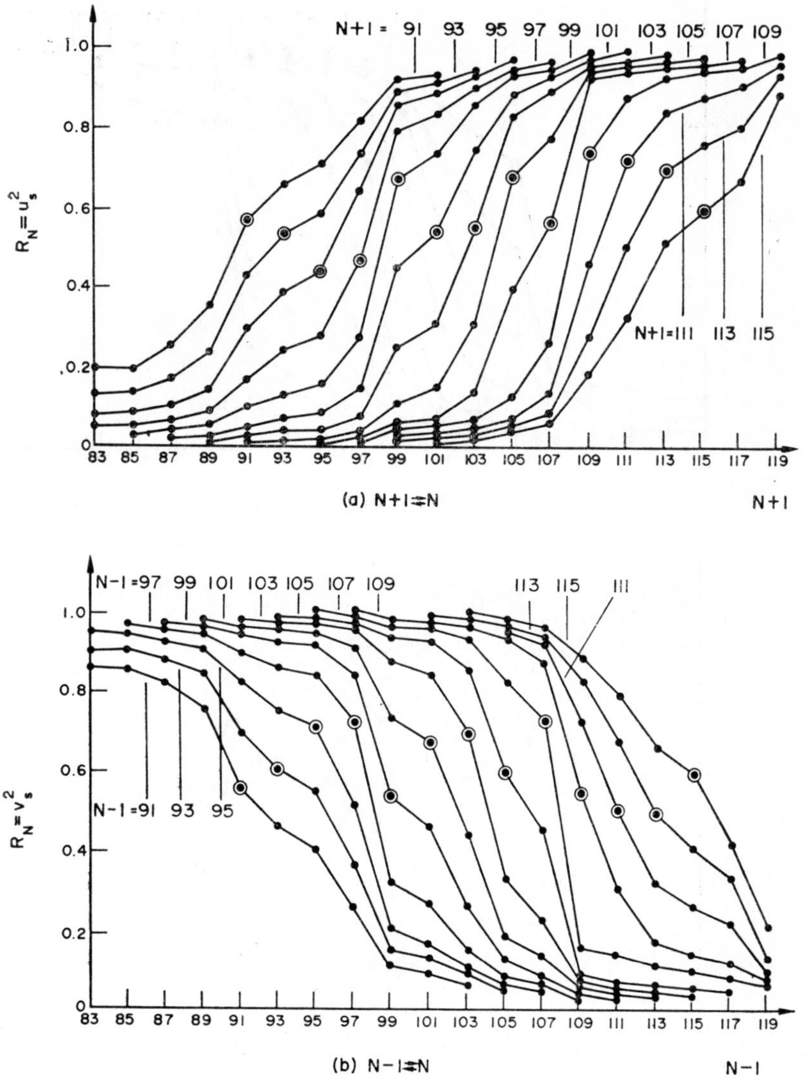

FIG. 6.4. The pairing correction factors R_N of the β-transition probabilities. The neutron systems with $90 \leqslant N \leqslant 116$ and the transitions connecting the quasiparticle vacuum with the one-quasiparticle states are considered. Graph (a) corresponds to the ($N+1 \rightleftarrows N$) transitions. The number of the paired neutrons is unchanged. Graph (b) corresponds to the ($N-1 \rightleftarrows N$) transitions; the number of pairs is changed by one pair. The single-particle states, enumerated according to the neutron number, are on the abscissa. The R_N values corresponding to the given state are on the ordinate. The curves connect the points corresponding to the same pair of nuclei ($N \rightleftarrows N \pm 1$). Each curve is labeled by the corresponding odd number of neutrons. The ground states are denoted by ◉. The points to the left of the ground state ◉ correspond to the hole transitions, while the points to the right correspond to the particle transitions.

We can see that the pairing corrections R_N and R_Z are smaller than 0.2 for the transitions to the hole states with $F-3$, $F-4$, ..., and an unchanged number of pairs and for the transitions to the particle states with $F+3$, $F+4$, ..., and a changed number of pairs. The inclusion of the pairing corrections R_N and R_Z is particularly important in these

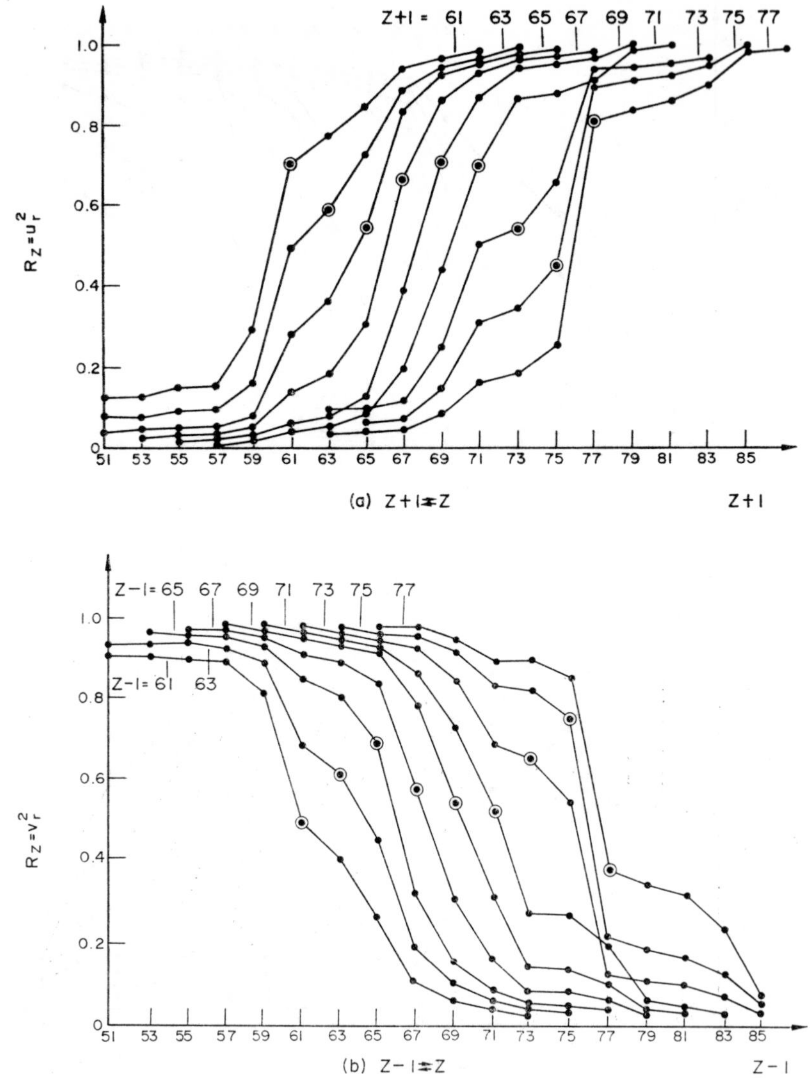

FIG. 6.5. Pairing correction factors R_Z of the β-transition probabilities. The proton systems with $60 \leqslant Z \leqslant 78$ and the transitions connecting the quasiparticle vacuum with one-quasiparticle states are considered. (Notation as Fig. 6.4.)

cases. Generally, the corrections R_N and R_Z are very useful in analysis of the β-decays between one-quasiparticle states in deformed nuclei.

Let us explain in detail how to find the R_N and R_Z values in Figs. 6.4–6.7. As an example we shall find the R_Z and R_N values for the β-decay of the $Fp523\uparrow$ state in ${}^{167}_{67}Ho_{100}$. The following final states in ${}^{167}_{68}Er_{99}$ will be considered: the ground state $Fn633\uparrow$, the hole state $F-1\ n523\downarrow$, and the particle state $F+2\ n512\uparrow$. The quantity R_Z is identical for all transitions, and it corresponds to the transition from the one-quasiparticle state $Fp523\uparrow$ of the system $Z-1 = 67$ to the quasiparticle vacuum of the $Z = 68$ system. From the $Z-1 = 67$ curve in Fig. 6.5(b) we find $R_Z = v_{523}^2 = 0.58$ corresponding to the point $Z-1 = 67$ on the

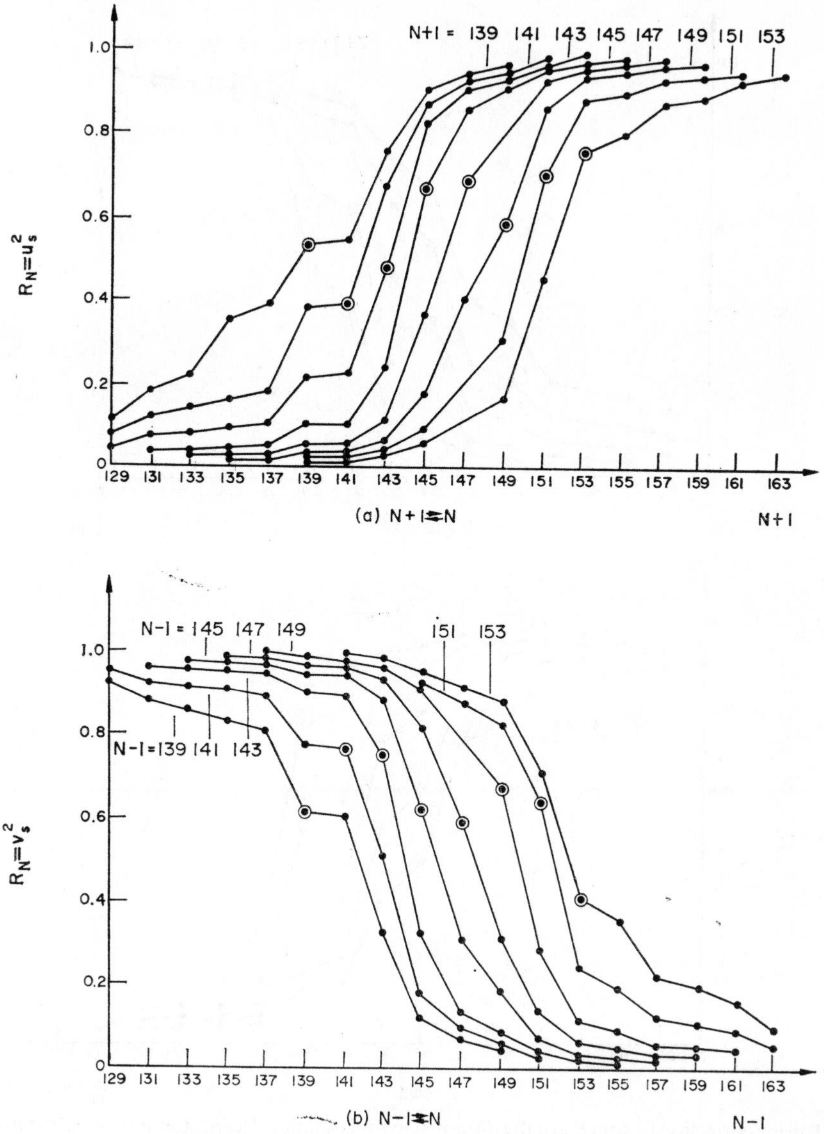

FIG. 6.6. Pairing correction factors R_N of the β-transition probabilities. The neutron systems with $138 \leqslant N \leqslant 154$ and the transitions connecting the quasiparticle vacuum with one-quasiparticle states are considered. (Notation as Fig. 6.4.)

abscissa. The R_N values are found on the $N-1 = 99$ curve in Fig. 6.4(b). The transition to the $Fn633\uparrow$ ground state corresponds to the abscissa $N-1 = 99$; the ordinate is $R_N = v_{633}^2 = 0.53$. The transition to the $F-1\,n523\downarrow$ state corresponds to the abscissa $N-1 = 97$; the ordinate is $R_N = v_{523}^2 = 0.84$. Finally, the transition to the $F+2\,n512\uparrow$ state corresponds to the abscissa $N-1 = 103$ and the ordinate is $R_N = v_{512}^2 = 0.25$.

The R_Z and R_N values from Figs. 6.4–6.7 may be used in the analysis of β-transitions between the two-quasiparticle states and the states without quasiparticles in even nuclei.

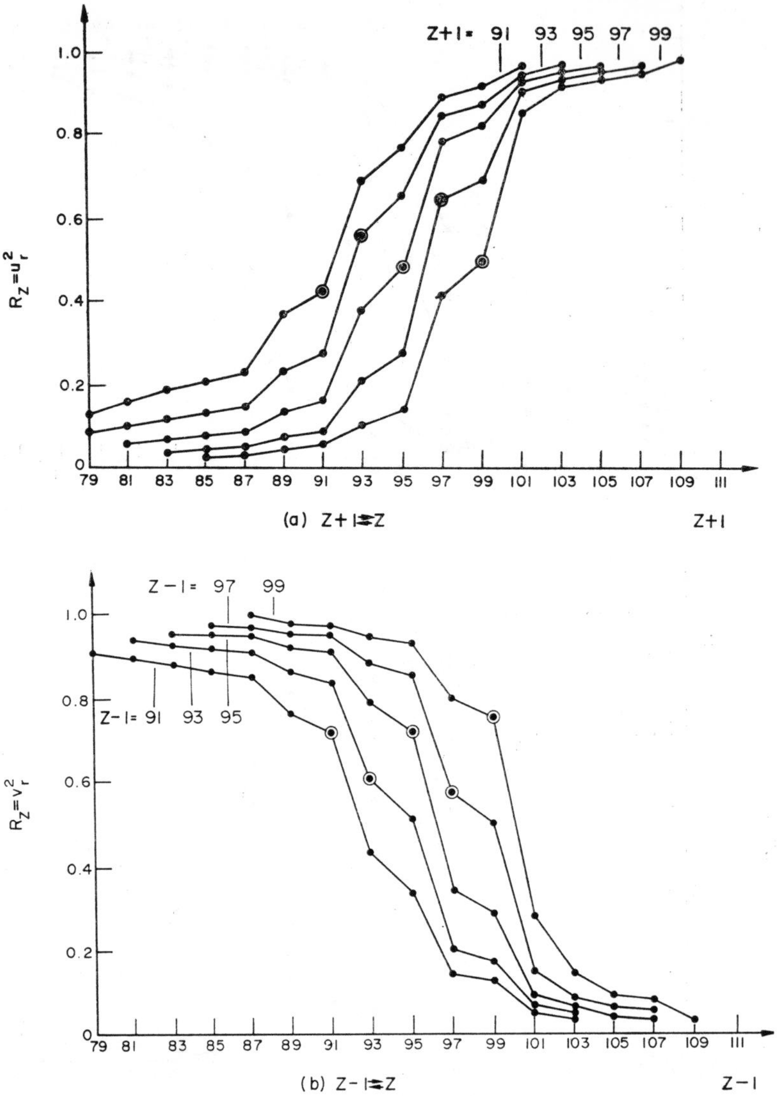

FIG. 6.7. Pairing correction factors R_N of the β-transition probabilities. The proton systems with $90 \leqslant Z \leqslant 100$ and the transitions connecting the quasiparticle vacuum with one-quasiparticle states are considered. (Notation as Fig. 6.4.)

They can be also used to analyze the transitions populating two-quasiparticle states when the specially calculated values of R_N and R_Z (with blocking effect included) are not available. The correction factors R_N and R_Z were calculated in ref. 296 using the projected wave functions. The improvement of the accuracy was not, however, very substantial. Similarly, the R_N and R_Z values calculated by means of the Woods–Saxon wave functions are very close to the values shown in Figs. 6.4–6.7.

Let us discuss the experimental information about the β-transitions between single-particle states in various nuclei. Similar analyses have been performed in refs. 46, 123, 147,

TABLE 6.7. *The au β-transitions between the* $p\frac{7}{2}^-$ [523] ⇄ $n\frac{5}{2}^-$ [523] *states*

$K\pi$	Transition	$K\pi$	log ft_e, experiment	$R\eta$	log ft_c, calculation	log $[ft_eR\eta]$	$\frac{ft_eR\eta}{(ft)_N}$	$\frac{ft_eR\eta}{(ft)_N(1-d_\beta^{(1)})^2}$	log $[ft_eR\eta C_\gamma C_\varrho]$
$\frac{7}{2}^-$	^{167}Tm ← ^{167}Yb	$\frac{5}{2}^-$	4.55	0.57	<u>4.6</u>	4.3	8.3	1.24	4.0
$\frac{7}{2}^-$	^{165}Tm ← ^{165}Yb	$\frac{5}{2}^-$	4.7	0.56	4.6	4.4	—	—	—
$\frac{7}{2}^-$	^{169}Ho → ^{169}Er	$\frac{5}{2}^-$	4.7	0.41	4.8	4.4	—	—	3.8
$\frac{7}{2}^-$	^{167}Ho → ^{167}Er	$\frac{5}{2}^-$	4.5	0.38	4.8	4.1	5.6	1.45	3.9
$\frac{7}{2}^-$	^{165}Ho ← ^{165}Er	$\frac{5}{2}^-$	4.6	0.42	4.7	4.2	7.7	1.06	4.0
$\frac{7}{2}^-$	^{163}Ho ← ^{163}Er	$\frac{5}{2}^-$	4.8	0.31	4.9	4.3	9.6	1.40	4.1
$\frac{7}{2}^-$	^{161}Ho → ^{161}Dy	$\frac{5}{2}^-$	4.8	0.32	4.9	4.3	7.0	0.9	4.1
$\frac{7}{2}^-$	^{159}Ho → ^{159}Dy	$\frac{5}{2}^-$	4.5	0.38	4.8	4.1	—	1.81	3.9
$\frac{7}{2}^-$	^{161}Tb ← ^{161}Gd	$\frac{5}{2}^-$	4.8	0.40	4.8	4.4	6.5	1.04	4.1
1^+	^{166}Tm ← ^{166}Yb	0^+	4.6	0.77	4.5	4.5	8.1	—	—
1^+	^{164}Tm ← ^{164}Yb	0^+	~5	1.0	4.3	~5	—	—	—
1^+	^{164}Tm → ^{164}Er	0^+	4.9	0.13	5.2	4.0	—	—	3.8
0^-	^{166}Ho → ^{166}Er	1^-	5.2	1.0	4.3	5.2	62	—	—
1^+	^{166}Ho ← ^{166}Dy	0^+	5.2	0.35	4.8	4.7	59	—	4.5
1^+	^{164}Ho → ^{164}Er	0^+	5.5	0.19	5.1	4.8	16	—	4.5
1^+	^{164}Ho → ^{164}Dy	0^+	⩽4.8	0.32	4.8	⩽4.3	—	—	4.1
1^+	^{162}Ho → ^{162}Dy	0^+	4.7	0.23	5.0	4.1	4.4	—	3.9
6^-	^{162}Ho → ^{162}Dy	5^-	4.6	0.3	4.9	4.1	3.0	—	—
5^+	^{160}Ho → ^{160}Dy	4^+	4.8	0.21	5.0	4.1	7.9	—	—
5^+	^{158}Ho → ^{158}Dy	4^+	⩽5.5	0.21	5.0	⩽4.8	—	—	—
1^-	^{162}Tb → ^{162}Dy	2^-	4.9	0.16	5.1	4.1	—	—	—
1^+	^{156}Eu ← ^{156}Sm	0^+	5.3	0.65	4.5	5.1	4.4	—	4.7

153, 196, and 297. Let us consider the *au* β-transition between the *p*523↑ and *n*523↓ states first. Table 6.7 shows the experimental log ft_e values for all such known transitions. The $R\eta$ values, with R_N and R_Z from Figs. 6.4 and 6.5, are shown in the fifth column. The sixth column shows the normalized values of log ft_c; the underlined experimental value for ^{167}Yb → ^{167}Tm transition was used for the normalization. The values $ft_eR\eta$ directly characterize the single-particle matrix elements; the effect of the pairing correlations and the statistical factor are excluded. The values of log $[ft_eR\eta]$ are shown in the seventh column. The eighth column includes the ratios between the $ft_eR\eta$ and the ft values calculated by means of the Nilsson model wave functions. The quantities in the last columns will be discussed later.

Table 6.7 clearly shows that the log $[ft_e R\eta]$ values for the β-transitions between one-quasiparticle states are very close to each other. The calculated $\log ft_c$ values are in a good agreement with the experimental values of $\log ft_e$. The spread of the log $[ft_e R\eta]$ values for the transitions in even nuclei is much larger than in odd-A nuclei. Thus the data again confirm that the Nilsson potential has a basic validity for the description of the average field in deformed nuclei.

The role of the pairing corrections R_N and R_Z is again illustrated in Table 6.8. The table contains some selected β-transitions in odd-A deformed nuclei. The supplementary classification is also shown: I denotes the transitions of the first group, and II denotes the transitions of the second group. The proton classification is shown to the left, followed by the neutron classification.

TABLE 6.8. *Beta-transitions in odd-A deformed nuclei*

β-transition	Classification	$\log ft_e$	$R\eta$	$\log ft_c$	$\log ft_e R\eta$	$\log [ft_e R\eta C_\nu C_\varrho]$
			$p532\uparrow \rightleftarrows n521\uparrow$			
$^{161}Ho \leftarrow {}^{161}Er$	*ah* I I	5.5	0.54	5.5	5.2	4.8
$^{159}Tb \leftarrow {}^{159}Gd$	*ah* II I	6.6	0.11	6.2	5.6	5.3
$^{157}Tb \leftarrow {}^{157}Dy$	*ah* I I	5.5	0.38	5.6	5.1	4.8
$^{155}Eu \leftarrow {}^{155}Sm$	*ah* II I	5.6	0.27	5.8	5.0	4.9
			$p411\downarrow \rightleftarrows n521\uparrow$			
$^{165}Tm \rightarrow {}^{165}Er$	1*u* I II	7.0	0.16	6.7	6.2	6.0
$^{163}Tm \rightarrow {}^{163}Er$	1*u* I II	6.4	0.30	6.4	5.9	5.7
$^{161}Ho \leftarrow {}^{161}Er$	1*u* II I	6.4	0.36	6.3	6.0	5.7
			$p404\downarrow \rightleftarrows n503\uparrow$			
$^{185}Ta \rightarrow {}^{185}W$	1*u* I II	6.5	0.20	6.5	5.9	5.8
$^{183}Ta \rightarrow {}^{183}W$	1*u* I II	6.7	0.11	6.8	5.8	5.7
$^{177}Ta \rightarrow {}^{177}Hf$	1*u* I I	6.4	0.48	6.2	6.1	5.8
$^{175}Ta \rightarrow {}^{175}Hf$	1*u* I II	6.2	0.50	6.2	5.9	5.7
			$p411\downarrow \rightleftarrows n510\uparrow$			
$^{183}Re \leftarrow {}^{183}Os$	1*u* I I	6.8	0.34	6.8	6.3	6.1
$^{181}Ta \leftarrow {}^{181}Hf$	1*u* II I	7.2	0.10	7.3	6.2	6.1

The importance of the pairing corrections R_N and R_Z for the second group transitions is evident. The inclusion of R explains relatively large variations of the ft_e values for identical single-particle states in different nuclei. The interval of the log $[ft_e R\eta]$ values is clearly narrower than the interval of the $\log ft_e$ values. The calculated relative $\log ft_c$ values (the values used for normalization and underlined) agree quite well with the experimental data.

The systematics of the experimental $\log ft_e$ values lead to the following division of the possible $\log ft_e$ values for the allowed and first-forbidden β-transitions between the quasiparticle states in deformed nuclei:

$$\left.\begin{array}{llll} 4.5 < \log ft_e < 5.5 & au, & 5.5 < \log ft_e < 8.0 & 1u, \\ 5.5 < \log ft_e < 8.0 & ah, & 7.5 < \log ft_e < 9.0 & 1h. \end{array}\right\} \quad (6.48)$$

As we have stated above, the pairing correlations strongly affect the β-decay probabilities. It is therefore appropriate to systemize the log $[ft_eR\eta]$ values instead of the log ft_e values. The log $[ft_eR\eta]$ values directly characterize the single-particle matrix elements; the effect of the pairing correlations and statistical factors is excluded. The corresponding intervals of the log $[ft_eR\eta]$ values are equal to

$$\left.\begin{array}{ll} 3.8 < \log[ft_eR\eta] < 4.6 \;\; au, & 6.0 < \log[ft_eR\eta] < 7.5 \;\; ah \;\; \Delta N = \pm 2, \\ 4.8 < \log[ft_eR\eta] < 6.0 \;\; ah \;\; \Delta N = 0, & 5.5 < \log[ft_eR\eta] < 7.5 \;\; 1u. \end{array}\right\} \quad (6.48')$$

The histograms in Figs. 6.8(a) and 6.9(a) show the distribution of the log $[ft_eR\eta]$ values for the β-transitions connecting one-quasiparticle states in odd-A deformed nuclei. Similar

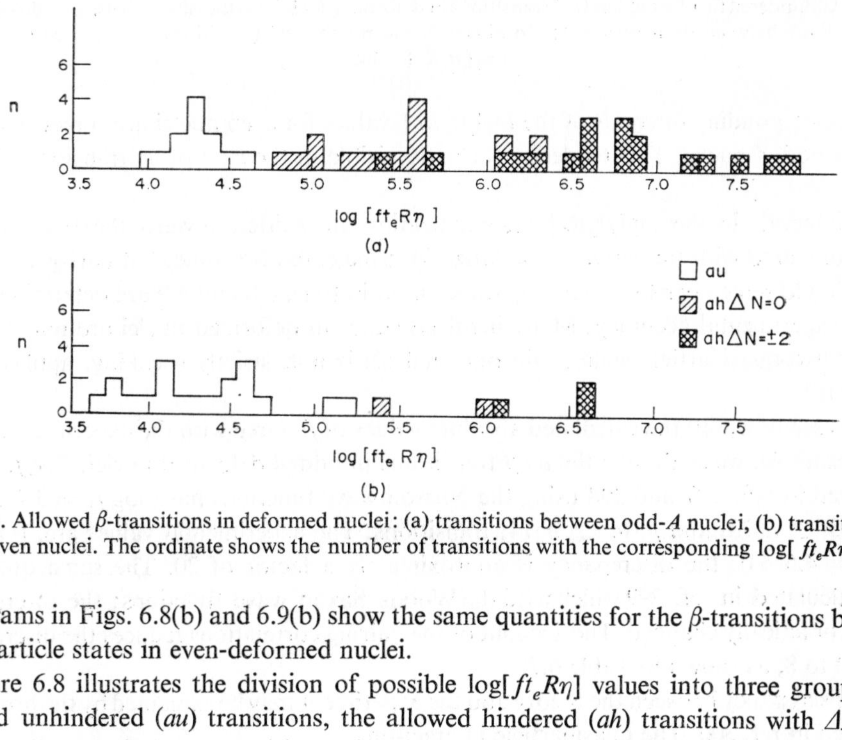

FIG. 6.8. Allowed β-transitions in deformed nuclei: (a) transitions between odd-A nuclei, (b) transitions between even nuclei. The ordinate shows the number of transitions with the corresponding log$[ft_eR\eta]$ value.

histograms in Figs. 6.8(b) and 6.9(b) show the same quantities for the β-transitions between quasiparticle states in even-deformed nuclei.

Figure 6.8 illustrates the division of possible log$[ft_eR\eta]$ values into three groups: the allowed unhindered (*au*) transitions, the allowed hindered (*ah*) transitions with $\Delta N = 0$, and the allowed hindered transitions with $\Delta N = \pm 2$. The transitions with $\Delta N = \pm 2$, i.e., with the change of the main oscillator quantum numbers by two units, must be considered separately. The fact that the log$[ft_eR\eta]$ values have such a characteristic distribution shows a deep physical importance of the asymptotic quantum numbers $[Nn_z\lambda]$.

The intervals of the log$[ft_eR\eta]$ values for the *ah* transitions and for the $1u$ transitions partially overlap. Thus the importance of the asymptotic quantum numbers is even increased.

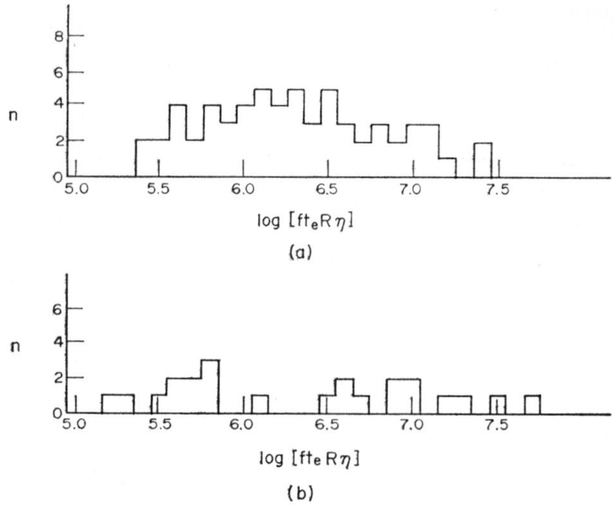

FIG. 6.9. Unhindered first forbidden $1u\beta$-transitions in deformed nuclei: (a) transitions between odd-A nuclei, (b) transitions between even nuclei. The ordinate is the number of transitions with the corresponding log $[ft_eR\eta]$ value.

The corresponding intervals of the log$[ft_eR\eta]$ values for even nuclei are somewhat wider than for odd-A nuclei. This tendency is clearly visible in the case of *au* transitions in Fig. 6.8.

The intervals in the log$[ft_eR\eta]$ systematization are shifted toward the smaller value when compared with the log ft_e systematization; they are also somewhat narrower.

We should remember that many ft_e values used in Figs. 6.8 and 6.9 are determined with a poor experimental accuracy. Many involved states in deformed nuclei are not the pure one- or two-quasiparticle states, and our analysis is not, strictly speaking, applicable to such states.

Let us see how well the calculated absolute values of ft agree with the experimental data. As an example, we shall take the *au* β-transitions in odd-A deformed nuclei. The ft values calculated in refs. 297 and 298 using the Nilsson wave functions have log ft = 3.4–3.6 for the $p523\uparrow \rightleftarrows n523\downarrow$ and $p514\uparrow \rightleftarrows n514\downarrow$ transitions. The experimental values are, however, log ft = 4.5–5.0; the discrepancy is approximately a factor of 20. The same quantities were calculated in ref. 299 which used the Woods–Saxon wave functions; the discrepancy has not practically changed. The account of the pairing correlations reduces the discrepancy from 20 to 8, as shown in Table 6.7.

The discrepancy between the theory and the experiment may be explained by the procedure suggested in ref. 300. The quasiparticle interaction

$$V_{12} = \frac{V}{A}(\boldsymbol{\tau}^{(1)}\boldsymbol{\tau}^{(2)})(\boldsymbol{\sigma}^{(1)}\boldsymbol{\sigma}^{(2)}), \tag{6.49}$$

with the interaction constant V was introduced in ref. 300. The interaction (6.49) is a part of the general quasiparticle interaction (3.84). It causes a spin polarization of the core, which in turn affects strongly the nuclear magnetic moments and the *au* β-transitions in odd-A nuclei.

Let us recapitulate the main results of Bochnacki and Ogaza[300]. The *ft* value of *au* transition based on the interaction (6.49) is equal to

$$(ft)_c = (ft)_{\mathrm{N}}\,(1-d_\beta^{(1)})^2\,R^{-1}\eta^{-1}, \tag{6.50}$$

where $(ft)_N$ denotes the single-particle value calculated by means of the Nilsson wave functions for the transitions between the $s_1\sigma_1$ and $r_2\sigma_2$ states. The $d^{(1)}$ is defined by

$$d_\beta^{(1)} = -\frac{2V}{A}\sum_{s,\,r}\sum_{\sigma=\pm 1}\Big\{[|\langle r+|\sigma^{(+)}\tau^{(-)}|s\sigma\rangle|^2+|\langle s+|\sigma^{(+)}\tau^{(+)}|r\sigma\rangle|^2]\times$$
$$\times\left(\frac{u_r^2v_s^2}{\varepsilon(r)+\varepsilon(s)\pm[\varepsilon(s_1)-\varepsilon(r_2)]}+\frac{u_s^2v_r^2}{\varepsilon(r)+\varepsilon(s)\mp[\varepsilon(s_1)-\varepsilon(r_2)]}\right)\Big\}, \tag{6.50'}$$

where the upper sign in denominators is used for the β-transitions with unchanged number of the proton and neutron pairs.

Table 6.7 contains in the ninth column the calculated values of $(ft)_eR\eta/(ft)_{\mathrm{N}}(1-d_\beta^{(1)})^2$. Evidently the quasiparticle interaction and the related core polarization remove the discrepancy between the experimental and calculated *ft* values for the *au* β-transitions in odd-*A* deformed nuclei.

The *ft* values of the *ah* β-transitions with $\Delta N = 0$ were calculated using both the Nilsson and the Woods–Saxon wave functions. The theoretical values are $\log ft \approx 4.0$–5.0 while the experimental values are $\log ft_e = 6.0$–7.0. The inclusion of the pairing correlations decreases the discrepancy, but approximately a factor of 10 remains unexplained.

The mentioned β-transitions with $\Delta N = 2$ are often called *N*-forbidden. When the interaction of the N and $N+2$ shells is not taken into account (as in refs. 41 and 42), the corresponding theoretical β-decay matrix elements vanish. Such a coupling was included in the calculations[55] using the Woods–Saxon potential. The corresponding wave functions with the asymptotic quantum number N then contain admixtures of components with $N\pm 2$.

The *ft* values of *ah* β-transitions with $\Delta N = 2$ were calculated using the mixed wave functions; the pairing correlations were naturally also included. The theoretical values are in the interval $\log ft_c = 6.5$–7.5,[301] while the experimental values are in the interval $\log ft_e = 6.5$–8.5. Thus the *ah* $\Delta N = 2$ transitions are 10–100 times slower than the *ah* $\Delta N = 0$ transitions. The discrepancy between the theoretical and calculated values is within one order of magnitude.

Calculations of the absolute *ft* values for the first-forbidden β-transitions and the comparison with the experimental data are more complicated than the same procedure for the allowed transitions. The calculations,[302] performed with the Nilsson and the Woods–Saxon wave functions, show similar discrepancies in the absolute *ft* values of the allowed and first-forbidden transitions.

5. Let us discuss the characteristic features of the β-decays in spherical nuclei. The Hamiltonian H will have the form (6.32′) and the matrix elements $\mathscr{M}(J_i \to J_F)$ as well as the quantities

$$S(J_i \to J_f)\eta = \frac{1}{2J_i+1}\sum_{M_i,\,M_f,\,\mu}|\mathscr{M}(J_i \to J_f)|^2 \tag{6.51}$$

will be calculated.

The β^--transitions between one-quasiparticle states in odd-A spherical nuclei will be treated first. The matrix element for the β-decay of an odd $N+1$ nucleus has the following form:

$$\mathcal{M}(j_1^n \to j_2^p) = \left(\Psi_0^*(j_2^p; Z+1)\,\Psi_0^*(N)\,H_\beta^\lambda \Psi_0(j_1^n; N+1)\,\Psi_0(Z)\right)$$
$$= \frac{1}{\sqrt{2\lambda+1}} \sum_{j^p m^p} \sum_{j^n m^n} (-1)^{j^n - m^n} \left(j^n m^n j^p - m^p \mid \lambda - \mu\right) \langle j^p | \Gamma^\lambda | j^n \rangle \times$$
$$\times \left(\Psi_0^*(j_2^p; Z+1)\, a^+_{j^p m^p} \Psi_0(Z)\right) \left(\Psi_0^*(N)\, a_{j^n m^n} \Psi_0(j_1^n; N+1)\right).$$

Using the one-quasiparticle wave function and the quasiparticle vacuum in the form (4.22″) and (4.33″),

$$\mathcal{M}(j_1^n \to j_2^p) = \frac{(-1)^{j_1^n - m^n}}{\sqrt{2\lambda+1}} \left(j_1^n m_1^n j_2^p - m_2^p \mid \lambda - \mu\right) \langle j_2^p | \Gamma^\lambda | j_1^n \rangle u_{j_1^n} u_{j_2^p}. \tag{6.52}$$

The number of pairs in the neutron and proton systems is unchanged in the $(N+1, Z) \to (N, Z+1)$ transition, hence $R = u_{j_1^n}^2 u_{j_2^p}^2$. Substituting (6.52) in (6.51) and using the properties of the Clebsch–Gordan coefficients, we write

$$S(j_1^n \to j_2^p)\eta = \frac{1}{2j_1^n+1} \left|\langle j_2^p | \Gamma^\lambda | j_1^n \rangle\right|^2 u_{j_1^n}^2 u_{j_2^p}^2. \tag{6.52′}$$

The statistical factor for the β-decay connecting one-quasiparticle states of odd-A nuclei equals

$$\eta = \frac{1}{2j_i+1}. \tag{6.52″}$$

Now we shall turn our attention to the β-decay of two-quasiparticle states in odd–odd nuclei to the ground state and to the excited two-quasiparticle states of even–even spherical nuclei. Let us calculate the matrix element of the β^+ decay $(N-1, Z+1 \to N, Z)$; the final state is a neutron two-quasiparticle state. The matrix element is equal to

$$\mathcal{M}(J_i M_i j_1^n j_2^p \to J_f M_f j_1^n j_3^n)$$
$$= \left(\Psi_0^*(Z)\,\Psi_0^*(J_f M_f | j_1^n, j_3^n; N)\,H_\beta^\lambda \Psi_0(J_i M_i | j_1^n, j_2^p; N-1, Z+1)\right)$$
$$= \frac{1}{\sqrt{2\lambda+1}} \sum_{j^p, m^p} \sum_{j^n, m^n} (-1)^{j^p - m^p} \left(j^p m^p j^n - m^n \mid \lambda - \mu\right) \times$$
$$\times \sum_{m_1^n, m_3^n} \sum_{m_2^n, m_2^p} \left(j_1^n m_1^n j_3^n m_3^n \mid J_f M_f\right) \left(j_1^n m_1^n j_2^p m_2^p \mid J_i M_i\right) \times$$
$$\times \langle j^n | \Gamma^\lambda | j^p \rangle \left(\Psi_0^*(j_1^n, j_3^n; N)\, a^+_{j^n m^n} \Psi_0(j_1^n, N-1)\right) \times$$
$$\times \left(\Psi_0^*(Z)\, a_{j^p m^p} \Psi_0(j_2^p; Z+1)\right). \tag{6.53}$$

The matrix element (6.53) fulfills the quasiparticle selection rules because the quasiparticle j_1^n is unchanged. The wave functions (4.31″), (4.31‴), and (4.49) are used in the further calculations.

If $j_1^n = j_3^n$, the matrix element (6.53) can be transformed into

$$\mathcal{M}(J_iM_ij_1^nj_2^p \to J_fM_fj_1^nj_1^n) = \langle j_1^n | \Gamma^\lambda | j_2^p \rangle u_{j_2^p} u_{j_1^n} \times$$
$$\times (1+(-1)^{J_f}) (J_iM_i\lambda\mu | J_fM_f) \, U(j_1^nj_2^pJ_f\lambda; J_ij_1^n); \quad (6.54)$$

the function U is defined in ref. 27.

After substituting (6.54) in (6.51), we find that

$$S(J_iM_ij_1^nj_2^p \to J_fM_fj_1^nj_1^n)\eta = \frac{2J_f+1}{2J_i+1} \left| \langle j_1^n | \Gamma^\lambda | j_2^p \rangle \right|^2 u_{j_2^p}^2 u_{j_1^n}^2 \times$$
$$\times U^2(j_1^nj_2^pJ_f\lambda; J_ij_1^n)\,(1+(-1)^{J_f}). \quad (6.54')$$

The statistical factor is more complicated, namely

$$\eta = \frac{2J_f+1}{2J_i+1} U^2(j_1^nj_2^pJ_f\lambda; J_ij_1^n). \quad (6.54'')$$

The expression (6.51) has the following form in the $j_1^n \neq j_3^n$ case:

$$S(J_iM_ij_1^nj_2^p \to J_fM_fj_1^nj_3^n)\eta = \left| \langle j_3^n | \Gamma^\lambda | j_2^p \rangle \right|^2 \times$$
$$\times u_{j_2^p}^2 u_{j_3^n}^2 \frac{2J_f+1}{2J_i+1} \frac{1}{2j_3^n+1} U^2(j_1^nj_2^pJ_f\lambda; J_ij_3^n), \quad (6.55)$$

where

$$\eta = \frac{2J_f+1}{2J_i+1} \frac{1}{2j_3^n+1} U^2(j_1^nj_2^pJ_f\lambda; J_ij_3^n). \quad (6.55')$$

Another possibility is the β^+ decay ($N-1$, $Z+1 \to N$, Z) from a two-quasiparticle state of an odd–odd nucleus to the ground state of an even–even nucleus.

$$S(J_iM_ij_1^nj_2^p \to J_i = 0, M_i = 0)\eta = \frac{1}{2\lambda+1} \left| \langle j_1^n | \Gamma^\lambda | j_2^p \rangle \right|^2 u_{j_2^p}^2 v_{j_1^n}^2, \quad (6.56)$$

where

$$\eta = \frac{1}{2\lambda+1}. \quad (6.56')$$

For the β-decays from the ground state of even–even nuclei $\eta = 1$.

The knowledge of the pairing correction factors R_N and R_Z is very useful for analysis of the β-transitions in spherical nuclei. Figures 6.10 and 6.11 are similar to Figs. 6.4–6.7; they show the R_N and R_Z values for the β-transitions between one-quasiparticle states of spherical nuclei with $100 < A < 150$. Figures 6.10 and 6.11 are used in exactly the same way as Figs. 6.4–6.7.

The allowed β-transitions in spherical nuclei can be further divided into three groups: the superallowed transitions, the unfavored transitions, and the l-forbidden transitions. In superallowed decays a neutron is transformed into a proton (or vice versa) in the same subshell; the corresponding ft values are $\log ft_e = 3$–4. Such transitions are observed in nuclei with $A \leq 40$; the β-decays of the mirror nuclei also belong to this group. For

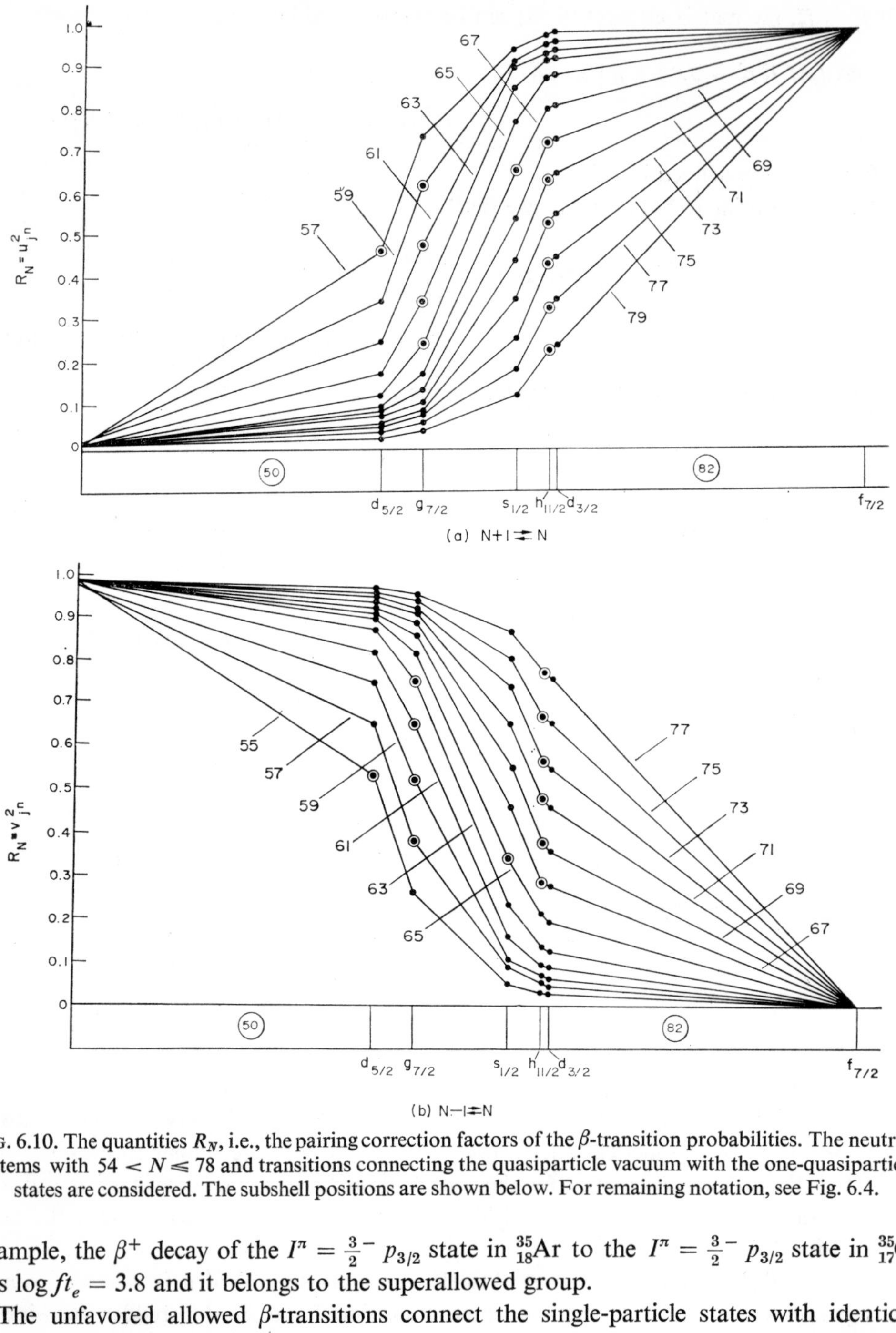

FIG. 6.10. The quantities R_N, i.e., the pairing correction factors of the β-transition probabilities. The neutron systems with $54 < N \leqslant 78$ and transitions connecting the quasiparticle vacuum with the one-quasiparticle states are considered. The subshell positions are shown below. For remaining notation, see Fig. 6.4.

example, the β^+ decay of the $I^\pi = \frac{3}{2}^- \; p_{3/2}$ state in $^{35}_{18}$Ar to the $I^\pi = \frac{3}{2}^- \; p_{3/2}$ state in $^{35}_{17}$Cl has $\log ft_e = 3.8$ and it belongs to the superallowed group.

The unfavored allowed β-transitions connect the single-particle states with identical l-values. For example, the transitions $p_{1/2} \rightleftarrows p_{3/2}$, $d_{3/2} \rightleftarrows d_{5/2}$, or $g_{7/2} \rightleftarrows g_{9/2}$ are unfavored allowed transitions. A majority of these transitions have ft values in the interval $\log ft_e = 4.5$–6.0.

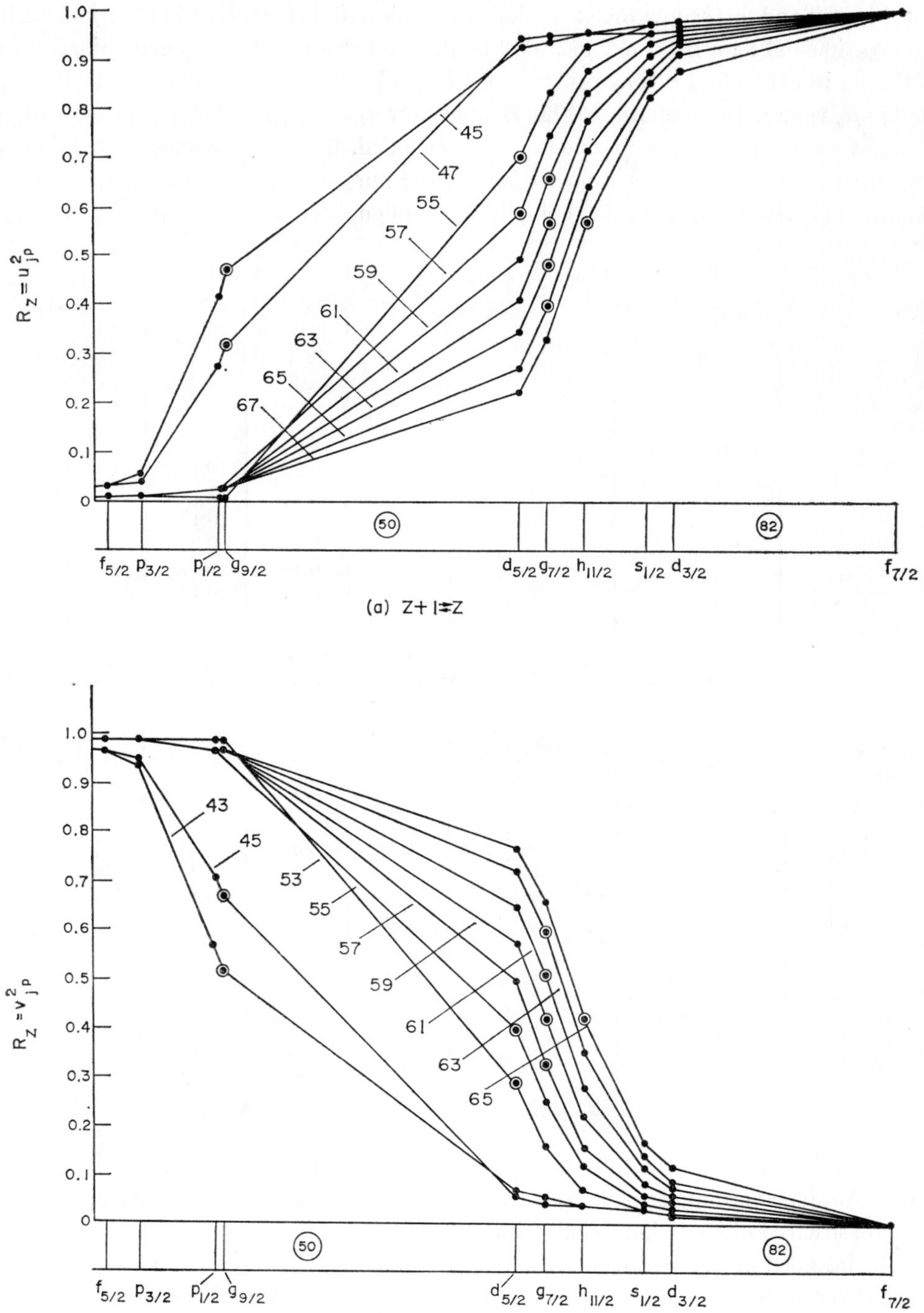

FIG. 6.11. The quantities R_Z, i.e., the pairing correction factors of the β-transition probabilities. The proton systems with $42 < Z \leqslant 46$ and $52 < Z < 68$ and the transitions connecting the quasiparticle vacuum with one-quasiparticle states are considered. The subshell positions are shown below. For remaining notation, see Fig. 6.4.

Tables 6.9 and 6.10 show the examples of the allowed β-transitions in spherical nuclei; the experimental data from ref. 172 were used. The importance of the corrections R and of the statistical factors η is evident; the log $[ft_e R\eta]$ values are considerably smaller than the log ft_e values. The scattering of the ft values for $g_{7/2} \rightleftarrows g_{9/2}$ and $d_{3/2} \rightleftarrows d_{5/2}$ transitions does not change when the corrections $R\eta$ are included. It should be remembered that the noncollective states in spherical nuclei are not the pure quasiparticle states; they contain important admixtures of the more complicated configurations.

TABLE 6.9. *Allowed β-transitions in odd-A spherical nuclei*

Parent nucleus	Transition	Daughter nucleus	$\log ft_e$	R	η	$\log[ft_e R\eta]$
$^{101}Tc_{53}$	$pg_{9/2} \rightleftarrows ng_{7/2}$	$^{101}Ru_{57}$	4.8	0.34	0.1	3.3
$^{101}Rh_{56}$		$^{101}Ru_{57}$	5.0	0.23	0.1	3.4
$^{131}Cs_{76}$		$^{131}Xe_{77}$	5.5	0.25	0.17	4.1
$^{133}Xe_{76}$	$pd_{5/2} \rightleftarrows nd_{3/2}$	$^{133}Cs_{78}$	5.7	0.17	0.25	4.3
$^{135}La_{78}$		$^{135}Ba_{79}$	5.7	0.14	0.17	4.1
$^{137}Pr_{78}$		$^{137}Ce_{79}$	5.4	0.12	0.17	3.7

TABLE 6.10. *Allowed β-transitions in even spherical nuclei*

Parent nucleus	Transition	Daughter nucleus	$\log ft_e$	R	η	$\log[ft_e R\eta]$
$^{104}Rh_{59}$		$^{104}Pd_{58}$	4.5	0.43	0.33	3.7
$^{106}Rh_{61}$	$pg_{9/2} \rightleftarrows ng_{7/2}$	$^{106}Pd_{60}$	5.2	0.32	0.33	4.2
$^{106}Ag_{61}$		$^{106}Pd_{60}$	6.0	0.17	0.33	3.8
$^{108}Ag_{61}$		$^{108}Pd_{62}$	4.5	0.21	0.33	3.4
$^{126}Ce_{71}$		$^{126}Xe_{72}$	5.0	0.32	0.33	4.0
$^{128}I_{75}$		$^{128}Xe_{74}$	6.0	0.12	0.33	4.6
$^{128}Ba_{72}$	$pd_{5/2} \rightleftarrows nd_{3/2}$	$^{128}Ce_{73}$	5.0	0.22	1	4.3
$^{130}Cs_{75}$		$^{130}Ba_{74}$	5.4	0.18	0.33	4.2
$^{134}La_{77}$		$^{134}Ba_{78}$	4.9	0.45	0.33	4.1

The transitions with $\Delta l = 2$, for example, the $p_{3/2} \rightleftarrows f_{5/2}$ or $d_{3/2} \rightleftarrows s_{1/2}$ transitions belong to the l-forbidden group. Their ft_e values are in the interval $\log ft_e = 6.0$–7.0. Such transitions are strictly forbidden in the independent particle model. The quasiparticle interaction must be included in any attempts to explain such transitions.

The effect of the pairing correlations on the β-decay probabilities in spherical nuclei has been often studied, for example, in refs. 303–306.

§ 3. Effect of Pairing Correlations on the Electromagnetic Transition Probabilities

1. We have already discussed several aspects of the nuclear electromagnetic transitions. The main characteristics of the electromagnetic radiation and of the transitions in the single-particle shell model have been treated in § 4 of Chapter 1. The single-particle transitions in deformed nuclei have been discussed in § 5 of the same chapter. The electromagnetic transitions in the rotational model have been studied in § 2, of Chapter 2, and the corrections caused by the Coriolis forces have been mentioned in § 4 of Chapter 4.

In this section we shall study the corrections of the reduced $B(E\lambda)$ and $B(M\lambda)$ transition probabilities caused by the superfluid pairing correlations. The electric and magnetic multipole operators will be expressed in terms of the quasiparticle operators, the explicit forms of $B(E\lambda)$ and $B(M\lambda)$ will be found, and the characteristic features of electromagnetic transitions between quasiparticle states will be discussed.

The probability of the electromagnetic transition is equal to

$$W(\lambda; I_i \to I_f) = 8\pi \frac{\lambda+1}{\lambda[(2\lambda+1)!!]^2} p^{2\lambda+1} B(\lambda), \tag{6.57}$$

and the reduced transition probability equals

$$B(\lambda; I_i \to I_f) = \frac{1}{2I_i+1} \sum_{M_i, M_f, \mu} |\langle I_f M_f | \mathfrak{M}(\lambda\mu) | I_i M_i \rangle|^2. \tag{6.58}$$

Let us find the explicit form of the multipole operators in deformed nuclei. The reduced probability of the electric transition is determined by

$$\begin{aligned} B(E\lambda; I_i \to I_f) = \frac{2\lambda+1}{4\pi} \left(\frac{1}{m\overset{\circ}{\omega}_0}\right)^2 &|(I_i K_i \lambda K_f - K_i | I_f K_f) \langle K_f | \mathfrak{M}(E\lambda; K_f - K_i) | K_i \rangle \\ &+ (-1)^{I_f - \frac{1}{2}} (I_i K_i \lambda - K_f - K_i | I_f - K_f) \langle -K_f | \mathfrak{M}(E\lambda; -K_f - K_i) | K_i \rangle |^2, \end{aligned} \tag{6.59}$$

where $\langle -K_f | = (-1)^{K_f + \frac{1}{2}} \varphi^*_{-K_f}$ for the Nilsson wave functions.

The electric transition operator in the body-fixed coordinate system is equal to

$$\mathfrak{M}(E\lambda) = \sum_{\substack{q, q', \\ \sigma, \sigma'}} \langle q\sigma | \Gamma(E\lambda) | q'\sigma' \rangle a^+_{q\sigma} a_{q'\sigma'}, \tag{6.60}$$

where

$$\Gamma(E\lambda) = \frac{1}{\sqrt{2}} e_{\text{eff}} e r^\lambda (Y_{\lambda\mu} + (-1)^\mu Y_{\lambda, -\mu})$$

and

$$e_{\text{eff}} = \left\{ \frac{1-\tau_z}{2} (1 + e_p^{(\lambda)}) + \frac{1+\tau_z}{2} e_n^{(\lambda)} \right\};$$

the $e_p^{(\lambda)}$ and $e_n^{(\lambda)}$ are the proton, respectively neutron effective charges for multipolarity λ.

The single-particle matrix elements obey the following symmetry relations:

$$\langle q+|\Gamma(E\lambda)|q'+\rangle = \langle q'+|\Gamma(E\lambda)|q+\rangle = \langle q-|\Gamma(E\lambda)|q'-\rangle, \tag{6.61}$$

$$\langle q+|\Gamma(E\lambda)|q'-\rangle = -\langle q-|\Gamma(E\lambda)|q'+\rangle = -\langle q'+|(E\lambda)|q-\rangle. \tag{6.61'}$$

Such relations are easy to prove; they follow from the hermiticity of the operator (6.60) and from the standard relations for the time reversal. Using (1.89) and (1.89″) and the invariance of the electric vector under the time reversal,

$$\mathfrak{T}\langle q+|\Gamma(E\lambda)|q'+\rangle = \langle (q'+)\mathfrak{T}^{-1}|\mathfrak{T}\Gamma(E\lambda)\mathfrak{T}^{-1}|\mathfrak{T}(q+)\rangle = \langle q'-|\Gamma(E\lambda)|q-\rangle,$$

$$\mathfrak{T}\langle q+|\Gamma(E\lambda)|q'-\rangle = \langle (q'-)\mathfrak{T}^{-1}|\mathfrak{T}\Gamma(E\lambda)\mathfrak{T}^{-1}|\mathfrak{T}(q+)\rangle = -\langle q'+|\Gamma(E\lambda)|q-\rangle.$$

Let us define new operators

$$\left.\begin{aligned} A(q,q') &= \frac{1}{\sqrt{2}}\sum_\sigma \sigma\alpha_{q'\sigma}\alpha_{q,-\sigma} = A(q',q), & A^+(q,q') &= \frac{1}{\sqrt{2}}\sum_\sigma \sigma\alpha^+_{q,-\sigma}\alpha^+_{q'\sigma}, \\ \bar{A}(q,q') &= \frac{1}{\sqrt{2}}\sum_\sigma \alpha_{q\sigma}\alpha_{q'\sigma} = -\bar{A}(q',q), & \bar{A}^+(q,q') &= \frac{1}{\sqrt{2}}\sum_\sigma \alpha^+_{q'\sigma}\alpha^+_{q\sigma}; \end{aligned}\right\} \tag{6.62}$$

$$\left.\begin{aligned} B(q,q') &= \sum_\sigma \alpha^+_{q\sigma}\alpha_{q'\sigma}, & B^+(q,q') &= B(q',q), \\ \bar{B}(q,q') &= \sum_\sigma \sigma\alpha^+_{q,-\sigma}\alpha_{q'\sigma}, & \bar{B}^+(q,q') &= -\bar{B}(q',q); \end{aligned}\right\} \tag{6.62'}$$

and the following quantities:

$$u_{qq'} = u_q v_{q'} + u_{q'} v_q, \quad u^{(-)}_{qq'} = u_q v_{q'} - u_{q'} v_q, \tag{6.63}$$

$$v_{qq'} = u_q u_{q'} - v_q v_{q'}, \quad v^{(+)}_{qq'} = u_q u_{q'} + v_q v_{q'}. \tag{6.63'}$$

Now we can express the $\mathfrak{M}(E\lambda)$ operators in terms of the quasiparticle operators and, by using (6.61) and (6.61′), in terms of the new operators (6.62) and (6.62′). As the result, we write:

$$\mathfrak{M}(E\lambda) = \sum_{q,q'}\Big\{\langle q+|\Gamma(E\lambda)|q'+\rangle\Big[v_{qq'}B(q,q') + \frac{1}{\sqrt{2}}u_{qq'}(A^+(q,q') + A(q,q'))\Big]$$
$$+\langle q+|\Gamma(E\lambda)|q'-\rangle\Big[v_{qq'}\bar{B}(q,q') + \frac{1}{\sqrt{2}}u_{qq'}(\bar{A}^+(q,q') + \bar{A}(q,q'))\Big]\Big\}. \tag{6.64}$$

By using the Nilsson wave functions for calculations of the single-particle matrix elements, we come to the final expressions

$$\mathfrak{M}(E\lambda; K_f - K_i) = e_{\text{eff}}e\sum_{q,q'}(G_{E\lambda})_{qq'}\Big\{v_{qq'}B(q,q') + \frac{1}{\sqrt{2}}u_{qq'}(A^+(q,q') + A(q,q'))\Big\}, \tag{6.64'}$$

$$\mathfrak{M}(E\lambda; -K_f - K_i) = e_{\text{eff}}e\sum_{q,q'}(b_{E\lambda}G_{E\lambda})_{qq'}\Big\{v_{qq'}\bar{B}(q,q') + \frac{1}{\sqrt{2}}u_{qq'}(\bar{A}(q,q') + \bar{A}^+(q,q'))\Big\}, \tag{6.64''}$$

where $G_{E\lambda}$ and $b_{E\lambda}$ were defined in (1.92′) and (1.92″).

Similar transformations are performed for the magnetic transitions. The reduced transition probability equals

$$B(M\lambda; I_i \to I_f) = \frac{1}{16\pi}\frac{2\lambda+1}{(m\mathring{\omega}_0)^{\lambda-1}} |(I_i K_i \lambda K_f - K_i | I_f K_f)\langle K_f | \mathfrak{M}(M\lambda; K_f - K_i) | K_i\rangle$$
$$+(-1)^{I_f - \frac{1}{2}}(I_i K_i \lambda - K_f - K_i | I_f - K_f)\langle -K_f | \mathfrak{M}(M\lambda; -K_f - K_i) | K_i\rangle|^2. \quad (6.65)$$

The magnetic transition operator has the form

$$\mathfrak{M}(M\lambda) = \sum_{q, q', \sigma, \sigma'} \langle q\sigma | \Gamma(M\lambda) | q'\sigma'\rangle a^+_{q\sigma} a_{q'\sigma'}, \quad (6.66)$$

where
$$\Gamma(M\lambda) = \frac{e}{2m}\left(g_s s + g_l \frac{2}{\lambda+1} l\right) \nabla \frac{1}{\sqrt{2}}\{r^\lambda (Y_{\lambda\mu} + (-1)^\mu Y_{\lambda, -\mu})\}, \quad (6.66')$$

and the symmetry relations of the matrix elements are

$$\langle q+ | \Gamma(M\lambda) | q'+\rangle = \langle q'+ | \Gamma(M\lambda) | q+\rangle = -\langle q- | \Gamma(M\lambda) | q'-\rangle, \quad (6.67)$$
$$\langle q+ | \Gamma(M\lambda) | q'-\rangle = \langle q'+ | \Gamma(M\lambda) | q-\rangle = \langle q'- | \Gamma(M\lambda) | q+\rangle. \quad (6.67')$$

By using the quasiparticle operators and the above-stated symmetry relations,

$$\mathfrak{M}(M\lambda) = \sum_{q, q'} \Big\{ \langle q+ | \Gamma(M\lambda) | q'+\rangle \left[v^{(+)}_{qq'} \mathfrak{B}(q, q') + \frac{1}{\sqrt{2}} u^{(-)}_{qq'} (\mathfrak{A}^+(q, q') + \mathfrak{A}(q, q'))\right]$$
$$+ \langle q+ | \Gamma(M\lambda) | q'-\rangle \left[v^{(+)}_{qq'} \overline{\mathfrak{B}}(q, q') - \frac{1}{\sqrt{2}} u^{(-)}_{qq'} (\overline{\mathfrak{A}}^+(q, q') + \overline{\mathfrak{A}}(q, q'))\right]\Big\}, \quad (6.68)$$

where

$$\left.\begin{aligned} \mathfrak{A}(q, q') &= \frac{1}{\sqrt{2}} \sum_\sigma \alpha_{q'\sigma} \alpha_{q, -\sigma} = -\mathfrak{A}(q', q), & \mathfrak{A}^+(q, q') &= \frac{1}{\sqrt{2}} \sum_\sigma \alpha^+_{q, -\sigma} \alpha^+_{q'\sigma}, \\ \overline{\mathfrak{A}}(q, q') &= \frac{1}{\sqrt{2}} \sum_\sigma \sigma \alpha_{q\sigma} \alpha_{q'\sigma} = -\overline{\mathfrak{A}}(q', q), & \overline{\mathfrak{A}}^+(q, q') &= \frac{1}{\sqrt{2}} \sum_\sigma \sigma \alpha^+_{q'\sigma} \alpha^+_{q\sigma}; \end{aligned}\right\} \quad (6.62'')$$

$$\left.\begin{aligned} \mathfrak{B}(q, q') &= \sum_\sigma \sigma \alpha^+_{q, -\sigma} \alpha_{q', -\sigma}, & \mathfrak{B}^+(q, q') &= \mathfrak{B}(q', q), \\ \overline{\mathfrak{B}}(q, q') &= \sum_\sigma \alpha^+_{q, -\sigma} \alpha_{q'\sigma}, & \overline{\mathfrak{B}}^+(q, q') &= \overline{\mathfrak{B}}(q', q). \end{aligned}\right\} \quad (6.62''')$$

The single-particle matrix elements can be calculated using the Nilsson model wave functions. The magnetic transition operators in (6.65) are expressed by

$$\mathfrak{M}(M\lambda; K_f - K_i) = \frac{e}{2m} \sum_{q, q'} (G_{M\lambda})_{qq'} \left\{ v^{(+)}_{qq'} \mathfrak{B}(q, q') + \frac{1}{\sqrt{2}} u^{(-)}_{qq'} (\mathfrak{A}^+(q, q') + \mathfrak{A}(q, q'))\right\}, \quad (6.68')$$

$$\mathfrak{M}(M\lambda; -K_f - K_i) = \frac{e}{2m} \sum_{q, q'} (G_{M\lambda} b_{M\lambda})_{qq'} \left\{ v^{(+)}_{qq'} \overline{\mathfrak{B}}(q, q') - \frac{1}{\sqrt{2}} u^{(-)}_{qq'} (\overline{\mathfrak{A}}^+(q, q') + \overline{\mathfrak{A}}(q, q'))\right\}; \quad (6.68'')$$

$G_{M\lambda}$ and $b_{M\lambda}$ were defined in (1.93′) and (1.93″).

2. Let us find the explicit form of the electric and magnetic multipole operators in spherical nuclei. The new operators are defined by:

$$A^+(j_1, j_2; \lambda\mu) = \sum_{m_1, m_2} (j_1m_1j_2m_2 \,|\, \lambda\mu)\alpha^+_{j_1m_1}\alpha^+_{j_2m_2}, \tag{6.69}$$

$$A(j_1, j_2; \lambda\mu) = \sum_{m_1, m_2} (j_1m_1j_2m_2 \,|\, \lambda\mu)\alpha_{j_2m_2}\alpha_{j_1m_1}, \tag{6.69'}$$

$$B(j_1, j_2; \lambda\mu) = \sum_{m_1, m_2} (-1)^{j_2+m_2}(j_1m_1j_2m_2 \,|\, \lambda\mu)\alpha^+_{j_1m_1}\alpha_{j_2, -m_2}. \tag{6.69''}$$

The reduced transition probability (6.58) is, after simple transformations, brought to the form

$$B(\lambda; I_i \to I_f) = \frac{1}{2I_i+1}|\langle I_f|\,\mathfrak{M}(\lambda)\,|I_i\rangle|^2.$$

The electric multipole operator equals

$$\mathfrak{M}(E\lambda) = \sum_{j_1, j_2} \sum_{m_1, m_2} \langle j_2m_2|\Gamma(E; \lambda\mu)\,|j_1m_1\rangle a^+_{j_2m_2}a_{j_1m_1}, \tag{6.70}$$

where

$$\Gamma(E; \lambda\mu) = e\left\{\frac{1-\tau_z}{2}(1+e_p^{(\lambda)})+\frac{1+\tau_z}{2}e_n^{(\lambda)}\right\}r^\lambda Y_{\lambda\mu}. \tag{6.70'}$$

The magnetic multipole operator equals

$$\mathfrak{M}(M\lambda) = \sum_{j_1, j_2} \sum_{m_1, m_2} \langle j_2m_2|\Gamma(M; \lambda\mu)\,|j_1m_1\rangle a^+_{j_2m_2}a_{j_1m_1}, \tag{6.71}$$

where

$$\Gamma(M; \lambda\mu) = \left(\frac{e}{2m}\right)\left(g_s\boldsymbol{s}+g_l\frac{2}{\lambda+1}\boldsymbol{l}\right)\nabla(r^\lambda Y_{\lambda\mu}). \tag{6.71'}$$

Let us use the quasiparticle operators and the symmetry relation (see ref. 27)

$$\langle j_1|\Gamma(E\lambda)\,|j_2\rangle = (-1)^{j_1-j_2-\lambda}\langle j_2|\Gamma(E\lambda)\,|j_1\rangle. \tag{6.72}$$

The electric transition operator is then of the form

$$\mathfrak{M}(E\lambda) = \frac{-1}{\sqrt{2\lambda+1}}\sum_{j_1, j_2}\langle j_2|\Gamma(E\lambda)\,|j_1\rangle(-1)^{j_1+j_2-\lambda}\times$$
$$\times\{v_{j_1j_2}B(j_1, j_2; \lambda\mu)+\tfrac{1}{2}u_{j_1j_2}[A^+(j_1, j_2; \lambda\mu)+(-1)^{\lambda-\mu}A(j_1, j_2; \lambda-\mu)]\}, \tag{6.73}$$

where $u_{j_1j_2}$ and $v_{j_1j_2}$ were defined in (6.63) and (6.63').

The symmetry rule of the magnetic transition operator is different, namely

$$\langle j_1|\Gamma(M\lambda)\,|j_2\rangle = (-1)^{j_1+j_2-\lambda}\langle j_2|\Gamma(M\lambda)\,|j_1\rangle, \tag{6.72'}$$

and after the quasiparticle transformation the resulting operator equals

$$\mathfrak{M}(M\lambda) = \frac{1}{\sqrt{2\lambda+1}}\sum_{j_1,j_2}\langle j_2|\Gamma(M\lambda)\,|j_1\rangle\times(-1)^{j_1+j_2-\lambda}\times$$
$$\times\{v^{(+)}_{j_1j_2}B(j_1, j_2; \lambda\mu)+\tfrac{1}{2}u^{(-)}_{j_1j_2}[A^+(j_1, j_2; \lambda\mu)-(-1)^{\lambda-\mu}A(j_1, j_2; \lambda-\mu)]\}; \tag{6.74}$$

the functions $u^{(-)}_{j_1j_2}$ and $v^{(+)}_{j_1j_2}$ were defined in (6.63) and (6.63').

3. Let us discuss the structure of the electric and magnetic multipole operators. We shall find the corresponding reduced transition probabilities in both spherical and deformed nuclei and study the role of the pairing corrections.

Equations (6.64), (6.68), (6.73), and (6.74) contain two types of terms. The terms containing the operators $B(q, q')$, $\bar{B}(q, q')$, $\mathfrak{B}(q, q')$, $\bar{\mathfrak{B}}(q, q')$, and $B(j_1, j_2; \lambda\mu)$ describe the transitions between the states with an equal number of quasiparticles. For example, γ-transitions between the one-quasiparticle states in odd-A nuclei belong to this group. On the other hand, the terms containing the operators $A(q, q')$, $\bar{A}(q, q')$, $\mathfrak{A}(q, q')$, $\bar{\mathfrak{A}}(q, q')$, and $A(j_1, j_2; \lambda\mu)$ and their Hermitian conjugated operators describe the transitions in which the number of quasiparticles was changed by two units. For example, the γ-transitions between the two-quasiparticle states and the ground state (i.e., quasiparticle vacuum) of even–even nuclei belong to this group.

Equations (6.64) (6.68), (6.73), and (6.74) define the explicit form of the electric and magnetic transition operators. They are easy to use, and it is simple to formulate the quasiparticle selection rules for the nuclear γ-transitions. The electromagnetic transitions can connect two states having the following properties: (a) the states have the same total number of quasiparticles and differ only in the state of a single quasiparticle; (b) one of the states has two quasiparticles more (or less) than the other one, and the remaining quasiparticles are in the same positions in both states.

We shall need several matrix elements for the calculations of the reduced electromagnetic transition probabilities. The wave function of the quasiparticle vacuum, one- and two-quasiparticle states are defined in (4.22), (4.31), and (4.33) for the deformed nuclei. Using these equations,

$$\left.\begin{aligned}
(\Psi_0^*\alpha_{q_1\sigma_1}B(q, q')\alpha^+_{q_2\sigma_2}\Psi_0) &= \delta_{\sigma_1\sigma_2}\delta_{q_1q}\delta_{q'q_2},\\
(\Psi_0^*\alpha_{q_1\sigma_1}\bar{B}(q, q')\alpha^+_{q_2\sigma_2}\Psi_0) &= \sigma_2\delta_{\sigma_1, -\sigma_2}\delta_{q_1q}\delta_{q'q_2},\\
(\Psi_0^*\alpha_{q_1\sigma_1}\mathfrak{B}(q, q')\alpha^+_{q_2\sigma_2}\Psi_0) &= -\sigma_1\delta_{\sigma_1, \sigma_2}\delta_{q_1q}\delta_{q'q_2},\\
(\Psi_0^*\alpha_{q_1\sigma_1}\bar{\mathfrak{B}}(q, q')\alpha^+_{q_2\sigma_2}\Psi_0) &= \delta_{\sigma_1, -\sigma_2}\delta_{q_1q}\delta_{q'q_2},
\end{aligned}\right\} \tag{6.75}$$

and

$$\left.\begin{aligned}
(\Psi_0[A^+(q, q')+A(q, q')]\alpha^+_{q_1\sigma_1}\alpha^+_{q_2\sigma_2}\Psi_0) &= \delta_{\sigma_2, -\sigma_1}\frac{\sigma_2}{\sqrt{2}}(\delta_{qq_1}\delta_{q'q_2}+\delta_{qq_2}\delta_{q'q_1}),\\
(\Psi_0^*[\bar{A}^+(q, q')+\bar{A}(q, q')]\alpha^+_{q_1\sigma_1}\alpha^+_{q_2\sigma_2}\Psi_0) &= \delta_{\sigma_1\sigma_2}\frac{1}{\sqrt{2}}(\delta_{q'q_1}\delta_{qq_2}-\delta_{qq_1}\delta_{q'q_2}),\\
(\Psi_0^*[\mathfrak{A}^+(q, q')+\mathfrak{A}(q, q')]\alpha^+_{q_1\sigma_1}\alpha^+_{q_2\sigma_2}\Psi_0) &= \delta_{\sigma_1, -\sigma_2}\frac{1}{\sqrt{2}}(\delta_{qq_1}\delta_{q'q_2}-\delta_{qq_2}\delta_{q'q_1}),\\
(\Psi_0^*[\bar{\mathfrak{A}}^+(q, q')+\bar{\mathfrak{A}}(q, q')]\alpha^+_{q_1\sigma_1}\alpha^+_{q_2\sigma_2}\Psi_0) &= \delta_{\sigma_1\sigma_2}\frac{\sigma_1}{\sqrt{2}}(\delta_{qq_2}\delta_{q'q_1}-\delta_{qq_1}\delta_{q'q_2}).
\end{aligned}\right\} \tag{6.75'}$$

The same wave functions for the spherical nuclei are defined in (4.22″), (4.31″), and (4.33″). They lead to

$$(\Psi_0^*\alpha_{j_1m_1}B(j, j'; \lambda\mu)\alpha^+_{j_2m_2}\Psi_0) = (-1)^{j_2-m_2}\delta_{j, j_1}\delta_{j'j_2}(j_1m_1j_2-m_2\,|\,\lambda\mu), \tag{6.76}$$

$$(\Psi_0^*A(j, j'; \lambda-\mu)\Psi_0(I_iM_i\,|j_1j_2)) = \delta_{\lambda I_i}\delta_{\mu, -M_i}(\delta_{jj_1}\delta_{j'j_2}-(-1)^{j_1+j_2-\lambda}\delta_{jj_2}\delta_{j'j_1}). \tag{6.76'}$$

The products of the form $\prod_{q\neq q_1, q_2 \ldots} (u_q u'_q + v_q v'_q)^2$ describe the difference in the superfluid characteristics of the initial and final states. Such terms have in most cases values close to unity, and thus can be omitted.

The reduced γ-transition probabilities in deformed nuclei can be calculated from (6.64′), (6.64″), (6.68′), (6.68″), (6.75), and (6.75′). The following expressions describe the transitions without change in the number of quasiparticles:

$$B(E\lambda;\, I_iK_iq_2 \to I_fK_fq_1) = \frac{2\lambda+1}{4\pi}\frac{e^2_{\text{eff}}e^2}{(m\overset{\circ}{\omega}_0)^2} v^2_{q_1q_2} |(I_iK_i\lambda K_f - K_i | I_fK_f)(G_{E\lambda})_{q_1q_2}$$
$$+(-1)^{I_f+K_f}(I_iK_i\lambda - K_f - K_i | I_f - K_f)(b_{E\lambda}G_{Fq})_{q_1q_2}|^2, \quad (6.77)$$

$$B(M\lambda;\, I_iK_iq_2 \to I_fK_fq_1) = \frac{1}{16\pi}\left(\frac{e}{2m}\right)^2 \frac{2\lambda+1}{(m\overset{\circ}{\omega}_0)^{\lambda-1}} (v^{(+)}_{q_1q_2})^2 \times$$
$$\times |(I_iK_i\lambda K_f - K_i | I_fK_f)(G_{M\lambda})_{q_1q_2} + (-1)^{I_f+K_f}\times$$
$$\times (I_iK_i\lambda - K_f - K_i | I_f - K_f)(b_{M\lambda}G_{M\lambda})_{q_1q_2}|^2. \quad (6.78)$$

These formulae describe the γ-transitions between one-quasiparticle states in odd-A nuclei, between two-quasiparticle states in even nuclei, etc. The presence of other quasiparticles, not participating in the γ-transition, causes blocking of the corresponding single-particle levels, i.e., the values of $v_{qq'}$ and $v^{(+)}_{qq'}$ are affected.

The reduced γ-transition probabilities with the quasiparticle number changed by two are equal to

$$B(E\lambda;\, I_iK_iq_1q_2q_3\ldots \to I_fK_fq_3\ldots) = \frac{2\lambda+1}{4\pi}\frac{e^2_{\text{eff}}e^2}{(m\overset{\circ}{\omega}_0)^2} u^2_{q_1q_2}\times$$
$$\times |(I_iK_i\lambda K_f - K_i | I_fK_f)(G_{E\lambda})_{q_1q_2} + (-1)^{I_f+K_f}\times$$
$$\times (I_iK_i\lambda - K_f - K_i | I_f - K_f)(b_{E\lambda}G_{E\lambda})_{q_1q_2}|^2, \quad (6.77')$$

$$B(M\lambda;\, I_iK_iq_1q_2q_3\ldots \to I_fK_fq_3\ldots) = \frac{1}{16\pi}\left(\frac{e}{2m}\right)^2 \frac{2\lambda+1}{(m\overset{\circ}{\omega}_0)^{\lambda-1}}\times$$
$$\times (u^{(-)}_{q_1q_2})^2 |(I_iK_i\lambda K_f - K_i | I_fK_f)(G_{M\lambda})_{q_1q_2} + (-1)^{I_f+K_f}\times$$
$$\times (I_iK_i\lambda - K_f - K_i | I_f - K_f)(b_{M\lambda}G_{M\lambda})_{q_1q_2}|^2. \quad (6.78')$$

These expressions describe the γ-transitions from the two-quasiparticle states to the ground states in even–even nuclei or from the three-quasiparticle states to the one-quasiparticle states in odd-A nuclei, etc.

The above expressions for $B(E\lambda)$ and $B(M\lambda)$ contain two terms. However, in a majority of cases one of them contributes overwhelmingly: if $\lambda \geqslant |K_i - K_f|$, then the first term is the contributing one, and if $\lambda \geqslant K_i + K_f$, then the second term contributes the most. The states with $K = \frac{1}{2}$ are exceptional; in the transitions involving such states, both above-mentioned terms contribute. The quantities $G_{E\lambda}$, $b_{E\lambda}$ and $G_{M\lambda}$, $b_{M\lambda}$ are usually calculated by means of the Nilsson wave functions. They can be replaced by the corresponding expressions using the Woods–Saxon wave functions.

The reduced γ-transition probabilities in spherical nuclei are even simpler; they follow from (6.73), (6.74), (6.76), and (6.76′). If the number of quasiparticles is unchanged and one quasiparticle from the subshell j_2 is transferred to the subshell j_1, the reduced transition probabilities equal

$$B(E\lambda; j_2 \to j_1) = \frac{1}{2j_2+1} |\langle j_2 | \Gamma(E\lambda) | j_1 \rangle|^2 \, v_{j_2 j_1}^2, \tag{6.79}$$

$$B(M\lambda; j_2 \to j_1) = \frac{1}{2j_2+1} |\langle j_2 | \Gamma(M\lambda) | j_1 \rangle|^2 \, (v_{j_1 j_2}^{(+)})^2. \tag{6.80}$$

These expressions describe the γ-transitions between the one-quasiparticle states in odd-A nuclei, between the two-quasiparticle states in even nuclei, etc.

In the transitions from the two-quasiparticle states to the ground states of even–even nuclei, the reduced γ-transition probability equals

$$B(E\lambda; I_i M_i j_1 j_2 \to I_f = 0) = \frac{\delta_{I_i\lambda}}{2I_i+1} u_{j_1 j_2}^2 |\langle j_2 | \Gamma(E\lambda) | j_1 \rangle|^2, \tag{6.79′}$$

$$B(M\lambda; I_i M_i j_1 j_2 \to I_f = 0) = \frac{\delta_{I_i\lambda}}{2I_i+1} (u_{j_1 j_2}^{(-)})^2 |\langle j_2 | \Gamma(M\lambda) | j_1 \rangle|^2. \tag{6.80′}$$

The same equations are applicable to the transitions from three-quasiparticle states to one-quasiparticle states in odd-A nuclei, or to the transitions from four-quasiparticle states to two-quasiparticle states in even nuclei.

Let us discuss the role of the pairing correlations in the nuclear γ-transitions. The effect is characterized by the correction factor R equal to:

(a)

$$\left.\begin{aligned} R_\gamma &= (u_{q_1}u_{q_2} - v_{q_1}v_{q_2})^2 \quad \text{for the } E\lambda \text{ transitions,} \\ R_\gamma &= (u_{q_1}u_{q_2} + v_{q_1}v_{q_2})^2 \quad \text{for the } M\lambda \text{ transitions,} \end{aligned}\right\} \tag{6.81}$$

when the number of quasiparticles is unchanged;

(b)

$$\left.\begin{aligned} R_\gamma &= (u_{q_1}v_{q_2} + u_{q_2}v_{q_1})^2 \quad \text{for the } E\lambda \text{ transitions,} \\ R_\gamma &= (u_{q_1}v_{q_2} - u_{q_2}v_{q_1})^2 \quad \text{for the } M\lambda \text{ transitions,} \end{aligned}\right\} \tag{6.81′}$$

when the number of quasiparticles is changed by two during the transition.

The quantum numbers q_1 and q_2 belong to the single-particle states where a quasiparticle has been formed or has disappeared. The labels q_1 and q_2 are replaced by j_1 and j_2 in the case of spherical nuclei.

The two parts of the pairing correction factor R_γ correspond to the direct and reverse processes. Their relative sign is determined by the time-reversal properties of the involved operators. The superfluid correction factors R_γ vary in an interval $10^{-5}-1$, i.e., the interval is much broader than in the case of the β-transitions discussed in the preceding section. The smallness of R_γ is sometimes caused by the small difference of the two almost equal terms. The accuracy of R_γ is insufficient in such a case; the R_γ values are extremely sensitive to the unknown exact position of the single-particle levels and to the equally poorly known admixtures of the more complicated (e.g., three-quasiparticle) configurations.

Without the pairing superfluid correlations, that is, if $G = 0$, we have two possibilities: $R_\gamma = 0$ or $R_\gamma = 1$. The particle–hole transitions with unchanged quasiparticle number and the transitions where two quasiparticles were formed or disappeared on two levels, both below or both above the Fermi level, have $R_\gamma = 0$ at $G = 0$. This means that such transitions are forbidden in the independent particle model.

The effect of the pairing correlations on the nuclear electromagnetic transitions has been discussed often, for example, in refs. 12, 146, 152, 296, 307, and 308.

4. Let us study the characteristic features of the electromagnetic transitions between the quasiparticle states in deformed and in spherical nuclei.

The experimental reduced γ-transition probabilities are deduced from the measured lifetimes of the excited states and from the cross-sections of the Coulomb excitation. The data are often presented in the form of the hindrance factors with respect to the Weisskopf single-particle values (1.67) and (1.74). The hindrance factor is the ratio

$$F_W = \frac{T_{(1/2)\gamma}(\text{exp})}{T_{(1/2)\gamma}(\text{W})} = \frac{B(\lambda)_{\text{s.p.}}}{B(\lambda)_{\text{exp}}}. \tag{6.82}$$

The other useful quantities are the hindrance factors F_N and F_{WS}. They express the ratio of the experimental lifetime to the single-particle value using the Nilsson wave functions, i.e.,

$$F_N = \frac{T_{(1/2)\gamma}(\text{exp})}{T_{(1/2)\gamma}(\text{N})} = \frac{B(\lambda)_N}{B(\lambda)_{\text{exp}}}, \tag{6.82'}$$

or the Woods–Saxon wave functions, i.e.,

$$F_{WS} = \frac{T_{(1/2)\gamma}(\text{exp})}{T_{(1/2)\gamma}(\text{WS})} = \frac{B(\lambda)_{WS}}{B(\lambda)_{\text{exp}}}. \tag{6.82''}$$

Several papers, for example, 299 and 307–317, were devoted to the systematization of the experimental data, to the calculation of the $B(E\lambda)$ and $B(M\lambda)$ values, and to the discussion of the pairing correction factors R_γ.

Let us treat the γ-transitions in deformed nuclei. The $E1$ transitions between one-quasiparticle states in odd-A nuclei are the most extensively studied type of the transitions. The F_W have the values 10^3–10^7 for these transitions. The quantities F_N are in the interval 0.1–10 when we consider only the $\Delta K = 0$ $E1$ transitions; the inclusion of the pairing correlations somewhat increases the discrepancy between the theory and the experimental data. However, $F_{WS} > F_N$ according to the results of ref. 299. In most cases the inclusion of the pairing correlations improves the agreement between the theory and the experimental data in the calculations using the Woods–Saxon wave functions. All these features are shown in Table 6.11. The table includes the hindrance factors F_W, F_N, and F_{WS} defined by (6.82), (6.82′), and (6.82″), and the quantities $R_\gamma F_N$ and $R_\gamma F_{WS}$. The $E1$ transitions with $\Delta K = \pm 1$ are also included in the table. Evidently, the agreement between the calculated values and the experimental data is not very good.

The hindrance factors F_N for the same single-particle transitions are often different in different nuclei. These variations are in many cases explained by the inclusion of the

TABLE 6.11. *The E1 transitions in odd-A deformed nuclei*

Nucleus	State Initial	State Final	E_γ (keV)	$T_{(1/2)\gamma}$ (s)	F_W	F_N	$R_\gamma F_N$	F_{WS}	$R_\gamma F_{WS}$
				$\Delta K = 0$					
^{173}Yb	514↓	633↑	286	8×10^{-10}	–	0.70	8×10^{-4}	30	0.06
^{161}Dy	523↓	642↑	25.7	9×10^{-8}	7×10^3	0.84	5×10^{-3}	22	2.4
^{169}Tm	523↑	404↓	63.1	1×10^{-7}	1×10^5	1.6	–	217	16
^{153}Eu	532↑	413↓	97	2×10^{-10}	8×10^2	0.01	9×10^{-4}	1.6	0.4
				$\Delta K = \pm1$					
^{177}Yb	514↓	624↑	104	7×10^{-9}	4×10^4	8	–	6.4	3.6
^{173}Yb	633↑	512↑	351	2×10^{-8}	5×10^6	880	500	73	29
^{169}Yb	512↑	633↑	191	4×10^{-9}	4×10^9	36	–	1.2	0.3
^{157}Gd	642↑	521↑	64	2×10^{-6}	2×10^6	920	–	36	0.2
^{179}Ta	514↑	404↓	30.7	7×10^{-6}	1×10^6	77	31	3.2	0.4
^{175}Lu	514↑	444↓	396	6×10^{-9}	2×10^6	182	36	6	2.9
^{159}Tb	532↑	411↑	363	1×10^{-10}	3×10^4	0.7	0.2	0.6	0.2
^{155}Eu	411↑	532↑	141	5×10^{-9}	3×10^3	1.8	1×10^{-2}	1.4	0.02

pairing correction factor R_γ. Table 6.12 shows as an example the $E1$ transitions between the $n743$↑ and $n622$↑ states in ^{234}U and ^{239}Pu. The ratio $R_\gamma F_N(^{239}\text{Pu})/R_\gamma F_N(^{234}\text{U})$ equals 0.8, i.e., the pairing correlations explain the variation of the F_N value between ^{239}Pu and ^{234}U.

TABLE 6.12. *The E1 transitions between the neutron n743↑ and n622↑ states in ^{234}U and ^{239}Pu*

Nucleus	Initial state $K\pi$	Initial state E (keV)	Final state $K\pi$	Final state E (keV)	Configuration Initial state	Configuration Final state	F_N	R_γ	$R_\gamma F_N$
^{234}U	5^+	1552	6^-	1421	$nF633↓$ $+n(F+2)$ 622↑	$nF633↑$ $+n(F+1)$ 743↑	0.27	0.61	0.16
^{239}Pu	$\frac{7}{2}^-$	392	$\frac{5}{2}^+$	286	$n(F-1)$ 743↑	$n(F+1)$ 622↑	13	0.009	0.12

The electric $E2$ transitions in deformed nuclei are divided into two groups. The first group contains the $\Delta K = 1$ transitions with $F_W = 0.1-60$ and $F_N = 10^{-5}-10^{-2}$. The enhancement of the $B(E2)$ values of these transitions is caused by the admixtures of collective states. The second group contains the $\Delta K = 2$ transitions; the F_N values are much larger than in the previous case, $F_N = 10^2-10^4$. The experimental data on $E3$ transitions are very rare; the known transitions have $F_N \approx 10^{-2}-10$.

The known magnetic dipole transitions in deformed nuclei have the hindrance factors of the following magnitude: $F_W = 10^2-10^5$, $F_N = 0.2-200$, and $F_{WS} = 0.2-20$. The calculated $B(M1)$ values depend on the collective gyromagnetic ratio g_R. The experimental data on the reduced transition probabilities of the $M2$ and $M3$ transitions are rare[311]; the known lifetimes give the following hindrance factors: $F_W = 10-100$, $F_N = 0.1-50$, and $F_{WS} = 0.1-5$

for the $M2$ transitions, and $F_W = 1\text{–}10^3$, $F_N = 0.01\text{–}100$, and $F_{WS} = 0.1\text{–}100$ for the $M3$ transitions.

As we have stated earlier, the large variations of the experimentally determined hindrance factors are caused by the impurities in the nuclear wave functions. The admixtures caused by the Coriolis coupling are particularly important.

We have shown in § 2 of Chapter 2 that the intensity ratios of the γ-transitions between two rotational bands in deformed nuclei should fulfill the Alaga rules. In many cases, these rules describe the experimental facts quite well. However, in a number of cases the deviations from the Alaga rules are substantial. Particularly large deviations are observed in the asymptotically forbidden single-particle transitions. The admixtures in the wave functions, caused by the Coriolis coupling, can remove the forbiddenness related to the asymptotic quantum numbers. Such effects are observed, for example, in the $\Delta K = \pm 1$ $E1$ transitions.(91)

The K-forbidden γ-transitions were discussed in §§ 2 and 4 of Chapter 4. We have stated there that the γ-transition probabilities are reduced approximately 100 times for each unit of the K-forbiddenness. Such transitions were studied in ref. 319, and the results confirm our previous conclusion.

The Λ-forbidden γ-transitions, analogous to the Λ-forbidden β-decays discussed in § 2 of this chapter, form another type of transition. An example of the twice Λ-forbidden $M2$ transition has been found in ref. 320 (the transition is called Ω-forbidden there). The measured hindrance factor $F_W = 800$ for the $M2$ transitions from the $K^\pi = 3^+$ $n633\uparrow - p411\downarrow$ state to the $K^\pi = 1^-$ $n521\downarrow + p411\downarrow$ state in ^{170}Tm.(320) The transition thus involves not only $n633\uparrow \to n521\downarrow$ but also the change in projection of the proton $p411\downarrow$ state. This fact explains the small rate of the observed transition.

We shall now discuss the γ-transitions between the quasiparticle states in spherical nuclei.

The $E1$ transitions in odd-A spherical nuclei are strongly hindered; they have $F_W = 2\times 10^4\text{–}10^6$ according to ref. 310. The $E1$ transition in Pb^{205} is, however, an exception; it has $F_W = 6$ only. Many features of the $E1$ transitions are caused by the relatively large admixtures to the quasiparticle states.

The $M4$ transitions in odd-A spherical nuclei were studied in ref. 296. These calculations show that the correction factors R_γ calculated using the quasiparticle wave functions (4.33″) and the projected wave functions (5.67) cannot explain the experimental $M4$ transition probabilities in nuclei with one closed shell. The calculated $B(M4)$ values differ from the experimental data by factors of 2–3.

In most cases it is useless to describe the γ-transitions in spherical nuclei in a framework of the independent quasiparticle model. The large admixtures to the quasiparticle states are too important, and thus they render the model inadequate. This statement is particularly true for the l-forbidden γ-transitions.

§ 4. Spectroscopic Factors of Direct Nuclear Reactions

1. In this section we shall formulate the basic expressions for the cross-sections of direct nuclear reactions. The reactions will be described by the distorted wave Born approximation, and the nuclear structure will be defined by the independent quasiparticle model. Our aim is to find out what structure information about the low-lying nuclear states is contained in the direct nuclear reactions at low energies.

The nuclear reactions are divided into two groups. The first type are the direct reactions which affect only a small fraction of the nuclear degrees of freedom. The second group contains the compound nuclear reactions in which many nuclear degrees of freedom are affected.

The most outstanding feature of the direct nuclear reactions is a small number of the nuclear degrees of freedom participating in the interaction of the incident particle with the nucleus. Consequently, the transition amplitude is determined by a relatively simple relation involving the wave function of the initial and final states. The Butler theory describing the stripping (*dp*) reaction in the plane wave Born approximation has been rather successful. In the next stage of the development it was necessary to overcome the mathematical difficulties related to the effect of the scattered waves on the reaction amplitude. The calculations have moved closer to the experimental data after the distorted waves had replaced the plane waves in the description of particle motion before and after the reaction. The agreement has improved in both the angular distribution and the absolute cross-section results. The distorted wave Born approximation gives a basically correct description of the direct nuclear reactions at low energies. However, a number of problems related to the reaction mechanism still remain open.

The theory has been explained in a number of publications, for example, in refs. 18–20, or 321–330.

The wave functions of the system in the initial or final states are represented in the distorted wave approximation as the products $\Phi\Psi$. The function Φ describes the relative motion of particles in the incoming and outgoing channels, while the function Ψ describes the intrinsic structure of the nuclear states. Such a factorization allows to separate the kinematic terms from the portion describing the nuclear structure. The cross-section of the direct nuclear reaction has the form

$$\frac{d\sigma}{d\Omega} = N \sum_{\xi} S_{\xi} \sigma_{\xi}(\theta), \tag{6.83}$$

where N is the normalization and statistical factor and ξ denotes the complete set of quantum numbers. The kinematic part of the cross-section $\sigma_{\xi}(\theta)$ depends on the energy and on the scattering angle. It contains the overlap integrals of the distorted waves and the bound state wave functions of the transferred particles. The quantity S_{ξ} is the spectroscopic factor in the nucleon transfer reactions, and it is replaced by the reduced probability $B(E\lambda)$ in the inelastic scattering. The quantities S_{ξ} or $B(E\lambda)$ are the primary sources of the nuclear structure information obtained from the direct nuclear reaction studies.

The direct nuclear reactions are treated as one-step processes. This means that the transfers of the angular momentum, of the linear momentum, and of the energy, are realized without additional nuclear intermediate states. Such a simple treatment explains the success of the distorted wave method. But the above described assumptions are invalid if the target nucleus has low-lying collective states with large excitation cross-section. The distorted wave approximation must be corrected in such cases. The method of coupled channels(329) offers such an improvement.

To conclude this general introduction, we shall enumerate the nuclear characteristics obtained in the direct nuclear reaction studies. They are: excitation energies, spins and parities of the excited states, and either the reduced $B(E\lambda)$ probabilities of the electric transitions or the spectroscopic factors S_ξ. In the following discussion we shall show which reactions excite the specific types of the nuclear states and how we may obtain some additional information (for example, additional quantum numbers) about the excited states.

2. Let us study the excitation of the low-lying collective states in inelastic scattering. The excitation mechanism is usually treated as a one-step process. The scattering amplitude can be factorized in such a case.

The cross-section of the inelastic scattering of the particle with charge Z_1e on the nucleus with charge Z_2e has in the distorted wave Born approximation the following form:

$$\frac{d\sigma(\theta)}{d\Omega} = \frac{2I_i+1}{2I_f+1} \sum_{\lambda,\mu} |\langle I_f|\, \mathfrak{M}_E^*(\lambda)\, |I_i\rangle\, \mathcal{L}^c_{\lambda\mu}(\theta) + \langle I_f|\, \mathfrak{M}^*_{\mathrm{nucl}}(\lambda)\, |I_i\rangle\, \mathcal{L}^{\mathrm{nucl}}_{\lambda\mu}(\theta)|^2. \tag{6.84}$$

The $\mathcal{L}^c_{\lambda\mu}(\theta)$ and $\mathcal{L}^{\mathrm{nucl}}_{\lambda\mu}(\theta)$ are the kinematical factors. The electric and nuclear transition operators are determined by

$$\mathfrak{M}_E(\lambda\mu) = \frac{3}{4\pi}(1+\tfrac{1}{2}\lambda\delta_{\lambda 3})e_{\mathrm{eff}}Z_2R_0^\lambda\alpha_{\lambda\mu}, \tag{6.85}$$

$$\mathfrak{M}_{\mathrm{nucl}}(\lambda\mu) = \alpha_{\lambda\mu}. \tag{6.86}$$

The parameters $\alpha_{\lambda\mu}$ describe the deviation of the nuclear surface from a sphere; they have been discussed in § 2 of Chapter 2. The parameters $\alpha_{\lambda\mu}$ enter in both the expressions (6.85) and (6.86). The nuclear transition operator is therefore proportional to $\mathfrak{M}_E(\lambda\mu)$, i.e.,

$$\mathfrak{M}_{\mathrm{nucl}}(\lambda\mu) = \frac{\mathfrak{M}_E(\lambda\mu)}{Z_2e_{\mathrm{eff}}R_0^\lambda(3/4\pi)\,(1+\frac{1}{2}\lambda\delta_{\lambda 3})}. \tag{6.86'}$$

Replacing the matrix element of $\mathfrak{M}_E(\lambda\mu)$ by the $B(E\lambda)$, we obtain the cross-section in the form

$$\frac{d\sigma(\theta)}{d\Omega} = \frac{2I_i+1}{2I_f+1}\sum_\lambda \frac{B(E\lambda;\, I_i \to I_f)}{e^2(\mathrm{barn})^\lambda}\,\sigma_\lambda(\theta) \tag{6.84'}$$

with

$$\sigma_\lambda(\theta) = \sum_\mu \left|\left(\Phi_f^{(-)*}q_\lambda(r)\, Y^*_{\lambda\mu}\Phi_i^{(+)}\right)\right|^2, \tag{6.87}$$

where $\Phi_f^{(-)}$ and $\Phi_i^{(+)}$ are the distorted waves in the outgoing and incoming channels, and

$$q_\lambda(r) = \frac{10^\lambda}{Z_2R_0^\lambda}\,\frac{1}{(3/4\pi)\,(1+\frac{1}{2}\lambda\delta_{\lambda 3})}\left\{\frac{3}{2\lambda+1}\,\frac{Z_1Z_2e^2}{r}\times\right.$$
$$\left.\times\left[\frac{1}{r}\left(\frac{R_c}{r}\right)^\lambda\theta(r-R_c)+\frac{1}{R_c}\left(\frac{r}{R_c}\right)^\lambda\theta(R_c-r)\right]+\frac{e}{e_{\mathrm{eff}}}R\frac{dU}{dr}\right\}. \tag{6.87'}$$

The $\theta(x)$ is the usual step function, $\theta(x) = 1$ for $x > 0$ and $\theta(x) = 0$ for $x < 0$. The R_c is the Coulomb radius. The last term in (6.87′) describes the nuclear scattering, and the remaining terms define the Coulomb excitation. The cross-section contains, naturally, the interference term of the two mechanisms.

The nuclear interaction is overwhelming if the incident particle has an energy above the Coulomb barrier or if the scattering angle is larger than the limiting classical Coulomb angle θ_c. The Coulomb interaction is usually neglected in these cases. The function $\sigma_\lambda(\theta)$ has a characteristic angular dependence on the transferred momentum. In the inelastic nuclear scattering, the $\sigma_2(\theta)$ is in phase and the $\sigma_3(\theta)$ is out of phase with the $\lambda = 0$ elastic scattering (Blair's phase rule).

The nuclear interaction in (6.87′) can be omitted at energies well below the Coulomb barrier because particles stay mostly outside the nucleus. We are then dealing with a pure Coulomb excitation; the cross-section increases with an increasing charge of the incident particle. The phase rule is not fulfilled in the Coulomb excitation and the cross-section does not have the characteristic dependence on the transferred angular momentum λ. The Coulomb excitation cannot be treated as a one-step process if the Coulomb interaction between the incident particle and the nucleus is sufficiently strong. For example, the heavy ion reactions belong to this category. The final state can be then excited via any of the close-lying states, and the corrections for such a mechanism must be included. The multiple Coulomb excitation theory[331] is designed for the treatment of these processes.

The quadrupole and octupole excitations can be separated according to their angular distributions when the energies above the Coulomb barrier are used. The energies of the quadrupole and octupole excited states are found from the energy spectra of the inelastically scattered particles, while the corresponding $B(E2)$ and $B(E3)$ values are found from the cross-section measurements. The absolute determination of the cross-section is often difficult and inaccurate. However, the relative measurements are more reliable, and the $B(E\lambda)$ values are often normalized in case one of them is known, for example, from its measured lifetime.

The (p, p'), (d, d'), (α, α') and heavier ion reactions are widely used for determination of the excitation energies and of the $B(E2)$ and $B(E3)$ values. These reactions excite preferably [90], [332–333] the rotational states in deformed nuclei and the vibrational states in deformed and spherical nuclei.

3. Let us discuss the single-nucleon transfer reactions on spherical nuclei. To this category belong, for example, the reactions (dp), (dt), (pd), $(^3\mathrm{He}\alpha)$, (dn), $(^3\mathrm{He}d)$, $(t\alpha)$, and others. The cross-section has the form

$$\frac{d\sigma}{d\Omega} = N \sum_{j,\lambda} S_{j\lambda}\sigma_\lambda(\theta). \tag{6.88}$$

Here

$$\sigma_\lambda(\theta) \sim \left|(\Phi_f^{(-)*} u_\lambda(r)\, Y_\lambda^*(\theta)\, \Phi_i^{(+)})\right|^2 \tag{6.88′}$$

is the differential cross-section calculated with the distorted waves $\Phi^{(\pm)}$. The distorted waves are calculated by the equation containing the optical potential. The $u_\lambda(r)$ is the radial part of the bound state wave function describing the transferred particle. The normalization factor N is different in different reactions; for example, in the (dp) reaction $N = (2J_f+1)/(2J_i+1)$.

The spectroscopic factor for the stripping reaction equals

$$S_j(J_i+j \rightarrow J_f) = \left| \sum_{m,\,M_i} (\Psi^*(J_fM_f;\, A+1)\,(J_iM_ijm\,|\,J_fM_f)\, a^+_{jm}\Psi(J_iM_i;\, A)) \right|^2, \qquad (6.89)$$

while for the pick-up reaction

$$S_j(J_i \rightarrow J_f+j) = \left| \sum_{m,\,M_i} (\Psi^*(J_fM_f;\, A)\,(J_fM_fjm\,|\,J_iM_i)\, a_{jm}\Psi(J_iM_i;\, A+1)) \right|^2. \qquad (6.89')$$

Several assumptions are implicitly contained in the cross-section formula (6.88). It is assumed that the distorted wave functions $\Phi^{(\pm)}$ and the nuclear wave functions Ψ are the eigenfunctions of the spherically symmetric potentials. It is also assumed that j and λ are good quantum numbers characterizing both the bound state of the transferred particle and the angular momentum transfer in the reaction.

Let us calculate the spectroscopic factors in the independent quasiparticle model. The spectroscopic factors for the even–even target nuclei with excitation of one-quasiparticle states are equal to

$$S_j = \delta_{jj_f} u^2_{j_f} \qquad (6.90)$$

for the (dp) reaction, and

$$S_j = \delta_{jj_f} v^2_{j_f} \qquad (6.90')$$

for the (dt) reaction.

This means that the particle states are excited in the stripping (dp) reaction, while the pick-up (dt) reaction excites the hole states.

Let us find out how to calculate the spectroscopic factors of a single-particle transfer reaction in the independent quasiparticle model.(334, 335) From (6.89) and (6.89′) it immediately follows that such reactions can excite only the states which have one quasiparticle more (or less) than the initial state; the positions of the remaining quasiparticles must be unchanged. If the number of paired nucleons is unchanged, then

$$S_j = u^2_j, \qquad (6.91)$$

and if the number of pairs changes by one pair, then

$$S_j = v^2_j, \qquad (6.91')$$

The quantum number j belongs to the subshell where a quasiparticle has been formed or has disappeared. The u^2_j and v^2_j are calculated by means of the correlation functions C and of the chemical potentials λ belonging to the nucleus with a lesser number of quasiparticles.

The single-particle levels are determined in the most direct way by the single-nucleon transfer reaction with the magic nuclei as the targets. The spectroscopic factors equal 1 or 0 in such a case. Any deviations from 1 or 0 signify either the admixtures in the single-particle states or the insufficiencies of the distorted wave Born approximation. The uncertainties in the spectroscopic factors caused by the inadequacy of the distorted wave approximation are very serious. The relative values of the spectroscopic factors are, however, free of these uncertainties. The neutron subshells were determined in the single neutron transfer reactions, as described in ref. 336. The proton subshells were determined from the (^{3}He, *d*), (*d*, ^{3}He), and (*dn*) reactions.[214]

The energies of the one-quasiparticle states in odd-A nuclei and the pairing factors v_j^2 and u_j^2 can be determined in the single-nucleon transfer reactions. The quantities u_j^2 and v_j^2 characterize the distribution of pairs in different subshells and depend on the corresponding single-particle energy. Alternatively, the same quantities may be determined in various nuclei, i.e., the dependence on the chemical potential λ can be studied. Figure 6.12 illustrates

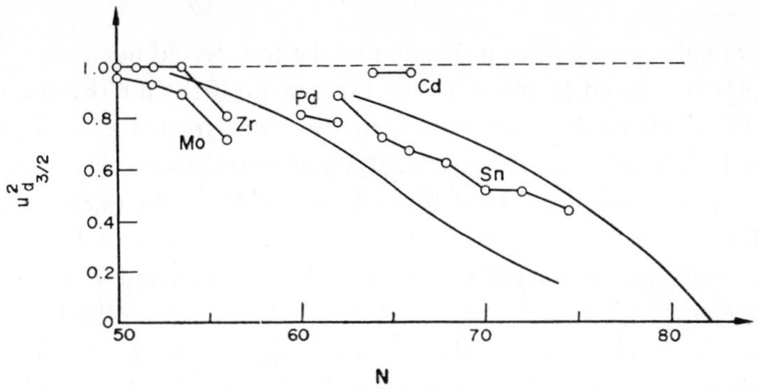

FIG. 6.12. The $u^2_{d_{3/2}}$ values of the neutron $d_{3/2}$ subshell. The data were obtained in the (*dp*) reactions on Zr, Mo, Pd, Cd, and Sn targets. The calculated $u^2_{d_{3/2}}$ values are also shown. The neutron number is plotted on the abscissa. The experimental $u^2_{d_{3/2}}$ values are denoted by points connected by the broken line. The calculated $u^2_{d_{3/2}}$ are denoted by the full curves. The left curve corresponds to the first third of the neutron $N = 50$–82 shell, the right curve corresponds to the last third of the same shell. The experimental data of ref. 214 were used.

such a procedure. It shows the u_j^2 value of the neutron $d_{3/2}$ subshell determined in the (*dp*) reaction with the even–even isotopes of zirconium, molybdenum, palladium, cadmium, and tin as targets. When the target mass increases, the $d_{3/2}$ is initially a particle state, then it becames the ground state, and, finally, a hole state. The experimental data in Fig. 6.12 reproduce rather well the expected behavior of the function $u^2_{d3/2}$.

4. Let us discuss the single-nucleon transfer reactions in deformed nuclei. The problem was often studied; e.g., in refs. 337–341.

The deformed nuclei have some peculiarities when compared to the spherical nuclei. They are related to the axial symmetry of the problem. The distorted waves $\Phi^{(\pm)}$ are the solutions of the Schrödinger equation containing the axially symmetric optical potential. The total angular momentum j and the orbital angular momentum l of the incident (or outgoing) particle are not conserved in the axially symmetric field. Thus, they are not identical with the conserved transferred momenta I and L. To include these features we have to treat the reaction as a many-step process. The preliminary (in the entrance channel) or subsequent (in the outgoing channel) excitations of the rotational and other low-lying states in the initial and final nuclei play an important role in these processes. The rotational states are probably the most important states. If we restrict our attention to the rotational states, we obtain the differential cross-section in the form

$$\frac{d\sigma(\theta)}{d\Omega} = N \sum_{I, l, n} \left| \beta_{Il}^{K, n} \mathcal{L}_l^{(n)}(\theta) \right|^2, \tag{6.92}$$

where n is a characteristic of the different channels. The statistical factors are equal to $N = (2I_f+1)/(2I_i+1)$ for the stripping reaction and $N = 1$ for the pick-up reaction. The following selection rules govern the reactions:

$$\left. \begin{array}{c} |I_f - I_i| \leq I \leq |I_i + I_f|, \\ K = K_f - K_i, \quad (-1)^l = \pi_i \pi_f, \end{array} \right\} \tag{6.93}$$

where K is the angular momentum projection of the transferred nucleon.

Formula (6.92) is reduced to the standard form (6.88) if the multistep excitation effects are small and the single-nucleon transfer reactions can be treated as a single-step process. This is the usual assumption used in the analysis of experimental data obtained from the single-nucleon transfer reactions in deformed nuclei. Formula (6.88) is used, and the characteristic features of the deformed field are included only in the calculations of spectroscopic factors. The kinematic part of the cross-section is calculated in the one-step approximation. The spectroscopic factors, on the other hand, depend on the Nilsson or Woods–Saxon wave functions. The transferred angular momenta I and L coincide in such a case with the particle characteristics j and l. The j and l characterize the spherical components in the expansion of the wave functions of deformed nuclei.

Formula (6.92) is simplified in case the above-described assumptions are used. It has now the form

$$\frac{d\sigma(\theta)}{d\Omega} = N \sum_l S_l \sigma_l(\theta), \tag{6.94}$$

where the spectroscopic factor is determined by

$$S_l = \frac{2I_i+1}{2I_f+1}(1+\delta_{K_i0}+\delta_{K_f0}) \sum_j \Big\{ (I_i K_i j K \,|\, I_f K_f)^2 \times \times \Big(\sum_\Lambda a_{l\Lambda}(l\Lambda s\Sigma \,|\, jK) \Big)^2 \times |(\Psi_{0f}^* a_{s\sigma}^+ (\text{or } a_{s\sigma}) \Psi_{0i})|^2 \Big\}. \tag{6.95}$$

The quantities $a_{l\Lambda}$ were defined in § 5 of Chapter 1 as the expansion coefficients of the Nilsson wave functions. The quasiparticle selection rules and the rules for the evaluation of the last factor in (6.95) are identical with the analogous rules in spherical nuclei.

When the independent quasiparticle model is used, and the neutron stripping on the even–even target is considered,

$$|(\Psi_f^* a_{s\sigma}^+ \Psi_i)|^2 = u_s^2, \tag{6.96}$$

while for the neutron pick-up

$$|(\Psi_f^* a_{s\sigma} \Psi_i)|^2 = v_s^2; \tag{6.96'}$$

the s corresponds to the state with a new quasiparticle.

Only one l-value contributes to the one-nucleon transfer reaction on an even–even target. This makes the cross-section formula particularly simple, i.e.,

$$\frac{d\sigma(\theta)}{d\Omega} = S_l \sigma_l(\theta), \tag{6.97}$$

where

$$S_l = 2\Big(\sum_{\Lambda} a_{l\Lambda}(l\Lambda \tfrac{1}{2}\Sigma \,|\, jK)\Big)^2 \,|(\Psi_f^* a_{s\sigma}^+(\text{or } a_{s\sigma})\Psi_i)\,|^2. \tag{6.97'}$$

For the (dp) reaction

$$S_l = 2\Big(\sum_{\Lambda} a_{l\Lambda}(l\Lambda \tfrac{1}{2}\Sigma \,|\, jK)\Big)^2 u_s^2, \tag{6.98}$$

and for the (dt) reaction

$$S_l = 2\Big(\sum_{\Lambda} a_{l\Lambda}(l\Lambda \tfrac{1}{2}\Sigma \,|\, jK)\Big)^2 v_s^2. \tag{6.98'}$$

The dependence of the functions u_s^2 and v_s^2 on the single-particle energy can be determined from the cross-section of the (dp) and (df) reactions. Figure 6.13 shows the cross-sections

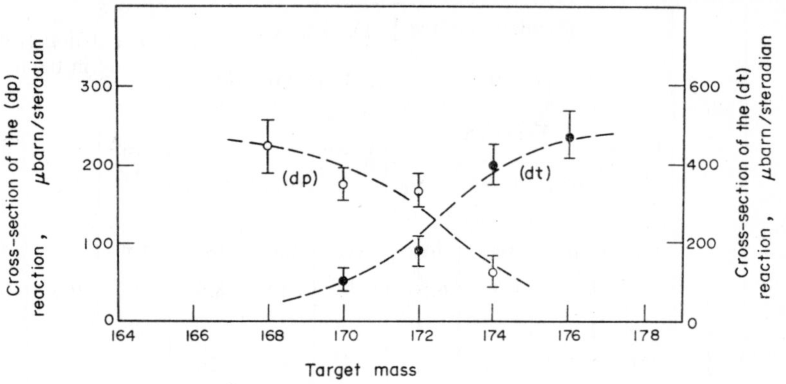

FIG. 6.13. Cross-section for excitation of the $\frac{7}{2}-\frac{5}{2}$ [512] state in (dp) and (dt) reactions on the even–even ytterbium isotopes (according to ref. 187).

for the excitation of the $\frac{7}{2}^-\frac{5}{2}[512]$ state in odd-A ytterbium isotopes. The independent quasiparticle model evidently works quite well, and the experimental points agree with the calculated u_s^2 and v_s^2. Figure 6.13 is similar to Fig. 6.12; again the increase in the target mass corresponds to the decrease in the single-particle energy relative to the chemical potential λ. By comparing eqns. (6.98) and (6.98′) we find that it is possible to distinguish between the hole-excited states and the particle-excited states when the absolute value of the cross-section ins known.

The population of the different rotational states in the same band depends not only on the K^π value of the band but also on the expansion coefficients $a_{l\Lambda}$. The intensities are distributed according to the $\Big(\sum_{\Lambda} a_{l\Lambda}(l\Lambda \tfrac{1}{2}\Sigma \,|\, jK)\Big)^2$ values, i.e., they depend on the contribution of the different spherical subshells. This feature can be used for the determination of the asymptotic quantum numbers $Nn_z\Lambda$ of the involved single-particle states. Table 6.13 demonstrates this fact. It shows the relative cross-sections at $\theta=90°$ and the $\left(\sum_\Lambda a_{l\Lambda}(l\Lambda\tfrac{1}{2}\Sigma|jK)\right)^2$ values for the rotational bands of the $\tfrac{1}{2}^-$[521] and $\tfrac{1}{2}^-$[510] states. (The data of refs. 187 and 342 were used.) The relative intensities are normalized to 100 for the $\tfrac{1}{2}^-\tfrac{1}{2}$ [521] and $\tfrac{3}{2}^-\tfrac{1}{2}$[510] states while the sum of the $\left(\sum_\Lambda a_{l\Lambda}(l\Lambda\tfrac{1}{2}\Sigma|jK)\right)^2$ values is equal to unity. The v_s^2 and u_s^2 values are identical for all members of the rotational band. When they are excluded, the pattern of the relative intensities is very similar in the (*dp*) and (*dt*) reactions. Table 6.13 clearly indicates that the distribution of intensities makes it possible to determine the asymptotic quantum numbers of the one-quasiparticle states. It is even possible to distinguish between the states with identical K^π values.

TABLE 6.13. *The relative intensities for the population of the rotational bands belonging to the* $\tfrac{1}{2}^-$ [521] *and* $\tfrac{1}{2}^-$ [510] *states*

Both the (*dp*) *and* (*dt*) *reactions are considered. The values of* $\left(\sum_\Lambda a_{l\Lambda}(l\Lambda\tfrac{1}{2}\Sigma|jK)\right)^2$ *are shown for comparison*

	Spin	Population of the $\tfrac{1}{2}^-$ [521] band							Population of the $\tfrac{1}{2}^-$ [510] band in the (*dp*) reaction			
		In (*dt*) reaction				In (*dp*) reaction						
		Theory	Experiment			Theory	Experiment		Theory	Experiment		
			^{171}Yb	^{173}Yb	^{175}Yb		^{159}Gd	^{161}Gd		^{171}Yb	^{173}Yb	^{175}Yb
Relative intensities at $\theta = 90°$	$\tfrac{1}{2}$	100	100	100	100	100	100	100	2.4	–	3.3	–
	$\tfrac{3}{2}$	9.3	24	6.3	6.5	9.8	12	8.8	100	100	100	100
	$\tfrac{5}{2}$	26		17	21	37	32	50	35	43	51	45
	$\tfrac{7}{2}$	31	32	27	27	47	65	50	25	16	25	20
	$\tfrac{9}{2}$	5.0	3	2.5	3.5	7	8	7.6	1.4	4.4	4.2	5.5
	$\tfrac{11}{2}$	0.7	1.4	1.7	1.2	1.2	–	–	0.2	–	–	–
$\left(\sum_\Lambda a_{l\Lambda}(l\Lambda\tfrac{1}{2}\Sigma\|jK)\right)^2$	$\tfrac{1}{2}$	0.249	0.284	0.300	0.282	0.25	0.23	0.25	0.01	–	0.01	–
	$\tfrac{3}{2}$	0.024	0.018	0.020	0.019	0.02	0.03	0.02	0.40	0.34	0.31	0.31
	$\tfrac{5}{2}$	0.182	0.142	0.142	0.165	0.18	0.15	0.24	0.29	0.31	0.33	0.29
	$\tfrac{7}{2}$	0.231	0.272	0.243	0.228	0.23	0.29	0.19	0.19	0.10	0.15	0.12
	$\tfrac{9}{2}$	0.269	0.185	0.164	0.216	0.27	0.30	0.29	0.09	0.24	0.21	0.28
	$\tfrac{11}{2}$	0.045	0.097	0.130	0.09	0.05	–	–	0.01	–	–	–

Many one-quasiparticle states in odd-*A* deformed nuclei were found in the (*dp*) and (*dt*) reactions on the even–even targets (see refs. 187, 191, 342–346). The same reactions were employed in studies of the two-quasiparticle neutron states in even–even deformed nuclei (see refs. 347–349). When the odd–odd targets are used, the (2*n*, *p*) three-quasiparticle states in odd-Z nuclei are excited (see ref. 206). While the cross-sections for excitation in the (dp) or (dt) reactions of the low-lying quasiparticle states are usually relatively large, the vibrational states have, as a rule, small cross–sections.

The experimental information describing the purity of the quasiparticle states in deformed nuclei is very important. The quasiparticle component of the wave function is found from the corresponding spectroscopic factors S_l, separated from the cross-section of the single-nucleon transfer reactions. The distorted wave Born approximation used in this procedure has, however, a number of serious shortcomings. Let us enumerate them: the nuclear deformation is included only approximately, the many-channel features of the rotational bands are neglected, and the finite-range interaction and the optical potential used in the $\sigma_l(\theta)$ calculation contain a number of uncertainties. Thus, remembering these facts, the accuracy of the absolute spectroscopic factors S_l is not better than a factor of 2. The accuracy of the relative S_l values is, however, considerably better.

The two-nucleon transfer reactions, for example, the (pt), (tp), or $(^3\text{He}, n)$ reactions represent a very important tool of the nuclear structure studies. Their cross-sections are, however, usually small, and their interpretation has several methodological difficulties. Consequently, the experimental data on the two-nucleon transfer reactions are relatively scant. More information about these reactions can be found in refs. 350–355.

The pairing superfluid correlations enhance the cross-section for the transfer of two nucleons in the relative S-state. This enhancement is analogous to the enhancement of the α-decay between the ground states of two even–even nuclei.

The direct nuclear reactions represent an important source of the nuclear structure information. They bring each year more and more valuable experimental information. The best way of studying the nuclear structure experimentally is to combine the direct nuclear reactions with the α-, β-, and γ-spectroscopical methods.

CHAPTER 7

NUCLEAR SHAPES AND MOMENTS OF INERTIA

§ 1. Ground State Equilibrium Deformations and Quadrupole Moments of Even–Even Nuclei

1. The equilibrium deformations were originally calculated by summing up the single-particle energies of the states occupied by the protons and neutrons. The equilibrium deformation β_0 was found from the minimum condition of the sum treated as a function of the deformation parameter β. This procedure predicted the nonvanishing values of β_0. These β_0 were, for nuclei far from the closed shells, identical with the deformations deduced experimentally from the $B(E2)$ values of the $E2$ transitions in the ground-state rotational bands of even–even nuclei.[43] However, we have noted earlier that the independent particle model does not generally predict the correct equilibrium deformations. The nuclear shape is a result of the competition between the quadrupole forces which tend to deform the nucleus and the pairing superfluid forces which tend to preserve the spherical nuclear form. Thus both forces must be included in reliable calculations of the equilibrium deformation. Such competition has been studied on a simple model (see ref. 10).

The quantitative calculations of the equilibrium deformations[356] include the pairing superfluid correlations and the Coulomb repulsion. They describe well the equilibrium deformations of the nuclei with $150 < A < 190$ and $230 < A < 250$. The method of calculations has been somewhat improved in ref. 357. The calculations were later generalized to include many elements, and thus a new region of the deformed nuclei with $50 < Z$, $N < 82$, has been predicted.[358]

The problem of the axial symmetry has been also studied. It was shown that a majority of deformed nuclei have the form of a prolate rotational ellipsoid.[49, 359]

The equilibrium deformations of the nuclei with $50 < Z$, $N < 82$, were calculated in refs. 360–362, while ref. 363 treated the nuclei with $28 < Z < 50$ and $50 < N < 82$. The equilibrium deformations of the light nuclei around ^{24}Mg were calculated in ref. 364.

The problem is formulated quite generally in ref. 365. The total nuclear energy was calculated as a function of the deformation parameters β and γ. The model of the "pairing correlations and quadrupole–quadrupole force" was applied and the inertial parameters B_2, the restoring constants C_2, and the zero-point vibrational energies were calculated. The problems of nuclear deformations are also discussed in refs. 366–369.

This section will explain the methods of calculations of the total nuclear energy as a function of the deformation parameters β and γ. The results of the calculations will be compared with the experimentally determined equilibrium deformation parameters β_0 and γ_0. The section will also include the studies of the deformation energies and of the ground-state quadrupole moments of even–even nuclei.

2. In calculations of the equilibrium deformations, the following important assumption is made: it is assumed that the average field potential has been selected in such a way that the density matrix $\varrho(q, q')$ is diagonal, and the correlation function $\Phi(q, q')$ has the canonical form at deformations close to equilibrium deformations of even–even nuclei (the β-stable nuclei are considered). In order to calculate the energy of the nucleus we have to average the Hamiltonian describing the interaction of nucleons in atomic nuclei over the ground-state wave function of an even–even nucleus. If the representation with the above-assumed properties is used, this average value consists of two terms in each of the (i.e., proton and neutron) systems. The first term is the sum of the single-particle energies multiplied by $\varrho(q, q)$. The second term is the average value of the pairing superfluid interaction. There is no additional quasiparticle interaction in the representation with diagonal density matrix $\varrho(q, q')$; only the pairing superfluid correlations should be included in calculations. Thus the assumption that the successful choice of the average field potential brought the function $\varrho(q, q')$ to the diagonal form is crucial for the equilibrium deformation calculation.

Taking into account the diagonal form of the density matrix $\varrho(q, q')$ just discussed we shall choose the nuclear interaction Hamiltonian in the simple form

$$H = H_{av} + H_{pair} = H_0(n) + H_0(p), \tag{7.1}$$

with

$$H_0 = \sum_{q,\sigma} E(q \mid \beta, \gamma)\, a_{q\sigma}^+ a_{q\sigma} - G \sum_{q,q'} a_{q+}^+ a_{q-}^+ a_{q'-} a_{q'+}\,. \tag{7.1'}$$

The single-particle energies $E(q \mid \beta, \gamma)$ depend on the deformation parameters β and γ. They are equal to the eigenvalues of the Schrödinger equation containing the Nilsson–Newton or Woods–Saxon potentials.

The calculations of the deformation energy and the search for the equilibrium deformation parameters involve an evaluation of the total nuclear energy. The total nuclear energy in such a procedure is equivalent, to a certain extent, to the sum of the kinetic and potential energies. Thus the potential energy is seemingly counted twice. However, the absolute value of the total energy is not needed; we are interested only in the energy dependence on the deformation parameters β and γ. Consequently, the incorrectness in handling of the potential energy does not cause large errors. This remarkable property is related to the conservation of the nuclear volume and to the short range of the residual interaction.

Let us describe how to find the dependence of the ground-state energy $\mathcal{E}_0(\beta, \gamma)$ on the parameters β and γ in even–even nuclei. We shall follow the method of ref. 356. To describe the deformation we shall use the parameter ε besides the usual parameter β; the relation between the two parameters is shown in Table 1.9.

The total ground-state energy of the even–even nucleus has the form

$$\mathcal{E}_0(\varepsilon, \gamma) = \mathcal{E}_0^p(\varepsilon, \gamma) + \mathcal{E}_0^n(\varepsilon, \gamma) + \mathcal{E}_c\,. \tag{7.2}$$

The proton energy $\mathcal{E}_0^p$ and the neutron energy $\mathcal{E}_0^n$ are determined according to (4.21). This means, for example,

$$\mathcal{E}_0^p(\varepsilon, \gamma) = \sum_r E(r \mid \varepsilon, \gamma) 2v_r^2 - \frac{C_p^2}{G_Z}. \tag{7.3}$$

The quantity $\mathcal{E}_c$ describes the nuclear Coulomb energy. This quantity is often expressed as the energy of a uniformly charged ellipsoid with a constant volume. Such an energy equals, for the axially symmetric ellipsoid,

$$\mathcal{E}_c(\varepsilon) = \mathcal{E}_c(0)\, g(\varepsilon), \tag{7.4}$$

where

$$\mathcal{E}_c(0) = 0.6\frac{Ze^2}{R_0}$$

is the Coulomb energy of a uniformly charged sphere. The function $g(\varepsilon)$ is, for $|\varepsilon| < 0.4$, equal to

$$g(\varepsilon) = \frac{(1-\frac{2}{3}\varepsilon)^{2/3}(1+\frac{1}{3}\varepsilon)^{1/3}}{\sqrt{2\varepsilon-\frac{1}{3}\varepsilon^2}} \ln \frac{1-\frac{2}{3}\varepsilon}{1+(\varepsilon/3)-\sqrt{2\varepsilon-\frac{1}{3}\varepsilon^2}} \tag{7.4$'$}$$

for $\varepsilon > 0$, and

$$g(\varepsilon) = \frac{(1-\frac{2}{3}\varepsilon)^{2/3}[1+(\varepsilon/3)]^{1/3}}{\sqrt{\frac{1}{3}\varepsilon^2-2\varepsilon}} \arctan \frac{\sqrt{(\varepsilon^2/3)-2\varepsilon}}{1+(\varepsilon/3)} \tag{7.4$''$}$$

for $\varepsilon < 0$.

The nuclear intrinsic quadrupole moments are calculated by two methods: the microscopic and the phenomenological. The mass quadrupole moment, according to the microscopic calculations, equals

$$Q_0^t = Q_0^p + Q_0^n, \tag{7.5}$$

where

$$Q_0^p = 2\sum_r \mathfrak{f}^{20}(r, r) v_r^2, \quad Q_0^n = 2\sum_s \mathfrak{f}^{20}(s, s) v_s^2, \tag{7.5$'$}$$

and the $\mathfrak{f}^{20}(q, q)$ are determined by the formula (3.65). The quadrupole moment of a uniformly charged ellipsoid with sharp surface is used in the phenomenological method. Such a moment equals

$$Q_0^{\text{unif}}(\varepsilon) = 0.4ZR_0^2\left(1-\frac{\varepsilon^2}{3}-\frac{2}{27}\varepsilon^3\right)^{2/3}\left\{\frac{1}{(1-\frac{2}{3}\varepsilon)^2}-\frac{1}{[1+(\varepsilon/3)]^2}\right\}. \tag{7.6}$$

Let us find the differences between the quadrupole moment calculated according to (7.5) and the quadrupole moment of the uniformly charged ellipsoid. We want also to see if the neutron and proton equilibrium deformations differ. Table 7.1 shows the ratios Q_0^t/Q_0^p calculated in ref. 362 for ^{126}Ba and the deformation interval $-0.4 \leqslant \varepsilon \leqslant 0.4$. The ratios are evidently almost constant (within approximately 5%). Hence, the proton and neutron equilibrium deformations coincide. Table 7.1 also demonstrates that the ratio Q_0^{unif}/Q_0^p is close to unity. This means that, for a majority of nuclei, the deformation parameter ε characterizing the deformation of the average field potential determines at the same time the density deformation of the neutron and proton systems.

Another modification of the above-explained method was suggested in ref. 357. The projected wave functions (5.67) replace the wave functions (4.22) in the energy calculations

TABLE 7.1. *The dependence of the quadrupole moment ratios Q_0^t/Q_0^p and Q_0^{unif}/Q_0^p on the deformation parameter*

ε	−0.4	−0.3	−0.2	−0.1	0.1	0.2	0.3	0.4
$Q_0^t/Q_0^{p(a)}$	2.28	2.36	2.42	2.49	2.43	2.41	2.35	2.36
Q_0^{unif}/Q_0^p	0.98	1.01	0.99	0.97	0.96	1.01	1.02	1.05

(a) Calculated microscopically.

according to (7.3). The expressions (4.21) for $\mathcal{E}_0^n$ and $\mathcal{E}_0^p$ are replaced at the same time by (5.68). This modification is unimportant in a majority of cases; however, it improves the calculation of transitional nuclei.

The ground-state wave functions of even–even deformed nuclei, used in the equilibrium deformation calculations, are not spherically symmetric. Reference 370 uses corrections related to the projection of the nuclear wave functions onto states with a fixed value of the angular momentum I and of its projection M. Such correction terms increase somewhat the deformation energy; the equilibrium deformation values are practically unchanged.

Strutinsky(371) suggested another method of finding the β and γ-dependence of the nuclear energy. Let us describe this method. The total ground-state energy of an even–even axially symmetric nucleus consists of two terms: the energy of a charged liquid drop and the so-called "shell correction" term. Thus,

$$\mathcal{E}_0^{\text{St}}(\beta) = \mathcal{E}_h(\beta) + \Delta\mathcal{E}_0(\beta). \tag{7.7}$$

The liquid-drop energy is determined by

$$\mathcal{E}_h(\beta) = \sigma_s A^{2/3}[0.18(1-x^2)\beta^2 - 0.011(1+2x)\beta^3 + \ldots], \tag{7.8}$$

where σ_s is the surface tension coefficient, $x = (Z^2/A)/(Z^2/A)_0$. The $(Z^2/A)_0$ is the critical value of Z^2/A determined in the fission theory; its numerical value is in the interval 45–50.

The shell correction is written as a sum of the proton and neutron contributions, i.e.,

$$\Delta\mathcal{E}_0(\beta) = \Delta\mathcal{E}_0^p(\beta) + \Delta\mathcal{E}_0^n(\beta). \tag{7.9}$$

If we consider, for example, the proton system,

$$\Delta\mathcal{E}_0^p(\beta) = \mathcal{E}_0^p(\beta) - \overline{\mathcal{E}}_0^p(\beta). \tag{7.10}$$

The $\mathcal{E}_0^p(\beta)$ is determined by (7.3) and the average energy is defined by

$$\overline{\mathcal{E}}_0^p(\beta) = \frac{2}{\sqrt{\pi}} \int_{-\infty}^{\lambda} dE \frac{E}{v} \sum_r \exp\left\{-\left(\frac{E-E(r\,|\,\beta)}{v}\right)^2\right\}. \tag{7.10'}$$

The chemical potential λ is found from the implicit equation

$$Z = \frac{2}{\sqrt{\pi}} \int_{-\infty}^{\lambda} \frac{dE}{v} \sum_r \exp\left\{-\left(\frac{E-E(r\,|\,\beta)}{v}\right)^2\right\} \tag{7.10''}$$

and the parameter ν is chosen close to the oscillator frequency $\mathring{\omega}_0$. The energy $\bar{\mathscr{E}}_0^p(\beta)$ is independent of the averaging interval ν if ν is larger than 5 MeV. The shell correction $\Delta\mathscr{E}_0(\beta)$ is thus equal to the energy difference of two superfluid systems. The first system uses the Nilsson or Woods–Saxon single-particle energies, while the other one has a uniform density of single-particle states.

The Strutinsky method directly relates the shell effects with the semiempirical mass formula based on the liquid-drop model.

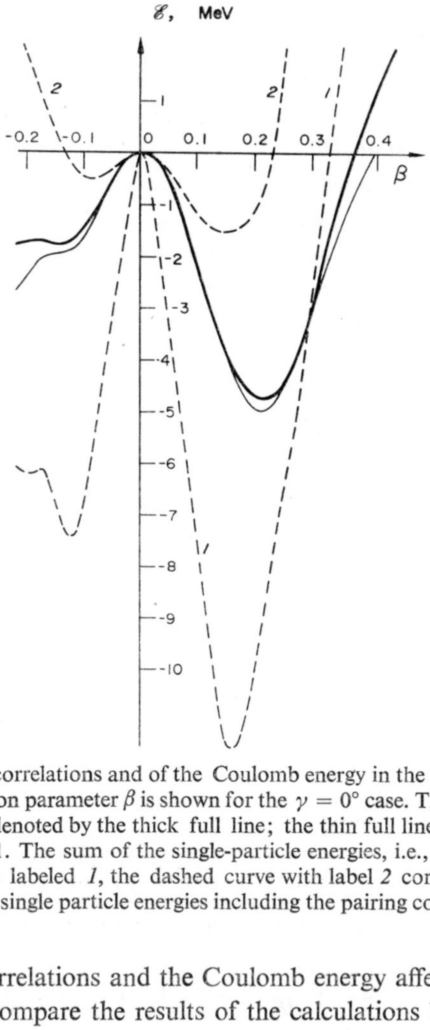

FIG. 7.1. Effect of the pairing correlations and of the Coulomb energy in the nucleus ^{238}U. The dependence of the energy on the deformation parameter β is shown for the $\gamma = 0°$ case. The $\mathscr{E}_0(\beta)$, i.e., the energy calculated according to ref. 356, is denoted by the thick full line; the thin full line shows $\mathscr{E}_0^{\mathrm{St}}(\beta)$, i.e., the energy calculated according to ref. 371. The sum of the single-particle energies, i.e., $\mathscr{E}_0^p(\beta, G_Z = 0)+\mathscr{E}_0^n(\beta, G_N = 0)$ is shown on the dashed curve labeled *1*, the dashed curve with label *2* corresponds to $\mathscr{E}_0^p(\beta)+\mathscr{E}_0^n(\beta)$, i.e., sum of the single particle energies including the pairing correlations.

How do the pairing correlations and the Coulomb energy affect the equilibrium deformations? We shall also compare the results of the calculations based on the methods of ref. 356 and on the Strutinsky method.[371] Figure 7.1 shows the dependence of the nuclear energy on the deformation parameter β (at $\gamma = 0°$). Four curves are plotted in the figure: $\mathscr{E}_0(\beta)$ is calculated according to ref. 356, $\mathscr{E}_0^{\mathrm{St}}(\beta)$ is calculated according to ref. 371, and the $\mathscr{E}_0^p+\mathscr{E}_0^n$ values calculated with and without pairing correlations are also shown. All energies are normalized to zero at $\beta = 0$. The $\mathscr{E}_0(\beta)$ and $\mathscr{E}_0^{\mathrm{St}}(\beta)$ values are close to each other in the $0-0.3$ interval of β-values. The two methods thus predict practically identical equilibrium deformations. At larger deformations, the Strutinsky method is more correct than the method of the ref. 356. Furthermore, Fig. 7.1 clearly demonstrates that the pairing correlations decrease the deformation energy by a large amount. The Coulomb energy makes the minimum at $\beta = \beta_0$ considerably deeper. The pairing correlations strongly affect the deformation energy but do not change the value of the equilibrium deformation in the case of strongly deformed nuclei. However, the situation is quite different in the case of transitional nuclei. The equilibrium deformation parameters β_0 are extremely sensitive to the variation of the pairing constants G_N and G_Z in the transitional region.

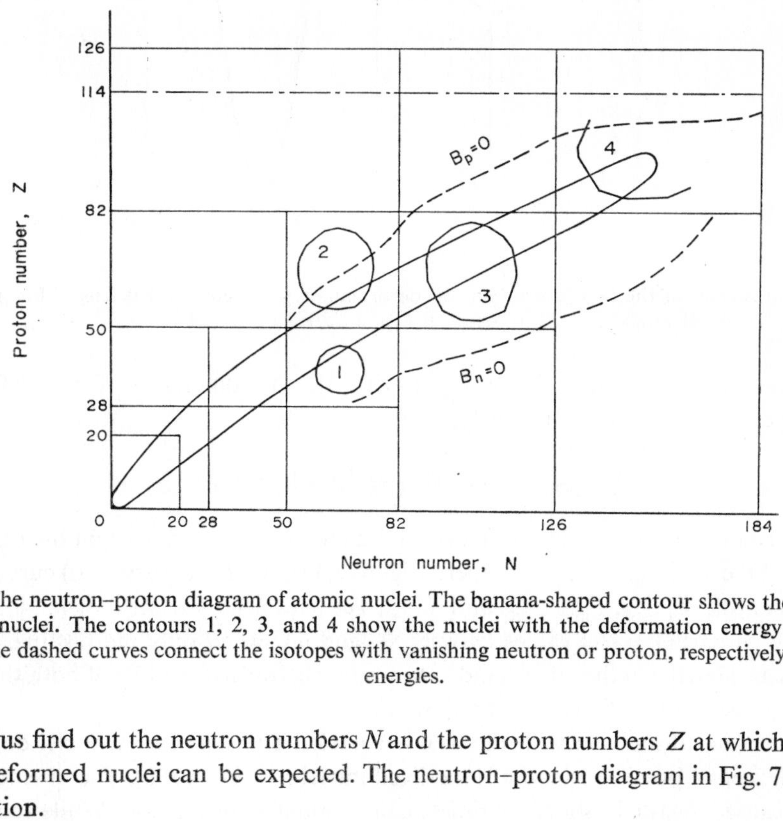

FIG. 7.2. The neutron–proton diagram of atomic nuclei. The banana-shaped contour shows the stable and long-lived nuclei. The contours 1, 2, 3, and 4 show the nuclei with the deformation energy larger than 2 MeV. The dashed curves connect the isotopes with vanishing neutron or proton, respectively, separation energies.

3. Let us find out the neutron numbers N and the proton numbers Z at which the existence of deformed nuclei can be expected. The neutron–proton diagram in Fig. 7.2 answers this question.

The nuclei with $150 < A < 190$ and $A > 226$ belong to the well-studied regions of strongly deformed nuclei. Some experimental data on the deformed nuclei within region 2 (of Fig. 7.2) have been already obtained, but the investigation of region 1 just begins. Several light nuclei are also deformed but we shall not discuss them in this section.

Before discussing the characteristic features of deformed nuclei in various regions, let us summarize our notations. Figure 7.3 shows a typical dependence of the ground state energy $\mathscr{E}_0(\varepsilon, \gamma = 0)$ on the deformation parameter ε. The curve describes ^{166}Er, i.e., the even–even nucleus, at $\gamma = 0°$. The $\mathscr{E}_0(\varepsilon, \gamma = 0)$ function has typically two minima. The deeper minimum corresponds to the positive deformation, it is denoted by ε_0^+. The shallower one corresponds to the negative deformation ε_0^-. The deformation energies of the minima are denoted by $\mathscr{E}_{\text{def}}^+$ and $\mathscr{E}_{\text{def}}^-$ (the deformation energy is equal to the difference of the energy

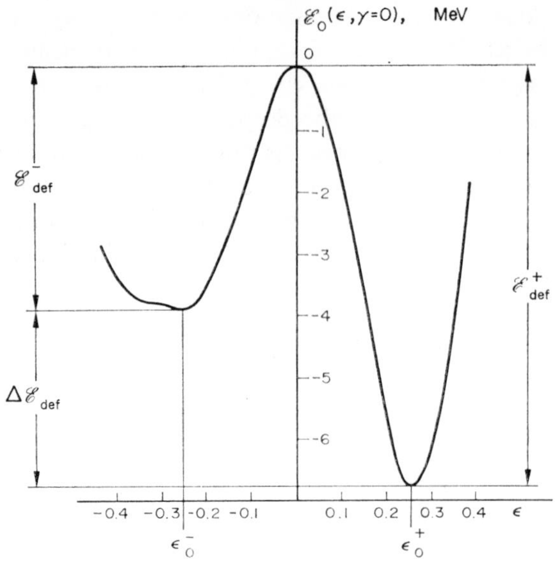

FIG. 7.3. Dependence of the total energy on the deformation parameter ε. Nucleus ^{166}Er, $\gamma = 0$ case is considered. $\mathcal{E}^+_{\text{def}} = \mathcal{E}_0(0) - \mathcal{E}_0(\varepsilon^+_0)$, $\mathcal{E}^-_{\text{def}} = \mathcal{E}_0(0) - \mathcal{E}_0(\varepsilon^-_0)$, $\Delta\mathcal{E}_{\text{def}} = \mathcal{E}^+_{\text{def}} - \mathcal{E}^-_{\text{def}}$.

at $\varepsilon = 0$ and at $\varepsilon = \varepsilon^{\pm}_0$, i.e., $\mathcal{E}^{\pm}_{\text{def}} = \mathcal{E}_0(0) - \mathcal{E}(\varepsilon^{\pm}_0)$). The difference of the deformation energies is denoted by $\Delta\mathcal{E}_{\text{def}}$. Obviously,

$$\Delta\mathcal{E}_{\text{def}} = \mathcal{E}^+_{\text{def}} - \mathcal{E}^-_{\text{def}} = \mathcal{E}_0(\varepsilon^-_0) - \mathcal{E}_0(\varepsilon^+_0). \tag{7.11}$$

Thus, if the minimum at ε^+_0 is deeper than the minimum at ε^-_0, $\Delta\mathcal{E}_{\text{def}} > 0$; on the other hand, $\Delta\mathcal{E}_{\text{def}} < 0$ if the minimum at ε^-_0 is deeper. Figure 7.3 shows the $\mathcal{E}_0(\varepsilon, \gamma = 0)$ curve together with the quantities ε^+_0, ε^-_0, $\mathcal{E}^+_{\text{def}}$, $\mathcal{E}^-_{\text{def}}$, and $\Delta\mathcal{E}_{\text{def}}$. The parameter γ is not shown. Note that the nonaxiality parameter γ has values between 0° and 60° when only the positive values of ε or β are considered. On the other hand, γ has values between 0° and 30° if both the positive and the negative values of ε (or β) are allowed.

Let us start our discussion with the heaviest nuclei. The superheavy nuclei do not exist in nature (except a possible island of long-lived nuclei with $A \approx 300$). The absence of such nuclei is caused by their short lifetime against spontaneous fission. While spontaneous fission is relatively unimportant in nuclei with $A < 230$, its role becomes noticeable when the atomic number A and, particularly, the nuclear charge increase. This fact is clearly demonstrated by the following sequence of the spontaneous fission half-lives: ^{230}Th, 10^{17} years; ^{240}Pu, 10^{11} years; ^{256}Fm, 2.1 h; ^{260}Ku, 0.1 s. The half-lives of the $Z = 106$ or 108 nuclei become extremely small and the nuclei are therefore unobservable. The importance of the α-decay also increases with increasing A.

The spontaneous fission half-lives of certain superheavy nuclei may be substantially longer than our estimate if spherically symmetric nuclei exist near the corresponding magic numbers.

The original calculations of the Nilsson single-particle levels[41] predicted the next proton magic number at $Z = 126$, i.e., the same magic number as the neutrons have. However, subsequent calculations,[42, 56, 372, 373] using the modified Nilsson potential or the Woods–Saxon potential, predict a different proton magic number. The magic number $Z = 114$ is caused by a shift of the $p_{3/2}$, $p_{1/2}$, and $f_{5/2}$ subshells. The $Z = 114$ magic number disappears at somewhat different values of the average potential parameters. The neutron magic number $N = 184$ is predicted for nuclei with $A \approx 300$.

The calculations of refs. 42 and 372–376 predict a spherically symmetric form of the doubly magic nucleus $Z = 114$, $N = 184$. The calculated lifetimes of α- and β-decays and of spontaneous fission are large, with values up to 10^8 years. The results, however, are very sensitive to the choice of parameters in the Nilsson or Woods–Saxon potentials. Many calculations predict the longlived nuclei around $Z = 126$. The experimental search for these islands of longlived superheavy nuclei is very intensive (see refs. 377–379).

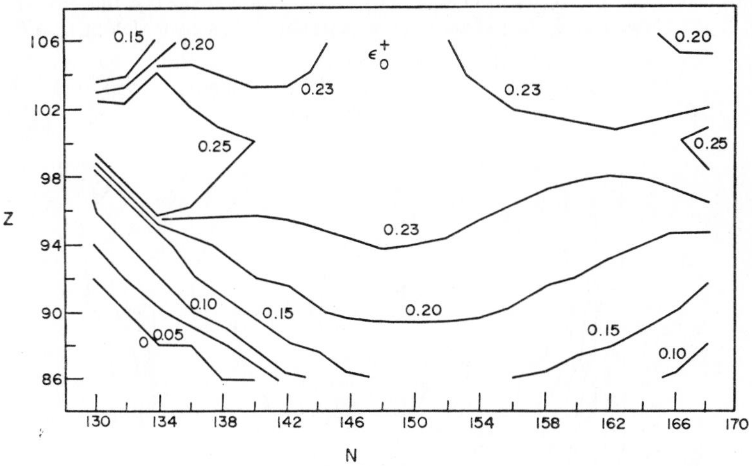

FIG. 7.4. The contour plot of the deformation parameter ε_0^+ and its dependence on the neutron number N and on the proton number Z. The curves are labeled by the corresponding ε_0^+ values.

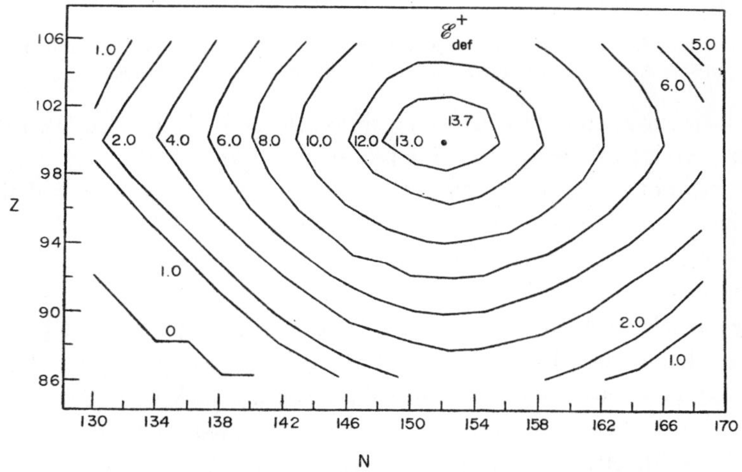

FIG. 7.5. Curves of the constant deformation energy $\mathscr{E}_{\text{def}}^+ = \mathscr{E}_0(\varepsilon = 0) - \mathscr{E}_0(\varepsilon_0^+)$. The dependence on the neutron number N and on the proton number Z is shown. The curves are labeled by the $\mathscr{E}_{\text{def}}^+$ values in MeV.

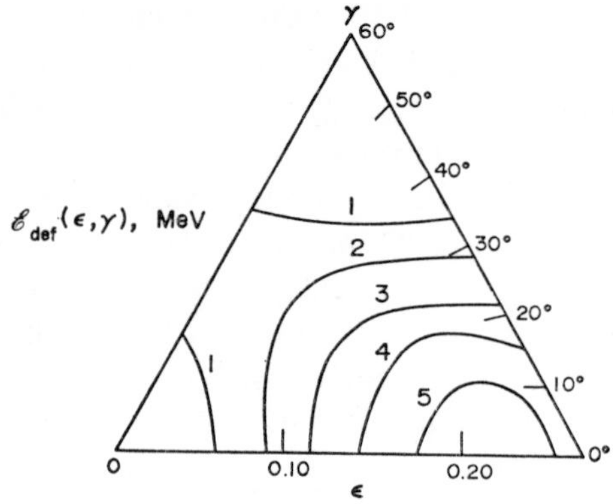

FIG. 7.6. Deformation energy $\mathscr{E}_{def}(\varepsilon, \gamma) = \mathscr{E}_0(\varepsilon = 0) - \mathscr{E}_0(\varepsilon, \gamma)$ for ^{238}U. The full curves correspond to $\mathscr{E}_{def} = 1, 2, 3, 4,$ or 5 MeV. The $\mathscr{E}_{def}(\varepsilon, \gamma)$ function has value approximately 6 MeV at $\varepsilon_0 = 0.25$ and $\gamma_0 = 0$.

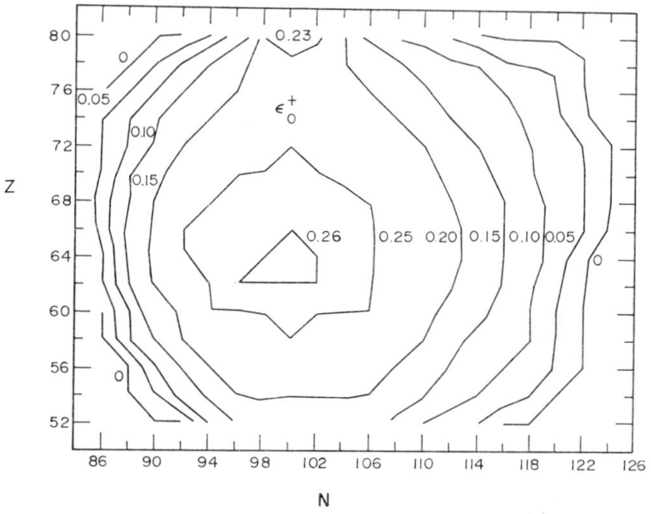

FIG. 7.7. The contour plot of the deformation parameter ε_0^+. The dependence on the neutron number N and on the proton number Z is shown. The curves are labeled by the corresponding ε_0^+ values.

4. Let us turn our attention to the equilibrium deformations, quadrupole moments, and deformation energies of the nuclei with $148 < A < 190$ and $220 < A < 260$. We shall consider the states of even–even nuclei in these regions. The results of calculations and the corresponding experimental data are collected in Tables 7.2 and 7.3 and Figs. 7.4–7.10. The tables consider the observable nuclei, while the figures show the general behavior for all nuclei within the discussed regions.

The single-particle level schemes of the modified Nilsson potential have been used in the presented calculations (see Figs. 1.5–1.8). The pairing constants G_Z and G_N determined in (4.52′) have been applied. Analogous calculations using the Woods–Saxon potential give essentially identical results. The ε_0^+ and E_{def}^+ values, calculated according to the methods of refs. 356 and 371, are close to each other.

Tables 7.2 and 7.3 demonstrate that the calculated ε_0^+ values reproduce the correspond-

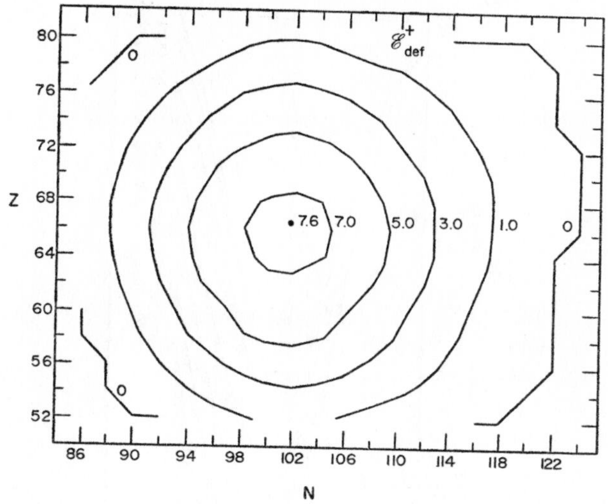

FIG. 7.8. The curves of constant $\mathscr{E}^+_{def}$ values. The dependence on N and Z is shown. The curves are labeled by the corresponding $\mathscr{E}^+_{def}$ value in MeV.

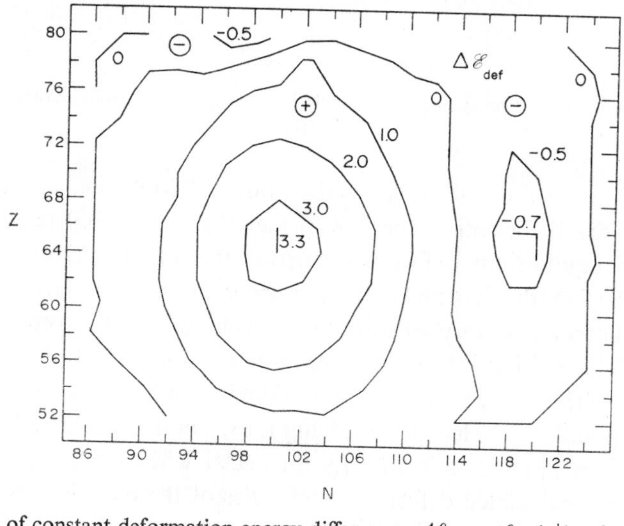

FIG. 7.9. The curves of constant deformation energy differences $\Delta\mathscr{E}_{def} = \mathscr{E}_{def}(\varepsilon_0^+) - \mathscr{E}_{def}(\varepsilon_0^-)$. The dependence on N and Z is shown. The thick full curve corresponds to $\Delta\mathscr{E}_{def} = 0$. Other curves are labeled by the corresponding $\Delta\mathscr{E}_{def}$ value in MeV. The region of the positive $\Delta\mathscr{E}_{def}$ is denoted by $\oplus$ and the region of the negative deformation energy difference is denoted by $\ominus$.

ing experimental data on the strongly deformed nuclei quite well. The equilibrium deformation parameter ε_0^+ has values $\varepsilon_0^+ = 0.23$–0.25 for nuclei with $220 < A < 260$ and the deformation energy has values up to 13 MeV. The observed deformations in the radon isotopes and in the light radium isotopes may be of a dynamic nature. The strongly deformed nuclei with $148 < A < 190$ have the ε_0^+ values between 0.20 and 0.26. The calculated ε_0^+ values are sometimes smaller than the experimental data; the discrepancy is, however, usually smaller than 0.03. Figure 7.8 shows that the deformation ener-

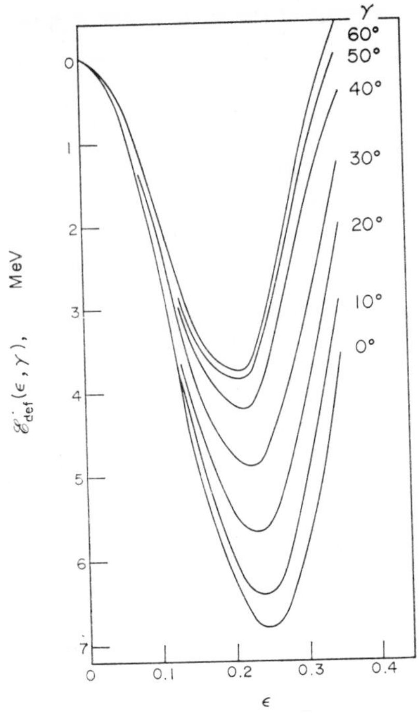

FIG. 7.10. The ε dependence of the total energy of ^{174}Yb at various γ values. The figure is reproduced from ref. 49.

gies of the $148 < A < 190$ nuclei have values only 5–8 MeV, i.e., they are approximately 5 MeV less than the deformation energies in the $220 < A < 260$ region.

The calculated equilibrium deformation parameters of the transitional nuclei, i.e., of radon, radium, the light thorium and uranium isotopes, the heavy tungsten and osmium isotopes and several others, are very sensitive to small changes in the average field potential or in the values of the pairing interaction constants G_N and G_Z. For example, the deeper energy minimum of the heavy tungsten and osmium isotopes can be shifted from the positive ε_0 values to the negative ε_0 values by changing the G_Z and G_N values. This means that the negative deformations predicted in Fig. 7.9 for nuclei with $N \geqslant 114$ are somewhat uncertain. Moreover, the calculated deformation energies of these nuclei are small and may be washed out by the zero-point vibrations.

Figure 7.3 shows the typical ε-dependence of the $\mathscr{E}_0(\varepsilon, \gamma = 0)$ function. We would like to know if the minimum at $\varepsilon < 0$ is a minimum or a saddle point. To answer this question we have to calculate the full energy surface $\mathscr{E}_0(\varepsilon, \gamma)$ for all γ-values. An example of such a surface is shown in Fig. 7.6 for ^{238}U; the Strutinsky method ([371]) was used in the calculation. The function $\mathscr{E}_{def}(\varepsilon, \gamma)$ in Fig 7.6 has only one minimum at $\varepsilon_0 > 0, \gamma = 0°$, the stationary point at $\gamma = 60°$ is a saddle point. Analogous results are shown in a somewhat different form in Fig. 7.10. The nucleus ^{174}Yb is considered and the method of ref. 356 is used.

Let us summarize the main characteristics of the strongly deformed nuclei in regions $148 < A < 190$ and $230 < A < 260$. (1) The prolate ellipsoid corresponds to the equilib-

TABLE 7.2. *The equilibrium deformation parameters* ε_0^+, *quadrupole moments* $Q_0(\varepsilon_0^+)$, *and deformation energies* $\mathcal{E}_{def}^+$ *for the nuclei with* $220 \leqslant A \leqslant 258$

Element	A	Experiment		Theory	
		ε_0^+	$Q_0(\varepsilon_0^+)$ (barn)	ε_0^+	$\mathcal{E}_{def}^+$ (MeV)
$_{86}$Rn	220	0.11	4.30±0.11	0	0
	222	0.13	4.78±0.15	0	0
$_{88}$Ra	222	0.17	6.63±0.26	0	0
	224	0.16	6.25±0.22	0	0
	226	0.18	7.21±0.36	0.05	0.04
	228	0.19	7.79±0.48	0.12	0.4
$_{90}$Th	226	0.20	8.25±0.46	0.10	0.1
	228	0.20	8.46±0.45	0.14	0.6
	230	0.21	8.91±0.45	0.16	1.5
	232	0.23	9.88±0.25	0.18	2.6
	234	0.21	8.97±0.56	0.20	3.7
$_{92}$U	230	0.22	9.46±0.69	0.17	1.7
	232	0.23	9.98±0.60	0.20	3.1
	234	0.23	10.03±0.40	0.20	4.4
	236	0.24	10.57±0.65	0.21	5.6
	238	0.25	11.26±0.27	0.22	6.7
	240	–	–	0.22	7.5
$_{94}$Pu	234	–	–	0.22	4.8
	236	–	–	0.22	6.2
	238	0.24	10.94±0.69	0.22	7.4
	240	0.25	11.30±0.18	0.23	8.5
	242	0.26	11.80±0.51	0.23	9.4
	244	–	–	0.23	9.9
	246	–	–	0.22	10.1
$_{96}$Cm	240	–	–	0.24	9.1
	242	–	–	0.24	10.2
	244	0.28	13.49±0.74	0.24	11.1
	246	–	–	0.24	11.6
	248	–	–	0.24	11.7
	250	–	–	0.23	11.5
$_{98}$Cf	246	–	–	0.25	12.3
	248	–	–	0.24	12.8
	250	–	–	0.24	12.9
	252	–	–	0.24	12.7
	254	–	–	0.24	12.4
$_{100}$Fm	250	–	–	0.25	13.6
	252	–	–	0.25	13.7
	254	–	–	0.24	13.4
	256	–	–	0.24	12.7
	258	–	–	0.24	12.0

TABLE 7.3. *The equilibrium deformation parameters* ε_0^+, *quadrupole moments* $Q_0(\varepsilon_0^+)$, *and the values* ε_0^+, ε_0^-, $\mathscr{E}_{def}^+$ *and* $\Delta\mathscr{E}_{def}^+$ *for nuclei with* $146 \leqslant A \leqslant 190$

Element	A	Experiment		Theory			
		ε_0^+	$Q_0(\varepsilon_0^+)$ (barn)	ε_0^+	ε_0^-	$\mathscr{E}_{def}^+$ (MeV)	$\Delta\mathscr{E}_{def}$ (MeV)
$_{60}$Nd	146	–	–	0	0	0	0
	148	–	–	0.10	0.10	0.20	0.10
	150	0.23	5.11±0.13	0.19	0.15	1.1	0.46
$_{62}$Sm	150	0.17	–	0.15	0.12	0.5	0.2
	152	0.26	5.96±0.11	0.21	0.28	1.7	0.6
	154	0.28	6.73±0.09	0.24	0.31	3.1	1.3
$_{64}$Gd	152	–	–	0.16	0.28	0.7	0.2
	154	0.26	6.18±0.11	0.23	0.29	2.1	0.6
	156	0.27	6.74±0.13	0.25	0.31	3.6	1.3
	158	0.29	7.24±0.13	0.26	0.32	4.9	2.0
	160	0.29	7.54±0.13	0.26	0.32	6.0	2.6
$_{66}$Dy	156	0.25	6.18±0.25	0.22	0.29	2.3	0.4
	158	0.27	6.85±0.30	0.25	0.31	3.8	1.3
	160	0.27	6.88±0.17	0.25	0.32	5.1	2.0
	162	0.27	7.12±0.09	0.26	0.32	6.2	2.6
	164	0.28	7.50±0.16	0.26	0.27	7.0	3.1
$_{68}$Er	162	0.26	7.02±0.18	0.25	0.32	4.8	1.8
	164	0.26	7.05±0.20	0.25	0.32	5.9	2.4
	166	0.28	7.58±0.14	0.26	0.26	6.7	2.8
	168	0.28	7.61±0.11	0 26	0.25	7.2	3.0
	170	0.27	7.43±0.10	0.26	0.25	7.3	2.9
$_{70}$Yb	168	0.26	7.43±0.18	0.25	0.25	6.1	2.5
	170	0.26	7.48±0.15	0.25	0.25	6.5	2.5
	172	0.27	7.73±0.13	0.25	0.25	6.6	2.4
	174	0.26	7.61±0.11	0.25	0.25	6.3	2.2
	176	0.26	7.45±0.11	0.24	0.25	5.7	1.7
$_{72}$Hf	174	0.25	7.28±0.24	0.25	0.25	5.6	2.0
	176	0.25	7.33±0.15	0.24	0.25	5.4	1.7
	178	0.23	6.80±0.11	0.22	0.24	4.8	1.4
	180	0.23	6.74±0.12	0.21	0.24	4.0	0.9
$_{74}$W	180	0.21	6.39±0.24	0.21	0.24	3.9	1.1
	182	0.21	6.46±0.11	0.20	0.23	3.2	0.8
	184	0.20	6.23±0.11	0.18	0.20	2.5	0.4
	186	0.19	5.95±0.12	0.15	0.19	1.9	0.1
$_{76}$Os	186	0.18	5.61±0.16	0.16	0.18	1.7	0.3
	188	0.17	5.22±0.20	0.15	0.17	1.2	0.1
	190	0.16	5.14±0.16	0.10	0.14	0.7	−0.1

rium shape, i.e., the energy has an absolute minimum at $\varepsilon_0 > 0$. (2) The $\mathscr{E}_0(\varepsilon, \gamma)$ function has only one minimum at $\gamma = 0°$. Thus, the equilibrium ellipsoid is axially symmetric and the nonaxial nuclei suggested by Davydov do not exist. (3) The difference in deformation energies $\Delta\mathscr{E}_{def}$ is relatively large, usually greater than 2 MeV. (4) The $\mathscr{E}_0(\varepsilon, \gamma = 0)$ curves have parabolic forms near the absolute minimum. This fact suggests a small anharmonicity of the surface vibrations.

5. We shall now discuss the so-called new deformation regions, i.e., the nuclei with $50 < Z, N < 82$ and $28 < Z < 50$, $50 < N < 82$.

Tables 7.4 and 7.5 and Figs. 7.11–7.18 contain the results of our calculations.[362, 363]

TABLE 7.4. *The calculated ground-state values of* $\varepsilon_0^{\pm}$, $Q_0(\varepsilon_0^{\pm})$, $\mathscr{E}_{def}^{+}$ *and* $\Delta\mathscr{E}_{def}$ *for even–even nuclei*

Element	A	ε_0^-	$Q_0(\varepsilon_0^-)$ (barn)	ε_0^+	$Q_0(\varepsilon_0^+)$ (barn)	$\mathscr{E}_{def}$ (MeV)	$\Delta\mathscr{E}_{def}$ (MeV)
$_{54}$Xe	114	−0.25	−3.1	0.21	3.2	1.1	+0.1
	116	−0.25	−3.1	0.25	3.9	1.4	0
	118	−0.25	−3.1	0.25	4.0	1.6	−0.2
	120	−0.24	−3.0	0.25	4.0	1.5	−0.4
	122	−0.22	−2.8	0.21	3.2	1.3	−0.6
	124	−0.21	−2.7	0.18	2.7	1.0	−0.5
	126	−0.17	−2.2	0.15	2.3	0.7	−0.2
$_{56}$Ba	116	−0.31	−4.0	0.31	5.5	2.8	+0.1
	118	−0.30	−3.9	0.30	5.3	3.2	+0.1
	120	−0.29	−3.8	0.29	5.1	3.3	0
	122	−0.27	−3.6	0.28	4.9	3.2	−0.2
	124	−0.26	−3.5	0.26	4.6	2.8	−0.6
	126	−0.25	−3.4	0.22	3.8	2.2	−0.7
	128	−0.23	−3.2	0.19	3.2	1.6	−0.4
	130	−0.17	−2.5	0.15	2.6	0.9	−0.1
$_{58}$Ce	122	−0.30	−4.2	0.31	5.9	5.1	+0.5
	124	−0.30	−4.2	0.30	5.7	5.0	+0.3
	126	−0.28	−4.0	0.29	5.6	4.5	−0.1
	128	−0.28	−4.0	0.26	5.0	3.6	−0.4
	130	−0.26	−3.8	0.22	4.2	2.6	−0.4
	132	−0.22	−3.4	0.18	3.4	1.7	−0.1
	134	−0.16	−2.6	0.15	2.9	0.7	−0.1
$_{60}$Nd	128	−0.30	−4.5	0.30	6.1	6.0	+0.4
	130	−0.29	−4.4	0.28	5.7	5.0	0
	132	−0.28	−4.3	0.25	5.1	3.7	−0.2
	134	−0.25	−4.0	0.21	4.3	2.5	−0.1
	136	−0.19	−3.2	0.16	3.3	1.2	−0.1
$_{62}$Sm	132	−0.30	−4.8	0.29	6.2	6.0	+0.3
	134	−0.29	−4.7	0.27	5.9	4.7	0
	136	−0.27	−4.5	0.23	5.0	3.2	0
	138	−0.20	−3.6	0.19	4.2	1.7	−0.1
	140	−0.15	−2.8	0.12	2.7	0.4	−0.2

TABLE 7.5. *The calculated ground-state values of $\varepsilon_0^{\pm}$, $Q_0(\varepsilon_0^-)$, $\mathscr{E}_{def}^-$ and $\Delta\mathscr{E}_{def}$ for even–even nuclei*

Element	A	ε_0^-	$Q_0(\varepsilon_0^-)$ (barn)	ε_0^+	$\mathscr{E}_{def}^-$ (MeV)	$\Delta\mathscr{E}_{def}$ (MeV)
$_{36}$Kr	94	−0.30	—	0.21	1.8	−0.7
	96	−0.32	−2.3	0.26	2.7	−1.0
	98	−0.32	−2.3	0.26	3.2	−1.2
$_{38}$Sr	94	−0.22	—	0.16	0.7	−0.3
	96	−0.28	—	0.23	1.6	−0.8
	98	−0.30	−2.3	0.30	2.4	−0.9
	100	−0.30	−2.3	0.30	2.9	−1.0
$_{40}$Zn	96	−0.12	—	0.21	1.4	−0.7
	98	−0.21	—	0.27	2.0	−0.9
	100	−0.28	−2.3	0.26	2.2	−0.9
	102	−0.28	−2.3	0.26	2.6	−0.9
$_{42}$Mo	98	−0.19	—	0.15	0.5	−0.1
	100	−0.24	—	0.19	1.2	−0.4
	102	−0.26	—	0.23	1.8	−0.7
	104	−0.27	−2.3	0.25	2.3	−0.8
	106	−0.26	−2.3	0.25	2.6	−0.9
	108	−0.25	−2.2	0.25	2.7	−1.1
$_{44}$Ru	102	−0.22	—	0.17	0.6	0.1
	104	−0.26	—	0.20	1.2	−0.2
	106	−0.26	—	0.22	1.7	−0.4
	108	−0.25	−2.3	0.23	2.0	−0.6
	110	−0.24	−2.2	0.22	2.1	−0.7
	112	−0.24	−2.3	0.21	2.0	−0.8
$_{46}$Pd	106	−0.17	—	0.16	0.3	0.3
	108	−0.23	—	0.19	0.7	0.1
	110	−0.23	—	0.20	1.0	0
	112	−0.22	—	0.20	1.2	−0.2
	114	−0.21	—	0.18	1.3	−0.2

They show the equilibrium deformation parameters $\varepsilon_0^{\pm}$, the quadrupole moments $Q_0(\varepsilon^{\pm})$, and the deformation energies. The single-particle level schemes of Figs. 1.10 and 1.11 and the pairing constants G_N and G_Z, determined in (4.53), have been used in the calculations.

The calculations predict prolate shapes for some even–even nuclei and oblate shapes for other even–even nuclei. Table 7.4 therefore states both deformation parameters ε_0^+ and ε_0^-. Figures 7.11–7.14 and 7.16–7.18 give an overall picture of the considered regions of atomic nuclei.

The deformation energies and the equilibrium deformation parameters of nuclei not included in the tables may be found in the corresponding figures. The procedure is a simple one: The $\Delta\mathscr{E}_{def}$ is found in Figs. 7.14 or 7.18. If $\Delta\mathscr{E}_{def} < 0$, the equilibrium form is oblate and the deformation parameter ε_0^- is found in Figs. 7.12 or 7.16. The $\mathscr{E}_{def}^-$ is found from

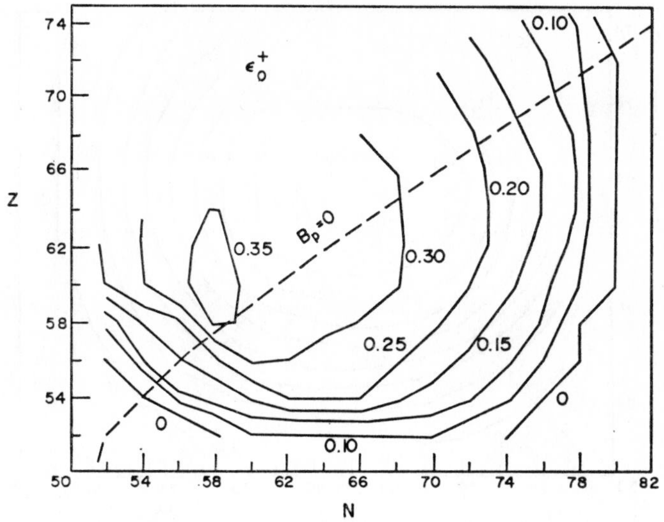

FIG. 7.11. The contour plot of the dependence of the deformation parameter ε_0^+ on the N- and Z-values. The dashed line $B_p = 0$ shows the vanishing proton separation energy in an even system. The curves are labeled by the corresponding ε_0^+ values. The figure is reproduced from ref. 362.

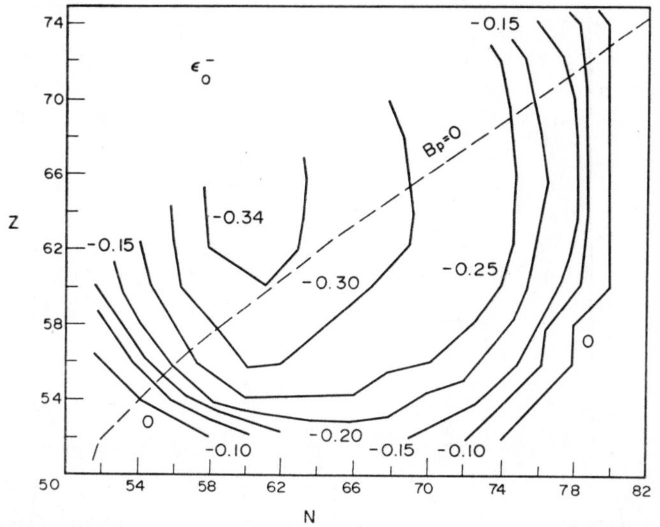

FIG. 7.12. The contour plot of the N- and Z-dependence of the ε_0^- deformation parameter. The curves are labeled by the corresponding ε_0^- values. The figure is reproduced from ref. 362.

Fig. 7.17 for a nucleus in the $28 < Z < 50$ and $50 < N < 82$ region. The $\mathcal{E}_{def}^-$ of other nuclei are calculated from the relation $\mathcal{E}_{def}^- = \mathcal{E}_{def}^+ - \Delta\mathcal{E}_{def}$; the $\mathcal{E}_{def}^+$ is plotted in Fig. 7.13.

The neutron-deficient nuclei in the $50 < Z$, $N < 82$ region and the neutron-rich nuclei in the $34 < Z < 46$ and $54 < N < 70$ region are transitional. Such nuclei are physically the most interesting. We have stated already earlier that the equilibrium deformations and the deformation energies of the transitional nuclei are very sensitive to the employed methods of calculations. Therefore the nuclei with $50 < Z$, $N < 82$ were calculated in ref. 361 by

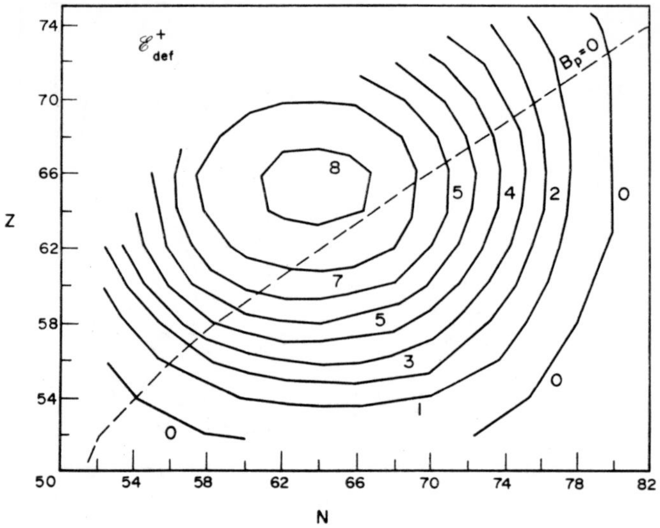

FIG. 7.13. The curves connect points with the constant deformation energy $\mathscr{E}^+_{def}$. The dependence on N and Z is shown. The curves are labeled by the $\mathscr{E}^+_{def}$ values in MeV. The figure is reproduced from ref. 362.

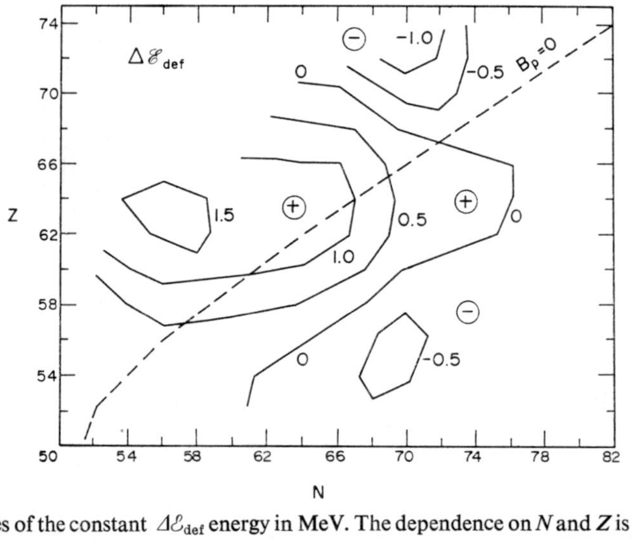

FIG. 7.14. The curves of the constant $\Delta\mathscr{E}_{def}$ energy in MeV. The dependence on N and Z is shown. The dashed line $B_p = 0$ shows the vanishing proton separation energy in an even system. Otherwise, the notation of Fig. 7.9 is used. The figure is reproduced from ref. 362.

both of the discussed methods, and three different single-particle level schemes were used. However, the results are always very close to each other. The sensitivity to the G_Z and G_N values was also tested in ref. 361 and it was found that all qualitative conclusions stay unchanged even if the pairing constants vary by up to 10%.

The nuclei in this region of atomic numbers have very small values of $\Delta\mathscr{E}_{def}$, i.e., the energy difference between the two minima corresponding to the prolate and oblate ellipsoids is very small. This main result of our studies is independent of the details of calcula-

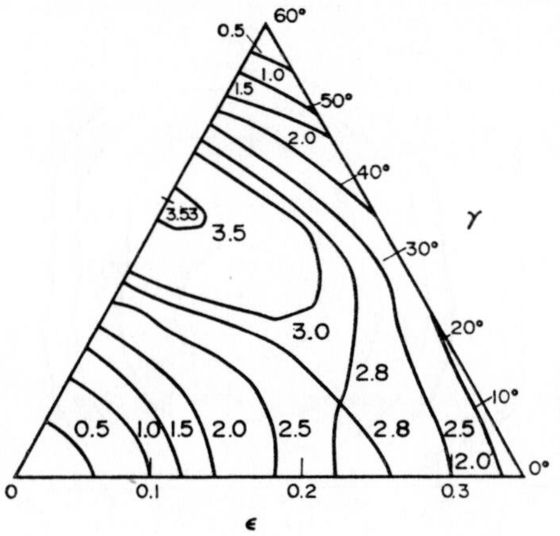

FIG. 7.15. The deformation energy $\mathscr{E}_{def}(\varepsilon, \gamma) = \mathscr{E}_0(\varepsilon = 0) - \mathscr{E}_0(\varepsilon, \gamma)$ of ^{124}Ba. The curves are labeled by the $\mathscr{E}_{def}$ values in MeV.

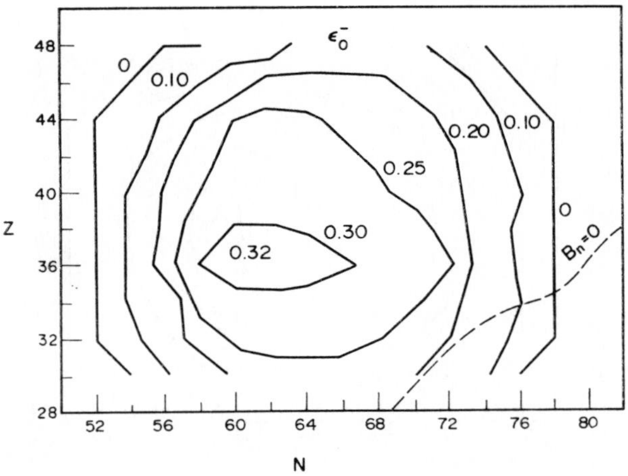

FIG. 7.16. The contour plot of the N- and Z-dependence of the deformation parameter ε_0^-. The curves are labeled by the ε_0^- values. The figure is reproduced from ref. 363.

tions. This feature makes the nuclei in these "new" deformation regions distinctly different from the nuclei in the "classical" deformation regions $150 < A < 190$ and $230 < A < 260$.

Tables 7.4 and 7.5 include the nuclei not far away from the line of β-stability. Such nuclei are available for experimental studies. The nuclei with a very small deformation energy (less than 1 MeV) may be treated as spherical nuclei.

Figure 7.14 shows the changes in nuclear shape when the isotopes further away from the line of β-stability are considered. Nuclei with $Z < 60$ and with small neutron deficiency are oblate. When the neutron number decreases and the $B_p = 0$ line (i. e., vanishing proton separation energy) becomes closer, the equilibrium shape is changed to the prolate ellipsoid. The exact position of the dividing line $\Delta\mathscr{E}_{def} = 0$ depends somewhat on the employed single-particle level scheme.

Figure 7.18 shows the situation for the nuclei with $28 < Z < 50$ and $50 < N < 82$.

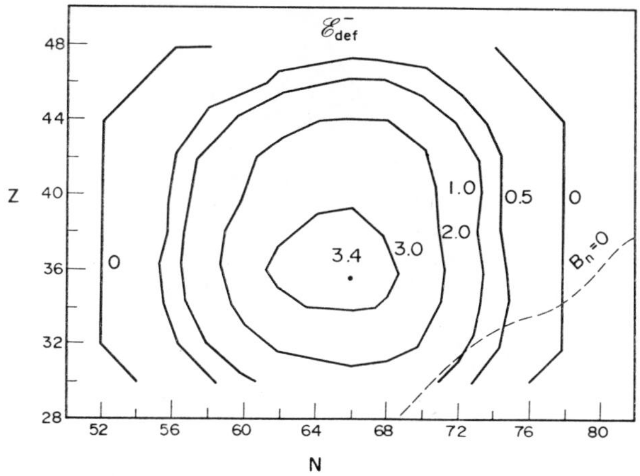

FIG. 7.17. The curves of the constant deformation energy $\mathscr{E}_{\overline{def}} = \mathscr{E}_0(\varepsilon = 0) - \mathscr{E}_0(\varepsilon_{\bar{0}})$. The N- and Z-dependence is shown. The curves are labeled by the $\mathscr{E}_{\overline{def}}$ values in MeV.

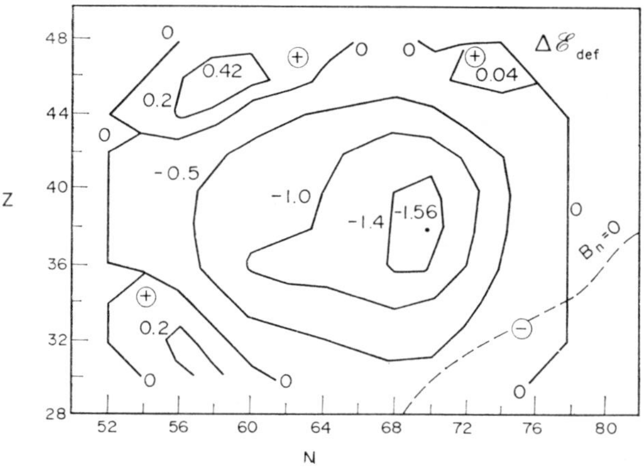

FIG. 7.18. The curves of constant deformation energy differences $\Delta\mathscr{E}_{def}$. The N- and Z-dependence is shown. The dashed $B_n = 0$ line denotes the vanishing neutron separation energy. (Notation as Fig. 7.9.)

Most of these nuclei have the oblate ellipsoidal form. All nuclei in this region have small deformation energies and they are "soft" against the β-vibrations.

Let us discuss the problem of the axial symmetry for these nuclei. Calculations of the γ-dependence of the total energy $\mathscr{E}_0(\varepsilon, \gamma)$ show that the energy minimum corresponds to the axially symmetric form. An example of such dependence is shown in Fig. 7.15. The deformation energy of ^{124}Ba is maximal at $\gamma = 60°$, i. e., it corresponds to the oblate ellipsoid. The energy surface thus contains a valley in the γ-direction. The local minimum at $\gamma = 0°$ is very shallow, its 0.1 MeV barrier is considerably lower than the energy of the zero-point motion. Consequently, the nucleus ^{124}Ba cannot have a prolate form in the ground state.

We have seen that all nuclei in the discussed region have rather flat energy dependence on the parameter γ; i.e., they are soft against the γ-deformation. The coupling of the rotational states with the γ-vibrational states should be, therefore, included in studies of the low-energy spectra.

6. We shall now discuss the problem of the higher multipole components of the static nuclear surface deformation. The radius describing the nuclear surface is chosen in the form

$$R(\theta) = R_0(1+\beta Y_{20}+\beta_4 Y_{40}+\beta_6 Y_{60}+\ \ldots). \tag{7.12}$$

Only the even multipoles (i.e., of the shape symmetrical around the equatorial plane) are included. Reference (380) has shown that the nuclei with a non-vanishing static octupole deformation do not exist in the $218 < A < 232$ region where the negative parity excited states have very low energies.

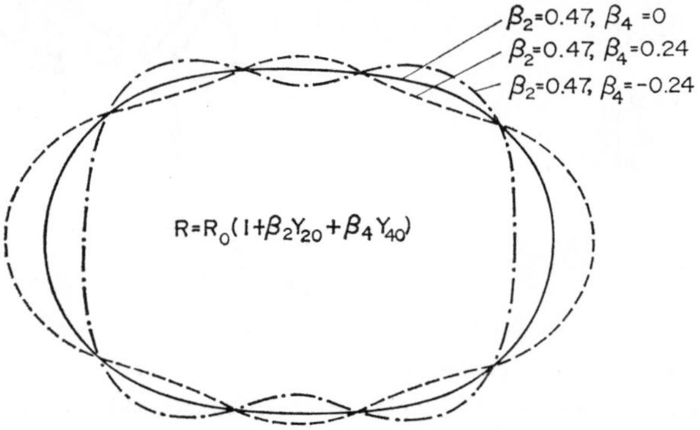

FIG. 7.19. The surface of an axially symmetric nucleus with different β_4 values.

There are convincing indications[381, 382] that several nuclei have $\beta_4 \neq 0$, i.e., that their surface has a nonvanishing static hexadecapole deformation. We shall restrict ourselves to the hexadecapole degree of freedom because there are no positive proofs that higher static multipole deformations really exist.

Figure 7.19 shows the nuclear shape with a nonzero value of the hexadecapole deformation parameter. Unrealistically large values $\beta_4 = \pm 0.24$ are chosen in Fig. 7.19 to stress the qualitative features of the corresponding surfaces. The appearance of a neck in the nuclear surface at $\beta_4 < 0$ is clearly visible.

The Nilsson or Woods–Saxon potentials must be modified in the $\beta_4 \neq 0$ case. An oscillator term $r^2\beta_4 Y_{40}$ is added to the Nilsson potential in such a case, while the nuclear surface in the Woods–Saxon potential is described by

$$R(\theta) = R_0(1+\beta Y_{20}(\theta)+\beta_4 Y_{40}(\theta)). \tag{7.12'}$$

The corresponding modified single-particle level schemes are then used in the calculations of the β and β_4 dependence of the energy surface $\mathcal{E}_0(\beta, \beta_4)$. The equilibrium deformation parameters β_0 and β_4^0 are again determined from the minimum condition of the function $\mathcal{E}_0(\beta, \beta_4)$. Such calculations using the Nilsson potential were performed in refs. 42 and 383, while the Woods–Saxon potential was used in ref. 384. The experimental[382] equilibrium β_4^0 values are compared with the theoretical predictions[383, 384] in Fig. 7.20. The agreement is evidently quite good. Particularly, the monotonic decrease in β_4^0 from the

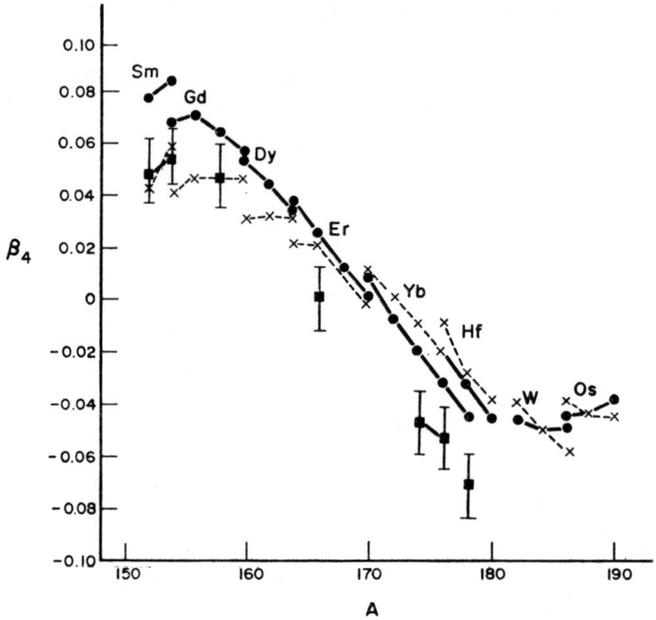

FIG. 7.20. Comparison of the experimental and calculated values of the hexadecapole deformation parameter β_4 in the $150 < A < 190$ region. ▪ denotes the experimental data from ref. 382. The black points connected by full lines are the calculations of ief. 383, while the crosses connected by dashed lines are the calculations of ref. 384.

positive $\beta_4^0 \approx 0.04$ values in the beginning of the $150 < A < 190$ to the vanishing $\beta_4^0 \approx 0.0$ values in the middle and to the negative $\beta_4^0 \approx -0.06$ values is clearly reproduced.

The nonvanishing equilibrium β_4^0 values affect the corresponding single-particle energies and wave functions. Several characteristics of the deformed nuclei are changed, e.g. the decoupling parameters a. The $\beta_4^0 \neq 0$ situation changes the mixing of the $N, N\pm 2$ components of the single-particle wave functions.[384] The interval $\Delta\beta$, where such mixing is appreciable, is wider in the $\beta_4^0 \neq 0$ case.

§ 2. Equilibrium Deformations of the Quasiparticle States

1. We shall discuss in this section the equilibrium deformations of the ground and low-lying excited states in odd-A and odd–odd nuclei, and excited states in even–even nuclei.

The β- and γ-dependence of the quasiparticle state energies can be easily calculated by using the formalism of the independent quasiparticle model described in Chapter 4. Thus, the energy of a one-quasiparticle state (the proton quasiparticle is in the r_0 state) has the familiar form

$$\mathscr{E}_p(r_0 \mid \varepsilon, \gamma) = E(r_0 \mid \varepsilon, \gamma) + 2 \sum_{r \neq r_0} E(r \mid \varepsilon, \gamma) v_r^2 - \frac{C_p^2(r_0)}{G_Z}. \tag{7.13}$$

Similarly, the energy of the two-quasiparticle state in the even proton system with the quasiparticles occupying levels r_1 and r_2 equals

$$\mathscr{E}_p(r_1, r_2 \mid \varepsilon, \gamma) = E(r_1 \mid \varepsilon, \gamma) + E(r_2 \mid \varepsilon, \gamma) + 2 \sum_{r \neq r_1, r_2} E(r \mid \varepsilon, \gamma) v_r^2 - \frac{C_p^2(r_1, r_2)}{G_Z}. \tag{7.14}$$

When calculating the energies (7.13) and (7.14) it is necessary to solve (4.37) and (4.38) or (4.46) and (4.47) for each state r_0 or r_1, r_2 and for each value of the deformation parameters β and γ. In other words, we need all values of the correlation functions $C_p(r_0)$ or $C_p(r_1, r_2)$ and of the chemical potentials $\lambda_p(r_0)$ or $\lambda_p(r_1, r_2)$.

The full energy of a one-quasiparticle state in an odd-Z nucleus is then equal to

$$\mathscr{E}(r_0 \mid \varepsilon, \gamma) = \mathscr{E}_p(r_0 \mid \varepsilon, \gamma) + \mathscr{E}_0^n(\varepsilon, \gamma) + \mathscr{E}_c, \tag{7.15}$$

where the $\mathscr{E}_p(r_0 \mid \varepsilon, \gamma)$ is determined according to (7.13). Similarly, the full energy of two-quasiparticle states in even–even nuclei is equal to

$$\mathscr{E}(s_1, s_2 \mid \varepsilon, \gamma) = \mathscr{E}_0^p(\varepsilon, \gamma) + \mathscr{E}_n(s_1, s_2 \mid \varepsilon, \gamma) + \mathscr{E}_c, \tag{7.16}$$

where $\mathscr{E}_n(s_1, s_2 \mid \varepsilon, \gamma)$ is determined by the formula (7.14). The quantities $\mathscr{E}_0^n(\varepsilon, \gamma)$, $\mathscr{E}_0^p(\varepsilon, \gamma)$, and $\mathscr{E}_c$ have been defined in (7.3) and (7.4). Finally, we need also a formula for an odd–odd nucleus. Its energy equals

$$\mathscr{E}(s_0, r_0 \mid \varepsilon, \gamma) = \mathscr{E}_p(r_0 \mid \varepsilon, \gamma) + \mathscr{E}_n(s_0 \mid \varepsilon, \gamma) + \mathscr{E}_c. \tag{7.17}$$

The energy of a quasiparticle state can be also found in the following way (equivalent to the method explained above): The energy $\mathscr{E}_0(\varepsilon, \gamma)$ of the quasiparticle vacuum is calculated first by application of the procedures of refs. 356 or 371. In the second step, the properly normalized difference

$$\mathscr{E}_{p,n}(q \mid \varepsilon, \gamma) - \mathscr{E}_0^{p,n}(\varepsilon, \gamma) \quad \text{or} \quad \mathscr{E}_{p,n}(q_1, q_2 \mid \varepsilon, \gamma) - \mathscr{E}_0^{p,n}(\varepsilon, \gamma)$$

is added. For example, we will calculate the energy $\mathscr{E}(r_0 \mid \varepsilon, \gamma)$ of the one-quasiparticle state in $^{165}_{67}$Ho. In order to do this we have to calculate the $\mathscr{E}_p(r_0 \mid \varepsilon, \gamma)$ value for the state r_0 and $Z = 67$ and then subtract the $\mathscr{E}_0^p(\varepsilon, \gamma)$ value for $Z = 66$, and finally add the result to the ground state $\mathscr{E}_0(\varepsilon, \gamma)$ value of ^{164}Dy. The normalization of the differences can be performed, for example, at ε_0, $\gamma = 0$.

The equilibrium deformations of the quasiparticle states are determined from the minimum conditions of the corresponding energies $\mathscr{E}(q \mid \varepsilon, \gamma)$, $\mathscr{E}(q_1, q_2 \mid \varepsilon, \gamma)$, or $\mathscr{E}(r_0, s_0 \mid \varepsilon, \gamma)$.

It should be remembered that the general behavior of the energy surface $\mathscr{E}(q_1, q_2, \ldots, \mid \varepsilon, \gamma)$ is, to a large extent, determined by the behavior of the total energy $\mathscr{E}_0(\varepsilon, \gamma)$ of the corresponding quasiparticle vacuum.

2. Let us study the energies of one- and two-quasiparticle states in deformed nuclei and find the equilibrium values ε_e, γ_e of the deformation parameters. The equilibrium deformations of the ground states in odd-A nuclei were calculated in ref. 385. The problem of the deformation of quasiparticle states is discussed in refs. 49, 335, 362, 386, and 387.

It is interesting to find under which circumstances can the equilibrium deformation ε_e of a one-quasiparticle state differ from the equilibrium deformation ε_0 of the core. This may obviously happen if the energy of the one-quasiparticle state decreases faster than the even–even core energy increases. The following simple rules may be formulated in the case of one-quasiparticle states: $\varepsilon_e > \varepsilon_0$ if the quasiparticle occupies either a hole state with a steeply increasing energy (as a function of increasing ε) or a particle state with a steeply decreasing energy. Similarly, $\varepsilon_e < \varepsilon_0$ if the quasiparticle occupies either a steeply descending hole level or a steeply ascending particle level.

However, a great majority of quasiparticle states have the same equilibrium deformation as the corresponding quasiparticle vacuum. For example, for the ground states of odd-A nuclei in the $161 < A < 181$ region, the differences between ε_e and ε_0 are less than ± 0.02. Only for few quasiparticle states is it meaningful to consider the differences between ε_0 and ε_0. Figure 7.21 demonstrates such a case. The figure shows the $\mathscr{E}(r_i\,|\,\varepsilon)$ curves for the

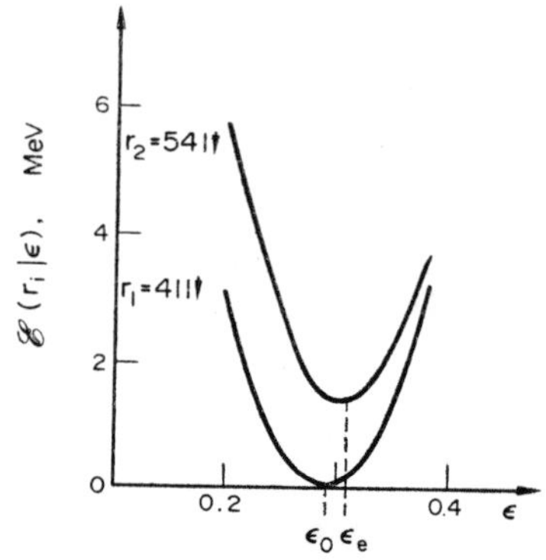

FIG. 7.21. The dependence of the total energy $\mathscr{E}(r_i\,|\,\varepsilon)$ on the deformation parameter ε. The curves are calculated for the 411↓ ground and 541↓ excited one-quasiparticle states in ^{169}Tm.

ground state 411↓ and for the excited state 541↓ in the nucleus ^{169}Tm. The equilibrium deformation ε_e of the 411↓ ground state is practically identical with the ε_0 value of the ^{168}Er core. However, the ε_e value of the 541↓ state is by 0.02 larger than ε_0. Such a difference is related to the steep decrease of the energy of the particle state 541↓ with increased ε. The corresponding level scheme is shown in Fig. 1.5.

Let us discuss the interesting nontrivial cases, i.e., the cases where a large difference between ε_e and ε_0 is expected. A number of such situations, with $|\varepsilon_0 - \varepsilon_e| \approx 0.02 - 0.05$, are collected in Table 7.6. In these cases the deviations are substantial and should be included in the calculations of the spectra of odd-A deformed nuclei. Several two-quasiparticle states with $|\varepsilon_0 - \varepsilon_e| > 0.03$ are shown in Table 7.7. The maximal deviations of ε_e from ε_0 are expected in odd–odd nuclei because it is easier to find two steeply ε-dependent single-particle levels in such a case. For example, the possible isomeric state $K^\pi = 12^-$ in ^{242}Am should have $\varepsilon_e - \varepsilon_0 = 0.05$. Such an increase in equilibrium deformation may decrease the corresponding spontaneous fission lifetime considerably. On the other hand, the expected isomeric states with $\varepsilon_e - \varepsilon_0 < 0$ in the odd–odd nuclei with $Z \geqslant 105$ should have spontaneous fission lifetimes larger than the ground states of the same nuclei. This fact may be of importance in the synthesis of heavy transuranium elements.

Another problem is a possible existence of the quasiparticle states with $\gamma_e \neq 0°$ and $\varepsilon_e > 0$. To illustrate this effect, Fig. 7.22 shows the deformation energy $\mathscr{E}_{\text{def}}(\varepsilon_0, \gamma)$ as a

TABLE 7.6. *The calculated differences $\varepsilon_e - \varepsilon_0$ of the equilibrium deformation parameters. The one-quasiparticle states in odd-A deformed nuclei are considered*

Nucleus	One-quasiparticle state	$\varepsilon_e - \varepsilon_0$
^{241}Am	$\frac{11}{2}-$ [505]	0.04
	$\frac{1}{2}+$ [651]	0.04
	$\frac{11}{2}+$ [615]	−0.03
^{241}Pu	$\frac{13}{2}+$ [606]	0.04
	$\frac{1}{2}-$ [761]	0.04
	$\frac{3}{2}-$ [761]	−0.03
	$\frac{7}{2}+$ [613]	−0.03
^{177}Hf	$\frac{1}{2}-$ [503]	−0.02
	$\frac{9}{2}-$ [505]	−0.02
^{175}Lu	$\frac{1}{2}-$ [541]	0.04
^{157}Gd	$\frac{1}{2}+$ [400]	0.04
	$\frac{11}{2}-$ [505]	−0.05

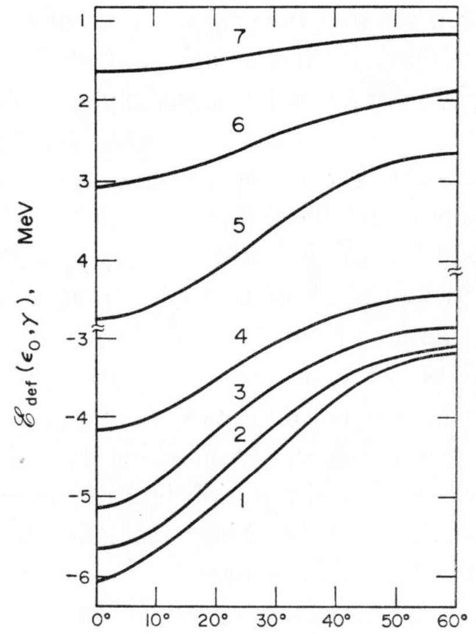

FIG. 7.22. The function $\mathscr{E}_{def}(\varepsilon_0, \gamma)$ (in MeV) calculated for several gadolinium and hafnium isotopes. The γ-dependence at $\varepsilon = \varepsilon_0$ is shown. The curves correspond to the following isotopes: 1, ^{174}Hf; 2, ^{176}Hf; 3, ^{178}Hf; 4, ^{180}Hf; 5, ^{156}Gd; 6, ^{154}Gd; 7, ^{152}Gd.

function of the nonaxiality parameter γ. The figure shows that the $\mathscr{E}_{def}(\varepsilon_0, \gamma)$ energy of the strongly deformed nuclei considered is an increasing function of γ. The γ-dependence of the single-particle energies has been shown in Fig. 1.9 (note that Fig. 1.9 contains the

TABLE 7.7. *The calculated differences* $\varepsilon_e - \varepsilon_0$ *of the equilibrium deformation parameters. The two-quasiparticle states of even-A deformed nuclei are considered*

Nucleus	K^π	Configuration	$\varepsilon_e - \varepsilon_0$
^{240}Pu	7^-, 6^-	$n606\uparrow$, $n761\downarrow$	0.03
	7^-, 6^-	$n606\uparrow$, $n501\downarrow$	0.03
	6^+, 7^+	$n606\uparrow$, $n880\uparrow$	0.03
	2^-, 5^-	$n613\uparrow$, $n761\uparrow$	−0.04
^{242}Am	12^-, 1^-	$p505\uparrow$, $n606\uparrow$	0.05
	11^-, 0^-	$p505\uparrow$, $n615\uparrow$	0.04
	5^+, 6^+	$p505\uparrow$, $n761\downarrow$	0.04
	4^+, 5^+	$p505\downarrow$, $n770\uparrow$	−0.04
	5^+, 4^+	$p505\downarrow$, $n761\downarrow$	−0.04
	1^-, 8^-	$p505\downarrow$, $n613\uparrow$	−0.03
	9^+, 2^+	$p615\uparrow$, $n613\uparrow$	−0.04
^{262}105	0^-, 9^-	$p505\downarrow$, $n604\uparrow$	−0.05
	10^-, 1^-	$p505\downarrow$, $n606\downarrow$	−0.04

quasicrossing levels). By comparing the behavior of the single-particle states and of the deformation energies, we can see that the γ-dependence of $\mathcal{E}_{def}(\varepsilon_0, \gamma)$ is, generally, overwhelming. Thus the results of the detailed calculations which show that the quasiparticle states in strongly deformed nuclei do not have the nonaxial $\gamma_e \neq 0$ forms are easily understood.

We have shown previously that the energy surface $\mathcal{E}_0(\varepsilon, \gamma)$ has a form of a valley with minimum at $\gamma = 0°$ and maximum at $\gamma = 60°$. Such energy surfaces are shown in Figs. 7.6 and 7.10. It is now legitimate to ask the following question: Could the decrease in single-particle energies compensate for the increase in $\mathcal{E}_0(\varepsilon, \gamma)$ with γ in such a way that the function $\mathcal{E}(q_1, q_2, \ldots, |\varepsilon, \gamma)$ has an absolute minimum at $\gamma = 60°$, i.e., it corresponds to the oblate shape? Our calculations have shown that such a compensation does not occur in the strongly deformed nuclei with $150 < A < 190$ and $A > 230$. This means that oblate quasiparticle states with $\gamma = 60°$ do not exist in these regions.

We have discussed the importance of the blocking effect in Chapter 4. However, the blocking effect is unimportant when the equilibrium deformation parameters of strongly deformed nuclei are calculated. But the blocking effect may somewhat change the calculated deformation parameters of the transitional nuclei.

The equilibrium deformation parameters ε_e and γ_e of the quasiparticle states in strongly deformed nuclei ($150 < A < 190$ and $230 < A < 256$) have the following characteristic properties:

(1) In a large majority of cases, the ε_e value is very close to the ε_0 value of the corresponding even–even core.

(2) In several cases, the equilibrium deformation parameters of special quasiparticle configurations differ by as much as $|\varepsilon_e - \varepsilon_0| \approx 0.05$ from the core equilibrium values.

(3) All quasiparticle states considered are prolate, i.e., they have $\gamma_e = 0°$. No states with $\gamma_e = 60°$ have been found.

3. In several regions of nuclei the density of the single-particle states decreases at large deformations. Such an effect was first recognized in ref. 388 and later studied in detail by Strutinsky.[371, 389] This decrease of the level density is sometimes characterized as a shell in the deformed nucleus; the deformation energy is increased at the corresponding values of the deformation parameter ε. In a number of actinide nuclei the increase of the deformation energy is so substantial that the full energy has a second minimum at certain value of ε. A typical example of the $\mathscr{E}_0(\varepsilon)$ function for the nuclei with two minima is shown in Fig. 7.23. The calculations predict such minima in many elements from uranium to californium. The second minimum is γ-stable, as shown, for example, in ref. 390.

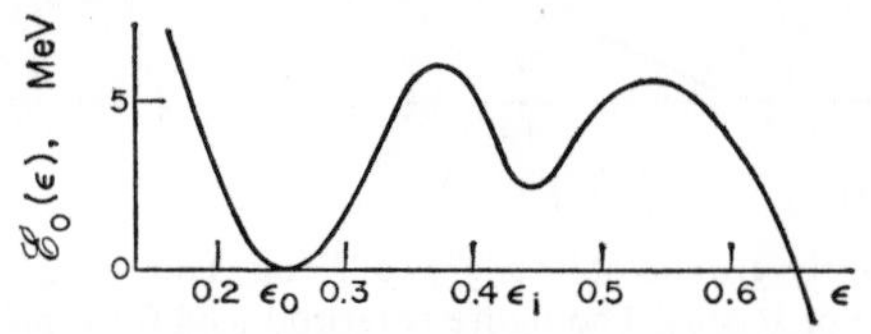

FIG. 7.23. The behavior of the $\mathscr{E}_0(\varepsilon)$ function at large ε and the second minimum.

The existence of the second minima of the function $\mathscr{E}_0(\varepsilon)$ causes several structural effects in the induced nuclear fission (see refs. 391–393).

The spontaneously fissioning isomers, discovered by Polikanov *et al.*,[394] are usually related to the second minima of the $\mathscr{E}_0(\varepsilon)$ curve.† It is generally believed that the observed spontaneously fissioning isomers are the shape isomers, i.e., the nuclear states concentrated in the second minimum. Such isomers are characterized by a large enhancement (up to 10^{20} times) of the spontaneous fission probability when compared with the ground states of the same nuclei. The first isomers were firmly established in the even Am isotopes: ^{240m}Am with $T_{1/2} = 0.0006$ s, ^{242m}Am with $T_{1/2} = 0.014$ s, and ^{244m}Am with $T_{1/2} = 0.001$ s. The interpretation of the spontaneously fissioning isomers was discussed in many papers, for example, in refs. 387, 392, and 395–397. The description of spontaneous fission contains, however, still a number of open problems (see ref. 398).

4. Let us turn our attention to the equilibrium deformations of the quasiparticle states in nuclei which cannot be considered as being strongly deformed. The transitional nuclei with relatively small deformation energies belong to such a group; however, the deformation energy is still larger than the energy associated with the zero-point motion. The region is characterized by the atomic numbers $220 < A < 230$, $184 < A < 192$, and $146 < A < 156$; the observable nuclei with $50 < N$, $Z < 82$, and $28 < Z < 50$, $50 < N < 82$ also belong there. Another property of such nuclei is their softness against the ε- and γ-deformations, i.e., the surfaces $\mathscr{E}_0(\varepsilon, \gamma)$ are relatively flat in the vicinity of the absolute minimum at ε_0 and γ_0.

Under these circumstances (i.e., because the function $\mathscr{E}_0(\varepsilon, \gamma)$ of the quasiparticle

† More recent development in this field is briefly discussed in the translator's addendum, p. 435.

vacuum weakly depends on ε and γ), the ε_e and γ_e values of many quasiparticle states substantially differ from the core values ε_0 and γ_0.

A majority of even–even nuclei with $184 < A < 192$ and $50 < N, Z < 82$ are, according to the calculations, soft against the γ-deformations and relatively hard against the ε-deformations. On the other hand, the nuclei with $28 < Z < 50$, $50 < N < 82$ are soft against both the ε- and the γ-deformations. A typical γ-dependence of the $\mathscr{E}_0(\varepsilon_0, \gamma)$ function is shown in Fig. 7.24 for ^{122}Ba; the $\mathscr{E}_0(\varepsilon_0, \gamma)$ energy increases by less than 300 keV for γ between

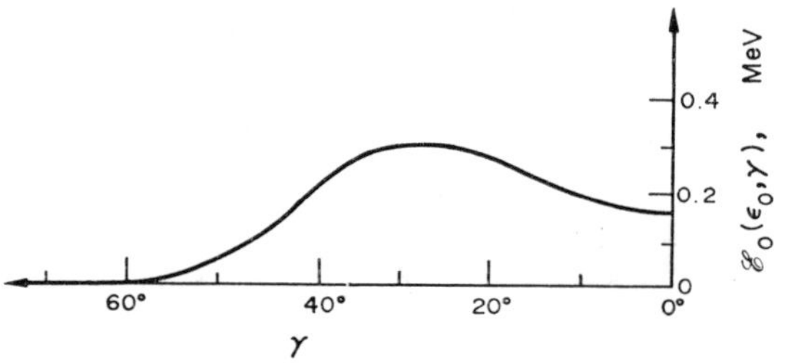

FIG. 7.24. The γ-dependence of the ground-state energy of ^{122}Ba. The curve was calculated for the minimum value of $\varepsilon = \varepsilon_0$. The figure is reproduced from ref. 362.

0° and 60°. The absolute minimum of the $\mathscr{E}_0(\varepsilon_0, \gamma)$ function may be shifted to $\gamma = 60°$ or to $\gamma = 0°$ by a modification of the pairing constants G_Z and G_N, or by a slight modification of the single-particle level scheme. The softness against the γ-deformation is, however, an essential feature present at all reasonable values of the parameters.

The $50 < N, Z < 82$ region illustrates, for example, the large variety of the equilibrium deformations of transitional nuclei. Table 7.8 shows the energies and the corresponding equilibrium deformations of several low-lying quasiparticle states in ^{127}Ba and ^{125}Cs calculated in ref. 362. The states with excitation energy up to 600 keV have a negative equilibrium deformation and $|\varepsilon_e - \varepsilon_0| \lesssim 0.03$. The states with a positive equilibrium deformation appear at excitation energies 600–650 keV while at higher energies both the $\varepsilon_e > 0$

TABLE 7.8. *Excitation energies and equilibrium deformations ε_e of several low-lying one-quasiparticle states in odd-A nuclei*

$^{127}_{56}Ba_{71}$			$^{125}_{55}Cs_{70}$		
State	Excitation energy (keV)	ε_e	State	Excitation energy (keV)	ε_e
503↓	0	−0.25	413↓	0	−0.24
503↑	200	−0.24	505↑	50	−0.23
501↑	330	−0.25	411↑	420	−0.25
431↑	410	−0.22	413↑	500	−0.22
431↓	450	−0.22	420↑	650	0.20
501↓	570	−0.25	330↑	770	−0.30
404↓	600	0.21	505↓	830	−0.23
402↑	700	0.22	411↓	830	−0.23
505↓	810	−0.15			

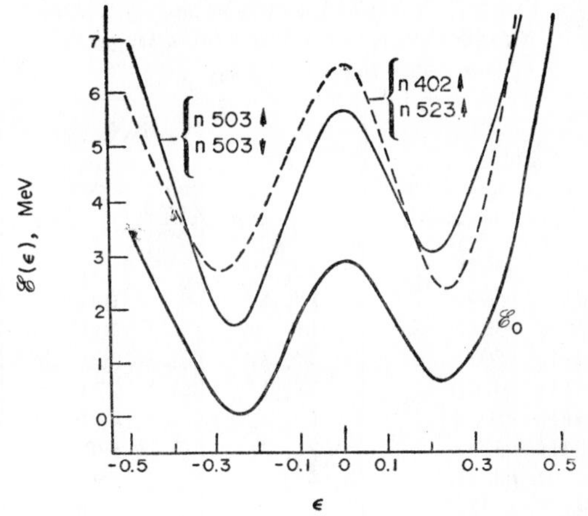

FIG. 7.25. The energies of the ground state (the thick curve) and of two two-quasiparticle states (the thin continuous curve and the dashed curve) in ^{126}Ba. The dependence of the deformation parameter ε is shown. The figure is reproduced from ref. 362.

and the $\varepsilon_e < 0$ states are present. The spectra of the two-quasiparticle states in even–even and odd–odd nuclei were also calculated in ref. 362. Tables 7.9 and 7.10 demonstrate that the spectra have similar properties, i.e., the quasiparticle energies have the minima either at $\gamma = 0°$ or at $\gamma = 60°$. Figure 7.25 shows the total ground-state energy $\mathscr{E}_0(\varepsilon, \gamma = 0°)$ for ^{126}Ba and the $\mathscr{E}(s_1, s_2 \,|\, \varepsilon, \gamma = 0°)$ energies for two two-quasiparticle states. The nucleus ^{126}Ba is oblate in the ground state and in several two-quasiparticle states (including the *nn*503↑, 503↓ state). However, the same nucleus is prolate in the *nn*523↑, 402↑ state and in several other two-quasiparticle states. Consequently, we expect to find many shape isomers in the nuclei of the region discussed.

The deformed nuclei in the $28 < Z < 50$, $50 < N < 82$ region are, as we have mentioned earlier, soft against both the ε- and the γ-deformations. Consequently, large deviations of ε_e and γ_e from the ε_0 and γ_0 values are expected in these nuclei.

TABLE 7.9. *Excitation energies and equilibrium deformations ε_e of the lowest two-quasiparticle states in* ^{126}Ba

State	K^π	Excitation energy (keV)	ε_e
*n*503↓ *n*503↑	6^+, 1^+	1700	−0.26
*n*501↑ *n*503↑	2^+, 5^+	2200	−0.25
*n*503↓ *n*420↑	3^-, 2^-	2220	−0.28
*n*431↑ *n*503↑	2^-, 5^-	2270	−0.22
*n*402↑ *n*523↑	1^-, 6^-	2340	0.24
*p*413↓ *p*411↑	4^+, 1^+	2400	−0.26
*p*505↑ *p*411↑	4^-, 7^-	2480	−0.24
*n*420↑ *n*503↑	3^-, 4^-	2480	−0.26
*p*413↓ *p*505↑	8^-, 3^-	2580	−0.23

TABLE 7.10. *Excitation energies and equilibrium deformations ε_e of the lowest two-quasiparticle states in* $^{126}_{57}La_{69}$

State	$K\pi$	Excitation energy (keV)	ε_e
$p411\uparrow$ $n503\uparrow$	5^-, 2^-	0	−0.27
$p413\downarrow$ $n503\uparrow$	1^-, 6^-	160	−0.26
$p505\downarrow$ $n503\uparrow$	1^+, 8^+	200	−0.27
$p411\downarrow$ $n503\uparrow$	3^-, 4^-	200	−0.27
$p411\uparrow$ $n420\uparrow$	2^+, 1^+	270	−0.28
$p411\uparrow$ $n503\downarrow$	1^-, 4^-	430	−0.28
$p541\uparrow$ $n523\uparrow$	5^+, 2^+	530	0.27
$p550\uparrow$ $n523\uparrow$	4^+, 3^+	530	0.25
$p420\uparrow$ $n523\uparrow$	4^-, 3^-	570	0.27
$p541\uparrow$ $n402\uparrow$	4^-, 1^-	580	0.25

The large variety of the equilibrium shapes of nuclei in the "new" deformation regions $28 < Z < 50$, $50 < N < 82$ and $50 < Z, N < 82$ is related not only to the considerable differences between the ε_e values of the excited states and the ε_0 core values but also to the possibility of the positive (prolate) and negative (oblate) deformations in the same nucleus.

There are the first experimental indications[(399–401)] confirming that the oblate ellipsoidal nuclei and the related shape isomers do, indeed, exist.

§ 3. Moments of Inertia

1. Moments of inertia are important characteristics of the ground and excited nuclear states. Studies of the rotational bands based on the nuclear ground and excited states are a source of a vast experimental material, and many moments of inertia $\mathcal{J}$ and collective gyromagnetic ratios g_R were obtained from these studies. It is very important to systemize the experimental material and to explain the basic trends in variations of the moments of inertia from one nucleus to another and from one nuclear state to another. The comparison of the calculated and experimental moments of inertia is an important test of any particular nuclear model. The problem of the nuclear moments of inertia was studied in many papers, for example, in 10, 45, 154, 155, and 402–407.

We shall discuss in this section the formula for the moment of inertia in the independent particle and independent quasiparticle models. We shall also compare the calculated and experimental values and study the effect of the nuclear rotation on the correlation functions. The conditions for disappearance of the pairing correlations will be discussed too.

The rotational energy in the nuclear rotational model is determined by (2.9), i.e.,

$$E_I = \frac{1}{2\mathcal{J}} I(I+1).$$

The coupling between the rotation and the intrinsic motion and a possible change in nuclear shape render this expression more complicated. Consequently, (2.9) is often replaced by (2.65)–(2.67) or by similar expressions.

The moment of inertia $\mathcal{J}$ in the nuclear liquid-drop model equals

$$\mathcal{J}_h = \frac{9}{8\pi} mAR^2\beta^2. \tag{7.18}$$

The experimental moments of inertia exceed, however, the hydrodynamic value by a factor of 2–5 times.

Let us derive the expression for the moment of inertia in the Inglis cranking model. We shall consider a rotation of an axially symmetric nucleus with the angular velocity Ω; the rotational axis $Ox = Ox'$ is perpendicular to the nuclear symmetry axis Oz'. The system is described by the Schrödinger equation

$$i\frac{\partial}{\partial t}\Psi(x, y, z, t) = H\Psi(x, y, z, t).$$

The standard quantum mechanics prescribes how to express the wave function $\Psi(x, y, z, t)$ in the laboratory coordinate system in terms of the wave function $\Psi'(x', y', z', t')$ in the body-fixed rotating system,

$$\Psi(x, y, z, t) = e^{-i\Omega R_x t}\Psi'(x', y', z', t),$$

where R_x is the x-component of the rotational operator. The wave function describes a stationary state, and, therefore,

$$i\frac{\partial}{\partial t}\Psi = E\Psi.$$

The eigenvalue equation is then of the form

$$(H+\Omega R_x)\Psi = E\Psi. \tag{7.19}$$

In other words, the rotating system is described by the time-independent Hamiltonian

$$H' = H+\Omega R_x. \tag{7.20}$$

Let us use the perturbation theory to calculate the energy corrections caused by the nuclear rotation. The unperturbed wave functions $|0\rangle$ and $|i'\rangle$ are, naturally, calculated at $\Omega = 0$. It is further assumed that H does not depend on the angular velocity of rotation. The first-order correction vanishes because $\langle 0| R_x |0\rangle = 0$.

The second-order correction to the ground-state energy equals

$$\Omega^2 \sum_{i'} \frac{\langle 0 | R_x | i'\rangle \langle i' | R_x | 0\rangle}{E(i')-E(0)},$$

where $E(0)$ and $E(i')$ are the energies of the ground and excited states. The next nonvanishing correction is of the fourth order, i.e., proportional to Ω^4. Using the hermiticity of the operator R_x, we obtain the energy of the rotating system in the form

$$E(\Omega) = E(0)+\Omega^2 \sum_{i'} \frac{|\langle 0 | R_x | i' \rangle|^2}{E(i')-E(0)} + \dots \, . \tag{7.21}$$

On the other hand, the rotational energy equals by definition $\frac{1}{2}\mathcal{J}\Omega^2$, and thus the moment of inertia is equal to

$$\mathcal{J} = 2 \sum_{i'} \frac{|\langle 0 | R_x | i' \rangle|^2}{E(i')-E(0)} \, . \tag{7.22}$$

Using further the relation $\mathcal{J}\Omega = \sqrt{I(I+1)}$, we finally obtain the familiar formula

$$E_I = E(0)+\frac{1}{2\mathcal{J}} I(I+1), \tag{7.21'}$$

Formula (7.22) can be calculated in a closed form if the harmonic oscillator potential (without the spin–orbit coupling) is used as the nuclear average field. Bohr and Mottelson(402) have shown that the rigid-body value of the moment of inertia is obtained in such a case, i.e.,

$$\mathcal{J} \approx \mathcal{J}_{\text{rig}} = \tfrac{2}{5} mAR^2 \, . \tag{7.23}$$

Thus we have now the expressions for the moment of inertia in two simple limiting cases: rigid-body and liquid-drop models. Their ratio equals

$$\mathcal{J}_h / \mathcal{J}_{\text{rig}} = 45\beta^2/16\pi = (\Delta R/R)^2 \, ,$$

where ΔR is the difference between the longer and shorter axes of the ellipsoid; it is related to β by $\Delta R/R = \frac{3}{4}(5/\pi)^{1/2}\beta$.

The experimentally determined nuclear moments of inertia have values between the two limits, i.e.,

$$\mathcal{J}_h < \mathcal{J}_{\text{exp}} < \mathcal{J}_{\text{rig}} \, . \tag{7.24}$$

2. Let us generalize the above-stated expressions and obtain the moment of inertia in the independent quasiparticle model. The corresponding formula was originally derived by Belyaev;(10) later it was modified several times and used in numerical calculations.(45, 87, 154–155, 403–405)

The Hamiltonian of the rotating nucleus is of the form

$$H_0' = H_0(n)+H_0(p)+\Omega \sum_{\substack{q,\, q' \\ \sigma,\, \sigma'}} \langle q\sigma | j_x | q'\sigma' \rangle a^+_{q\sigma} a_{q'\sigma'} \, , \tag{7.25}$$

where $H_0(n)$ and $H_0(p)$ were determined in (4.2). The R_x operator can be rewritten in the second quantization form as

$$R_x = \sum_{\substack{q,\, q' \\ \sigma,\, \sigma'}} \langle q\sigma | j_x | q'\sigma' \rangle a^+_{q\sigma} a_{q'\sigma'} \, . \tag{7.26}$$

In the next step we shall exclude the sign of the projection σ, i.e., the quantum numbers σ,

σ'. Note that if q and q' have $K \neq \frac{1}{2}$ or $K' \neq \frac{1}{2}$, then $\sigma = \sigma'$ and thus

$$\langle q\sigma | j_x | q'\sigma' \rangle = \delta_{\sigma\sigma'} \langle q\sigma | j_x | q'\sigma \rangle; \tag{7.27}$$

while if $K = K' = \frac{1}{2}$, then $\sigma = -\sigma'$, and thus

$$\langle q\sigma | j_x | q'\sigma' \rangle = \delta_{\sigma, -\sigma'} \langle q\sigma | j_x | q', -\sigma \rangle. \tag{7.27'}$$

Further, we shall utilize the hermiticity of the operator j_x, the fact that the matrix elements $\langle q\sigma | jx | q'\sigma' \rangle$ are real, and the standard properties of the time-reversal operator. Under the time-reversal operation, obviously

$$\mathfrak{T}\langle q+ | j_x | q'+ \rangle = \langle (q'+) \mathfrak{T}^{-1} | \mathfrak{T} j_x \mathfrak{T}^{-1} | \mathfrak{T}(q+) \rangle$$

and, remembering that

$$\mathfrak{T} j_x \mathfrak{T}^{-1} = -j_x, \quad \mathfrak{T} | q+ \rangle = | q- \rangle, \quad \mathfrak{T} | q- \rangle = -| q+ \rangle,$$

we obtain the symmetry relation

$$\langle q+ | j_x | q'+ \rangle = -\langle q'- | j_x | q- \rangle. \tag{7.28}$$

Further

$$\begin{aligned} \mathfrak{T}\langle q+ | j_x | q'- \rangle &= \langle (q+) \mathfrak{T}^{-1} | \mathfrak{T} j_x \mathfrak{T}^{-1} | \mathfrak{T}(q'-) \rangle^* \\ &= \langle q- | -j_x | (-1) q'+ \rangle^* = \langle q- | j_x | q'+ \rangle. \end{aligned} \tag{7.28'}$$

The moment of inertia in the independent particle model is equal to

$$\mathcal{J} = 2 \sum_{i'} \frac{|\langle 0 | R_x | i' \rangle|^2}{E(i') - E(0)}.$$

The excited states $|i'\rangle$ are the particle–hole states because in (7.25) the operator $a_{q'\sigma'}$ creates a hole while the operator $a^+_{q\sigma}$ creates a particle.

How does the analogous expression appear in the independent quasiparticle model? We shall start with the ground states of even–even nuclei; the blocking effect will be neglected. The two-quasiparticle states with the wave functions (4.31), i.e.,

$$\alpha^+_{q_1\sigma_1} \alpha^+_{q_2\sigma_2} \Psi_0$$

(Ψ_0 is the ground-state wave function) are the only candidates for the corresponding excited states. As a next step we have to calculate the matrix element $\langle 0 | R_x | i' \rangle$ using the wave functions (4.31) and (4.22). The energy difference $E(i') - E(0)$ in the denominator must be replaced by the energy difference between the two-quasiparticle state and the quasiparticle vacuum which is, according to (4.31), equal to

$$\mathcal{E}_0(q_1, q_2) - \mathcal{E}_0 = \varepsilon(q_1) + \varepsilon(q_2).$$

The corresponding matrix element equals

$$\langle | R_x \alpha^+_{q_1\sigma_1} \alpha^+_{q_2\sigma_2} | \rangle_0 = \sigma_2 \langle q_2\sigma_2 | j_x | q_1, -\sigma_1 \rangle u_{q_1} v_{q_2} - \sigma_1 \langle q_1\sigma_1 | j_x | q_2, -\sigma_2 \rangle u_{q_2} v_{q_1}. \tag{7.29}$$

If $K_1 \neq \frac{1}{2}$ or $K_2 \neq \frac{1}{2}$, then $\sigma_1 = -\sigma_2$ and by using (7.28) we can transform (7.29) into

$$\langle | R_x \alpha^+_{q_1\sigma_1} \alpha^+_{q_2\sigma_2} | \rangle_0 = \delta_{\sigma_1, -\sigma_2} \sigma_1 \langle q_1\sigma_1 | j_x | q_2\sigma_1 \rangle (u_{q_1} v_{q_2} - u_{q_2} v_{q_1}).$$

On the other hand, if $K_1 = K_2 = \frac{1}{2}$, then $\sigma_1 = \sigma_2$, and, according to (7.28′), we write

$$\langle | R_x \alpha^+_{q_1\sigma_1} \alpha^+_{q_2\sigma_2} | \rangle_0 = \sigma_1 \delta_{\sigma_1\sigma_2} \langle q_1\sigma_1 | j_x | q_2 - \sigma_1 \rangle (u_{q_1} v_{q_2} - u_{q_2} v_{q_1}).$$

Finally, we shall replace the energy expression in (7.22), use the matrix element (7.29), and perform the summation over σ. As a result we obtain the following moment of inertia:

$$\mathcal{J} = 4 \sum_{q_1, q_2} \frac{|\langle q_1 + | j_x | q_2 \pm \rangle|^2}{\varepsilon(q_1) + \varepsilon(q_2)} (u_{q_1} v_{q_2} - u_{q_2} v_{q_1})^2 . \tag{7.30}$$

The upper sign in the matrix element corresponds to the $K_1 \neq \frac{1}{2}$ or $K_2 \neq \frac{1}{2}$ case, while the lower sign corresponds to the $K_1 = K_2 = \frac{1}{2}$ case. The pairing factor $(u_{q_1} v_{q_2} - u_{q_2} v_{q_1})^2$ shows that the particle–hole states give the largest contributions. The total moment of inertia equals

$$\mathcal{J} = \mathcal{J}_p + \mathcal{J}_n, \tag{7.31}$$

i.e., it contains the contributions of protons and neutrons.

The expression (7.30) is somewhat modified when the blocking effect is taken into account. The moment of inertia is then equal to

$$\mathcal{J} = 4 \sum_{q_1 q_2} \frac{|\langle q_1 + | j_x | q_2 \pm \rangle|^2}{\mathcal{E}_0(q_1, q_2) - \mathcal{E}_0} (u_{q_1} v_{q_2} - u_{q_2} v_{q_1})^2 \prod_{q \neq q_1, q_2} (u_q u_q(q_1 q_2) + v_q v_q(q_1 q_2))^2 , \tag{7.32}$$

where $\mathcal{E}_0(q_1, q_2)$ and $\mathcal{E}_0$ were defined in (4.48) or (4.21), respectively. The product $\prod_{q \neq q_1, q_2} (u_q u_q(q_1 q_2) + v_q v_q(q_1 q_2))^2$ has a value close to unity, and it is therefore usually omitted. The blocking effect is not very important in calculations of the moment of inertia of the quasiparticle vacuum.

Inclusion of the pairing correlations decreases the theoretical moment of inertia which is then substantially smaller than the rigid-body value. Such a decrease has two reasons: the pairing factor $(u_{q_1} v_{q_2} - u_{q_2} v_{q_1})^2$ is quite small when both q_1 and q_2 are close to the Fermi level. Thus the nucleons strongly affected by the pairing correlations, i.e., those in the "superfluid" state, contribute little to the nuclear moment of inertia. The second reason for the decreased $\mathcal{J}$-value is related to the existence of the gap in spectra of even–even nuclei. With the nonvanishing gap all energy denominators in (7.30) or (7.32) are larger than the corresponding independent particle values; consequently, the resulting moment of inertia is smaller.

The decrease in the moment of inertia caused by the pairing correlations is illustrated in Figs. 7.26 and 7.27. The figures show the experimental ground-state moments of inertia of even–even nuclei with $150 < A < 190$ and $228 < A < 254$; the independent particle model (i.e., the rigid-body values) moments of inertia and the results of calculations in the nuclear superfluid model are also shown. It is apparent that the rigid-body values exceed the experimental values 2–3 times. On the other hand, the moments of inertia calculated in the nuclear superfluid model agree reasonably well with the experiment; a small systematic discrepancy (the experimental values are typically somewhat larger than the theoretical values) is, however, present. The discrepancy could be probably removed by improvement of the calculational methods. Particularly, the Woods–Saxon single-particle

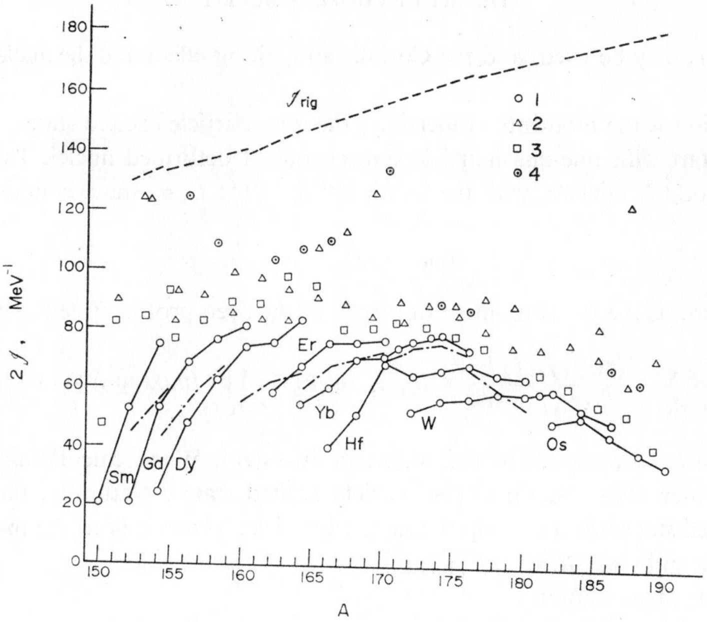

FIG. 7.26. Moments of inertia of the deformed nuclei in the 150 < A < 190 region. The experimental moments of inertia are denoted in the following way: 1 (empty circles) connected by full lines denote the even–even nuclei; 2 (triangles) denote the odd-Z nuclei; 3 (squares) denote the odd-N nuclei; 4 (circles with a dot) denote the odd–odd nuclei. The dashed line shows the rigid-body values. The calculated values in the nuclear superfluid model[45] are connected by the broken lines.

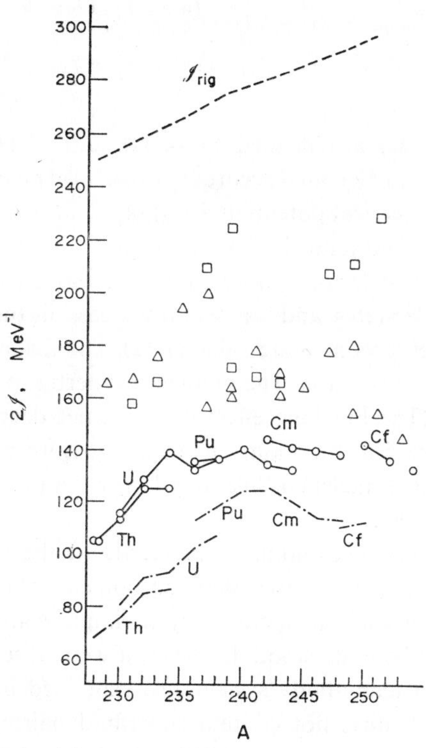

FIG. 7.27. Moments of inertia of the deformed nuclei in the 228 < A < 254 region. (Notation as Fig. 7.26.)

wave functions may be used, and the Coriolis antipairing effect and the nuclear stretching included.

3. Let us discuss the moments of inertia of the quasiparticle excited states.

We shall start with one-quasiparticle states in odd-A deformed nuclei. The moment of inertia of an odd-N nucleus, with the quasiparticle in the s_0, σ_0 state, contains two terms, i.e.,

$$\mathcal{J}(s_0) = \mathcal{J}_p + \mathcal{J}_n(s_0).$$

The first term is the usual moment of inertia of the even proton system, i.e.,

$$\mathcal{J}_p = 4 \sum_{r_1, r_2} \frac{|\langle r_1 + | j_x | r_2 \pm \rangle|^2}{\mathcal{E}_0(r_1, r_2) - \mathcal{E}_0} (u_{r_1} v_{r_2} - u_{r_2} v_{r_1})^2 \prod_{r \neq r_1, r_2} (u_r u_r(r_1 r_2) + v_r v_r(r_1 r_2))^2 . \quad (7.33)$$

The second term corresponds to the odd neutron system. When calculating the $\mathcal{J}_n(s_0)$ we have to consider one- and three-quasiparticle excited states. Naturally, the three-quasiparticle excited states must contain the s_0 quasiparticle. Thus we need the matrix elements $\langle | \alpha_{s_0\sigma_0} R_x \alpha^+_{s\sigma} | \rangle_0$ and $\langle | \alpha_{s_0\sigma_0} R_x \alpha^+_{s_0\sigma_0} \alpha^+_{s_1\sigma_1} \alpha^+_{s_2\sigma_2} | \rangle_0$.

After simple manipulations,

$$\mathcal{J}_n(s_0) = 2 \sum_{s \neq s_0} \frac{|\langle s_0 \sigma_0 | j_x | s \pm \sigma_0 \rangle|^2}{\mathcal{E}_0(s) - \mathcal{E}_0(s_0)} (u_s(s_0)\, u_{s_0}(s) + v_{s_0}(s)\, v_s(s_0))^2 \times$$
$$\times \prod_{s' \neq s_0, s} (u_{s'}(s_0)\, u_{s'}(s) + v_{s'}(s_0)\, v_{s'}(s))^2$$
$$+ 4 \sum_{s_1, s_2} \frac{|\langle s_2 + | j_x | s_1 \pm \rangle|^2}{\mathcal{E}_0(s_0, s_1, s_2) - \mathcal{E}_0(s_0)} (u_{s_1}(s_0)\, v_{s_2}(s_0) - u_{s_2}(s_0)\, v_{s_1}(s_0))^2 \times$$
$$\times \prod_{s' \neq s_0, s_1, s_2} (u_{s'}(s_0)\, u_{s'}(s_0, s_1, s_2) + v_{s'}(s_0)\, v_{s'}(s_0, s_1, s_2))^2 , \quad (7.34)$$

where $\mathcal{E}_0(s)$ and $\mathcal{E}_0(s_0, s_1, s_2)$ are defined in (4.39) and (4.44). The $u_s(s_0)$, $v_s(s_0)$, and $u_s(s_0, s_1, s_2)$, $v_s(s_0, s_1, s_2)$ are calculated according to (4.16). The correlation functions $C_n(s_0)$ and $C_n(s_0, s_1, s_2)$ and the chemical potentials $\lambda_n(s_0)$, $\lambda_n(s_0, s_1, s_2)$ should be used.

The second term in the righthand side of (7.34) has the same structure as the analogous expression in (7.32). The first term, however, is a new term. It corresponds to the contribution of one-quasiparticle states and causes an increase in the moment of inertia (7.33) compared to the even–even ground-state value (7.32). The magnitude of this term strongly depends on the quantum numbers s_0. The moment of inertia of an odd-A nucleus is larger also for another reason. The blocking effect causes a net decrease in the nuclear superfluidity and thus an increase in the moment of inertia. Both effects make the theoretical moments of inertia of odd-A nuclei considerably larger than the moments of inertia of the neighboring even–even nuclei.

The experimentally determined moments of inertia of the odd-A deformed nuclei are larger (almost twice as large in many cases) than the moments of inertia of the even–even core. It is impossible to explain such an increase by a simple contribution of one additional nucleon. The increase can be understood, however, if we realize the basic differences in the characteristics of the two many-body systems, as explained in the preceding discussion. Any nuclear model which does not contain superfluid pairing correlations fails in this respect. Even if it is sometimes possible in such model to explain (by an adjustment of the parameters) the even–even moments of inertia, it is impossible to understand the change caused by the addition of one nucleon to the system.

Figures 7.26 and 7.27 illustrate the increase in the moments of inertia of the odd-A nuclei compared to the ground state moments of inertia of the even–even nuclei; the strong variations of the odd-A moments of inertia are also apparent. Tables 7.11 and 7.12 demonstrate this fact again.

TABLE 7.11. *Moments of inertia of the one-quasiparticle states in odd-N nuclei*

Single-particle state	Nucleus	Energy (keV)	$\mathcal{J}_{A+1}(q)$ (MeV^{-1})	$\delta\mathcal{J}_{A+1}(q)$ (MeV^{-1})
$\frac{3}{2}^-$ [521]	^{155}Sm	0	46.7	9.4
	^{155}Gd	0	41.3	11.3
	^{157}Gd	0	45.8	9.6
	^{159}Dy	0	44.2	10.0
	^{161}Dy	75	43.2	5.8
	^{163}Dy	421	45.7	6.0
$\frac{5}{2}^-$ [523]	^{157}Gd	437	43.5	7.3
	^{157}Gd	146	43.8	4.4
	^{159}Dy	310	41.0	6.8
	^{161}Dy	26	45.2	7.8
	^{163}Dy	0	47.2	7.5
	^{163}Er	0	41.6	10.0
	^{165}Er	0	45.1	9.3
$\frac{7}{2}^+$ [633]	^{165}Dy	0	53.7	12.0
	^{167}Er	0	56.7	18.8
	^{169}Er	244	60.2	22.0
	^{169}Yb	0	63.3	27.7
	^{171}Yb	95	61.3	24.3

Tables 7.11 and 7.12 show the increase in the moments of inertia $\delta\mathcal{J}_{A+1}(q)$. This quantity is defined as $\delta\mathcal{J}_{A+1}(q) = \mathcal{J}_{A+1}(q) - \frac{1}{2}(\mathcal{J}_A + \mathcal{J}_{A+2})$, i.e., it is equal to the difference between the moment of inertia of the one-quasiparticle state and the average moment of inertia of the neighboring even–even nuclei. All $\delta\mathcal{J}_{A+1}(q)$ values are positive, thus proving that the moment of inertia has indeed increased. The strong variation is also apparent; while the $\delta\mathcal{J}_{A+1}(q)$ are relatively small for the $\frac{1}{2}^+$ [411] and $\frac{7}{2}^+$ [404] states, they become very important for the $\frac{7}{2}^+$ [633] states. The odd–even differences $\delta\mathcal{J}_{A+1}(q)$ belonging to the same one-quasiparticle state q but to the different nuclei are not constant. Such fluctuations can be related to the Coriolis effect and, to a certain extent, to the quasiparticle–phonon interaction.

The calculated moments of inertia of the odd-A nuclei[60, 87, 405] agree reasonably well with the experimental data.

Moments of inertia of the two-quasiparticle states in odd-odd deformed nuclei have the form

$$\mathcal{J}(r_0, s_0) = \mathcal{J}_p(r_0) + \mathcal{J}_n(s_0), \tag{7.35}$$

TABLE 7.12. *Moments of inertia of the one-quasiparticle states in odd-Z nuclei*

Single-particle state	Nucleus	Energy (keV)	$\mathcal{J}_{A+1}(q)$ (MeV^{-1})	$\delta\mathcal{J}_{A+1}(q)$ (MeV^{-1})
$\frac{3}{2}+$ [411]	^{155}Eu	246	39.0	3.7
	^{157}Tb	0	41.3	8.5
	^{159}Tb	0	44.6	7.0
	^{161}Tb	0	44.2	3.0
	^{165}Ho	362	42.9	7.1
$\frac{7}{2}-$ [523]	^{163}Ho	0	49.0	13.2
	^{165}Ho	0	47.6	11.8
	^{169}Tm	379	48.0	11.2
$\frac{1}{2}+$ [411]	^{167}Tm	0	40.3	3.6
	^{169}Tm	0	40.6	3.8
	^{171}Tm	0	41.6	3.2
	^{171}Lu	382	37.0	2.9
	^{173}Lu	425	39.0	3.2
	^{177}Lu	459	37.6	2.8
$\frac{7}{2}+$ [404]	^{169}Tm	316	38.4	0
	^{171}Lu	0	37.0	2.9
	^{175}Lu	0	39.3	2.3
	^{177}Lu	0.	37.3	2.5
	^{181}Ta	0	33.1	1.7

where the $\mathcal{J}_p(r_0)$ and $\mathcal{J}_n(s_0)$ are the expressions analogous to (7.34). The moments of inertia of the odd–odd nuclei can be then estimated using the experimental $\mathcal{J}_p(r_0)$ and $\mathcal{J}_n(r_0)$ of the neighboring odd-A nuclei.

Several moments of inertia of the two-quasiparticle states in odd–odd nuclei are shown in Table 7.13. The corresponding increase defined by

$$\delta\mathcal{J}_{Z+1,\,N-1}(q_1, q_2) = \mathcal{J}(q_1, q_2) - \tfrac{1}{2}(\mathcal{J}_{Z,\,N} + \mathcal{J}_{Z+2,\,N-2}) \tag{7.36}$$

is also listed. The quantity $\delta\mathcal{J}(q_1, q_2) > 0$, i.e., it is positive for all states, except for the

TABLE 7.13. *Moments of inertia of the two-quasiparticle states in even deformed nuclei*

Nucleus	K^π	Configuration	Energy (keV)	$\mathcal{J}_{(q_1, q_2)}$ (MeV^{-1})	$\delta\mathcal{J}_{(q_1, q_2)}$ (MeV^{-1})
^{238}Np	3+	$n631\downarrow$, $p642\uparrow$	26	113.6	45.1
	3−	$n631\downarrow$, $p523\downarrow$	135	92.6	24.1
^{186}Re	1−	$n512\downarrow$, $p402\uparrow$	0	34.2	10.2
^{182}W	4−	$n624\uparrow$, $n510\uparrow$	1554	51.0	20.7
^{176}Lu	7−	$n514\downarrow$, $p404\downarrow$	0	28.7	−7.0
^{172}Yb	3+	$n512\uparrow$, $n521\downarrow$	1174	44.2	5.8
^{170}Tm	1−	$n521\downarrow$, $p411\downarrow$	0	43.7	6.6
^{166}Ho	0−	$n633\uparrow$, $p523\uparrow$	0	55.4	17.5
	3+	$n521\downarrow$, $p523\uparrow$	191	57.2	19.3

$K^{\pi} = 7^-$ state in ^{176}Lu. The fluctuations of the quantity $\delta\mathcal{J}(q_1, q_2)$ are substantially larger than the similar fluctuations of the $\delta\mathcal{J}_{A+1}(q)$ in the odd-A nuclei.

Let us, for completeness, give the formula for the moment of inertia of a two-quasiparticle state in an even–even nucleus. If we suppose that we have two neutron quasiparticles $(\varrho_1\sigma_1^0, \varrho_2\sigma_2^0)$, then

$$\mathcal{J}(\varrho_1, \varrho_2) = \mathcal{J}_p + \mathcal{J}_n(\varrho_1, \varrho_2). \tag{7.37}$$

If, for simplicity, the product $\prod(u_s u_s' + v_s v_s')^2$ is omitted, we obtain $\mathcal{J}_n(\varrho_1, \varrho_2)$ of the form

$$\mathcal{J}_n(\varrho_1, \varrho_2) = 4 \sum_{s_1, s_2} \frac{|\langle s_1 + | j_x | s_2 \pm \rangle|^2}{\mathcal{E}_0(s_1, s_2, \varrho_1, \varrho_2) - \mathcal{E}_0(\varrho_1, \varrho_2)} \times$$
$$\times (u_{s_1}(\varrho_1, \varrho_2)\, v_{s_2}(\varrho_1, \varrho_2) - u_{s_2}(\varrho_1, \varrho_2)\, v_{s_1}(\varrho_1, \varrho_2))^2$$
$$+ 2 \sum_{s_1} \frac{|\langle \varrho_1\sigma_1^0 | j_x | s_1 \pm \sigma_1^0 \rangle|^2}{\mathcal{E}_0(s_1, \varrho_2) - \mathcal{E}_0(\varrho_1, \varrho_2)} (u_{s_1}(\varrho_1, \varrho_2)\, u_{\varrho_1}(s_1, \varrho_2) + v_{s_1}(\varrho_1, \varrho_2)\, v_{\varrho_1}(s_1, \varrho_2))^2$$
$$+ 2 \sum_{s_2} \frac{|\langle \varrho_2\sigma_2^0 | j_x | s_2 \pm \sigma_2^0 \rangle|^2}{\mathcal{E}_0(\varrho_1, s_2) - \mathcal{E}_0(\varrho_1, \varrho_2)} (u_{s_2}(\varrho_1, \varrho_2)\, u_{\varrho_2}(\varrho_1, s_2) + v_{s_2}(\varrho_1, \varrho_2)\, v_{\varrho_2}(\varrho_1, s_2))^2$$
$$+ 2 \frac{|\langle \varrho_1\sigma_1^0 | j_x | \varrho_2\sigma_2^0 \rangle|^2}{\mathcal{E}_0 - \mathcal{E}_0(\varrho_1, \varrho_2)} (u_{\varrho_1} v_{\varrho_2} - u_{\varrho_2} v_{\varrho_1})^2. \tag{7.38}$$

Thus we have the contributions of the two-quasiparticle and four-quasiparticle states.

The experimental moments of inertia of two-quasiparticle states are, generally, larger than the ground state moments of inertia. Table 7.13 shows two examples of this effect, the moments of inertia (and the corresponding differences $\delta\mathcal{J}$) for ^{172}Yb and ^{182}W. The strong fluctuations are quite apparent.

We have shown in § 2 of Chapter 4 that the calculated correlation functions $C(q_1, q_2)$ of several two-quasiparticle states are very small. Thus we may expect a large increase in the moment of inertia caused by the near vanishing of the correlation function $C(q_1, q_2)$. The moments of inertia of many two-quasiparticle states were calculated in refs. 404 and 405. A reasonably good agreement between the experimental and theoretical values was found; even in the limiting case $C(q_1, q_2) = 0$, the calculated moments of inertia are considerably smaller than the rigid-body limit.

4. There is an analogy between the effect of a magnetic field on a superconductor and the effect of a rotation on the superfluid characteristics of the atomic nuclei. The analogy is actually quite deep—the superconductivity disappears at certain critical magnetic field and, similarly, the pairing correlations disappear at certain critical angular momentum I_c.

Let us study in detail the effect of the nuclear rotation on the correlation function values. Such an effect is often referred to as the "Coriolis antipairing" effect. The effect has been discussed in a number of original studies, e.g. in refs. 408–415.

Let us derive an equation for the correlation function including the nuclear rotation. In order to do this we use the results of § 2 of Chapter 3. We are considering the case $\Phi^*(f, f') = -\Phi(f', f)$ using the matrix form of the equations. The matrix $F(f, f')$ has the form (3.19′), i.e.,

$$F = \begin{vmatrix} \varrho & -\Phi \\ \Phi & 1-\varrho \end{vmatrix};$$

it fulfills the projection condition (3.19″), $F^2 = F$.

We shall define another matrix L:

$$L = \begin{vmatrix} \xi-\Omega j_x & -C \\ C & -(\xi-\Omega j_x) \end{vmatrix}, \tag{7.39}$$

where, according to (3.18) and (3.17),

$$C_{f,f'} = \tfrac{1}{2}\sum_{f_1', f_2'} G(f, f'; f_2', f_1')\,\Phi(f_2', f_1'),$$
$$\xi(f, f') = T'(f, f') - \sum_{f_1, f_2} G(f, f_1; f_2, f')\,\varrho(f_1, f_2).$$

Equation

$$[L, F] = 0 \tag{7.40}$$

is equivalent to (3.12′) with the functions $\xi(f,f')$ replaced by $\xi(f, f')-\Omega\langle f\,|\,j_x\,|\,f'\rangle$.

Because $R_x = \mathcal{J}\Omega$, we obtain the following identity:

$$\mathcal{J}\Omega = 2\,\mathrm{Sp}\, j_x\varrho, \tag{7.41}$$

where Sp denotes the trace of the matrix in the space of the quantum numbers q. The factor 2 in (7.41) is related to the summation over $\sigma = \pm 1$ because $f = (q, \sigma)$. Let us denote the trace of the second order matrix (L or F) by Tr. We obtain

$$\mathrm{Sp\,Tr}\,\frac{\partial L}{\partial \Omega} F = \mathrm{Sp\,Tr} \begin{vmatrix} -j_x & 0 \\ 0 & j_x \end{vmatrix} \begin{vmatrix} \varrho & -\Phi \\ \Phi & 1-\varrho \end{vmatrix} = -2\,\mathrm{Sp}\, j_x\varrho,$$

because $\mathrm{Sp}\, j_x = 0$. Using (7.41), we can write the moment of inertia as

$$\mathcal{J} = -\frac{1}{\Omega}\,\mathrm{Sp\,Tr}\,\frac{\partial L}{\partial \Omega} F. \tag{7.41′}$$

We have, further,

$$\mathrm{Sp\,Tr}\,\frac{\partial L}{\partial C} F = \mathrm{Sp\,Tr} \begin{vmatrix} 0 & -1 \\ -1 & 0 \end{vmatrix} \begin{vmatrix} \varrho & -\Phi \\ \Phi & 1-\varrho \end{vmatrix} = -2\,\mathrm{Sp}\,\Phi. \tag{7.42}$$

It is easy to see that if $F = F^2$, then $\delta F = F\,\delta F + \delta F F$ and $F\,\delta F F = F\,\delta F F^2 + F^2\,\delta F F = 2F\,\delta F F = 0$. Using this result and (7.40),

$$\mathrm{Sp}\, L\,\delta F = \mathrm{Sp}\, L(\delta F F + F\,\delta F) = 2\,\mathrm{Sp}\, LF\,\delta F F = 0,$$

and, consequently,

$$\mathrm{Sp\,Tr}\, L\frac{\partial F}{\partial \Omega} = 0, \quad \mathrm{Sp\,Tr}\, L\frac{\partial F}{\partial C} = 0. \tag{7.43}$$

The expressions (7.41′) and (7.42) are transformed into the form

$$\mathcal{J} = -\frac{1}{\Omega}\frac{\partial}{\partial \Omega} \text{Sp Tr } LF, \tag{7.41″}$$

$$\text{Sp } \Phi = -\frac{1}{2}\frac{\partial}{\partial C} \text{Sp Tr } LF. \tag{7.42′}$$

Let us find the relation between (7.41″) and (7.42′). We shall do the following steps:

$$\frac{1}{4}\times\frac{1}{C}\frac{\partial}{\partial C}\mathcal{J} = \frac{1}{2}\frac{\partial}{\partial C^2}\mathcal{J} = -\frac{\partial}{\partial C^2}\frac{\partial}{\partial \Omega^2} \text{Sp Tr } LF,$$

$$\frac{1}{C}\frac{\partial}{\partial \Omega^2} \text{Sp } \Phi = -\frac{\partial}{\partial C^2}\frac{\partial}{\partial \Omega} \text{Sp Tr } LF$$

and obtain

$$\frac{1}{2}\times\frac{\partial}{\partial C^2}\mathcal{J} = \frac{1}{C}\times\frac{\partial}{\partial \Omega^2} \text{Sp } \Phi. \tag{7.44}$$

By integrating this equation over Ω^2 within the 0 and Ω^2 limits, we finally obtain

$$\frac{1}{C} \text{Sp } \Phi = \frac{1}{C}(\text{Sp } \Phi)_{\Omega=0} + \frac{1}{2}\Omega^2\frac{\partial \mathcal{J}}{\partial C^2}. \tag{7.45}$$

Let us consider our usual approximation G = const and $\Phi(f, f') = \Phi(q)\delta_{q, q'}\delta_{\sigma, -\sigma'}$. In such an approximation, (3.18) is rewritten as

$$C = \tfrac{1}{2}G\sum_{q, \sigma}\Phi(q) = G \text{ Sp } \Phi. \tag{7.46}$$

From (3.36′) and (3.46′) it follows that

$$\frac{1}{C}(\text{Sp } \Phi)_{\Omega^2=0} = \frac{1}{C}\sum_q u_q v_q = \frac{1}{2}\sum_q\frac{1}{\varepsilon(q)}. \tag{7.46′}$$

Substituting (7.46) and (7.46′) in (7.45),

$$\frac{2}{G} = \sum_q\frac{1}{\varepsilon(q)} + \Omega^2\frac{\partial \mathcal{J}}{\partial C^2}, \tag{7.47}$$

and, using $\Omega\mathcal{J} = \sqrt{I(I+1)}$, we can rewrite (7.47) in the form

$$\frac{2}{G} = \sum_q\frac{1}{\varepsilon(q)} + \frac{I(I+1)}{\mathcal{J}^2}\frac{\partial \mathcal{J}}{\partial C^2}. \tag{7.47′}$$

Thus we have now the equation which includes the dependence of the correlation function on the nuclear angular momentum I. The correlation function C obviously decreases when the nuclear angular momentum increases. Equation (7.47′) should be solved together with (4.20) for each value of the angular momentum I. The neutron and proton systems are still treated separately. The critical angular momentum I_c is determined from the condition of the vanishing values C_p or C_n. The numerical calculations[410, 414] gave $I_c = 16$–20 for the rare-earth nuclei and somewhat larger values of I_c for the actinide nuclei. The neutron correlation functions vanish generally at lower critical angular momenta I_c than the proton correlation functions.

However, we should remember that the calculation of the critical values I_c is rather crude. Actually, when the correlation functions C are small (which is the case of I being close to I_c), the approximate equations (4.19) or (7.47) become inaccurate. In such cases it is necessary to use more accurate calculations, for example, those discussed in § 3 of Chapter 5, where the average energy calculated by means of the projected wave functions is minimized.

The calculated critical angular momenta I_c in odd-A nuclei strongly fluctuate. They are, however, systematically smaller than the similar quantities in even–even nuclei.[411]

The simple formula (2.9), i.e.,

$$E_I = \frac{I(I+1)}{2\mathcal{J}},$$

describes the rotational states with $I < 8$ with a reasonable accuracy. When the states with a higher angular momentum values are considered, several corrections become important. These corrections include the change in nuclear superfluid characteristics by rotation (Coriolis antipairing), change in nuclear shape by rotation (nuclear centrifugal stretching), and the direct Coriolis effects.

The "long" rotational bands may be described in two somewhat different ways.

In the first approach we use the simple formula

$$E_I = \frac{I(I+1)}{2\mathcal{J}_I^{\text{eff}}}, \tag{7.48}$$

where $\mathcal{J}_I^{\text{eff}}$ is an effective moment of inertia explicitly I-dependent. The effective moment of inertia should be approximately constant at small values of I, and it should slowly and smoothly grow for larger values of I. All difficulties of the description of the long rotational bands are concentrated in the effective moment of inertia in such an approach.

Another, more traditional, approach consists of the expansion of energy in powers of $I(I+1)$, i.e.,

$$E_I = \frac{I(I+1)}{2\mathcal{J}} + \mathfrak{B}I^2(I+1)^2 + \dots \tag{7.49}$$

[see eqns. (2.65) and (2.67)]. The quantity $\mathfrak{B}$ is divided into several terms in microscopic calculations:[414, 415]

$$\mathfrak{B} = \mathfrak{B}_{\text{cor}} + \mathfrak{B}_\lambda + \mathfrak{B}_\Delta + \mathfrak{B}_\beta + \mathfrak{B}_\gamma. \tag{7.50}$$

The $\mathfrak{B}_{\text{cor}}$ is related to the direct Coriolis interaction. It is calculated in the perturbation theory using the model of independent quasiparticles as a basis. The chemical potentials λ_p, λ_n vary slightly from one rotational state to another. These variations lead to the term $\mathfrak{B}_\lambda$ which is generally quite small and can be usually neglected. The third term in (7.50), the $\mathfrak{B}_\Delta$, is related to the change of the correlation functions C_p and C_n caused by the nuclear rotation. Using the same approximation as in (7.47),

$$\mathfrak{B}_\Delta = -\frac{1}{8\mathcal{J}^4}\left\{\frac{\left(\dfrac{\partial \mathcal{J}}{\partial C_n}\right)^2}{\sum\limits_s \dfrac{C_n^2}{\varepsilon^3(s)}} + \frac{\left(\dfrac{\partial \mathcal{J}}{\partial C_p}\right)^2}{\sum\limits_r \dfrac{C_p^2}{\varepsilon^3(r)}}\right\}. \tag{7.51}$$

The last two terms in (7.50) are related to the centrifugal stretching of the nucleus. They are of the form

$$\mathfrak{B}_\beta = -\frac{1}{8\mathcal{J}^4}\frac{1}{C_\beta}\left(\frac{\partial \mathcal{J}}{\partial \beta}\right)^2, \tag{7.52}$$

$$\mathfrak{B}_\gamma = -\frac{1}{8\mathcal{J}^4}\frac{1}{C_\gamma}\left(\frac{\partial \mathcal{J}}{\partial \gamma}\right)^2, \tag{7.52'}$$

where C_β and C_γ are the restoring force parameters of the β- or γ-vibrations, respectively (see § 3 of Chapter 2).

When the above-discussed effects were included, the description of the rotational bands has considerably improved. The whole complex of related problems is discussed in a number of publications, for example, in refs. 60, 63, 412, and 414–416.

CHAPTER 8

SEMIMICROSCOPIC THEORY OF NUCLEAR VIBRATIONS

§ 1. The Selfconsistent Field Method†

1. In this chapter we shall discuss the mathematical methods used in developing the semimicroscopic theory of nuclear vibrations. The classical perturbation theory, very important in the solution of various physical problems, cannot properly describe the quasiparticle interaction. Other methods, including the description of the coherent collective effects, must be therefore used.

There are three different formal methods applicable in nuclear theory; all of them satisfy the above-stated condition, i.e., they contain description of the coherent effects. We shall discuss these methods in this chapter, and we shall reserve one section to each method. The selfconsistent field method will be discussed in § 1; the method of approximate second quantization in § 2, and the Green function method in § 3.

These mathematical methods are not exclusive to nuclear physics; they are quite general, and often used in other theoretical physics applications. Let us enumerate the basic references. The necessary modifications of the general methods were performed in refs. 150 and 417–419. The semimicroscopic description of the vibrational states in spherical nuclei was developed in refs. 420–423 and the same problem has been solved in refs. 424–433 for deformed nuclei. Many monographs, textbooks, and review articles describe this subject, among others refs. 22, 74, 128, 129, 140, 158, 335, 434, and 435. This chapter will discuss the general mathematical formalisms, then the basic equations will be written in easily numerically solvable form in § 4, and, finally, in § 5, we shall discuss the 0^+ excited states differing in several aspects from the other vibrational states.

This section will explain the selfconsistent field method in the Bogolyubov formulation.[120] We shall derive the general form of the secular equation and discuss several special problems.

2. To derive our basic equations, we shall start with (5.37) in the form

$$i\frac{\partial}{\partial t}\varrho(f_1, f_2) = \mathfrak{B}(f_1, f_2), \quad i\frac{\partial}{\partial t}\Phi(f_1, f_2) = \mathfrak{A}(f_1, f_2).$$

The functions $\mathfrak{A}(f_1, f_2)$, $\mathfrak{B}(f_1, f_2)$ are defined by (3.15′) or (3.16′), respectively. Let us show

† This section can be omitted at the first reading.

again their explicit form:

$$\mathfrak{A}(f_1, f_2) = \sum_f \{\xi(f_1, f)\,\Phi(f, f_2)+\Phi(f_1, f)\,\xi(f_2, f)\}$$
$$-\tfrac{1}{2}\sum_{f, f_1', f_2'} \Phi(f_2', f_1')\{G(f_1, f; f_2', f_1')\,\varrho(f, f_2)+G(f, f_2; f_2', f_1')\,\varrho(f, f_1)\}$$
$$+\tfrac{1}{2}\sum_{f_1', f_2'} \Phi(f_2', f_1')\,G(f_1, f_2; f_2', f_1');$$

$$\mathfrak{B}(f_1, f_2) = \sum_f \{\xi(f_2, f)\,\varrho(f_1, f)-\xi(f, f_1)\,\varrho(f, f_2)\}$$
$$+\tfrac{1}{2}\sum_{f, f_1', f_2'} \{\Phi^*(f_1, f)\,G(f_2, f; f_2', f_1')\,\Phi(f_2', f_1')$$
$$-\Phi(f_2, f)\,G(f_1, f; f_2', f_1')\,\Phi^*(f_2', f_1')\},$$

where
$$\xi(f, f') = T'(f, f')-\sum_{f_1', f_2'} G(f, f_1'; f_2', f)\,\varrho(f_1', f_2');$$

the remaining notation has been explained in § 2 of Chapter 3. We must remember that the correlation function $\Phi(f_1, f_2)$ and the density matrix $\varrho(f_1, f_2)$, defined in (3.14), are not independent; they are related by (3.20), i.e.,

$$\varrho(f_1, f_2) = \sum_{f'} \{\varrho(f_1, f')\,\varrho(f', f_2)+\Phi^*(f', f_1)\,\Phi(f', f_2)\},$$
$$\sum_{f'} \{\varrho(f_1, f')\,\Phi(f', f_2)+\varrho(f_2, f')\,\Phi(f', f_1)\} = 0.$$

The quasiparticle vacuum is the ground state of even–even nuclei when the quasiparticle interaction can be neglected. The functions $\varrho(f_1, f_2)$ and $\Phi(f_1, f_2)$ corresponding to the quasiparticle vacuum (i.e., the implicit averaging was performed over the vacuum wave function) are denoted by $\varrho_0(f_1, f_2)$ and $\Phi_0(f_1, f_2)$. With proper choice of the nuclear average field, the function $\varrho_0(f_1, f_2)$ becomes diagonal and $\Phi_0(f_1, f_2)$ has the canonical form. This property has been already discussed earlier. By adding small deviations from the vacuum values to the functions ϱ_0 and Φ_0, we shall describe the vibrations of the nuclear surface. Thus,

$$\left.\begin{aligned}\varrho(f_1, f_2) &= \varrho_0(f_1, f_2)+\delta\varrho(f_1, f_2),\\ \Phi(f_1, f_2) &= \Phi_0(f_1, f_2)+\delta\Phi(f_1, f_2).\end{aligned}\right\} \tag{8.1}$$

Now we have to find the equations of motion for the quantities $\delta\varrho$ and $\delta\Phi$.

Substituting (8.1) in (5.37),

$$i\frac{\partial}{\partial t}\,\delta\varrho(f_1, f_2) = \delta\mathfrak{B}(f_1, f_2), \tag{8.2}$$

$$i\frac{\partial}{\partial t}\,\delta\Phi(f_1, f_2) = \delta\mathfrak{A}(f_1, f_2). \tag{8.3}$$

The quantities $\delta\varrho$ and $\delta\Phi$ are also connected by the additional relations

$$\left.\begin{aligned}\delta\Big\{\varrho(f_1, f_2)-\sum_{f'}\varrho(f_1, f')\,\varrho(f', f_2)-\sum_{f'}\Phi^*(f', f_1)\,\Phi(f', f_2)\Big\} &= 0,\\ \delta\Big\{\sum_{f'}\varrho(f_1, f')\,\Phi(f', f_2)+\sum_{f'}\varrho(f_2, f')\,\Phi(f', f_1)\Big\} &= 0.\end{aligned}\right\} \tag{8.4}$$

Later, we shall introduce new functions replacing $\delta\varrho$ and $\delta\Phi$ in such a way that (8.4) will be automatically valid.

With the use of the canonical transformation (3.5), the functions ϱ and Φ can be rewritten in terms of the quasiparticle operators:

$$\varrho(f_1, f_2) = \langle | a_{f_1}^{+} a_{f_2} | \rangle = \varrho_0(f_1, f_2) + \delta\varrho(f_1, f_2) = \sum_g v^*(f_1 g)\, v(f_2 g)$$
$$+ \sum_{g_1, g_2} \{u^*(f_1 g_1)\, u(f_2 g_2) \langle | \alpha_{g_1}^{+} \alpha_{g_2} | \rangle - v^*(f_1 g_1)\, v(f_2 g_2) \langle | \alpha_{g_2}^{+} \alpha_{g_1} | \rangle$$
$$+ u^*(f_1 g_1)\, v(f_2 g_2) \langle | \alpha_{g_1}^{+} \alpha_{g_2}^{+} | \rangle + v^*(f_1 g_1)\, u(f_2 g_2) \langle | \alpha_{g_1} \alpha_{g_2} | \rangle\},$$
$$\Phi(f_1, f_2) = \langle | a_{f_1} a_{f_2} | \rangle = \Phi_0(f_1, f_2) + \delta\Phi(f_1, f_2) = \sum_g u(f_1 g)\, v(f_2 g)$$
$$+ \sum_{g_1, g_2} \{v(f_1 g_1)\, u(f_2 g_2) \langle | \alpha_{g_1}^{+} \alpha_{g_2} | \rangle - u(f_1 g_1)\, v(f_2 g_2) \langle | \alpha_{g_2}^{+} \alpha_{g_1} | \rangle$$
$$+ u(f_1 g_1)\, u(f_2 g_2) \langle | \alpha_{g_1} \alpha_{g_2} | \rangle + v(f_1 g_1)\, v(f_2 g_2) \langle | \alpha_{g_1}^{+} \alpha_{g_2}^{+} | \rangle\}.$$

When the quasiparticle interaction is included, the ground state is not identical with the quasiparticle vacuum. However, we shall assume that the average number of quasiparticles in the ground state is, nevertheless, small. This basic assumption can be expressed mathematically as the condition

$$\langle | \alpha_{g_1}^{+} \alpha_{g_2} | \rangle = 0. \tag{8.5}$$

The deviation of the ground state from the quasiparticle vacuum will be characterized by the new function

$$\mu(g_1, g_2) = \langle | \alpha_{g_1} \alpha_{g_2} | \rangle, \tag{8.6}$$

which must be, naturally, antisymmetric,

$$\mu(g_1, g_2) = -\mu(g_2, g_1). \tag{8.6'}$$

Using (8.5) and (8.6), we can express $\delta\varrho$ and $\delta\Phi$ in terms of μ and μ^*:

$$\delta\varrho(f_1, f_2) = \sum_{g_1, g_2} \{v^*(f_1 g_1)\, u(f_2 g_2)\, \mu(g_1, g_2) + u^*(f_1 g_1)\, v(f_2 g_2)\, \mu^*(g_2, g_1)\}, \tag{8.7}$$
$$\delta\Phi(f_1, f_2) = \sum_{g_1, g_2} \{u(f_1 g_1)\, u(f_2 g_2)\, \mu(g_1, g_2) + v(f_1 g_1)\, v(f_2 g_2)\, \mu^*(g_2, g_1)\}. \tag{8.7'}$$

Now we have to find the inverse transformation, i.e., to express the function μ in terms of $\delta\varrho$ and $\delta\Phi$. In order to do that, we shall multiply (8.7) by $v(f_1 g)$ and (8.7′) by $u^*(f_1 g)$. Further, we shall add the resulting expressions and make the summation over f_1. By using the orthogonality relations (3.6′), we finally obtain

$$\sum_{f_1} \{v(f_1 g)\, \delta\varrho(f_1, f_2) + u^*(f_1 g)\, \delta\Phi(f_1, f_2)\}$$
$$= \sum_{f_1, g_1, g_2} \{[v(f_1 g)\, v^*(f_1 g_1) + u^*(f_1 g)\, u(f_1 g_1)]\, u(f_2, g_2)\, \mu(g_1, g_2)$$
$$+ [v(f_1 g)\, u^*(f_1 g_1) + u^*(f_1 g)\, v(f_1 g_1)]\, v(f_2 g_2)\, u^*(g_2, g_1)\}$$
$$= \sum_{g_2} u(f_2 g_2)\, \mu(g, g_2). \tag{8.8}$$

Similarly, by multiplying (8.7) by $u(f_1g)$ and (8.7′) by $v^*(f_1g)$, adding, and summing over f_1 we shall obtain another relation, namely,

$$\sum_{f_1}\{u(f_1g)\,\delta\varrho(f_1,f_2)+v^*(f_1g)\,\delta\Phi(f_1,f_2)\} = \sum_{g_2} v(f_2g_2)\,\mu^*(g_2,g) \tag{8.8'}$$

or

$$\sum_{f_1}\{u^*(f_1g)\,\delta\varrho^*(f_1,f_2)+v(f_1g)\,\delta\Phi^*(f_1,f_2)\} = -\sum_{g_2} v^*(f_2g_2)\,\mu(g,g_2).$$

The same procedure has to be used once more. By multiplying (8.8) by $u^*(f_2g')$ and (8.8′) by $v(f_2g)$, subtracting the resulting expressions, and summing over f_2, we finally obtain

$$\mu(g,g') = \sum_{f_1,f_2}\{v(f_1,g)\,u^*(f_2g')\,\delta\varrho(f_1,f_2)+u^*(f_1g)u^*(f_2g')\,\delta\Phi(f_1,f_2)$$
$$-u^*(f_1g)\,v(f_2g')\,\delta\varrho^*(f_1,f_2)-v(f_1g)\,v(f_2g')\,\delta\Phi^*(f_1,f_2)\}.$$

Now it is easy to find the equation of motion for the function $\mu(g_1,g_2)$. We have to take the time derivative of the above expression and use (8.2) and (8.3). Thus

$$i\frac{\partial}{\partial t}\mu(g_1,g_2) = \sum_{f_1,f_2}\{u^*(f_1g_1)\,u^*(f_2g_2)\,\delta\mathfrak{A}(f_1,f_2)+v(f_1g_1)\,v(f_2g_2)\,\delta\mathfrak{A}^*(f_1,f_2)$$
$$+v(f_1g_1)\,u^*(f_2g_2)\,\delta\mathfrak{B}(f_1,f_2)+u^*(f_1g_1)\,v(f_2g_2)\,\delta\mathfrak{B}^*(f_1,f_2)\}. \tag{8.9}$$

Let us find the explicit form of (8.9). From (3.15′) and (3.16′) it easily follows that

$$\delta\mathcal{U}(f_1,f_2) = \sum_{f}\{\delta\xi(f_1,f)\,\Phi_0(f,f_2)+\xi_0(f_1,f)\,\delta\Phi(f,f_2)+\delta\Phi(f_1,f)\,\xi_0(f_2,f)$$
$$+\Phi_0(f_1,f)\,\delta\xi(f_1,f)-\tfrac{1}{2}\sum_{f,f',f_2}\delta\Phi(f_2',f_1')\times$$
$$\times\{G(f_1,f;f_2',f_1')\,\varrho_0(f,f_2)+G(f,f_2';f_2',f_1')\,\varrho_0(f,f_1)\}$$
$$-\tfrac{1}{2}\sum_{f,f_1',f_2'}\Phi_0(f_2',f_1')\{G(f_1,f;f_2',f_1')\,\delta\varrho(f,f_2)+G(f,f_2;f_2',f_1')\,\delta\varrho(f,f_1)\}$$
$$+\tfrac{1}{2}\sum_{f_1',f_2'}\delta\Phi(f_2',f_1')\,G(f_1,f_2;f_2',f_1') \tag{8.10}$$

$$\delta\mathfrak{B}(f_1,f_2) = \sum_{f}\{\delta\xi(f_2,f)\,\varrho_0(f_1,f)+\xi_0(f_2,f)\,\delta\varrho(f_1,f)$$
$$-\delta\xi(f,f_1)\,\varrho_0(f,f_2)-\xi_0(f,f_1)\,\delta\varrho(f,f_2)\}$$
$$+\tfrac{1}{2}\sum_{f,f_1',f_2'}G(f_2,f;f_2',f_1')\{\delta\Phi^*(f_1,f)\,\Phi_0(f_2',f_1')+\Phi_0^*(f_1,f)\,\delta\Phi(f_2',f_1')\}$$
$$-\tfrac{1}{2}\sum_{f,f_1',f_2'}G(f_1,f;f_2',f_1')\{\delta\Phi(f_2,f)\,\Phi_0^*(f_2',f_1')+\Phi_0(f_2,f)\,\delta\Phi^*(f_2',f_1')\}, \tag{8.10'}$$

where

$$\delta\xi(f,f') = -\sum_{f_1',f_2'}G(f,f_1';f_2',f')\,\delta\varrho(f_1',f_2')$$

and
$$\xi_0(f,f') = T(f,f')-\lambda\delta_{f,f'}-\sum_{f_1',f_2'}G(f,f_1';f_2',f')\,\varrho_0(f_1',f_2').$$

We have to substitute $\delta\mathfrak{A}$ and $\delta\mathfrak{B}$ in (8.9) and replace $\delta\varrho$ and $\delta\Phi$ by μ and μ^*. After the rather tedious formal transformations have been performed, the basic equations can be written in the following form:

$$i\frac{\partial}{\partial t}\mu(g_1, g_2) = \sum_{g'}\{\Omega(g_2, g')\,\mu(g_1, g')-\Omega(g_1, g')\,\mu(g_2, g')\}$$
$$+\sum_{g_1', g_2'}\{X(g_1, g_2; g_1', g_2')\,\mu(g_1', g_2')+Y(g_1, g_2; g_1', g_2')\,\mu^*(g_2', g_1')\}, \quad (8.11)$$

$$-i\frac{\partial}{\partial t}\mu^*(g_1, g_2) = \sum_{g'}\{\Omega^*(g_2, g')\,\mu^*(g_1, g')-\Omega^*(g_1, g')\,\mu^*(g_2, g')\}$$
$$+\sum_{g_1', g_2'}\{X^*(g_1, g_2; g_1', g_2')\,\mu^*(g_1', g_2')+Y^*(g_1, g_2; g_1', g_2')\,\mu(g_2', g_1')\}. \quad (8.12)$$

The second equation is the complex conjugate of the first one. The quasiparticle vacuum correlation function has been defined in (3.18) as

$$C^0_{f_1, f_2} = \tfrac{1}{2}\sum_{f_1', f_2'} G(f_1, f_2; f_2', f_1')\,\Phi_0(f_2', f_1').$$

Using this definition we shall rewrite the implicit expressions in (8.11) and (8.12):

$$\Omega(g, g') = \sum_{f, f'}\xi_0(f, f')\{u^*(fg)\,u(f'g')-v^*(fg)\,v(f'g')\}$$
$$-\sum_{f_1, f_2}\{C^0_{f_1, f_2}u^*(f_1g)\,v^*(f_2g')+C^{0*}_{f_1, f_2}\,v(f_2g)\,u(f_1g')\}, \quad (8.13)$$

$$X(g_1, g_2; g_1', g_2') = -\tfrac{1}{2}\sum_{f_1, f_2 f_1', f_2'} G(f_1, f_2; f_2', f_1')\{u(f_1\,g_2)\,u(f_2g_1)\,u(f_1'g_2')\,u(f_2'g_1')$$
$$+v(f_1'g_1)\,v(f_2'g_2)\,v(f_1g_1')\,v(f_2g_2')+[v(f_1'g_1)\,u(f_1g_2)$$
$$-u(f_1g_1)\,v(f_1'g_2)]\,[v(f_2g_1')\,u(f_2'g_2')-v(f_2g_2')\,u(f_2'g_1')]\}, \quad (8.14)$$

$$Y(g_1, g_2; g_1', g_2') = -\tfrac{1}{2}\sum_{f_1, f_2, f_1', f_2'} G(f_1, f_2; f_2'\,f_1')\{u(f_2g_1)\,u(f_1g_2)\,v(f_2'g_1')\,v(f_1'g_2')$$
$$+v(f_1'g_1)\,v(f_2'g_2)\,u(f_2\,g_2')\,u(f_1g_1')+[v(f_1'g_1)\,u(f_1g_2)$$
$$-v(f_1'g_2)\,u(f_1g_1)]\,[u(f_2g_1')\,v(f_2'\,g_2')-u(f_2g_2')\,v(f_2'g_1')]\}. \quad (8.15)$$

The solutions of the homogenous equations (8.11) and (8.12) can be expanded into a superposition of the normal vibrational modes:†

$$\mu(g_1, g_2) = \sum_{\omega} e^{-i\omega t}\xi_\omega(g_1, g_2), \qquad \mu^*(g_1, g_2) = \sum_{\omega} e^{-i\omega t}\eta_\omega(g_1, g_2). \quad (8.16)$$

Obviously, after the complex conjugation of the first equation (8.16),

$$\mu^*(g_1, g_2) = \sum_{\omega} e^{+i\omega t}\xi^*_\omega(g_1, g_2) = \sum_{\omega} e^{-i\omega t}\xi^*_{-\omega}(g_1, g_2),$$

or, in other words,

$$\eta_\omega(g_1, g_2) = \xi^*_{-\omega}(g_1, g_2).$$

The functions $\xi_\omega(g_1, g_2)$ and $\eta_\omega(g_1, g_2)$ are antisymmetric, i.e.,

$$\xi_\omega(g_1, g_2) = -\xi_\omega(g_2, g_1), \qquad \eta_\omega(g_1, g_2) = -\eta_\omega(g_2, g_1). \quad (8.17)$$

† This method was developed in ref. 436.

We may substitute (8.16) in (8.11) and (8.12) and obtain the secular equations in the form

$$\omega\xi_\omega(g_1, g_2) = \sum_{g'}\{\Omega(g_2, g')\,\xi_\omega(g_1, g') - \Omega(g_1, g')\,\xi_\omega(g_2, g')\}$$
$$+ \sum_{g_1', g_2'}\{X(g_1, g_2; g_1', g_2')\,\xi_\omega(g_1', g_2') - Y(g_1, g_2; g_1', g_2')\,\eta_\omega(g_1', g_2')\}. \quad (8.11')$$

$$-\omega\eta_\omega(g_1, g_2) = \sum_{g'}\{\Omega^*(g_2, g')\,\eta_\omega(g_1, g') - \Omega^*(g_1, g')\,\eta_\omega(g_2, g')\}$$
$$+ \sum_{g_2', g_1'}\{X^*(g_1, g_2; g_1', g_2')\,\eta_\omega(g_1', g_2') - Y^*(g_1, g_2; g_1', g_2')\,\xi_\omega(g_1', g_2')\}. \quad (8.12')$$

It is useful to remember that if ξ_ω, η_ω, and ω are the solutions of (8.11′) and (8.12′), then the transformation

$$\omega \to -\omega, \qquad \xi_\omega \to \eta_\omega^*, \qquad \eta_\omega \to \xi_\omega^*$$

leads again to the solution of the same system of equations.

3. Let us simplify the system of (8.11′) and (8.12′). We shall make the following approximations: the functions $u(fg)$ and $v(fg)$ will be diagonal and real, i.e.,

$$\left.\begin{aligned} v(fg) &= v_f\delta_{f,\,g}, & v_{-f} &= -v_f, & v_f^* &= v_f,\\ u(fg) &= u_f\delta_{f,\,-g}, & u_{-f} &= u_f, & u_f^* &= u_f, \end{aligned}\right\} \quad (8.18)$$

and the function $\xi_0(f_1, f_2)$ will be diagonal. In other words, the single-particle energies of the average field will be of the form

$$\xi_0(f_1, f_2) = \xi(f_1)\delta_{f_1, f_2} \equiv (E(f_1) - \lambda)\delta_{f_1, f_2}. \quad (8.18')$$

In such an approximation

$$C^0_{f_1, f_2} = C^{0*}_{f_1, f_2} = C_f\delta_{f,\,-f_2},$$
$$C_f = \tfrac{1}{2}\sum_{f'} G(f, -f; -f', f')u_{f'}v_{f'}. \quad (8.19)$$

The functions Ω, X, and Y can be further simplified. Thus we can write

$$\Omega(g, g') = \sum_{f, f'}\xi(f)\,\delta_{f,f'}\{u_f\delta_{f,\,-g}u_{f'}\delta_{f',\,-g'} - v_f\delta_{f,\,g}v_{f'}\delta_{f',\,g'}\}$$
$$-\sum_{f_1, f_2} C_f\delta_{f_1,\,-f_2}\{u_{f_1}\delta_{f_2,\,-g}v_{f_2}\delta_{f_2,\,g'} + u_{f_1}\delta_{f_1,\,-g'}v_{f_2}\delta_{f_2,\,g}\}$$
$$= \delta_{g,\,g'}\{\xi(g)\,(u_g^2 - v_g^2) + 2C_g u_g v_g\}.$$

Using (3.46) and (3.46′),

$$\Omega(g, g') = \delta_{g,\,g'}\varepsilon(g), \quad (8.13')$$

where

$$\varepsilon(g) = \sqrt{C_g^2 + \xi(g)^2}\,.$$

Similarly,

$$X(g_1, g_2; g_1', g_2') = -\tfrac{1}{2}G(g_1, g_2; g_2', g_1')u_{g_1}u_{g_2}u_{g_1'}u_{g_2'} - \tfrac{1}{2}G(-g_1, -g_2; -g_2', -g_1')\times$$
$$\times v_{g_1}v_{g_2}v_{g_1'}v_{g_2'} - \tfrac{1}{2}G(g_1, -g_2'; g_1', -g_2)u_{g_1}v_{g_2}u_{g_1'}v_{g_2'}$$
$$-\tfrac{1}{2}G(-g_1, g_2'; -g_1', g_2)v_{g_1}u_{g_2}v_{g_1'}u_{g_2'} + \tfrac{1}{2}G(g_1, -g_1'; g_2', -g_2)\times$$
$$\times u_{g_1}v_{g_2}u_{g_2'}v_{g_1'} + \tfrac{1}{2}G(-g_1, g_1'; -g_2', g_2)v_{g_1}u_{g_2}v_{g_2'}u_{g_1'}, \quad (8.14')$$

$$
\begin{aligned}
Y(g_1, g_2; g_1', g_2') = &-\tfrac{1}{2}G(g_1, g_2; g_2', g_1')u_{g_1}u_{g_2}v_{g_1}v_{g_2'} - \tfrac{1}{2}G(-g_1, -g_2; -g_2', -g_1')\times \\
&\times v_{g_1}v_{g_2}u_{g_1'}u_{g_2'} + \tfrac{1}{2}G(g_1, -g_2'; g_1', -g_2)u_{g_1}v_{g_2}u_{g_2'}v_{g_1'} \\
&+\tfrac{1}{2}G(-g_1, g_2'; -g_1', g_2)v_{g_1}u_{g_2}u_{g_1'}v_{g_2'} - \tfrac{1}{2}G(g_1, -g_1'; g_2', -g_2)\times \\
&\times u_{g_1}v_{g_2}u_{g_1'}v_{g_2'} - \tfrac{1}{2}G(-g_1, g_1'; -g_2, g_2)v_{g_1}u_{g_2}v_{g_1'}u_{g_2'}, \qquad (8.15')
\end{aligned}
$$

where $X^* = X$, $Y^* = Y$.

Equations (8.11′) and (8.12′) have the following form in the approximation discussed:

$$
\begin{aligned}
\omega\xi_\omega(g_1, g_2) = &[\varepsilon(g_1)+\varepsilon(g_2)]\,\xi_\omega(g_1, g_2) + \sum_{g_1', g_2'}\{X(g_1, g_2; g_1', g_2')\,\xi_\omega(g_1', g_2') \\
&-Y(g_1, g_2; -g_1', -g_2')\eta_\omega(-g_1', -g_2')\}, \qquad (8.11'')
\end{aligned}
$$

$$
\begin{aligned}
-\omega\eta_\omega(-g_1, -g_2) = &[\varepsilon(g_1)+\varepsilon(g_2)]\,\eta_\omega(-g_1, -g_2) + \sum_{g_1', g_2'}\{X(-g_1, -g_2; -g_1', -g_2')\times \\
&\times\eta(-g_1', -g_2') - Y(-g_1, -g_2; g_1', g_2')\,\xi_\omega(g_1', g_2')\}. \qquad (8.12'')
\end{aligned}
$$

Let us introduce a new combination of the unknown functions, namely

$$
\nu^{(\pm)}(g_1, g_2) = \tfrac{1}{2}\{\xi_\omega(g_1, g_2) \pm \eta_\omega(-g_1, -g_2)\},
$$

Equations (8.11″) and (8.12″) can be now written in the unified form:

$$
\begin{aligned}
\omega\nu^{(\mp)}(g_1, g_2) = &[\varepsilon(g_1)+\varepsilon(g_2)]\,\nu^{(\pm)}(g_1, g_2) + \tfrac{1}{2}\sum_{g_1', g_2'}\{[X(g_1, g_2; g_1', g_2') \\
&+X(-g_1, -g_2; -g_1', -g_2')] \mp [Y(g_1, g_2; -g_1', -g_2') \\
&+Y(-g_1, -g_2; g_1', g_2')]\}\,\nu^{(\pm)}(g_1', g_2') + \tfrac{1}{2}\sum_{g_1', g_2'}\{[X(g_1, g_2; g_1', g_2') \\
&-X(-g_1, -g_2; -g_1', -g_2')] \pm [Y(g_1, g_2; -g_1', -g_2') \\
&-Y(-g_1, -g_2; g_1', g_2')]\}\,\nu^{(\mp)}(g_1', g_2'), \qquad (8.20)
\end{aligned}
$$

where
$$
\begin{aligned}
&\{X(g_1, g_2; g_1', g_2') + X(-g_1, -g_2; -g_1', -g_2')\} \mp \{Y(g_1, g_2; -g_1', -g_2') \\
&+Y(\cdot g_1, -g_2; g_1', g_2')\} = -\tfrac{1}{2}\{G(g_1, g_2; g_2', g_1') \\
&+G(-g_1, -g_2; -g_2', -g_1';)\}v^{(\mp)}_{g_1g_2}v^{(\mp)}_{g_1'g_2'} - \tfrac{1}{2}\{[G(g_1, -g_2'; g_1', -g_2) \\
&\mp G(g_1, -g_1'; g_2', -g_2)] + [G(-g_1, g_2'; -g_1', g_2) \\
&\mp G(-g_1, g_1'; -g_2', g_2)]\}u^{(\pm)}_{g_1g_2}u^{(\pm)}_{g_1'g_2'}, \qquad (8.21)
\end{aligned}
$$

$$
\begin{aligned}
&\{X(g_1, g_2; g_1', g_2') - X(-g_1, -g_2; -g_1', -g_2')\} \pm \{Y(g_1, g_2; -g_1', -g_2') \\
&-Y(-g_1, -g_2; g_1', g_2')\} = -\tfrac{1}{2}\{G(g_1, g_2; g_2', g_1') \\
&-G(-g_1, -g_2; -g_2', -g_1')\}v^{(\mp)}_{g_1g_2}v^{(\pm)}_{g_1'g_2'} - \tfrac{1}{2}\{[G(g_1, -g_2'; g_1', -g_2) \\
&\pm G(g_1, -g_1', g_2', -g_2)] - [G(-g_1, g_2'; -g_1'\ g_2) \\
&\pm G(-g_1, g_1'; -g_2', g_2)]\}u^{(\pm)}_{g_1g_2}u^{(\mp)}_{g_1'g_2'}, \qquad (8.21')
\end{aligned}
$$

and

$$
u^{(\pm)}_{g_1g_2} = u_{g_1}v_{g_2} \pm u_{g_2}v_{g_1}, \quad v^{(\pm)}_{g_1g_2} = v_{g_1}u_{g} \pm v_{g_1}v_{g_2}. \qquad (8.22)
$$

The functions $u^{\pm}$, $v^{\pm}$ have the following symmetry properties:

$$u^{(\pm)}_{-g_1 g_2} = u^{(\mp)}_{g_1 g_2}, \qquad u^{(\pm)}_{g_1, -g_2} = -u^{(\mp)}_{g_1 g_2},$$
$$v^{(\pm)}_{g_1, -g_2} = v^{(\mp)}_{g_1 g_2}, \qquad v^{(\pm)}_{-g_1 g_2} = v^{(\mp)}_{g_1 g_2}.$$

Let us give, for completeness, the final form of the basic equations

$$\begin{aligned}
&[\varepsilon(g_1)+\varepsilon(g_2)]\, \nu^{(\pm)}(g_1, g_2) - \omega \nu^{(\mp)}(g_1, g_2) - \tfrac{1}{4}\sum_{g_1', g_2'} \{G(g_1, g_2; g_2' g_1') \\
&\quad + G(-g_1, -g_2; -g_2', -g_1')\}\, v^{(\mp)}_{g_1 g_2} v^{(\mp)}_{g_1' g_2'} \nu^{(\pm)}(g_1', g_2') \\
&\quad - \tfrac{1}{4}\sum_{g_1', g_2'} \{[G(g_1, -g_2'; g_1', -g_2) \mp G(g_1, -g_1'; g_2', -g_2)] \\
&\quad + [G(-g_1, g_2'; -g_1', g_2) \mp G(-g_1, g_1'; -g_2', g_2)]\} u^{(\pm)}_{g_1 g_2} u^{(\pm)}_{g_1' g_2'} \nu^{(\pm)}(g_1', g_2') \\
&\quad - \tfrac{1}{4}\sum_{g_1', g_2'} \{G(g_1, g_2; g_2', g_1') - G(-g_1, -g_2; -g_2', -g_1')\} v^{(\mp)}_{g_1 g_2} v^{(\pm)}_{g_1' g_2'} \nu^{(\mp)}(g_1' g_2') \\
&\quad - \tfrac{1}{4}\sum_{g_1', g_2'} \{[G(g_1, -g_2'; g_1', -g_2) \pm G(g_1, -g_1'; g_2', -g_2)] \\
&\quad - [G(-g_1, g_2'; -g_1', g_2) \pm G(-g_1, g_1'; -g_2', g_2)]\} u^{(\pm)}_{g_1 g_2} u^{(\mp)}_{g_1' g_2'} \nu^{(\mp)}(g_1', g_2') = 0. \qquad (8.20')
\end{aligned}$$

4. Equation (8.20′) can be further transformed. We shall use the quantum numbers $q\sigma$ instead of g, and exclude the σ-dependence according to ref. 437.

We shall use the properties of the time-reversal operator as we did in the similar problem before. It is known that the operator $a^{+}_{q\sigma}$ is transformed as

$$\mathfrak{T}^{-1} a^{+}_{q\sigma} \mathfrak{T} = \gamma_\sigma a_{q, -\sigma},$$

where γ_σ obeys the relations

$$\gamma_\sigma \gamma_{-\sigma} = -1, \quad \gamma_\sigma^2 = 1, \quad \gamma_{-\sigma} = -\gamma_\sigma.$$

The following relation is a direct consequence of the invariance of the Hamiltonian (3.1) under the time conjugation:

$$G(q_1\sigma_1, q_2\sigma_2; q_2'\sigma_2', q_1'\sigma_1') = \gamma_{\sigma_1}\gamma_{\sigma_2}\gamma_{\sigma_2'}\gamma_{\sigma_1'} G(q_1, -\sigma_1, q_2, -\sigma_2; q_2', -\sigma_2', q_1', -\sigma_1'). \qquad (8.23)$$

Equations (8.18) may be supplemented by several additional relations, namely,

$$\left.\begin{aligned}
&u_f = u_{-f} = u_{q\sigma} = u_q, \\
&v_f = v_{q\sigma} = \gamma_\sigma v_q, \quad v_{-f} = v_{q, -\sigma} = \gamma_{-\sigma} v_q = -\gamma_\sigma v_q.
\end{aligned}\right\} \qquad (8.23')$$

Functions (8.22) therefore obey the relations

$$\left.\begin{aligned}
u^{(\pm)}_{q\sigma, q'\sigma} &= \gamma_\sigma (u_q v_{q'} \pm u_{q'} v_q) \equiv \gamma_\sigma u^{(\pm)}_{qq'}, \\
u^{(\pm)}_{q'\sigma, q', -\sigma} &= \gamma_{-\sigma} (u_q v_{q'} \mp u_{q'} v_q) \equiv \gamma_{-\sigma} u^{(\mp)}_{qq'}, \\
v^{(\pm)}_{q\sigma, q'\sigma} &= u_q u_{q'} \pm v_q v_{q'} \equiv v^{(\pm)}_{qq'}, \\
v^{(\pm)}_{q\sigma, q', -\sigma} &= u_q u_{q'} \mp v_q v_{q'} \equiv v^{(\mp)}_{qq'}.
\end{aligned}\right\} \qquad (8.22')$$

Equation (8.20′) can be now expressed in terms of the variables $q\sigma$ [using relations (8.23)]:

$$
\begin{aligned}
&[\varepsilon(q_1)+\varepsilon(q_2)]\, \nu^{(\pm)}(q_1\sigma_1, q_2\sigma_2)-\omega\nu^{(\mp)}(q_1\sigma_1, q_2\sigma_2)\\
&\quad -\tfrac{1}{4}\sum_{q_1'\sigma_1',\, q_2'\sigma_2'} G(q_1\sigma_1, q_2\sigma_2; q_2'\sigma_2', q_1'\sigma_1')\,(1+\gamma_{\sigma_1}\gamma_{\sigma_2}\gamma_{\sigma_2'}\gamma_{\sigma_1})\times\\
&\quad \times v^{(\mp)}_{q_1\sigma_1,\, q_2\sigma_2} v^{(\mp)}_{q_1'\sigma_1',\, q_2'\sigma_2'}\nu^{(\pm)}(q_1'\sigma_1', q_2'\sigma_2')-\tfrac{1}{4}\sum_{q_1'\sigma_1',\, q_2\sigma_2'}\{G(q_1\sigma_1, q_2', -\sigma_2'; q_1'\sigma_1', q_2, -\sigma_2)\\
&\quad \mp G(q_1\sigma_1, q_1', -\sigma_1'; q_2'\sigma_2', q_2, -\sigma_2)\}\times(1+\gamma_{\sigma_1}\gamma_{\sigma_2}\gamma_{\sigma_2'}\gamma_{\sigma_1'})\, u^{(\pm)}_{q_1\sigma_1,\, q_2\sigma_2}u^{(\pm)}_{q_1'\sigma_1',\, q_2'\sigma_2'}\nu^{(\pm)}(q_1\, \sigma_1', q_2'\sigma_2')\\
&\quad -\tfrac{1}{4}\sum_{q_1'\sigma_1',\, q_2'\sigma_2'} G(q_1\sigma_1, q_2\sigma_2; q_2'\sigma_2', q_1'\sigma_1')\,(1-\gamma_{\sigma_1}\gamma_{\sigma_2}\gamma_{\sigma_2'}\gamma_{\sigma_1'})\times\\
&\quad \times v^{(\mp)}_{q_1\sigma_1,\, q_2\sigma_2} v^{(\pm)}_{q_1'\sigma_1',\, q_2'\sigma_2'}\nu^{(\mp)}(q_1'\sigma_1', q_2'\sigma_2')-\tfrac{1}{4}\sum_{q_1'\sigma_1',\, q_2'\sigma_2'}\{G(q_1\sigma_1, q_2', -\sigma_2'; q_1'\sigma_1', q_2, -\sigma_2)\\
&\quad \pm G(q_1\sigma_1, q_1', -\sigma_1'; q_2'\sigma_2', q_2, -\sigma_2)\}\,(1-\gamma_{\sigma_1}\gamma_{\sigma_2}\gamma_{\sigma_2'}\gamma_{\sigma_1'})\times\\
&\quad \times u^{(\pm)}_{q_1\sigma_1,\, q_2\sigma_2} u^{(\mp)}_{q_1'\sigma_1',\, q_2'\sigma_2'}\nu^{(\mp)}(q_1'\sigma_1'; q_2'\sigma_2') = 0.
\end{aligned}
$$

The products of several γ-factors can be excluded if we realize that, for example, for $\sigma_1 = \sigma_2$

$$
1+\gamma_{\sigma_1}\gamma_{\sigma_2}\gamma_{\sigma_2'}\gamma_{\sigma_1'} = 1+\gamma_{\sigma_2'}\gamma_{\sigma_1} = \begin{cases} 2, & \sigma_1' = \sigma_2'; \\ 0, & \sigma_1' = -\sigma_2'. \end{cases}
$$

Similar relations are also valid for the $\sigma_1 = -\sigma_2$ case. Thus we shall rewrite our equation again:

$$
\begin{aligned}
&[\varepsilon(q_1)+\varepsilon(q_2)]\, \nu^{(\pm)}(q_1\sigma, q_2\sigma)-\omega\nu^{(\mp)}(q_1\sigma, q_2\sigma)\\
&\quad -\tfrac{1}{2}\sum_{q_1,\, q_2',\, \sigma'} G(q_1\sigma, q_2\sigma; q_2'\sigma', q_1'\sigma')\, \bar{v}^{(\mp)}_{q_1,\, q_2} v^{(\mp)}_{q_1'q_2'}\nu^{(\pm)}(q_1'\sigma', q_2'\sigma')\\
&\quad -\tfrac{1}{2}\sum_{q_1',\, q_2',\, \sigma'}\{G(q_1\sigma, q_2', -\sigma'; q_1'\sigma', q_2, -\sigma)\mp G(q_1\sigma, q_1', -\sigma'; q_2'\sigma', q_2, -\sigma)\}\times\\
&\quad \times\gamma_\sigma\gamma_{\sigma'} u^{(\pm)}_{q_1q_2}u^{(\pm)}_{q_1'q_2'}\nu^{(\pm)}(q_1'\sigma', q_2'\sigma')-\tfrac{1}{2}\sum_{q_1',\, q_2',\, \sigma'} G(q_1\sigma, q_2\sigma; q_2', -\sigma', q_1'\sigma')\times\\
&\quad \times v^{(\mp)}_{q_1q_2}v^{(\mp)}_{q_1'q_2'}\nu^{(\mp)}(q_1'\sigma', q_2', -\sigma')-\tfrac{1}{2}\sum_{q_1',\, q_2',\, \sigma'}\{G(q_1\sigma, q_2'\sigma'; q_1'\sigma', q_2, -\sigma)\\
&\quad \pm G(q\sigma, q_1', -\sigma' : , q_2', -\sigma', q_2, -\sigma)\}\gamma_\sigma\gamma_{-\sigma'} u^{(\pm)}_{q_1q_2}u^{(\pm)}_{q_1'q_2'}\nu^{(\mp)}(q_1'\sigma', q_2', -\sigma') = 0, \qquad (8.20'')
\end{aligned}
$$

$$
\begin{aligned}
&[\varepsilon(q_1)+\varepsilon(q_2)]\, \nu^{(\mp)}(q_1\sigma, q_2, -\sigma)-\omega\nu^{(\pm)}(q_1\sigma, q_2, -\sigma)\\
&\quad -\tfrac{1}{2}\sum_{q_1',\, q_2',\, \sigma'} G(q_1\sigma, q_2, -\sigma; q_2', -\sigma', q_1'\sigma')\, v^{(\mp)}_{q_1q_2}v^{(\mp)}_{q_1'q_2'}\nu^{(\mp)}(q_1'\sigma', q_2', -\sigma')\\
&\quad -\tfrac{1}{2}\sum_{q_1',\, q_2',\, \sigma'}\{G(q_1\sigma, q_2'\sigma'; q_1'\sigma', q_2\sigma)\pm G(q_1\sigma, q_1', -\sigma'; q_2', -\sigma', q_2\sigma)\}\times\\
&\quad \times\gamma_\sigma\gamma_{\sigma'} u^{(\pm)}_{q_1q_2}u^{(\pm)}_{q_1'q_2'}\nu^{(\mp)}(q_1'\sigma', q_2', -\sigma')-\tfrac{1}{2}\sum_{q_1',\, q_2',\, \sigma'} G(q_1\sigma, q_2, -\sigma; q_2'\sigma', q_1'\sigma')\times\\
&\quad \times v^{(\mp)}_{q_1q_2}v^{(\mp)}_{q_1'q_2'}\nu^{(\pm)}(q_1'\sigma', q_2'\sigma')-\tfrac{1}{2}\sum_{q_1',\, q_2',\, \sigma'}\{G(q_1\sigma, q_2', -\sigma'; q_1'\sigma', q_2\sigma)\\
&\quad \mp G(q_1\sigma, q_1', -\sigma'; q_1'\sigma', q_2\sigma)\}\, \gamma_{-\sigma}\gamma_{\sigma'} u^{(\pm)}_{q_1q_2}u^{(\pm)}_{q_1q_2'}\nu^{(\pm)}(q_1'\sigma', q_2'\sigma') = 0. \qquad (8.20''')
\end{aligned}
$$

Let us introduce the following notation:

$$\left.\begin{aligned}
G^{\xi}(q_1+, q_2+; q_2'+, q_1'+) &\equiv \tfrac{1}{2}\sum_{\sigma} G(q_1\sigma, q_2\sigma; q_2'\sigma', q_1'\sigma'),\\
G^{\xi}(q_1+, q_2+; q_2'-, q_1'+) &\equiv \tfrac{1}{2}\sum_{\sigma} G(q_1\sigma, q_2\sigma; q_2', -\sigma', q_1'\sigma')\gamma_{\sigma'},\\
G^{\xi}(q_1+, q_2-; q_2'+, q_1'+) &\equiv \tfrac{1}{2}\sum_{\sigma} G(q_1\sigma, q_2, -\sigma; q_2'\sigma', q_1'\sigma')\gamma_{\sigma},\\
G^{\xi}(q_1+, q_2-; q_2'-, q_1'+) &\equiv \tfrac{1}{2}\sum_{\sigma} G(q_1\sigma, q_2, -\sigma; q_2', -\sigma', q_1'\sigma')\gamma_{\sigma}\gamma_{\sigma'},\\
G^{\omega}(q_1+, q_2+; q_2'+, q_1'+) &\equiv \tfrac{1}{2}\sum_{\sigma} G(q_1\sigma, q_2'\sigma'; q_1'\sigma', q_2\sigma),\\
G^{\omega}(q_1+, q_2+; q_2'-, q_1'+) &\equiv \tfrac{1}{2}\sum_{\sigma} G(q_1\sigma, q_2', -\sigma'; q_1'\sigma', q_2\sigma)\gamma_{-\sigma'},\\
G^{\omega}(q_1+, q_2-; q_2'+, q_1'+) &\equiv \tfrac{1}{2}\sum_{\sigma} G(q_1\sigma, q_2'\sigma'; q_1'\sigma', q_2, -\sigma)\gamma_{-\sigma},\\
G^{\omega}(q_1+, q_2-; q_2'-, q_1'+) &\equiv \tfrac{1}{2}\sum_{\sigma} G(q_1\sigma, q_2', -\sigma'; q_1'\sigma', q_2, -\sigma)\gamma_{-\sigma}\gamma_{-\sigma'}.
\end{aligned}\right\} \tag{8.24}$$

Such a notation is rigorous because the righthand side expressions do not depend on the quantum number σ'. Let us prove this statement for one case:

$$\begin{aligned}
\sum_{\sigma}\gamma_{-\sigma'}G(q_1\sigma, q_2', -\sigma'; q_1'\sigma', q_2\sigma) &= \sum_{\sigma}\gamma_{-\sigma'}\gamma_{\sigma}\gamma_{-\sigma'}\gamma_{\sigma'}\gamma_{\sigma}G(q_1, -\sigma, q_2'\sigma'; q_1', -\sigma', q_2, -\sigma)\\
&= \sum_{\sigma}\gamma_{\sigma'}G(q_1\sigma, q_2'\sigma'; q_1', -\sigma', q_2\sigma),
\end{aligned}$$

i.e., the expression does not change when the sign of σ' changes. The other relations (8.24) can be proven in an analogous way.

Now we can perform the summation over σ in (8.20″). Equation (8.20‴) is first multiplied by γ_σ and then summed over σ. After that,

$$\begin{aligned}
&[\varepsilon(q_1)+\varepsilon(q_2)]\sum_{\sigma}\nu^{(\pm)}(q_1\sigma, q_2\sigma) - \omega\sum_{\sigma}\nu^{(\mp)}(q_1\sigma, q_2\sigma)\\
&\quad - \sum_{q_1', q_2'} G^{\xi}(q_1+, q_2+; q_2'+, q_1'+)v^{(\mp)}_{q_1q_2}v^{(\mp)}_{q_1'q_2'}\sum_{\sigma}\nu^{(\pm)}(q_1'\sigma, q_2'\sigma)\\
&\quad - \sum_{q_1', q_2'} G^{\xi}(q_1+, q_2+; q_2'-, q_1'+)v^{(\mp)}_{q_1q_2}v^{(\mp)}_{q_1'q_2'}\sum_{\sigma}\gamma_{\sigma}\nu^{(\mp)}(q_1'\sigma, q_2', -\sigma)\\
&\quad - \sum_{q_1', q_2'}\{G^{\omega}(q_1+, q_2-; q_2'-, q_1'+)\mp G^{\omega}(q_1+, q_2-; q_1'-, q_2'+)\}\times\\
&\quad \times u^{(\pm)}_{q_1q_2}u^{(\pm)}_{q_1'q_2'}\sum_{\sigma}\nu^{(\pm)}(q_1'\sigma, q_2'\sigma) - \sum_{q_1', q_2'}\{G^{\omega}(q_1+, q_2-; q_2'+, q_1'+)\\
&\quad \pm G^{\omega}(q_1+, q_2-; q_1'+, q_2'+)\}u^{(\pm)}_{q_1q_2}u^{(\pm)}_{q_1'q_2'}\sum_{\sigma}\gamma_{\sigma}\nu^{(\mp)}(q_1'\sigma, q_2', -\sigma) = 0;
\end{aligned}$$

$$\begin{aligned}
&[\varepsilon(q_1)+\varepsilon(q_2)]\sum_{\sigma}\gamma_{\sigma}\nu^{(\mp)}(q_1\sigma, q_2, -\sigma) - \omega\sum_{\sigma}\gamma_{\sigma}\nu^{(\pm)}(q_1\sigma, q_2, -\sigma)\\
&\quad - \sum_{q_1', q_2'} G^{\xi}(q_1+, q_2-; q_2'-, q_1'+)v^{(\mp)}_{q_1q_2}v^{(\mp)}_{q_1'q_2'}\sum_{\sigma}\gamma_{\sigma}\nu^{(\mp)}(q_1'\sigma, q_2', -\sigma)\\
&\quad - \sum_{q_1', q_2'} G^{\xi}(q_1+, q_2-; q_2'+, q_1'+)v^{(\mp)}_{q_1q_2}v^{(\mp)}_{q_1'q_2'}\sum_{\sigma}\nu^{(\pm)}(q_1'\sigma, q_2'\sigma)\\
&\quad - \sum_{q_1', q_2'}\{G^{\omega}(q_1+, q_2+; q_2'+, q_1'+)\pm G^{\omega}(q_1+, q_2+; q_1'+, q_2'+)\}\times\\
&\quad \times u^{(\pm)}_{q_1q_2}u^{(\pm)}_{q_1'q_2'}\sum_{\sigma}\gamma_{\sigma}\nu^{(\mp)}(q_1'\sigma, q_2', -\sigma) - \sum_{q_1', q_2'}\{G^{\omega}(q_1+, q_2+; q_2'-, q_1'+)\\
&\quad \mp G^{\omega}(q_1+, q_2+; q_1'-, q_2'+)\}u^{(\pm)}_{q_1q_2}u^{(\pm)}_{q_1'q_2'}\sum_{\sigma}\nu^{(\pm)}(q_1'\sigma, q_2'\sigma) = 0.
\end{aligned}$$

The expression $\sum_\sigma \nu^{(\pm)}(q\sigma, q', -\sigma)$ is antisymmetric when q and q' are exchanged. On the other hand, the expression $\sum_\sigma \gamma_\sigma \nu^{(\pm)}(q\sigma, q', -\sigma)$ is symmetric in q and q'. Actually,

$$\sum_\sigma \gamma_\sigma \nu^{(\pm)}(q'\sigma, q, -\sigma) = -\sum_\sigma \gamma_\sigma \nu^{(\pm)}(q, -\sigma, q'\sigma)$$
$$= -\sum_\sigma \gamma_{-\sigma} \nu^{(\pm)}(q\sigma, q', -\sigma) = \sum_\sigma \gamma_\sigma \nu^{(\pm)}(q\sigma, q', -\sigma).$$

For each type of nuclear excitation, and for fixed values of q and q', only one of the two combinations

$$\sum_\sigma \nu^{(\pm)}(q\sigma, q'\sigma) \quad \text{and} \quad \sum_\sigma \gamma_\sigma \nu^{(\pm)}(q\sigma, q', -\sigma),$$

is nonvanishing. Consequently, we can use their sum instead of either of them. Under such circumstances, the basic equations can be further simplified, i.e.,

$$\begin{aligned}
&[\varepsilon(q_1)+\varepsilon(q_2)]\sum_\sigma \nu^{(\pm)}(q_1\sigma, q_2\sigma) - \omega\sum_\sigma \nu^{(\mp)}(q_1\sigma, q_2\sigma)\\
&\quad - \sum_{q_1', q_2'} \{G^\xi(q_1+, q_2+; q_2'+, q_1'+) + G^\xi(q_1+, q_2+; q_2'-, q_1'+)\}\times\\
&\quad \times v^{(\mp)}_{q_1q_2} v^{(\mp)}_{q_1'q_2'} \left\{\sum_\sigma \nu^{(\pm)}(q_1'\sigma, q_2'\sigma) + \sum_\sigma \gamma_\sigma \nu^{(\mp)}(q_1'\sigma, q_2', -\sigma)\right\}\\
&\quad -2\sum_{q_1', q_2} \{G^\omega(q_1+, q_2-; q_2'-, q_1'+) + G^\omega(q_1+, q_2-; q_2'+, q_1'+)\}\times\\
&\quad \times u^{(\pm)}_{q_1q_2} u^{(\pm)}_{q_1'q_2'} \left\{\sum_\sigma \nu^{(\pm)}(q_1'\sigma, q_2'\sigma) + \sum_\sigma \gamma_\sigma \nu^{(\mp)}(q_1'\sigma, q_2', -\sigma)\right\} = 0,
\end{aligned} \tag{8.25}$$

$$\begin{aligned}
&[\varepsilon(q_1)+\varepsilon(q_2)]\sum_\sigma \gamma_\sigma \nu^{(\mp)}(q_1\sigma, q_2, -\sigma) - \omega\sum_\sigma \gamma_\sigma \nu^{(\pm)}(q_1\sigma, q_2, -\sigma)\\
&\quad - \sum_{q_1', q_2'} \{G^\xi(q_1+, q_2-; q_2'-, q_1'+) + G^\xi(q_1+, q_2-; q_2'+, q_1'+)\}\times\\
&\quad \times v^{(\mp)}_{q_1q_2} v^{(\mp)}_{q_1'q_2'} \left\{\sum_\sigma \nu^{(\pm)}(q_1'\sigma, q_2'\sigma) + \sum_\sigma \gamma_\sigma \nu^{(\pm)}(q_1'\sigma, q_2', -\sigma)\right\}\\
&\quad -2\sum_{q_1', q_2'} \{G^\omega(q_1+, q_2+; q_2'+, q_1'+) + G^\omega(q_1+, q_2+; q_2'-, q_1'+)\}\times\\
&\quad \times u^{(\pm)}_{q_1q_2} u^{(\pm)}_{q_1'q_2'} \left\{\sum_\sigma \nu^{(\pm)}(q_1'\sigma, q_2'\sigma) + \sum_\sigma \gamma_\sigma \nu^{(\mp)}(q_1'\sigma, q_2', -\sigma)\right\} = 0.
\end{aligned} \tag{8.25'}$$

It is useful to introduce another unknown function defined by

$$Z^{(\pm)}(q_1, q_2) = \sum_\sigma \nu^{(\pm)}(q_1\sigma, q_2\sigma) + \sum_\sigma \gamma_\sigma \nu^{(\mp)}(q_1\sigma, q_2, -\sigma).$$

The following notation is also useful:

$$\left.\begin{aligned}
G^\xi(q_1, q_2; q_2', q_1') \equiv{}& G^\xi(q_1+, q_2+; q_2'+, q_1'+) + G^\xi(q_1+, q_2+; q_2'-, q_1'+)\\
&+ G^\xi(q_1+, q_2-; q_2'+, q_1'+) + G^\xi(q_1+, q_2-; q_2'-, q_1'+),\\
G^\omega(q_1, q_2; q_2', q_1') \equiv{}& G^\omega(q_1+, q_2+; q_2'+, q_1'+) + G^\omega(q_1+, q_2+; q_2'-, q_1'+)\\
&+ G^\omega(q_1+, q_2-; q_2'+, q_1'+) + G^\omega(q_1+, q_2-; q_2'-, q_1'+).
\end{aligned}\right\} \tag{8.24'}$$

Now we can take the final step, i.e., add (8.25) and (8.25′) together and use the new function $Z^{(\pm)}$. After simple manipulations we obtain the basic equations in the form

$$[\varepsilon(q_1)+\varepsilon(q_2)]\,Z^{(\pm)}(q_1,q_2)-\omega Z^{(\mp)}(q_1,q_2)-\sum_{q_1',q_2'} G^{\xi}(q_1,q_2;q_2',q_1')v^{(\mp)}_{q_1q_2}v^{(\mp)}_{q_1'q_2'}Z^{(\pm)}(q_1',q_2')$$
$$-2\sum_{q_1',q_2'} G^{\omega}(q_1,q_2;q_2',q_1')u^{(\pm)}_{q_1q_2}u^{(\pm)}_{q_1'q_2'}Z^{(\pm)}(q_1',q_2')=0. \quad (8.26)$$

Thus we have two independent equations for the unknown functions $Z^{(+)}(q_1, q_2)$ and $Z^{(-)}(q_1, q_2)$. The quasiparticle interaction is used in its most general form. Equations (8.26) contain the particle–hole interaction (terms with $u^{(\pm)}_{q_1q_2}$ factors) and the particle–particle interaction (terms with $v^{(\pm)}_{q_1q_2}$ factors). Equation (8.26) was originally formulated in ref. 437.

When studying the low-lying states in medium and heavy nuclei it is important to realize that the general interaction $G(g_1, g_2; g_2', g_1')$ affects two distinct regions of the momenta of interacting particles. Part of the observed collective effects is related to the particle–hole correlations of the quadrupole, octupole, etc., type. They are determined by the low-momentum transfer part of the interaction which is denoted by $G^{\omega}(q_1, q_2; q_2', q_1')$ in our equations. The other part of the collective effects is related to the pairing superfluid correlations. These effects correspond to the interaction with large momentum transfer (zero total momentum); the corresponding interaction is denoted by $G^{\xi}(q_1, q_2; q_2', q_1')$. For practical calculations, the interactions G^{ω} and G^{ξ} may be treated as independent.

5. Let us discuss in detail several important particular forms of (8.26).

First case

It is well known that the vibrational excitations are related to the particle–hole interaction. The largest coherent effect is caused by the interaction term proportional to $u^{(+)}_{qq'}$. Let us consider a special case and assume that the particle–particle interaction does not affect the vibrational states. Consequently, we shall keep only the terms proportional to $u^{(+)}_{qq'}$ in (8.26). Thus our equations have the form

$$[\varepsilon(q_1)+\varepsilon(q_2)]\,Z^{(+)}(q_1,q_2)-\omega Z^{(-)}(q_1,q_2)-2\sum_{q_1',q_2'} G^{\omega}(q_1,q_2;q_2',q_1')\,u^{(+)}_{q_1'q_2'}Z^{(+)}(q_1',q_2')=0, \quad (8.27)$$

$$[\varepsilon(q_1)+\varepsilon(q_2)]\,Z^{(-)}(q_1,q_2)-\omega Z^{(+)}(q_1,q_2)=0. \quad (8.27')$$

We shall substitute $Z^{(-)}(q_1, q_2)$ from (8.27′) in (8.27) and obtain the single equation

$$\left(\varepsilon(q_1)+\varepsilon(q_2)-\frac{\omega^2}{\varepsilon(q_1)+\varepsilon(q_2)}\right)Z^{(+)}(q_1,q_2)=u^{(+)}_{q_1q_2}D^{(+)}(q_1,q_2), \quad (8.27'')$$

where

$$D^{(+)}(q_1,q_2)=2\sum_{q_1',q_2'} G^{\omega}(q_1,q_2;q_2',q_1')\,u^{(+)}_{q_1'q_2'}Z^{(+)}(q_1',q_2'). \quad (8.28)$$

Substituting for the $Z^{(+)}(q_1, q_2)$ from (8.27″), we obtain the following integral equation:

$$D^{+}(q_1,q_2)=2\sum_{q_1',q_2'}\frac{G^{\omega}(q_1,q_2;q_2',q_1')\,(u^{(+)}_{q_1'q_2'})^2\,[\varepsilon(q_1')+\varepsilon(q_2')]}{[\varepsilon(q_1')+\varepsilon(q_2')]^2-\omega^2}D^{(+)}(q_1',q_2'). \quad (8.29)$$

We shall further assume that the interaction $G^{\omega}(q_1, q_2; q_2', q_1')$ can be factorized. (This is actually true in the case of the multipole–multipole interaction.) Thus, we assume that

$$G^{\omega}(q_1, q_2; q_2', q_1') = \varkappa f(q_1, q_2) f(q_1', q_2'). \tag{8.30}$$

We shall denote the functions $\mathfrak{f}^{\lambda\mu}(q, q')$ and $\bar{\mathfrak{f}}^{\lambda\mu}(q, q')$ determined in (3.65), (3.65'), and (3.65'') by $f(q, q')$. The expression (8.28) then has the form

$$D^{(+)}(q_1, q_2) = 2\varkappa f(q_1, q_2) \sum_{q_1', q_2'} f(q_1', q_2') u^{(+)}_{q_1'q_2'} Z^{(+)}(q_1', q_2') \equiv \varkappa f(q_1, q_2) D_0. \tag{8.28'}$$

Let us use (8.28') and (8.30) in (8.29) and divide the equation by D_0. Consequently, we obtain the secular equation

$$1 = 2\varkappa \sum_{q_1, q_2} \frac{(f(q_1, q_2)\, u^{(+)}_{q_1q_2})^2 [\varepsilon(q_1)+\varepsilon(q_2)]}{[\varepsilon(q_1)+\varepsilon(q_2)]^2-\omega^2}. \tag{8.29'}$$

The energies of the vibrational states are equal to the solutions of (8.29'). This particular form of the secular equation is often used in the calculations of the vibrational states in even–even nuclei.

Second case

Let us assume that the particle–hole interaction consists of two terms: the multipole–multipole interaction and the spin–multipole interaction, i.e.,

$$G^{\omega}(q_1, q_2; q_2', q_1') = \varkappa f(q_1, q_2) f(q_1', q_2') + \varkappa_{\sigma} t(q_1, q_2)\, t(q_1', q_2'). \tag{8.30'}$$

The functions $t^{2\mu}(q, q')$ and $\bar{t}^{2\mu}(q, q')$ determined in (3.74), (3.74'), and (3.74'') are denoted by $t(q, q')$. We shall again neglect the particle–particle interaction. Later, we shall show that the particle–hole part of the multipole moment operator (3.63) contains the pairing factor $u^{(+)}_{qq'}$, while the particle–hole part of the spin–multipole moment operator is proportional to $u^{(-)}_{qq'}$. Consequently, eqns. (8.26) have the form

$$[\varepsilon(q_1)+\varepsilon(q_2)]\, Z^{(+)}(q_1, q_2) - \omega Z^{(-)}(q_1, q_2) - 2\varkappa f(q_1 q_2) u^{(+)}_{q_1q_2} \sum_{q_1', q_2'} f(q_1', q_2') u^{(+)}_{q_1'q_2'} Z^{(+)}(q_1', q_2') = 0, \tag{8.31}$$

$$[\varepsilon(q_1)+\varepsilon(q_2)]\, Z^{(-)}(q_1, q_2) - \omega Z^{(+)}(q_1, q_2) - 2\varkappa_{\sigma} t(q_1, q_2) u^{(-)}_{q_1q_2} \sum_{q_1', q_2'} t(q_1', q_2')\, u^{(-)}_{q_1'q_2'} Z^{(-)}(q_1', q_2') = 0. \tag{8.31'}$$

To simplify the equations further we shall use the constant D_0 of (8.28') and another constant R defined by

$$R = 2 \sum_{q_1', q_2'} u^{(-)}_{q_1'q_2'} t(q_1', q_2')\, Z^{(-)}(q_1', q_2').$$

Equations (8.31) and (8.31') will now have the form

$$[\varepsilon(q_1)+\varepsilon(q_2)] Z^{(+)}(q_1, q_2) - \omega Z^{(-)}(q_1, q_2) - \varkappa f(q_1, q_2) u^{(+)}_{q_1q_2} D_0 = 0,$$
$$[\varepsilon(q_1)+\varepsilon(q_2)]\, Z^{(-)}(q_1, q_2) - \omega Z^{(+)}(q_1, q_2) - \varkappa_{\sigma} t(q_1, q_2) u^{(-)}_{q_1q_2} R = 0,$$

and, consequently,

$$Z^{(+)}(q_1, q_2) = \frac{\varkappa f(q_1, q_2)[\varepsilon(q_1)+\varepsilon(q_2)]u^{(+)}_{q_1q_2}D_0+\varkappa_\sigma t(q_1, q_2)\omega u^{(-)}_{q_1q_2}R}{[\varepsilon(q_1)+\varepsilon(q_2)]^2-\omega^2},$$

$$Z^{(-)}(q_1, q_2) = \frac{\varkappa f(q_1, q_2)\omega u^{(+)}_{q_1q_2}D_0+\varkappa_\sigma t(q_1, q_2)[\varepsilon(q_1)+\varepsilon(q_2)]u^{(-)}_{q_1q_2}R}{[\varepsilon(q_1)+\varepsilon(q_2)]^2-\omega^2}.$$

By substituting the above expressions in the definitions of the quantities D_0 and R we obtain two homogenous equations. The system has a nontrivial solution if its determinant equals zero. Such a condition then leads to the secular equation

$$\left\{1-2\varkappa\sum_{q,q'}\frac{f^2(q,q')(u^{(+)}_{qq'})^2[\varepsilon(q)+\varepsilon(q')]}{[\varepsilon(q)+\varepsilon(q')]^2-\omega^2}\right\}\times\left\{1-2\varkappa_\sigma\sum_{q,q'}\frac{t^2(q,q')(u^{(-)}_{qq'})^2[\varepsilon(q)+\varepsilon(q')]}{[\varepsilon(q)+\varepsilon(q')]^2-\omega^2}\right\}$$
$$= 4\varkappa\varkappa_\sigma\left(\sum_{q,q'}\frac{f(q,q')\,t(q,q')u^{(+)}_{qq'}u^{(-)}_{qq'}\omega}{[\varepsilon(q)+\varepsilon(q')]^2-\omega^2}\right)^2. \qquad (8.32)$$

An equation of this type was discussed in the study on quadrupole states in deformed nuclei.(131)

The secular equation (8.29′) is a particular form of the more general equation (8.32). It is obtained when $\varkappa_\sigma = 0$. As we have mentioned before, (8.29′) was often discussed in detail, for example, in refs. 129 and 435.

Third case

Let us consider the collective states determined primarily by the particle–particle correlations. Consequently, we shall assume that

$$G^\omega(q_1, q_2; q_2', q_1') = 0,$$
$$G^\xi(q_1, q_2; q_2', q_1') = G\delta_{q_1q_2}\delta_{q_1q_2}, \qquad (8.30'')$$

and that (8.26) will have the form

$$2\varepsilon(q)Z^{(\pm)}(q,q)-\omega Z^{(\mp)}(q,q)-Gv^{(\mp)}_{qq}\sum_{q'}v^{(\mp)}_{q'q'}Z^{(\pm)}(q',q') = 0. \qquad (8.33)$$

Let us denote by $d^{(\pm)}$ the quantities

$$d^{(\pm)} = G\sum_{q'}v^{(\mp)}_{q'q'}Z^{(\pm)}(q',q') \qquad (8.33')$$

and rewrite (8.33) as

$$2\varepsilon(q)Z^{(\pm)}(q,q)-\omega Z^{(\mp)}(q,q)-v^{(\mp)}_{qq}d^{(\pm)} = 0.$$

Therefore

$$Z^{(\pm)}(q,q) = \frac{2\varepsilon(q)v^{(\mp)}_{qq}d^{(\pm)}+\omega v^{(\pm)}_{qq}d^{(\mp)}}{4\varepsilon^2(q)-\omega^2}.$$

By substituting the expressions for $Z^{(+)}(q, q)$ in the definition of $d^{(\pm)}$ we obtain again two homogenous equations. The determinant of the system vanishes if

$$\left\{\sum_q \frac{2\varepsilon(q)\,(v_{qq}^{(-)})^2}{4\varepsilon^2(q)-\omega^2}-\frac{1}{G}\right\}\left\{\sum_q \frac{2\varepsilon(q)}{4\varepsilon^2(q)-\omega^2}-\frac{1}{G}\right\}=\omega^2\left\{\sum_q \frac{v_{qq}^{(-)}}{4\varepsilon^2(q)-\omega^2}\right\}^2, \qquad (8.33'')$$

which is the secular equation determining the energies of the collective states.

Such equations were derived in ref. 438 for the spherical nuclei and in ref. 431 for the deformed nuclei. They are often used in the theory of pairing vibrations.[439]

Fourth case

We shall show later that the inclusion of the particle–particle interaction excludes the spurious states in calculations of the 0^+ excited states. This is the case when it is necessary to include both the particle–hole and particle–particle interactions. We shall choose the interaction in the form

$$\left.\begin{aligned} G^{\omega}(q_1, q_2; q_2', q_1') &= \varkappa f(q_1, q_2) f(q_1', q_2'), \\ G^{\xi}(q_1, q_2; q_2', q_1') &= G\delta_{q_1q_2}\delta_{q_1'q_2'}, \end{aligned}\right\} \qquad (8.30''')$$

i.e., a combination of the first and third cases.

Equations (8.26) are now more complicated:

$$\begin{aligned} &[\varepsilon(q_1)+\varepsilon(q_2)]\,Z^{(+)}(q_1, q_2)-\omega Z^{(-)}(q_1, q_2)-Gv_{q_1q_2}^{(-)}\delta_{q_1q_2}\sum_{q_1'} v_{q_1'q_1'}^{(-)}Z^{(+)}(q_1', q_1') \\ &\quad -2\varkappa f(q_1, q_2)u_{q_1q_2}^{(+)}\sum_{q_1', q_2'} f(q_1', q_2')\,u_{q_1'q_2'}^{(+)}Z^{(+)}(q_1', q_2') = 0, \end{aligned} \qquad (8.34)$$

$$[\varepsilon(q_1)+\varepsilon(q_2)]\,Z^{(-)}(q_1, q_2)-\omega Z^{(+)}(q_1, q_2)-Gv_{q_1q_2}^{(+)}\delta_{q_1q_2}\sum_{q_1'} v_{q_1'q_1'}^{(+)}Z^{(-)}(q_1', q_2') = 0. \quad (8.34')$$

We shall use the constants D_0 and $d^{(\pm)}$ defined earlier in (8.28') and (8.33'), and obtain

$$\begin{aligned} &[\varepsilon(q_1)+\varepsilon(q_2)]\,Z^{(+)}(q_1, q_2)-\omega Z^{(-)}(q_1, q_2)-v_{q_1q_1}^{(-)}\delta_{q_1q_2}d^{(+)}-\varkappa f(q_1, q_2)u_{q_1q_2}^{(+)}D_0 = 0, \\ &[\varepsilon(q_1)+\varepsilon(q_2)]\,Z^{(-)}(q_1, q_2)-\omega Z^{(+)}(q_1, q_2)-v_{q_1q_1}^{(+)}\delta_{q_1q_2}d^{(-)} = 0. \end{aligned}$$

Therefore

$$\left.\begin{aligned} Z^{(+)}(q_1, q_2) &= \frac{\varkappa f(q_1, q_2)\,[\varepsilon(q_1)+\varepsilon(q_2)]u_{q_1q_2}^{(+)}D_0+2\varepsilon(q_1)\delta_{q_1q_2}v_{q_1q_1}^{(-)}d^{(+)}+\omega\delta_{q_1q_2}v_{q_1q_2}^{(+)}d^{(-)}}{[\varepsilon(q_1)+\varepsilon(q_2)]^2-\omega^2}, \\ Z^{(-)}(q_1, q_2) &= \frac{\varkappa f(q_1, q_2)\omega u_{q_1q_2}^{(+)}D_0+\omega\delta_{q_1q_2}v_{q_1q_1}^{(-)}d^{(+)}+2\varepsilon(q_1)\delta_{q_1q_2}v_{q_1q_1}^{(+)}d^{(-)}}{[\varepsilon(q_1)+\varepsilon(q_2)]^2-\omega^2}. \end{aligned}\right\} \qquad (8.34'')$$

By substituting the $Z^{(\pm)}(q_1, q_2)$ functions in the relations

$$D_0 = 2\sum_{q_1, q_2} f(q_1, q_2)\,u_{q_1q_2}^{(+)}Z^{(+)}(q_1, q_2),$$

$$d^{(\pm)} = G\sum_{q'} v_{q'q'}^{(\mp)}Z^{(\pm)}(q', q'),$$

we obtain a system of three homogenous equations in the unknown quantities D_0, $d^{(+)}$ and $d^{(-)}$. The determinant of the system must vanish, and hence we obtain the secular equation in the form

$$\begin{vmatrix} 2\sum_{q,q'} \frac{[f(q,q')u_{qq'}^{(+)}]^2[\varepsilon(q)+\varepsilon(q')]}{[\varepsilon(q)+\varepsilon(q')]^2-\omega^2} - \frac{1}{\varkappa} & \sum_q \frac{f(q,q)u_{qq}^{(+)}\omega}{4\varepsilon^2(q)-\omega^2} & \sum_q \frac{2f(q,q)u_{qq}^{(+)}v_{qq}^{(-)}\varepsilon(q)}{4\varepsilon^2(q)-\omega^2} \\ \sum_q \frac{2f(q,q)u_{qq}^{(+)}\omega}{4\varepsilon^2(q)-\omega^2} & \sum_q \frac{2\varepsilon(q)}{4\varepsilon^2(q)-\omega^2} - \frac{1}{G} & \sum_q \frac{\omega v_{qq}^{(-)}}{4\varepsilon^2(q)-\omega^2} \\ \sum_q \frac{4f(q,q)u_{qq}^{(+)}v_{qq}^{(-)}\varepsilon(q)}{4\varepsilon^2(q)-\omega^2} & \sum_q \frac{\omega v_{qq}^{(-)}}{4\varepsilon^2(q)-\omega^2} & \sum_q \frac{2(v_{qq}^{(-)})^2\varepsilon(q)}{4\varepsilon^2(q)-\omega^2} - \frac{1}{G} \end{vmatrix} = 0. \tag{8.35}$$

The equation of this type was originally suggested in ref. 150; it was often used in numerical calculations, for example, in refs. 424 and 429.

Equations (8.26) are very general. By a particular choice of the interactions G^{ξ} and G^{ω}, we can obtain different secular equations applicable to the studies of collective states in medium and heavy nuclei.

6. Let us generalize (8.11) and (8.12) and allow a weak external potential. In order to do this, we shall add the term

$$\sum_{f,f'} \delta I(f,f')a_f^+ a_{f'} \tag{8.36}$$

to the Hamiltonian (3.1). The function $\delta I(f,f') = \delta I^*(f',f)$ characterizes the external potential. When such an additional term is present, the functions $\mathfrak{A}(f_1,f_2)$ and $\mathfrak{B}(f_1,f_2)$ will contain besides the original terms with $T'(f_1,f_2)$ the terms proportional to the $\delta I(f,f')$. The increments $\delta\mathfrak{A}(f_1,f_2)$ defined in (8.10) and $\delta\mathfrak{B}(f_1,f_2)$ defined in (8.10′) are supplemented by the following additional terms:

$$\delta\mathfrak{A}_{\text{ex}}(f_1,f_2) = \sum_f \{\delta I(f_1,f)\,\Phi(f,f_2) + \delta I(f_2,f)\,\Phi(f_1,f)\},$$

$$\delta\mathfrak{B}_{\text{ex}}(f_1,f_2) = \sum_f \{\delta I(f_2,f)\,\varrho(f_1,f) - \delta I(f,f_1)\,\varrho(f,f_2)\}.$$

Using the Fourier transformations

$$\delta I(f,f') = \sum_\omega e^{-i\omega t}\delta I_\omega(f,f'),$$

$$\delta I^*(f,f') = \sum_\omega e^{-i\omega t}\delta I^*_{-\omega}(f,f'), \tag{8.36′}$$

we can rewrite (8.11′) and (8.12′) in a form which includes the external potential:

$$\begin{aligned}\omega\xi_\omega(g_1,g_2) = &\sum_{g'} \{\Omega(g_2,g')\,\xi_\omega(g_1,g') - \Omega(g_1,g')\,\xi_\omega(g_2,g')\} \\ &+ \sum_{g_1',g_2'} \{X(g_1,g_2;g_1',g_2')\,\xi_\omega(g_1',g_2') - Y(g_1,g_2;g_1',g_2')\,\eta_\omega(g_1,g_2')\} \\ &+ \sum_{f,f'} \{v(f'g_1)\,u^*(fg_2) - u^*(fg_1)\,v(f'g_2)\}\,\delta I_\omega(f,f'),\end{aligned} \tag{8.37}$$

$$-\omega\eta_\omega(g_1, g_2) = \sum_{g'}\{\Omega^*(g_2, g')\eta_\omega(g_1, g')-\Omega^*(g_1, g')\eta_\omega(g_2, g')\}$$
$$+\sum_{g_1', g_2'}\{X^*(g_1, g_2; g_1', g_2')\eta_\omega(g_1', g_2')-Y^*(g_1, g_2; g_1', g_2')\,\xi_\omega(g_1', g_2')\}$$
$$+\sum_{f, f'}\{v^*(f'g_1)\,u(fg_2)-u(fg_1)\,v^*(f'g_2)\,\delta I^*_{-\omega}(f, f'). \tag{8.37'}$$

Such equations were obtained originally in ref. 120.

Let us express the general equations (8.37) and (8.37′) in the approximation defined by (8.18) and (8.18′). By introducing the function $\nu^{(\pm)}(g_1, g_2)$ and by repeating all steps leading to (8.20), we obtain two new equations:

$$[\varepsilon(g_1)+\varepsilon(g_2)]\,\nu^{(\pm)}(g_1, g_2)-\omega\nu^{(\mp)}(g_1, g_2)+\sum_{g_1', g_2'}\{X(g_1, g_2; g_1', g_2')$$
$$\mp Y(g_1, g_2; -g_1', -g_2')\}\,\nu^{(\pm)}(g_1, g_2)-\tfrac{1}{2}u^{(\pm)}_{g_1, g_2}[\delta I^*_{-\omega}(g_2, -g_1)$$
$$\mp\delta I_\omega(-g_2, g_1)] = 0. \tag{8.37''}$$

As before, we shall use the quantum numbers $q\sigma$ instead of g, and exclude the σ-dependence. The derivation is analogous to the derivation of (8.26). The resulting system has the form

$$[\varepsilon(q_1)+\varepsilon(q_2)]\,Z^{(\pm)}(q_1, q_2)-\omega Z^{(\mp)}(q_1, q_2)-\sum_{q_1', q_2'}G^\xi(q_1, q_2; q_2', q_1')\,v^{(\mp)}_{q_1q_2}v^{(\mp)}_{q_1'q_2'}Z^{(\pm)}(q_1', q_2')$$
$$-2\sum_{q_1', q_2'}G^\omega(q_1, q_2; q_2', q_1')\,u^{(\pm)}_{q_1q_2}u^{(\pm)}_{q_1'q_2'}Z^{(\pm)}(q_1', q_2') = \tfrac{1}{2}u^{(\pm)}_{q_1q_2}[\delta I_\omega(q_1, q_2)\pm\delta I^*_{-\omega}(q_1, q_2)], \tag{8.38}$$

where

$$\delta I_\omega(q_1, q_2) = \sum_\sigma\{\delta I_\omega(q_1\sigma, q_2\sigma)-\gamma_\sigma\delta I_\omega(q_1\sigma, q_2, -\sigma)\},$$
$$\delta I^*_{-\omega}(q_1, q_2) = \sum_\sigma\{\delta I^*_{-\omega}(q_1\sigma, q_2\sigma)-\gamma_\sigma\delta I^*_{-\omega}(q_1\sigma, q_2, -\sigma)\}.$$

Thus we have obtained a system of two independent equations describing the vibrations of the system in a weak external field. These equations were discussed in ref. 437.

7. Let us transform (4.38) and discuss their relation to the equations of the finite Fermi system theory.[13, 14] We shall introduce the new unknown functions

$$d^{(\pm)}(q_1, q_2) = \sum_{q_1', q_2'}G^\xi(q_1, q_2; q_2', q_1')\,v^{(\mp)}_{q_1'q_2'}Z^{(\pm)}(q_1', q_2'), \tag{8.39}$$

$$V^{(\pm)}(q_1, q_2) = 2\sum_{q_1', q_2'}G^\omega(q_1, q_2; q_2', q_1')\,u^{(\pm)}_{q_1'q_2'}Z^{(\pm)}(q_1', q_2')+V_0^{(\pm)}(q_1, q_2), \tag{8.39'}$$

$$V_0^{(\pm)}(q_1, q_2) = \tfrac{1}{2}[\delta I_\omega(q_1, q_2)\pm\delta I^*_{-\omega}(q_1, q_2)]. \tag{8.39''}$$

When these functions are used, equations (8.38) are transformed into the form

$$[\varepsilon(q_1)+\varepsilon(q_2)]\,Z^{(\pm)}(q_1, q_2)-\omega Z^{(\mp)}(q_1, q_2)-v^{(\mp)}_{q_1q_2}d^{(\pm)}(q_1, q_2)-u^{(\pm)}_{q_1q_2}V^{(\pm)}(q_1, q_2) = 0. \tag{8.38'}$$

Consequently,

$$Z^{(\pm)}(q_1, q_2) = \frac{1}{[\varepsilon(q_1)+\varepsilon(q_2)]^2-\omega^2}\{[\varepsilon(q_1)+\varepsilon(q_2)]\,[u^{(\pm)}_{q_1q_2}V^{(\pm)}(q_1, q_2)$$
$$+v^{(\mp)}_{q_1q_2}d^{(\pm)}(q_1, q_2)]+\omega[u^{(\mp)}_{q_1q_2}V^{(\mp)}(q_1, q_2)+v^{(\pm)}_{q_1q_2}d^{(\mp)}(q_1, q_2)]\}. \tag{8.38''}$$

By substituting (8.38″) in the definitions (8.39) and (8.39′) we obtain a system of four equations for the determination of the quantities $d^{(\pm)}$ and $V^{(\pm)}$. They have the form

$$V^{(\pm)}(q_1, q_2) = V_0^{(\pm)}(q_1, q_2) + 2 \sum_{q_1', q_2'} G^{\omega}(q_1, q_2; q_2', q_1') \frac{u^{(\pm)}_{q_1'q_2'}}{[\varepsilon(q_1') + \varepsilon(q_2')]^2 - \omega^2} \times$$
$$\times \{[\varepsilon(q_1') + \varepsilon(q_2')] [u^{(\pm)}_{q_1'q_2'} V^{(\pm)}(q_1', q_2') + v^{(\mp)}_{q_1'q_2'} d^{(\pm)}(q_1', q_2')]$$
$$+ \omega [u^{(\mp)}_{q_1'q_2'} V^{(\mp)}(q_1', q_2') + v^{(\pm)}_{q_1'q_2'} d^{(\mp)}(q_1', q_2')]\}, \tag{8.40}$$

$$d^{(\pm)}(q_1, q_2) = \sum_{q_1', q_2'} G^{\xi}(q_1, q_2; q_2', q_1') \frac{v^{(\mp)}_{q_1'q_2'}}{[\varepsilon(q_1') + \varepsilon(q_2')]^2 - \omega^2} \times$$
$$\times \{[\varepsilon(q_1') + \varepsilon(q_2')] [u^{(\pm)}_{q_1'q_2'} V^{(\pm)}(q_1', q_2') + v^{(\mp)}_{q_1'q_2'} d^{(\pm)}(q_1', q_2')]$$
$$+ \omega [u^{(\mp)}_{q_1'q_2'} V^{(\mp)}(q_1', q_2') + v^{(\pm)}_{q_1'q_2'} d^{(\mp)}(q_1', q_2')]\}. \tag{8.40′}$$

The basic equations of the finite Fermi-system theory,[13, 14] are identical with (8.40) and (8.40′) except for the interaction G^{ω} being replaced by the quantity $\mathcal{F}^{\omega}$ and for the interaction G^{ξ} being replaced by the quantity $\mathcal{F}^{\xi}$. The equations in the finite Fermi-system theory were obtained by the Green function method. In our case, the equations were derived in the framework of the selfconsistent field method. The identical form of the equations in both methods is a consequence of the general theorem derived by Bogolyubov.[228] The theorem relates the variations of the correlation functions (for example, our functions $\varrho(g_1, g_2)$ and $\Phi(g_1, g_2)$ to the corresponding Green functions.

The system of four equations (8.40) and (8.40′) can be solved by various approximate methods in the finite Fermi-system theory. It should be remembered, however, that only two of the equations are independent. As a matter of fact, if the two equations (8.38) are solved and the solutions $Z^{(\pm)}$ found, the unknown functions $d^{(\pm)}$ and $V^{(\pm)}$ are simply calculated by using definitions (8.39) and (8.39′).

The local quasiparticle interaction is not calculable in the finite Fermi-system theory. Instead, its parameters are fitted by comparing the corresponding measured and calculated quantities. It is assumed that the parameters are universal, i.e., that they have identical values in all medium and heavy nuclei.

The interaction $\mathcal{F}^{\omega}$ is believed to be density dependent. Consequently, its values are determined by (3.82) and they are different inside the nucleus and on the nuclear periphery. The interaction $\mathcal{F}^{\omega}$ is responsible for the description of the processes with low momentum transfer. Therefore it is assumed that $\mathcal{F}^{\omega}$ does not depend on the momentum transfer and can be described by the formula (3.80), i.e.,

$$\mathcal{F}^{\omega} = \frac{\pi^2}{m^* p_F} \{f + g\sigma^{(1)} . \sigma^{(2)} + [f' + g'\sigma^{(1)} . \sigma^{(2)}]\tau^{(1)} . \tau^{(2)}\}.$$

The quantities f, f', g, and g' depend on the quasiparticle momentum. But the considered transitions all occur near the Fermi level, and hence all momenta may be replaced by the Fermi momentum p_F. The interaction $\mathcal{F}^{\omega}$ depends on the angle between the incoming momenta. However, usually only the first term in the Legendre polynomial expansion (3.81) is included. The other interaction channel $\mathcal{F}^{\xi}$ is taken as a δ-function.

We would like to compare the multipole–multipole interaction with the effective interaction in the Fermi-system theory. In order to do that, we shall calculate the effective forces between particles in one partially filled shell (see refs. 440 and 441). The effect of the particle–particle interaction on the particle–hole excitation will be neglected, and the effective field equations will have the form

$$V^{(\pm)}(q_1, q_2) = V_0^{(\pm)}(q_1, q_2) + \sum_{q_1', q_2'} \mathcal{F}^{\omega}(q_1, q_2; q_1', q_2')\, \mathcal{L}^{(\pm)}(q_1', q_2')\, V^{(\pm)}(q_1', q_2'), \tag{8.41}$$

where

$$\mathcal{L}^{(\pm)}(q_1, q_2) = u_{q_1 q_2}^{(\pm)} \frac{[\varepsilon(q_1)+\varepsilon(q_2)]u_{q_1q_2}^{(\pm)} + \omega u_{q_1q_2}^{(\mp)}\hat{P}}{[\varepsilon(q_1)+\varepsilon(q_2)]^2 - \omega^2},$$

$$\hat{P}V^{(\pm)}(q_1, q_2) = V^{(\mp)}(q_1, q_2).$$

The function $\mathcal{L}^{(\pm)}(q_1, q_2)$ is divided into two terms

$$\mathcal{L}^{(\pm)}(q_1, q_2) = \mathcal{L}_1^{(\pm)}(q_1, q_2) + \mathcal{L}_2^{(\pm)}(q_1, q_2).$$

The function $\mathcal{L}_1^{(\pm)}(q_1, q_2)$ is nonvanishing only if both q_1 and q_2 belong to the partially filled shell. At the same time, the function $\mathcal{L}_2^{(\pm)}(q_1, q_2)$ vanishes if both q_1 and q_2 belong to the same partially filled shell. We shall use the definition

$$\begin{aligned}\mathcal{F}'(q, q'; q_1, q_2) &= \mathcal{F}^{\omega}(q, q'; q_1, q_2)\\ &\quad + \sum_{q_1', q_2'} \mathcal{F}'(q, q'; q_1', q_2')\, \mathcal{L}_2^{(\pm)}(q_1', q_2')\, \mathcal{F}^{\omega}(q_1', q_2'; q_1, q_2).\end{aligned} \tag{8.41'}$$

By multiplying (8.41) by $\mathcal{F}'(q, q'; q_1, q_2)\, \mathcal{L}^{(\pm)}(q_1, q_2)$ and by summing over q_1, q_2

$$\begin{aligned}&\sum_{q_1, q_2} \mathcal{F}'(q, q'; q_1, q_2)\, \mathcal{L}_2^{(\pm)}(q_1, q_2)\, V^{(\pm)}(q_1, q_2)\\ &\quad = \sum_{q_1, q_2} \mathcal{F}'(q, q'; q_1, q_2)\, \mathcal{L}_2^{(\pm)}(q_1, q_2)\, V_0^{(\pm)}(q_1, q_2)\\ &\quad\quad + \sum_{q_1, q_2, q_1', q_2'} \mathcal{F}'(q, q'; q_1, q_2)\, \mathcal{L}_2^{(\pm)}(q_1, q_2)\, \mathcal{F}^{\omega}(q_1, q_2; q_1', q_2') \times\\ &\quad\quad \times \{\mathcal{L}_1^{(\pm)}(q_1', q_2') + \mathcal{L}_2^{(\pm)}(q_1', q_2')\}\, V^{(\pm)}(q_1', q_2').\end{aligned}$$

Adding this equation to (8.41), we can write

$$\begin{aligned}&V^{(\pm)}(q, q') + \sum_{q_1, q_2} \mathcal{F}'(q, q'; q_1, q_2)\, \mathcal{L}_2^{(\pm)}(q_1, q_2)\, V^{(\pm)}(q_1, q_2)\\ &\quad = \Delta I_{\omega}^{(\pm)}(q, q') + \sum_{q_1, q_2} \Big\{ \mathcal{F}^{\omega}(q, q'; q_1, q_2) + \sum_{q_1', q_2'} \mathcal{F}'(q, q'; q_1', q_2') \times\\ &\quad\quad \times \mathcal{L}_2^{(\pm)}(q_1', q_2')\, \mathcal{F}^{\omega}(q_1', q_2'; q_1, q_2) \Big\} \{\mathcal{L}_1^{(\pm)}(q_1, q_2) + \mathcal{L}_2^{(\pm)}(q_1, q_2)\}\, V^{(\pm)}(q_1, q_2),\end{aligned}$$

where

$$\Delta I_{\omega}^{(\pm)}(q, q') = V_0^{(\pm)}(q, q') + \sum_{q_1, q_2} \mathcal{F}'(q, q'; q_1, q_2)\, \mathcal{L}_2^{(\pm)}(q_1, q_2)\, V_0^{(\pm)}(q_1, q_2).$$

Now, by using definition (8.41′) we derive the final equation

$$V^{(\pm)}(q, q') = \Delta I_{\omega}^{(\pm)}(q, q') + \sum_{q_1, q_2} \mathcal{F}'(q, q'; q_1, q_2) \mathcal{L}_1^{(\pm)}(q_1, q_2) V^{(\pm)}(q_1, q_2). \quad (8.42)$$

Thus, the function $\mathcal{F}(q, q'; q_1, q_2)$ obviously acts as the effective interaction in the unfilled shell.

Let us consider the quadrupole excited states. Naturally, the function $\mathcal{F}'$ is not reduced to the matrix elements of the quadrupole interaction. However, according to refs. 440 and 441, the ratio

$$\frac{\mathcal{F}'(q_1, q_1; q_2, q_2')}{\varkappa^{(2)} f^{\lambda=2}(q_1, q_1') f^{\lambda=2}(q_2, q_2')}$$

fluctuates around the average value equal to unity. The fluctuations are quite large in isolated cases. But the calculated characteristics of the collective quadrupole states involve summations over many states. Consequently, the fluctuations are largely averaged out. This means that the $\mathcal{F}'$ interaction and the quadrupole–quadrupole interaction predict the similar collective vibrational quadrupole states.

§ 2. The Approximate Second Quantization Method

1. The method of the approximate second quantization was originally suggested by Bogolyubov[4]. Later it was further developed in ref. 442 and applied to the solution of the electron gas problem in ref. 443. The method is often used in the solutions of various many-body problems. There are two basic variants of the approximate second quantization method: the Tamm–Dancoff (TD) method and the Random Phase Approximation (RPA).

The TD method was originally formulated by Tamm[444] in the quantum field theory; later this method was developed independently by Dancoff.[445] The mathematical basis for the method was developed by Fock[446] even earlier when he formulated the quantum electrodynamics by the method of functionals. The TD method (without pairing superfluid correlations) is often used in studies of light nuclei[447, 448] and medium and heavy nuclei.

The TD method takes into account the quasiparticle interaction in excited states, but the interaction does not affect the ground state. The ground state of an even–even nucleus is, therefore, the quasiparticle vacuum. The diagrams shown in Fig. 8.1(a) are included in the TD method. The main shortcoming of the TD method is the asymmetric treatment of the ground and excited states.

This defect is corrected in the RPA which includes the quasiparticle interaction in all states. The RPA method uses a wider class of diagrams in the summation, namely those shown in Fig. 8.1(b). The RPA variant of the approximate second quantization will be discussed in this section. The corresponding formulae of the TD variant will be also shown.

The approximate second quantization method has certain advantages when compared to the selfconsistent field method. Namely, the normalization condition related to the commutation relations of the former method gives explicitly the wave functions of the vibrational states.

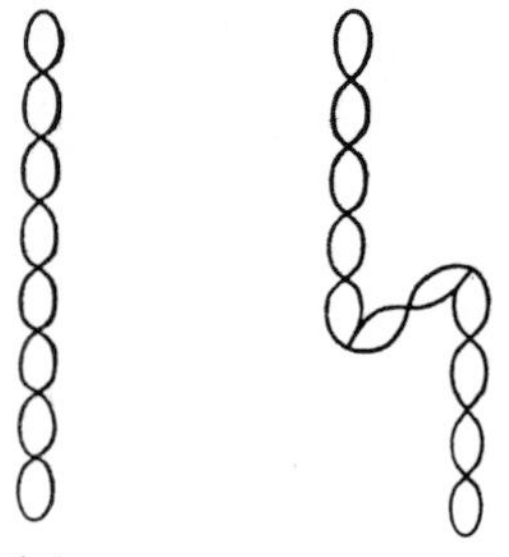

FIG. 8.1. The diagrams included in the approximate second quantization method. (a) The TD method; (b) the RPA.

Equations (8.11′) and (8.12′) may be derived by using the approximate second quantization method. When the approximation (8.18) is used we obtain equations (8.26) determining the frequencies of the collective vibrational states. It is unnecessary to derive the most general equations (8.11′), (8.12′), and (8.26) again. Therefore we shall derive in this section the equations in the particular form applicable to the description of the vibrational states in complex nuclei.

Let us discuss the main principles of the approximate second quantization method. We shall use the quasiparticle pair operators $A(q, q')$ defined in (6.62) or (6.69). The operators consist of the products $\alpha_{q\sigma}\alpha_{q'\sigma'}$ and obey the following commutation relations

$$[A(q, q'), A^{+}(q_2, q_2')] = \delta_{qq_2}\delta_{q'q_2'} + \delta_{qq_2'}\delta_{q'q_2} + \sum_{\substack{q'', \sigma''; \\ q''', \sigma'''}} l(q, q'; q_2, q_2; q'', q''')\alpha^{+}_{q''\sigma''}\alpha_{q'''\sigma'''}, \tag{8.43}$$

$$[A(q, q'), A(q_2, q_2')] = [A^{+}(q_2, q_2'), A^{+}(q, q')] = 0, \tag{8.43'}$$

where l is an expression containing the Kronecker δ-function.

The quasiparticle interaction affects the ground states of even–even nuclei. The ground-state wave function is not equal to the quasiparticle vacuum; it contains small components with different (but always even) number of quasiparticles. We shall consider, however, only such situations where the average number of quasiparticles in the ground state is small. This is our main assumption. Mathematically, it coincides with the condition (8.5), i.e.,

$$\langle|\alpha^{+}_{q\sigma}\alpha_{q'\sigma'}|\rangle = 0.$$

Thus we are using the same basic assumption as in the selfconsistent field method. When using our basic assumption, we may omit the $\alpha^{+}_{q\sigma}\alpha_{q'\sigma'}$ terms in the commutation relation (8.43). Thus the operators $A(q, q')$ are effectively described by the boson commutation relations. Hence, the method is often called "the quasiboson" approximation. In the following discussion we shall replace the commutation relation (8.43) by

$$[A(q, q'), A^{+}(q_2, q_2')] = \delta_{qq_2}\delta_{q'q_2'} + \delta_{qq_2'}\delta_{q'q_2}. \tag{8.44}$$

The new operators, the phonons, are introduced by

$$Q_i = \tfrac{1}{2}\sum_{q, q'}\{\psi^{i}_{qq'}A(q, q') - \varphi^{i}_{qq'}A^{+}(q, q')\}, \tag{8.45}$$

$$Q_i^{+} = \tfrac{1}{2}\sum_{q, q'}\{\psi^{i}_{qq'}A^{+}(q, q') - \varphi^{i}_{qq'}A(q, q')\}. \tag{8.45'}$$

The pairs (q, q') are the pairs of the single-particle states connected by certain selection rules. The index $i = 1, 2, 3, \ldots$, determines the sequence of the one-phonon states. Naturally, the number of pairs (q, q') and the number of the states i are identical. Hence, the matrices $\psi^i_{qq'}$, and $\varphi^i_{qq'}$ are the square matrices.

The ground state of an even–even nucleus will be defined as a phonon vacuum, i.e., by the condition

$$Q_i \Psi = 0 \tag{8.46}$$

valid for all i. The excited states are then the one-phonon states

$$Q_i^+ \Psi, \tag{8.47}$$

two-phonon states

$$Q_i^+ Q_{i'}^+ \Psi, \tag{8.47'}$$

etc.

To ensure the orthonormality of the wave functions corresponding to the ground state and to the excited states, the phonon operators must fulfill Bose-type commutation relations

$$[Q_i, Q_{i'}^+] = \delta_{ii'}, \tag{8.48}$$

$$[Q_i, Q_{i'}] = [Q_i^+, Q_{i'}^+] = 0. \tag{8.48'}$$

(We are assuming that $A(q, q)$ and $A^+(q_2, q_2')$ satisfy (8.43′) and (8.44)). From the compatibility of the conditions (8.48), (8.48′) and (8.43′), (8.44) we obtain the following equations for the unknown matrices $\psi^i_{qq'}$ and $\varphi^i_{qq'}$:

$$\sum_{q,q'} (\psi^i_{qq'}\psi^{i'}_{qq'} - \varphi^i_{qq'}\varphi^{i'}_{qq'}) = 2\delta_{ii'}, \tag{8.49}$$

$$\sum_{q,q'} (\psi^i_{qq'}\varphi^{i'}_{qq'} - \psi^{i'}_{qq'}\varphi^i_{qq'}) = 0, \tag{8.49'}$$

$$\sum_i (\psi^i_{qq'}\psi^i_{q_2q_2'} - \varphi^i_{qq'}\varphi^i_{q_2q_2'}) = \delta_{qq_2}\delta_{q'q_2'} + \delta_{qq_2'}\delta_{q'q_2}. \tag{8.50}$$

Now it is easy to find the inverse transformation, i.e., to express the operators $A(q, q')$ and $A^+(q, q')$ using the phonon operators

$$A(q, q') = \sum_i \{\psi^i_{qq'}Q_i + \varphi^i_{qq'}Q_i^+\}, \tag{8.51}$$

$$A^+(q, q') = \sum_i \{\psi^i_{qq'}Q_i^+ + \varphi^i_{qq'}Q_i\}. \tag{8.51'}$$

Let us find the explicit form of the phonon vacuum (ground state) Ψ. It is advantageous to use a matrix form of (8.45) and (8.45′). Thus

$$\begin{pmatrix} Q \\ Q^+ \end{pmatrix} = \begin{pmatrix} D & -F \\ -F & D \end{pmatrix} \begin{pmatrix} A \\ A^+ \end{pmatrix}, \tag{8.45''}$$

where

$$D = \tfrac{1}{2}\,|\psi^i_{qq'}|, \quad F = \tfrac{1}{2}\,|\varphi^i_{qq'}|$$

and Q and A denote the corresponding column vectors. The quasiparticle vacuum is determined by the condition

$$A\Psi_0 = 0. \tag{8.52}$$

The wave function Ψ will be found from the condition (8.46), i.e., $Q\Psi = 0$.

From (8.45″) it follows that

$$Q = DA - FA^+$$

and thus

$$\Gamma = A - D^{-1}FA^+ = A + 2LA^+,$$

where $\Gamma = D^{-1}Q$ and $2L = -D^{-1}F$.

Let us define the operator

$$\mathcal{M} = \frac{1}{\sqrt{N}} e^{-(1/2)A^+LA^+},$$

where $\frac{1}{\sqrt{N}}$ is (undetermined) normalizing factor. The operator A^+ clearly commutes with $\mathcal{M}$ and therefore

$$\mathcal{M}^{-1}\Gamma\mathcal{M} = \mathcal{M}^{-1}A\mathcal{M} + 2LA^+.$$

Let us find the operator $\mathcal{M}^{-1}A\mathcal{M}$; obviously

$$A\mathcal{M} = A(q, q') \sum_{n=0}^{\infty} \frac{1}{n!} \left(-\tfrac{1}{2} \sum_{\substack{q_1, q_1'; \\ q_2, q_2'}} A^+(q_1, q_1') L(q_1, q_1'; q_2, q_2') A^+(q_2, q_2') \right)^n$$

and, besides,

$$\begin{aligned} A(q, q') \left(-\tfrac{1}{2}\right) \sum_{\substack{q_1, q_1'; \\ q_2, q_2'}} A^+(q_1, q_1') L(q_1, q_1'; q_2, q_2') A^+(q_2, q_2') \\ = -\tfrac{1}{2} \sum_{\substack{q_1, q_1'; \\ q_2, q_2'}} A^+(q_1, q_1') L(q_1, q_1'; q_2, q_2') A^+(q_2, q_2') A(q, q') \\ -2 \sum_{q_2, q_2'} L(q, q'; q_2, q_2') A^+(q_2, q_2'). \end{aligned}$$

After some further manipulations,

$$\mathcal{M}^{-1}A\mathcal{M} = A - 2LA^+, \tag{8.53}$$

and, therefore,

$$\mathcal{M}^{-1}\Gamma\mathcal{M} = A. \tag{8.53'}$$

By multiplying (8.46) from left by D^{-1}, we obtain

$$\Gamma\Psi = 0, \tag{8.54}$$

and, therefore,

$$\mathcal{M}^{-1}\Gamma\mathcal{M}\mathcal{M}^{-1}\Psi = 0. \tag{8.54'}$$

With use of (8.53′), (8.54′) is rewritten as

$$\mathcal{M}^{-1}\Gamma\mathcal{M}\mathcal{M}^{-1}\Psi = A\mathcal{M}^{-1}\Psi = 0. \tag{8.54''}$$

But, according to (8.52),

$$A\Psi_0 = 0,$$

and, consequently,

$$\mathcal{M}^{-1}\Psi = \Psi_0$$

or

$$\Psi = \mathcal{M}\Psi_0 = \frac{1}{\sqrt{N}} e^{-(1/2)A^+LA^+}\Psi_0 . \tag{8.55}$$

Thus we have shown that the ground state of an even–even system has the form

$$\Psi = \frac{1}{\sqrt{N}} \exp\left\{\tfrac{1}{4} \sum_{\substack{q_1, q_1' \\ q_2, q_2'}} (\psi^{-1})^i_{q_1, q_1'} \varphi_{q_2, q_2'} A^+(q_1, q_1')\, A^+(q_2, q_2')\right\}\Psi_0 \tag{8.55'}$$

(the explicit form of the normalizing factor $1/\sqrt{N}$ can be found in ref. 431). The relation (8.55′) shows that the ground-state wave function contains a term without quasiparticles, and terms with four quasiparticles, eight quasiparticles, etc. Note that in this approximation the wave-function Ψ does not contain the two-quasiparticle components.

2. Let us consider the particle–hole interactions leading to the coherent effects. The full nuclear Hamiltonian, expressed in terms of the quasiparticle operators, is of the form (5.7′). As before, the particle–hole part of the Hamiltonian (5.7′) contains the terms proportional to the pairing factors

$$u^{(+)}_{qq'} = u_q v_{q'} + u_{q'} v_q \equiv u_{qq'}, \quad u^{(-)}_{qq'} = u_q v_{q'} - u_{q'} v_q .$$

Let us isolate from (5.7′) the part

$$-\tfrac{1}{4} \sum_{q_1, q_1'; q_2, q_2'} G(q_1, q_2; q_2', q_1')\, u_{q_1q_1'} u_{q_2q_2'} [A(q_1, q_1') + A^+(q_1, q_1')]\, [A(q_2, q_2') + A^+(q_2, q_2')],$$

which contains the symmetric particle–hole interaction proportional to the $u_{qq'}$ function For simplicity, we are considering the terms with the identical σ values.

We have to add a term describing the unperturbed quasiparticle energy. Such a term, according to (5.7′) and (5.17), is equal to $\sum_q \varepsilon(q)\, B(q, q)$. The operator B is defined by the relation

$$B(q, q') = \sum_q \alpha^+_{q\sigma} \alpha_{q'\sigma} .$$

Using (8.51) and (8.51′) we can express the operators $A(q, q')$, $A^+(q, q')$ in terms of the phonon operators. It is useful to define new unknown functions $Z^{(\pm)}$

$$Z^{i(\pm)}_{qq'} = \psi^i_{qq'} \pm \varphi^i_{qq'} . \tag{8.56}$$

Finally, the Hamiltonian describing the particle–hole interaction has the form

$$H_1 = \sum_q \varepsilon(q)\, B(qq) - \tfrac{1}{4} \sum_{q_1, q_1'; q_2, q_2'} G(q_1, q_2; q_2', q_1') u_{q_1q_1'} u_{q_2q_2'} \sum_{i, i'} Z^{i(+)}_{q_1q_1'} Z^{i'(+)}_{q_2q_2'} (Q^+_i + Q_i)\, (Q^+_{i'} + Q_{i'}). \tag{8.57}$$

Let us find the expectation value of H_1 in the one-phonon state (8.47):

$$\begin{aligned}\langle Q_i H_1 Q^+_i\rangle = {} & \tfrac{1}{4} \sum_{q, q'} [\varepsilon(q) + \varepsilon(q')]\, \{(Z^{i(+)}_{qq'})^2 + (Z^{i(-)}_{qq'})^2\} \\ & - \tfrac{1}{2} \sum_{q_1, q_1'; q_2, q_2'} G(q_1, q_2; q_2', q_1')\, u_{q_1q_1'} u_{q_2q_2'} \{Z^{i(+)}_{q_1q_1'} Z^{i(+)}_{q_2q_2'} + \tfrac{1}{2} \sum_{i'} Z^{i'(+)}_{q_1q_1'} Z^{i'(+)}_{q_2q_2'}\}\end{aligned} \tag{8.58}$$

The phonon–vacuum expectation value equals

$$\langle H_1\rangle = -\tfrac{1}{4}\sum_{q_1,\, q_1';\, q_2,\, q_2'} G(q_1, q_2; q_2', q_1')\, u_{q_1q_1'}u_{q_2q_2'}\sum_{i'} Z^{i'(+)}_{q_1q_1'}Z^{i'(+)}_{q_2q_2'}. \tag{8.58'}$$

Here and in the further discussion, the symbol $\langle\ldots\rangle$ denotes the averaging over one of the states (8.55′). The condition (8.49) can be written in the form

$$\sum_{q,\, q'} Z^{i(+)}_{qq'}Z^{i(-)}_{qq'} - 2 = 0.$$

The energies and amplitudes $Z^{i(\pm)}_{qq'}$ will be found by using the variational principle

$$\delta\left\{\langle Q_iH_1Q_i^+\rangle - \langle H_1\rangle - \frac{\omega_i}{2}\left[\sum_{q,\, q'} Z^{i(+)}_{qq'}Z^{i(-)}_{qq'} - 2\right]\right\} = 0. \tag{8.59}$$

The Lagrange multiplier ω_i ensures the fulfillment of the supplementary condition (8.49). Physically, the ω_i equals to the energy of the one-phonon state $Q_i^+\Psi$. We are considering one-phonon states with a particular value of i; therefore $i = i'$, the condition (8.49′) is fulfilled automatically, and the condition (8.50) is inapplicable. After the variations $\delta Z^{i(+)}_{q_1q_1}$ and $\delta Z^{i(-)}_{q_1q}$ are excluded, the following equations are obtained:

$$\tfrac{1}{2}[\varepsilon(q_1)+\varepsilon(q_1')]Z^{i(+)}_{q_1,\, q_1'} - u_{q_1q_1'}\sum_{q_2,\, q_2'} G(q_1, q_2; q_2', q_1')u_{q_2q_2'}Z^{i(+)}_{q_2q'_2} - \frac{\omega_i}{2} Z^{i(-)}_{q_1q_1'} = 0,$$

$$\tfrac{1}{2}[\varepsilon(q_1)+\varepsilon(q_1')]Z^{i(-)}_{q_1q_1'} - \frac{\omega_i}{2} Z^{i(+)}_{q_1q_1'} = 0.$$

The above-stated equations coincide with (8.27) and (8.27′) obtained in the selfconsistent field method. From the second equation

$$Z^{i(-)}_{q_1q_1'} = \frac{\omega_i}{\varepsilon(q_1)+\varepsilon(q_1')} Z^{i(+)}_{q_1q_1'},$$

and, substituting in the first equation, we write

$$\left[\varepsilon(q_1)+\varepsilon(q_1') - \frac{\omega_i^2}{\varepsilon(q_1)+\varepsilon(q_1')}\right]Z^{i(+)}_{q_1q_1'} - 2u_{q_1q_1'}\sum_{q_2,\, q_2'} G(q_1, q_2; q_2', q_1')u_{q_2q_2'}Z^{i(+)}_{q_2q_2'} = 0. \tag{8.60}$$

As before, a special notation is used for the sum, namely,

$$D^{(+)}(q_1, q_1') = 2\sum_{q_2,\, q_2'} G(q_1, q_2; q_2', q_1')u_{q_2q_2'}Z^{i(+)}_{q_2q_2'}, \tag{8.61}$$

and (8.60) is rewritten as

$$\left[\varepsilon(q_1)+\varepsilon(q_1') - \frac{\omega_i^2}{\varepsilon(q_1)+\varepsilon(q_1')}\right]Z^{(+)}_{q_1q_1'} = u_{q_1q_1'}D^{(+)}(q_1, q_1'). \tag{8.60'}$$

Consequently,

$$Z^{i(+)}_{q_1q_1'} = \frac{D^{(+)}(q_1, q_1')\, u_{q_1q_1'}[\varepsilon(q_1)+\varepsilon(q_1')]}{[\varepsilon(q_1)+\varepsilon(q_1')]^2 - \omega_i^2}. \tag{8.62}$$

The function $D^{(+)}$ will be therefore determined by the following integral equation:

$$D^{(+)}(q_1, q_1') = 2\sum_{q_2,\, q'} \frac{G(q_1, q_2; q_2', q_1')\, u^2_{q_2q_2'}[\varepsilon(q_2)+\varepsilon(q_2')]}{[\varepsilon(q_2)+\varepsilon(q_2')]^2-\omega_i^2} D^{(+)}(q_2, q_2'). \tag{8.63}$$

The equivalent equation (8.29) was again derived in the selfconsistent field method.

3. Let us use the factorized form of the quasiparticle interaction, i.e., let us assume that

$$G(q_1, q_2; q_2', q_1') = \varkappa f(q_1, q_1') f(q_2, q_2'). \tag{8.64}$$

Expression (8.61) will be simplified in such a case, namely,

$$D^{(+)}(q, q') = 2\varkappa f(q, q') \sum_{q_2\, q_2'} f(q_2, q_2') u_{q_2q_2'} Z^{i(+)}_{q_2q_2'} \equiv f(q, q') D_0. \tag{8.61'}$$

Substituting (8.61') in (8.63) and dividing by (8.63), we obtain the following secular equation:

$$1 = 2\varkappa \sum_{q,\, q'} \frac{f^2(q, q')\, u^2_{qq'}[\varepsilon(q)+\varepsilon(q')]}{[\varepsilon(q)+\varepsilon(q')]^2-\omega_i^2}. \tag{8.65}$$

The roots $\omega_1, \omega_2, \ldots,$ of (8.65) are the energies of the one-phonon states; they coincide with the solutions of (8.29') previously obtained in identical approximation by the selfconsistent field method.

Let us find the explicit form of the function $Z^{i(\pm)}_{qq'}$. The quantity $D^{(+)}(q, q')$ in the relation (8.61') is substituted in (8.62). Thus

$$Z^{i(+)}_{qq'} = D_0 \frac{f(q, q')\, u_{qq'}[\varepsilon(q)+\varepsilon(q')]}{[\varepsilon(q)+\varepsilon(q')]^2-\omega_i^2}.$$

The unknown constant D_0 is found from the normalization condition (8.49). Let us denote $D_0 = 2/Y^i$, then

$$Z^{i(+)}_{qq'} = \sqrt{\frac{2}{Y^i}}\, \frac{f(q, q')\, u_{qq'}[\varepsilon(q)+\varepsilon(q')]}{[\varepsilon(q)+\varepsilon(q')]^2-\omega_i^2}, \tag{8.66}$$

$$Z^{i(-)}_{qq'} = \sqrt{\frac{2}{Y^i}}\, \frac{f(q, q')\, u_{qq'}\omega_i}{[\varepsilon(q)+\varepsilon(q')]^2-\omega_i^2}, \tag{8.66'}$$

and

$$Y^i = \sum_{q,\, q'} \frac{f^2(q, q')\, u^2_{qq'}\omega_i[\varepsilon(q)+\varepsilon(q')]}{\{[\varepsilon(q)+\varepsilon(q')]^2-\omega_i^2\}^2}. \tag{8.67}$$

The original unknown amplitudes are found from (8.56). After simple manipulation,

$$\psi^i_{qq'} = \frac{1}{\sqrt{2Y^i}}\, \frac{f(q, q')u_{qq'}}{\varepsilon(q)+\varepsilon(q')-\omega_i}, \tag{8.68}$$

$$\varphi^i_{qq'} = \frac{1}{\sqrt{2Y^i}}\, \frac{f(q, q')u_{qq'}}{\varepsilon(q)+\varepsilon(q')+\omega_i}. \tag{8.68'}$$

Thus our problems is solved and the wave functions of the one-phonon excitations found.

Let us check if the quantity ω_i really determines the energy of the $Q_i^+\Psi$ state. In order to prove it we have to calculate the quantity

$$\langle Q_i H_1 Q_i^+\rangle - \langle H_1\rangle = \tfrac{1}{2}\sum_{q,q'}[\varepsilon(q)+\varepsilon(q')]\{(Z^{i(+)}_{q,q'})^2+(Z^{i(-)}_{q,q'})^2\} - \frac{\varkappa}{2}\Big(\sum_{q,q'} f(q,q')u_{qq'}Z^{i(+)}_{q,q'}\Big)^2 .$$

Substituting the expressions (8.66) and (8.66′) and using (8.65),

$$\begin{aligned}\langle Q_i H_1 Q_i^+\rangle - \langle H_1\rangle &= \frac{1}{2Y^i}\sum_{q,q'} f^2(q,q')\,u^2_{qq'}\,[\varepsilon(q)+\varepsilon(q')]\,\frac{[\varepsilon(q)+\varepsilon(q')]^2+\omega_i^2}{\{[\varepsilon(q)+\varepsilon(q')]^2-\omega_i^2\}^2}\\ &\quad -2\varkappa\sum_{q,q'} f^2(q,q')u^2_{qq'}\frac{\varepsilon(q)+\varepsilon(q')}{[\varepsilon(q)+\varepsilon(q')]^2-\omega_i^2}\times\\ &\quad \times\frac{1}{2Y^i}\sum_{q,q'} f^2(q,q')u^2_{qq'}\frac{\varepsilon(q)+\varepsilon(q')}{[\varepsilon(q)+\varepsilon(q')]^2-\omega_i^2}\\ &= \frac{1}{2Y^i}\sum_{q,q'} f^2(q,q')\,u^2_{qq'}[\varepsilon(q)+\varepsilon(q')]\times\\ &\quad \times\frac{[\varepsilon(q)+\varepsilon(q')]^2+\omega_i^2-[\varepsilon(q)+\varepsilon(q')]^2+\omega_i^2}{\{[\varepsilon(q)+\varepsilon(q')]^2-\omega_i^2\}^2} = \omega_i .\end{aligned}$$

Let us discuss some other characteristic features of the secular equation (8.65). They are related to the properties of the function $X(\omega)$ defined by

$$\frac{1}{\varkappa} = 2\sum_{q,q'}\frac{f^2(q,q')\,u^2_{qq'}[\varepsilon(q)+\varepsilon(q')]}{[\varepsilon(q)+\varepsilon(q')]^2-\omega^2} \equiv X(\omega). \tag{8.65′}$$

Figure 8.2 shows as an example the function $X(\omega)$ of the $K^\pi = 2^+$ state in ^{164}Dy. The function $X(\omega)$ has simple poles at the points $\omega = \varepsilon(q)+\varepsilon(q')$, i.e., at the values of the correspond-

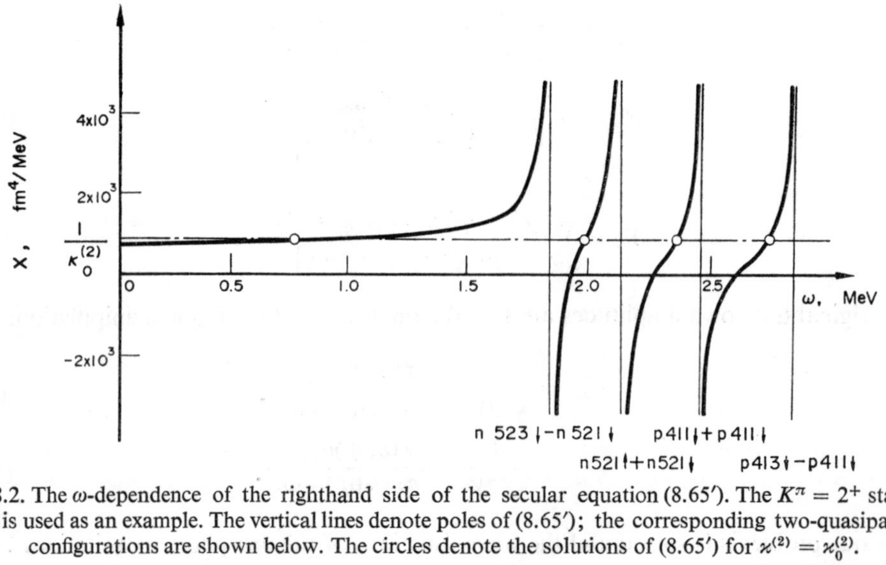

FIG. 8.2. The ω-dependence of the righthand side of the secular equation (8.65′). The $K^\pi = 2^+$ state in ^{164}Dy is used as an example. The vertical lines denote poles of (8.65′); the corresponding two-quasiparticle configurations are shown below. The circles denote the solutions of (8.65′) for $\varkappa^{(2)} = \varkappa_0^{(2)}$.

ing two-quasiparticle energies. The solutions $\omega_1, \omega_2, \ldots$, etc., are determined by the intersection of the horizontal line $1/\varkappa$ with the curve $X(\omega)$. The solution of (8.65′) with an energy below the first pole exists if

$$\varkappa < \varkappa_{\text{cr}} = \left[2 \sum_{q,\, q'} \frac{f^2(q, q')u_{qq'}^2}{\varepsilon(q)+\varepsilon(q')}\right]^{-1}. \tag{8.69}$$

If $\varkappa > \varkappa_{\text{cr}}$, the corresponding first solution is imaginary and the system is unstable. At higher energies there is always one solution between two subsequent poles.

When the one-phonon state energy is close to the pole value, the curve $X(\omega)$ and the horizontal line $1/\varkappa$ intersect at almost right angles. Such a situation corresponds to the third and fourth solutions in Fig. 8.2; the corresponding states are then only weakly collective.

The lowest solution of (8.65) is strongly collective if its energy is substantially lower than the first pole. The $1/\varkappa$ line and the $X(\omega)$ curve intersect at a very small angle (see the case in Fig. 8.2). The ω-value in such a case is very sensitive to the strength $\varkappa$; a small change in $\varkappa$ will change the ω_1 value appreciably.

In the extreme situation of the very collective states, i.e., of very small values of ω_1, the quasi-boson approximation is insufficiently accurate. In such a situation, the quasiparticle density $\langle \alpha_{q\sigma}^+ \alpha_{q\sigma} \rangle$ is not small and cannot be neglected. The same problem arises in the self-consistent field method. Interesting attempts to improve the quasi-boson approximation in this case were discussed in refs. 449 and 450.

4. Let us study the properties of the one-phonon state in two limiting cases: (1) if the solution ω_i is close to the corresponding pole and (2) if the solution ω_1 has much smaller value than the first pole.

The wave function of the one-phonon state has the form (8.45′), i.e.,

$$Q_{i_0}^+ \Psi = \tfrac{1}{2} \sum_{q,\, q'} \{\psi_{qq'}^{i_0} A^+(q, q') - \varphi_{qq'}^{i_0} A(q, q')\} \Psi.$$

Let us find the limit of the wave function $Q_i^+ \Psi$ for ω_i converging to the pole value $\varepsilon(q_1)+\varepsilon(q_2)$ with the corresponding matrix element $f(q_1, q_2)$:

$$\begin{aligned}
Q_{i_0}^+ \Psi \,|_{\varepsilon(q_1)+\varepsilon(q_2)-\omega_{i_0} \to 0} &= \left\{ \frac{1}{2\sqrt{2}} \frac{1}{\sqrt{Y^{i_0}}} \sum_{q,\, q'} \left[\frac{f(q, q')u_{qq'}}{\varepsilon(q)+\varepsilon(q')-\omega_{i_0}} A^+(q, q') \right.\right. \\
&\quad \left.\left. - \frac{f(q, q')u_{qq'}}{\varepsilon(q)+\varepsilon(q')+\omega_{i_0}} A(q, q') \right] \Psi \right\}_{\varepsilon(q_1)+\varepsilon(q_2)-\omega_{i_0} \to 0} \\
&= \frac{1}{2\sqrt{2}} \left\{ \left[2 \frac{f^2(q_1, q_2)\, u_{q_1 q_2}^2 \omega_{i_0} [\varepsilon(q_1)+\varepsilon(q_2)]}{\{[\varepsilon(q_1)+\varepsilon(q_2)]^2 - \omega_{i_0}^2\}^2} + \ldots \right]^{-1/2} \times \right. \\
&\quad \times \left[2 \frac{f(q_1, q_2)u_{q_1 q_2}}{\varepsilon(q_1)+\varepsilon(q_2)-\omega_{i_0}} A^+(q_1, q_2) - 2 \frac{f(q_1, q_2)u_{q_1 q_2}}{\varepsilon(q_1)+\varepsilon(q_2)+\omega_{i_0}} \times \right. \\
&\quad \left.\left. \times A(q_1, q_2) + \ldots \right] \Psi \right\}_{\varepsilon(q_1)+\varepsilon(q_2)-\omega_{i_0} \to 0} = \frac{1}{2} \frac{\varepsilon(q_1)+\varepsilon(q_2)+\omega_{i_0}}{\sqrt{\omega_{i_0}[\varepsilon(q_1)+\varepsilon(q_2)]}} \times \\
&\quad \times A^+(q_1, q_2)\, \Psi = A^+(q_1, q_2)\, \Psi = \frac{1}{\sqrt{2}} \left(\alpha_{q_1 -}^+ \alpha_{q_2 +}^+ - \alpha_{q_1 +}^+ \alpha_{q_2 -}^+ \right) \Psi.
\end{aligned} \tag{8.70}$$

Thus, when the phonon energy is close to a pole, the corresponding wave function is converging to the wave function of the corresponding two-quasiparticle state.

The method therefore describes the collective and two-quasiparticle states at the same time. The one-phonon states are simple superpositions of the two-quasiparticle states.

Let us consider the other limiting case when the energy of the first solution is very small, i.e.,

$$\omega_1 \ll \min\{\varepsilon(q)+\varepsilon(q')\}. \tag{8.71'}$$

Such a case corresponds to the adiabatical limit, when $\omega_{\text{vib}} \ll \omega_{\text{in}}$. This approximation forms the basis of the phenomenological description of the nuclear vibrations. The secular equation (8.65′) can be rewritten as

$$\frac{1}{\varkappa} = 2\sum_{q,q'} \frac{f^2(q,q')u^2_{qq'}}{\varepsilon(q)+\varepsilon(q')} \frac{1}{1-\dfrac{\omega_1^2}{[\varepsilon(q)+\varepsilon(q')]^2}}. \tag{8.65'}$$

Expanding the denominator and keeping only the first and second terms,

$$\frac{1}{\varkappa} = 2\sum_{q,q'} \frac{f^2(q,q')u^2_{qq'}}{\varepsilon(q)+\varepsilon(q')} + 2\omega_1^2 \sum_{q,q'} \frac{f^2(q,q')u^2_{qq'}}{[\varepsilon(q)+\varepsilon(q')]^3} + \dots .$$

Thus

$$\omega_1^2 = \frac{\dfrac{1}{\varkappa} - 2\sum\limits_{q,q'} \dfrac{f^2(q,q')u^2_{qq'}}{\varepsilon(q)+\varepsilon(q')}}{2\sum\limits_{q,q'} \dfrac{f^2(q,q')u^2_{qq'}}{[\varepsilon(q)+\varepsilon(q')]^3}}. \tag{8.72}$$

Thus the equations of the approximate second quantization method contain the formulae of the adiabatic approximation as a limiting case. (Remember that in Chapter 2 we have shown the following relation for the adiabatic approximation):

$$\omega_1 = \sqrt{C/B}\,.$$

5. Let us use the secular equation (8.65) and transform the interaction Hamiltonian. We remember from (8.57) that

$$H_1 = \sum_q \varepsilon(q)\, B(q,q) - \frac{\varkappa}{4} \sum_{q_1, q_1'} \sum_{q_2', q_2'} \sum_{i,i'} f(q_1, q_1') f(q_2, q_2')\, u_{q_1q_1'} u_{q_2q_2'} \times$$
$$\times Z^{i(+)}_{q_1q_1'} Z^{i'(+)}_{q_2q_2'} (Q_i^+ Q_{i'} + Q_{i'}^+ Q_i + Q_i Q_{i'} + Q_i^+ Q_{i'}^+). \tag{8.73}$$

It is necessary to prove that the Hamiltonian H_1 does not contain terms corresponding to the creation or annihilation of two phonons from the phonon vacuum.

When calculating (8.58) we have used the following commutation relation:

$$[B(q,q), Q_i^+] = \sum_{i',q'} \{(\psi^i_{qq'}\psi^{i'}_{qq'} + \varphi^i_{qq'}\varphi^{i'}{}_{q'})\, Q^+_{i'} + (\psi^i_{qq'}\varphi^{i'}_{qq'} + \varphi^i_{qq'}\psi^{i'}_{qq'}) Q_{i'}\}. \tag{8.74}$$

Consequently, the operator $B(q,q)$ can be expressed in terms of the phonon operators as

$$B(q,q) = \sum_{i,i',q'} \{(\psi^i_{qq'}\psi^{i'}_{qq'} + \varphi^i_{qq'}\varphi^{i'}_{qq'})\, Q_i^+ Q_{i'} + \tfrac{1}{2}(\psi^i_{qq'}\varphi^{i'}_{qq'} + \varphi^{i'}_{qq'}\psi^i_{qq'})\,(Q_i Q_{i'} + Q^+_{i'} Q^+_i)\}. \tag{8.75}$$

This formula is exact if the corresponding averaging is performed over the one-phonon states. Therefore the expression $\sum_q \varepsilon(q)\,B(q,q)$ in the Hamiltonian may be replaced by

$$\sum_q \varepsilon(q)B(q,q) = \sum_{i,i'}\sum_{q,q'}[\varepsilon(q)+\varepsilon(q')]\{\tfrac{1}{2}(\psi^i_{qq'}\psi^{i'}_{qq'}+\varphi^i_{qq'}\varphi^{i'}_{qq'})Q^+_iQ_{i'} + \tfrac{1}{4}(\psi^i_{qq'}\varphi^{i'}_{qq'}+\psi^{i'}_{qq'}\varphi^i_{qq'})(Q_iQ_{i'}+Q^+_{i'}Q^+_i)\}. \quad (8.75')$$

The second term in (8.73) can be also transformed. We shall substitute $Z^{i(+)}_{q,q'}$ from (8.66) and obtain

$$-\frac{\varkappa}{2}\sum\frac{1}{\sqrt{Y^iY^{i'}}}\sum_{q,q'}\frac{f^2(q,q')\,u^2_{qq'}[\varepsilon(q)+\varepsilon(q')]}{[\varepsilon(q)+\varepsilon(q')]^2-\omega_i^2}\sum_{q_2,q_2'}\frac{f^2(q_2,q_2')\,u^2_{q_2q_2'}[\varepsilon(q_2)+\varepsilon(q_2')]}{[\varepsilon(q_2)+\varepsilon(q_2')]^2-\omega_{i'}^2}\times$$
$$\times(Q^+_iQ_{i'}+Q^+_{i'}Q_i+Q_iQ_{i'}+Q^+_{i'}Q^+_i).$$

Using (8.65) we obtain the final form of this term, i.e.,

$$-\frac{1}{8\varkappa}\sum_{i,i'}\frac{1}{\sqrt{Y^iY^{i'}}}(Q^+_iQ_{i'}+Q^+_{i'}Q_i+Q_iQ_{i'}+Q^+_{i'}Q^+_i).$$

Thus, the Hamiltonian H_1 can be written in the form

$$H_1 = \sum_{i,i'}\sum_{q,q'}[\varepsilon(q)+\varepsilon(q')]\{\tfrac{1}{2}(\psi^i_{qq'}\psi^{i'}_{qq'}+\varphi^i_{qq'}\varphi^{i'}_{qq'})Q^+_iQ_{i'}+\tfrac{1}{4}(\psi^i_{qq'}\varphi^{i'}_{qq'}+\psi^{i'}_{qq'}\varphi^i_{qq'})\times$$
$$\times(Q_iQ_{i'}+Q^+_{i'}Q^+_i)\}-\frac{1}{8\varkappa}\sum_{i,i}\frac{1}{\sqrt{Y^iY^{i'}}}(Q^+_iQ_{i'}+Q^+_{i'}Q_i+Q_iQ_{i'}+Q^+_{i'}Q^+_i).$$

Let us find the explicit form of the corresponding coefficients. The relation (8.49′) can be rewritten as

$$\sum_{q,q'}(\psi^i_{qq'}\varphi^{i'}_{qq'}-\psi^{i'}_{qq'}\varphi^i_{qq'}) = \frac{\omega_i-\omega_{i'}}{\sqrt{Y^iY^{i'}}}\sum_{q,q'}\frac{f^2(q,q')\,u^2_{qq'}[\varepsilon(q)+\varepsilon(q')]}{\{[\varepsilon(q)+\varepsilon(q')]^2-\omega_i^2\}\{[\varepsilon(q)+\varepsilon(q')]^2-\omega_{i'}^2\}} = 0.$$

The coefficient at $Q^+_iQ_{i'}$ equals

$$\tfrac{1}{2}\sum_{q,q'}[\varepsilon(q)+\varepsilon(q')](\psi^i_{qq'}\psi^{i'}_{qq'}+\varphi^i_{qq'}\varphi^{i'}_{qq'})-\frac{1}{4\varkappa}\frac{1}{\sqrt{Y^iY^{i'}}}$$
$$= \frac{1}{4}\frac{1}{\sqrt{Y^iY^{i'}}}\left\{\sum_{q,q'}\frac{f^2(q,q')\,u^2_{qq'}[\varepsilon(q)+\varepsilon(q')]\,2\{[\varepsilon(q)+\varepsilon(q')]^2+\omega_i\omega_{i'}\}}{\{[\varepsilon(q)+\varepsilon(q')]^2-\omega_i^2\}\{[\varepsilon(q)+\varepsilon(q')]^2-\omega_{i'}^2\}}-\frac{1}{\varkappa}\right\}$$
$$= \frac{1}{4}\frac{(\omega_i+\omega_{i'})^2}{\sqrt{Y^iY^{i'}}}\sum_{q,q'}\frac{f^2(q,q')\,u^2_{qq'}[\varepsilon(q)+\varepsilon(q')]}{\{[\varepsilon(q)+\varepsilon(q')]^2-\omega_i^2\}\{[\varepsilon(q)+\varepsilon(q')]^2-\omega_{i'}^2\}}.$$

This expression vanishes if $i\neq i'$, and it equals ω_i for $i=i'$. The coefficient of $(Q_iQ_{i'}+Q^+_{i'}Q^+_i)$ equals

$$\tfrac{1}{4}\sum_{q,q'}[\varepsilon(q)+\varepsilon(q')](\psi^i_{qq'}\varphi^{i'}_{qq'}+\psi^{i'}_{qq'}\varphi^i_{qq'})-\frac{1}{8\varkappa}\frac{1}{\sqrt{Y^iY^{i'}}}$$
$$= \frac{1}{8}\frac{1}{\sqrt{Y^iY^{i'}}}\left\{2\sum_{b,q'}\frac{f^2(q,q')\,u^2_{qq'}[\varepsilon(q)+\varepsilon(q')]\{[\varepsilon(q)+\varepsilon(q')]^2-\omega_i\omega_{i'}\}}{\{[\varepsilon(q)+\varepsilon(q')]^2-\omega_i^2\}\{[\varepsilon(q)+\varepsilon(q')]^2-\omega_{i'}^2\}}-\frac{1}{\varkappa}\right\}$$
$$= \frac{1}{8}\frac{(\omega_i-\omega_{i'})^2}{\sqrt{Y^iY^{i'}}}\sum_{q,q'}\frac{f^2(q,q')\,u^2_{qq'}[\varepsilon(q)+\varepsilon(q')]}{\{[\varepsilon(q)+\varepsilon(q')]^2-\omega_i^2\}\{[\varepsilon(q)+\varepsilon(q')]^2-\omega_{i'}^2\}} = 0.$$

Therefore the phonon vacuum cannot create or annihilate a phonon pair.

Thus, finally, the Hamiltonian H_1 was brought to the form

$$H_1 = \sum_i \omega_i Q_i^+ Q_i, \tag{8.76}$$

i.e., it describes the independent phonons. (Note that (8.76) actually describes the excitation energy, the constant term $\langle H_1 \rangle$ has been subtracted.)

6. We have mentioned already earlier that the analogous secular equations and one-phonon excited states can be obtained in the framework of the Tamm–Dancoff (TD) method. In such a method, the ground state of an even–even system is simply the quasi-particle vacuum with the obvious property

$$A(q, q')\Psi_0 = 0.$$

The phonon operator is then

$$\bar{Q}_i = \frac{1}{2} \sum_{q, q'} \psi^i_{qq'} A(q, q') \tag{8.77}$$

and the excited one-phonon states are described by the wave function $\bar{Q}_i^+ \Psi_0$.

The equations of the TD method may be obtained from the more general equations of the random phase approximation by assuming that

$$\varphi^i_{qq'} = 0. \tag{8.78}$$

Nevertheless, we shall proceed and derive the TD secular equation directly. The linearization method, often used in solving similar problems, will be used. The Hamiltonian H_1 has the form

$$H_1 = \sum_q \varepsilon(q)\, B(q, q) - \frac{\varkappa}{4} \left(\sum_{q, q'} f(q, q')\, u_{qq'} [A(q, q') + A^+(q, q')] \right)^2 . \tag{8.79}$$

Let $\mathcal{E}_0$ be the ground state energy and $\mathcal{E}_0 + \omega_i$ the excited state energy. Then

$$H_1 \Psi_0 = \mathcal{E}_0 \Psi_0,$$
$$H_1 \bar{Q}_i^+ \Psi_0 = (\mathcal{E}_0 + \omega_i) \bar{Q}_i^+ \Psi_0.$$

We shall subtract

$$\bar{Q}_i^+ H_1 \Psi_0 = \mathcal{E}_0 \bar{Q}_i^+ \Psi_0$$

from the above-stated expressions and obtain

$$(H_1 \bar{Q}_i^+ - \bar{Q}_i^+ H_1)\Psi_0 = \omega_i \bar{Q}_i^+ \Psi_0, \tag{8.80}$$

or, in other words, we have obtained the operator equation

$$[H_1, \bar{Q}_i^+] = \omega_i \bar{Q}_i^+ . \tag{8.80'}$$

By calculating the commutator and comparing the coefficients at $A^+(q, q')$ we obtain the following equation:

$$[\varepsilon(q) + \varepsilon(q') - \omega_i]\, \psi^i_{qq'} = \varkappa f(q, q') u_{qq'} \sum_{q_2, q_2'} f(q_2, q_2') u_{q_2 q_2'} \psi^i_{q_2 q_2'} . \tag{8.80''}$$

Defining the constant D^i by

$$D^i = \varkappa \sum_{q_2, q_2'} f(q_2, q_2')u_{q_2q_2'}\psi^i_{q_2q_2'}, \tag{8.81}$$

we find that

$$\psi^i_{qq'} = D^i \frac{f(q, q')u_{qq'}}{\varepsilon(q)+\varepsilon(q')-\omega_i},$$

and substituting $\psi^i_{qq'}$ in (8.81) we can finally write the secular equation

$$1 = \varkappa \sum_{q, q'} \frac{f^2(q, q')u^2_{qq'}}{\varepsilon(q)+\varepsilon(q')-\omega_i}. \tag{8.82}$$

The normalization condition has the form

$$\sum_{q, q'} \psi^i_{qq'}\psi^{i'}_{qq'} = 2\delta_{ii'},$$

and consequently

$$(D^i)^{-2} = \frac{1}{2} \sum_{q, q'} \frac{f^2(q, q')\, u^2_{qq'}}{[\varepsilon(q)+\varepsilon(q')-\omega_i]^2}. \tag{8.81'}$$

The energies of the one-phonon states are the solutions $\omega_1, \omega_2, \ldots,$ of the secular equation (8.82). The equation has a solution with an energy below the first pole if

$$\varkappa \sum_{q, q'} \frac{f^2(q, q')u^2_{qq'}}{\varepsilon(q)+\varepsilon(q')} < 1.$$

Other solutions are always sandwiched between the corresponding poles of (8.82).

The wave function of the one-phonon state has the form

$$Q_i^+\Psi_0 = \frac{1}{2\sqrt{D^i}} \sum_{q, q'} \frac{f(q, q')u_{qq'}}{\varepsilon(q)+\varepsilon(q')-\omega_i} A^+(q, q')\Psi_0. \tag{8.83}$$

It is easy to show[451] that the solutions of (8.65) and (8.82) practically coincide for small $\varkappa$, i.e., when ω_i is close to the corresponding pole.

§ 3. The Green Function Method†

1. The Green function method is one of the most powerful tools in quantum field theory and in statistical physics (see, for example, refs. 28, 124, 228, and 452–463). The method is also often used in nuclear theory (e.g., in refs. 13, 22, 154–156, and 252). The Green function in many-body theory has somewhat different meaning than the same term used in mathematical theory of linear differential equations. Although the Green function satisfies an equation containing the δ-function on the righthand side, in the many-body problem such an equation may be nonlinear. The Green function, by its physical meaning, is a generalization of the correlation function.

† Th section may be omitted at the first reading.

The Green functions are used in nuclear theory in studies of the nuclear matter, in calculations of the nuclear excitation energies and transition probabilities, and in other similar problems. The zero temperature Green functions are, as a rule, used in nuclear theory.

In this section, we shall first define the Green functions and find the corresponding determining equations. Later, we shall discuss their application to the system of noninteracting fermions and describe the application of the Green functions in studies of the pairing superfluid correlations and in calculations of the nuclear vibrational states. We shall quote only the most important results directly related to the description of the ground and low-lying nuclear states. More detailed description of the Green function theory can be found in special publications, for example, in refs. 13 and 458–460. The diagram technique described in detail by Migdal(13) will be omitted completely.

Let us define the two-time-point Green function. We shall consider three types of the Green functions: the causal function $G_c(t, t')$, the retarded function $G_r(t, t')$, and the advanced function $G_a(t, t')$. They are defined by the relations:

$$G_c(t, t') \equiv \langle\langle F_1(t) \mid F_2(t')\rangle\rangle_c = -i\langle T\{F_1(t), F_2(t')\}\rangle, \tag{8.84}$$

$$G_r(t, t') \equiv \langle\langle F_1(t) \mid F_2(t')\rangle\rangle_r = -i\theta(t-t')\langle [F_1(t), F_2(t')]\rangle, \tag{8.85}$$

$$G_a(t, t') \equiv \langle\langle F_1(t) \mid F_2(t')\rangle\rangle_a = i\theta(t'-t)\langle [F_1(t), F_2(t')]\rangle, \tag{8.86}$$

where $\langle \ldots \rangle$ denotes the expectation value in some state. The symbols $F_1(t)$, $F_2(t)$ denote operators in Heisenberg representation; these operators obey the commutation relation

$$[F_1, F_2] = F_1F_2 - \eta F_2F_1, \quad \eta = \pm 1.$$

The quantity T is the chronological operator. It orders the operators from left to right according to the decreasing value of the time variable, i.e.,

$$T\{F_1(t)\, F_2(t')\} = \theta(t-t')\, F_1(t)\, F_2(t') + \eta\theta(t'-t)\, F_2(t')\, F_1(t).$$

Finally, $\theta(t)$ is the step function

$$\theta(t) = \begin{cases} 1, & \text{if} \quad t > 0, \\ 0, & \text{if} \quad t < 0, \end{cases}$$

or, in the integral representation,

$$\theta(t) = \int_{-\infty}^{t} \delta(t')\, dt'.$$

The Green functions (8.84)–(8.86) can be rewritten as

$$G_c(t, t') = -i\theta(t-t')\langle F_1(t)\, F_2(t')\rangle - i\eta\theta(t'-t)\langle F_2(t')\, F_1(t)\rangle, \tag{8.84'}$$

$$G_r(t, t') = -i\theta(t-t')\{\langle F_1(t)\, F_2(t')\rangle - \eta\langle F_2(t')\, F_1(t)\rangle\}, \tag{8.85'}$$

$$G_a(t, t') = i\theta(t'-t)\{\langle F_1(t)\, F_2(t')\rangle - \eta\langle F_2(t')\, F_1(t)\rangle\}. \tag{8.86'}$$

In the absence of external field, the Green functions depend only on the time difference $t-t'$.

Let us discuss several examples. The one-particle Green function is defined by

$$G_c(p, t) = -i\langle 0| T\{a_p(t)\, a_p^+(0)\}\, |0\rangle. \tag{8.87}$$

The quantity $|0\rangle$ denotes the ground state of the system; thus $H|0\rangle = \mathscr{E}_0|0\rangle$. In the Heisenberg representation, $a_p(t) = e^{iHt}a_p(0)e^{-iHt}$, and, therefore,

$$\left.\begin{aligned} G_c(p, t) &= -i\langle 0|\, a_p e^{-iHt} a_p^+ \,|0\rangle e^{i\mathscr{E}_0 t} \qquad t > 0,\\ G_c(p, t) &= i\langle 0|\, a_p^+ e^{iHt} a_p \,|0\rangle e^{-i\mathscr{E}_0 t} \qquad t < 0. \end{aligned}\right\} \tag{8.87'}$$

The Fourier transform $G(p, t)$ is defined in the usual way, i.e.,

$$G(p, t) = \frac{1}{2\pi}\int_{-\infty}^{\infty} dE\, G(p, E)e^{-iEt}. \tag{8.88}$$

Let us find the Green function of a free particle. The $|0\rangle$ state is the vacuum state, and from (8.87′) it follows that

$$G(p, t) = \begin{cases} -ie^{-iT(p)t} & \text{for} \quad t > 0,\\ 0 & \text{for} \quad t < 0, \end{cases} \tag{8.89}$$

where $T(p) = p^2/2m$ is the kinetic energy. The Fourier transform of $G(p, t)$ equals

$$\begin{aligned} G(p, E) &= \int_{-\infty}^{\infty} dt e^{iEt} G(p, t) = (-i)\int_0^{\infty} e^{iEt - iT(p)t - \varepsilon t}\, dt\\ &= (-1)\frac{e^{-\varepsilon t}}{E - T(p) + i\varepsilon}\bigg|_0^{\infty} = \frac{1}{E - T(p) + i\varepsilon}; \end{aligned} \tag{8.90}$$

the infinitesimal quantity ε determines the integration path. Consequently

$$G(p, t) = \frac{1}{2\pi}\int_{-\infty}^{\infty} dE \frac{e^{-iEt}}{E - T(p) + i\varepsilon}; \tag{8.89'}$$

The integration contour is shown in Fig. 8.3. At $t < 0$ the contour must be closed in the upper half-plane and the Green function vanishes. For $t > 0$ the contour is closed in the lower half-plane. The corresponding residuum equals $-2\pi i e^{-iT(p)t}$ and we obtain the result (8.89).

We shall often use in calculations with the Green function the following formula:

$$\int_{-\infty}^{\infty} \mathscr{F}(t) \frac{dt}{t + i\varepsilon} = \int \mathscr{F}(t)\frac{dt}{t} - i\pi\mathscr{F}(0), \tag{8.91}$$

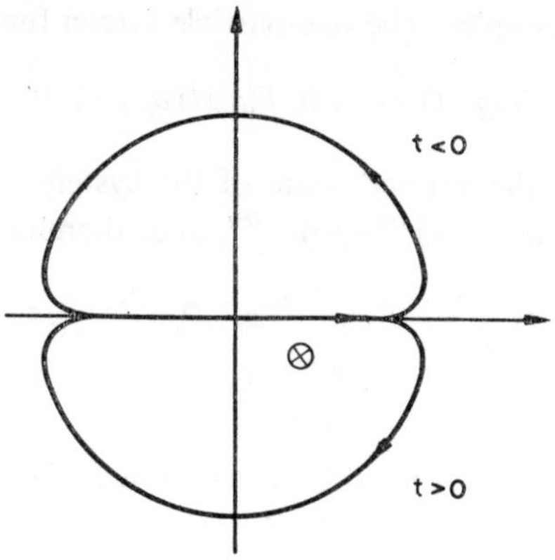

FIG. 8.3. The integration path. $\otimes$ denotes the pole $T(p)-i\varepsilon$.

where $\mathcal{F}(t)$ is an arbitrary function without singularities, and $⨍$ denotes the principal value of the integral. Equation (8.91) can be rewritten in the symbolical form as

$$\frac{1}{t+i\varepsilon}=\frac{1}{t}-i\pi\delta(t). \tag{8.91'}$$

Let us consider a system of noninteracting fermions. The ground state of such a system obeys the following conditions:

$$\begin{aligned} n_p &= 1, \quad \text{if} \quad p<p_F, \qquad a_p^+ |0\rangle = 0 \quad \text{when} \quad p<p_F,\\ n_p &= 0, \quad \text{if} \quad p>p_F, \qquad a_p |0\rangle = 0 \quad \text{when} \quad p>p_F, \end{aligned}$$

where p_F is the Fermi momentum. Besides, the following equations are also valid:

$$\begin{aligned} Ha_p^+ |0\rangle &= (E(p)+\mathcal{E}_0)a_p^+ |0\rangle,\\ Ha_p |0\rangle &= (\mathcal{E}_0-E(p))a_p |0\rangle. \end{aligned}$$

Using (8.87'), we obtain the one-particle Green function

$$G(p,t)=\begin{cases} -i(1-n_p)e^{-iE(p)t} & \text{if} \quad t>0,\\ in_pe^{-iE(p)t} & \text{if} \quad t<0, \end{cases} \tag{8.92}$$

with the Fourier transform

$$G(p,E)=\int_{-\infty}^{\infty} dtG(p,t)e^{iEt}=i\int_{-\infty}^{0} n_pe^{it(E-E(p)-i\varepsilon)}\,dt$$

$$-i(1-n_p)\int_{0}^{\infty} e^{it(E-E(p)+i\varepsilon)}\,dt=\frac{n_p}{E-E(p)-i\varepsilon}+\frac{1-n_p}{E-E(p)+i\varepsilon}.$$

The two above-stated terms can be combined and the one-particle Green function in the noninteracting system will have the form

$$G(p,E)=\frac{1}{E-E(p)+i\varepsilon\,\mathrm{sign}(|p_F|-p)}, \tag{8.92'}$$

where sign $(|p_F|-p)$ means the sign of the quantity $(|p_F|-p)$. The single-particle energies coincide with the poles of the Green function. The propagation of a hole is described by $G(p, F)$ with $t < 0$.

2. Let us find the equation determining the two-time-point Green function. The equation of motion of an operator $F(t)$ has the form

$$i\frac{dF(t)}{dt} = FH - HF. \tag{8.93}$$

By differentiating the Green functions (8.84), (8.85), and (8.86) over t, we obtain

$$i\frac{d}{dt}\langle\langle F_1(t)\,|\,F_2(t')\rangle\rangle = \frac{d\theta(t-t')}{dt}\langle[F_1(t), F_2(t')]\rangle + \left\langle\left\langle i\frac{dF_1(t)}{dt}\,\middle|\,F_2(t')\right\rangle\right\rangle.$$

Such an equation has the same form for the causal, retarded, and advanced Green functions because

$$\frac{d}{dt}\theta(-t) = -\frac{d}{dt}\theta(t).$$

After simple manipulations,

$$i\frac{d}{dt}\langle\langle F_1(t)\,|\,F_2(t')\rangle\rangle = \delta(t-t')\langle[F_1(t), F_2(t')]\rangle + \langle\langle\{F_1(t)H - HF_1(t)\}\,|\,F_2(t')\rangle\rangle. \tag{8.94}$$

The righthand side of (8.94) contains the Green functions of higher order than the original equations. These functions can be also described by the equations similar to (8.94). Thus we shall obtain an infinite system of coupled equations for the Green functions. In order to obtain a solvable system we have to break the chain of equations by using some appropriate approximation.

The equations determining the Green functions can be brought to a closed form, to the so-called Dyson equation. The Dyson equation has the form

$$G(x_1, x_2) = G_0(x_1, x_2) + \int dx_3\, dx_4 G_0(x_1, x_3)\,\Sigma(x_3, x_4)\,G(x_4, x_2), \tag{8.95}$$

where $G_0(x_1, x_2)$ denotes the Green function of the free particle and $\Sigma(x, x')$ is called either "self-energy term" or "mass operator." The self-energy term can be expressed in terms of the functional derivatives or in terms of the vertex functions. The diagram technique is particularly suitable for handling (8.95).

Let us find the spectral representation of the time-correlation functions and of the advanced Green functions. Let ψ_n, E_n denote the eigenfunctions and eigenvalues of the operator $H-\lambda N$. The step function $\theta(t)$ is represented by

$$\theta(t) = \int_{-\infty}^{t} e^{\varepsilon t}\,\delta(t)\,dt \quad \text{for} \quad \varepsilon \to 0, \quad \varepsilon > 0.$$

Using the δ-function of the form

$$\delta(t) = \frac{1}{2\pi}\int_{-\infty}^{\infty} e^{-ixt}\,dx,$$

we obtain

$$\theta(t) = \frac{i}{2\pi} \int_{-\infty}^{\infty} \frac{e^{-ixt}}{x+i\varepsilon}\, dx. \tag{8.96}$$

Let us consider the correlation functions $\langle F_1(t)\, F_2(t')\rangle$ and $\langle F_2(t')\, F_1(t)\rangle$ averaged over the ground state with the energy E_0. Using the completeness of the system ψ_n,

$$\langle F_1(t)\, F_2(t')\rangle = \sum_n (\psi_0^* F_1(t)\psi_n)\, (\psi_n^* F_2(t')\psi_0) = \sum_n (\psi_0^* F_1(0)\psi_n)\, (\psi_n^* F_2(0)\psi_0) e^{-i(E_n-E_0)\,(t-t')}$$

$$= \int_0^{\infty} d\omega e^{-i\omega(t-t')} \sum_n (\psi_0^* F_1(0)\psi_n)\, (\psi_n^* F_2(0)\psi_0)\, \delta(E_n - E_0 - \omega);$$

$$\langle F_2(t')\, F_1(t)\rangle = \sum_n (\psi_0^* F_2(0)\psi_n)\, (\psi_n^* F_1(0)\psi_0) e^{-i(E_0-E_n)\,(t-t')}$$

$$= \int_{-\infty}^{0} d\omega e^{-i\omega(t-t')} \sum_n (\psi_0^* F_2(0)\psi_n)\, (\psi^* F_1(0)\psi_0)\, \delta(E_0 - E_n - \omega).$$

The spectral distribution function is defined by

$$I(\omega) = \sum_n (\psi_0^* F_1(0)\psi_n)\, (\psi_n^* F_2(0)\psi_0)\, \delta(E_n - E_0 - \omega) + \sum_n (\psi_0^* F_2(0)\psi_n)\, (\psi_0^* F_1(0)\psi_0)\, \delta(E_0 - E_n - \omega) \tag{8.97}$$

and the correlation functions are related to $I(\omega)$ by the equations

$$\langle F_1(t)\, F_2(t')\rangle = \int_0^{\infty} d\omega e^{-i\omega(t-t')} I(\omega) \tag{8.98}$$

and

$$\langle F_2(t')\, F_1(t)\rangle = \int_{-\infty}^{0} d\omega e^{-i\omega(t-t')} I(\omega). \tag{8.98'}$$

What is the spectral representation of the retarded and advanced Green functions? The Fourier transform of the retarded function is defined as

$$G_r(t-t') = \int_{-\infty}^{\infty} G_r(E) e^{-iE(t-t')}\, dE. \tag{8.99}$$

The quantity $G_r(E)$ equals

$$G_r(E) = \frac{1}{2\pi} \int_{-\infty}^{\infty} G_r(t-t')\, dt e^{iE(t-t')}$$

$$= \frac{1}{2\pi i} \int_{-\infty}^{\infty} dt e^{iE(t-t')} \theta(t-t') \{\langle F_1(t)\, F_2(t')\rangle - \eta \langle F_2(t')\, F_1(t)\rangle\}$$

$$= \frac{1}{2\pi i} \int_{-\infty}^{\infty} I(\omega)\, d\omega \{\theta(+\omega) - \eta\theta(-\omega)\} \int_{-\infty}^{\infty} dt \theta(t) e^{i(E-\omega)t}.$$

Using (8.96) we find that

$$\frac{1}{2\pi i}\int_{-\infty}^{\infty} dt\theta(t)e^{i(E-\omega)t} = \frac{1}{2\pi}\,\frac{1}{E-\omega+i\varepsilon},$$

and, consequently,

$$G_r(E) = \frac{1}{2\pi}\int_{-\infty}^{\infty} I(\omega)\,d\omega\,\frac{\theta(+\omega)-\eta\theta(-\omega)}{E-\omega+i\varepsilon}. \tag{8.100}$$

Similarly,

$$G_a(E) = \frac{1}{2\pi}\int_{-\infty}^{\infty} I(\omega)\,d\omega\,\frac{\theta(+\omega)-\eta\theta(-\omega)}{E-\omega-i\varepsilon}. \tag{8.100'}$$

When the operators F_1 and F_2 obey the Fermi statistics, i.e., $\eta = -1$, we write

$$G_r(E) = \frac{1}{2\pi}\int_{-\infty}^{\infty} d\omega\,\frac{I(\omega)}{E-\omega+i\varepsilon}, \tag{8.101}$$

$$G_a(E) = \frac{1}{2\pi}\int_{-\infty}^{\infty} d\omega\,\frac{I(\omega)}{E-\omega-i\varepsilon}. \tag{8.101'}$$

The δ-function can be represented by

$$\delta(x) = \frac{1}{2\pi i}\left(\frac{1}{x-i\varepsilon}+\frac{1}{x+i\varepsilon}\right) \tag{8.102}$$

and, therefore,

$$G_a(E)-G_r(E) = i\int_{-\infty}^{\infty} I(\omega)\,\delta(E-\omega)\,d\omega = iI(E). \tag{8.103}$$

Thus, when the Green functions are known it is easy to use the spectral representation and to find the correlation functions $\langle F_1F_2\rangle$.

3. Let us rederive the basic equations of the pairing theory using the Green function method. To obtain closed system of equations we shall utilize the procedure described in § 2 of Chapter 5. The auxiliary Hamiltonian H' is defined by

$$H' = \sum_{q\sigma} T(q)a^+_{q\sigma}a_{q\sigma} - G\sum_{q,\,q'}\{\Phi^*(q)a_{q'-}a_{q'+}+a^+_{q+}a^+_{q-}\Phi(q')-\Phi^*(q)\Phi(q)\}, \tag{8.104}$$

where $\Phi(q) = \langle a_{q-}a_{q+}\rangle$.

The Green function in this particular case satisfies the equation

$$i\frac{d}{dt}\langle\langle a_{q\sigma}(t)\,|\,a^+_{q\sigma}(t')\rangle\rangle = \delta(t-t')+T(q)\langle\langle a_{q\sigma}(t)\,|\,a^+_{q\sigma}(t')\rangle\rangle - G\sum_{q'}\Phi(q')\,\sigma\langle\langle a^+_{q,\,-\sigma}(t)\,|\,a^+_{q\sigma}(t')\rangle\rangle. \tag{8.105}$$

The system of equations will be closed by using the equation for $\langle\langle a^+_{q,-\sigma}(t)\,|\,a^+_{q\sigma}(t')\rangle\rangle$, i.e.,

$$i\frac{d}{dt}\langle\langle a^+_{q,-\sigma}(t)\,|\,a^+_{q\sigma}(t')\rangle\rangle = -T(q)\langle\langle a^+_{q,-\sigma}(t)\,|\,a^+_{q\sigma}(t')\rangle\rangle - G\sum_{q'}\Phi^*(q')\,\sigma\langle\langle a_{q\sigma}(t)\,|\,a^+_{q\sigma}(t')\rangle\rangle. \tag{8.106}$$

Using the Fourier transforms and the shorthand notation

$$C = G\sum_{q'}\Phi(q') = G\sum_{q'}\Phi^*(q'),$$

we obtain

$$(E-T(q))\langle\langle a_{q\sigma}\,|\,a^+_{q\sigma}\rangle\rangle_E = \frac{1}{2\pi} - C\sigma\langle\langle a^+_{q,-\sigma}\,|\,a^+_{q\sigma}\rangle\rangle_E, \tag{8.105'}$$

$$(E+T(q))\langle\langle a^+_{q,-\sigma}\,|\,a^+_{q\sigma}\rangle\rangle_E = -C\sigma\langle\langle a_{q\sigma}\,|\,a^+_{q\sigma}\rangle\rangle_E. \tag{8.106'}$$

After simple manipulations we can write

$$\langle\langle a_{q\sigma}\,|\,a^+_{q\sigma}\rangle\rangle_{E\pm i\varepsilon} = \frac{1}{2\pi}\left\{\frac{u_q^2}{E-\varepsilon(q)\pm i\varepsilon} + \frac{v_q^2}{E+\varepsilon(q)\pm i\varepsilon}\right\}, \tag{8.107}$$

$$\langle\langle a^+_{q,-\sigma}\,|\,a^+_{q\sigma}\rangle\rangle_{E\pm i\varepsilon} = -\frac{\sigma C}{4\pi\varepsilon(q)}\left\{\frac{1}{E-\varepsilon(q)\pm i\varepsilon} - \frac{1}{E+\varepsilon(q)\pm i\varepsilon}\right\}, \tag{8.108}$$

where

$$\varepsilon(q) = \sqrt{C^2+T^2(q)}, \quad u_q^2 = \frac{1}{2}\left\{1+\frac{T(q)}{\varepsilon(q)}\right\}, \quad v_q^2 = \frac{1}{2}\left\{1-\frac{T(q)}{\varepsilon(q)}\right\}.$$

The spectral function $I(E)$ is obtained from (8.102) and (8.103), namely,

$$I_1(E) = -i\langle\langle a_{q\sigma}\,|\,a^+_{q\sigma}\rangle\rangle_{E-i\varepsilon} + i\langle\langle a_{q\sigma}\,|\,a^+_{q\sigma}\rangle\rangle_{E+i\varepsilon} = u_q^2\delta(E-\varepsilon(q)) + v_q^2\delta(E+\varepsilon(q)), \tag{8.109}$$

$$I_2(E) = -i\langle\langle a^+_{q,-\sigma}\,|\,a^+_{q\sigma}\rangle\rangle_{E-i\varepsilon} + i\langle\langle a^+_{q,-\sigma}\,|\,a^+_{q\sigma}\rangle\rangle_{E+i\varepsilon} = -\frac{\sigma}{2}\frac{C}{\varepsilon(q)}\{\delta(E-\varepsilon(q)) - \delta(E+\varepsilon(q))\}. \tag{8.109'}$$

Now it is easy to use (8.98) and find the correlations functions. At $t = t' = 0$

$$\langle a_{q\sigma}a^+_{q\sigma}\rangle = \int_0^\infty I_1(E)\,dE = u_q^2, \tag{8.110}$$

$$\langle a^+_{q\sigma}a_{q\sigma}\rangle = 1 - \langle a_{q\sigma}a^+_{q\sigma}\rangle = v_q^2, \tag{8.110'}$$

$$\langle a^+_{q,-\sigma}a^+_{q\sigma}\rangle = \int_0^\infty I_2(E)\,dE = -\frac{\sigma}{2}\frac{C}{\varepsilon(q)}. \tag{8.111}$$

Let us use the definition of C and the formula (8.111):

$$C = G\sum_q \Phi^*(q) = G\sum_q \langle a^+_{q+}a^+_{q-}\rangle = \frac{G}{2}\sum_q \frac{C}{\varepsilon(q)}.$$

Consequently,

$$1 = \frac{G}{2}\sum_q \frac{1}{\varepsilon(q)}. \tag{8.112}$$

The total number of particles equals

$$N = \sum_{q\sigma} \langle a^+_{q\sigma} a_{q\sigma} \rangle = 2 \sum_q v_q^2. \tag{8.113}$$

Thus we have demonstrated that the Green function method leads to the usual basic equations of the pairing theory.

Let us show how the quasiparticles discussed earlier in Chapter 4 can be introduced in the framework of this method.

The Green function $G(p, t)$ describes the propagation of a particle with momentum p. The Fourier transform of the causal Green function equals

$$G_c(E) = \frac{1}{2\pi} \int_{-\infty}^{\infty} G_c(t) e^{iEt}.$$

Using (8.84′), (8.98), and (8.98′) we can write

$$G_c(E) = \frac{1}{2\pi} \int_{-\infty}^{\infty} I(\omega)\, d\omega \left\{ \frac{\theta(+\omega)}{E-\omega+i\varepsilon} - \frac{\eta\theta(-\omega)}{E-\omega-i\varepsilon} \right\}. \tag{8.114}$$

The propagation of a particle is determined by the spectral function $I(\omega)$. In the case of noninteracting particles, such a function is a δ-function, and we obtain (8.92′) for the Fourier transform. When the pairing correlations are present, the spectral function $I_1(\omega)$ has the form (8.109). The causal Green function then equals, obviously,

$$\langle\langle a_{q\sigma} | a^+_{q\sigma} \rangle\rangle_c = \frac{1}{2\pi} \left\{ \frac{u_q^2}{E-\varepsilon(q)+i\varepsilon} + \frac{v_q^2}{E+\varepsilon(q)-i\varepsilon} \right\}. \tag{8.115}$$

The poles of the Green function determine the quasiparticle energies. The simple form of (8.115) is related to the fact that such a function describes noninteracting quasiparticles. The $I(\omega)$ function becomes more complicated in the case of interacting quasiparticles.

The system with pairing correlations is characterized by two Green functions; the usual function corresponds to the normal pairing, and the Green function $\langle\langle a^+_{q,-\sigma} | a^+_{q\sigma} \rangle\rangle_c$, corresponds to the anomalous pairing. Such Green function is easily calculated from (8.114) using the spectral function $I_2(\omega)$ from (8.109′). The resulting formula can be written in the form

$$\langle\langle a^+_{q,-\sigma} | a^+_{q\sigma} \rangle\rangle_c = -\frac{\sigma}{4\pi} \frac{C}{\varepsilon(q)} \left\{ \frac{1}{E-\varepsilon(q)+i\varepsilon} - \frac{1}{E+\varepsilon(q)-i\varepsilon} \right\}. \tag{8.116}$$

Without pairing correlations the quantity $C = 0$ and, consequently, the Green function (8.116) vanishes as well.

Gorkov[463] derived a system of equations determining the Green function. This method is often used in studies of systems with pairing correlations.

4. The apparatus of the two-particle Green functions may be used for the description of the nuclear vibrational states. The two-particle Green functions are again defined by

(8.84)–(8.86). The corresponding operators can change the state of two particles at the same time, hence the name "two-particle function."

The general causal two-particle Green functions equals

$$G_c(\boldsymbol{r}_1t_1, \boldsymbol{r}_2t_2, \boldsymbol{r}_3t_3, \boldsymbol{r}_4t_4) = -i\langle T\{\psi(\boldsymbol{r}_1t_1)\,\psi^+(\boldsymbol{r}_2t_2)\,\psi(\boldsymbol{r}_3t_3)\,\psi^+(\boldsymbol{r}_4t_4)\}\rangle. \tag{8.117}$$

Very often the Green function uses only two points of time, e.g., $t_1 = t_2$, $t_3 = t_4$, and, therefore,

$$G_c(\boldsymbol{r}_1t, \boldsymbol{r}_2t, \boldsymbol{r}_3t', \boldsymbol{r}_4t') = -i\langle T\{\psi(\boldsymbol{r}_1t)\,\psi^+(\boldsymbol{r}_2t)\,\psi(\boldsymbol{r}_3t')\,\psi^+(\boldsymbol{r}_4t')\}\rangle. \tag{8.117'}$$

The creation and annihilation operators are in the Heisenberg representation; their explicit form has been shown in (1.7).

Let us derive the secular equation (8.65) using the two-particle Green functions. The system Hamiltonian (8.79) will be used.

First, we need the equation determining the Green functions

$$\langle\langle A(q, q'; t)\,|\,A^+(q_2, q_2'; t_2)\rangle\rangle \quad \text{and} \quad \langle\langle A^+(q, q'; t)\,|\,A^+(q_2, q_2', t_2)\rangle\rangle.$$

By using the exact commutation relations we obtain an infinite system of coupled equations. However, when the approximate commutation relations (8.44) are used (i.e., the quasiboson approximation), the resulting system of equations is closed.

The quasiboson commutation relations have the form

$$[B(q, q'), A^+(q_2, q_2')] = \delta_{q_2q'}A^+(q, q_2')+\delta_{q_2'q'}A^+(q, q_2), \tag{8.118}$$

$$[A(q_2, q_2'), B(q, q')] = \delta_{qq_2'}A(q', q_2)+\delta_{qq_2}A(q', q_2'). \tag{8.118'}$$

Let (q, q') and (q_2, q_2') be different sets of quantum numbers, i.e., $q \neq q_2$, $q' \neq q_2'$. Using (8.94), we obtain the following equations:

$$\begin{aligned} i\frac{d}{dt}\langle\langle A(q, q'; t)\,|\,A^+(q_2, q_2'; t_2)\rangle\rangle &= [\varepsilon(q)+\varepsilon(q')]\times \\ &\times\langle\langle A(q, q'; t)\,|\,A^+(q_2, q_2'; t_2)\rangle\rangle - \varkappa f(q, q')u_{qq'}\sum_{q_3, q_3'} f(q_3, q_3')u_{q_3q'}\times \\ &\times\{\langle\langle A(q_3, q_3'; t)\,|\,A^+(q_2, q_2'; t_2)\rangle\rangle + \langle\langle A^+(q_3, q_3'; t)\,|\,A^+(q_2, q_2'; t_2)\rangle\rangle\}, \end{aligned} \tag{8.119}$$

$$\begin{aligned} i\frac{d}{dt}\langle\langle A^+(q, q'; t)\,|\,A^+(q_2, q'; t_2)\rangle\rangle &= -[\varepsilon(q)+\varepsilon(q')]\times \\ &\times\langle\langle A^+(q, q'; t)\,|\,A^+(q_2, q_2'; t_2)\rangle\rangle + \varkappa f(q, q')u_{qq'}\times \\ &\times\sum_{q_3, q_3} f(q_3, q_3')\,u_{q_3q_3'}\{\langle\langle A^+(q_3, q_3'; t)\,|\,A^+(q_2, q'; t_2)\rangle\rangle \\ &+\langle\langle A(q_3, q_3'; t)\,|\,A^+(q_2, q_2'; t_2)\rangle\rangle\}. \end{aligned} \tag{8.120}$$

In Fourier transform, the above written equations have the form

$$\begin{aligned} \{\omega-[\varepsilon(q)+\varepsilon(q')]\}\langle\langle A(q, q')\,|\,A^+(q_2, q_2')\rangle\rangle_\omega &= -\varkappa f(q, q')u_{qq'}\sum_{q_3, q_3'} f(q_3, q_3')\,u_{q_3q_3'}\langle\langle A(q_3, q_3') \\ &+A^+(q_3, q_3')\,|\,A^+(q_2, q_2')\rangle\rangle_\omega, \end{aligned} \tag{8.119'}$$

$$\{\omega+[\varepsilon(q)+\varepsilon(q')]\}\langle\langle A^+(q,q')\,|\,A^+(q_2,q_2')\rangle\rangle_\omega$$
$$=\varkappa f(q,q')u_{qq'}\sum_{q_3,q_3'} f(q_3,q_3')\,u_{q_3q_3'}\langle\langle A(q_3,q_3')+A^+(q_3,q_3')\,|\,A^+(q_2,q_2')\rangle\rangle_\omega. \qquad (8.120')$$

Equation (8.119′) will be multiplied by $\omega+[\varepsilon(q)+\varepsilon(q')]$, and (8.120′) by $\omega-[\varepsilon(q)+\varepsilon(q')]$; after simple transformations we find that

$$\{[\varepsilon(q)+\varepsilon(q')]^2-\omega^2\}\langle\langle A(q,q')+A^+(q,q')\,|\,A^+(q_2,q_2')\rangle\rangle_\omega$$
$$=2\varkappa f(q,q')u_{qq'}[\varepsilon(q)+\varepsilon(q')]\sum_{q_3,q_3'} f(q_3,q_3')u_{q_3q_3'}\langle\langle A(q_3,q_3')+A^+(q_3\,q_3')\,|\,A^+(q_2,q_2')\rangle\rangle. \qquad (8.121)$$

Using a shorthand notation,

$$D=2\sum_{q_3,q_3'} f(q_3,q_3')\,u_{q_3q_3'}\langle\langle A(q_3,q_3')+A^+(q_3,q_3')\,|\,A^+(q_2,q_2')\rangle\rangle_\omega \qquad (8.122)$$

we obtain

$$\langle\langle A(q,q')+A^+(q,q')\,|\,A^+(q_2,q_2')\rangle\rangle_\omega=\frac{\varkappa f(q,q')\,u_{qq'}D[\varepsilon(q)+\varepsilon(q')]}{[\varepsilon(q)+\varepsilon(q')]^2-\omega_i^2}.$$

Finally, by substituting the above-written expression in (8.122), we come to the familiar secular equation

$$1=2\varkappa\sum_{q,q'}\frac{f^2(q,q')\,u_{qq'}^2[\varepsilon(q)+\varepsilon(q')]}{[\varepsilon(q)+\varepsilon(q')]^2-\omega^2}.$$

Thus, the secular equation determining the energies of the vibrational states was obtained by using the two-particle Green function technique. The same equation (8.65) was derived earlier in the random phase approximation.

The secular equation (8.65) can be also derived from another type of the two-particle Green functions, namely,

$$\langle\langle \alpha_{q_1\sigma_1}(t)\,\alpha_{q_2\sigma_2}(t)\,|\,\alpha^+_{q_3\sigma_3}(t')\,\alpha^+_{q_4\sigma_4}(t')\rangle\rangle.$$

However, it is necessary to find such a reduction of the higher-order Green functions which exactly corresponds to the quasiboson approximation used in our discussion.

§ 4. The Multipole–Multipole Interaction

1. The multipole–multipole interaction is often used in calculations of the properties of vibrational states in spherical and deformed nuclei. The spin–multipole interaction is also occasionally employed. In this section we shall discuss the essential features of these interactions; we shall transform the Hamiltonian containing the pairing and multipole forces to the form accessible to the numerical calculations, and derive the corresponding secular equation. We shall consider only harmonic vibrations; the vibration–rotation coupling will not be discussed. The 0^+ one-phonon states will be discussed separately in the next section.

To describe the one-phonon states in even–even spherical and deformed nuclei, we shall use the interaction Hamiltonian of the following form:

$$H = H_0(n)+H_0(p)+H_Q+H_\sigma, \tag{8.123}$$

where, according to (4.2), (4.2″), (3.62), and (3.71), the $H_0(n)$ of the deformed nuclei equals:

$$H_0(n) = \sum_{s,\,\sigma}\{E_0(s)-\lambda_n\}a^+_{s\sigma}a_{s\sigma}-G_N\sum_{s,\,s'}a^+_{s+}a^+_{s-}a_{s'-}a_{s'+},$$

while in the case of spherical nuclei

$$H_0(n) = \sum_{j,\,m}\{E_0(j)-\lambda_n\}a^+_{jm}a_{jm}-\frac{G_N}{4}\sum_{j,\,j'}\sum_{m,\,m'}(-1)^{j-m}(-1)^{-j'+m'}a^+_{jm}a^+_{j,\,-m}a_{j',\,-m'}a_{j'm'}.$$

The remaining parts are defined by

$$H_Q = -\tfrac{1}{2}\sum_{\lambda=2,3}\sum_{\mu\geq 0}\{\varkappa_n^{(\lambda)}Q^+_{\lambda\mu}(n)\,Q_{\lambda\mu}(n)+\varkappa_p^{(\lambda)}Q^+_{\lambda\mu}(p)\,Q_{\lambda\mu}(p)$$

$$+\varkappa_{np}^{(\lambda)}(Q^+_{\lambda\mu}(n)\,Q_{\lambda\mu}(p)+Q^+_{\lambda\mu}(p)\,Q_{\lambda\mu}(n))\},$$

$$H_\sigma = -\tfrac{1}{2}\sum_{\lambda=1,\,2\,\ldots}\sum_{\mu=-\lambda}^{\lambda}\varkappa_\sigma^\lambda\sum_{n=\lambda,\,\lambda\pm 1}T^+_\sigma(\lambda\mu n)\,T_\sigma(\lambda\mu n).$$

As the next step, we have to replace the particle operators $a_{q\sigma}$ by the quasiparticle operators. The canonical transformations (4.5) or (4.5″) will be used for this purpose. Because the operators $Q_{\lambda\mu}$ and $T(\lambda\mu n)$ are somewhat different in spherical and deformed nuclei, we shall divide our task into two steps. First, the Hamiltonian (8.123) will be transformed for the deformed nuclei. Then, the similar transformation will be performed for the spherical nuclei.

2. The Hamiltonian (8.123) for deformed nuclei will be transformed to the quasiparticle representation.[129] In § 1 of Chapter 5 we have shown that the term $H_0(n)$ equals [see (5.21)]

$$H_0(n) = H_0^0(n)+H_0^\beta(n)+H_0'(n)+H_0''(n),$$

where, according to (5.27),

$$H_0^0(n) = \sum_s 2\{E(s)-\lambda_n\}v_s^2-\frac{C_n^2}{G_N}+\sum_s \varepsilon(s)\,B(s,s),$$

and the remaining terms are defined in (5.28), (5.29), and (5.30).

The multipole moment operator was found in (3.63) as

$$Q_{\lambda\mu}(n) = \sum_{\substack{s,\,s'\\ \sigma,\,\sigma'}}\langle s\sigma|\mathfrak{f}^{\lambda\mu}|s'\sigma'\rangle a^+_{s\sigma}a_{s'\sigma'} = \sum_{s,\,s',\,\sigma}\{\mathfrak{f}^{\lambda\mu}(s,s')a^+_{s\sigma}a_{s'\sigma}+\sigma\bar{\mathfrak{f}}^{\lambda\mu}(s,s')a^+_{s\sigma}a_{s',\,-\sigma}\},$$

and the functions $\mathfrak{f}^{\lambda\mu}(s,s')$ and $\bar{\mathfrak{f}}^{\lambda\mu}(s,s')$ were defined in (3.65), (3.65′), and (3.65″). There are two distinct matrix elements for each multipolarity λ and $\mu\neq 0$. The $\mathfrak{f}^{\lambda\mu}(q,q')$ element obeys the angular momentum projection selection rule $K\pm\mu = K'$, while the $\bar{\mathfrak{f}}^{\lambda\mu}(q,q')$ matrix element obeys the selection rule $K+K' = \pm\mu$. For a given pair of states (q,q'), only one of the matrix elements can be nonvanishing. Occasionally, when the difference between the matrix elements $\bar{\mathfrak{f}}^{\lambda\mu}(q, q')$ and $\mathfrak{f}^{\lambda\mu}(q, q')$ is unimportant, we shall denote them by the common notation $f^{\lambda\mu}(q, q')$.

Using (6.62) and (6.62′) we shall introduce the quasiparticle pair operators $A(q, q')$, $\bar{A}(q, q')$, $B(q, q')$, and $\bar{B}(q, q')$. We should make, however, some additional comments: the operators $A(q, q')$ or $\bar{A}(q_2, q_2')$, respectively, are defined in such a way that the whole set of pairs of states q, q', corresponding to the multipolarity λ and projection μ, is divided into two nonoverlapping subsets. The subset (q, q') corresponds to the operator $A(q, q')$, while the subset (q_2, q_2') corresponds to the operator $\bar{A}(q_2, q_2')$. Similar division exists for the operators $B(q, q')$ or $\bar{B}(q_2, q_2')$, respectively.

The operators $A(q, q')$, $\bar{A}(q_2, q_2')$, $B(q, q')$, and $\bar{B}(q_2, q_2')$ obey the following exact commutation rules:

$$[A(q, q'), A^+(q_3, q_3')] = \delta_{qq_3}\delta_{q'q_3'}+\delta_{qq_3'}\delta_{q'q_3}-\tfrac{1}{2}\{\delta_{qq_3}B(q_3', q')+\delta_{q'q_3'}B(q_3, q) + \delta_{qq_3'}B(q_3, q')+\delta_{q'q_3}B(q_3', q)\}, \quad (8.124)$$

$$[\bar{A}(q, q'), \bar{A}^+(q_2, q_2')] = \delta_{qq_2}\delta_{q'q_2'}-\delta_{qq_2'}\delta_{q'q_2}-\tfrac{1}{2}\{\delta_{qq_2}B(q_2', q')+\delta_{q'q_2'}B(q_2, q)-\delta_{qq_2'}B(q_2, q') - \delta_{q'q_2}B(q_2', q)\}, \quad (8.124')$$

$$[B(q, q'), B(q_2, q_2')] = \delta_{q_2q'}B(q, q_2')-\delta_{qq_2'}B(q_2, q'). \quad (8.124'')$$

The remaining commutation rules have been given in (8.118), (8.118′), and in ref. 129.

The multipole moment operator expressed in terms of the above-discussed quasiparticle pair operators has the following form:

$$Q_{\lambda\mu}(n) = \frac{1}{\sqrt{2}}\sum_{s,s'} u_{ss'}\{\mathfrak{f}^{\lambda\mu}(s, s')[A(s, s')+A^+(s, s')]+\bar{\mathfrak{f}}^{\lambda\mu}(s, s')[\bar{A}(s, s')+\bar{A}^+(s, s')]\} + \sum_{s,s'} v_{ss'}\{\mathfrak{f}^{\lambda\mu}(s, s')B(s, s')+\bar{\mathfrak{f}}^{\lambda\mu}(s, s')\bar{B}(s, s')\}+2\sum_{s}\mathfrak{f}^{\lambda\mu}(s, s)v_s^2, \quad (8.125)$$

where $v_{ss'} \equiv v_{ss'}^{(-)} = u_s u_{s'} - v_s v_{s'}$. The expression $Q_{\lambda\mu}^+(n)\, Q_{\lambda\mu}(n)$ can be divided into three parts (the constant part of (8.125) is again omitted):

$$Q_{\lambda\mu}^+(n)\, Q_{\lambda\mu}(n) = h_1^{(\lambda\mu)}(n)+h_2^{(\lambda\mu)}(n)+h_3^{(\lambda\mu)}(n), \quad (8.126)$$

where

$$h_1^{(\lambda\mu)}(n) = \tfrac{1}{2}\Big(\sum_{s,s'} u_{ss'}\{\mathfrak{f}^{\lambda\mu}(s, s')[A(s, s')+A^+(s, s')]+\bar{\mathfrak{f}}^{\lambda\mu}(s, s')[\bar{A}(s, s')+\bar{A}^+(s, s')]\}\Big)^2, \quad (8.126')$$

$$h_2^{(\lambda\mu)}(n) = \Big(\sum_{s,s'} v_{ss'}\{\mathfrak{f}^{\lambda\mu}(s, s')B(s, s')+\bar{\mathfrak{f}}^{\lambda\mu}(s, s')\bar{B}(s, s')\}\Big)^2, \quad (8.126'')$$

$$h_3^{(\lambda\mu)}(n) = \frac{1}{\sqrt{2}}\sum_{s,s'}\sum_{s_2,s_2'} u_{ss'}v_{s_2s_2'}(\{\mathfrak{f}^{\lambda\mu}(s, s')[A(s, s')+A^+(s, s')] + \bar{\mathfrak{f}}^{\lambda\mu}(s, s')[\bar{A}(s, s')+\bar{A}^+(s, s')]\}\{\mathfrak{f}^{\lambda\mu}(s_2, s_2')B(s_2, s_2')+\bar{\mathfrak{f}}^{\lambda\mu}(s_2, s_2')\bar{B}(s_2, s_2')\} + \{\mathfrak{f}^{\lambda\mu}(s_2, s_2')B(s_2, s_2')\}+\bar{\mathfrak{f}}^{\lambda\mu}(s_2, s_2')\bar{B}(s_2, s_2')\}\{\mathfrak{f}^{\lambda\mu}(s, s')[A(s, s')+A^+(s, s')] + \bar{\mathfrak{f}}^{\lambda\mu}(s, s')[\bar{A}(s, s')+\bar{A}^+(s, s')]\}). \quad (8.126''')$$

The phonon operators are defined in the usual way, i.e.,

$$Q_i(\lambda\mu) = \frac{1}{2}\sum_{q,\,q'}\{\psi_{qq'}^{\lambda\mu i}A(q,q')-\varphi_{qq'}^{\lambda\mu i}A^+(q,q')+\bar{\psi}_{qq'}^{\lambda\mu i}\bar{A}(q,q')-\bar{\varphi}_{qq'}^{\lambda\mu i}\bar{A}^+(q,q')\}, \tag{8.127}$$

where

$$\psi_{qq'}^{\lambda\mu i} = \psi_{q'q}^{\lambda\mu i}, \quad \varphi_{qq'}^{\lambda\mu i} = \varphi_{q'q}^{\lambda\mu i}, \quad \bar{\psi}_{qq'}^{\lambda\mu i} = -\bar{\psi}_{q'q}^{\lambda\mu i}, \quad \bar{\varphi}_{qq'}^{\lambda\mu i} = -\bar{\varphi}_{q'q}^{\lambda\mu i},$$

and the summation over q, q' includes both neutron and proton states. The exact commutation relations (8.124) are replaced by the approximate commutation relations (8.44) in the approximate second quantization method. In the quasiboson approximation, the phonon operators obey the following relations.

$$[Q_i(\lambda\mu), \quad Q_{i'}^+(\lambda'\mu')] = \delta_{ii'}\delta_{\lambda\lambda'}\delta_{\mu\mu'}, \tag{8.128}$$

$$[Q_i(\lambda\mu), \quad Q_i(\lambda'\mu')] = [Q_{i'}^+(\lambda'\mu'), \quad Q_i^+(\lambda\mu)] = 0. \tag{8.128'}$$

To fulfill these, the functions ψ and φ are restricted by the equations

$$\sum_{q,\,q'}\{\psi_{qq'}^{\lambda\mu i}\psi_{qq'}^{\lambda'\mu' i'}-\varphi_{qq'}^{\lambda\mu i}\varphi_{qq'}^{\lambda'\mu' i'}+\bar{\psi}_{qq'}^{\lambda\mu i}\bar{\psi}_{qq'}^{\lambda'\mu' i'}-\bar{\varphi}_{qq'}^{\lambda\mu i}\bar{\varphi}_{qq'}^{\lambda'\mu' i'}\} = 2\delta_{\lambda\lambda'}\delta_{\mu\mu'}\delta_{ii'}, \tag{8.129}$$

$$\sum_{q,\,q'}\{\psi_{qq'}^{\lambda\mu i}\varphi_{qq'}^{\lambda'\mu' i'}-\psi_{qq'}^{\lambda'\mu' i'}\varphi_{qq'}^{\lambda\mu i}+\bar{\psi}_{qq'}^{\lambda\mu i}\bar{\varphi}_{qq'}^{\lambda'\mu' i'}-\bar{\psi}_{qq'}^{\lambda'\mu' i'}\bar{\varphi}_{qq'}^{\lambda\mu i}\} = 0. \tag{8.129'}$$

According to (8.51) and (8.51'), the inverse transformation, i.e., the transformation from the phonon operators to the operators $A(q, q')$, $A^+(q, q')$, has the form

$$A(q,q') = \sum_i \{\psi_{qq'}^{\lambda\mu i}Q_i(\lambda\mu)+\varphi_{qq'}^{\lambda\mu i}Q_i^+(\lambda\mu)\}.$$

Such relations put new restrictions on the functions ψ and φ (as a consequence of the commutation relations (8.43'), (8.44)), namely,

$$\sum_i (\psi_{qq'}^{\lambda\mu i}\psi_{q_2q_2'}^{\lambda\mu i}-\varphi_{qq'}^{\lambda\mu i}\varphi_{q_2q_2'}^{\lambda\mu i}) = \delta_{qq_2}\delta_{q'q_2'}+\delta_{qq_2'}\delta_{q'q_2}, \tag{8.130}$$

$$\sum_i (\psi_{qq'}^{\lambda\mu i}\varphi_{q_2q_2'}^{\lambda\mu i}-\psi_{q_2q_2'}^{\lambda\mu i}\varphi_{qq'}^{\lambda\mu i}) = 0 \tag{8.130'}$$

(the remaining conditions are listed in ref. 129).

We shall use the combined functions

$$g_{qq'}^{\lambda\mu i} = \psi_{qq'}^{\lambda\mu i}+\varphi_{qq'}^{\lambda\mu i}, \quad \bar{g}_{qq'}^{\lambda\mu i} = \bar{\psi}_{qq'}^{\lambda\mu i}+\bar{\varphi}_{qq'}^{\lambda\mu i},$$

$$w_{qq'}^{\lambda\mu i} = \psi_{qq'}^{\lambda\mu i}-\varphi_{qq'}^{\lambda\mu i}, \quad \bar{w}_{qq'}^{\lambda\mu i} = \bar{\psi}_{qq'}^{\lambda\mu i}-\bar{\varphi}_{qq'}^{\lambda\mu i}.$$

The condition (8.129) is now expressed as

$$\sum_{q,\,q'}\{g_{qq'}^{\lambda\mu i}w_{qq'}^{\lambda\mu i}+\bar{g}_{qq'}^{\lambda\mu i}\bar{w}_{qq'}^{\lambda\mu i}\} = 2. \tag{8.129''}$$

The Hamiltonian H_Q can be transformed to the form which explicitly includes the phonon operators. After some manipulations, we obtain the following expression:

$$H_Q = -\frac{1}{2}\sum_{\lambda=2,3\ldots}\sum_{\mu\geqslant 0}\{\varkappa_n^{(\lambda)}[h_1^{(\lambda\mu)}(n)+h_2^{(\lambda\mu)}(n)+h_3^{(\lambda\mu)}(n)]+\varkappa_p^{(\lambda)}[h_1^{(\lambda\mu)}(p)+h_2^{(\lambda\mu)}(p)+h_3^{(\lambda\mu)}(p)]$$
$$+2\varkappa_{np}^{(\lambda)}[h_1^{(\lambda\mu)}(np)+h_2^{(\lambda\mu)}(np)+h_3^{(\lambda\mu)}(np)]\}. \tag{8.131}$$

The operator $h_2^{(\lambda\mu)}$ was defined in (8.126″) and

$$h_1^{(\lambda\mu)}(n) = \frac{1}{2}\sum_{i, i'} \sum_{s, s', s_2, s_2'} u_{ss'}u_{s_2s_2}[\mathfrak{f}^{\lambda\mu}(s, s')g_{ss'}^{\lambda\mu i}+\bar{\mathfrak{f}}^{\lambda\mu}(s, s')\bar{g}_{ss'}^{\lambda\mu i}][\mathfrak{f}^{\lambda\mu}(s_2, s_2')g_{s_2s_2'}^{\lambda\mu i'}+\bar{\mathfrak{f}}^{\lambda\mu}(s_2, s_2')\bar{g}_{s_2s_2'}^{\lambda\mu i'}]\times$$
$$\times[Q_i^+(\lambda\mu)Q_{i'}(\lambda\mu)+Q_{i'}^+(\lambda\mu)Q_i^+(\lambda\mu)+Q_i^+(\lambda\mu)Q_{i'}^+(\lambda\mu)+Q_{i'}(\lambda\mu)Q_i(\lambda\mu)], \qquad (8.131')$$

$$h_3^{(\lambda\mu)}(n) = \sum_i \frac{1}{\sqrt{2}} \sum_{s, s', s_2, s_2'} u_{ss'}v_{s_2s_2'}[\mathfrak{f}^{\lambda\mu}(s, s')g_{ss'}^{\lambda\mu i}+\bar{\mathfrak{f}}^{\lambda\mu}(s, s')\bar{g}_{ss'}^{\lambda\mu i}]\{[Q_i^+(\lambda\mu)+Q_i(\lambda\mu)]\times$$
$$\times[\mathfrak{f}^{\lambda\mu}(s_2, s_2')B(s_2, s_2')+\bar{\mathfrak{f}}^{\lambda\mu}(s_2, s_2')\bar{B}(s_2, s_2')]+[\mathfrak{f}^{\lambda\mu}(s_2, s_2')B(s_2, s_2')$$
$$+\bar{\mathfrak{f}}^{\lambda\mu}(s_2, s_2')\bar{B}(s_2, s_2')][Q_i^+(\lambda\mu)+Q_i(\lambda\mu)]\}. \qquad (8.131'')$$

The expressions (8.131′) and (8.126′) are very similar; their difference is a constant independent of the phonon operators.

3. Let us find the secular equation using the approximate second quantization method. As we have stated earlier, we are considering the one-phonon states in even–even deformed nuclei. In the quasiboson approximation, our Hamiltonian has the form

$$H_v = \sum_q \varepsilon(q)B(q, q) - \frac{1}{2}\sum_{\lambda=2, 3\ldots}\sum_{\mu\geq 0}\{\varkappa_n^{(\lambda)}h_1^{(\lambda\mu)}(n)+\varkappa_p^{(\lambda)}h_1^{(\lambda\mu)}(p)+\varkappa_{np}^{(\lambda)}h_1^{(\lambda\mu)}(np)\}, \qquad (8.132)$$

with $h_1^{(\lambda\mu)}$ defined in (8.131′). As before in § 2, we shall use the variational principle, i.e., we shall demand that

$$\delta\left\{\langle Q_i(\lambda\mu)H_vQ_i^+(\lambda\mu)\rangle - \frac{\omega_i^{\lambda\mu}}{2}\left[\sum_{q, q'}(g_{qq'}^{\lambda\mu i}w_{qq'}^{\lambda\mu i}+\bar{g}_{qq'}^{\lambda\mu i}\bar{w}_{qq'}^{\lambda\mu i})-2\right]\right\} = 0.$$

The Lagrange multiplier $\omega_i^{\lambda\mu}$ determines the energy of the one phonon state with one particular value of $\lambda\mu$.

Assuming that the variations $\delta g_{qq'}^{\lambda\mu i}$ and $\delta w_{qq'}^{\lambda\mu i}$ are independent, we obtain two equations

$$\{[\varepsilon(s)+\varepsilon(s')]^2-(\omega_i^{\lambda\mu})^2\}g_{ss'}^{\lambda\mu i} = \mathfrak{f}^{\lambda\mu}(s, s')u_{ss'}(\varkappa_n^{(\lambda)}D^{\lambda\mu i}(n)+\varkappa_{np}^{(\lambda)}D^{\lambda\mu i}(p))(\varepsilon(s)+\varepsilon(s')),$$
$$w_{qq'}^{\lambda\mu i} = \frac{\omega_i^{\lambda\mu}}{\varepsilon(q)+\varepsilon(q')}g_{qq'}^{\lambda\mu i}$$

(similar equations, naturally, are valid for the quantities $\bar{g}^{\lambda\mu i}$ and $\bar{w}^{\lambda\mu i}$). The constants D are defined by

$$D^{\lambda\mu i}(n) = 2\sum_{s, s'} u_{ss'}(\mathfrak{f}^{\lambda\mu}(s, s')g_{ss'}^{\lambda\mu i}+\bar{\mathfrak{f}}^{\lambda\mu}(s, s')\bar{g}_{ss'}^{\lambda\mu i}). \qquad (8.133)$$

By solving the equations for $g_{ss'}^{\lambda\mu i}$ and $\bar{g}_{ss'}^{\lambda\mu i}$, and substituting the result in (8.133), we obtain the system

$$\left.\begin{aligned} D^{\lambda\mu i}(n) &= X^{\lambda\mu i}(n)\{\varkappa_n^{(\lambda)}D^{\lambda\mu i}(n)+\varkappa_{np}^{(\lambda)}D^{\lambda\mu i}(p)\}, \\ D^{\lambda\mu i}(p) &= X^{\lambda\mu i}(p)\{\varkappa_p^{(\lambda)}D^{\lambda\mu i}(p)+\varkappa_{np}^{(\lambda)}D^{\lambda\mu i}(n)\}, \end{aligned}\right\} \qquad (8.133')$$

where, for example,

$$X^{\lambda ni}(n) = 2\sum_{s, s'}\frac{(f^{\lambda\mu}(s, s')^2u_{ss'}^2[\varepsilon(s)+\varepsilon(s')]}{[\varepsilon(s)+\varepsilon(s')]^2-(\omega_i^{\lambda\mu})^2}. \qquad (8.133'')$$

The system of (8.133′) has a nontrivial solution if its determinant vanishes, i.e.,

$$\begin{vmatrix} \varkappa_n^{(\lambda)} X^{\lambda\mu i}(n) - 1 & \varkappa_{np}^{(\lambda)} X^{\lambda\mu i}(n) \\ \varkappa_{np}^{(\lambda)} X^{\lambda\mu i}(p) & \varkappa_p^{(\lambda)} X^{\lambda\mu i}(p) - 1 \end{vmatrix} = 0. \tag{8.134}$$

Thus, the phonons exist only for the $\omega_i^{\lambda\mu}$ values which are the solutions of the secular equation (8.134). Let us, for completeness, show its explicit form

$$1 = 2\varkappa_n^{(\lambda)} \sum_{s, s'} \frac{(f^{\lambda\mu}(s, s'))^2 u_{ss'}^2 [\varepsilon(s) + \varepsilon(s')]}{[\varepsilon(s) + \varepsilon(s')]^2 - (\omega_i^{\lambda\mu})^2} + 2\varkappa_p^{(\lambda)} \sum_{r, r'} \frac{(f^{\lambda\mu}(r, r'))^2 u_{rr'}^2 [\varepsilon(r) + \varepsilon(r')]}{[\varepsilon(r) + \varepsilon(r')]^2 - (\omega_i^{\lambda\mu})^2}$$

$$+ 4[(\varkappa_{np}^{(\lambda)})^2 - \varkappa_n^{(\lambda)} \varkappa_p^{(\lambda)}] \sum_{s, s'} \frac{(f^{\lambda\mu}(s, s'))^2 u_{ss'}^2 [\varepsilon(s) + \varepsilon(s')]}{[\varepsilon(s) + \varepsilon(s')]^2 - (\omega_i^{\lambda\mu})^2} \sum_{r, r'} \frac{(f^{\lambda\mu}(r, r'))^2 u_{rr'}^2 [\varepsilon(r) + \varepsilon(r')]}{[\varepsilon(r) + \varepsilon(r')]^2 - (\omega_i^{\lambda\mu})^2}. \tag{8.134′}$$

Equation (8.134′) is a generalization of the simple equation (8.65); the system consisting of neutrons and protons is treated in this case.

Equation (8.134′) is simpler in the symmetric case of $\varkappa_n^{(\lambda)} = \varkappa_p^{(\lambda)} = \varkappa_{np}^{(\lambda)} \equiv \varkappa^{(\lambda)}$,

$$1 = 2\varkappa^{(\lambda)} \sum_{q, q'} \frac{(f^{\lambda\mu}(q, q'))^2 u_{qq'}^2 [\varepsilon(q) + \varepsilon(q')]}{[\varepsilon(q) + \varepsilon(q')]^2 - (\omega_i^{\lambda\mu})^2}. \tag{8.134″}$$

By following the steps leading to the expressions (8.66)–(8.68) we can find the wave functions of the one-phonon states:

$$g_{qq'}^{\lambda\mu i} = \frac{\sqrt{2}}{\sqrt{Y^i(\lambda\mu)}} \frac{\mathfrak{f}^{\lambda\mu}(q, q') u_{qq'} [\varepsilon(q) + \varepsilon(q')]}{[\varepsilon(q) + \varepsilon(q')]^2 - (\omega_i^{\lambda\mu})^2}, \tag{8.135}$$

$$\psi_{qq'}^{\lambda\mu i} = \frac{1}{\sqrt{2Y^i(\lambda\mu)}} \frac{\mathfrak{f}^{\lambda\mu}(q, q') u_{qq'}}{\varepsilon(q) + \varepsilon(q') - \omega_i^{\lambda\mu}}, \tag{8.135′}$$

$$\varphi_{qq'}^{\lambda\mu i} = \frac{1}{\sqrt{2Y^i(\lambda\mu)}} \frac{\mathfrak{f}^{\lambda\mu}(q, q') u_{qq'}}{\varepsilon(q) + \varepsilon(q') + \omega_i^{\lambda\mu}}, \tag{8.135″}$$

where

$$Y^i(\lambda\mu) = \sum_{q, q'} \frac{(f^{\lambda\mu}(q, q'))^2 u_{qq'}^2 [\varepsilon(q) + \varepsilon(q')] \omega_i^{\lambda\mu}}{\{[\varepsilon(q) + \varepsilon(q')]^2 - (\omega_i^{\lambda\mu})^2\}^2}. \tag{8.136}$$

The expressions for the quantities $\bar{g}_{qq'}^{\lambda\mu i}$, $\bar{\psi}_{qq'}^{\lambda\mu i}$, and $\bar{\varphi}_{qq'}^{\lambda\mu i}$ are obtained from the above relations by replacing the matrix element $\mathfrak{f}^{\lambda\mu}(q, q')$ by the matrix element $\bar{\mathfrak{f}}^{\lambda\mu}(q, q')$.

Let us conclude: The energies $\omega_1^{\lambda\mu}, \omega_2^{\lambda\mu}, \ldots$, of the one-phonon states are the solutions of the secular equations (8.134′) or (8.134″). The corresponding one-phonon wave functions are calculated from (8.135′) and (8.135″). The properties of (8.134″) were discussed in § 2 of this chapter. Note that the poles of (8.134) are the two-quasiparticle energies in the neutron or proton, respectively, systems.

The generalized equations of the Tamm–Dancoff method are easily obtained by following the steps leading to (8.82) and (8.83).

4. Now let us turn our attention to the problem of spin–quadrupole interaction. In the charge symmetric case $\varkappa_\sigma^{(2)} \equiv \varkappa_\sigma^{(2)}(n) = \varkappa_\sigma^{(2)}(p) = \varkappa_\sigma^{(2)}(np)$, the spin–quadrupole interaction (3.71) has the form

$$-\tfrac{1}{2}\varkappa_\sigma^{(2)} T_\sigma^+(2\mu 2) T_\sigma(2\mu 2),$$

where

$$T_\sigma(2\mu 2) = \sum_{q, q', \sigma} \{t^{2\mu}(q, q')a^+_{q\sigma}a_{q'\sigma} + \sigma \bar{t}^{2\mu}(q, q')a_{q\sigma}a_{q', -\sigma}\},$$

and the functions $t^{2\mu}(q, q')$ and $\bar{t}^{2\mu}(q, q')$ are calculated according to (3.74), (3.74′), and (3.74″).

Let us rewrite the operator $T_\sigma(2\mu 2)$ in terms of the quasiparticle pair operators. After the canonical transformation,

$$T_\sigma(2\mu 2) = \frac{1}{\sqrt{2}} \sum_{q, q'} u^{(-)}_{qq'}\{t^{2\mu}(q, q')[A^+(q, q') - A(q, q')] + \bar{t}^{2\mu}(q, q')[\bar{A}^+(q, q') - \bar{A}(q, q')]\}$$
$$+ \sum_{q, q'} v^{(+)}_{qq'}\{t^{2\mu}(q, q')B(q, q') + \bar{t}^{2\mu}(q, q')\bar{B}(q, q')\}. \tag{8.137}$$

Now, we shall use the phonon operators $Q_i(2\mu)$, defined in (8.127), and obtain

$$T_\sigma(2\mu 2) = \frac{1}{\sqrt{2}} \sum_i \sum_{q, q'} u^{(-)}_{qq'}\{t^{2\mu}(q, q')w^{2\mu i}_{qq'} + \bar{t}^{2\mu}(q, q')\bar{w}^{2\mu i}_{qq'}\}[Q^+_i(2\mu) - Q_i(2\mu)]$$
$$+ \sum_{q, q'} v^{(+)}_{qq'}\{t^{2\mu}(q, q')B(q, q') + \bar{t}^{2\mu}(q, q')\bar{B}(q, q')\}. \tag{8.137′}$$

The Hamiltonian describing the one-phonon states in even–even deformed nuclei will now consist of two parts: the original Hamiltonian (8.132) and the spin–quadrupole interaction chosen as

$$H^{2\mu}_\sigma = -\frac{\varkappa^{(2)}_\sigma}{2} \sum_{i, i'} \sum_{q, q'} u^{(-)}_{qq'}\{t^{2\mu}(q, q')w^{2\mu i}_{qq'} + \bar{t}^{2\mu}(q, q')\bar{w}^{2\mu i}_{qq'}\} \times$$
$$\times \sum_{q_2, q_2'} u^{(-)}_{q_2 q_2'}\{t^{2\mu}(q_2, q_2')w^{2\mu i'}_{q_2 q_2'} + \bar{t}^{2\mu}(q_2, q_2')\bar{w}^{2\mu i'}_{q_2 q_2'}\} \times$$
$$\times [Q^+_i(2\mu)Q_{i'}(2\mu) + Q^+_{i'}(2\mu)Q_i(2\mu) - Q^+_i(2\mu)Q^+_{i'}(2\mu) - Q_{i'}(2\mu)Q_i(2\mu)]. \tag{8.138}$$

Let us, for example, consider the quadrupole $\mu = 2$ states. By using the variational principle we obtain two equations:

$$\left.\begin{aligned} [\varepsilon(q) + \varepsilon(q')]g^{2i}_{qq'} - \omega^{22}_i w^{22i}_{qq'} - \varkappa^{(2)} u_{qq'} \bar{t}^{22}(q, q')D^{22i} &= 0, \\ [\varepsilon(q) + \varepsilon(q')]w^{22i}_{qq'} - \omega^{22}_i g^{22i}_{qq'} - \varkappa^{(2)}_\sigma u^{(-)}_{qq'} t^{22}(q, q')R^{22i} &= 0, \end{aligned}\right\} \tag{8.139}$$

where the constant D^{22i} is determined by (8.133) and

$$R^{22i} = 2\sum_{q, q'} u^{(-)}_{qq'}\{t^{22}(q, q')w^{22i}_{qq'} + \bar{t}^{22}(q, q')\bar{w}^{22i}_{qq'}\}. \tag{8.140}$$

From (8.139) we obtain the solutions

$$\left.\begin{aligned} g^{22i}_{qq'} &= \frac{\varkappa^{(2)}[\varepsilon(q) + \varepsilon(q')]\, u_{qq'} \bar{t}^{22}(q, q')D^{22i} + \varkappa^{(2)}_\sigma \omega^{22}_i u^{(-)}_{qq'} t^{22}(q, q')R^{22i}}{[\varepsilon(q) + \varepsilon(q')]^2 - (\omega^{22}_i)^2}, \\ w^{22i}_{qq'} &= \frac{\varkappa^{(2)}_\sigma[\varepsilon(q) + \varepsilon(q')]\, u^{(-)}_{qq'} t^{22}(q, q')R^{22i} + \varkappa^{(2)} \omega^{22}_i u_{qq'} \bar{t}^{22}(q, q')D^{22i}}{[\varepsilon(q) + \varepsilon(q')]^2 - (\omega^{22}_i)^2}. \end{aligned}\right\} \tag{8.139′}$$

Hereafter we shall use a special notation:

$$\left.\begin{aligned} X^{2\mu i} &= X^{2\mu i}(n)+X^{2\mu i}(p), \\ S^{2\mu i} &= 2\sum_{q,\,q'}\frac{(t^{2\mu}(q,q'))^2\,(u^{(-)}_{qq'})^2\,[\varepsilon(q)+\varepsilon(q')]}{[\varepsilon(q)+\varepsilon(q')]^2-(\omega_i^{2\mu})^2}, \\ W^{2\mu i} &= 2\sum_{q,\,q'}\frac{t^{2\mu}(q,q')\,\bar{t}^{2\mu}(q,q')\,u_{qq'}u^{(-)}_{qq'}\omega_i^{2\mu}}{[\varepsilon(q)+\varepsilon(q')]^2-(\omega_i^{2\mu})^2}, \end{aligned}\right\} \tag{8.141}$$

and the quantities $X^{2\mu i}(n)$ are determined according to (8.133''). The summations in (8.141') and (8.141'') include both the $t^{2\mu}(q, q')$ and $\bar{t}^{2\mu}(q, q')$. Substituting the expressions (8.139') in the definitions of the quantities D^{22i} and R^{22i}, and using the notation of (8.141), (8.141'), and (8.141''), we obtain the system of equations

$$\left.\begin{aligned} (1-\varkappa^{(2)}X^{22i})D^{22i}-\varkappa^{(2)}W^{22i}R^{22i} &= 0, \\ -\varkappa^{(2)}W^{22i}D^{22i}+(1-\varkappa_\sigma^{(2)}S^{22i})R^{22i} &= 0. \end{aligned}\right\} \tag{8.139''}$$

The secular equation is then obtained from the condition that the determinant of system (8.139'') vanishes. The resulting secular equation has the form

$$\begin{vmatrix} 1-\varkappa^{(2)}X^{22i} & -\varkappa_\sigma^{(2)}W^{22i} \\ -\varkappa^{(2)}W^{22i} & 1-\varkappa_\sigma^{(2)}S^{22i} \end{vmatrix} = 0, \tag{8.142}$$

or we can write it in the form

$$(1-\varkappa^{(2)}X^{22i})\,(1-\varkappa_\sigma^{(2)}S^{22i}) = \varkappa^{(2)}\varkappa_\sigma^{(2)}(W^{22i})^2. \tag{8.142'}$$

Using the normalization condition (8.129''), we obtain the following expressions for the wave functions:

$$g^{22i}_{qq'} = \sqrt{\frac{2}{Z^i(22)}}\,\frac{1}{[\varepsilon(q)+\varepsilon(q')]^2-(\omega_i^{22})^2}\{\bar{t}^{22}(q,q')\,u_{qq'}[\varepsilon(q)+\varepsilon(q')]+\varkappa_\sigma^{(2)}L^{22i}\omega_i^{22}u^{(-)}_{qq'}t^{22}(q,q')\}, \tag{8.143}$$

$$w^{22i}_{qq'} = \sqrt{\frac{2}{Z^i(22)}}\,\frac{1}{[\varepsilon(q)+\varepsilon(q')]^2-(\omega_i^{22})^2}\{\bar{t}^{22}(q,q')u_{qq'}\omega_i^{22}+\varkappa_\sigma^{(2)}L^{22i}[\varepsilon(q)+\varepsilon(q')]u^{(-)}_{qq'}t^{22}(q,q')\}, \tag{8.143'}$$

where

$$L^{22i} = \frac{W^{22i}}{1-\varkappa_\sigma^{(2)}S^{22i}}, \tag{8.144}$$

$$Z^i(22) = \frac{1}{4}\left\{\frac{\partial}{\partial\omega_i^{22}}X^{22i}+2\varkappa_\sigma^{(2)}L^{22i}\frac{\partial}{\partial\omega_i^{22}}W^{22i}+(\varkappa_\sigma^{(2)}L^{22i})^2\frac{\partial}{\partial\omega_i^{22}}S^{22i}\right\}. \tag{8.144'}$$

Naturally, when $\varkappa_\sigma^{(2)} = 0$, we obtain again our original equations (8.134)–(8.136).

Let us discuss the characteristic features of the secular equation (8.142'). The quantity W^{22i} contains positive and negative contributions. Therefore, it is substantially (usually 20–40 times) smaller than the quantities X^{22i} and S^{22i} which contain only the coherent contributions. By assuming that $W^{22i} = 0$, eqn. (8.141') is split into two independent equations. The first equation includes only the pure quadrupole interaction, while the second one includes only the spin–quadrupole interaction.

Let us rewrite (8.142′) in a slightly different form:

$$\frac{1}{\varkappa_\sigma^{(2)}} = P^{22i}(\omega, \varkappa_\sigma^{(2)}), \tag{8.142''}$$

where

$$P^{22i}(\omega, \varkappa_\sigma^{(2)}) = X^{22i} + \varkappa_\sigma^{(2)} \frac{(W^{22i})^2}{1-\varkappa_\sigma^{(2)} S^{22i}}\,.$$

It is interesting to see how the function $P^{22}(\omega, \varkappa_\sigma^{(2)})$ behaves in the vicinity of the first pole $\varepsilon(q_1)+\varepsilon(q_2)$ of the X^{22i} function. Such a behavior is clearly visible if we consider the limit

$$P^{22i}(\omega, \varkappa_\sigma^{(2)})\Big|_{\varepsilon(q_1)+\varepsilon(q_2)-\omega_1^{22}\to 0} = 2\frac{(\mathfrak{f}^{22}(q_1, q_2))^2 u_{q_1q_2}^2}{\varepsilon(q_1)+\varepsilon(q_2)}\,.$$

Thus, the function P^{22} has no poles at the ω^{22} values corresponding to the two-quasiparticle states. Instead, the poles of P^{22} are determined by the equation

$$1-\varkappa_\sigma^{(2)} S^{22} = 0. \tag{8.145}$$

Figure 8.4 shows the behavior of the functions P^{22}, X^{22}, and S^{22} for the nucleus ^{164}Er. The spin–quadrupole interaction shifts, in this case, the first pole toward the lower energies. The solutions of (8.142″) have, generally, somewhat lower energies than the solutions

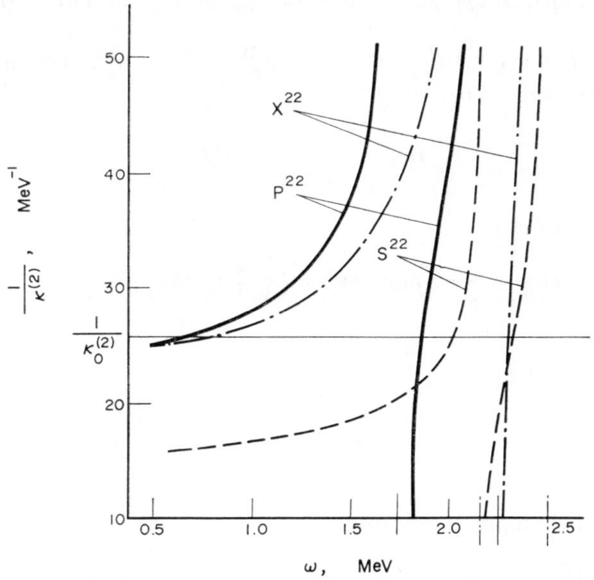

FIG. 8.4. The behavior of the functions $P^{22}(\omega)$, $X^{22}(\omega)$, and $S^{22}(\omega)$ of the $K^\pi = 2^+$ states in ^{164}Er. The interaction constants are $\varkappa_\sigma^{(2)} = 5.7A^{-4/3}\,\mathring{\omega}_0$ and $\varkappa_0^{(2)} = 4.5A^{-4/3}\,\mathring{\omega}_0$. The solutions of the secular equation (8.142″) are determined by the intersection of the curve $P^{22}(\omega)$ and horizontal line $1/\varkappa_0^{(2)}$. The full vertical lines denote the first and second poles of the function $P^{22}(\omega)$, while the dot-and-dash vertical lines denote the first and second two-quasiparticle states. The corresponding calculations were performed in ref. 131.

of the usual equation (8.134″). Such a tendency is more pronounced in the case of the second solutions.

5. Let us find the secular equation and the one-phonon wave function for the vibrational states in spherical nuclei. We shall start with the transformation of the Hamiltonian (8.123). The $H_0(n)$ term will have the form (5.21), with

$$H_0^0(n) = \sum_j (2j+1)\{E(j)-\lambda_n\}v_j^2 - \frac{C_n^2}{G_N} + \sum_j \varepsilon(j) \sum_m \alpha_{jm}^+ \alpha_{jm}. \tag{8.146}$$

The quasiparticle pair operators (6.69), (6.69′), and (6.69″) will be used in the form

$$A(j_1, j_2; \lambda\mu) = \sum_{m_1, m_2} (j_1 m_1 j_2 m_2 \mid \lambda\mu) \alpha_{j_2 m_2} \alpha_{j_1 m_1},$$
$$A^+(j_1, j_2; \lambda\mu) = \sum_{m_1, m_2} (j_1 m_1 j_2 m_2 \mid \lambda\mu) \alpha_{j_1 m_1}^+ \alpha_{j_2 m_2}^+,$$
$$B(j_1, j_2; \lambda\mu) = \sum_{m_1, m_2} (-1)^{j_2+m_2} (j_1 m_1 j_2 m_2 \mid \lambda\mu) \alpha_{j_1 m_1}^+ \alpha_{j_2, -m_2}.$$

They obey the symmetry relations

$$A(j_2, j_1; \lambda\mu) = (-1)^{j_1+j_2+1-\lambda} A(j_1, j_2; \lambda\mu), \tag{8.147}$$
$$B^+(j_1, j_2; \lambda\mu) = (-1)^{j_2-j_1+\mu} B(j_2, j_1; \lambda-\mu). \tag{8.147′}$$

We shall again use to quasiboson approximation. The operators A and A^+ then obey the following commutation relations:

$$[A(j, j'; \lambda\mu), A^+(j_2, j_2'; \lambda_2\mu_2)] = \delta_{\lambda\lambda_2}\delta_{\mu\mu_2}(\delta_{jj_2}\delta_{j'j_2'} + (-1)^{j_2'-j_2+\lambda}\delta_{j_2 j'}\delta_{j' j}). \tag{8.148}$$

Let us consider the symmetric case $\varkappa_n^{(\lambda)} = \varkappa_p^{(\lambda)} = \varkappa_{np}^{(\lambda)} \equiv \varkappa^{(\lambda)}$. The multipole–multipole interaction H_Q is then equal to

$$H_Q = -\tfrac{1}{2} \sum_{\lambda, \mu} \varkappa^{(\lambda)} Q_{\lambda\mu}^+ Q_{\lambda\mu}. \tag{8.149}$$

The multipole operator equals

$$Q_{\lambda\mu} = \sum_{\substack{j, j' \\ m, m'}} \langle j'm' \mid i^\lambda Y_{\lambda, -\mu} (\sqrt{m\mathring{\omega}_0}\, r)^\lambda \mid jm \rangle a_{j'm'}^+ a_{jm}, \tag{8.150}$$

and its complex conjugate

$$Q_{\lambda\mu}^+ = (-1)^{\lambda-\mu} \sum_{\substack{j, j', \\ m, m'}} \langle j'm' \mid i^\lambda Y_{\lambda\mu} (\sqrt{m\mathring{\omega}_0}\, r)^\lambda \mid jm \rangle a_{j'm'}^+ a_{jm}. \tag{8.150′}$$

Using the quasiparticle representation,

$$Q_{\lambda\mu}^+ = \frac{(-1)^{\lambda-\mu}}{\sqrt{2\lambda+1}} \sum_{j, j'} f^\lambda(j', j) \Big\{ \tfrac{1}{2} u_{jj'} [A^+(j', j; \lambda\mu) + (-1)^{-\lambda+\mu} A(j', j; \lambda-\mu)]$$
$$+ v_{jj'} B(j', j; \lambda\mu) + \sum_m (-1)^{j-m} \delta_{jj'} v_j^2 \Big\}, \tag{8.151}$$

where

$$\mathfrak{f}^{\lambda}(j', j) = \langle j' | i^{\lambda} Y_{\lambda} (\sqrt{m\mathring{\omega}_0} r)^{\lambda} | j \rangle,$$

The explicit form of matrix element will be discussed in Chapter 10.

The spherical phonon operators are defined by the relation

$$Q_i(\lambda\mu) = \tfrac{1}{2} \sum_{j, j'} \{\psi_{jj'}^{\lambda i} A(j, j'; \lambda\mu) - (-1)^{\lambda-\mu} \varphi_{jj'}^{\lambda i} A^{+}(j, j'; \lambda-\mu)\}. \qquad (8.152)$$

The operators satisfy the commutation relations (8.128) and (8.128′). Consequently, the functions $\psi_{jj'}^{\lambda i}$ and $\varphi_{jj'}^{\lambda i}$ are restricted by the orthogonality relations (8.129), (8.129′), (8.130), and (8.130′). The inverse transformation to (8.152) has the form

$$A(j, j'; \lambda\mu) = \sum_{i} \{\psi_{jj'}^{\lambda i} Q_i(\lambda\mu) + (-1)^{\lambda-\mu} \varphi_{jj'}^{\lambda i} Q_i^{+}(\lambda-\mu)\}, \qquad (8.153)$$

$$A^{+}(j, j'; \lambda\mu) = \sum_{i} \{\psi_{jj'}^{\lambda i} Q_i^{+}(\lambda\mu) + (-1)^{\lambda-\mu} \varphi_{jj'}^{\lambda i} Q_i(\lambda-\mu)\}. \qquad (8.153')$$

The Hamiltonian in the harmonic approximation contains only quadratic terms in Q, Q^{+}, i.e.,

$$H_v = \sum_{j} \varepsilon(j) \sum_{m} \alpha_{jm}^{+} \alpha_{jm} - \tfrac{1}{2} \sum_{\lambda=2, 3, \ldots} \frac{\varkappa^{(\lambda)}}{2\lambda+1} \tfrac{1}{4} \sum_{j, j', j_2, j_2'} \mathfrak{f}^{\lambda}(j', j) \times$$
$$\times \mathfrak{f}^{\lambda}(j_2', j_2) \sum_{i, i'} u_{jj'} u_{j_2 j_2'} \sum_{\mu} (-1)^{\lambda-\mu} g_{j'j}^{\lambda i} g_{j_2' j_2}^{\lambda i'} \{Q_i^{+}(\lambda\mu) Q_{i'}^{+}(\lambda-\mu) + Q_{i'}(\lambda-\mu) \times$$
$$\times Q_i(\lambda\mu) + (-1)^{\lambda-\mu} Q_i^{+}(\lambda\mu) Q_{i'}(\lambda\mu) + (-1)^{\lambda-\mu} Q_{i'}^{+}(\lambda-\mu) Q_i(\lambda-\mu)\}, \qquad (8.154)$$

where

$$g_{j'j}^{\lambda i} = \psi_{j'j}^{\lambda i} + \varphi_{j'j}^{\lambda i}, \quad w_{j'j}^{\lambda i} = \psi_{j'j}^{\lambda i} - \varphi_{j'j}^{\lambda i}. \qquad (8.155)$$

The particle–phonon interaction term is also contained in our Hamiltonian:

$$H_3 = -\tfrac{1}{4} \sum_{\lambda=2, 3, \ldots} \frac{\varkappa^{(\lambda)}}{2\lambda+1} \sum_{j, j', j_2, j_2'} \mathfrak{f}^{\lambda}(j', j) \mathfrak{f}^{\lambda}(j_2', j_2) u_{jj'} v_{j_2 j_2'} \times$$
$$\times \sum_{i} \sum_{\mu} (-1)^{\lambda-\mu} g_{j'j}^{\lambda i} \{[Q_i^{+}(\lambda\mu) + (-1)^{\lambda-\mu} Q_i(\lambda-\mu)] \times$$
$$\times B(j_2', j_2; \lambda-\mu) + B(j_2', j_2; \lambda\mu) [Q_i^{+}(\lambda-\mu) + (-1)^{\lambda+\mu} Q_i(\lambda\mu)]\}. \qquad (8.156)$$

The secular equation is, as before, obtained in the framework of the approximate second quantization method. We shall use the usual variational principle

$$\delta \left\{ \langle Q_i(\lambda\mu) H_v Q_i^{+}(\lambda\mu) \rangle - \frac{\omega_i^{\lambda}}{2} \left[\sum_{j, j'} g_{j'j}^{\lambda i} w_{j'j}^{\lambda i} - 2 \right] \right\} = 0,$$

where

$$\langle Q_i(\lambda\mu) H_v Q_i^{+}(\lambda\mu) \rangle = \tfrac{1}{4} \sum_{j, j'} [\varepsilon(j) + \varepsilon(j')] [(g_{jj'}^{\lambda i})^2 + (w_{jj'}^{\lambda i})^2] - \frac{1}{4} \frac{\varkappa^{(\lambda)}}{2\lambda+1} \left(\sum_{j, j'} \mathfrak{f}^{\lambda}(j', j) u_{jj'} g_{j'j}^{\lambda i} \right)^2. \qquad (8.157)$$

Treating the variations δg and δw as independent,

$$[\varepsilon(j)+\varepsilon(j')]g^{\lambda i}_{j'j}-\omega^{\lambda}_{i}w^{\lambda i}_{j'j} = \frac{\varkappa^{(\lambda)}}{2\lambda+1}\mathfrak{f}^{\lambda}(j',j)u_{j'j}D^{\lambda i}, \tag{8.158}$$

$$w^{\lambda i}_{j'j} = \frac{\omega^{\lambda}_{i}}{\varepsilon(j)+\varepsilon(j')}g^{\lambda i}_{j'j}, \tag{8.158'}$$

where

$$D^{\lambda i} = \sum_{j_2, j_2'} \mathfrak{f}^{\lambda}(j_2', j_2)u_{j_2j_2'}g^{\lambda i}_{j_2'j_2}. \tag{8.158''}$$

After some simple manipulations, we finally obtain the familiar secular equation

$$1 = \frac{\varkappa^{(\lambda)}}{2\lambda+1}\sum_{j,j'}\frac{(\mathfrak{f}^{\lambda}(j,j'))\,u^2_{jj'}[\varepsilon(j)+\varepsilon(j')]}{[\varepsilon(j)+\varepsilon(j')]^2-(\omega^{\lambda}_{i})^2}. \tag{8.159}$$

Using the normalization condition, we can calculate the one-phonon wave functions:

$$g^{\lambda i}_{jj'} = \sqrt{\frac{2}{Y^i(\lambda)}}\frac{\mathfrak{f}^{\lambda}(j,j')\,u_{jj'}[\varepsilon(j)+\varepsilon(j')]}{[\varepsilon(j)+\varepsilon(j')]^2-(\omega^{\lambda}_{i})^2}, \tag{8.160}$$

$$\psi^{\lambda i}_{jj'} = \frac{1}{\sqrt{2Y^i(\lambda)}}\frac{\mathfrak{f}^{\lambda}(j,j')u_{jj'}}{\varepsilon(j)+\varepsilon(j')-\omega^{\lambda}_{i}}, \tag{8.161}$$

$$\varphi^{\lambda i}_{jj'} = \frac{1}{\sqrt{2Y^i(\lambda)}}\frac{\mathfrak{f}^{\lambda}(j,j')u_{jj'}}{\varepsilon(j)+\varepsilon(j')+\omega^{\lambda}_{i}}, \tag{8.161'}$$

where

$$Y^i(\lambda) = \sum_{j,j'}\frac{(\mathfrak{f}^{\lambda}(j,j'))^2u^2_{jj'}\omega^{\lambda}_{i}[\varepsilon(j)+\varepsilon(j')]}{\{[\varepsilon(j)+\varepsilon(j')]^2-(\omega^{\lambda}_{i})^2\}^2}. \tag{8.162}$$

The formulae derived in this section will be applied in Chapters 9 and 10 to the actual case of the vibrational states in deformed and spherical nuclei.

§ 5. The Structure of the Excited 0^+ States

1. The excited $I^\pi = 0^+$ states have a special position in nuclear theory. Mathematical difficulties of the approximate methods employed are quite apparent when these states are calculated. The $I^\pi = 0^+$ states have different structure; some are related to the pairing vibrations, others have the properties of β-vibrations, yet others are related to the spin–quadrupole forces or to the two-phonon states, etc. Many papers are dealing with the different aspects of this problem, for instance, refs. 123, 129, 131, 150, 153, 424, 429–431, 438, 439, 464–470. In this section, we shall discuss the structure of the excited 0^+ states using the approximate second quantization method.

Let us start with the 0^+ states in even–even nuclei and use the framework of the independent quasiparticle model. The ground state is the quasiparticle vacuum; it has the quantum numbers $I^\pi=0^+$. Let us suppose that we are dealing with Ω doubly degenerate average field levels. Consequently, we shall expect Ω two-quasiparticle excited 0^+ states, with both quasiparticle on the same average field level. Besides, other 0^+ states have four, six, or more, quasiparticles, with the pairs of quasiparticles on several levels. Thus, we have the ground state, plus Ω two-quasiparticle 0^+ states, $\Omega+1$ states altogether, while we would expect only Ω 0^+ states. This means that there is one superfluous state, so called "spurious," among the two-quasiparticle states. Similar spurious states exist among the four-, six-, and more, quasiparticle states. Particularly, all states having more quasiparticles than the available number of particles in the system are spurious.

The wave function of a two-quasiparticle 0^+ state (with the blocking effect included) has the form

$$\Psi_0(q_1, q_1) = \{u_{q_1}(q_1, q_1)a^+_{q_1+}a^+_{q_1-} - v_{q_1}(q_1, q_1)\}\times \times \prod_{q\neq q_1} (u_q(q_1, q_1) + v_q(q_1, q_1)a^+_{q+}a^+_{q-})\Psi_{00}. \quad (8.163)$$

The wave functions (8.163) are not orthogonal to each other or to the ground state (4.22). Actually, the overlap equals

$$(\Psi_0^*(q_1, q_1)\,\Psi_0(q_2, q_2)) \equiv \langle F\pm i_1, F\pm i_1 | F\pm i_2, F\pm i_2\rangle = \pm[u_{q_1}(q_1, q_1)\, v_{q_1}(q_2, q_2) - u_{q_1}(q_2, q_2)\, u_{q_1}(q_1, q_2)]\times \times[u_{q_2}(q_1, q_1)\, v_{q_2}(q_2, q_2) - u_{q_2}(q_2, q_2)\, v_{q_2}(q_1, q_1)]\times \times \prod_{q\neq q_1, q_2} [u_q(q_1, q_1)\, u_q(q_2, q_2) + v_q(q_1, q_1)\, v_q(q_2, q_2)]. \quad (8.164)$$

Let us analyze the dependence of the nonorthogonality between the state (8.163) and the ground state (4.22) on the pairing constant G. Table (8.1) shows the matrix elements simi-

TABLE 8.1. *Nonorthogonality of the 0^+ states*

State	G/G_0				
	0.6	0.8	1.0	1.2	1.4
$\langle 0 \mid F-1, F-1\rangle$	10^{-5}	0.13	0.18	0.12	0.07
$\langle 0 \mid F, F\rangle$	10^{-5}	0.27	0.24	0.11	0.05
$\langle 0 \mid F+1, F+1\rangle$	0.02	0.31	0.24	0.09	0.05
$\langle 0 \mid F+2, F+2\rangle$	0.02	0.09	0.16	0.12	0.07
$\langle F, F \mid F+1, F+1\rangle$	0.92	0.52	0.20	0.02	0.01
$\langle F, F \mid F+2, F+2\rangle$	0.01	0.07	0.07	0.02	0.01
$\langle F-1, F-1 \mid F+1, F+1\rangle$	0.02	0.10	0.07	0.02	0.01

lar to (8.164) for different ratios G/G_0; the $G = G_0$ value corresponds to the realistic nuclear forces. The symbol $|0\rangle$ denotes the ground state, while the two-quasiparticle states are denoted by $|F\pm i, F\pm i\rangle$ (the quasiparticles occupy the levels $F\pm i$, counted from the Fermi level). From Table 8.1 it follows that the $|F, F\rangle$ and $|F+1, F+1\rangle$ states form one state when $G\to 0$; at the same time, the other states become orthogonal to each other. The quantity $\langle F, F|F+1, F+1\rangle$ decreases with increasing G, and at the same time the nonorthogonality of the low-lying 0^+ states increases. The nonorthogonality again decreases at very large values of G because the number of the nonorthogonal 0^+ states increases.

2. The spurious state can be excluded when the quasiparticle interaction is taken into account. We shall explain this procedure using the even–even deformed nuclei; the secular equation and the wave functions of the one-phonon 0^+ states will be obtained. We shall use the pairing plus quadrupole–quadrupole force model, and employ the approximate second quantization method. The rotation–vibration coupling and the quasiparticle–phonon interaction will be neglected.

To exclude the spurious state, we have to consider the terms $H_0^\beta(n)$ and $H_0^\beta(p)$ in the Hamiltonians $H_0(n)$ and $H_0(p)$. According to (5.28), we shall write

$$H_0^\beta(n) = -\frac{G_N}{2}\sum_{s,\,s'}[u_s^2A^+(s,s)-v_s^2A(s,s)]\,[u_{s'}^2A(s',s')-v_{s'}^2A^+(s',s')].$$

For simplicity, we shall assume that $\varkappa_n^{(2)} = \varkappa_p^{(2)} = \varkappa_{np}^{(2)} = \varkappa^{(2)}$ (the general case is discussed in ref. (431). The indexes (20) will be omitted in the matrix elements and in the functions ψ, φ, g, and w.

The phonon operators $Q_i(20)$ are defined in (8.127) with $\bar{\psi}_{qq}^i = \bar{\varphi}_{qq}^i = 0$. The H_0^β can be then expressed in terms of the phonon operators. The Hamiltonion of the system has the form

$$\begin{aligned}H_v^\beta = \sum_q \varepsilon(q)\,B(q,q) - \tfrac{1}{4}\sum_{i,\,i'}\Big\{&G_N\sum_{s,\,s'}[(u_s^2-v_s^2)(u_{s'}^2-v_{s'}^2)g_{ss}^i g_{s's'}^{i'}+w_{ss}^i w_{s's'}^{i'}]\\ &+G_Z\sum_{r,\,r'}[(u_r^2-v_r^2)(u_{r'}^2-v_{r'}^2)g_{rr}^i g_{r'r'}^{i'}+w_{rr}^i w_{r'r'}^{i'}]\Big\}Q_i^+(20)\,Q_{i'}(20)\\ &-\frac{\varkappa^{(2)}}{4}\sum_{i,\,i'}\sum_{\substack{q_1,\,q_1',\\ q_2,\,q_2'}} u_{qq'}u_{q_2q_2'}\mathfrak{f}(q,q')\,g_{qq'}^i\mathfrak{f}(q_2,q_2')\,g_{q_2q_2'}^{i'}\{Q_i^+(20)\,Q_{i'}(20)\\ &+Q_{i'}^+(20)\,Q_i(20)+Q_i^+(20)\,Q_{i'}^+(20)+Q_{i'}(20)\,Q_i(20)\}.\end{aligned}\tag{8.165}$$

The expectation value of H_v^β in the one-phonon state equals

$$\begin{aligned}\langle Q_i(20)\,H_v^\beta Q_i^+(20)\rangle = \tfrac{1}{4}&\sum_{q,\,q'}[\varepsilon(q)+\varepsilon(q')]\,[(g_{qq'}^i)^2+(w_{qq'}^i)^2]\\ &-\frac{G_N}{4}\sum_{s,\,s'}[(u_s^2-v_s^2)(u_{s'}^2-v_{s'}^2)g_{ss}^i g_{s's'}^i+w_{ss}^i w_{s's'}^i]\\ &-\frac{G_Z}{4}\sum_{r,\,r'}[(u_r^2-v_r^2)(u_{r'}^2-v_{r'}^2)g_{rr}^i g_{r'r'}^i+w_{rr}^i w_{r'r'}^i]\\ &-\frac{\varkappa^{(2)}}{2}\Big(\sum_{q,\,q'}\mathfrak{f}(q,q')u_{qq'}g_{qq'}^i\Big)^2.\end{aligned}\tag{8.166}$$

Let us use the variational principle in the form (8.59). Varying over δg and δw, we obtain the equations

$$[\varepsilon(s)+\varepsilon(s')]g_{ss'}^i = \omega_i w_{ss'}^i+\delta_{ss'}G_N(u_s^2-v_s^2)\sum_{s''}(u_{s''}^2-v_{s''}^2)g_{s''s''}^i+2\varkappa^{(2)}\mathfrak{f}(s,s')u_{ss'}\sum_{q,\,q'}\mathfrak{f}(q,q')u_{qq'}g_{qq'}^i,$$

$$[\varepsilon(s)+\varepsilon(s')]w_{ss'}^i = \omega_i g_{ss'}^i+\delta_{ss'}G_N\sum_{s''}w_{s''s''}^i,$$

and similar equations for $g_{rr'}^i$ and $w_{rr'}^i$.

$$\begin{vmatrix} X^i(n)+X^i(p)-\dfrac{1}{\varkappa^{(2)}} & V^i(n) & W^i(n) & V^i(p) & W^i(p) \\ V^i(n) & \dfrac{2}{\omega_i^2}\left\{\sum_s \dfrac{2\varepsilon(s)}{4\varepsilon^2(s)-\omega_i^2}-\dfrac{1}{G_N}\right\} & 2\sum_s \dfrac{u_s^2-v_s^2}{4\varepsilon^2(s)-\omega_i^2} & 0 & 0 \\ W^i(n) & 2\sum_s \dfrac{u_s^2-v_s^2}{4\varepsilon^2(s)-\omega_i^2} & \sum_s \dfrac{4\varepsilon(s)\,(u_s^2-v_s^2)^2}{4\varepsilon^2(s)-\omega_i^2}-\dfrac{2}{G_N} & 0 & 0 \\ V^i(p) & 0 & 0 & \dfrac{2}{\omega_i^2}\left\{\sum_r \dfrac{2\varepsilon(r)}{4\varepsilon^2(r)-\omega_i^2}-\dfrac{1}{G_Z}\right\} & 2\sum_r \dfrac{u_r^2-v_r^2}{4\varepsilon^2(r)-\omega_i^2} \\ W^i(p) & 0 & 0 & 2\sum_r \dfrac{u_r^2-v_r^2}{4\varepsilon^2(r)-\omega_i^2} & \sum_r \dfrac{4\varepsilon(r)\,(u_r^2-v_r^2)}{4\varepsilon^2(r)-\omega_i^2}-\dfrac{2}{G_Z} \end{vmatrix} = 0. \tag{8.167}$$

$$\begin{vmatrix} X^i(n)+X^i(p)-\dfrac{1}{\varkappa^{(2)}} & V^i(n) & W^i(n) & V^i(p) & W^i(p) \\ V_i(n) & \sum_s \dfrac{1}{\varepsilon(s)\,[4\varepsilon^2(s)-\omega_i^2]} & 2\sum_s \dfrac{u_s^2-v_s^2}{4\varepsilon^2(s)-\omega_i^2} & 0 & 0 \\ W^i(n) & 2\sum_s \dfrac{u_s^2-v_s^2}{4\varepsilon^2(s)-\omega_i^2} & \sum_s \dfrac{\omega_i^2-4C_n^2}{\varepsilon(s)\,[4\varepsilon^2(s)-\omega_i^2]} & 0 & 0 \\ V^i(p) & 0 & 0 & \sum_r \dfrac{1}{\varepsilon(r)\,[4\varepsilon^2(r)-\omega_i^2]} & 2\sum_r \dfrac{u_r^2-v_r^2}{4\varepsilon^2(r)-\omega_i^2} \\ W^i(p) & 0 & 0 & 2\sum_r \dfrac{u_r^2-v_r^2}{4\varepsilon^2(r)-\omega_i^2} & \sum_r \dfrac{\omega_i^2-4C_p^2}{\varepsilon(r)\,[4\varepsilon^2(r)-\omega_i^2]} \end{vmatrix} = 0. \qquad (8.167')$$

After some straightforward but cumbersome manipulations, the secular equation (8.167) is obtained (see p. 341). The equation has, besides the solutions $\omega_1, \omega_2, \ldots$, also the value $\omega_0 = 0$ as a solution. For a system of particles of only one type (neutrons or protons), (8.167) has the same form as (8.35) obtained earlier by the selfconsistent field method.

Assuming that the correlation functions C_n and C_p satisfy the usual equations (4.19), we can write (8.167) as (8.167′) (see p. 342).

The function $X^i(n)$ was defined in (8.133″), and

$$V^i(n) = 2\sum_s \frac{\mathfrak{f}(s, s)C_n}{\varepsilon(s)\,[4\varepsilon^2(s)-\omega_i^2]}, \tag{8.168}$$

$$W^i(n) = \sum_s \frac{\mathfrak{f}(s, s)\,4C_n[E(s)-\lambda_n]}{\varepsilon(s)\,[4\varepsilon^2(s)-\omega_i^2]}. \tag{8.168′}$$

The expressions for $V^i(p)$, $W^i(p)$ are, naturally, analogous.

The wave functions are found with the use of the normalization condition (8.129″). The result is

$$\psi_{ss'}^{20i} = \frac{1}{\sqrt{2Y^i(20)}}\left\{\frac{\mathfrak{f}(s, s')u_{ss'}}{\varepsilon(s)+\varepsilon(s')-\omega_i} - \delta_{ss'}\frac{C_n}{\varepsilon(s)}\frac{1}{\gamma_n^i}\left[\frac{\Gamma_n^i(s)}{2\varepsilon(s)-\omega_i}+\frac{\Xi_n^i}{\omega_i}\right]\right\}, \tag{8.169}$$

$$\varphi_{ss'}^{20i} = \frac{1}{\sqrt{2Y^i(20)}}\left\{\frac{\mathfrak{f}(s, s')u_{ss'}}{\varepsilon(s)+\varepsilon(s')+\omega_i} - \delta_{ss'}\frac{C_n}{\varepsilon(s)}\frac{1}{\gamma_n^i}\left[\frac{\Gamma_n^i(s)}{2\varepsilon(s)+\omega_i}-\frac{\Xi_n^i}{\omega_i}\right]\right\}, \tag{8.169′}$$

where

$$Y^i(20) = \sum_{q\neq q'} \frac{\mathfrak{f}^2(q, q')u_{qq'}^2\omega_i[\varepsilon(q)+\varepsilon(q')]}{\{[\varepsilon(q)+\varepsilon(q')]^2-\omega_i^2\}^2}+y_n^i+y_p^i, \tag{8.170}$$

$$\gamma_n^i = \sum_{s,\, s'} \frac{4C_n^2-\omega_i^2+4[E(s)-\lambda_n]\,[E(s')-\lambda_n]}{\varepsilon(s)\,[4\varepsilon^2(s)-\omega_i^2]\,\varepsilon(s')\,[4\varepsilon^2(s')-\omega_i^2]}, \tag{8.171}$$

$$\Gamma_n^i(s) = \sum_{s_2,\, s_2'} \frac{\mathfrak{f}(s_2, s_2)}{\varepsilon(s_2)\,[4\varepsilon^2(s_2)-\omega_i^2]\,\varepsilon(s_2')\,[4\varepsilon^2(s_2')-\omega_i^2]}\{4[E(s_2')-\lambda_n]\,[E(s_2)-\lambda_n]$$
$$-4[E(s)-\lambda_n]\,[E(s_2)-\lambda_n]+4[E(s)-\lambda_n]\,[E(s_2')-\lambda_n]+4C_n^2-\omega_i^2\}, \tag{8.171′}$$

$$\Xi_n^i = \sum_{s,\, s'} \frac{\mathfrak{f}(s, s)}{\varepsilon(s)\,[4\varepsilon^2(s)-\omega_i^2]}\;\frac{4C_n^2-\omega_i^2+4[E(s)-\lambda_n]\,[E(s')-\lambda_n]}{\varepsilon(s')\,[4\varepsilon^2(s')-\omega_i^2]}. \tag{8.171″}$$

In (8.170),

$$y_n^i = \sum_s \frac{2\omega_i C_n^2}{\varepsilon(s)\,[4\varepsilon^2(s)-\omega_i^2]^2}\left(\mathfrak{f}(s, s) - \frac{\Gamma_n^i(s)}{\gamma_n^i}\right). \tag{8.170′}$$

The quantity y_p^i is determined by a similar expression.

The energies $\omega_1, \omega_2, \ldots$, of the one-phonon 0^+ states are the solutions of (8.167′). Substituting the calculated values in (8.169) and (8.169′), we obtain the corresponding wave functions. The spurious state has vanishing energy $\omega_0 = 0$, and it is orthogonal to all other solutions.

Let us discuss the properties of eqn. (8.167′). In order to do that, we shall calculate the determinant and rewrite the equation in the form

$$\frac{1}{\varkappa^{(2)}} = F^{20}(\omega), \tag{8.167″}$$

where

$$F^{20}(\omega) = 2\sum_{q,q'} \frac{\mathfrak{f}^2(q,q')\,u_{qq'}^2[\varepsilon(q)+\varepsilon(q')]}{[\varepsilon(q)+\varepsilon(q')]^2-\omega^2} - \frac{4C_n^2}{\gamma_n}\sum_s \frac{\mathfrak{f}(s,s)\,\Gamma_n(s)}{\varepsilon(s)\,[4\varepsilon^2(s)-\omega^2]} - \frac{4C_p^2}{\gamma_p}\sum_r \frac{\mathfrak{f}(r,r)\,\Gamma_p(r)}{\varepsilon(r)\,[4\varepsilon^2(r)-\omega^2]}\,. \tag{8.172}$$

The function $F^{20}(\omega)$ has no singularities at $\omega = 2\varepsilon(q)$, it has first order poles at $\omega = \varepsilon(q)+\varepsilon(q')$, $(q \neq q')$, and at

$$\gamma_n = 0, \quad \gamma_p = 0. \tag{8.173}$$

The pole values (8.173) are independent of the quadrupole constant $\varkappa^{(2)}$. There is always one solution of the secular equation (8.167′) between two poles of the function $F^{20}(\omega)$.

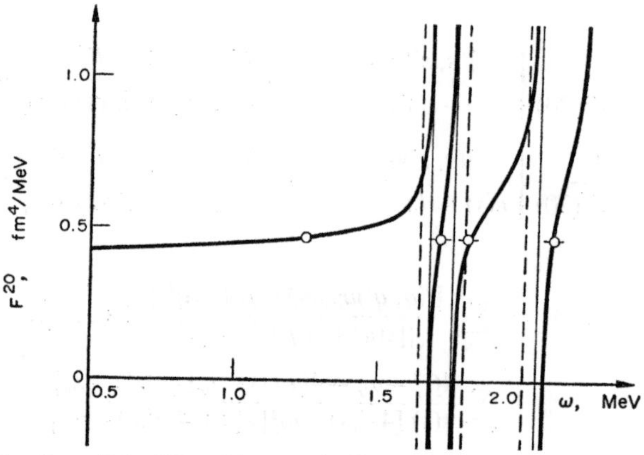

FIG. 8.5. The $F^{20}(\omega)$ function of the $K^\pi = 0^+$ states in ^{164}Er. The vertical dashed lines denote the two-quasiparticle energies $2\varepsilon(q)$; the vertical thin lines show the $\gamma_n = 0$ and $\gamma_p = 0$ poles. The circles denote the solutions of the secular equation (8.167′).

The function $F^{20}(\omega)$ for the nucleus ^{164}Er is shown in Fig. 8.5. The regularity at $\omega = 2\varepsilon(q)$ and the poles at $\gamma_n = 0$ and $\gamma_p = 0$ are apparent. The exclusion of the spurious state somewhat shifted the poles.

Let us analyze a simplified model where all diagonal quadrupole moment matrix elements are equal, i.e., $\mathfrak{f}(q,q) = \mathfrak{f}_0$. In such a model,

$$\Gamma_n^i(s) = \mathfrak{f}_0\gamma_n^i, \quad \Gamma_p^i(r) = \mathfrak{f}_0\gamma_p^i, \quad \Xi_n^i = \mathfrak{f}_0\gamma_n^i, \quad \Xi_p^i = \mathfrak{f}_0\gamma_p^i\,.$$

The secular equation (8.167′) has the form

$$\left\{\sum_{q\neq q'} \frac{f^2(q,q')\,u_{qq'}^2[\varepsilon(q)+\varepsilon(q')]}{[\varepsilon(q)+\varepsilon(q')]^2-\omega_i^2} - \frac{1}{2\varkappa^{(2)}}\right\}\gamma_n^i\gamma_p^i = 0. \tag{8.174}$$

The energies of the 0^+ states are then independent of the $\mathfrak{f}_0$ value.

We have two types of the excited 0^+ states in our $\mathfrak{f}(q, q) = \mathfrak{f}_0$ model. One type is determined by the nondiagonal matrix elements of the quadrupole moment operator. The other type of the 0^+ state does not depend on the quadrupole–quadrupole interaction; it is determined by the pairing interaction H_{pair}. Thus, we are dealing with two distinct and independent collective states.

In real atomic nuclei, the diagonal matrix elements $\mathfrak{f}(q, q)$ are not equal. Consequently, the two collective 0^+ modes are coupled.

3. The spin–quadrupole interaction strongly affects the structure of the excited 0^+ states if the interaction constants $\varkappa_\sigma^{(2)}$ and $\varkappa_0^{(2)}$ are close to each other. Let us show the secular equation for such a case. It was derived in ref. 131 and it is valid in the case of even–even deformed nuclei.

The spin–quadrupole forces are described by an additional term in the Hamiltonian (8.165). According to (8.138), this term equals

$$H_\sigma^{20} = -\frac{\varkappa_\sigma^{(2)}}{4}\sum_{i,i'}\sum_{q,q'} t^{20}(q,q')\,u_{qq'}^{(-)}w_{qq'}^{i}\sum_{q_2,q_2'} t^{20}(q_2,q_2')\,u_{q_2q_2'}^{(-)}w_{q_2q_2'}^{i}\times$$
$$\times\{Q_i^+(20)Q_{i'}(20)+Q_{i'}^+(20)Q_i(20)-Q_i^+(20)Q_{i'}^+(20)-Q_{i'}(20)Q_i(20)\}. \tag{8.175}$$

The secular equation is obtained by the above-described procedure. The expectation value of $H_v^\beta + H_\sigma^{20}$ in the one-phonon state is calculated and minimized. The resulting secular equation can be brought to the form

$$(1-\varkappa^{(2)}F^{20i}(\omega))\,(1-\varkappa_\sigma^{(2)}S^{20i}) = \varkappa^{(2)}\varkappa_\sigma^{(2)}(W^{20i})^2, \tag{8.176}$$

the functions $F^{20i}(\omega)$, $S^{20i}(\omega)$, and $W^{20i}(\omega)$ were defined in (8.172), (8.141′), and (8.141″).

Equation (8.176) can be rewritten in the standardized form

$$\frac{1}{\varkappa^{(2)}} = P^{20i}(\omega, \varkappa_\sigma^{(2)}), \tag{8.176′}$$

where

$$P^{20i}(\omega, \varkappa_\sigma^{(2)}) = F^{20i}(\omega)+\varkappa_\sigma^{(2)}\frac{(W^{20i}(\omega))^2}{1-\varkappa_\sigma^{(2)}S^{20i}}\,. \tag{8.177}$$

As we have mentioned earlier, the function $F^{20i}(\omega)$ has simple poles at values satisfying the relations $\gamma_n^i = 0$, $\gamma_p^i = 0$, and at $\omega = \varepsilon(q)+\varepsilon(q')$, $(q \neq q')$. It is easy to show that the more complicated function $P^{20i}(\omega, \varkappa_\sigma^{(2)})$ does not have the poles at $\omega = \varepsilon(q)+\varepsilon(q')$; instead, its poles are the solutions of the equations

$$\gamma_n^i = 0, \quad \gamma_p^i = 0, \tag{8.178}$$
$$1-\varkappa_\sigma^{(2)}S^{20i}(\omega) = 0. \tag{8.178′}$$

Thus, the spin–quadrupole interaction replaces the poles $\omega = \varepsilon(q)+\varepsilon(q')$ by the solutions of (8.178′) while leaving the poles (8.178) unchanged. The downward shift of the new poles causes an increase in the density of low-lying 0^+ states. Such an effect can even bring two or more 0^+ states inside the energy gap.

4. Let us discuss now the properties of the 0^+ states using a model including only the pairing superfluid interaction H_{pair}. This means that we have to use $\varkappa^{(2)} = 0$ in the secular

equation (8.167). Consequently, the equation disintegrates into two equations; each of them describes one type of particles (neutrons or protons). The spurious solution $\omega = 0$ is, however, still present.

The secular equation of one system, for example, of the neutron system, has the form

$$\left\{\sum_s \frac{2\varepsilon(s)}{4\varepsilon^2(s)-\omega_i^2} - \frac{1}{G_N}\right\} \left\{\sum_s \frac{2\varepsilon(s)\,(u_s^2-v_s^2)^2}{4\varepsilon^2(s)-\omega_i^2} - \frac{1}{G_N}\right\} - \omega_i^2 \left(\sum_s \frac{u_s^2-v_s^2}{4\varepsilon^2(s)-\omega_i^2}\right)^2 = 0. \quad (8.179)$$

The correlation function is not determined by the usual equation (4.19) in such a case.

The last term in (8.179) vanishes if the single-particle levels are symmetrically distributed around the Fermi level. Thus, in a certain approximation, the last term can be neglected. Assuming further that the correlation function nevertheless satisfies (4.19), i.e.,

$$\frac{1}{G_N} = \sum_s \frac{2}{\varepsilon(s)},$$

(8.179) simply means that

$$(\omega_i^2-4C^2)\sum_s \frac{1}{\varepsilon(s)\,[4\varepsilon^2(s)-\omega_i^2]} = 0. \quad (8.180)$$

When the last term in (8.179) is included and (4.19) is still valid, the resulting more complicated equation is

$$\gamma_n^i \equiv \sum_{s,\,s'} \frac{4[E(s)-\lambda_n]\,[E(s')-\lambda_n]+4C_n^2-\omega_i^2}{\varepsilon(s)\,\varepsilon(s')\,[4\varepsilon^2(s)-\omega_i^2]\,[4\varepsilon^2(s')-\omega_i^2]} = 0, \quad (8.179')$$

or, in a slightly different form,

$$4\left(\sum_s \frac{E(s)-\lambda_n}{\varepsilon(s)\,[4\varepsilon^2(s)-\omega_i^2]}\right)^2 = (\omega_i^2-4C_n^2)\left(\sum_s \frac{1}{\varepsilon(s)\,[4\varepsilon^2(s)-\omega_i^2]}\right)^2.$$

Finally, we can compute the square root of the above-written expression and obtain[431]

$$\sum_s \frac{\sqrt{\omega_i^2-4C_n^2} \pm 2[E(s)-\lambda_n]}{\varepsilon(s)\,[4\varepsilon^2(s)-\omega_i^2]} = 0. \quad (8.181)$$

The wave function of the 0^+ state is equal to

$$\tfrac{1}{2}\sum_s \{\psi_{ss}^i A^+(s,s) - \varphi_{ss}^i A(s,s)\}\Psi, \quad (8.182)$$

where

$$\left.\begin{aligned} A(s,s) &= \sqrt{2}\alpha_{s+}\alpha_{s-}, \\ \psi_{ss}^i &= \frac{1}{\sqrt{Y_n^i}}\,\frac{\varepsilon(s)\sqrt{\omega_i^2-4C_n^2}-\omega_i[E(s)-\lambda_n]}{\omega_i\varepsilon(s)\,[2\varepsilon(s)-\omega_i]}, \\ \varphi_{ss}^i &= \frac{1}{\sqrt{Y_n^i}}\,\frac{\varepsilon(s)\sqrt{\omega_i^2-4C_n^2}-\omega_i[E(s)-\lambda_n]}{\omega_i\varepsilon(s)\,[2\varepsilon(s)+\omega_i]}, \end{aligned}\right\} \quad (8.182')$$

$$Y_n^i = \sum_s \frac{\omega_i(\sqrt{\omega_i^2-4C_n^2}-2[E(s)-\lambda_n])^2}{\varepsilon(s)\,[4\varepsilon^2(s)-\omega_i^2]^2}. \quad (8.182'')$$

Thus, we have seen the consequencies of excluding the spurious states. When the single-particle levels are symmetrically distributed around λ_n, (8.181) shows that the lowest 0^+ state has the energy $\omega_1 = 2C_n$, i.e., it has the same value as the energy gap. The spurious states in spherical nuclei are excluded in a similar way.[438]

In spherical nuclei with one closed shell, the 0^+ states can be calculated by direct diagonalization. The G = constant approximation was used, and the number of subshells was truncated. For example, calculations[464] of the tin isotopes use all neutron subshells between $N = 50$ and $N = 82$; 105 seniority zero configurations form the basis.

Comparison between the exact diagonalization and the results of approximate calculations shows that the latter are sometimes insufficiently accurate. Particularly, the situations with low density of single-particle levels near the Fermi level are not adequately described.

It is interesting to know how important is the approximation G = constant in the case of the 0^+ states. The δ-forces and the finite range forces have, typically, the diagonal matrix elements larger than the nondiagonal elements. The energies of the 0^+ states depend not only on the strength, but also on the form of the residual interaction. In diagonalization of the Gaussian forces in ref. 464, the density of the low-lying 0^+ states was larger than in similar calculations with G = constant. The most striking feature of the excited 0^+ states is the large cross-section of the two nucleon transfer reactions. Similarly, the α-transitions between ground states of even–even nuclei have enhanced probabilities.

5. Using this property, Bohr[439] has suggested the existence of a new mode of the collective motion, the so called "pairing vibrations." Each collective mode is characterized by a certain operator $\mathcal{F}$. For example, the quadrupole moment operator is the characteristic operator of the quadrupole vibrations. For the pairing vibrations, the characteristic operator is usually chosen as

$$A^+ = \sum_s a_{s+}^+ a_{s-}^+ , \tag{8.183}$$

i.e., it is a creation or annihilation operator of a particle pair on one level.

The pairing forces

$$-GA^+A \tag{8.184}$$

determine the properties of the collective pairing vibrations in the same way as the quadrupole–quadrupole forces determine the quadrupole collective states. The analogy goes even further. The nuclei with strong pairing correlations are described by the potential

$$U_{\text{pair}} = -C^*A - CA^+ , \tag{8.185}$$

with $C = G\langle|A|\rangle$.

The average potential in the permanently deformed nuclei has violated the angular momentum conservation. Similarly, the potential (8.185) violates the particle number conservation law.

The operator A is not Hermitian; the correlation function is, therefore, generally complex. The phase angle φ, related to the particle number operator [see (5.59)] is usually chosen $\varphi = 0$; such a choice gives real values of the correlation functions C_n and C_p.

The correlation function of magic nuclei fluctuates around zero average value. The corresponding excited states have, therefore, a vibrational character. When the average value of the correlation function C becomes larger than the fluctuations, we are dealing with a system described by the static pairing field. The nuclear motion is split into intrinsic motion related to the quasiparticles, and the collective motion analogous to the rotational or vibrational states of deformed nuclei. The equation determining the correlation function is then similar to the corresponding cranking model formula.

The vibrational 0^+ states are related to the quasiparticle interaction (5.28), i.e.,

$$H_0^\beta(n) = -\frac{G_N}{2}\sum_s [u_s^2 A^+(s,s) - v_s^2 A(s,s)] \sum_{s'} [u_{s'}^2 A(s',s') - v_{s'}^2 A^+(s',s')].$$

This expression can be modified in the following way:

$$H_0^\beta(n) = \frac{G_N}{8}\left\{\sum_s [A^+(s,s) - A(s,s)]\right\}^2 - \frac{G_N}{8}\left\{\sum_s (u_s^2 - v_s^2)[A^+(s,s) - A(s,s)]\right\}^2; \qquad (8.186)$$

the first term is important for excluding the spurious state, and the second term describes pairing vibrational states.

The neutron pairing vibrational states in vicinity of the doubly magic nucleus ^{208}Pb were extensively studied. The two-neutron transfer reactions (t, p) and (p, t) on ^{208}Pb target strongly populate the ground states of ^{210}Pb or ^{206}Pb, respectively. Such an enhanced cross-section is related to the neutron pairing correlations. Consequently, the ^{210}Pb and ^{206}Pb ground states can be treated as collective excitations of the nucleus ^{208}Pb. The excitations are characterized by the pair of quantum numbers (n_1, n_2); n_1 denotes the number of correlated pairs removed from ^{208}Pb, n_2 denotes the number of correlated pairs added to ^{208}Pb. Thus, the ground states are characterized by the quantum numbers (0, 0) for ^{208}Pb, (1, 0) for ^{206}Pb, (0, 1) for ^{210}Pb, etc. The energies of the pairing vibrations are simply equal to $(n_1+n_2)\omega_{\text{pair}}$ in the harmonic approximation.

Figure 8.6 shows schematically the pairing vibrational states based on ^{208}Pb. The arrows connect the states with the enhanced two-neutron transfer reaction cross-section. Such reactions have the following obvious selection rules in the harmonic approximation:

$$\left.\begin{aligned} \Delta n_1 &= \pm 1, & \Delta n_2 &= 0, \\ \Delta n_1 &= 0, & \Delta n_2 &= \pm 1. \end{aligned}\right\} \qquad (8.187)$$

Most of the states shown in Fig. 8.6 were actually experimentally observed.[354] Their energies and cross-sections are close to the predictions of the harmonic approximation.

Deviations from the harmonic theory (8.187) are related to the anharmonicity of the vibrations.[471, 472]

The pairing collective states form rotation-like bands in nonmagic nuclei. The energies of the involved states are described by the usual rigid rotor formula. The states of such a type have been discussed in ref. 439.

6. The effective interaction should fulfill all requirements following from the general laws. The principle of gradient invariance is one of them. According to this principle,[467, 468] the effective interaction should be (at least approximately) invariant when the wave functions are transformed by

$$\varphi_q(r) \to e^{i\Lambda(r)}\varphi_q(r), \qquad (8.188)$$

where $\Lambda(r)$ is an arbitrary function.

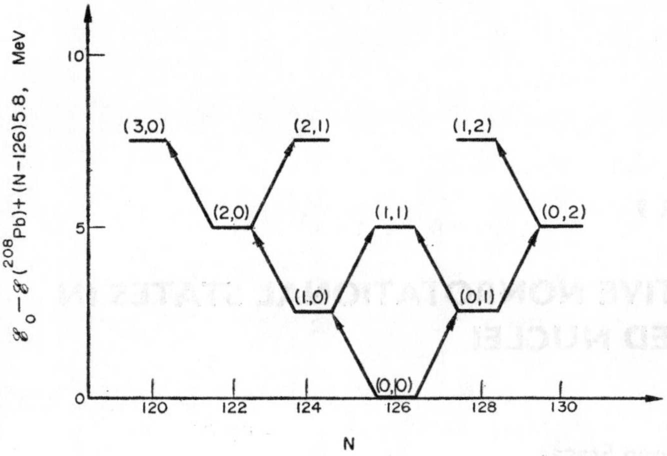

FIG. 8.6. Neutron pairing vibrations in vicinity of ^{208}Pb. (n_1, n_2) are the pairing vibration quantum numbers; n_1 is the number of removed pairs, n_2 is the number of added pairs. The arrows denote the enhanced two-neutron transfer reactions.

The condition of gradient invariance means that the particle–particle interaction G should satisfy the following relation:

$$\langle q|\Lambda|q'\rangle = \tfrac{1}{2}\sum_{q_2,\, q_2'} \langle q+, q'-|G|q_2'-, q_2+\rangle \frac{\varepsilon(q_2)+\varepsilon(q_2')}{2\varepsilon(q_2)\,\varepsilon(q_2')} \langle q_2|\Lambda|q_2'\rangle. \tag{8.189}$$

The usually employed approximation (3.54), i.e., $G = \text{const}$, or

$$\langle q+, q'-|G|q_2'-, q_2+\rangle = G\delta_{qq'}\delta_{q_2q_2'},$$

obviously violates the condition (8.189).

Reference 467 suggests a modification of the interaction (4.2) by addition of several terms. Thus, we have a new interaction

$$\langle q+, q'-|G|\, q_2'-, q_2+\rangle = G\Big\{\delta_{qq'}\delta_{q_2q_2'} + \sum_{n=1,\,2,\,\ldots} \langle q+|\mathfrak{f}_n|q'-\rangle\langle q_2+|\mathfrak{f}_n|q_2'-\rangle^*\Big\}, \tag{8.190}$$

where $\langle q+|\mathfrak{f}_n|q'-\rangle$ are the single-particle matrix elements of a certain function $\mathfrak{f}(r)$. The condition (8.189) is valid if the functions $\mathfrak{f}_n$ satisfy the orthogonality conditions

$$G\sum_{q,\, q'} \frac{\varepsilon(q)+\varepsilon(q')}{4\varepsilon(q)\,\varepsilon(q')} \langle q+|\mathfrak{f}_n|q'-\rangle\langle q'-|\mathfrak{f}_{n'}|q+\rangle = \delta_{nn'}. \tag{8.191}$$

In such an approach, the gradient invariance of the theory is achieved if the interaction is modified in the prescribed way. The modification is, however, ambiguous; the exact form and the number of the functions $\mathfrak{f}_n$ are not determined. Such terms were included in ref. 467. It was found that many excited 0^+ states are sensitive to such a modification.

In this chapter we have discussed the main mathematical methods used for analyzing the vibrational states of even–even nuclei. In the next two chapters we shall show how to improve these methods by including the quasiparticle–phonon interaction.

CHAPTER 9

COLLECTIVE NONROTATIONAL STATES IN DEFORMED NUCLEI

§ 1. One-phonon States

1. This chapter will discuss the vibrational states in deformed nuclei. The experimental data on collective nonrotational states will be systemized and compared with the results of calculations. We shall use the model with pairing and multipole–multipole interaction, and study the role of the spin–quadrupole and surface δ-interactions. The quasiparticle–phonon interaction will also be discussed.

Let us start with one-phonon states in even–even deformed nuclei. We shall use the mathematical apparatus developed in the preceding chapter, but only the harmonic vibrations will be treated (the anharmonic corrections will be discussed in § 5).

The vibrational states in even–even nuclei are interpreted as one-phonon states. The nuclear model Hamiltonian which includes the pairing superfluid interactions and the multipole–multipole forces has the form

$$H_v = \sum_q \varepsilon(q)\, B(q,q) - \sum_{\lambda=2,3} \frac{\varkappa^{(\lambda)}}{4} \sum_{\mu\geq 0} \sum_{i,i'} \sum_{q,q',q_2,q_2'} u_{qq'} u_{q_2q_2'} \times$$
$$\times \left[\mathfrak{f}^{\lambda\mu}(q,q') g_{qq'}^{\lambda\mu i} + \bar{\mathfrak{f}}^{\lambda\mu}(q,q') \bar{g}_{qq'}^{\lambda\mu i}\right] \left[\mathfrak{f}^{\lambda\mu}(q_2,q_2') g_{q_2q_2'}^{\lambda\mu i'} + \bar{\mathfrak{f}}^{\lambda\mu}(q_2,q_2') \bar{g}_{q_2q_2'}^{\lambda\mu i'}\right] \times$$
$$\times \left[Q_i^+(\lambda\mu)\, Q_{i'}(\lambda\mu) + Q_{i'}^+(\lambda\mu)\, Q_i(\lambda\mu) + Q_i^+(\lambda\mu)\, Q_{i'}^+(\lambda\mu) + Q_{i'}(\lambda\mu)\, Q_i(\lambda\mu)\right] \tag{9.1}$$

(we have assumed that $\varkappa^{(\lambda)} \equiv \varkappa_n^{(\lambda)} = \varkappa_p^{(\lambda)} = \varkappa_{np}^{(\lambda)}$). All notation was explained in § 4 of Chapter 8. The calculations performed with the Hamiltonian (9.1) in the framework of the approximate second quantization method (random phase approximation) will be characterized by the symbol $P+QQ$.

Let us repeat the main formulae related to the one-phonon states. The ground state of an even–even nucleus is the phonon vacuum determined by the equation

$$Q_i(\lambda\mu)\Psi = 0. \tag{9.2}$$

The formula (8.55′) shows the explicit form of Ψ. The one-phonon state wave function is defined by

$$Q_i^+(\lambda\mu)\Psi \tag{9.3}$$

with the phonon creation operator

$$Q_i^+(\lambda\mu) = \tfrac{1}{2} \sum_{q,q'} \{\psi_{qq'}^{\lambda\mu i} A^+(q,q') - \varphi_{qq'}^{\lambda\mu i} A(q,q') + \bar{\psi}_{qq'}^{\lambda\mu i} \bar{A}^+(q,q') - \bar{\varphi}_{qq'}^{\lambda\mu i} \bar{A}(q,q')\}. \tag{9.4}$$

The orthogonality of various one-phonon states and their orthogonality to the ground state can be easily proven using the definitions (9.2) and (9.3) and the commutation relations (8.128). The function $\psi_{qq'}^{\lambda\mu i}$, $\bar{\psi}_{qq'}^{\lambda\mu i}$, $\varphi_{qq'}^{\lambda\mu i}$, and $\bar{\varphi}_{qq'}^{\lambda\mu i}$ are determined by the formulae (8.135), (8.135′), (8.135″), (8.136), (8.169), and (8.169′).

The energies of one-phonon states, except for $K^\pi = 0^+$, are the solutions of the secular equation

$$\frac{1}{\varkappa_0^{(\lambda)}} = 2 \sum_{q,q'} \frac{(f^{\lambda\mu}(q,q')u_{qq'})^2 [\varepsilon(q)+\varepsilon(q')]}{[\varepsilon(q)+\varepsilon(q')]^2 - (\omega_i^{\lambda\mu})^2} \equiv X^{\lambda\mu}(\omega). \tag{9.5}$$

The summation in (9.5) involves the pairs of states (q, q') allowed by the selection rules of the matrix elements $\mathfrak{f}^{\lambda\mu}(q, q')$ and $\bar{\mathfrak{f}}^{\lambda\mu}(q, q')$. Each pair (q, q') contributes to (9.5) by

$$X^{\lambda\mu i}(q,q') = 4\varkappa_0^{(\lambda)} \frac{(f^{\lambda\mu}(q,q')u_{qq'})^2 [\varepsilon(q)+\varepsilon(q')]}{[\varepsilon(q)+\varepsilon(q')]^2 - (\omega_i^{\lambda\mu})^2}, \tag{9.6}$$

i.e., (9.5) can be rewritten as

$$1 = \sum_{q>q'} X^{\lambda\mu i}(q,q'). \tag{9.5′}$$

Many states participate in the righthand side of (9.5′) if the corresponding state is collective. Equation (9.5) was discussed in detail in the second section of the preceding chapter.

The secular equation (8.167′) of the one-phonon 0^+ states is more complicated. The summation in (8.167′) includes the diagonal $q = q'$ terms and the nondiagonal $q \neq q'$ terms satisfying the selection rules of matrix elements $\mathfrak{f}^{20}(q, q')$. The form of the individual contributions is apparent from (8.172). The whole problem was discussed in detail in the last section of the preceding chapter.

The one-phonon state wave function $Q_i^+(\lambda\mu)\Psi$ contains contributions of many two-quasiparticle states. The weighing factors of various states can be calculated as follows: the normalization condition has the form

$$(\Psi^* Q_i(\lambda\mu) Q_i^+(\lambda\mu)\Psi) = 1, \tag{9.7}$$

or, in a slightly different notation,

$$\frac{1}{Y^i(\lambda\mu)} \sum_{q,q'} y_i^{\lambda\mu}(q,q') = 1, \tag{9.7′}$$

where

$$\frac{y_i^{\lambda\mu}(q,q')}{Y^i(\lambda\mu)} = \frac{1}{2}\{(\psi_{qq'}^{\lambda\mu i})^2 - (\varphi_{qq'}^{\lambda\mu i})^2\}. \tag{9.7″}$$

The quantity $y_i^{\lambda\mu}(q, q')$ is simply (except for $\lambda = 2$, $\mu = 0$) equal to

$$y_i^{\lambda\mu}(q,q') = \frac{(f^{\lambda\mu}(q,q')u_{qq'})^2 [\varepsilon(q)+\varepsilon(q')]\omega_i^{\lambda\mu}}{\{[\varepsilon(q)+\varepsilon(q')]^2 - (\omega_i^{\lambda\mu})^2\}^2}. \tag{9.8}$$

In the $\lambda = 2$, $\mu = 0$ case, the function $y_i^{\lambda\mu}(q, q')$ is determined by (9.8) for $q \neq q'$. If $q = q'$, the formula becomes more complicated:

$$y_i^{20}(s, s) = 2\frac{C_n^2\omega_i^{20}}{\varepsilon(s)} \frac{\mathfrak{f}^{20}(s, s) - \dfrac{\Gamma_n^i(s)}{\gamma_n^i}}{[4\varepsilon^2(s)-(\omega_i^{20})^2]^2} \left\{ \mathfrak{f}^{20}(s, s) - \frac{\Gamma_n^i(s)}{\gamma_n^i} - \frac{\Xi_n^i}{\gamma_n^i} \frac{4\varepsilon^2(s)-(\omega_i^{20})^2}{(\omega_i^{20})^2} \right\}. \quad (9.8')$$

(the notation was explained in § 5 of Chapter 8).

We have stated earlier that the two-quasiparticle states with relatively large energies contribute non-negligibly to the secular equation. On the other hand, the same states contribute much less to the normalization (9.8). For example, the distant states contributing 1% to the normalization change the energy ω_1 of the first one-phonon state by 10%.

Not all solutions of the secular equation are collective. We shall characterize a state as a two-quasiparticle state if one of the components contributes more than 98% to its normalization.

2. Let us now discuss the results of numerical calculations and the choice of the corresponding parameters.

The particle number is not a conserved quantity in the approximate second quantization method. Consequently, we have to assure that the average neutron and proton numbers have correct values. This problem was previously solved by the blocking effect method. We have demonstrated the related improvements earlier in our discussion of the quasiparticle excited states.

However, the correct handling of the blocking effect in the one-phonon states is quite difficult. Thus, we shall follow ref. 431, and include the blocking effect only approximately. The task is accomplished by two operations. Firstly, the secular equations (9.5) and (8.167') are supplemented by the conditions

$$\begin{aligned} N &= \left\langle Q_i(\lambda\mu) \sum_{s,\sigma} a_{s\sigma}^{+} a_{s\sigma} Q_i^{+}(\lambda\mu) \right\rangle, \\ Z &= \left\langle Q_i(\lambda\mu) \sum_{r,\sigma} a_{r\sigma}^{+} a_{r\sigma} Q_i^{+}(\lambda\mu) \right\rangle, \end{aligned} \quad (9.9)$$

and the new values of chemical potentials λ_n, λ_p are found for each one-phonon state. Secondly, for several lower states, the two-quasiparticle energies $\varepsilon(q)+\varepsilon(q')$ are replaced by the corresponding values $\mathscr{E}_0(q, q')-\mathscr{E}_0$ calculated according to (4.21) and (4.48).

The shifts in the λ_n and λ_p values are, generally, quite small. Their inclusion changes the one-phonon state energies only by 50–100 keV. On the other hand, the change in the pole energies, i.e., the replacement of $\varepsilon(q)+\varepsilon(q')$ by $\mathscr{E}_0(q, q')-\mathscr{E}_0$ may strongly affect certain one-phonon states, particularly those with a predominantly two-quasiparticle structure.

The one-phonon states in even–even deformed nuclei with $150 < A < 190$ and $228 < A < 254$ are calculated using the single-particle energies of the Nilsson or Woods–Saxon potentials. The calculations based on the Nilsson potential were performed in refs. 426–433 and 473–476. Here we shall use the results of papers 474–476 which use the Nilsson model parameters of Table 1.4. These parameters will be denoted by the symbol N in the following discussion.

The Woods–Saxon potential contains an essential A-dependence. To account for it, we have divided the whole region of deformed nuclei into several groups. In refs. 474–477, the $150 < A < 190$ region is divided into four parts: $A = 155$ ($\beta_0 = 0.31$), $A = 165$ ($\beta_0 = 0.31$), $A = 173$ ($\beta_0 = 0.26$), and $A = 181$ ($\beta_0 = 0.23$). The $234 < A < 254$ region is divided into two parts: $A = 237$ ($\beta_0 = 0.20$, $\beta_4 = 0.08$) and $A = 247$ ($\beta_0 = 0.23$, $\beta_4 = 0.06$). In the following tables and figures the corresponding results are denoted by letters WS. The single-particle energies and the wave functions of refs. 55 and 56 are used.

The pairing interaction constants G_N and G_Z of the formulae (4.52′) were employed.

The multipole–multipole interaction constants $\varkappa^{(\lambda)}$ are determined empirically. They are chosen by demanding the best overall agreement between the experimental data and the calculated energies of the one-phonon states. As we have discussed earlier, the actual values of $\varkappa^{(\lambda)}$ strongly depend on the number of states (q, q') included in (9.5) and (8.167′).

This important point may be understood in the following way. Let us rewrite the secular equation (9.5) approximately as

$$\frac{1}{\varkappa^{(\lambda)}} = 2 \sum_{q,q' \in \theta_1} \frac{(f^{\lambda\mu}(q, q')u_{qq'})^2 [\varepsilon(q)+\varepsilon(q')]}{[\varepsilon(q)+\varepsilon(q')]^2-(\omega_i^{\lambda\mu})^2} + 2 \sum_{q,q' \in \theta_2} \frac{(f^{\lambda\mu}(q, q'))^2 u_{qq'}^2}{\varepsilon(q)+\varepsilon(q')}. \tag{9.10}$$

We have divided the set (q, q') into two groups. One-phonon states in different nuclei are described correctly if the group (θ_2) contains only the states which satisfy

$$\varepsilon(q)+\varepsilon(q') \gg \omega_1^{\lambda\mu}, \tag{9.11}$$

and if neither q nor q' come close to the Fermi surface in the considered region of nuclei.

Under these circumstances, the interaction constant $\varkappa^{(\lambda)}$ can be renormalized using the relation

$$\frac{1}{\varkappa_0^{(\lambda)}} = \frac{1}{\varkappa^{(\lambda)}} - 2 \sum_{q,q' \in \theta_2} \frac{(f^{\lambda\mu}(q, q')u_{qq'})^2}{\varepsilon(q)+\varepsilon(q')}. \tag{9.12}$$

After this normalization, only the (q, q') pairs belonging to the region θ_1 should be included in the secular equation (9.5). Obviously the constant is larger when the size of θ_1 is decreased. Different authors use different sets of states (q, q') in their calculations. Consequently, it is quite difficult to compare the resulting $\varkappa^{(\lambda)}$ values.

Another problem is the A-dependence on the constants $\varkappa^{(2)}$ and $\varkappa^{(3)}$. To determine this dependence we shall use the general properties of the short-range forces; we have stated already that the matrix elements of such forces are inversely proportional to the volume of the system, i.e., they are proportional to A^{-1}. This conclusion may be easily verified in the extreme case of the δ-forces, assuming that the single-particle wave functions have a constant value inside the nucleus and vanish outside. The same conclusion is valid for noninteracting particles confined in the nuclear volume. Thus, for the multipole–multipole interaction, we shall assume that

$$\varkappa^{(\lambda)}(\langle r^\lambda \rangle)^2 \sim A^{-1}.$$

Remembering that $\langle r^\lambda \rangle \approx A^{\lambda/3}$,

$$\varkappa^{(2)} = \frac{k^{(2)}}{A^{7/3}}, \quad \varkappa^{(3)} = \frac{k^{(3)}}{A^3}, \tag{9.13}$$

with $k^{(2)}$, $k^{(3)}$ independent of A.[435]

The calculations using the Nilsson potential use the dimensionless matrix element proportional to $A^{-\lambda/3}$. Therefore, in such a case,

$$\varkappa_0^{(2)} = \frac{k_N^{(2)}}{A^{5/3}}\,\text{MeV}, \quad \varkappa_0^{(3)} = \frac{k_N^{(3)}}{A^2}\,\text{MeV}. \tag{9.14}$$

The following values of the parameters were used in the numerical calculations:(129, 433, 474–476)

for $150 < A < 190$:

$$\left.\begin{aligned} k_N^{(2)} &\approx 250\ \text{MeV}, & \varkappa_0^{(2)} &\approx 50\ \text{keV},\\ k_N^{(3)} &\approx 100\ \text{MeV}, & \varkappa_0^{(3)} &\approx 4\ \ \text{keV}; \end{aligned}\right\} \tag{9.14'}$$

for $228 < A < 254$:

$$\left.\begin{aligned} k_N^{(2)} &\approx 330\ \text{MeV}, & \varkappa_0^{(2)} &\approx 40\ \ \text{keV},\\ k_N^{(3)} &\approx 130\ \text{MeV}, & \varkappa_0^{(3)} &\approx 2.5\ \text{keV}. \end{aligned}\right\} \tag{9.14''}$$

In the calculations using the Woods–Saxon potential, the A-dependence of $\varkappa^{(\lambda)}$ is determined by (9.13). The constant $\varkappa^{(\lambda)}$ has then the dimensionality MeV/(fm)$^{2\lambda}$. In refs. 474–477 the number of pairs (q, q') in the actinide $228 < A < 254$ region was considerably larger than the number of the states in the rare earths $150 < A < 190$ region. This means that the renormalization of $\varkappa^{(\lambda)}$ is different in the two regions. The resulting constants have the following values:

for $150 < A < 190$:

$$\left.\begin{aligned} k^{(2)} &\approx 400\,\frac{\text{MeV}}{(\text{fm})^4}, & \varkappa_0^{(2)} &\approx 2.5\,\frac{\text{keV}}{(\text{fm})^4},\\ k^{(3)} &\approx 220\,\frac{\text{MeV}}{(\text{fm})^6}, & \varkappa_0^{(3)} &\approx 0.04\,\frac{\text{keV}}{(\text{fm})^6}; \end{aligned}\right\} \tag{9.13'}$$

for $228 < A < 254$:

$$\left.\begin{aligned} k^{(2)} &\approx 310\,\frac{\text{MeV}}{(\text{fm})^4}, & \varkappa_0^{(2)} &\approx 1\,\frac{\text{keV}}{(\text{fm})^4},\\ k^{(3)} &\approx 180\,\frac{\text{MeV}}{(\text{fm})^6}, & \varkappa_0^{(3)} &\approx 0.01\,\frac{\text{keV}}{(\text{fm})^6}. \end{aligned}\right\} \tag{9.13''}$$

The experimental data of refs. 171–173, 258, 332–333, 348–349, and 478–494 were used in the tables of this and subsequent sections.

3. We shall turn our attention to the one-phonon $\lambda = 2$, $\mu = 2$ states. The lowest state $(i = 1)$ is usually identified as the γ-vibrational state.

Let us see how well the lowest solutions ω^{22} of (9.5) agree with the experimental $K^\pi = 2^+$ energies. We shall denote the value of $\varkappa_0^{(2)}$ resulting in the correct ω_1^{22} energy by $\varkappa_{\text{exp}}^{(2)}$. Instead of comparing the experimental and calculated energies, we shall compare the $\varkappa_{\text{exp}}^{(2)}$ values with the expected values (9.13) and (9.14). Figure 9.1 shows such a comparison; the quantity $\varkappa_{\text{exp}}^{(2)} A^{7/3}$ is plotted for the nuclei with $150 < A < 190$ and for calculations using the Woods-

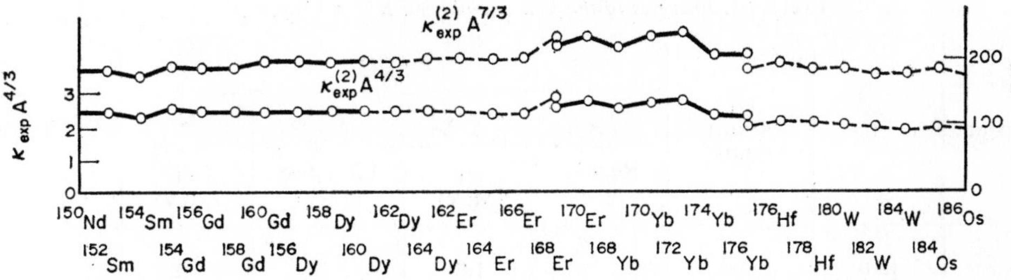

FIG. 9.1. The quadrupole–quadrupole interaction constants $\varkappa^{(2)}_{exp}A^{7/3}$ and $\varkappa^{(2)}_{exp}A^{4/3}$. The calculated and experimental energies of the lowest $K^\pi = 2^+$ states coincide when the $\varkappa^{(2)}_{exp}$ value is used. The full curves denote the nuclei in the vicinity of $A = 155$ or $A = 173$, respectively; the dashed curves show the nuclei around $A = 165$ or $A = 181$, respectively. The calculations were performed in ref. 474 using the Woods–Saxon potential.

Saxon potential. The quantity $\varkappa^{(2)}_{exp}A^{7/3}$ is apparently constant in the considered region, except for the slight fluctuations between ^{168}Er and ^{176}Yb. Note that the A-dependence of $\varkappa^{(2)}$ is not quite uniquely determined. Figure 9.1 shows also another curve, $\varkappa^{(2)}_{exp}A^{4/3}$ which is still reasonably close to a horizontal line. Similar calculations using the Nilsson model potential give somewhat larger fluctuations of the quantity $\varkappa^{(2)}_{exp}A^{5/3}$.

Thus, this comparison shows that our theory describes quite well the energies of the first one-phonon 2^+ states. It is also apparent that the calculations using Woods–Saxon potential are somewhat better than the calculations using the Nilsson potential.

The detailed comparison between the experimental and calculated $K^\pi = 2^+$ energies is recorded in Tables 9.1 and 9.2. Many nuclei with $150 < A < 190$ and $228 < A < 254$ are included. The calculations clearly reproduce the main characteristic features of the ω_1^{22} energies. Particularly, the minimum observed in dysprosium and erbium isotopes is correctly reproduced by the calculations.

Let us discuss the structure of the $K^\pi = 2^+$ states. As an example we shall consider the first and second $K^\pi = 2^+$ states in ^{166}Er. Table 9.3 shows the two-quasiparticle components contributing substantially to the $Q_1^+(22)\Psi$ and $Q_2^+(22)\Psi$ states. The values of the matrix elements $f^{22}(q, q')$ and of the poles $\mathscr{E}_0(q, q')-\mathscr{E}_0$ are also shown. The contributions are characterized by the quantities $X^{221}(q, q')$ and $X^{222}(q, q')$, defined in (9.6), and by the normalization contributions (9.7′) and (9.7″). Both considered states have clearly a collective structure; many two-quasiparticle states contribute to the righthand side of the secular equation (9.5).

A majority of the lowest $K^\pi = 2^+$ states are collective and contain many two-quasiparticle components. This fact is clearly visible in Table 9.4 which shows the structure of such states in ^{154}Sm, ^{158}Gd, ^{164}Dy, ^{178}Hf, and ^{182}W.

The first $K^\pi = 2^+$ state in ^{172}Yb is somewhat exceptional. The two-quasiparticle neutron $nn512\uparrow$–$521\downarrow$ state contributes 98.8% to the normalization. The function $X^{22}(\omega)$ for this nucleus is shown in Fig. 9.2. The first 2^+ state has almost pure two-quasiparticle structure while the second state is collective. However, if a larger $\varkappa_0^{(2)}$ value is used, the situation is reversed. The first state becomes collective and the second state acquires the two-quasiparticle structure. This example clearly demonstrates the interrelation between the two-quasiparticle and collective states.

TABLE 9.1. *Energies of the first one-phonon $K^{\pi} = 2^{+}$ states*

Element	A	ω_1 (exp.) (MeV)	*P+QQ*				*SDI*
			WS			N	N
			Region of A	ω_1 (MeV)	First pole (MeV)	ω_1 (MeV)	ω_1 (MeV)
Nd	150	1.060		1.0	2.2n	1.1	–
Sm	152	1.088		1.0	2.2n	1.1	1.72
	154	1.437		1.3	2.3n	1.4	1.73
	156	–		1.2	2.3n	1.4	–
Gd	154	0.998		1.0	2.2n	0.9	1.46
	156	1.155	155	1.2	2.3n	1.2	1.45
	158	1.185		1.2	2.3n	1.2	1.39
	160	0.988		1.0	1.9n	1.2	1.22
Dy	156	0.890		0.92	2.2n	–	–
	158	0.948		1.1	2.3n	0.84	1.05
	160	0.966		1.0	2.3n	0.87	0.99
Dy	162	0.890		0.83	2.2n	0.83	0.83
	164	0.761		0.80	1.8n	0.67	0.65
Er	160	–		0.95	2.0p	0.96	–
	162	0.897		0.85	2.0p	0.98	–
	164	0.858		0.86	2.0p	0.96	0.93
	166	0.786	165	0.80	1.8n	0.79	0.77
	168	0.821		1.1	1.8n	1.1	0.97
	170	0.931		1.3	1.4n	1.4	1.19
Yb	166	–		1.3	2.2n	–	–
	168	0.986		0.96	2.8n	–	1.19
	170	1.140		1.4	1.8n	–	–
Yb	172	1.486		1.5	1.5n	–	1.59
	174	1.630		1.6	1.9n	1.6	1.53
	176	1.254		1.1	2.1n	1.5	1.08
Hf	170	–	173	1.0	2.0n	–	–
	172	–		1.5	1.7n	–	–
	174	1.227		1.2	1.5n	–	–
	176	1.343		1.5	2.0n	1.6	1.51
Yb	178	–		1.1	2.1n	1.3	–
Hf	178	1.174		1.4	2.0p	1.5	–
	180	1.2		1.2	2.0p	1.3	–
W	178	–		1.7	2.0p	1.5	–
	180	1.7	181	1.3	2.0p	1.4	–
	182	1.222		1.2	2.0p	1.2	–
	184	0.904		0.8	1.6n	1.0	–
	186	0.730		0.7	1.5n	1.1	–
Os	182	–		1.4	2.3n	1.3	–
	184	0.940		1.1	2.1n	1.1	–
	186	0.768		0.7	1.6n	0.8	–
	188	0.633		0.6	1.5n	1.0	–

TABLE 9.2. *Energies of the first $K^\pi = 2^+$ and 0^+ states*

Element	A	ω_1 (MeV)					
		$K^\pi = 2^+$			$K^\pi = 0^+$		
		Experimental	Calculated		Experimental	Calculated	
			WS	N		WS	N
Th	228	0.969	0.9	1.0	0.825	0.83	0.82
	230	0.783	0.8	0.7	0.636	0.63	0.65
	232	0.788	0.8	0.8	0.725	0.70	0.72
	234		1.0	0.9			1.2
U	232	0.893	0.9	0.9	0.694	0.60	0.69
	234	0.927	0.9	0.9	0.810	0.81	0.78
	236		1.1	1.0	0.910	0.90	0.90
	238	1.061	1.2	1.0	0.925	0.85	1.1
Pu	238	1.030	1.1	1.0	0.943	1.05	0.96
	240	0.945	1.0	1.0	0.863	1.0	0.87
	242		1.4	1.2		0.6	0.90
Cm	242		1.3	1.0		0.92	1.0
	244		1.4	1.3		0.85	1.0
	246	1.126	1.1	1.2		1.2	1.2
	248			1.1			1.2
Cf	248		1.2	1.1		1.0	1.2
	250	1.032	1.0	1.0		0.9	1.3
	252		0.8	0.8		1.1	0.92
Fm	252			1.0		1.1	1.1
	254	0.692	0.8	0.8		1.1	0.88

The second states ω_2^{22} have the energies between the first and second poles of the secular equation (9.5). In many cases, the two poles are close to each other, and, consequently, they determine the position of the second state. Many such states have a collective structure; occasionally, the first and second poles overwhelmingly contribute to the state. Several second $K^\pi = 2^+$ states have a purely two-quasiparticle structure. The varying structure of the second $K^\pi = 2^+$ states is demonstrated in Table 9.5. Some second $K^\pi = 2^+$ states contain noticeable two-phonon components.

4. In this paragraph we shall discuss the one-phonon $\lambda = 2$, $\mu = 0$ states. The 0^+ states primarily related to the quadrupole–quadrupole forces are identified as the β-vibrations, and the states primarily related to the pairing forces are identified as the pairing vibrations. The energies ω_i^{20} are obtained by solving the secular equation (8.167′).

Tables 9.2 and 9.6 show the calculated energies of the first excited $K^\pi = 0^+$ states (in $150 < A < 190$ region, the second states are also shown). The $\varkappa_0^{(2)}$ values used in the calculations are close to the values used in similar calculations of the γ-vibrational states. The agreement between the experimental and theoretical ω_1^{20} energies is worse than in the ω_1^{22} case. The function $\varkappa_{\text{exp}}^{(2)} A^{7/3}$ is quite far from being constant. Nevertheless, the calculations

TABLE 9.3. *Structure of the* $K^{\pi} = 2^{+}$ *states in* ^{166}Er ($\omega_1^{22} = 0.86$ MeV, $\omega_2^{22} = 1.9$ MeV)

Two quasi-particle configuration	$f^{22}(q, q')$	Poles $\varepsilon(q)+\varepsilon(q')$ (MeV)	$X^{221}(q, q')\times 100$	$y_1^{22}(q, q')$ (%)	$X^{222}(q, q')\times 100$	$y_2^{22}(q, q')$ (%)
			Neutron states			
523↓−521↓	6.7	1.8	15.4	40.0	−100.2	55.9
521↑+521↓	7.3	2.1	14.1	26.1	73.6	23.1
633↑−651↑	−4.3	2.5	3.1	3.6	6.4	0.5
624↑−642↑	−3.4	3.2	1.8	1.2	2.5	0.1
642↑−660↑	−5.0	3.5	1.1	0.7	1.6	–
651↑+660↑	5.5	4.0	0.7	0.3	1.0	–
402↓+400↑	−9.5	4.6	1.5	0.5	1.9	–
512↑−510↑	9.7	4.8	1.9	0.5	2.0	–
505↑−503↑	8.5	5.8	5.8	1.1	6.3	–
514↑−512↑	6.6	5.9	3.5	0.7	3.8	–
			Proton states			
411↑+411↓	6.0	2.5	9.7	18.3	55.9	19.9
413↓−411↓	−4.9	3.0	3.7	3.0	5.8	0.2
532↑−530↑	3.8	6.2	1.1	0.2	1.2	–
404↑−402↑	6.5	6.4	3.3	0.5	3.5	–
523↑−521↑	5.1	6.8	1.6	0.2	1.7	–

TABLE 9.4. *The two-quasiparticle components of the first one-phonon* $K^{\pi} = 2^{+}$ *states. The quantity* y_1^{22} *in % is shown*

Two quasi-particle configuration	$f^{22}(q, q')$	^{154}Sm	^{158}Gd	^{164}Dy	^{172}Yb	^{178}Hf	^{182}W
			Neutron states				
633↑−651↑	−4.3	4.3	4.0	4.0	0.04	–	–
523↓−521↓	6.7	9.7	24.9	41.0	0.2	–	–
642↑−660↑	−5.0	22.0	9.6	0.8	0.01	–	–
505↑−503↑	8.5	1.0	1.1	1.2	0.04	1.5	1.0
512↑−521↓	0.4	–	–	–	98.8	0.1	0.03
512↑−510↑	9.7	–	–	–	–	42.6	24.7
514↓−512↓	−8.2	–	–	–	–	28.8	25.5
402↓+400↑	−9.5	10.2	3.6	0.6	–	–	–
512↓+510↑	10.3	–	–	–	–	10.2	35.0
651↑+660↑	5.5	11.8	2.9	0.4	0.01	–	–
521↑+521↓	7.3	21.9	36.8	28.5	0.1	–	–
			Proton states				
523↑−541↑	−2.8	1.6	1.1	0.4	–	–	–
413↓−411↓	−4.9	1.6	3.1	3.3	0.03	0.6	0.1
532↑−550↑	−3.5	2.0	0.5	–	–	–	–
404↑−402↑	6.5	0.8	0.7	0.5	0.01	0.5	0.3
404↓−402↓	7.3	–	–	–	–	1.4	1.6
402↑−400↑	9.5	–	–	–	–	1.2	2.7
411↑+411↓	6.0	1.8	4.1	10.4	0.08	2.1	0.4

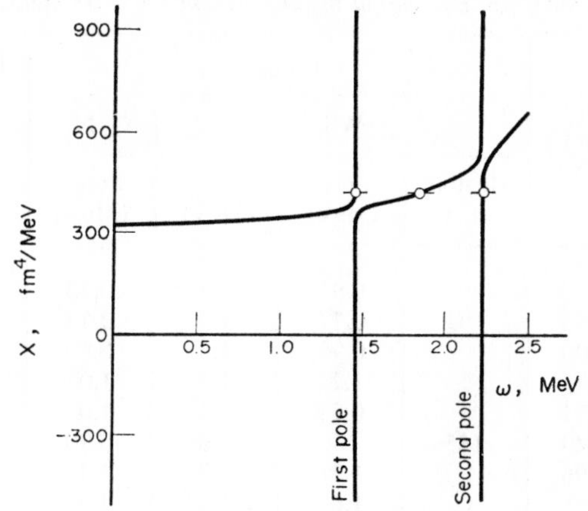

FIG. 9.2. The functon $X^{22}(\omega)$ of the 2^+ states in ^{172}Yb. The empty circles denote the solutions of the secular equation (9.5) at $\varkappa^{(2)} = \varkappa_0^{(2)}$.

TABLE 9.5. *Energies and main components of the second $K^\pi = 2^+$ states*

Nucleus	Energy (MeV)		Poles (MeV)		Structure (%)
	Experimental	Theoretical	First	Second	
^{174}Yb	2.189	2.0	1.99	2.26	$nn512\uparrow\ 521\downarrow - 99.7$
^{172}Yb	1.599	1.8	1.50	2.26	$nn512\uparrow\ 510\uparrow - 30.8$; $nn523\downarrow\ 521\downarrow - 15.1$; $nn521\downarrow\ 521\uparrow - 11.6$; $nn624\uparrow\ 642\uparrow - 8.3$; $pp411\downarrow\ 411\uparrow - 6.5$; $nn514\downarrow\ 512\downarrow - 4.9$;
^{164}Dy	1.987	1.9	1.77	2.05	$nn523\downarrow\ 521\downarrow - 49.4$; $nn521\downarrow\ 521\uparrow\ 46.5$; $pp411\downarrow\ 411\uparrow - 2.5$

describe the overall behavior of the 0^+ states correctly. For example, the relative ordering of $K^\pi = 2^+$ and $K^\pi = 0^+$ states is sufficiently explained.

Let us analyze the structure of the 0^+ states. Table 9.7 shows as an example the components of the first and second excited 0^+ states in ^{164}Er. The main contribution to the wave function $Q_i^+(20)\Psi$ comes from the diagonal terms. The nondiagonal terms, related to the matrix elements $\mathfrak{f}^{20}(q, q')$, $q \neq q'$, contribute usually less than 10% to the wave function normalization. This behavior has two reasons. Firstly, the diagonal matrix elements are usually larger than the nondiagonal elements. Secondly, most of the low-lying poles of (8.167′) are the $\gamma_n^i = 0$ or $\gamma_p^i = 0$ poles, i.e., the poles related to the diagonal matrix elements. We should remember, however, that the nondiagonal matrix elements contribute relatively more to the secular equation. On the other hand, the distant poles give only a small contribution to the normalization.

The solutions of the secular equation (8.167′) have a simple two-quasiparticle structure only when ω^{20} is close to the $\varepsilon(q)+\varepsilon(q')$, $(q \neq q')$ pole. On the other hand, the structure of the 0^+ state remains complex even if the energy ω^{20} is close to the $2\varepsilon(q)$ or to the $\gamma^1 = 0$ pole. This fact is demonstrated in Table 9.7 where the third and fifth columns show the solution for $\varkappa^{(2)} = 0$, i.e., when the quadrupole–quadrupole forces are completely turned off.

TABLE 9.6. *Energies of the one-phonon $K^{\pi} = 0^{+}$ states*

Element	A	ω_1 (exp.) (MeV)	$P+QQ$				$P+QQ+T_{\sigma}T_{\sigma}$	*SDI*
			WS			N	N	N
			Region of A	ω_1 (MeV)	ω_2 (MeV)	ω_1 (MeV)	ω_1 (MeV)	ω_1 (MeV)
Nd	150	0.687		0.63	2.1	0.78	0	–
Sm	152	0.685		0.77	1.9	0.75	0.67	–
	154	1.100		0.96	1.8	0.80	0.96	–
	156			1.2	1.7	1.0	–	–
Gd	154	0.681		0.74	2.1	1.0	0.91	0.656
	156	1.049	155	1.2	1.8	0.98	1.13	0.823
	158	1.196		1.2	1.7	1.1	1.39	–
	160			1.6	1.7	1.6	1.28	–
Dy	156	0.674		0.80	2.1	1.2	–	–
	158	0.993		1.1	1.8	1.1	1.23	0.969
	160	1.263		1.3	1.7	1.2	1.45	–
Dy	162			1.1	1.8	1.6	1.34	–
	164			1.5	2.3	1.8	0.97	–
Er	160			1.3	1.7	1.0	–	–
	162			1.1	1.7	1.1	–	–
	164	1.246	165	1.2	1.7	1.4	1.34	1.425
	166	1.460		1.6	1.7	1.7	0.98	1.410
	168	1.215		1.4	1.5	1.8	1.23	–
	170			1.0	1.7	1.4	1.38	–
Yb	166			1.0	1.8	–	–	–
	168	1.191		1.3	1.8	–	1.16	–
	170	1.069		1.2	1.3	–	1.36	–
Yb	172	1.043		1.0	1.7	–	1.42	–
	174			1.3	1.7	1.3	1.43	–
	176			1.4	1.7	1.5	–	–
Hf	170		173	0.8	1.7	–	–	–
	172			0.9	1.3	–	–	–
	174	0.827		0.9	1.6	–	–	1.252
	176	1.250		1.3	1.5	1.2	–	1.327
Yb	178			1.2	1.5	1.4	–	–
Hf	178	1.197		1.4	1.8	1.5	–	1.265
	180			1.3	1.9	1.4	–	–
W	178		181	1.2	2.1	1.1	–	–
	180		181	1.4	1.8	1.5	–	1.233
	182			1.3	1.9	1.4	–	–
	184			1.1	1.8	1.6	–	–
	186			1.4	1.8	1.4	–	–
Os	182			0.8	1.8	1.1	–	–
	184			0.8	1.8	1.2	–	–
	186			0.8	1.8	1.1	–	–
	188	1.086		0.9	1.8	0.8	–	1.111

TABLE 9.7. *Contribution of various two-quasiparticle states to the first and second excited* 0^+ *states in* ^{164}Er, (%)

Model	$P+QQ$				$P+QQ+T_\sigma T_\sigma$			
Potential	WS				N			
	First state $\omega_1^{20} = 1.20$ MeV		Second state $\omega_2^{20} = 1.67$ MeV		First state $\omega_1^{20} = 1.25$ MeV		Second state $\omega_2^{20} = 1.7$ MeV	
					$\varkappa^{(2)} = 5.3A^{-4/3}\mathring{\omega}_0$		$\varkappa^{(2)} = 5.3A^{-4/3}\mathring{\omega}_0$	
Two-quasi-particle configuration	$\varkappa^{(2)} = \varkappa_0^{(2)}$	$\varkappa^{(2)} = 0$	$\varkappa^{(2)} = \varkappa_0^{(2)}$	$\varkappa^{(2)} = 0$	$\varkappa_\sigma^{(2)} = 8.8A^{-4/3}\mathring{\omega}_0$	$\varkappa_\sigma^{(2)} = 0$	$\varkappa_\sigma^{(2)} = 8.8A^{-4/3}\mathring{\omega}_0$	$\varkappa_\sigma^{(2)} = 0$
				Neutron states				
633↑−633↑	2.1	2.2	2.2	1.1	0.8	3.2	0.8	0.9
523↓−523↓	5.4	34.8	35.2	60.9	3.9	20.5	23.5	25.5
642↑−642↑	36.3	39.7	39.6	15.6	2.3	12.2	32.6	35.3
505↑−505↑	37.8	12.2	12.0	6.6	9.1	47.0	8.7	9.4
651↑−651↑	3.5	0.6	0.6	0.5	0.2	1.0	0.6	0.7
402↓−402↓	1.6	0.2	0.2	0.4	—	—	—	—
512↑−512↑	0.4	0.1	0.1	0.4	0.6	3.6	0.8	0.9
521↑−521↑	0.4	1.4	1.5	10.9	2.2	4.2	0.5	0.5
541↓−510↑	—	—	—	—	10.8	0.1	1.3	—
521↑−512↓	—	—	—	—	5.1	—	0.6	—
523↓−512↑	0.2	—	0.1	—	6.2	0.1	1.0	—
660↑−400↑	0.7	—	0.1	—		—	—	—
				Proton states				
411↓−411↓	1.9	3.1	3.5	—	0.1	0.3	4.9	5.1
523↑−523↑	1.6	2.4	2.6	—	1.7	8.0	10.5	11.6
411↑−402↓	—	—	—	—	3.7	—	0.5	—
411↓−420↑	—	—	—	—	3.7	—	0.5	—
541↑−532↓	—	—	—	—	3.9	0.2	0.2	—

The calculations[469] have shown that the first 0^+ states in rare-earth region have a β-vibrational character in the beginning of the deformed region and a pairing vibrational character in the center of the deformed region.

In the interval 1–2 MeV, three or more excited states were observed[482, 488, 495] in a number of even–even nuclei. Such a large density of excited 0^+ states is difficult to explain in our $P+QQ$ model. The first O^+ states in thorium and uranium isotopes contain noticeable two-phonon components.

5. We shall now discuss the octupole one-phonon $\lambda = 3$, $\mu = 0$, 1, 2, and 3 states, i.e., the lowest $K^\pi = 0^-$, 1^-, 2^-, and 3^- states.

The calculations using both the Nilsson and Saxon–Woods potentials predict the first ω_1^{33} energies very close to the corresponding poles of the secular equation (9.5). Thus, most $K^\pi = 3^-$ states have the two-quasiparticle structure.

Let us analyze the structure of the octupole one-phonon states with $\mu = 0$, 1, and 2. According to the calculations,[129] a majority of the lowest $K^\pi = 0^-$ states has a collective structure, while the $K^\pi = 2^-$ and particularly $K^\pi = 1^-$ are less collective. In the $K^\pi = 1^-$

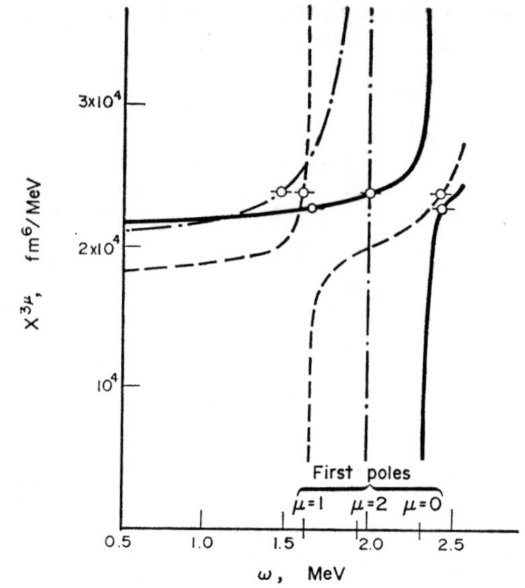

FIG. 9.3. The function $X^{3\mu}(\omega)$ ($\mu = 0, 1, 2$) for ^{166}Er. Full line: $\mu = 0$, $K^\pi = 0^-$. Dot-and-dash line: $\mu = 1$, $K^\pi = 1^-$. Dashed line: $\mu = 2$, $K^\pi = 2^-$. Empty circles denote the solutions of the secular equations; the $\varkappa_0^{(3)}$ for $\mu = 0$ is 1.05 larger than the $\varkappa_0^{(3)}$ for $\mu = 1$ and 2. The calculations used the Woods–Saxon potential.

states, the components corresponding to the first poles contribute 70–95% to the normalization of the one-phonon wave function. The decrease in collectivity with the increasing μ is related to the following fact. The sum of squares of all matrix elements included in the calculations decreases with increasing μ. Reference[129] contains calculations of these sums; for $150 < A < 190$, we obtained 186 ($\mu = 0$), 165 ($\mu = 1$), 173 ($\mu = 2$), and 143 ($\mu = 3$), while for $228 < A < 254$ we obtained 427 ($\mu = 0$), 230 ($\mu = 1$,) 249 ($\mu = 2$), and 173 ($\mu = 3$).

The calculations based on the Woods–Saxon potential predict a stronger collectivity of the $K^\pi = 1^-$ and 2^- states than the calculations based on the Nilsson potential. Figure 9.3 shows as an example the $X^{3\mu}(\omega)$ function in ^{166}Er. Obviously, the first $\mu = 0$ and 2 states are collective; they are strongly shifted from the corresponding poles. On the other hand, the $\mu = 1$ state is very close to its pole, and it is, consequently, noncollective. The octupole states in the beginning of the $150 < A < 190$ region are generally more collective than the states in the middle and at the end of the rare-earth region. The structure of many $K^\pi = 0^-, 1^-$, and 2^- states is shown in Tables 9.8, 9.9, and 9.10. The tables show the largest two-quasiparticle components calculated with the Woods–Saxon potential. The experimental data qualitatively agree with the above conclusions.

The experimental and calculated (using the Woods–Saxon potential) energies of the first one-phonon octupole states with $K^\pi = 0^-, 1^-, 2^-$ are collected in Tables 9.11 and 9.12. The first table contains the nuclei with $150 < A < 190$; the second table contains the actinide nuclei with $228 < A < 254$. It is obvious that the theory well describes the energies of the first octupole states. The first octupole states in both regions are strongly affected by the Coriolis coupling.[496]

TABLE 9.8. *Two-quasiparticle components of the first one-phonon $K^{\pi} = 0^{-}$ states (%)*

Two-quasiparticle configuration	$f^{30}(q, q')$	^{154}Sm	^{162}Dy	^{164}Er	^{166}Er	^{172}Yb	^{178}Hf
			Neutron states				
615↑ − 505↑	−4.8	3.2	6.4	5.8	10.8	3.6	5.4
402↓ − 512↓	−3.9	4.0	4.6	4.2	7.0	2.0	2.8
400↑ − 510↑	−4.0	5.0	4.6	4.2	6.8	2.0	3.2
633↑ − 514↓	−0.9	0.03	0.4	0.4	1.4	72.6	38.6
404↓ − 514↓	−3.8	2.0	2.0	1.8	2.8	0.8	0.3
642↑ − 512↑	0.8	0.2	1.4	1.2	4.4	4.0	2.4
402↑ − 512↑	−3.9	4.0	4.0	3.6	5.4	1.0	0.3
400↑ − 521↓	−1.3	3.2	2.6	2.4	3.8	0.2	0.03
642↑ − 523↓	1.1	10.2	38.0	42.8	15.0	0.2	0.2
402↑ − 523↓	−1.8	1.8	0.4	0.4	0.3	0.01	0.003
651↑ − 521↑	1.3	20.2	4.6	4.4	2.2	0.1	0.2
633↑ − 503↑	0.6	—	0.02	0.02	0.05	0.1	2.8
			Proton states				
651↑ − 541↑	4.8	2.0	3.0	2.8	4.2	1.2	2.8
660↑ − 550↑	6.2	3.2	4.6	4.4	6.4	2.0	4.6
411↓ − 541↓	1.2	0.02	0.2	1.0	1.6	1.0	8.8
431↓ − 541↓	−3.6	2.0	1.0	1.0	1.4	0.4	1.6
404↑ − 514↑	−3.5	3.8	4.4	4.2	6.4	1.2	1.5
413↓ − 532↑	−0.5	8.0	0.2	0.02	0.04	0.002	0.02

TABLE 9.9. *Two-quasiparticle components of the first one-phonon $K^{\pi} = 1^{-}$ states (%)*

Two-quasiparticle configuration	$f^{31}(q, q')$	^{154}Sm	^{160}Dy	^{164}Er	^{168}Er	^{174}Yb
			Neutron states			
624↑ − 514↓	−0.6	2×10^{-3}	2×10^{-3}	2×10^{-3}	10^{-3}	99.6
633↑ − 523↓	1.3	0.7	1.1	30.0	98.2	10^{-3}
651↑ − 523↓	−2.1	7.6	4.6	4.7	0.08	10^{-4}
642↑ − 521↑	2.6	40.3	71.2	51.6	0.2	10^{-3}
400↑ − 521↑	1.4	1.7	0.7	0.2	0.01	10^{-5}
624↑ − 505↑	3.4	0.9	0.9	1.7	0.2	0.01
402↓ − 521↓	−2.1	1.9	1.3	1.4	0.1	10^{-3}
633↑ − 514↑	4.0	1.4	0.8	0.6	0.05	10^{-4}
660↑ − 521↑	−1.7	1.7	0.6	0.6	0.02	10^{-4}
			Proton states			
642↑ − 523↑	3.6	0.03	0.1	0.3	0.03	4×10^{-3}
402↑ − 523↑	1.6	0.02	0.1	0.2	0.02	0.01
413↓ − 541↑	1.3	1.7	0.1	0.01	10^{-3}	10^{-5}
411↑ − 532↑	2.5	20.1	1.0	0.1	0.02	10^{-3}
404↑ − 523↑	3.0	2.4	0.8	0.2	0.02	10^{-4}

TABLE 9.10. *Two-quasiparticle components of the first one-phonon $K^\pi = 2^-$ states (%)*

Two-quasiparticle configuration	$f^{32}(q, q')$	^{158}Gd	^{162}Dy	^{166}Er	^{174}Yb	^{175}Hf	^{182}W
			Neutron states				
402↓ − 514↓	3.9	1.5	1.3	1.3	0.5	0.1	0.01
624↑ − 512↑	4.3	0.3	0.4	1.0	89.8	68.2	1.3
624↑ − 523↓	2.7	1.3	1.5	2.8	1.1	0.3	0.02
400↑ − 512↑	4.7	4.2	2.9	2.9	0.3	0.1	0.07
633↑ − 521↑	3.9	25.5	24.2	65.1	0.4	0.5	0.03
633↑ − 532↓	2.7	2.4	1.2	0.9	0.05	0.03	0.01
642↑ − 521↓	− 1.2	2.1	1.7	5.3	0.07	0.02	10^{-3}
642↑ − 530↑	3.1	3.5	0.8	0.3	0.02	0.04	3×10^{-3}
400↑ + 512↓	4.9	1.2	0.9	0.9	0.4	0.4	0.06
402↑ − 510↑	5.1	1.9	1.3	1.3	0.6	0.2	0.03
651↑ + 521↓	− 1.4	1.7	0.9	0.0	0.05	0.03	2×10^{-3}
			Proton states				
402↑ − 514↑	3.9	0.04	0.05	0.1	1.8	22.1	96.7
651↑ − 523↑	− 2.2	0.08	0.1	0.3	0.1	0.2	0.04
411↑ − 523↑	3.9	36.3	51.4	6.3	0.4	0.3	0.01

The $K^\pi = 0^-$ states in thorium, uranium, and plutonium isotopes are strongly collective. Their energies are smaller than the energies of the β- and γ-vibrations in the same nuclei. The first $I^\pi K = 1^-0$ states in the lightest nuclei included in Table 9.12 have the energies 564 keV in ^{232}U, 508 keV in ^{230}Th, and 328 keV in ^{228}Th. These states have even lower energies in the lighter thorium isotopes and in the isotopes of radium, namely, 231 keV in ^{226}Th, 246 keV in ^{224}Th, 253 keV in ^{226}Ra, 216 keV in ^{224}Ra, and 242 keV in ^{222}Ra. This is a very exceptional situation; the nonrotational states in even–even nuclei with such a low energy cannot be found anywhere else. These states have the energies considerably smaller than the energy gap value.

The low energies of the octupole 1^-0 states in the beginning of the actinide region are related to the behavior of the single-particle levels. The secular equation for these nuclei contains many low-lying poles with large matrix elements. The octupole $K^\pi = 1^-$ and 2^- states are also collective. Generally, the collectivity of the octupole states in the $228 < A < 254$ decreases with the increasing mass number A.

6. Let us compare the above results with a more general theory allowing $\varkappa_{np}^{(\lambda)} \neq \varkappa_n^{(\lambda)}$, $\varkappa_{np}^{(\lambda)} \neq \varkappa_p^{(\lambda)}$. In this paragraph, we shall also discuss the results obtained in the Tamm–Dancoff approximation.

It is easy to solve the secular equation (8.134′) for $\varkappa_{np}^{(2)} \neq \varkappa^{(2)} \equiv \varkappa_n^{(2)} = \varkappa_p^{(2)}$. Such an equation was analyzed in ref. 431. The analysis has shown that the energies equal to the previously calculated energies of the first $K^\pi = 0^+$ and 2^+ states can be obtained for $\varkappa_{np}^{(2)} < \varkappa^{(2)}$ if a somewhat larger $\varkappa^{(2)}$ value is chosen, and vice versa. For example, very similar results have been obtained using $\varkappa^{(2)} = 9.5A^{-4/3}\mathring{\omega}_0$, $\varkappa_{np}^{(2)} = \varkappa^{(2)}$ on the one hand, and $\varkappa^{(2)} = 8.2A^{-4/3}\mathring{\omega}_0$, $\varkappa_{np}^{(2)} = 1.38\varkappa^{(2)}$, or $\varkappa_p^{(2)} = 11.5A^{-4/3}\mathring{\omega}_0$, $\varkappa_{np}^{(2)} = 0.7\varkappa^{(2)}$ on the other

TABLE 9.11. *Energies of the first one-phonon $K^\pi = 0^-, 1^-$, and 2^- states*

Element	A	Region of A	ω_1 (MeV)					
			$K^\pi = 0^-$		$K^\pi = 1^-$		$K^\pi = 2^-$	
			Experimental	Calculated	Experimental	Calculated	Experimental	Calculated
Nd	150			0.8		1.4		1.6
Sm	152		0.966	1.0	1.511	1.4		1.6
	154		0.920	1.0		1.4		1.8
	156			1.0		1.5		1.4
Gd	154		1.241	1.1	1.509	1.4	1.720	1.8
	156	155	1.242	1.1		1.4		1.8
	158		1.263	1.1	0.977	1.1	1.791	1.7
	160			1.2		1.7		1.2
Dy	156			1.2		1.5		1.3
	158			1.2		1.5		1.3
	160			1.5	1.285	1.2	1.264	1.2
Dy	162		1.275	1.3		1.5	1.148	1.0
	164			1.6		1.6	0.977	1.1
Er	160			1.3		1.6		1.9
	162			1.3		1.6		1.6
	164	165	1.386	1.4		1.6		1.5
	166		1.663	1.6	1.824	1.6	1.460	1.5
	168		1.354	1.5		1.6	1.569	1.8
	170			1.8		1.4		1.8
Yb	166			1.4		1.6		1.5
	168			1.5		1.4		1.5
	170		1.364	1.4		1.5		1.9
Yb	172		1.600	1.6		1.2		1.6
	174			1.6		1.6	1.321	1.3
	176			2.1		1.5		1.3
Hf	170	173		1.6		1.5		1.2
	172			1.7		1.5		1.4
	174			1.6		1.4		1.4
	176		1.722	1.7		1.5	1.280	1.2
Yb	178			1.9		1.6		2.0
Hf	178			2.0		1.3	1.250	1.3
	180			2.2		1.4		1.7
W	178			1.6		1.8		0.9
	180	181		2.0		1.3	1.006	1.0
	182			2.2		1.6	1.289	1.3
	184			2.4		1.7		1.3
	186			2.4		2.0		1.0
Os	182			2.0		1.3		1.4
	184			2.3		1.6		1.7
	186			2.4		1.7		1.7
	188			2.5		2.0	1.450	1.3

TABLE 9.12. *Energies of the one-phonon $K^\pi = 0^-, 1^-$, and 2^- states*

Element	A	ω_1 (MeV)					
		$K^\pi = 0^-$		$K^\pi = 1^-$		$K^\pi = 2^-$	
		Experimental	Calculated	Experimental	Calculated	Experimental	Calculated
Th	228	0.328	0.5		0.7	1.123	1.3
	230	0.508	0.5	0.954	1.0	1.080	1.3
	232	1.045	0.7		1.0		1.2
	234		0.7		1.0		1.3
U	232	0.564	0.6	1.146	0.9	1.018	1.1
	234	0.787	0.7	1.438	1.1	0.989	1.1
	236	0.687	0.7		1.0	0.688	0.9
	238	0.680	0.8	0.931	0.9	1.129	1.1
Pu	238	0.605	0.7		1.0		1.3
	240	0.610	0.7		0.9		1.0
	242		1.0		0.9		1.0
Cm	242		0.9		0.9		1.2
	244		1.0		0.9	0.950	1.0
	246		1.4	1.080	0.9	0.842	0.9
	248		1.3		0.8		1.2
Cf	248		1.1		0.8		0.7
	250		1.3		1.0	0.850	0.9

hand. Naturally, in such a case we have essentially one additional parameter, and hence the agreement between the theory and the experiment is improved. However, the improvement is quite small, and thus it is better to use only one parameter, i.e., use the condition $\varkappa_0^{(2)} = \varkappa_{np}^{(2)} = \varkappa_p^{(2)} = \varkappa_n^{(2)}$.

The secular equation in the Tamm–Dancoff approximation (§ 2 of Chapter 8) has the form

$$1 = \varkappa^{(\lambda)} \sum_{q,q'} \frac{(f^{\lambda\mu}(q, q')u_{qq'})^2}{\varepsilon(q)+\varepsilon(q')-\omega_i^{\lambda\mu}} . \tag{9.15}$$

Let us see how this (less accurate) equation describes the first $K^\pi = 0^+$, 2^+, and 0^- states in even–even deformed nuclei. Equation (9.15) was solved in ref. 431; the Nilsson potential was used there. The solutions of (9.15) are not as close to the experimental energies as the solutions of the secular equation (9.5). Besides, the solutions of the Tamm–Dancoff equation (9.15) change less than the solutions of (9.5) when the different nuclei are considered.

Consequently, the more accurate approximate second quantization method using the random phase approximation is clearly preferable compared to the Tamm–Dancoff approximation.

7. Let us discuss the effect of the spin–quadrupole forces on the properties of one-

phonon states. We shall also briefly mention the results obtained from calculations including the surface δ-interaction.

As we have mentioned earlier, the more complicated equation (8.142) describes the spin–quadrupole forces acting together with the usual multipole–multipole forces. Let us see when the spin–quadrupole forces can essentially change our previous conclusions. The results of the calculations treating the Hamiltonian (8.138) will be denoted by $P+QQ+T_\sigma T_\sigma$.

The spin–quadrupole forces cause substantial changes in the structure of the 0^+ one-phonon states. The corresponding secular equation has the form (8.176). The poles $\varepsilon(q)+\varepsilon(q')$ with $q \neq q'$ are replaced now by the solutions of (8.178′), i.e.,

$$1-\varkappa_\sigma^{(2)} S^{20i}(\omega) = 0 .$$

However, the $\gamma_i = 0$ poles are unchanged. Figure 9.4 shows various functions entering (8.176′). We can see that, if $\varkappa_\sigma^{(2)} > \varkappa_0^{(2)}$, then the first 0^+ state has a lower energy but the next three 0^+ states are practically unchanged.

Table 9.7 shows the main components of the first and second 0^+ states in ^{164}Er. Results of both calculations $P+QQ$ and $P+QQ+T_\sigma T_\sigma$ are stated. The spin–quadrupole forces clearly enhance the effect of the nondiagonal matrix elements.

The energies of the first 0^+ states in $150 < A < 190$ region are correctly reproduced if $\varkappa_\sigma^{(2)}$ is chosen 30% larger than the $\varkappa_0^{(2)}$. This conclusion is demonstrated in Table 9.6 showing

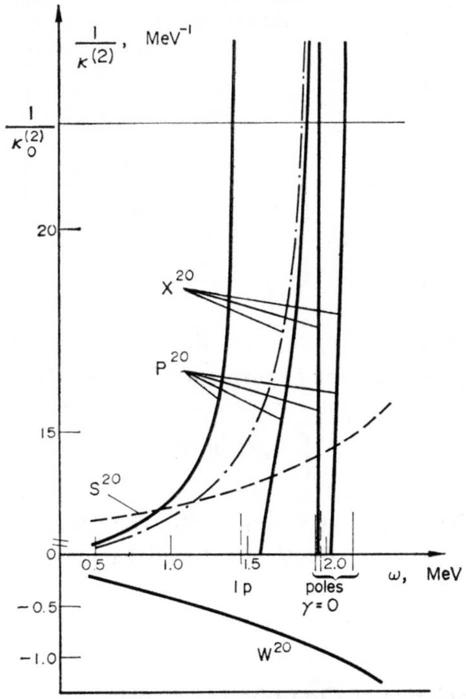

FIG. 9.4. The functions $P^{20}(\omega)$, $F^{20}(\omega)$, $S^{20}(\omega)$, and $W^{20}(\omega)$ for the $K^\pi = 0^+$ states in ^{164}Er and $\varkappa_\sigma^{(2)} = 8.8A^{-4/3}\mathring{\omega}_0$. The short vertical line $1p$ denotes the first pole; the full vertical lines denote the $\gamma = 0$ poles. The functions X^{20} and P^{20} are identical for ω larger than the first $\gamma = 0$ pole. The solutions of (8.176′) are the crossing points of the horizontal line $1/\varkappa_0^{(2)}$ ($\varkappa_0^{(2)} = 5.3A^{-4/3}\mathring{\omega}_0$) and the P^{20} curve. The results of ref. 131, using the Nilsson potential, were used in the calculations.

the results of the $P+QQ+T_\sigma T_\sigma$ calculations performed in ref. 131. The spin–quadrupole part of the wave function $Q_i^+(20)\Psi$ contributes up to 50% to the normalization. The inclusion of the additional parameter $\varkappa_\sigma^{(2)}$ undoubtedly improves the description of the first 0^+ states.

The spin-quadrupole forces, on the other hand, have a little effect on the $K^\pi = 2^+$ states. The corresponding part contributes not more than 10–20% to the normalization. Certain higher lying $K^\pi = 2^+$ states can be, however, strongly affected.

The problem of the spin–quadrupole forces has not been satisfactorily solved as yet. The $\varkappa_\sigma^{(2)}$ values, at which the forces have substantial effect, seem unrealistically large. The stated results for 0^+ states are extremely sensitive to the $\varkappa_\sigma^{(2)}$ value. There is no convincing experimental evidence proving the large role of the spin–quadrupole forces.

The one-phonon energies of the quadrupole and octupole states were also calculated using the surface δ-interaction.[136] The results are denoted by *SDI*; some of them are shown in Tables 9.1 and 9.6. The quadrupole and octupole interaction constants are close to the corresponding value in the $P+QQ$ model. The *SDI* model predicts the energies of the one-phonon states close to the energies calculated in $P+QQ$.

Thus two forces with a different radial dependence describe the one-phonon states equally well. Table 9.13 explains this fact; it shows the average values of the radial matrix elements

TABLE 9.13. *Average radial matrix elements of the multitpole–multipole interaction and surface δ-interaction*

Region	λ	$P+QQ$ $\varkappa^{(\lambda)}(f^{\lambda}_{(q,q')})^2$	*SDI* F
$150 \leqslant A \leqslant 190$	2	29.6	27.5
$150 \leqslant A \leqslant 190$	3	36.0	34.0
$228 \leqslant A \leqslant 254$	2	34.5	29.4
$228 \leqslant A \leqslant 254$	3	29.0	33.5

in both models. The F values of (3.76) and the average $\varkappa^{(\lambda)}(f^{\lambda\mu}(q, q'))^2$ are apparently quite close to each other. This is not so surprising because the integrand in

$$\int_0^\infty u_1(r)r^\lambda u_2(r)r^2\,dr$$

has a maximum near the nuclear surface $r = R_0$. The energies of the one-phonon states therefore depend only weakly on the detailed radial dependence of the employed interaction.[496]

2. Integral Characteristics of One-phonon States

1. The agreement between the calculated and experimental one-phonon state energies is insufficient test of our model; the structure of the states must be adequately described too. The structure is best defined by two types of the characteristics. The integral characteristics

test the collectivity of the involved states, while the differential characteristics probe the individual two-quasiparticle components.

The reduced probabilities $B(\lambda)$ of the electromagnetic multipole radiation and the reduced α-transition probabilities are the integral characteristics of the one-phonon states. The electromagnetic transition probabilities are directly related to the coherent character of the corresponding states; consequently, the enhancement of the reduced probability $B(E\lambda)$ (when compared to the single-particle values) is a commonly used criterion of collectivity.

In this section, we shall derive the formulae for the reduced probabilities of the $E0$, $E2$, and $E3$ transitions involving one-phonon states. The experimental data will be systemized and compared with the results of calculations. The reduced α-transition probabilities will be also discussed.

Let us now derive the formulae for the $E2$ and $E3$ reduced transition probabilities related to the Coulomb excitation of the quadrupole and octupole one-phonon states. Such a reduced transition probability has the form

$$B(E\lambda_0;\, 0^+0 \to I_f^{\pi f} K_f) = (00\lambda_0\mu_0 \,|\, I_f K_f)^2 \mathcal{M}^2. \tag{9.16}$$

where $(\lambda\mu f)$ denotes the quantum numbers of the one-phonon state. We shall use the single-particle units (1.67).

The matrix element $\mathcal{M}$ is easily found from the expression (6.64) for the $E\lambda$ transition operator and

$$A^+(q, q') + A(q, q') = \sum_i g_{qq'}^{\lambda\mu i}(Q_i^+(\lambda\mu) + Q_i(\lambda\mu)). \tag{9.17}$$

Thus, using the approximation $\varkappa^{(\lambda)} = \varkappa_n^{(\lambda)} = \varkappa_p^{(\lambda)} = \varkappa_{np}^{(\lambda)}$, we obtain for the matrix element $\mathcal{M}$ connecting the ground state Ψ and the one-phonon state $Q_i^+(\lambda\mu)\Psi$ the following formula:

$$\mathcal{M} = \frac{e}{\sqrt{2}}\left\{(1+e_q^{(\lambda_0)}) \sum_{r, r'} g_{rr'}^{\lambda\mu i} p^{\lambda_0\mu_0}(r, r') u_{rr'} + e_n^{(\lambda_0)} \sum_{s, s'} g_{ss'}^{\lambda\mu i} p^{\lambda_0\mu_0}(s, s') u_{ss'}\right\}. \tag{9.18}$$

Here the quantity $p^{\lambda_0\mu_0}(q, q')$ is the single-particle matrix element of the electric transition operator. The functions $g^{\lambda\mu i}$ are determined in (8.135), except for the $\lambda = 2$, $\mu = 0$ case when $g_{qq'}^{20i} = \psi_{qq'}^{20i} + \varphi_{qq'}^{20i}$, and the $\psi_{qq'}^{20i}$ and $\varphi_{qq'}^{20i}$ are defined in (8.169) and (8.169′). The matrix element (9.18) depends on the number of two-quasiparticle states participating in the summation. This dependence, similar to the renormalization (9.12) of the interaction constants $\varkappa^{(\lambda)}$, is expressed here in terms of the effective charges $e_p^{(\lambda)}$ and $e_n^{(\lambda)}$. To decrease the number of free parameters we shall usually assume that $e_n^{(\lambda)} = e_p^{(\lambda)} = e_{\text{eff}}^{(\lambda)}$.

Let us show here the explicit form of the matrix element $\mathcal{M}$ (the $\lambda = 2$, $\mu = 0$ case is not included):

$$\mathcal{M} = \frac{e}{\sqrt{Y^i(\lambda\mu)}}\left\{(1+e_{\text{eff}}^{(\lambda_0)}) \sum_{r, r'} \frac{f^{\lambda\mu}(r, r')\, p^{\lambda_0\mu_0}(r, r')\, u_{rr'}^2 [\varepsilon(r)+\varepsilon(r')]}{[\varepsilon(r)+\varepsilon(r')]^2 - (\omega_i^{\lambda\mu})^2}\right.$$
$$\left. + e_{\text{eff}}^{(\lambda_0)} \sum_{s, s'} \frac{f^{\lambda\mu}(s, s')\, p^{\lambda_0\mu_0}(s, s')\, u_{ss'}^2 [\varepsilon(s)+\varepsilon(s')]}{[\varepsilon(s)+\varepsilon(s')]^2 - (\omega_i^{\lambda\mu})^2}\right\}. \tag{9.19}$$

The quantity $\omega_i^{\lambda\mu}$ is the one-phonon energy, $Y^i(\lambda\mu)$ is defined in (8.136). Note that the expression (9.19) does not depend explicitly on the interaction strength $\varkappa^{(\lambda)}$.

TABLE 9.14. *Reduced probabilities $B(E2; 0^+0 \rightarrow 2^+2)$ for transitions from the ground states to the first $K^\pi = 2^+$ states*

Element	A	$B(E2)_{s.p.u.}$ experiment	$P+QQ$, $e_{eff}^{(2)} = 0.2$			SDI, $e_{eff}^{(2)} = 0.4$
			WS		N	N
			Region of A	$B(E2)_{s.p.u.}$	$B(E2)_{s.p.u}$	$B(E2)_{s.p.u.}$
Nd	150	–		4.3	9.2	–
Sm	152	3.4		3.7	9.7	2.6
	154	2.7		3.0	7.8	2.8
	156	–		2.5	6.8	–
Gd	154	4.7		5.9	12.7	3.6
	156	4.0	155	5.2	9.8	3.7
	158	2.7		3.7	8.6	3.9
	160	2.8		2.1	8.0	4.0
Dy	156	8.7		4.7	–	–
	158	6.3		5.7	13.6	5.0
	160	5.7		6.2	12.1	5.3
Dy	162	4.8		5.5	11.3	5.9
	164	4.4		6.1	11.9	6.7
Er	160	–		6.8	12.7	–
	162	7.1		5.0	11.3	–
	164	7.1	165	8.0	10.5	5.2
	166	5.8		5.9	10.6	5.6
	168	5.8		4.7	8.3	4.7
	170	4.0		5.5	5.8	3.8
Yb	166	–		1.7	–	–
	168	4.5		3.9	–	3.1
	170	2.8		4.0	–	–
Yb	172	1.3		0.04	–	0.6
	174	1.4		2.6	1.2	2.7
	176	2.1	173	1.9	1.9	3.7
Hf	170			2.8	–	–
	172			1.8	–	–
	174			1.0	–	0.4
	176			1.5	1.0	2.8
Yb	178			2.6	2.0	–
Hf	178			2.0	1.8	2.1
	180			2.2	2.0	–
W	178			2.4	2.2	–
	180		181	2.2	2.7	2.8
	182			2.3	3.0	3.4
	184			3.0	3.2	4.2
	186			2.1	3.0	4.6
Os	182			3.3	4.9	–
	184			3.4	5.4	5.4
	186			4.6	6.2	7.6
	188			3.3	6.0	10.1

In the more general case $\varkappa_n^{(\lambda)} \neq \varkappa_p^{(\lambda)} \neq \varkappa_{np}^{(\lambda)}$, the formula (9.19) becomes more complicated, namely,

$$\mathcal{M} = e(1+e_p^{(\lambda_0)}) \frac{1-\varkappa_n^{(\lambda)} X^{\lambda\mu i}(n)}{([1-\varkappa_n^{(\lambda)} X^{\lambda\mu i}(n))^2 Y^{\lambda\mu i}(p) + (\varkappa_{np}^{(\lambda)} X^{\lambda\mu i}(p))^2 Y^{\lambda\mu i}(n)]^{1/2}} \times$$
$$\times \sum_{r,r'} \frac{f^{\lambda\mu}(r,r')\, p^{\lambda_0\mu_0}(r,r')\, u_{rr'}^2 [\varepsilon(r)+\varepsilon(r')]}{[\varepsilon(r)+\varepsilon(r')]^2-(\omega_i^{\lambda\mu})^2}$$
$$+ee_n^{(\lambda_0)} \frac{1-\varkappa_p^{(\lambda)} X^{\lambda\mu i}(p)}{[(1-\varkappa_p^{(\lambda)} X^{\lambda\mu i}(p))^2 Y^{\lambda\mu i}(n) + (\varkappa_{np}^{(\lambda)} X^{\lambda\mu i}(n))^2 Y^{\lambda\mu i}(p)]^{1/2}} \times$$
$$\times \sum_{s,s'} \frac{f^{\lambda\mu}(s,s')\, p^{\lambda_0\mu_0}(s,s')\, u_{ss'}^2 [\varepsilon(s)+\varepsilon(s')]}{[\varepsilon(s)+\varepsilon(s')]^2-(\omega_i^{\lambda\mu})^2}, \tag{9.19'}$$

where $X^{\lambda\mu i}(n)$ is defined in (8.133″) and

$$Y^{\lambda\mu i}(n) = \frac{1}{4}\,\frac{\partial}{\partial\omega_i^{\lambda\mu}}\, X^{\lambda\mu i}(n).$$

For the $E2$ excitation of the quadrupole states and $E3$ excitation of the octupole states we have, obviously, $\lambda = \lambda_0$, $\mu = \mu_0$ and $f^{\lambda\mu}(q, q') = p^{\lambda\mu}(q, q')$. Thus all terms in the summation (9.18), 9.19), and (9.19′) give coherent positive contributions in the $i = 1$ case. Therefore, the $E2$ or $E3$ transitions have the enhanced rates. When the energy $\omega_i^{\lambda\mu}$ is close to the pole value $\varepsilon(q_1)+\varepsilon(q_2)$, the matrix element (9.19) converges to the single-particle value $f^{\lambda\mu}(q_1, q_2)u_{q_1q_2}$.

The $B(E\lambda)$ value increases with decreasing energy $\omega_1^{\lambda\mu}$. This means that $B(E\lambda)$ increases when the interaction strength $\varkappa^{(\lambda)}$ is increased. The dependence of $B(E2)$ on $\varkappa^{(2)}$ is illustrated in Fig. 9.5; note the sharp increase at large $\varkappa^{(2)}$ values.

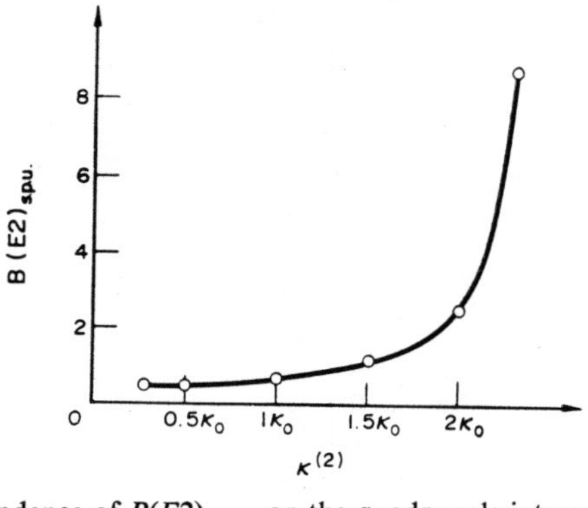

FIG. 9.5. The dependence of $B(E2)_{\text{s.p.u.}}$ on the quadrupole interaction constant $\varkappa^{(2)}$.

The enhancement in Fig. 9.5 amounts to only a few single-particle units. We must remember, however, that the single-particle transition probabilites calculated using the Nilsson or Woods–Saxon wave functions are in many cases 100–1000 times hindered when compared to the $B(E\lambda)_{\text{s.p.u.}}$ values of (1.67′) and (1.67″). Thus, the $B(E\lambda)_{\text{s.p.u.}}$ values close to unity, as shown in Fig. 9.5 for $\varkappa^{(2)} = \varkappa_0^{(2)}$, actually represent a strong enhancement; the corresponding states are, therefore, collective.

2. Let us discuss the $B(E2)$ transition probabilities for the excitation of the $I^\pi K = 2^+2$ states.

The experimental data and the calculated $B(E2)_{\text{s.p.u.}}$ values are collected in Table 9.14 for the nuclei with $150 \leq A \leq 190$ and in Table 9.15 for the $228 \leq A \leq 254$ nuclei. The experimental $B(E2)_{\text{s.p.u.}}$ values clearly show the collective character of the lowest $K^\pi = 2^+$ states. The correlation between the large $B(E2)$ values and small ω_1^{22} values (see Table 9.1) is apparent in the $150 \leq A \leq 190$ region.

The $B(E2)$ values in the $P+QQ$ model were calculated in refs. 129, 430, 432, and 497 using the Nilsson potential, while the Woods–Saxon potential was used in refs. 474–477. The calculations with surface δ-interaction[136] used the Nilsson potential. The calculations contain an additional free parameter, the effective charge $e_{\text{eff}}^{(2)}$; the values $e_{\text{eff}}^{(2)} = 0.2$–$0.5$ were used.

TABLE 9.15. *Reduced probabilities $B(E2)$ and $B(E3)$ for transitions from the ground states to the first $I^\pi K = 2^+2$, 2^+0, and 3^-0 states*

$(\varkappa^{(\lambda)} = \varkappa_{\text{exp.}}^{(\lambda)}, e_{\text{eff}}^{(2)} = 0.2, e_{\text{eff}}^{(3)} = 0.2)$.

Element	A	$B(E2; 0^+0 \to 2^+2)_{\text{s.p.u.}}$ Experimental	Calculated WS	Calculated N	$B(E2; 0^+0 \to 2^+0)_{\text{s.p.u.}}$ Calculated WS	$B(E3; 0^+0 \to 3^-0)_{\text{s.p.u.}}$ Experimental	Calculated
Th	228		6.7	8.3	9.2		28.1
	230		4.3	10.0	18.2		23.1
	232	10.5	3.6	6.6	18.1	12	8.9
	234		8.7	4.4	16.0		17.5
U	232		3.4	7.1	16.4		16.9
	234		4.7	4.4	17.3		15.2
	236		3.5	3.0	10.3		17.3
	238	2	4.6	0.5	6.7	20	17.0
Pu	238		3.8	2.0	7.6		27.2
	240		2.2	0.2	6.5		16.6
	242		0.1	0.8	10.7		5.6
Cm	242		0.2	0.2	5.6		9.0
	244		2.1	0.4	5.8		10.0
	246		3.3	2.1			14.1
	248		3.4	2.6			12.0
Cf	248		0.1	2.7			6.4
	250		10.4	3.3			13.0
	252		6.0	3.6			16.0
Fm	252		0.1	3.4			1.8
	254		5.0	3.6			11.0

The Nilsson and Woods–Saxon potentials do not predict the same $B(E2)$ values. The transition probabilities for the rare-earth nuclei calculated with the Nilsson model wave functions are larger.

Table 9.14 contains also $B(E2)_{\text{s.p.u.}}$ values using the surface δ-interaction (SDI) in calculations. By comparing these values with the experimental data we can conclude that both of the interactions, i.e., $P+QQ$ and SDI, describe the ω_1^{22} and $B(E2)$ values quite well.

3. Let us now discuss the $E2$ and $E0$ transitions involving the $\lambda = 2$, $\mu = 0$ one-phonon states. As we have stated earlier, the wave function $Q_i^+(20)\Psi$ is quite complicated. Consequently, the electromagnetic transition matrix element is also more involved. The $E2$ matrix element has the form (9.18), with both the diagonal $q = q'$ and nondiagonal $q \neq q'$ pairs participating in the summation. Using (8.169)–(8.171′) we find that

$$\sum_{s,s'} g_{ss'}^{20i} \mathfrak{f}^{20}(s,s') u_{ss'} = \frac{1}{\sqrt{2Y^i(20)}} \sum_{s \neq s'} \frac{(\mathfrak{f}^{20}(s,s')\, u_{ss'})^2\, [\varepsilon(s)+\varepsilon(s')]}{[\varepsilon(s)+\varepsilon(s')]^2 - (\omega_i^{20})^2}$$

$$- \frac{4}{\sqrt{2Y^i(20)}} \frac{C_n^2}{\gamma_n^2} \sum_s \frac{\Gamma_n^i(s)\, \mathfrak{f}^{20}(s,s)}{\varepsilon(s)\,[4\varepsilon^2(s) - (\omega_i^{20})^2]}.$$

The first sum in the above-written expression contains coherent contributions, i.e., it renders the main part of the $B(E2)$ value. The β-vibrational 0^+ states therefore have large $B(E2)$ values, while the pairing vibrational states have small $B(E2)$ values contributed mostly by the second sum.

If the spin–quadrupole forces are included together with the quadrupole–quadrupole forces, the function $g_{ss'}^{20i}$, in (9.18) has the form

$$g_{ss'}^{20i} = \sqrt{\frac{2}{Z^i(20)}} \left\{ \frac{\mathfrak{f}^{20}(s,s')\, u_{ss'}[\varepsilon(s)+\varepsilon(s')]}{[\varepsilon(s)+\varepsilon(s')]^2 - (\omega_i^{20})^2} - 2\delta_{ss'} \frac{C_n}{\gamma_n^i} \frac{\Gamma_n^i(s)}{4\varepsilon^2(s) - (\omega_i^{20})^2} \right.$$
$$\left. + \varkappa_\sigma^{(2)} \frac{W^{20i}}{1 - \varkappa_\sigma^{(2)} S^{20i}} \frac{t^{20}(s,s')\, u_{ss'}^{(-)} \omega_i^{20}}{[\varepsilon(s)+\varepsilon(s')]^2 - (\omega_i^{20})^2} \right\}, \tag{9.20}$$

where $t^{20}(q, q')$ is the matrix element (3.74) and

$$Z^i(20) = Y^i(20) + \frac{\varkappa_\sigma^{(2)}}{2} \frac{W^{20i}}{1 - \varkappa_\sigma^{(2)} S^{20i}} \frac{\partial}{\partial \omega_i^{20}} W^{20i} + \frac{1}{2} \left(\frac{\varkappa_\sigma^{(2)} W^{20i}}{1 - \varkappa_\sigma^{(2)} S^{20i}} \right)^2 \frac{\partial}{\partial \omega_i^{20}} S^{20i}.$$

The functions S^{20i}, W^{20i}, Y^{20i}, and $\Gamma_n^i(s)$ are defined in (8.141′), (8.141″), (8.170), and (8.171′). The spin–quadrupole forces have small effect on the $B(E2)$ values.

Different nuclear $I^\pi = 0^+$ states can be connected by the monopole $E0$ transition. Such a transition is caused by the interaction between the nuclear charge and the atomic electrons. The most probable $E0$ process is an emission of one conversion electron. The probabilities of the $E0$ transitions depend on the properties of the corresponding nuclear states. The general theory of monopole transitions was explained in ref. 498.

The reduced $E0$ probability is expressed in terms of the dimensionless quantity $\varrho(E0)$ defined in ref. 499 by

$$\varrho(E0) = \int \Psi_f^* \sum_p \left[\left(\frac{r_p}{R_0}\right)^2 - \zeta \left(\frac{r_p}{R_0}\right)^4 + \ldots \right] \Psi_i\, dr, \tag{9.21}$$

where Ψ_i and Ψ_f are the initial and final nuclear states and r_p is the proton radius vector. The quantity ζ depends on the nuclear charge distribution; it is small and often neglected. The $E0$ transition matrix element equals

$$\mathcal{M}(E0) = \varrho(E0)R_0^2, \tag{9.22}$$

and the probability of the $E0$ internal conversion is

$$W_e(E0) = \varrho^2(E0)\Omega_e.$$

The quantity Ω_e is fully determined by the atomic electron wave functions. Its explicit form can be found in ref. 498.

The 0^+ states in deformed nuclei are characterized by a relative quantity x introduced in ref. 500. It is defined by

$$x = \frac{e^2 R_0^4 \varrho^2(E0)}{B(E2; 0^+0 \to 2^+0)}. \tag{9.23}$$

The parameter x can be calculated in simple situations, for example, in a uniformly charged ellipsoid

$$x = 4\beta^2. \tag{9.23'}$$

In the model of independent particles with coherent contribution of all protons, we obtain

$$x \approx 9\beta^2. \tag{9.23''}$$

Using our wave functions we can easily find the quantity $\varrho^2(E0)$ for the transition between the ground state and the one-photon 0^+ state. Such a quantity is equal to

$$\varrho_i^2(E0) = \frac{e^2}{2R_0^4}\left| (1+e_{\text{eff}}^{(0)}) \sum_{r,r'} \langle r|r^2|r'\rangle g_{rr'}^{20i} u_{rr'} + e_{\text{eff}}^{(0)} \sum_{s,s'} \langle s|r^2|s'\rangle g_{ss'}^{20i} u_{ss'} \right|^2, \tag{9.24}$$

where $\langle q|r^2|q'\rangle$ is the single-particle r^2 matrix element. The effective charge parameter $e_{\text{eff}}^{(0)}$ is usually chosen equal to $e_{\text{eff}}^{(2)}$.

Let us take a look at the calculated $B(E2)$, $\varrho(E0)$, and x-values for the excited 0^+ states. The effective charge $e_{\text{eff}}^{(2)}$ was fixed in the analysis of the $K^\pi = 2^+$ states. Consequently, we do not have free parameters in our analysis of the $K^\pi = 0^+$ states.

The calculations[466] using the Nilsson model wave functions give the following results: the $B(E2)_{\text{s.p.u.}}$ have values 1–4 for the β-vibrational states. On the other hand, for the pure pairing vibrations (i.e., when $\varkappa^{(2)} = \varkappa_\sigma^{(2)} = 0$) we obtain $B(E2)_{\text{s.p.u.}} = 10^{-4}$–$10^{-1}$. The quantity $\varrho(E0) \approx 0.01$ and the quantity x strongly fluctuates. For the pure spin–quadrupole states, i.e., when $\varkappa^{(2)} = 0$, we obtain the following values: $B(E2)_{\text{s.p.u.}} \sim 10^{-2}$, $\varrho(E0) \sim 0.01$, $x \sim 0.11$–0.14. Thus, the $E2$ and $E0$ transition probabilities are quite small in the last two situations.

The experimental and calculated[466] $B(E2)$, $\varrho(E0)$, and x-values are collected in Table 9.16. The varying structure of the one-phonon 0^+ states is clearly born out. Several 0^+ states in samarium and gadolinium isotopes have the properties of β-vibrations. In other cases, for example, in ^{164}Er and ^{170}Yb, the 0^+ states have the properties of pairing vibrations. It is obvious that we need more experimental data in order to recognize the character of all excited 0^+ states.

4. Let us now discuss the $E3$ transition probabilities involved in the excitation of the octupole $I^{\pi}K = 3^{-}K$ states. Tables 9.15 and 9.16 contain the theoretical $B(E3; 0^{+}0 \to 3^{-}K)_{\text{s.p.u.}}$ values. The Woods–Saxon potential was used in the calculations.[475–477] The effective charge $e_{\text{eff}}^{(3)} = 0.2$ appears in formula (9.19). Both the first and the second solutions ($i = 1, 2$) of the secular equation are included in Table 9.17.

The lowest $K^{\pi} = 0^{-}$ states are collective in all $150 \leqslant A < 170$ nuclei; the $B(E3)_{\text{s.p.u.}}$ have values up to 10 units there. The $K^{\pi} = 0^{-}$ states are closer to the two-quasiparticle states in the heavier rare-earth nuclei. Several of the second $K^{\pi} = 0^{-}$ states are collective, too.

The first $K^{\pi} = 1^{-}$ states are collective in the nuclei with $150 \leqslant A \leqslant 160$. Later, the collectivity decreases. Many second $K^{\pi} = 1^{-}$ states are even more collective than the first states.

The first $K^{\pi} = 2^{-}$ states are collective in practically all nuclei with $150 \leqslant A \leqslant 190$. However, the first half of this region is somewhat more collective than the second one.

The first and second $K^{\pi} = 3^{-}$ states have a two-quasiparticle structure in all nuclei with $150 \leqslant A \leqslant 190$ and $228 \leqslant A \leqslant 254$; the corresponding $B(E3)_{\text{s.p.u.}}$ values are typically equal to 0.01.

The wave functions of the octupole states, examples of which we have seen in Tables 9.8, 9.9, and 9.10 support these conclusions.

The $K^{\pi} = 0^{-}$ states in the actinide $228 \leqslant A \leqslant 254$ region are strongly collective; the $B(E3)_{\text{s.p.u.}}$ have values up to 20 units. The $K^{\pi} = 1^{-}$ and 2^{-} states are collective in many nuclei of this region. However, the collectivity decreases with increasing atomic number A.

The octupole states calculated with the Woods–Saxon potential are usually more collective than the states calculated in the framework of the Nilsson model. This tendency is particularly strong in the $150 \leqslant A \leqslant 160$ region and for the $K^{\pi} = 1^{-}$ and 2^{-} states.

Many $B(E3)$ values were experimentally measured. It is, however, somewhat difficult[496] to compare these values directly with the calculated values. There are two reasons for this: firstly, the $K^{\pi} = 0^{-}$, 1^{-}, and 2^{-} octupole states are strongly mixed by the Coriolis interaction,[496, 501] especially in the lighter rare-earth nuclei. Secondly, the K quantum numbers of the experimentally observed states are seldom known. To overcome these difficulties,[496] we have listed the experimental $B(E3)_{\text{s.p.u.}}$ values for two lowest $I^{\pi} = 3^{-}$ states in Table 9.17. Figure 9.6 compares the experimental and calculated $B(E3)_{\text{s.p.u.}}$ for the lowest $I^{\pi} = 3^{-}$ states, irrespective of their K-value. The good overall agreement is apparent.

Another problem are the reduced $E1$ transition probabilities connecting the nuclear ground states and the octupole $I^{\pi}K = 1^{-}0$ states. To calculate these probabilities we shall use formula (9.19) and $p^{\lambda_0\mu_0}(q, q') = \mathfrak{f}^{10}(q, q')$. The summation in (9.19) thus contains the terms $\mathfrak{f}^{30}(q, q')\mathfrak{f}^{10}(q, q')$ acquiring both positive and negative values. Different terms have a tendency to cancel each other and, consequently, the $E1$ rates are small. The calculations using the effective charges $e_{p}^{(1)} = N/A$, $e_{n}^{(1)} = -Z/A$ result in $B(E1)_{\text{s.p.u.}} = (4\text{–}8) \times 10^{-2}$. Such values do not contradict the scant experimental information.

Let us conclude: we have seen that the calculations based on the superfluid nuclear model basically correctly reproduce the $E2$ and $E3$ transition probabilities connecting the ground and one-phonon quadrupole or octupole states.

TABLE 9.16. *Reduced transition probabilities $B(E2)$, $\varrho(E0)$, and x-values for the γ-transitions from $K\pi = 0^+$ excited states*[a]

Nucleus	Energy ω_1^{20} (experimental) (MeV)	$B(E2; 0^+0 \rightarrow 2^+0)_{s.p.u.}$			$\varrho(E0)$			x	
		Experimental	Calculated		Experimental	Calculated		Experimental	Calculated
			$\varkappa_\sigma^{(2)} = 0$	$\varkappa_\sigma^{(2)} = 8.2A^{-4/3}\mathring{\omega}_0$		$\varkappa_\sigma^{(2)} = 0$	$\varkappa_\sigma^{(2)} = 8.2A^{-4/3}\mathring{\omega}_0$		
^{152}Sm	0.68	1.0	4.1	4.0	0.2	0.35	0.34	0.3	0.21
^{154}Sm	1.10	1.1	2.6	2.4	—	0.28	0.26	—	0.21
^{154}Gd	0.68	4.8	2.7	2.6	0.4	0.28	0.28	0.1	0.20
^{156}Gd	1.05	2.8	1.8	1.8	0.4	0.23	0.22	0.5	0.21
^{164}Er	1.25		0.8	0.4		0.16	0.09	0.15	0.24
	1.70		0.02	0.3		0.05	0.13	0.39	0.85
	1.77		0.08	0.02		0.03	0.05	0.78	0.05
	2.17		0.22	0.07		0.03	0.03	1.7	0.02
	2.18		10^{-3}	0.21		−0.09	0.03	5.5	50.0
^{170}Yb	1.07		1.4	1.3		0.18	0.16	—	0.15
	1.23		0.2	0.1		0.02	0.06	0.13	0.03
	1.48		10^{-2}	0.2		0.06	0.04	0.94	4.4
	1.57		—	—		—	—	0.5	—
^{178}Hf	1.20		3.8	2.6		0.33	0.26	0.2	0.20
	1.43		0.1	0.6		0.05	0.17	0.1	0.15
	1.44		0.3	0.4		−0.02	0.11	0.4	0.01

[a] The Nilsson model wave functions were used in the calculations and $\varkappa^{(2)} = \varkappa^{(2)}_{exp}$, $e^{(2)}_{eff} = 0.2$.

TABLE 9.17. *The reduced transition probabilities $B(E3, 0^+0 \to 3^-K)$ for the octupole $K^\pi = 0^-, 1^-$, and 2^- states*[a]

Element	A	$B(E3)_{s.p.u.}$ experimental		$B(E3)_{s.p.u.}$ calculated					
				$K^\pi = 0^-$		$K^\pi = 1^-$		$K^\pi = 2^-$	
		$I_1^\pi = 3^-$	$I_2^\pi = 3^-$	$i = 1$	$i = 2$	$i = 1$	$i = 2$	$i = 1$	$i = 2$
Nd	150			10.7	0.2	9.9	0.3	7.0	0.03
Sm	152	14.7	8.2	10.7	0.1	9.4	0.1	6.9	0.04
	154	8.1	6.0	8.6	0.1	6.9	0.1	3.9	0.01
	156			9.5	0.02	4.4	0.3	6.4	0.01
Gd	154	16	4.5	9.3	0.06	8.7	0.002	4.8	0.02
	156	11	4.2	9.1	0.001	6.3	0.01	4.1	0.2
	158	6.6	2.2	8.3	0.6	3.9	0.3	5.0	0.1
	160	6.9	2.0	6.9	1.3	2.8	0.002	6.2	0.02
Dy	156	18		7.8	0.2	7.5	0.2	6.2	0.2
	158	15		7.6	0.001	8.2	0.06	4.5	0.4
	160	11		2.8	0.8	2.9	0.01	6.6	0.1
Dy	162	8.7	2.5	7.8	2.5	5.9	0.002	5.9	0.1
	164	6.0		8.7	0.6	1.2	0.02	4.8	0.2
Er	160			8.2	0.01	4.2	0.4	4.7	0.01
	162	12	2.1	7.6	1.0	2.4	0.6	4.1	0.001
	164	8.1	3.6	6.4	2.6	1.6	0.04	3.3	0.01
	166	6.1	3.0	4.6	0.7	1.2	1.9	2.4	0.03
	168	3.3	3.0	3.3	0.2	0.5	0.02	2.2	0.07
	170	1.1	3.6	5.4	0.3	1.2	0.01	1.7	0.6
Yb	166			5.7	0.8	2.0	0.04	3.3	0.01
	168	7.3	3.7	5.9	0.3	0.3	2.4	2.3	0.003
	170	6.2		5.1	1.2	0.6	0.03	2.0	0.1
Yb	172	2.6	4.8	1.7	1.2	1.3	0.05	2.1	0.3
	174	4.0	4.8	0.7	2.3	0.2	0.3	1.4	1.5
	176	2.6		0.5	1.2	0.1	0.2	1.6	0.4
Hf	170			5.5	0.1	0.01	0.9	5.2	0.1
	172			3.8	0.08	0.01	0.6	4.2	0.03
	174			1.3	0.9	0.3	0.03	3.4	0.2
	176			0.6	3.3	0.1	0.03	2.4	0.6
Yb	178			4.2	0.9	0.4	0.02	2.3	0.001
Hf	178			0.1	0.8	0.3	0.03	3.2	0.2
	180			1.0	0.2	0.1	0.4	2.2	0.4
W	178			4.3	2.2	2.4	0.04	6.4	0.1
	180			0.4	0.8	0.2	0.4	4.1	0.1
	182			0.9	0.2	0.5	0.1	2.6	0.6
	184			1.5	0.5	0.4	0.1	2.9	0.03
	186			0.6	0.8	0.4	0.1	1.6	0.4
Os	182			2.1	0.9	0.2	3.0	2.6	0.001
	184			2.0	0.3	0.6	0.2	0.1	2.8
	186			0.9	0.5	0.5	0.1	1.5	0.002
	188			0.4	0.5	0.4	0.06	0.9	0.005

[a] The Woods–Saxon potential wave functions were used and $e_{eff}^{(3)} = 0.2$, $\varkappa_0^{(3)} = \varkappa_{exp}^{(3)}$.

5. Let us study the excitation of the one-phonon states in the nuclear α-decay.

As a first step, we shall obtain the expression for the α-decay probability. Our model will use the quasiparticle operators $\alpha_{q\sigma}$ and phonon operators $Q_i(\lambda\mu)$. Such operators are different in different nuclei. In processes involving the α- or β-decays of various nuclei, we can use the particle operators $a_{q\sigma}$ independent of the proton or neutron numbers.

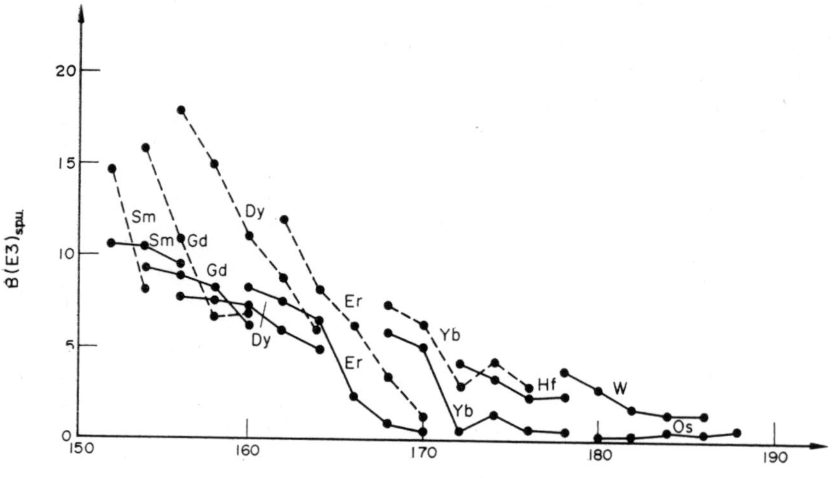

FIG. 9.6. The $B(E3)_{\text{s.p.u.}}$ values for the lowest $I^\pi = 3^+$ states. The dashed curves show the experimental data and the full curve shows the calculated values. The Woods–Saxon potential and parameters $\varkappa^{(3)} = \varkappa^{(3)}_{\text{exp}}$ and $e^{(3)}_{\text{eff}} = 0.2$ were used in the calculations.

This program can be carried out in Tamm–Dancoff approximation because the corresponding ground state is simply the quasiparticle vacuum. On the other hand, the ground state in the random phase approximation theory is the phonom vacuum, and it is very difficult to rewrite it in terms of the particle operators. To avoid these problems we shall make a number of simplifying asumptions when calculating the probabilities of α- or β-decays. We will assume that the phonom vacuum of the parent and daughter nuclei is identical. In the case of an α-decay, this means that

$$\left.\begin{aligned} Q_i(\lambda\mu; Z, N) &= Q_i(\lambda\mu; Z-2, N-2) \equiv Q_i(\lambda\mu), \\ Q_i(\lambda\mu)\Psi_i &= Q_i(\lambda\mu)\Psi_f = 0. \end{aligned}\right\} \tag{9.25}$$

The reduced probability of the α-decay from the ground state of the parent nucleus [wave function $\Psi_i(N)\Psi_i(Z)$] to the one-phonon state of the daughter nucleus [wave function $Q_i^+(\lambda\mu)\Psi_f$] has the form

$$\gamma^2_{li}(Q) = \left|\left(\Psi_f^* Q_i(\lambda\mu)\frac{1}{4}\sum_{\substack{r,\, r' \\ \sigma_1,\, \sigma_1'}}\sum_{\substack{s,\, s' \\ \sigma_2,\, \sigma_2'}} W^{II}_{\sigma_1\sigma_1';\, \sigma_2\sigma_2'}(rr' \mid ss') a_{r\sigma_1} a_{r'\sigma_1'} a_{s\sigma_2} a_{s'\sigma_2'} \Psi_i(N)\Psi_i(Z)\right)\right|^2. \tag{9.26}$$

The functions W describe the formation of the α-particle; they satisfy the relations (6.8′).

Using the approximation (9.25),[129]

$$\gamma_{ll}(Q) = -\frac{1}{\sqrt{2}}\sum_{r_1, r_2, s} W^{ll}_{+-;\,+-}(r_1 r_2 \mid ss)\,[\psi^{\lambda\mu i}_{r_1 r_2} v_{r_1}(Z-2)\, v_{r_2}(Z-2) - \varphi^{\lambda\mu i}_{r_1 r_2} u_{r_1}(Z-2) \times$$

$$\times u_{r_2}(Z-2)]\,\xi_s(N) + \frac{1}{\sqrt{2}}\sum_{r_1, r_2, s} W^{ll}_{++;\,+-}(r_1 r_2 \mid ss)\,[\bar{\psi}^{\lambda\mu i}_{r_1 r_2} v_{r_1}(Z-2)\, v_{r_2}(Z-2)$$

$$-\bar{\varphi}^{\lambda\mu i}_{r_1 r_2} u_{r_1}(Z-2)\, u_{r_2}(Z-2)]\,\xi_s(N) - \frac{1}{\sqrt{2}}\sum_{s_1, s_2 r} W^{ll}_{+-,\,+-}(rr \mid s_1 s_2)\,[\psi^{\lambda\mu i}_{s_1 s_2} v_{s_1}(N-2)\, v_{s_2}(N-2)$$

$$-\varphi^{\lambda\mu i}_{s_1 s_2} u_{s_1}(N-2)\, u_{s_2}(N-2)]\,\xi_r(Z) + \frac{1}{\sqrt{2}}\sum_{s_1, s_2, r} W^{ll}_{+-;\,++}(rr \mid s_1 s_2)\,[\bar{\psi}^{\lambda\mu i}_{s_1 s_2} v_{s_1}(N-2) \times$$

$$\times v_{s_2}(N-2) - \bar{\varphi}^{\lambda\mu i}_{s_1 s_2} u_{s_1}(N-2)\, u_{s_2}(N-2)]\,\xi_r(Z), \qquad (9.27)$$

the quantity $\xi_s(N)$ is defined by (6.10′) and the functions $\psi^{\lambda\mu i}_{qq'}$ and $\varphi^{\lambda\mu i}_{qq'}$ are determined by (8.169) and (8.169′) in the $\lambda = 2$, $\mu = 0$ case, and by (8.135′) and (8.135″) in all other cases.

Formula (9.27) contains summation over all quasiparticle pairs (q, q'), and the reduced widths $\gamma_{ll}(Q)$ belong, therefore, among the integral characteristics of the one-phonon states. The reduced widths $\gamma_{ll}(Q)$ are smaller than the ground state width γ_{00} because (9.27) contains only the pairs of states which form the phonon, and each pair has a weighing factor smaller than unity.

The hindrance factor $F_\alpha(Q)$ for the α-decay exciting the one-phonon state has the form (see 6.6)

$$F_\alpha(Q) = \frac{(\gamma^2_{00})_g}{\gamma^2_{ll}(Q)}\,. \qquad (9.28)$$

From the condition $\gamma_{ll}(Q) < \gamma_{00}$ it follows that $F_\alpha(Q) > 1$. On the other hand, by comparing (9.27) with $\gamma_{ll}(q_1, q_2)$ for the purely two-quasiparticle states, we see that

$$\gamma_{ll}(Q) > \gamma_{ll}(q_1, q_2), \qquad (9.29)$$

and, consequently,

$$1 < F_\alpha(Q) < F_\alpha(q_1, q_2). \qquad (9.30)$$

The enhancement of the reduced α-width of one-phonon states is a very characteristic feature of the collective states.

The experimentally determined hindrance factors of α-decays involving the quadrupole and octupole phonon states are collected in Table 9.18. The inequality (9.30) is obviously valid for all these factors.

The hindrance factors $F_\alpha(Q)$ were calculated using (9.27) and (9.28). Satisfactory agreement with the experimental data was usually found.[502]

The integral characteristics of the quadrupole and octupole one-phonon states clearly show the coherent contribution of many nucleons. This coherence causes enhancement of the $E2$ and $E3$ reduced transition probabilities and of the reduced α-widths when compared with typical single-particle values. The superfluid nuclear model containing the pairing and multipole–multipole interactions describes adequately the integral characteristics of the one-phonon states.

TABLE 9.18. *Alpha-transitions between the ground states and lowest $I^{\pi}K = 2^{+}2, 0^{+}0$, and $1^{-}0$ states in deformed even–even nuclei*

α-Transition	Final state		Hindrance factor $F_{\alpha}(Q)$
	K^{π}	Energy (MeV)	
$^{254}Fm \rightarrow {}^{250}C$	2^{+}	1.032	18
$^{244}Cm \rightarrow {}^{240}Pu$	0^{-}	0.600	240
	0^{+}	0.861	2.1
$^{242}Cm \rightarrow {}^{238}Pu$	0^{-}	0.609	190
	0^{+}	0.940	19
$^{238}Pu \rightarrow {}^{234}U$	0^{-}	0.795	50
	0^{+}	0.810	6
	2^{+}	0.922	>20
$^{236}Pu \rightarrow {}^{232}U$	0^{-}	0.564	160
	0^{+}	0.693	12
	2^{+}	0.893	>50
$^{234}U \rightarrow {}^{230}Th$	0^{-}	0.510	300
	0^{+}	0.640	28
	2^{+}	0.780	>20
$^{232}U \rightarrow {}^{228}Th$	0^{-}	0.328	270
	0^{+}	0.825	10

§ 3. Differential Characteristics of One-phonon States

1. In the semiempirical method, the two-quasiparticle states are considered belonging to the same group as the one-phonon states. The wave functions of the one-phonon states are simply the superpositions of the two-quasiparticle states. The majority of the lowest quadrupole and octupole states has a clearly pronounced collective structure, i.e., the states contain a large number of different two-quasiparticle components.

In this section, we shall discuss the nuclear processes giving an information about the individual components of the collective states. These processes include the β-decay exciting the one-phonon states, γ-transitions from the higher lying states and the one-nucleon transfer reactions. Only a few quasiparticle components of the collective state participate in these processes. In particular cases it is even possible to extract from the experimental data the information about the individual two-quasiparticle components of one-phonon states.

Let us find the nuclear matrix element of the β-decay from the two-quasiparticle state of an odd–odd nucleus (the odd proton is in the r_1 state, the odd neutron is in the s_1 state) to the one-phonon state $Q_i(\lambda\mu)\Psi_f$ in an even–even nucleus. The wave function of the initial state has the form

$$\frac{1}{\sqrt{2}}(\alpha_{s_1-}^+\alpha_{r_1+}^+ + \alpha_{s_1+}^+\alpha_{r_1-}^+)\Psi_i, \quad \text{if} \quad \pm K = K_n - K_p,$$

$$\frac{1}{\sqrt{2}}(\alpha_{s_1+}^+\alpha_{r_1+}^+ + \alpha_{s_1-}^+\alpha_{r_1-}^+)\Psi_i \quad \text{if} \quad \pm K = K_n + K_p.$$

We shall now make the assumption which we have already used in the α-decay case, namely, we shall assume that the quasiparticle and phonon operators are identical in the parent and daughter nuclei, i.e.,

$$\left.\begin{array}{l} Q_i(\lambda\mu; N, Z) = Q_i(\lambda\mu; N\pm 1, Z\mp 1) \equiv Q_i(\lambda\mu), \\ Q_i(\lambda\mu)\Psi_i = Q_i(\lambda\mu)\Psi_f = 0. \end{array}\right\} \tag{9.31}$$

Besides, we shall also assume that the superfluid characteristics of the parent and daughter nuclei are identical.

The β-decay matrix element in our case has the form

$$\mathcal{M} = (\Psi_f^*[Q_i(\lambda\mu), H_\beta\alpha_{r_1\sigma_1'}^+\alpha_{s_1\sigma_1}^+]\Psi_i),$$

where H_β is defined by (6.32). Using the approximations (9.31) and (9.31') we obtain after some simple arithmetic the following formulae:

$$\beta^- \text{ decay}, \quad \pm K = K_n - K_p$$

$$\mathcal{M}\begin{pmatrix} Z-1 \to Z \\ N+1 \to N \end{pmatrix} = -\sum_r \{\langle s_1 + |\Gamma_\beta| r + \rangle\psi_{rr_1}^{\lambda\mu i} + \langle s_1 - |\Gamma_\beta| r + \rangle\overline{\psi}_{rr_1}^{\lambda\mu i}\} u_r u_{s_1}$$
$$+ \sum_s \{\langle s + |\Gamma_\beta| r_1 + \rangle\psi_{ss_1}^{\lambda\mu i} - \langle s - |\Gamma_\beta| r_1 + \rangle\overline{\psi}_{ss_1}^{\lambda\mu i}\} v_s v_{r_1}; \tag{9.32}$$

$$\beta^- \text{ decay}, \quad \pm K = K_n + K_p$$

$$\mathcal{M}\begin{pmatrix} Z-1 \to Z \\ N+1 \to N \end{pmatrix} = -\sum_r \{\langle s_1 - |\Gamma_\beta| r + \rangle\psi_{rr_1}^{\lambda\mu i} - \langle s_1 + |\Gamma_\beta| r + \rangle\overline{\psi}_{rr_1}^{\lambda\mu i}\} u_r u_{s_1}$$
$$+ \sum_s \{\langle s - |\Gamma_\beta| r_1 \rangle\psi_{ss_1}^{\lambda\mu i} + \langle s + |\Gamma_\beta| r_1 + \rangle\overline{\psi}_{ss_1}^{\lambda\mu i}\} v_s v_{r_1}; \tag{9.32'}$$

$$\beta^+ \text{ decay}, \quad \pm K = K_n - K_p$$

$$\mathcal{M}\begin{pmatrix} Z+1 \to Z \\ N-1 \to N \end{pmatrix} = \sum_r \{\langle s_1 + |\Gamma_\beta| r + \rangle\psi_{rr_1}^{\lambda\mu i} + \langle s_1 - |\Gamma_\beta| r + \rangle\overline{\psi}_{rr_1}^{\lambda\mu i}\} v_r v_{s_1}$$
$$- \sum_s \{\langle s + |\Gamma_\beta| r_1 + \rangle\psi_{ss_1}^{\lambda\mu i} - \langle s - |\Gamma_\beta| r_1 + \rangle\overline{\psi}_{ss_1}^{\lambda\mu i}\} u_{r_1} u_s; \tag{9.32''}$$

$$\beta^+ \text{ decay}, \quad \pm K = K_n + K_p$$

$$\mathcal{M}\begin{pmatrix} Z+1 \to Z \\ N-1 \to N \end{pmatrix} = \sum_r \{\langle s_1 - |\Gamma_\beta| r + \rangle\psi_{rr_1}^{\lambda\mu i} - \langle s_1 + |\Gamma_\beta| r + \rangle\overline{\psi}_{rr_1}^{\lambda\mu i}\} v_r v_{s_1}$$
$$- \sum_s \{\langle s - |\Gamma_\beta| r_1 + \rangle\psi_{ss_1}^{\lambda\mu i} + \langle s + |\Gamma_\beta| r + \rangle\overline{\psi}_{ss_1}^{\lambda\mu i}\} u_{r_1} u_s. \tag{9.32'''}$$

Here we have denoted the single-particle β-transition matrix element by $\langle s\sigma | \Gamma_\beta | r\sigma' \rangle$. Such matrix elements were extensively discussed in § 2 of Chapter 6. The proton (neutron) numbers $Z(N)$ are even; the quantities u_q, v_q belong to the ground state of the daughter nucleus. The functions $\psi_{qq'}^{\lambda\mu i}$ have been often discussed. They are determined by (8.169) for $\lambda = 2$, $\mu = 0$ and by (8.135′) in all other cases.

The summations in (9.32)–(9.32″) are very restricted. Only two-neutron states containing the fixed orbit r_1 can participate. Moreover, the remaining summation is further restricted by the selection rules of the single-particle β-transition matrix element. The β-decay to a collective one-phonon state is slower than the β-decay to a pure two-quasiparticle state. Hence, we expect that

$$\log (ft)_Q > \log (ft)_{q_1 q_2}. \tag{9.33}$$

This conclusion is obvious; the corresponding two-quasiparticle states enter the phonon wave function with a weighing factor smaller than unity.

The examples of the β-transitions populating the one-phonon states $K^\pi = 2^+$ and 0^+ are shown in Tables 9.19 and 9.20. The inhibition of the β-transitions is quite apparent. We must remember that the β-decays populating 0^+ states in ^{178}Hf and ^{164}Er are caused predominantly by the *au* type transitions and therefore $\log ft_e = 5$–6.

TABLE 9.19. *The experimental* $\log ft_e$ *values for the* β*-decays populating the lowest* $I^\pi K = 2^+2$ *states*

Initial state		β-Transition	Final state energy (MeV)	$\log ft_e$
K^π	Configuration			
		Allowed transitions		
2^+	$p523\downarrow - n734\uparrow$	^{246}Am → ^{246}Cm	1.126	7.8
3^+	$p411\downarrow - n633\uparrow$	^{168}Tm → ^{168}Er	0.822	8.8
2^+	$p411\downarrow - n642\uparrow$	^{166}Tm → ^{166}Er	0.786	8.0
1^+	$p523\uparrow - n523\downarrow$	^{164}Tm → ^{164}Er	0.858	7.6
		First forbidden transitions		
2^-	$p633\uparrow - n725\uparrow$	^{254}Es → ^{254}Fm	0.693	6.8
2^-	$p521\uparrow - n624\downarrow$	^{246}Bk → ^{246}Cm	1.126	6.6
0^-	$p523\downarrow - n622\uparrow$	^{240}Np → ^{240}Pu	0.945	6.9
2^-	$p402\downarrow + n510\uparrow$	^{188}Ir → ^{188}Os	0.633	8.0
1^-	$p402\uparrow - n512\downarrow$	^{188}Re → ^{188}Os	0.633	9.4
1^-	$p402\uparrow - n512\downarrow$	^{186}Re → ^{186}Os	0.767	8.5
3^-	$p404\downarrow - n510\uparrow$	^{182}Ta → ^{182}W	1.222	10.2
2^-	$p411\downarrow - n512\uparrow$	^{172}Tm → ^{172}Yb	1.468	7.0
1^-	$p411\uparrow - n523\downarrow$	^{162}Tb → ^{162}Dy	0.888	6.4
3^-	$p411\uparrow + n521\uparrow$	^{160}Tb → ^{160}Dy	0.966	8.9
3^-	$p411\uparrow + n521\uparrow$	^{156}Tb → ^{156}Gd	1.154	7.8
3^-	$p411\uparrow + n521\uparrow$	^{154}Eu → ^{154}Gd	0.996	11.0

The quantities $\psi_{qq'}^{\lambda\mu i}$ may be, in certain cases, extracted from the experimental $\log ft_e$ values. This is possible if one of the larger components is populated by an *au* transition.

We have seen in § 2 of Chapter 6 that the calculated quantities $\log ft_c$ for the transitions between the quasiparticle states are 10–100 times smaller than the corresponding experi-

TABLE 9.20. *The experimental log ft_e values for the β-decays populating the lowest $I^{\pi}K = 0^{+}0$ states*

Initial state		β-Transition	Final state energy (MeV)	log ft_e
K^{π}	Configuration			
		Allowed transitions		
2^+	$p642\uparrow - n631\downarrow$	$^{238}Np \rightarrow {}^{238}Pu$	1.030	6.2
1^+	$p514\uparrow - n514\downarrow$	$^{178}Ta \rightarrow {}^{178}Hf$	1.197	5.9
			1.434	5.2
			1.444	5.1
1^+	$p523\uparrow - n523\downarrow$	$^{164}Tm \rightarrow {}^{164}Er$	1.246	5.4
			1.698	5.6
			1.766	5.4
		First forbidden transitions		
0^-	$p523\downarrow - n622\uparrow$	$^{240}Np \rightarrow {}^{240}Pu$	0.863	6.9
0^-	$p530\uparrow - n631\downarrow$	$^{234}Pa \rightarrow {}^{234}U$	0.810	7.0
			1.045	6.7
1^-	$p402\uparrow - n512\downarrow$	$^{198}Re \rightarrow {}^{188}Os$	1.086	9.1
0^-	$p523\uparrow - n633\uparrow$	$^{166}Ho \rightarrow {}^{166}Er$	1.460	7.3

mental data. At the same time the relative log *ft* values (i.e., the ratios with the appropriate normalization) are in a good agreement with the experiment. But it is difficult to calculate the relative log *ft* values in the case of collective final states because several two-quasi-particle states participate in the decay.

The absolute and relative log *ft* values for the β-decay leading to the one-phonon states were calculated in refs. 503 and 504. The relative log *ft* values have the same order of magnitude as the corresponding experimental data, while the absolute *ft* quantities are 10–100 smaller than the measured values. Thus this discrepancy does not represent any failure in our description of the one-phonon states.

2. Another source of the information about the individual components of one-phonon states are the γ-transitions from the higher two- or four-quasiparticle states.

Let us find the $E\lambda$ matrix element connecting the two-quasiparticle state $\alpha^{+}_{q_1\sigma_1}\alpha^{+}_{q_2\sigma_2}\Psi$, and the one-phonon state $Q_i^{+}(\lambda\mu)\Psi$. The $E\lambda$ transition operator has the form (6.64), and consequently

$$\begin{aligned}\mathcal{M}(q_1q_2 \rightarrow Q_i(\lambda\mu)) &= (\Psi^* Q_i(\lambda\mu)\,\mathfrak{M}(E\lambda)\alpha^{+}_{q_1\sigma_1}\alpha^{+}_{q_2\sigma_2}\Psi)\\ &= -\frac{\delta_{\sigma_1,-\sigma_2}}{\sqrt{2}}\sum_q \{\sigma_1\langle q+|\Gamma(E\lambda)|q_1+\rangle\,\psi^{\lambda\mu i}_{qq_2}v_{qq_1}\\ &\quad+\sigma_1\langle q+|\Gamma(E\lambda)|q_2+\rangle\,\psi^{\lambda\mu i}_{qq_1}v_{qq_2}+\langle q+|\Gamma(E\lambda)|q_1-\rangle\bar{\psi}^{\lambda\mu i}_{qq_2}v_{qq_1}\\ &\quad+\langle q+|\Gamma(E\lambda)|q_2-\rangle\bar{\psi}^{\lambda\mu i}_{q_1q}v_{qq_2}\}-\frac{\delta_{\sigma_1\sigma_2}}{\sqrt{2}}\sum_q\{\langle q+|\Gamma(E\lambda)|q_1+\rangle\times\\ &\quad\times\bar{\psi}^{\lambda\mu i}_{qq_2}v_{qq_1}+\langle q+|\Gamma(E\lambda)|q_2+\rangle\,\bar{\psi}^{\lambda\mu i}_{q_1q}v_{qq_2}-\sigma_1\langle q+|\Gamma(E\lambda)|q_1-\rangle\times\\ &\quad\times\psi^{\lambda\mu i}_{qq_2}v_{qq_1}+\sigma_1\langle q+|\Gamma(E\lambda)|q_2-\rangle\psi^{\lambda\mu i}_{q_1q}v_{qq_2}\},\end{aligned} \tag{9.34}$$

where $\langle q\sigma|\Gamma(E\lambda)|q'\sigma'\rangle$ represents the single-particle matrix element (6.60)or (6.61).

The first term in (9.34) is nonvanishing if $\sigma_1 = -\sigma_2$, while the second enters if $\sigma_1 = \sigma_2$. The formula (9.34) clearly shows that the γ-transition selects the individual components of the one-phonon state. An analogous formula can be derived for the transitions starting from a four-quasiparticle state. The individual $\psi^{\lambda\mu i}_{qq'}$ values can be found from the measured $B(E\lambda)$ values if one term in the formula (9.34) contributes overwhelmingly. We expect that a γ-transition probability will be lower if the final state is collective and higher in the purely two-quasiparticle case.

Naturally, the $M\lambda$ transition probabilities have analogous properties.

3. The direct one-nucleon transfer reactions offer the most promising way of extracting the individual components of the collective states from the experimental data.

We shall use the approximations which were introduced in the derivation of the one-nucleon transfer cross-section in (6.94). After simple transformations, we obtain the following formula for the one-nucleon transfer cross-section:

$$\frac{d\sigma}{d\Omega} = N \sum_{j,\,i} |(I_i K_i j K | I_f \mu) \sum_q \left(\sum_\Lambda a^q_{l\Lambda} (l \Lambda s \Sigma | jK) \right) u_q \psi^{\lambda\mu i}_{qq_0} + (I_i - K_i j K | I_f \mu) \times$$

$$\times \sum_q \left(\sum_\Lambda a^q_{l\Lambda} (l \Lambda s \Sigma | jK) \right) u_q \overline{\psi}^{\lambda\mu i}_{qq_0} |^2 \sigma_l(\theta); \qquad (9.35)$$

the parameters $a^q_{l\Lambda}$ are the expansion coefficients of the corresponding Nilsson or Woods–Saxon single-particle wave functions; the remaining notation is explained in § 4 of Chapter 6. In (9.35) we are considering the reactions (dp) or $(^3\text{He}, d)$ on the odd-A deformed target; the odd nucleon is in the state q_0. The final one-phonon state has the wave function $Q_i^+(\lambda\mu)\Psi$.

The individual functions $\psi^{\lambda\mu i}_{qq_0}$ clearly determine the cross-section (9.35). The selection rules of the matrix elements $\mathfrak{f}^{\lambda\mu}(q, q_0)$ restrict the number of possible components to two, four, and, in several cases, to a single one. The function $\psi^{\lambda\mu i}_{qq_0}$ giving an overwhelming contribution to (9.35) can be, naturally, determined from the experimental cross-section.

Several $\psi^{221}_{qq_0}$ values were determined from the excitation of γ-vibrational states in the (dp) reaction.[505] The corresponding cross-sections were normalized with respect to the pure two-quasiparticle states in the same final nuclei. The experimental cross-sections[505] are compared with the calculated values in Table 9.21. The table also contains the calculated values using the Nilsson model wave functions N I (see ref. 505) or N II (calculated in ref. 473) and the same quantities calculated in ref. 474 using the Woods–Saxon wave functions. The calculated and experimental cross-sections involving large $\psi^{221}_{qq_0}$ values agree well with each other. On the other hand, the restriction to one component is too severe in the case of small $\psi^{221}_{qq_0}$ values as, e.g., for ^{162}Dy or ^{168}Er.

The contributions of the individual components in one-phonon states have been often discussed.[348, 349] For example, the experimentally determined contribution of the nn $633\uparrow - 521\uparrow$ state to the lowest $K^\pi = 2^-$ in ^{168}Er is about 50%, while the calculations result in 35%.

The whole complex of experimental information about differential characteristics of one-phonon states shows that the calculations in our model correctly describe the large, most important components.

The approximate second quantization method based on the superfluid nuclear model

TABLE 9.21. *The excitation cross-section of the $K^\pi = 2^+$ states in (dp) reaction. The calculated quantities $\psi^{221}_{qq_0}$ are also shown*

Initial state	Final nucleus	Configuration	Excitation cross-section at 65° (μbarn/steradian)		$\psi^{221}_{qq_0}$, calculated		
			Experimental	Calculated N I	N I	N II	WS
521↑	^{156}Gd	521↑+521↓	–	21.6	0.41	0.44	0.60
521↑	^{158}Gd	521↑+521↓	40±6	32.5	0.52	0.60	0.80
642↑	^{162}Dy	642↑–640↑	5±2	1.8	0.10	0.11	0.04
		642↑–660↑		–	–	0.28	0.24
523↓	^{164}Dy	523↓–521↓	18±2	18.1	0.52	0.71	0.75
633↑	^{168}Er	633↑–631↑	1.4±0.5	0.5	0.12	–	–
		633↑–651↑		–	–	0.16	0.15
512↑	^{174}Yb	512↑–510↑	–	36	0.56	1.1	1.1
514↓	^{178}Hf	514↓–512↓	–	17	0.60	0.81	0.75
510↑	^{184}W	510↑+512↓	–	25	0.76	0.97	1.1

gives a correct general description of the vibrational states in deformed even–even nuclei. This general success should be, however, supplemented by further improvements. The most urgent, and also the most difficult part of such an effort, is the development of the methods free from the restrictions of the quasiboson approximation.

§ 4. Quasiparticle–Phonon Interaction in Odd-*A* Deformed Nuclei

1. The nonrotational excited states of deformed nuclei are described in terms of quasiparticle and phonon operators. However, the real nuclear states are neither the pure quasiparticle states nor the pure phonon states.

Two components of the nuclear Hamiltonian cause the mixing of the quasiparticle and phonon states. One of them is the coupling of the intrinsic and rotational motions, and the other is the quasiparticle–phonon interaction. In this section we shall discuss the effect of the quasiparticle–phonon interaction on the structure of nonrotational states in odd-*A* deformed nuclei. We shall derive the equations describing the interaction and analyze the nuclear processes giving an information about the structure of involved states. The changing structure of nuclear states at different excitation energies will be also discussed.

The odd-*A* nucleus consists of one quasiparticle in addition to the phonons and quasiparticles of an even–even nucleus. In principle, this additional quasiparticle changes the involved phonons. However, such changes are usually small and can be neglected. Similarly to the previous discussions, we shall assume that the phonons of an odd nucleus $A+1$ are identical with the phonons of the even–even nucleus A.

This means that we shall treat the quasiparticle and phonon operators as fully determined entities. In other words, (4.19) and (4.20) fully describe the superfluid characteristics of our system, and the secular equations (9.5) and (8.167′) determine the phonon energies and wave functions $\psi^{\lambda\mu i}_{qq'}$ and $\varphi^{\lambda\mu i}_{qq'}$. The multipole–multipole interaction constants $\varkappa^{(\lambda)}$ have been fixed in studies of even–even nuclei. Thus, there are no free parameters at our disposal in the analysis of odd-A nuclei.

The Hamiltonian describing the quasiparticle–phonon interaction is, according to (5.27), (8.131), (8.131′), and (8.131″), of the form

$$H_{vq} = \sum_q \varepsilon(q)\,B(q,q) - \sum_{\lambda=2,3}\frac{\varkappa^{(\lambda)}}{2}\sum_{\mu\geq 0}\Big\{\tfrac{1}{2}\sum_{i,i'}\sum_{q,q',q_2,q_2'} u_{qq'}u_{q_2q_2'}[\mathfrak{f}^{\lambda\mu}(q,q')\,g^{\lambda\mu i}_{qq'}+\bar{\mathfrak{f}}^{\lambda\mu}(q,q')\,\bar{g}^{\lambda\mu i}_{qq'}]\times$$
$$\times[\mathfrak{f}^{\lambda\mu}(q_2,q_2')\,g^{\lambda\mu i'}_{q_2q_2'}+\bar{\mathfrak{f}}(q_2,q_2')\bar{g}^{\lambda\mu i'}_{q_2q_2'}][Q_i^+(\lambda\mu)\,Q_{i'}(\lambda\mu)+Q_{i'}^+(\lambda\mu)\,Q_i(\lambda\mu)+Q_i^+(\lambda\mu)\,Q_{i'}^+(\lambda\mu)$$
$$+Q_{i'}(\lambda\mu)\,Q_i(\lambda\mu)]+\frac{1}{\sqrt{2}}\sum_i\sum_{q,q',q_2,q_2'} u_{q_2q_2'}v_{qq'}[\mathfrak{f}^{\lambda\mu}(q_2,q_2')g^{\lambda\mu i}_{q_2q_2'}+\bar{\mathfrak{f}}^{\lambda\mu}(q_2,q_2')\bar{g}^{\lambda\mu i}_{q_2q_2'}]\times$$
$$\times[[Q_i^+(\lambda\mu)+Q_i(\lambda\mu)]\,[\mathfrak{f}^{\lambda\mu}(q,q')\,B(q,q')+\bar{\mathfrak{f}}^{\lambda\mu}(q,q')\,\bar{B}(q,q')]+[\mathfrak{f}^{\lambda\mu}(q,q')\,B(q,q')$$
$$+\bar{\mathfrak{f}}^{\lambda\mu}(q,q')\,\bar{B}(q,q')]\,[Q_i^+(\lambda\mu)+Q(\lambda\mu)]]\Big\}. \tag{9.36}$$

Using the expression (8.135) for $g^{\lambda\mu i}_{qq'}$ and (9.5), we find that

$$\varkappa^{(\lambda)}\sum_{q,q'} u_{qq'}[\mathfrak{f}^{\lambda\mu}(q,q')\,g^{\lambda\mu i}_{qq'}+\bar{\mathfrak{f}}^{\lambda\mu}(q,q')\,\bar{g}^{\lambda\mu i}_{qq'}] = \frac{2\varkappa^{(\lambda)}}{\sqrt{2Y^i(\lambda\mu)}}\sum_{q,q'}\frac{(f^{\lambda\mu}(q,q')u_{qq'})^2[\varepsilon(q)+\varepsilon(q')]}{[\varepsilon(q)+\varepsilon(q')]^2-(\omega_i^{\lambda\mu})^2}$$
$$= \frac{1}{\sqrt{2Y^i(\lambda\mu)}}. \tag{9.37}$$

Consequently, the Hamiltonian (9.36) can be rewritten in the more compact form, i.e.,

$$H_{vq} = \sum_q \varepsilon(q)\,B(q,q) - \tfrac{1}{4}\sum_{\lambda=2,3}\sum_{\mu\geq 0}\sum_{i,i'}\frac{1}{2\varkappa^{(\lambda)}}\frac{1}{\sqrt{Y^i(\lambda\mu)\,Y^{i'}(\lambda\mu)}}[Q_i^+(\lambda\mu)\,Q_{i'}(\lambda\mu)$$
$$+Q_{i'}^+(\lambda\mu)\,Q_i(\lambda\mu)+Q_i^+(\lambda\mu)\,Q_{i'}^+(\lambda\mu)+Q_{i'}(\lambda\mu)\,Q_i(\lambda\mu)]$$
$$-\tfrac{1}{4}\sum_{\lambda=2,3}\sum_{\mu\geq 0}\sum_i\frac{1}{\sqrt{Y^i(\lambda\mu)}}\sum_{q,q'} v_{qq'}\{[Q_i^+(\lambda\mu)+Q_i(\lambda\mu)]\,[\mathfrak{f}^{\lambda\mu}(q,q')\,B(q,q')$$
$$+\bar{\mathfrak{f}}^{\lambda\mu}(q,q')\,\bar{B}(q,q')]+[\mathfrak{f}^{\lambda\mu}(q,q')\,B(q,q')+\bar{\mathfrak{f}}^{\lambda\mu}(q,q')\,\bar{B}(q,q')]\times$$
$$\times[Q_i^+(\lambda\mu)+Q_i(\lambda\mu)]\}, \tag{9.38}$$

where $Y^i(\lambda\mu)$ are determined by (8.136) and (8.170).

The additional interaction (5.28) assures the exclusion of the spurious state in the $\lambda = 2$, $\mu = 0$ case. The analogous terms from the Hamiltonian $H_0(n)$ and $H_0(p)$ should be included in the treatment of the quasiparticle–phonon interaction. This additional interaction (5.30) can be rewritten for our purposes as

$$H_0''(n) = -\frac{G_N}{\sqrt{2}}\sum_{s,s'}(u_s^2-v_s^2)u_{s'}v_{s'}\sum_i\{[\psi^{20i}_{ss'}Q_i^+(20)+\varphi^{20i}_{ss'}Q_i(20)]\,B(s,s')+B(s,s')\times$$
$$\times[\psi^{20i}_{ss'}Q_i(20)+\varphi^{20i}_{ss'}Q_i(20)]\}. \tag{9.39}$$

The Coriolis interaction is not included in our present model. Hence, the angular momentum projection on the nuclear symmetry axis K is a conserved quantity. The quasiparticle–phonon interaction is treated in the quasiboson approximation. This means that the quasiparticle operators $\alpha_{q\sigma}$ which do not enter in the operators $B(q, q')$ or $B(q, q')$ commute with the phonons, i.e.,

$$[\alpha_{q\sigma} Q_i(\lambda\mu)] = 0. \tag{9.40}$$

The phonon–phonon scattering is also neglected.

2. Let us find the secular equation determining the energies of nonrotational states in odd-A nuclei.

The wave function of the considered K^π state is assumed in the form

$$\Psi_n(K^\pi; \varrho) = C_\varrho^n \frac{1}{\sqrt{2}} \sum_\sigma \left\{ \alpha_{\varrho\sigma}^+ + \sum_{\lambda\mu i} \sum_s D_{\varrho s\sigma}^{\lambda\mu i n} \alpha_{s\sigma}^+ Q^+(\lambda\mu) \right\} \Psi \tag{9.41}$$

for an odd-N nucleus and

$$\Psi_n(K^\pi; v) = C_v^n \frac{1}{\sqrt{2}} \sum_\sigma \left\{ \alpha_{v\sigma}^+ + \sum_{\lambda\mu i} \sum_r D_{vr\sigma}^{\lambda\mu i n} \alpha_{r\sigma}^+ Q_i^+(\lambda\mu) \right\} \Psi \tag{9.41'}$$

for an odd-Z nucleus.

The function Ψ represents the phonon vacuum (ground state) of the corresponding even–even nucleus; it is determined by (9.2). The quantum numbers ϱ or v denote the single-particle level with the considered values K^π. The integer n denotes different solutions of our equation. The wave function (9.41) is properly normalized if

$$(C_\varrho^n)^2 \left\{ 1 + \tfrac{1}{2} \sum_{\lambda\mu i} \sum_{s\sigma} (D_{\varrho s\sigma}^{\lambda\mu i n})^2 \right\} = 1. \tag{9.42}$$

The functions (9.41) with $n \neq n'$ are orthogonal.

Let us find the expectation value of H_{vq} in the state (9.41):

$$(\Psi_n^*(K^\pi; \varrho)\, H_{vq} \Psi_n(K^\pi; \varrho) = (C_\varrho^n)^2 \Bigg\{ \varepsilon(\varrho) + \tfrac{1}{2} \sum_{\lambda\mu i} \sum_{s\sigma} [\varepsilon(s) - \omega_i^{\lambda\mu}] (D_{\varrho s\sigma}^{\lambda\mu i n})^2$$
$$- \tfrac{1}{2} \sum_{\lambda\mu i} \frac{1}{\sqrt{Y^i(\lambda\mu)}} \sum_{s\sigma} D_{s\sigma}^{\lambda\mu i} v_{\varrho s} [\mathfrak{f}^{\lambda\mu}(\varrho s) - \sigma \bar{\mathfrak{f}}^{\lambda\mu}(\varrho s)] \Bigg\}.$$

The functions C_ϱ^n and $D_{\varrho s\sigma}^{\lambda\mu i n}$ are determined from the variational equation, i.e., from the minimum energy condition. The variational principle is symbolically formulated in the usual way, i.e.,

$$\delta \left\{ (\Psi_n^*(K^\pi; \varrho)\, H_{vq} \Psi(K^\pi; \varrho)) - \eta_n \left[(C_\varrho^n)^2 \left(1 + \tfrac{1}{2} \sum_{\lambda\mu i} \sum_{s\sigma} (D_{\varrho s\sigma}^{\lambda\mu i n})^2 \right) - 1 \right] \right\} = 0, \tag{9.43}$$

where the Lagrange multiplier η_n assures the validity of the normalization condition (9.42). It is equal to the energy of the resulting state in the odd-A nucleus.

As a result of simple transformations (see ref. 506 for details), we obtain the secular equation of the form

$$-P(\eta) \equiv \varepsilon(\varrho) - \eta_n - \tfrac{1}{4} \sum_{\lambda\mu i} \sum_s \frac{v_{\varrho s}^2}{Y^i(\lambda\mu)} \frac{(f^{\lambda\mu}(\varrho s))^2}{\varepsilon(s) + \omega_i^{\lambda\mu} - \eta_n} = 0. \tag{9.44}$$

The solutions η_n, enumerated by the integer n ($n = 1, 2, 3, \ldots$), determine the energy of the state (9.41). Equation (9.44) has poles at $\varepsilon(s)+\omega_i^{\lambda\mu}$, i.e., at the sum of quasiparticle and phonon energies. Another important point is the point $\eta = \varepsilon(\varrho)$, where the first term of the sum changes the sign. The pairing factor $v_{\varrho s}^2$ in (9.44) suggests that the interaction has a predominantly particle–particle character. Equation (9.44) is formally independent of $\varkappa^{(\lambda)}$; it does not contain any free parameters.

The resulting energies are shifted with respect to the values $\varepsilon(\varrho)$ and $\varepsilon(s)+\omega_i^{\lambda\mu}$; such shifts are caused by the quasiparticle–phonon interaction. The magnitude of shift is related to the strength of interaction.

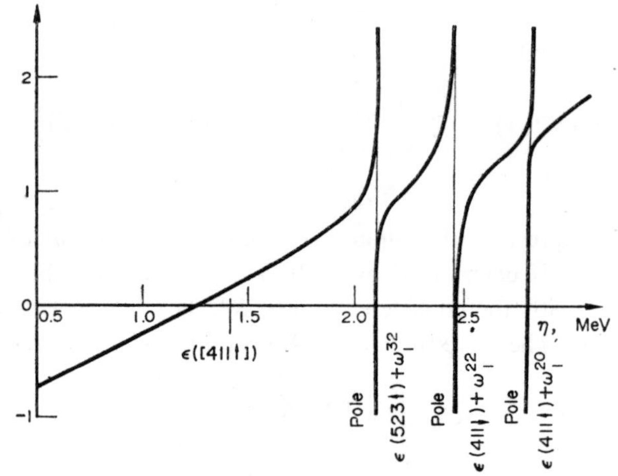

FIG. 9.7. The function $P(\eta)$ of (9.44) for the $K^\pi = \frac{3}{2}^+$, $\varrho = 411\uparrow$ states in ^{165}Ho. The vertical lines show the poles of (9.44); their characteristics are also shown; the one-quasiparticle energy is $\varepsilon(411\uparrow) = 1430$ keV.

Let us explain the whole concept with one example. Figure 9.7 shows the function $P(\eta)$ for the $K^\pi = \frac{3}{2}^+$ state in ^{165}Ho, with $\varrho = 411\uparrow$. The solutions η_n correspond to the points $P(\eta) = 0$. The first solution η_1 is 200 keV less than the quasiparticle energy $\varepsilon(411\uparrow)$. The second solution practically coincides with the pole $\varepsilon(523\uparrow)+\omega_1^{32}$. The resulting state has quasiparticle + octupole phonon structure. The third solution is between the second and third poles of (9.44). This solution has predominantly the structure quasiparticle + γ-vibrational phonon.

The functions C_ϱ^n and $D_{\varrho s\sigma}^{\lambda\mu i n}$ are found from (9.44) and from the normalization condition (9.42). They are equal to

$$(C_\varrho^n)^{-2} = 1+\tfrac{1}{4}\sum_{\lambda\mu i}\sum_s \frac{v_{\varrho s}^2}{Y^i(\lambda\mu)} \frac{(f^{\lambda\mu}(\varrho v))^2}{[\varepsilon(s)+\omega_i^{\lambda\mu}-\eta_n]^2}, \tag{9.45}$$

$$D_{\varrho s\sigma}^{\lambda\mu i n} = \tfrac{1}{2}\frac{v_{\varrho s}}{\sqrt{Y^i(\lambda\mu)}} \frac{f^{\lambda\mu}(\varrho s)}{\varepsilon(s)+\omega_i^{\lambda\mu}-\eta_n}. \tag{9.45'}$$

The quantity $(C_\varrho^n)^2$ determines the contribution of the one-quasiparticle component in the wave function (9.41). The value $(C_\varrho^n)^2 \to 1$ if the energy η_n converges to the quasiparticle value $\varepsilon(\varrho)$; the wave function has a purely one-quasiparticle character in such a case. The quantity

$$d_n^{\lambda\mu i}(\varrho, s) = \tfrac{1}{2}(C_\varrho^n)^2 \sum_\sigma (D_{\varrho s\sigma}^{\lambda\mu i n})^2 \tag{9.46}$$

characterizes the contribution of the quasiparticle s+phonon $(\omega_i^{\lambda\mu})$ in the state (9.41). In the limit $\eta_n \to \varepsilon(s)+\omega_i^{\lambda\mu}$,

$$\Psi_n(K^\pi; \varrho)\Big|_{\varepsilon(s)+\omega_{i_0}^{\lambda_0\mu_0}-\eta_n \to 0} = \frac{1}{\sqrt{2}} \sum_\sigma \alpha_{s\sigma}^+ Q_{i_0}^+(\lambda_0\mu_0)\Psi, \tag{9.47}$$

and $d_n^{\lambda\mu i}(\varrho, s) \to 1$. The state (9.47) is usually characterized as a vibrational state in an odd-A nucleus. If $\lambda_0 = 2$, $\mu_0 = 2$, the corresponding state is γ-vibrational. If $\lambda_0 = 3$, then the state is an octupole state, etc. For example, the second solution in Fig. 9.7 is an octupole state; its wave function has the form (9.47).

The spectra and the wave functions of odd-A deformed nuclei are found in the following way. Equation (9.44) is solved for every value of quantum numbers K^π; the corresponding solutions represent the energies

$$\eta_n = (\Psi_n^*(K^\pi; \varrho)\, H_{vq}\Psi_n(K^\pi; \varrho)). \tag{9.48}$$

The ground-state energy η_1^F is equal to the smallest η_1 value. The excitation energies of nonrotational states are simply the differences

$$\eta_n - \eta_1^F. \tag{9.48'}$$

The secular equation (9.44) contains a summation over different phonons $(\lambda\mu i)$. Calculations[140, 475, 507–510] included the $i = 1$ and 2 phonons with $(\lambda\mu)$=(2, 0), (2, 2), (3, 0), (3, 1), and (3, 2). The $\lambda = 2$, $\mu = 2$, $i = 1$ were found to be the most important phonon states in these calculations. However, the $(\lambda\mu) = (2, 0)$ and (3, 0) phonons also played an important role. The truncation to $\lambda < 4$ phonons and $i < 3$ solutions is sufficient in a majority of cases. We should realize that the quantity $Y^i(\lambda\mu)$ diverges for the purely two-quasiparticle states. Consequently, the contribution of such states in (9.44) is converging to zero.

3. The basic equation (9.44) can be modified and the additional terms can be added. Let us discuss briefly the possible improvement and show when such improvements are needed.

The first possibility is to include the blocking effect. Even an approximate account of the blocking effect improves the description of energies. The blocking effect can be approximately included if the energies $\varepsilon(\varrho)$ are calculated using the $C(\varrho)$ and $\lambda(\varrho)$ of the odd-A system. Another possibility is to replace the $\varepsilon(\varrho)$ in (9.44) by a difference between the energy of the $N+1$ system (counted from $\lambda(\varrho)$) and the N system (counted from λ). Such a modification was used in refs. 475 and 507–510.

The second modification was suggested in ref. 511. In this case, the phonons of an even–even nucleus are not introduced explicitly. Instead, the interaction between the quasiparticles with a certain multipolarity is taken into account. The phonon operator of the form

$$\Omega = \sum_s \{R_s B(s, \varrho) - R_s' B(\varrho, s) + S_s \alpha_{\varrho+}\alpha_{s-} - S_s' \alpha_{s+}^+ \alpha_{\varrho-}^+\} + \sum_{q,q'} \{\mathcal{L}_{qq'}\alpha_{q'-}\alpha_{q+} - \mathcal{L}_{qq'}' \alpha_{q+}^+ \alpha_{q'-}^+\} \tag{9.49}$$

acts on the wave function $\alpha^+_{\varrho\sigma}\Psi$ of an odd-A nucleus. The resulting secular equation then has the form

$$\frac{1}{\varkappa^{(\lambda)}_2} = \sum_{q,\, q' \neq \varrho} \frac{(f^{\lambda\mu}(q,q')u_{qq'})^2[\varepsilon(q)+\varepsilon(q')]}{[\varepsilon(q)+\varepsilon(q')]^2-\eta_n^2}$$
$$+\sum_{\substack{s\\ K_s-K_\varrho=\mu}} (f^{\lambda\mu}(\varrho s))^2 \left\{\frac{u^2_{s\varrho}[\varepsilon(s)+\varepsilon(\varrho)]}{[\varepsilon(s)+\varepsilon(\varrho)]^2-\eta_n^2}+\frac{v^2_{s\varrho}[\varepsilon(s)-\varepsilon(\varrho)]}{[\varepsilon(s)-\varepsilon(\varrho)]^2-\eta_n^2}\right\}. \tag{9.50}$$

Such an equation treats the Pauli exclusion principle more consistently than (9.44). However, the quasiparticle interaction is restricted to only one $(\lambda\mu)$ combination.

Both (9.44) and (9.50) give similar results in the cases where the interaction of quasiparticles with γ-vibrational phonons is more important than any other type of interaction.

The third possible improvement of our basic equation (9.44) is based on the inclusion of the quasiparticle–phonon interaction in the nuclear ground state. The wave function of an odd-Z nucleus has the form

$$O^+(K^\pi)\Psi', \tag{9.51}$$

where

$$O^+(K^\pi) = C_\nu \frac{1}{\sqrt{2}}\left\{\sum_\sigma \alpha^+_{\nu\sigma}+\sum_{\lambda\mu i}\sum_{r\sigma}[D^{\lambda\mu i}_{\nu r\sigma}\alpha^+_{r\sigma}Q^+_i(\lambda\mu)-\Delta^{\lambda\mu i}_{\nu r\sigma}Q_i(\lambda\mu)\alpha_{r,\,-\sigma}]\right\},$$

and $O(K^\pi)\Psi' = 0$.

The normalization condition is also modified, namely

$$C^2_\nu\left\{1+\tfrac{1}{2}\sum_{\lambda\mu i}\sum_{r\sigma}[(D^{\lambda\mu i}_{\nu r\sigma})^2+(\Delta^{\lambda\mu i}_{\nu r\sigma})^2]\right\} = 1. \tag{9.51'}$$

Using our usual variational method we can obtain the secular equation which contains an additional term when compared with (9.44). The additional term is regular, i.e., it has no poles for $\eta > 0$. The secular equation is

$$\varepsilon(\nu)-\eta_n-\tfrac{1}{4}\sum_{\lambda\mu i}\sum_r \frac{(f^{\lambda\mu}(\nu r))^2}{Y^i(\lambda\mu)}\left\{\frac{v^2_{\nu r}}{\varepsilon(r)+\omega^{\lambda\mu}_i-\eta_n}-\frac{u^2_{\nu r}}{\varepsilon(r)+\omega^{\lambda\mu}_i+\eta^n}\right\} = 0. \tag{9.52}$$

The amplitude $D^{\lambda\mu i}_{r\nu\sigma}$ has the form (9.45′) and

$$\Delta^{\lambda\mu i}_{\nu r\sigma} = -\frac{\sigma}{2}\frac{u_{\nu r}}{\sqrt{Y^i(\lambda\mu)}}\frac{f^{\lambda\mu}(\nu r)}{\varepsilon(r)+\omega^{\lambda\mu}_i+\eta_n}.$$

The solutions η_n of (9.52) are closer to the one-quasiparticle energy $\varepsilon(\nu)$ than the solutions of (9.44). However, the change is more or less the same for all solutions close to the one-quasiparticle states. Hence, the corresponding excitation energies are close to each other if (9.52) or (9.44) are used. The main effect of the modification (9.52) is a small (less than 10%) increase in energies of vibrational states.

The fourth modification is used if the interaction with the β-vibrational phonons is important. The additional interaction (9.39) helps to reduce the errors related to the nonconservation of the particle number. The secular equation including the additional interaction (9.39) is more complex, namely

$$\varepsilon(\varrho)-\eta_n-\tfrac{1}{4}\sum_{\lambda\mu i}\sum_{s\neq\nu}\frac{v_{\varrho s}^2}{Y^i(\lambda\mu)}\frac{(f^{\lambda\mu}(\varrho,s))^2}{\varepsilon(s)+\omega_i^{\lambda\mu}-\eta_n}$$
$$-\tfrac{1}{4}\sum_i\frac{1}{Y^i(20)}\frac{1}{\varepsilon^2(\varrho)\,[\varepsilon(\varrho)+\omega_i^{20}-\eta_n]}\Big\{\mathfrak{f}^{20}(\varrho,\varrho)\,[E(\varrho)-\lambda]$$
$$-\frac{4C_n^2}{\gamma_n^i}\sum_{s,s'}\frac{\mathfrak{f}^{20}(s,s)}{\varepsilon(s)\,[4\varepsilon^2(s)-(\omega_i^{20})^2]}\frac{E(s)-E(s')}{\varepsilon(s')\,[4\varepsilon^2(s')-(\omega_i^{20})^2]}\Big\}^2=0, \qquad (9.53)$$

the quantities $Y^i(20)$ and γ_n^i were determined in (8.170) and (8.171). However, if the effect of the $\lambda = 2$, $\mu = 0$ terms is not substantial, we can still use the simpler equation (9.44).

Finally, the fifth and last modification includes simultaneously several one-quasiparticle states with identical K^π values. The single-particle spectrum contains many states with identical quantum numbers K^π, some of them are even relatively close to each other. Suppose that we want to include $\varrho_1, \varrho_2, \varrho_3, \ldots,$ states with the same K^π. The trial wave function is then chosen as

$$\Psi_n(K^\pi;\varrho_1,\varrho_2,\varrho_3,\ldots)=N(\varrho_1,\varrho_2,\varrho_3,\ldots)\frac{1}{\sqrt{2}}\sum_\sigma\Big\{C'^n_{\varrho_1}\alpha^+_{\varrho_1\sigma}$$
$$+C'^n_{\varrho_2}\alpha^+_{\varrho_2\sigma}+\ldots+\sum_{\lambda\mu i}\sum_s D^{\lambda\mu i n}_{\varrho_1\varrho_2\varrho_3\ldots s\sigma}\alpha^+_{s\sigma}Q_i^+(\lambda\mu)\Big\}\Psi, \qquad (9.54)$$

where $N(\varrho_1, \varrho_2, \varrho_3, \ldots,)$ represents the normalization factor. The general secular equation was derived in ref. 506; the particular case of two states was discussed in refs. 507, 510, and 512.

Let us restrict our attention to the description of two one-quasiparticle states in odd-N nucleus. The trial wave function is chosen as

$$\Psi_n(K^\pi;\varrho_1,\varrho_2)=N_n(\varrho_1,\varrho_2)\frac{1}{\sqrt{2}}\sum_\sigma\Big\{C'^n_{\varrho_1}\alpha^+_{\varrho_1\sigma}+C'^n_{\varrho_2}\alpha^+_{\varrho_2\sigma}+\sum_{\lambda\mu i}\sum_s D^{\lambda\mu i n}_{\varrho_1\varrho_2 s\sigma}\alpha^+_{s\sigma}Q_i^+(\lambda\mu)\Big\}\Psi, \quad (9.55)$$

with the normalization

$$N_n^2(\varrho_1,\varrho_2)\Big\{(C'^n_{\varrho_1})^2+(C'^n_{\varrho_2})^2+\tfrac{1}{2}\sum_{s\sigma}(D^{\lambda\mu i n}_{\varrho_1\varrho_2 s\sigma})^2\Big\}=1. \qquad (9.55')$$

The secular equation is found by the variational method. It has the form

$$P(\eta)\equiv\begin{vmatrix}V_n(\varrho_1,\varrho_1)-[\varepsilon(\varrho_1)-\eta_n] & V_n(\varrho_1,\varrho_2)\\ V_n(\varrho_1,\varrho_2) & V_n(\varrho_2,\varrho_2)-[\varepsilon(\varrho_2)-\eta_n]\end{vmatrix}=0, \qquad (9.56)$$

where

$$V_n(\varrho,\varrho')=\tfrac{1}{4}\sum_{\lambda\mu i s}\frac{v_{\varrho s}v_{\varrho' s}}{Y^i(\lambda\mu)}\frac{f^{\lambda\mu}(\varrho s)\,f^{\lambda\mu}(\varrho' s)}{\varepsilon(s)+\omega_i^{\lambda\mu}-\eta_n}.$$

Equation (9.56) contains only the first-order poles; the second-order poles exactly cancel each other. The functions $C'^n_{\varrho_1}$, $C'^n_{\varrho_2}$ may be chosen in the symmetric form

$$\left.\begin{aligned} C'^n_{\varrho_1} &= 1-\frac{V_n(\varrho_1,\varrho_2)}{V_n(\varrho_1,\varrho_1)-[\varepsilon(\varrho_1)-\eta_n]},\\ C'^n_{\varrho_2} &= 1-\frac{V_n(\varrho_1,\varrho_2)}{V_n(\varrho_2,\varrho_2)-[\varepsilon(\varrho_2)-\eta_n]}. \end{aligned}\right\} \tag{9.57}$$

In such a case

$$D^{\lambda\mu i n}_{\varrho_1\varrho_2 s\sigma} = C'^n_{\varrho_1} D^{\lambda\mu i n}_{\varrho_1 s\sigma} + C'^n_{\varrho_2} D^{\lambda\mu i n}_{\varrho_2 s\sigma}, \tag{9.57'}$$

$$\left(N_n(\varrho_1,\varrho_2)\right)^{-2} = (C'^n_{\varrho_1})^2+(C'^n_{\varrho_2})^2+\tfrac{1}{2}\sum_{\lambda\mu i}\sum_{s\sigma}(D^{\lambda\mu i n}_{\varrho_1\varrho_2 s\sigma})^2, \tag{9.57''}$$

with $D^{\lambda\mu i n}_{\varrho s\sigma}$ determined by (9.45′). Each of the one-quasiparticle states is characterized by its contribution $(N_n(\varrho_1,\varrho_2)C'^n_{\varrho_1})^2$. The quasiparticle s+phonon $(\lambda\mu i)$ contributes the amount

$$d^{\lambda\mu i}_n(\varrho_1,\varrho_2;s) = \tfrac{1}{2}(N_n(\varrho_1,\varrho_2))^2\sum_\sigma(D^{\lambda\mu i n}_{\varrho_1\varrho_2 s\sigma})^2. \tag{9.58}$$

Let us compare the solutions of (9.44) and (9.56). We shall use the $K^\pi = \frac{5}{2}^-$ states in ^{167}Yb as an example and display the corresponding functions in Fig. 9.8. The figure shows

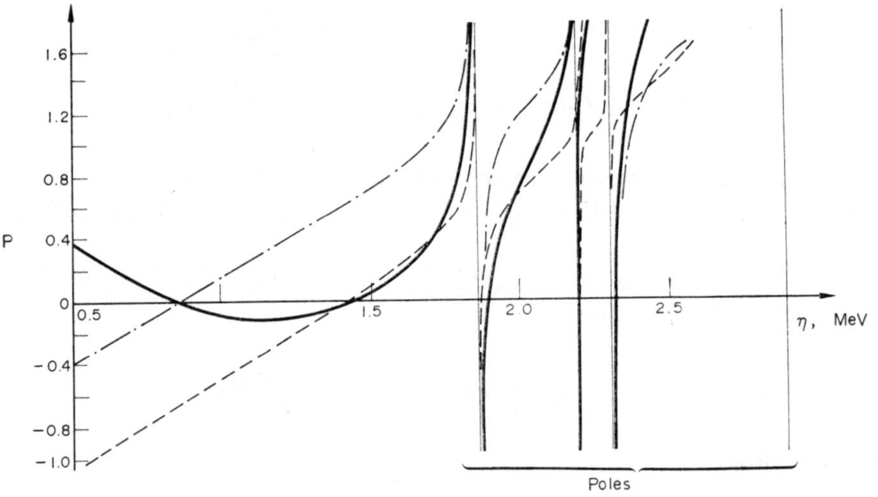

FIG. 9.8. The function $P(\eta)$ of (9.44) for $\varrho_1 = 523\downarrow$ (dot-and-dash curve) and for $\varrho_2 = 512\uparrow$ (dashed curve). The function $P(\eta)$ of (9.56) for $\varrho_1+\varrho_2 = 523\downarrow+512\uparrow$ is shown by the full curve. The calculations were performed for the $K^\pi = \frac{5}{2}^-$ states in ^{167}Yb. The vertical lines show the positions of the poles in (9.44) and (9.56).

the functions $P(\eta)$ of (9.44) for $\varrho_1 = 523\downarrow$ and $\varrho_2 = 512\uparrow$. At the same time, the function $P(\eta)$ of (9.56) is shown for $\varrho_1+\varrho_2 = 523\downarrow+512\uparrow$. Obviously, the solutions for the two equations are quite close. By solving (9.58) we have obtained the following structure of the two lowest states: $\eta_1 = 0.86$ MeV, $523\downarrow-90\%$, $512\uparrow-2\%$; $\eta_2 = 1.436$ MeV, $523\downarrow-2\%$, $512\uparrow-76\%$. The third solution at 1.9 MeV is collective; it contains 2% of the $523\downarrow$ state and 3.6% of $512\uparrow$ state. Thus, this state contains both of the one-quasiparticle states in comparable amounts. The main advantage of (9.44) is the more correct number of states in the former case. Equation (9.56) should be used whenever two one-quasiparticle states have the energies $\varepsilon(\varrho_1)$, $\varepsilon(\varrho_2)$ close to each other. Such calculations were performed in refs. 475 and 507–510.

4. We have seen that the structure of nonrotational states in odd-A deformed nuclei is strongly affected by the quasiparticle–phonon interaction. The individual components of the wave function of such states can be experimentally determined by complex studies combining the methods of traditional α-, β-, and γ-spectroscopy with nuclear reactions. Let us discuss the individual nuclear processes giving information about different parts of the nuclear wave function.

1. *Reduced probabilities of the electric transitions*

The reduced probability of the $E\lambda$ transition equals, according to (6.59),

$$B(E\lambda; I_0 \to I_f) = \frac{2\lambda+1}{4\pi}\left(\frac{1}{m\mathring{\omega}_0}\right)^2 \times \\ \times |(I_0K_0\lambda K_f-K_0 \,|\, I_fK_f)\langle K_f \,|\, \mathfrak{M}(E\lambda; K_f-K_0) \,|\, K_0\rangle + (-1)^{I_f-1/2} \times \\ \times (I_0K_0\lambda -K_f-K_0 \,|\, I_f-K_f)\langle -K_f \,|\, \mathfrak{M}(E\lambda; -K_f-K_0) \,|\, K_0\rangle|^2 \quad (9.59)$$

(for notation, see § 3 of Chapter 6). The matrix element $\langle K_f \,|\, \mathfrak{M}(E2); K_f-K_0) \,|\, K_0\rangle$ for an $E2$ transition between two states described by the wave functions $\Psi_{n_0}(K_0^{\pi_0}; \varrho_0)$ and $\Psi_{n_f}(K_f^{\pi_f}; \varrho_f)$ (see (9.41)) consists of four terms:

$$\langle K_f \,|\, \mathfrak{M}(E2; K_f-K_0) \,|\, K_0\rangle = C^{n_0}_{\varrho_0} C^{n_f}_{\varrho_f} \Big\{ \sum_i D^{22in_f}_{\varrho_f\varrho_0+} \mathscr{M}((\varrho_0 \to \varrho_0 + Q_i(22)) \\ + \sum_i D^{22in_0}_{\varrho_0\varrho_f+} \mathscr{M}_i(\varrho_f + Q_i(22) \to \varrho_f) + e_{\text{eff}} e f^{22}(\varrho_0, \varrho_f) v_{\varrho_0\varrho_f} \\ + e_{\text{eff}} e \sum_{\lambda\mu i} \sum_{s,\, s'} D^{\lambda\mu in}_{\varrho_f s+} D^{\lambda\mu in}_{\varrho_0 s'+} f^{22}(s, s') v_{s's} \Big\}. \quad (9.60)$$

The quantity $\mathscr{M}_i(\varrho_0 \to \varrho_0 + Q_i(22))$ is the $E2$ matrix element connecting the ground and one-phonon ($22i$) states in the corresponding even–even nucleus; this quantity is determined according to (9.18). All terms in (9.60) have an intuitive physical meaning. The first term describes a transition from the one-quasiparticle component in initial state to the γ-vibrational phonon plus the same quasiparticle in final state. The second term describes the identical process with the initial and final states interchanged. The third term describes the transition between the one-quasiparticle components in both states and the last term describes a transition between the quasiparticle plus phonon components in both states. In the particular case of a transition from a purely one-quasiparticle state to a purely quasiparticle plus γ-phonon state, we expect to observe $B(E2)$ of the even–even nucleus. The effective charge $e^{(2)}_{\text{eff}}$ should be identical in odd-A and even–even nuclei.

In many cases, several terms in (9.60) give important contributions to the total $E2$ matrix element. For example, the first, second and third terms add coherently if η_1 is below the lowest pole in (9.44). The enhancement of $B(E2)$ value, compared to the $E2$ transition between one-quasiparticle states, is a clearly recognizable feature of the states having large component quasiparticle plus γ-vibrational phonon. Similarly, the enhancement of $B(E3)$ will signal a large octupole component of the wave function.

2. *Alpha-decay hindrance factor* F_α

Let us consider the α-decay from the predominantly one-quasiparticle state $\Psi_{n_0}(K_0^{\pi_0}; \varrho_0)$ to the predominantly quasiparticle ϱ_0+ phonon $(\lambda_0\mu_0 i)$ state $\Psi_{n_f}(K_f^{\pi_f}; \varrho_f)$.

The reduced width of such a transition equals

$$\gamma_{II}^2(\varrho_0 \to \varrho_0+Q_{i_0}(\lambda_0\mu_0)) = (C_{\varrho_0}^{n_0})^2\, d_{n_f}^{\lambda_0\mu_0 i_0}(\varrho_f, \varrho_0)\gamma_{II}^2(Q), \tag{9.61}$$

where $\gamma_{II}^2(Q)$ is the reduced width (9.27) of the α-decay populating one-phonon state in corresponding even–even nucleus. The quantity $d_{n_f}^{\lambda_0\mu_0 i_0}(\varrho_f, \varrho_0)$ is determined by (9.46). The hindrance factor is then equal to

$$F_\alpha(\varrho_0 \to \varrho_0+Q_{i_0}(\lambda_0\mu_0)) = \frac{F_\alpha(Q_{i_0}(\lambda_0\mu_0))}{(C_{\varrho_0}^{n_0})^2\, d_{n_f}^{\lambda_0\mu_0 i_0}(\varrho_f, \varrho_0)}, \tag{9.62}$$

with $F_\alpha(Q_{i_0}(\lambda_0\mu_0))$ denoting the hindrance factor (9.28) in even–even nucleus. The hindrance factors in odd-A nuclei can be calculated if the $F_\alpha(Q)$ values are known.[508] The reduced width (9.61) describes a favored α-decay; the hindrance factors are small in such a case. On the other hand, when the state of the odd quasiparticle is changed ($s \neq \varrho_0$), we are dealing with an unfavored α-decay. The hindrance factors of such decays have been discussed in § 1 of Chapter 6.

3. *ft values in* β*-decay*

Let us take as an example the β^+ decay from a one-quasiparticle state ν_0 of an odd-Z nucleus to a state $\Psi_{n_f}(K_f^{\pi_f}; \varrho_f)$ in odd-N nucleus. The transition matrix element is equal to

$$\mathcal{M}\begin{pmatrix} Z+1 \to Z \\ N \to N+1 \end{pmatrix} = \tfrac{1}{2}C_{\nu_0}^{n_0}C_{\varrho_f}^{n_f} \sum_{\sigma,\sigma'} \Big\{ \langle \varrho_f\sigma | \Gamma_\beta | \nu_0\sigma' \rangle u_{\varrho_f} u_{\nu_0} - \frac{1}{\sqrt{2}} \sum_{\lambda\mu i} \sum_{s,r} \langle s\sigma | \Gamma_\beta | r\sigma' \rangle u_s v_r D_{\varrho_0 s, -\sigma}^{\lambda\mu i n_f} (\psi_{r\nu_0}^{\lambda\mu i} + \sigma\varphi_{\nu r_0}^{\lambda\mu i}) \Big\}. \tag{9.63}$$

Similarly, the β^+ decay of an odd-N nucleus is described by

$$\mathcal{M}\begin{pmatrix} Z \to Z-1 \\ N+1 \to N+2 \end{pmatrix} = -\tfrac{1}{2}C_{\nu_f}^{n_f}C_{\varrho_0}^{n_0} \sum_{\sigma,\sigma'} \Big\{ \sigma\sigma' \langle \varrho_0\sigma | \Gamma_\beta | \nu_f\sigma' \rangle v_{\nu_f} v_{\varrho_0} + \frac{\sigma'}{\sqrt{2}} \sum_{\lambda\mu i} \sum_{s,r} \langle s\sigma | \Gamma_\beta | r\sigma' \rangle v_r u_s D_{\nu_f r\sigma'}^{\lambda\mu i n_f} (\psi_{s\varrho_0}^{\lambda\mu i}\sigma - \varphi_{s\varrho_0}^{\lambda\mu i}) \Big\}. \tag{9.63'}$$

The first term in the above expressions describes β-transitions between one-quasiparticle states; such transitions are slower by the factor $(C_\varrho^n C_\nu^{n'})^2$ than the similar transitions calculated without the quasiparticle-phonon interaction. The remaining terms in (9.63) and (9.63′) describe β-transitions involving various three-quasiparticle components; two of the three quasiparticles involved always form one component of the phonon. The β-decay to the state quasiparticle plus phonon is somewhat slower than the similar β-decay populating one-phonon state.

In favorable cases we can extract from the experimental $\log ft$ values the information about some particular three-quasiparticle component of (9.41). For example, if the final state contains a relatively large component populated by the *au* transition, we can neglect all remaining terms and determine the quantity $(u_s v_r C_{\nu_0}^{n_0} d_{nf}^{\lambda\mu i}(\varrho_f, s))^2$. The ft value is, in such a case, considerably smaller than the typical *ah* β-transitions between one-quasiparticle states.

Fast β-transitions from $p411\uparrow$ level in ^{163}Tb populate several levels in ^{163}Dy.(181) They can be explained as the transitions to the three-quasiparticle component $n523\downarrow$, $p523\uparrow$, $p411\uparrow$; the combination $pp523\uparrow$–$411\uparrow$ forms a part of the $\lambda = 3$, $\mu = 2$, $i = 1$ phonon. The *au* transition transfers the $n523\downarrow$ neutron into the $p523\uparrow$ proton; the $\log ft$ value for such transitions equals 5.2–6.0.

4. *One-nucleon transfer reactions*

The cross-section of one-nucleon transfer reaction on an even–even deformed target equals, according to (6.97),

$$\frac{d\sigma}{d\Omega} = S_l \sigma_l(\theta), \tag{9.64}$$

all notation was explained in § 3 of Chapter 6. The spectroscopic factors of (*dp*) and (*dt*) reactions equal

$$S_l = 2\Big(\sum_\Lambda a_{l\Lambda}(l\Lambda \tfrac{1}{2}\Sigma \mid jK_f)\Big)^2 (\Psi_{n_f}^*(K_f^{\pi_f};\varrho) a_{\varrho\sigma}^+ \Psi), \tag{9.65}$$

$$S_l = 2\Big(\sum_\Lambda a_{l\Lambda}(l\Lambda \tfrac{1}{2}\Sigma \mid jK_f)\Big)^2 (\Psi_n^*(K_f^{\pi_f};\varrho) a_{\varrho\sigma} \Psi). \tag{9.65′}$$

In the above expressions, Ψ represents the phonon vacuum of the even–even nucleus. Substituting $\Psi_{n_f}(K_f^{\pi_f}, \varrho)$ in the form (9.41),

$$S_l = 2\Big(\sum_\Lambda a_{l\Lambda}(l\Lambda \tfrac{1}{2}\Sigma \mid jK_f)\Big)^2 (u_\varrho C_\varrho^{n_f})^2 \tag{9.66}$$

for (*dp*) reaction, and

$$S_l = 2\Big(\sum_\Lambda a_{l\Lambda}(l\Lambda \tfrac{1}{2}\Sigma \mid jK_f)\Big)^2 (v_\varrho C_\varrho^{n_f})^2 \tag{9.66′}$$

for the (*dt*) reaction.

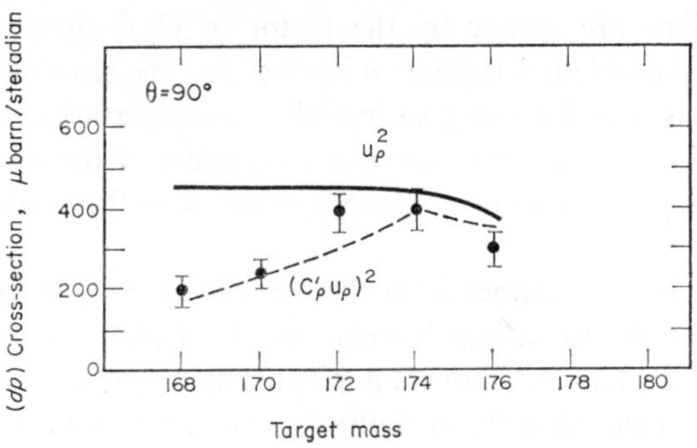

FIG. 9.9. Excitation cross-section of the $\frac{3}{2}-\frac{1}{2}$ [510] state in ytterbium isotopes and the quantities u_ϱ^2 and $(C'_\varrho u_\varrho)^2$. The experimental data with the error bars are reproduced from ref. 187. The calculated $(C'_\varrho u_\varrho)^2$ curve was normalized to the experiment at $A = 174$.

Thus, the cross-section of the one-nucleon transfer reaction depends only on the one-quasiparticle components of the excited states.

We may be able to determine the one-quasiparticle components by measuring the absolute cross-section (9.66) or (9.66′). Figure 9.9 shows an example of such a procedure; the $\frac{3}{2}-\frac{1}{2}$ [510] state was excited in *(dp)* reaction in a number of ytterbium isotopes. The calculated values of u_ϱ^2 and $(C'_\varrho u_\varrho)^2$ are also shown. Note that the $(C'_\varrho u_\varrho)^2$ value was normalized to the experimental value of ^{174}Yb target. The cross-section should remain constant in ^{171}Yb and ^{169}Yb according to the independent quasiparticle model. In reality, the cross-section decreases for the lighter ytterbium isotopes. Such a decrease is easily explained by the decrease of the one-quasiparticle component $\varrho = 510\uparrow$ in the $K^\pi = \frac{1}{2}^-$ wave function. Our model and the experimental data are in a perfect agreement. On the other hand, the decrease in the cross-section in ^{177}Yb is caused by decreasing the u_ϱ^2 value, and it is independent of the quasiparticle–phonon interaction. This example thus shows how valuable are the direct one-nucleon transfer reactions for determining one-quasiparticle components in the wave function (9.41).

5. *The decoupling parameter a of* $K = \frac{1}{2}$ *bands*

The decoupling parameter of the $K = \frac{1}{2}$ band is easily calculated with the wave functions (9.41). As the result,

$$a = (C_\varrho^n)^2 \left\{ a_{\varrho\varrho}^N + \sum_{s,s'} \sum_i a_{ss'}^N (D_{\varrho s+}^{20in} D_{\varrho s'+}^{20in} - D_{\varrho s+}^{30in} D_{\varrho s'+}^{30in}) \right\}, \tag{9.67}$$

where $a_{\varrho\varrho}^N$, a_{ss}^N are the single-particle decoupling parameters (2.12) calculated with the Nilsson or Woods–Saxon wave functions. The second and third terms in (9.67) give usually a negligible contribution to *a*.

Thus, the decoupling parameter of a predominantly one-quasiparticle state simply equals $a_{\varrho\varrho}^N$. On the other hand, *a* vanishes in the purely collective vibrational states with $\mu \neq 0$. The decoupling parameters often offer a convenient way of determining the one-quasiparticle component of the wave function. For example, the 521↓ quasiparticle forms the ground state of ^{169}Er, ^{171}Yb, and ^{173}Hf; the value of $(C'_{\varrho})^2$ equals 0.96–0.98 in this case, and the experimental decoupling parameters are close to the single-particle value a^N. The contribution of the quasiparticle 521↓ decreases in lighter nuclei and correspondingly, the experimental decoupling parameter is also smaller. The value a in ^{161}Dy and ^{163}Er is, in accordance with the theoretical expectation, only a half of the a^N value.

5. Let us discuss generally the structure of nonrotational states in odd-A deformed nuclei. Such states can be approximately divided into three groups: predominantly one-quasiparticle states, predominantly collective (vibrational) states, and the states with a complex structure. There are, naturally, higher lying states with purely three, five, and more, quasiparticles. Tables 9.22, 9.23, and 9.24 contain examples of the three groups. The experimental data of refs. 169, 170, 175–187, 206, 345, and 513–523 are used in the tables.

The predominantly one-quasiparticle states have energies close to but smaller than the $\varepsilon(\varrho)$ value. Such states usually satisfy the condition

$$\varepsilon(\varrho) \ll \min\{\varepsilon(s)+\omega_1^{\lambda\mu}\}, \tag{9.68}$$

i.e., the quasiparticle energy is considerably smaller than the first pole in (9.44). However, there are some predominantly one-quasiparticle states which do not satisfy (9.68); particularly the states with large K-values belong to this group.

Table 9.22 shows several predominantly one-quasiparticle states. The quasiparticle–

TABLE 9.22. *The predominantly one-quasiparticle states*

Nucleus	K^π	Energy (keV)		ϱ	$(C_{\varrho}^1)^2$ (%)
		Experimental	Calculated		
^{181}Re	$\frac{5}{2}+$	0	0	402↑	94
^{177}Hf	$\frac{7}{2}-$	0	0	514↓	97
	$\frac{9}{2}+$	324	300	624↑	99
^{177}Yb	$\frac{9}{2}+$	0	0	624↑	99
^{175}Yb	$\frac{7}{2}-$	0	0	514↓	98
^{169}Yb	$\frac{7}{2}+$	0	0	633↑	97
	$\frac{1}{2}-$	24	86	521↓	96
^{165}Er	$\frac{5}{2}-$	0	0	523↓	96
^{163}Dy	$\frac{5}{2}-$	0	0	523↓	94

phonon interaction affects the high K ($K = \frac{9}{2}, \frac{11}{2}$, etc.) states relatively weakly. The effect on the states with a smaller K-value is somewhat larger. Consequently, the resulting energies are shifted from the $\varepsilon(\varrho)$ value by unequal amounts and the resulting sequence of states can be different from the sequence of single-particle states. Hence, we can explain why the $K = \frac{9}{2}, \frac{11}{2}$ states do not form ground states in deformed nuclei when the single-particle levels with smaller K-values have the energies close to the energy values of these states. For example, the 505↑ state is close to the Fermi level in odd-N nuclei with $N = 91$–97, but it does not form the ground state in any of these nuclei. The quasiparticle–phonon interaction can also cause changes of energy levels in different isotopes with the same odd proton number Z and in different isotones with the same odd neutron number N.

The predominantly vibrational states have the energies η_n close to the pole value $\varepsilon(s) + \omega_{i_0}^{\lambda_0 \mu_0}$. Such states usually exist if

$$\varepsilon(s) + \omega_{i_0}^{\lambda_0 \mu_0} \ll \varepsilon(\varrho). \tag{9.69}$$

However, vibrational states can also exist among the higher excited states. We have already stated that the $K = \frac{9}{2}, \frac{11}{2}$ etc., states are weakly affected by the quasiparticle–phonon interaction. Hence, such states often have predominantly vibrational character and are described by the wave function (9.47). The states with $K = \frac{1}{2}, \frac{3}{2}$ usually have a complex structure.

TABLE 9.23. *States with a large γ-vibrational component*

Nucleus	K^π	Energy (keV)		$B(E2)_{\text{s.p.u.}}$		Structure (%)
		Experimental	Calculated	Experimental	Calculated	
^{187}Re	$\frac{1}{2}^+$	511	710	3.8	2.0	402↑+Q_1 (22)−85; 400↑−14
	$\frac{9}{2}^+$	840	840	2.5	2.0	402↑+Q_1 (22)−99
^{185}Re	$\frac{1}{2}^+$	645	660	3.6	2.3	402↑+Q_1 (22)−80; 400↑−18
	$\frac{9}{2}^+$	966	860	2.6	2.7	402↑+Q_1 (22)−99
^{181}Re	$\frac{1}{2}^+$	932	930	−	1.2	402↑+Q_1 (22)−58; 400↑−38
^{177}Lu	$\frac{11}{2}^+$	1306	1360	−	2.3	404↓+Q_1 (22)∼100
^{169}Tm	$\frac{3}{2}^+$	570	1074	1.5	2.9	411↓+Q_1 (22)−79; 411↑−20
	$\frac{5}{2}^+$	1190	1100	1.3	2.6	411↓+Q_1 (22)−72; 413↓−2
^{159}Tb	$\frac{1}{2}^+$	580	600	1.5	2.0	411↓−64; 411↑+Q_1 (22)−27
	$\frac{7}{2}^+$	1270	1400	2.0	2.5	411↑+Q_1 (22)−100
^{171}Yb	$\frac{3}{2}^-$	902	920	−	1.0	521↓+Q_1 (22)−44; 521↑−55
^{169}Er	$\frac{3}{2}^-$	714	730	−	1.5	521↓+Q_1 (22)−56; 521↑−39
^{165}Er	$\frac{1}{2}^-$	920	930	−	2.5	523↓+Q_1 (22)−66; 510↑−17

Papers 169, 170, and 524 contain the analysis of the experimental information about the vibrational states in odd-A deformed nuclei. The vibrational states in odd-A nuclei appear if the corresponding even–even nuclei have the low-lying one-phonon states. For example, the γ-vibrational states in dysprosium and erbium isotopes have energies below 1 MeV. Correspondingly, the low-lying states with large γ-vibrational components are observed in the neighboring odd-A nuclei.

Table 9.23 shows several states of the γ-vibrational type. The large $B(E2)_{\text{s.p.u.}}$ values suggest a large quasiparticle plus γ-vibrational phonon component of these states.

Two γ-vibrational states, $K = K_0-2$ and $K = K_0+2$, can be built on the $K = K_0$ ground state of an odd-A deformed nucleus. If the energy $\varepsilon(\varrho)$ is lower than the first pole, we expect that

$$\mathscr{E}_0(K_0-2) < \mathscr{E}_0(K_0+2). \tag{9.70}$$

The validity of (9.70) is related to the smaller number of levels s in the equation (9.44) for the larger K-value.

The prevalent case in real nuclei is an intermediate situation such as

$$\varepsilon(\varrho) \approx (0.5-2.0) \min\{\varepsilon(s)+\omega_1^{\lambda\mu}\}. \tag{9.71}$$

The solutions of (9.44) are maximally shifted in such a case with respect to both the $\varepsilon(\varrho)$ value and the first pole. The resulting states have a complex structure. The one-quasiparticle component of such states gives 30–80% of the normalization. Many states with different quantum numbers s and $\lambda\mu i$ contribute to the righthand side of the secular equation (9.44). Table 9.24 shows the examples of the states with a complex structure.

TABLE 9.24. *States with a complex structure*

Nucleus	K^π	Energy (keV) Experimental	Energy (keV) Calculated	Structure (%)		
^{181}Re	$\frac{1}{2}+$	826	940	$411\downarrow-83$;	$411\uparrow+Q_1\,(22)-10$;	$402\uparrow+Q_1\,(22)-5$
	$\frac{3}{2}+$	1060	1140	$411\uparrow-23$;	$411\downarrow+Q_1\,(22)-76$	
^{165}Ho	$\frac{1}{2}+$	428	490	$411\downarrow-88$;	$411\uparrow+Q_1\,(22)-10$;	$521\uparrow+Q_1\,(32)-0.5$
^{175}Yb	$\frac{3}{2}-$	809	730	$512\downarrow-79$;	$510\uparrow+Q_1\,(22)-14$;	$514\downarrow+Q_1\,(22)-4$
^{169}Yb	$\frac{3}{2}-$	657	590	$521\uparrow-87$;	$521\uparrow+Q_1\,(20)-5.9$;	$521\downarrow+Q_1\,(22)-2.5$
^{167}Er	$\frac{3}{2}-$	750	600	$521\uparrow-87$;	$521\downarrow+Q_1\,(22)-6.8$;	$633\uparrow+Q_1\,(32)-1.2$
	$\frac{1}{2}-$	767	660	$510\uparrow-41$;	$512\uparrow+Q_1\,(22)-50$;	$512\downarrow+Q_1\,(22)-5$
^{165}Er	$\frac{5}{2}-$	477	390	$512\uparrow-85$;	$510\uparrow+Q_1\,(22)-9.3$;	$624\uparrow+Q_1\,(32)-3.1$
^{165}Dy	$\frac{1}{2}-$	570	550	$510\uparrow-42$;	$512\uparrow+Q_1\,(22)-47$;	$523\downarrow+Q_1\,(22)-2.4$

6. As a consequence of the quasiparticle–phonon interaction, the one-quasiparticle, three-quasiparticle, etc., components are spread over several excited states of odd-A deformed nuclei. The structure of states becomes more complex at higher excitation energies. Let us discuss this structure and the distribution of the individual components.[512]

The ground states and a few low-lying states are weakly affected by the quasiparticle–phonon interaction; they have a predominantly one-quasiparticle character. The amount of admixtures grows with the increasing excitation energy. The single-particle states of the Nilsson or Woods–Saxon potentials become distributed over several nonrotational states. Table 9.25 shows several examples of such a process; the Woods–Saxon potential was used in the calculations. The number of nonrotational states with a certain K^π value excited by the (dp) and (dt) reactions is larger than the number of such states in the single-particle level scheme. However, the total strength agrees with the estimate based on the Nilsson potential. Such experimental material thus supports our conclusion about the distribution of one-quasiparticle components over several nonrotational states.

This tendency should be more pronounced in the three-quasiparticle states. Only the states with large K-values or the states which cannot be combined as one quasiparticle plus

TABLE 9.25. *Distribution of one-quasiparticle components*

Nucleus	K^π	ϱ	$\varepsilon(\varrho)$ (MeV)	n	η_n (MeV)	$(C_\varrho^n)^2$ (%)	Poles of (9.44) (MeV)
^{167}Yb	$\frac{7}{2}+$	633↑	1.17	1	1.00	94.0	2.34
				2	2.38	2.3	2.59
	$\frac{1}{2}-$	521↓	1.30	1	1.17	93.0	1.06
				2	2.10	2.6	2.18
				3	2.19	1.3	2.46
				4	2.48	2.5	2.59
	$\frac{5}{2}-$	512↑	2.06	1	1.42	79.0	1.87
				2	1.88	2.8	2.21
	$\frac{9}{2}+$	624↑	2.38	1	1.94	59.0	2.15
				2	2.22	15.0	2.51
				3	2.56	8.0	2.98
				4	2.99	0.3	3.37
^{157}Dy	$\frac{3}{2}-$	532↓	2.06	1	0.97	61.0	1.67
				2	1.72	1.5	1.75
				3	1.79	3.6	1.82
^{155}Gd	$\frac{5}{2}+$	642↑	1.28	1	0.94	86.0	2.02
				2	2.09	4.5	2.34
				3	—	—	2.41
				4	2.44	1.6	2.71

$\lambda = 2$ or $\lambda = 3$ phonon remain relatively pure. The *au* β-decay can help to discover the distribution of various three-quasiparticle components over several levels.

Let us see how two different states with identical K^π values ϱ_1 and ϱ_2 mix together. By solving (9.56) we can calculate the mixing of the state ϱ_1 and state ϱ_2. The calculations show that the corresponding admixtures are relatively small. Large mixing of the one-quasiparticle states ϱ_1 and ϱ_2 is possible only in two cases: if the energies $\varepsilon(\varrho_1)$ and $\varepsilon(\varrho_2)$ are close to each other, or if either of the $\varepsilon(\varrho_1)$ or $\varepsilon(\varrho_2)$ is larger than the first pole of (9.56).

Several examples of ϱ_1 and ϱ_2 mixing are shown in Table 9.26. The quantities $(NC_{\varrho_1}^n)^2$ and $(NC_{\varrho_2}^n)^2$ determine the contributions of the one-quasiparticle states ϱ_1 and ϱ_2 in the resulting state with the energy η_n. The table shows several states with appreciable admixtures of both components. If ϱ_1 is a hole state and ϱ_2 is a particle state, and if both are relatively far from the Fermi level, the resulting state can be populated in both the (*dp*) and the (*dt*) reaction. The (*dp*) reaction will populate the ϱ_2 component and the (*dt*) reaction will populate the ϱ_1 component in such a case. The experimental evidence[187, 191, 345] suggests the existence of such states.

TABLE 9.26. *Distribution and mixing of one-quasiparticle components*

Nucleus	K^π	ϱ_1	$\varepsilon(\varrho_1)$ (MeV)	ϱ_2	$\varepsilon(\varrho_2)$ (MeV)	η_n (MeV)	$(NC^n_{\varrho_1})^2$ (%)	$(NC^n_{\varrho_2})^2$ (%)
^{177}Yb	$\frac{1}{2}-$	510↑	1.11	521↓	2.34	0.93	91.0	0.001
						2.12	0.002	88.0
						2.92	0.1	8.2
^{167}Yb	$\frac{5}{2}-$	523↓	1.13	512↑	2.06	0.86	90.0	2.0
						1.44	2.4	76.0
						1.90	2.0	3.6
^{165}Dy	$\frac{5}{2}-$	523↓	1.34	512↑	1.65	0.98	25.0	58.0
						1.17	69.0	23.0
						3.14	0.1	4.9
	$\frac{3}{2}-$	521↑	1.64	532↓	3.35	1.23	79.0	1.0
						1.85	2.3	0.1
						2.00	12.5	4.5
						2.65	0.4	51.0
^{161}Dy	$\frac{5}{2}-$	523↓	1.42	512↑	3.05	0.85	72.0	4.0
						1.74	11.0	14.0
						1.98	0.8	2.0

We have already discussed the mixing of one-particle states with the identical K^π values and different main oscillator numbers N and $N+2$. Using the Woods–Saxon potential we expect such a mixing in a very narrow interval ($\Delta\beta \approx 0.001$) near the quasicrossing of the corresponding single-particle levels. The quasiparticle–phonon interaction makes this interval considerably larger. The observation of $\Delta N = \pm 2$ mixing is, naturally, possible if the interval $\Delta\beta$ is of such a magnitude that it includes the equilibrium deformation of the considered odd-*A* nucleus. The experimental evidence[170, 206] proves the mixing of N and $N\pm 2$ states. Such experiments thus show that the interval $\Delta\beta$ with an appreciable mixing is, in reality, quite wide.

The tendency to distribute the one- and three-quasiparticle strength over several nonrotational states in odd-*A* deformed nuclei should be at 2–3 MeV excitation energy stronger than at lower excitation energies. The experimental data about such high excitation states are very scarce. The studies of the β-decay in nuclei far off the line of β-stability[525, 526] support our qualitative conclusions.

Including the quasiparticle–phonon interaction in calculations helps one to understand the basic characteristics of nonrotational states in odd-*A* deformed nuclei. The calculations including such interaction give clearly a better description of energies and other experimental features than the calculations based on the model of independent quasiparticles.

§ 5. Quasiparticle–Phonon Interaction in Even Deformed Nuclei

1. We have discussed in the preceding section the departure from a purely quasiparticle or a purely phonon structure of the nonrotational nuclear states. These effects are caused by the quasiparticle–phonon interaction in odd-A deformed nuclei.

In this section, we shall discuss the effect of the quasiparticle–phonon interaction on the one- and two-phonon states in even–even nuclei and on the two-quasiparticle states in both even–even and odd–odd deformed nuclei.

For such a purpose we shall use the Hamiltonian (9.38) and employ the approximate second quantization method. As in the preceding section, we shall assume that the additional quasiparticles do not change the phonon operators.

First, let us study the admixtures of two-phonon states to the lowest one-phonon states in even–even deformed nuclei.[527] The trial wave function of a state with considered K^π values (corresponding to certain quantum numbers λ_0, μ_0, i_0) is chosen as

$$\Psi_{i_0}(K^\pi) = C_{i_0}(\lambda_0\mu_0)\left\{Q_{i_0}^+(\lambda_0\mu_0)+\sum_{\substack{\lambda\mu i\\ \lambda'\mu' i'}} \Delta_{\lambda'\mu' i'}^{\lambda\mu i}(\lambda_0\mu_0 i_0)\, Q_i^+(\lambda\mu)\, Q_{i'}^+(\lambda'\mu')\right\}\Psi, \tag{9.72}$$

with the normalization condition

$$C_{i_0}^2(\lambda_0\mu_0)\left\{1+2\sum_{\substack{\lambda\mu i\\ \lambda'\mu' i'}}\left(\Delta_{\lambda'\mu' i'}^{\lambda\mu i}(\lambda_0\mu_0 i_0)\right)^2\right\} = 1. \tag{9.72$'$}$$

Let us find the expectation value of H_{vq}:

$$\left(\Psi_{i_0}^*(K^\pi)\, H_{vq}\Psi_{i_0}(K^\pi)\right) = C_{i_0}^2(\lambda_0\mu_0)\left\{\omega_{i_0}^{\lambda_0\mu_0}+2\sum_{\substack{\lambda\mu i\\ \lambda'\mu' i'}}\left(\omega_i^{\lambda\mu}+\omega_{i'}^{\lambda'\mu'}\right)\left(\Delta_{\lambda'\mu' i'}^{\lambda\mu i}(\lambda_0\mu_0 i_0)\right)^2\right.$$
$$\left. -2\sum_{\substack{\lambda\mu i\\ \lambda'\mu' i'}}\Delta_{\lambda'\mu' i'}^{\lambda\mu i}(\lambda_0\mu_0 i_0)\, U_{\lambda'\mu' i'}^{\lambda\mu i}(\lambda_0\mu_0 i_0)\right\}, \tag{9.73}$$

and determine the unknown quantities $C_{i_0}(\lambda_0\mu_0)$ and $\Delta_{\lambda'\mu' i'}^{\lambda\mu i}(\lambda_0\mu_0 i_0)$ from the minimum condition of the energy (9.73). The resulting energy ζ_n is equal to the solution of the secular equation

$$\omega_{i_0}^{\lambda_0\mu_0}-\zeta_n = \frac{1}{2}\sum_{\substack{\lambda\mu i\\ \lambda'\mu' i'}}\frac{[U_{\lambda'\mu' i'}^{\lambda\mu i}(\lambda_0\mu_0 i_0)]^2}{\omega_i^{\lambda\mu}+\omega_{i'}^{\lambda'\mu'}-\zeta_n}, \tag{9.74}$$

where

$$U_{\lambda'\mu' i'}^{\lambda\mu i}(\lambda_0\mu_0 i_0) = \frac{1}{2}\sum_{q,\, q',\, q_2} v_{qq'}\left\{\frac{f^{\lambda\mu}(q, q')}{\sqrt{Y^i(\lambda\mu)}}\left(\psi_{q_2 q}^{\lambda_0\mu_0 i_0}\psi_{q_2 q'}^{\lambda'\mu' i'}+\varphi_{q_2 q}^{\lambda_0\mu_0 i_0}\varphi_{q_2 q'}^{\lambda'\mu' i'}\right)\right.$$
$$\left.+\frac{f^{\lambda'\mu'}(q, q')}{\sqrt{Y^i(\lambda'\mu')}}\left(\psi_{q_2 q}^{\lambda_0\mu_0 i_0}\psi_{q_2 q'}^{\lambda\mu i}+\varphi_{q_2 q}^{\lambda_0\mu_0 i_0}\varphi_{q_2 q'}^{\lambda\mu i}\right)+\frac{f^{\lambda_0\mu_0}(q, q')}{\sqrt{Y^{i_0}(\lambda_0\mu_0)}}\left(\psi_{q_2 q'}^{\lambda\mu i}\varphi_{q_2 q}^{\lambda'\mu' i'}+\varphi_{q_2 q'}^{\lambda\mu i}\psi_{q_2 q}^{\lambda'\mu' i'}\right)\right\}. \tag{9.75}$$

The quantities C and Δ are determined from

$$C_{i_0}^{-2}(\lambda_0\mu_0) = 1+\frac{1}{2}\sum_{\substack{\lambda\mu i\\ \lambda'\mu' i'}} \frac{[U^{\lambda\mu i}_{\lambda'\mu' i'}(\lambda_0\mu_0 i_0)]^2}{(\omega_i^{\lambda\mu}+\omega_{i'}^{\lambda'\mu'}-\zeta_n)^2}, \tag{9.76}$$

$$\Delta^{\lambda\mu i}_{\lambda'\mu' i'}(\lambda_0\mu_0 i_0) = \frac{1}{2}\,\frac{U^{\lambda\mu i}_{\lambda'\mu' i'}(\lambda_0\mu_0 i_0)}{\omega_i^{\lambda\mu}+\omega_{i'}^{\lambda'\mu'}-\zeta_n}. \tag{9.77}$$

The shifts of the lowest one-phonon states and the components C have been calculated in ref. 527.

The calculations based on the Nilsson potential single-particle wave functions show that the admixtures of the two-phonon components to the one-phonon states are small. However, there are some exceptions from this rule, in particular, if the one-phonon and two-phonon states have almost degenerate energies. For example, the one-phonon state (201) in ^{166}Er has an energy almost equal to the two-phonon pole (221) (221); thus the two $K^\pi = 0^+$ states mix.

The generally small mixing of two-phonon and one-phonon states means that the anharmonic effects are small. This conclusion qualitatively agrees with the dependence of the total nuclear energy on the deformation parameters β and γ as discussed in § 1 of Chapter 7. For example, the ε-dependence of the total energy of ^{174}Yb, shown in Fig. 7.10 for different γ-values, has a clearly parabolic form in the vicinity of minima.

The anharmonic effects become more important in the transitional nuclei. For example, the γ-vibrational states in ^{152}Sm and ^{150}Nd contain very small admixtures. However, the β-vibrational states in these nuclei have large admixtures. The amount of mixing sharply increases from ^{158}Dy to the isotopes ^{152}Sm and particularly ^{150}Nd. This effect is apparently related to the sudden change of the deformation between $N = 90$ and $N = 88$.

A different situation is expected for the isotopes of tungsten and osmium. The one-phonon $K^\pi = 0^+$ states remain relatively pure. However, the lowest one-phonon $K^\pi = 2^+$ states contain larger admixtures, and the amount of mixing increases with the increasing neutron and proton numbers. Thus, the anharmonicity of γ-vibrations in tungsten and osmium isotopes increases as the corresponding nuclei become closer and closer to the upper end of the deformed region.

2. Let us now discuss the two-phonon states in even–even deformed nuclei.

Many quadrupole two-phonon states have been observed in spherical even–even nuclei. On the other hand, in the deformed nuclei the experimental evidence is very scarce.[528, 529] The nature of vibrational states in spherical and deformed nuclei is identical. Thus, there is no doubt that the two-phonon states also exist in deformed nuclei. The large difference in experimentally observed two-phonon states is related to the differing spectra of spherical and deformed nuclei. The two-phonon states in spherical nuclei have energies below the energy gap, i.e., smaller than the two-quasiparticle states. Usually, no other states are in their vicinity. The situation in deformed nuclei is rather different. The two-phonon states have larger energies and, moreover, the energies are larger than the lowest two-quasiparticle states. We expect the two-phonon states in deformed nuclei at energies 1.5–2.5 MeV. Many two-quasiparticle states, first and second quadrupole $K^\pi = 0^+$, 2^+ states and octupole $K^\pi = 0^-, 1^-, 2^-$ states, together with the corresponding rotational bands, are in the same energy interval. Thus, the large number of other states makes the identification of the two-phonon states in even–even deformed nuclei more difficult.

We shall denote two-phonon states by $K^{\pi}(\lambda\mu i)(\lambda'\mu' i')$, i.e., by the characteristics of the corresponding phonons. The most important feature of two-phonon states is the enhanced $E\lambda$ transition probability connecting them with one-phonon states (and their rotational bands). To find, for example, the $K^{\pi} = 2^{+}$ or 2^{-} two-phonon states, we have to look for the states having $B(E2)$ or $B(E3)$ values of the transitions to the one-phonon states larger than the transitions to the ground state band.[530]

Microscopically, the two-phonon states are the superpositions of four-quasiparticle states. We have shown earlier that such states in even–even nuclei can be of the ($4n$), ($4p$), or ($2n$, $2p$) type; all of them can participate in a two-phonon state. The two-phonon state $(\lambda\mu i)(\lambda'\mu' i')$ has two components with $K = |\mu - \mu'|$ and $K = \mu + \mu'$. It is interesting to find the splitting of this doublet.

The lowest two-phonon states in deformed even–even nuclei have relatively small admixtures of one-phonon states. The states without one-phonon analog, i.e., $K^{\pi}_{i} = 1^{+}, 3^{+}, 4^{-}, 4^{+}$, etc., should remain quite pure. Apparently, the energies of such states are equal to the sum of the energies of corresponding phonons. Similarly, because the anharmonicity of one-phonon states is small, we expect a small splitting of two-phonon states in deformed nuclei. The structure of all states in deformed nuclei becomes more complex at higher excitation energies. Thus, we expect that the higher lying two-phonon states will contain some important admixtures.

The two-phonon states can be populated in β-decay. Let us take, as an example, the β^{+} decay of an odd–odd nucleus with quasiparticles s_0, r_0, and $K_{00} = K_{s_0} + K_{r_0}$. The matrix element for the decay populating the two-phonon state $(\lambda\mu i)(\lambda'\mu' i')$ equals

$$\mathcal{M}\begin{pmatrix} Z-1 \to Z \\ N+1 \to N \end{pmatrix}; s_0, r_0 \Big) = \frac{1}{\sqrt{2}} \sum_{s,r} u_s v_r \{\langle s+|\Gamma_\beta|r-\rangle \times$$
$$\times \left(\psi^{\lambda\mu i}_{ss_0}\psi^{\lambda'\mu' i'}_{rr_0} + \psi^{\lambda'\mu' i'}_{ss_0}\psi^{\lambda\mu i}_{rr_0} + \bar{\varphi}^{\lambda\mu i}_{ss_0}\bar{\varphi}^{\lambda'\mu' i'}_{rr_0} + \bar{\varphi}^{\lambda'\mu' i'}_{ss_0}\bar{\varphi}^{\lambda\mu i}_{rr_0}\right)$$
$$+ \langle s+|\Gamma_\beta|r+\rangle \left(\psi^{\lambda\mu i}_{ss_0}\bar{\varphi}^{\lambda'\mu' i'}_{rr_0} + \psi^{\lambda'\mu' i'}_{ss_0}\bar{\varphi}^{\lambda\mu i}_{rr_0} + \bar{\varphi}^{\lambda\mu i}_{ss_0}\psi^{\lambda'\mu' i'}_{rr_0} + \bar{\varphi}^{\lambda'\mu' i'}_{ss_0}\psi^{\lambda\mu i}_{rr_0}\right)\}, \quad (9.78)$$

where $\langle s+|\Gamma_\beta|r\pm\rangle$ is the single particle β-decay matrix element. If $K_{00} = K_{s_0} - K_{r_0}$, the matrix elements $\langle s+|\Gamma|r-\rangle$ and $\langle s+|\Gamma|r+\rangle$ in (9.78) have to be interchanged. For the β^{-} decay we have to replace the quantities $u_s v_r$ by $u_r v_s$. The numerical estimate based on (9.78) shows that such a β-decay proceeds 1–2 orders of magnitude slower than the similar β-decays between one-quasiparticle states. However, in certain cases the inhibition can be smaller.

Let us take, for example, the β-decay of the $K^{\pi} = 5^{+}$ ($p411\uparrow + n633\uparrow$) state in ^{164}Tb to the $K^{\pi} = 4^{-}$ state in ^{164}Dy; experimentally $\log ft = 6.2$ (see ref. 531). Assuming that the $K^{\pi} = 4^{-}$ state is the (221) (321) two-phonon state we obtain a number of possible final four-quasiparticle components. One of the components is the combination $n633\uparrow - n521\uparrow + p411\uparrow + p411\downarrow$. Let us estimate the corresponding matrix element. From (9.78) it follows that

$$\mathcal{M}_1 = \frac{1}{\sqrt{2}} v_{n521\uparrow} u_{p411\uparrow} \langle n521\uparrow|\Gamma_\beta|p411\uparrow\rangle \bar{\varphi}^{221}_{p411\uparrow + p411\downarrow} \psi^{321}_{n633\uparrow - n521\uparrow}.$$

According to[473]

$$\psi^{221}_{p411\uparrow+p411\downarrow} = 0.87, \quad \psi^{321}_{n633\uparrow-n521\uparrow} = 0.3; \quad v^2_{n521\uparrow} = 0.83, \quad u^2_{p411\uparrow} = 0.84,$$

and, consequently,

$$\mathcal{M}_1 = \langle n521\uparrow |\Gamma_\beta| p411\uparrow\rangle 0.2.$$

Taking into account the fact that the typical β-decays to one-quasiparticle levels contain u_s^2 or v_s^2 equal approximately 0.5 (for the states near the Fermi level) and that our β-decay can proceed also via other four-quasiparticle components, we conclude that the β-decay to the 4^- two-phonon state in ^{164}Dy is approximately five times hindered compared to the one-particle decay connecting the $p411\uparrow$ and $n521\uparrow$ states.

The experimental identification of the two-phonon states, determination of the doublet splitting energy, etc., are very important for a better understanding of the structure of deformed nuclei.

3. Let us turn our attention to the effect of quasiparticle–phonon interaction on the two-quasiparticle states in even–even deformed nuclei. We want to see how the structure of such states changes with an increasing excitation energy.

The form of the wave function (9.72) is quite general. Therefore, we can consider the limiting case when one or two of the participating phonons are replaced by the two-quasiparticle operators. If the phonons are restricted to $\lambda = 2$, $\mu = 0, 2$, and $\lambda = 3$, $\mu = 0, 1, 2$, we obtain the resulting wave function such as:[512]

$$\Psi_n(K^\pi; \varrho_1, \varrho_2) = C^n_{\varrho_1\varrho_2}\Big\{\frac{1}{\sqrt{2}}\sum_{\sigma_1, \sigma_2} \theta_{\sigma_1, \sigma_2}\Big[\alpha^+_{\varrho_1\sigma_1}\alpha^+_{\varrho_1\sigma_2} + \sum_{\lambda\mu i}\sum_{q_1, q_2} D^{\lambda\mu in}_{q_1q_2\sigma_1\sigma_2}(\varrho_1, \varrho_2)\alpha^+_{q_1\sigma_1}\alpha^+_{q_2\sigma_2}Q_i^+(\lambda\mu)\Big] + \sum_{\lambda\mu i}\sum_{\lambda'\mu' i'} \Delta^{\lambda\mu i}_{\lambda'\mu' i'}(\varrho_1, \varrho_2)\, Q_i^+(\lambda\mu)Q_{i'}^+(\lambda'\mu')\Big\}\,\Psi, \tag{9.79}$$

where $\psi^{\lambda\mu i}_{\varrho_1\varrho_2} = 0$ and $\psi^{\lambda\mu i}_{q_1q_2} = 0$. Besides, $\theta_{\sigma_1\sigma_2} = \delta_{\sigma_1\sigma_2}$ if $K = K_{\varrho_1}+K_{\varrho_2}$ and $\theta_{\sigma_1\sigma_2} = -\sigma_1\delta_{\sigma_1, -\sigma_2}$ if $K = |K_{\varrho_1}-K_{\varrho_2}|$.

The normalization condition for such a wave function has the form

$$1 = (C^n_{\varrho_1\varrho_2})^2\Big\{1+\sum_{\lambda\mu i}\sum_{q_1, q_2, \sigma}[D^{\lambda\mu in}_{q_1q_2\sigma,\pm\sigma}(\varrho_1, \varrho_2)]^2 + 2\sum_{\lambda\mu i}\sum_{\lambda'\mu' i'}[\Delta^{\lambda\mu i}_{\lambda'\mu' i'}(\varrho_1, \varrho_2)]^2\Big\}. \tag{9.79'}$$

Let us find the expectation value of H_{vq} in the state (9.79):

$$\big(\Psi^*(K^\pi; \varrho_1, \varrho_2)\, H_{vq}\Psi(K^\pi; \varrho_1, \varrho_2)\big) = (C^n_{\varrho_1\varrho_2})^2\Big\{\varepsilon(\varrho_1)+\varepsilon(\varrho_2)+\sum_{\lambda\mu i}\sum_{q_1, q_2\, \sigma}[\varepsilon(q_1)+\varepsilon(q_2)+\omega_i^{\lambda\mu}]\times$$
$$\times(D^{\lambda\mu in}_{q_1q_2\sigma, \pm\sigma}(\varrho_1, \varrho_2))^2 - \sum_{\lambda\mu i}\frac{1}{\sqrt{Y^i(\lambda\mu)}}\sum_{q, \sigma}[v_{\varrho_1 q}D^{\lambda\mu in}_{q\varrho_2\sigma, \pm\sigma}f^{\lambda\mu}(\varrho_1, q) + v_{\varrho_2 q}D^{\lambda\mu in}_{q\varrho_1\sigma, \pm\sigma}f^{\lambda\mu}(\varrho_2, q)]$$
$$+2\sum_{\lambda\mu i}\sum_{\lambda'\mu' i'}(\omega_i^{\lambda\mu}+\omega_{i'}^{\lambda'\mu'})(\Delta^{\lambda\mu i}_{\lambda'\mu' i'}(\varrho_1, \varrho_2))^2 - 2\sum_{\lambda\mu i}\sum_{\lambda'\mu' i'}\Delta^{\lambda\mu i}_{\lambda'\mu' i'}(\varrho_1, \varrho_2)\, U^{\lambda\mu i}_{\lambda'\mu' i'}(\varrho_1, \varrho_2)\Big\}, \tag{9.80}$$

where

$$U^{\lambda\mu i}_{\lambda'\mu' i'}(\varrho_1, \varrho_2) = \frac{1}{2\sqrt{2}} \frac{1}{\sqrt{Y^i(\lambda\mu)\,Y^{i'}(\lambda'\mu')}} \sum_q \left\{ v_{\varrho_2 q} u_{\varrho_1 q} \times \right.$$

$$\times \left[\frac{f^{\lambda\mu}(\varrho_2, q)\, f^{\lambda'\mu'}(\varrho_1, q)}{\varepsilon(\varrho_1)+\varepsilon(q)-\omega_{i'}^{\lambda'\mu'}} + \frac{f^{\lambda'\mu'}(\varrho_2, q)\, f^{\lambda\mu}(\varrho_1, q)}{\varepsilon(\varrho_1)+\varepsilon(q)-\omega_i^{\lambda\mu}} \right]$$

$$\left. + v_{\varrho_1 q} u_{\varrho_2 q} \left[\frac{f^{\lambda\mu}(\varrho_1, q)\, f^{\lambda'\mu'}(\varrho_2, q)}{\varepsilon(\varrho_2)+\varepsilon(q)-\omega_{i'}^{\lambda'\mu'}} + \frac{f^{\lambda'\mu'}(\varrho_1, q)\, f^{\lambda\mu}(\varrho_2, q)}{\varepsilon(\varrho_2)+\varepsilon(q)-\omega_i^{\lambda\mu}} \right] \right\}. \quad (9.81)$$

From the usual variational procedure we obtain the following secular equation:

$$\varepsilon(\varrho_1)+\varepsilon(\varrho_2)-\zeta_n - \frac{1}{2} \sum_{\lambda\mu i} \sum_q \frac{1}{Y^i(\lambda\mu)} \left\{ \frac{(f^{\lambda\mu}(\varrho_1, q))^2 v^2_{\varrho_1 q}}{\varepsilon(q)+\varepsilon(\varrho_2)+\omega_i^{\lambda\mu}-\zeta_n} \right.$$

$$\left. + \frac{(f^{\lambda\mu}(\varrho_2, q))^2 v^2_{\varrho_2 q}}{\varepsilon(\varrho_1)+\varepsilon(q)+\omega_i^{\lambda\mu}-\zeta_n} \right\} - \frac{1}{2} \sum_{\lambda\mu i} \sum_{\lambda'\mu' i'} \frac{(U^{\lambda\mu i}_{\lambda'\mu' i'}(\varrho_1, \varrho_2))^2}{\omega_i^{\lambda\mu}+\omega_{i'}^{\lambda'\mu'}-\zeta_n} = 0, \quad (9.82)$$

and from the normalization condition we find that

$$(C^n_{\varrho_1\varrho_2})^{-2} = 1 + \sum_{\lambda\mu i} \sum_{q_1, q_2, \sigma} (D^{\lambda\mu i n}_{q_1 q_2 \sigma, \pm\sigma}(\varrho_1, \varrho_2))^2 + 2 \sum_{\lambda\mu i} \sum_{\lambda'\mu' i'} (\Delta^{\lambda\mu i}_{\lambda'\mu' i'}(\varrho_1, \varrho_2))^2, \quad (9.83)$$

$$D^{\lambda\mu i n}_{q_1 q_2 \sigma, \pm\sigma}(\varrho_1, \varrho_2) = \frac{1}{\sqrt{4Y^i(\lambda\mu)}} \frac{\delta_{\varrho_2 q_2} v_{\varrho_1 q_1} f^{\lambda\mu}(\varrho_1, q_1) + \delta_{\varrho_1 q_1} v_{\varrho_2 q_2} f^{\lambda\mu}(\varrho_2, q_2)}{\varepsilon(q_1)+\varepsilon(q_2)+\omega_i^{\lambda\mu}-\zeta_n}, \quad (9.84)$$

$$\Delta^{\lambda\mu i}_{\lambda'\mu' i'}(\varrho_1, \varrho_2) = \frac{1}{2} \frac{U^{\lambda\mu i}_{\lambda'\mu' i'}(\varrho_1, \varrho_2)}{\omega_i^{\lambda\mu}+\omega_{i'}^{\lambda'\mu'}-\zeta_n}. \quad (9.85)$$

The quantity $U^{\lambda\mu i}_{\lambda'\mu' i'}(\varrho_1, \varrho_2) = 0$ for $K^\pi = 1^+$ and $K \geqslant 3$ states. Equation (9.82) is simpler in such a case, namely,

$$\varepsilon(\varrho_1)+\varepsilon(\varrho_2)-\zeta_n - \frac{1}{2} \sum_{\lambda\mu i} \sum_q \frac{1}{Y^i(\lambda\mu)} \times$$

$$\times \left\{ \frac{(f^{\lambda\mu}(\varrho_1, q) v_{\varrho_1 q})^2}{\varepsilon(q)+\varepsilon(\varrho_2)+\omega_i^{\lambda\mu}-\zeta_n} + \frac{(f^{\lambda\mu}(\varrho_2, q) v_{\varrho_2 q})^2}{\varepsilon(q)+\varepsilon(\varrho_1)+\omega_i^{\lambda\mu}-\zeta_n} \right\} = 0. \quad (9.86)$$

Tables 9.27 and 9.28 show the examples of the solutions of (9.86). The Nilsson model single-particle wave functions were used in the calculations.

Table 9.27 suggests that the low-lying two-quasiparticle states remain essentially pure. The admixtures are typically only about 1–5%. On the other hand, the two-quasiparticle strength is distributed among several levels if the energy $\varepsilon(\varrho_1)+\varepsilon(\varrho_2)$ is close to the first pole or larger than the energy of the first pole. The states of such a type are shown in Table 9.28; they have the energies 3 MeV or more in nuclei with $150 < A < 190$.

The distribution of the two-quasiparticle strength can be experimentally observed in nuclear β-decays and in direct nuclear reactions. It would be necessary to show that the number of populated states with a certain K^π value is larger than the number of possible

TABLE 9.27. *The structure of two-quasiparticle states*

Nucleus	K^π	ϱ_1	ϱ_2	$\varepsilon(\varrho_1)+\varepsilon(\varrho_2)$ (MeV)	n	ζ_n (MeV)	$(C^n_{\varrho_2\varrho_1})^2$ (%)	Poles (9.86) (MeV)
^{178}Hf	8−	$n514\downarrow$	$n624\uparrow$	1.88	1	1.86	99.0	3.10
	8−	$p514\uparrow$	$p404\downarrow$	1.82	1	1.80	99.8	3.10
^{170}Yb	5−	$p411\downarrow$	$p514\uparrow$	2.33	1	2.24	94.1	2.91
					2	2.93	2.6	3.12
	5+	$p411\uparrow$	$p404\downarrow$	2.84	1	2.77	94.6	3.35
					2	3.37	3.6	3.75
	3+	$n512\downarrow$	$n512\uparrow$	2.57	1	2.54	96.6	3.04
					2	3.05	0.3	3.38
					3	3.40	1.8	3.42
	5+	$n523\downarrow$	$n512\uparrow$	2.95	1	2.92	97.3	3.53
					2	3.54	0.2	3.64
					3	3.65	1.2	—
^{166}Er	5−	$n642\uparrow$	$n523\downarrow$	2.26	1	2.22	97.3	3.34
					2	3.35	0.1	3.36
					3	3.37	0.8	3.53
^{164}Dy	6+	$n642\uparrow$	$n633\uparrow$	2.57	1	2.52	96.8	3.37
					2	3.39	1.3	3.50
					3	3.51	0.4	3.63
^{160}Dy	4−	$n521\uparrow$	$n642\uparrow$	2.37	1	2.30	97.6	3.21
					2	3.22	0.2	3.34
					3	3.35	0.5	3.48

TABLE 9.28. *Distribution of the two-quasiparticle strength on different levels*

Nucleus	K^π	ϱ_1	ϱ_2	$\varepsilon(\varrho_1)+\varepsilon(\varrho_2)$ (MeV)	n	ζ_n (MeV)	$(C^n_{\varrho_1\varrho_2})^2$ (%)	Poles (9.86) (MeV)
^{170}Yb	3+	$p411\downarrow$	$p402\uparrow$	2.78	1	2.65	78.3	2.91
					2	2.96	13.0	3.12
					3	3.28	8.2	—
	7−	$n633\uparrow$	$n514\downarrow$	3.19	1	2.94	24.0	3.10
					2	3.23	73.8	3.56
	3+	$n512\downarrow$	$n514\downarrow$	3.37	1	3.23	51.8	3.04
					2	3.42	1.8	3.43
					3	3.48	13.4	3.51
					4	3.54	32.0	—
^{166}Er	5+	$n523\downarrow$	$n512\uparrow$	3.02	1	2.95	88.0	3.14
					2	3.25	0.9	3.52
					3	3.59	10.4	—
^{160}Dy	5−	$n521\uparrow$	$n633\uparrow$	3.30	1	3.17	61.4	3.21
					2	3.23	18.5	3.25
					3	3.37	15.9	3.49

two-quasiparticle configurations. It seems that such an effect indeed exists in a number of cases. For example, several $K^\pi = 4^+$ states have been observed in ^{158}Dy, several $K^\pi = 1^+$ and 1^- states in ^{160}Dy, and many $K^\pi = 1^+$, 1^-, and 0^- states in ^{170}Yb.[480–482, 495, 532] The observed enhancement of $\log ft$ values, when compared with typical *ah* and 1*u* values, is an additional argument for such an interpretation.

A majority of two-quasiparticle states, with the excitation energy up to 2 MeV, have a pure two-quasiparticle configuration. However, at higher excitation energies, the structure of these states becomes more complex. The states with large admixtures are common, and the two-quasiparticle strength is distributed among several states.

4. Let us now discuss the two-quasiparticle states in odd–odd deformed nuclei.[533] The trial wave function has the form

$$\Psi_n(K^\pi; \varrho, \nu) = C^n_{\varrho\nu}\frac{1}{2}\sum_{\sigma,\sigma'}\delta_{\sigma,\pm\sigma'}\left\{\alpha^+_{\varrho\sigma}\alpha^+_{\nu\sigma'}+\sum_{\lambda\mu i}\sum_{s,r}D^{\lambda\mu in}_{s\sigma r\sigma'}(\varrho,\nu)\alpha^+_{s\sigma}\alpha^+_{r\sigma'}Q^+_i(\lambda\mu)\right\}\Psi \tag{9.87}$$

with the normalization condition

$$\frac{1}{4}(C^n_{\varrho\nu})^2\sum_{\sigma,\sigma'}\delta_{\sigma,\pm\sigma'}\left\{1+\sum_{\lambda\mu i}\sum_{s,r}(D^{\lambda\mu in}_{s\sigma r\sigma'}(\varrho,\nu))^2\right\} = 1. \tag{9.87'}$$

The quantity $(C^n_{\varrho\nu})^2$ determines the two-quasiparticle component.

The continuation is straightforward. The expectation value of H_{vq} in the state $\Psi_n(K^\pi; \varrho, \nu)$ equals

$$\begin{aligned}(\Psi^*(K^\pi; \varrho, \nu)\, H_{vq}\Psi(K^\pi; \varrho, \nu))\\ = \frac{1}{2}(C^n_{\varrho\nu})^2\sum_{\sigma,\sigma'}\delta_{\sigma,\pm\sigma}\Bigg\{\varepsilon(\varrho)+\varepsilon(\nu)+\sum_{\lambda\mu i}\sum_{s,r}[\varepsilon(s)+\varepsilon(r)+\omega^{\lambda\mu}_i]\,(D^{\lambda\mu in}_{s\sigma r\sigma'}(\varrho\nu))^2\\ -\sum_{\lambda\mu i}\frac{1}{\sqrt{Y^i(\lambda\mu)}}\Bigg[\sum_s D^{\lambda\mu in}_{s\sigma\nu\sigma'}(\varrho,\nu)\,f^{\lambda\mu}(\varrho,s)v_{\varrho s}\\ +\sum_r D^{\lambda\mu in}_{\varrho\sigma r\sigma'}(\varrho,\nu)\,f^{\lambda\mu}(\nu,r)v_{\nu r}\Bigg]\Bigg\}.\end{aligned} \tag{9.88}$$

The minimum condition of (9.88) means that

$$\begin{aligned}\delta\Bigg\{(\Psi^*_n(K^\pi; \varrho, \nu)\, H_{vq}\Psi_n(K^\pi; \varrho, \nu))\\ -\zeta_n\left[\frac{1}{4}(C^n_{\varrho\nu})^2\sum_{\sigma,\sigma'}\delta_{\sigma\sigma'}\left[1+\sum_{\lambda\mu i}\sum_{s,r}(D^{\lambda\mu in}_{s\sigma r\sigma'}(\varrho,\nu))^2\right]-1\right]\Bigg\} = 0,\end{aligned} \tag{9.89}$$

and the resulting secular equation determining the energies ζ_n has the form

$$\varepsilon(\varrho)+\varepsilon(\nu)-\zeta_n-\frac{1}{2}\sum_{\lambda\mu i}\frac{1}{Y^i(\lambda\mu)}\left\{\sum_s\frac{(f^{\lambda\mu}(\varrho,s)v_{\varrho s})^2}{\varepsilon(s)+\varepsilon(\nu)+\omega^{\lambda\mu}_i-\zeta_n}+\sum_r\frac{(f^{\lambda\mu}(\nu,r)v_{\nu r})^2}{\varepsilon(\varrho)+\varepsilon(r)+\omega^{\lambda\mu}_i-\zeta_n}\right\} = 0. \tag{9.90}$$

Equation (9.90) has the same form for both $\sigma = \sigma'$ and $\sigma = -\sigma'$ situations. This means that the quasiparticle–phonon interaction does not break the spin degeneracy. Using the normalization condition (9.87′) we find that

$$(C^n_{\varrho\nu})^{-2} = 1+\frac{1}{4}\sum_{\lambda\mu i}\frac{1}{Y^i(\lambda\mu)}\left\{\sum_s \frac{(f^{\lambda\mu}(\varrho, s)v_{\varrho s})^2}{[\varepsilon(s)+\varepsilon(\nu)+\omega_i^{\lambda\mu}-\zeta_n]^2}+\sum_r \frac{(f^{\lambda\mu}(\nu, r)v_{\nu r})^2}{[\varepsilon(\varrho)+\varepsilon(r)+\omega_i^{\lambda\mu}-\zeta_n]^2}\right\}, \quad (9.91)$$

$$(D^{\lambda\mu in}_{s\sigma r\sigma'}(\varrho, \nu))^2 = \frac{1}{4Y^i(\lambda\mu)}\,\frac{\delta_{\nu r}(f^{\lambda\mu}(\varrho, s)v_{\varrho s})^2+\delta_{\varrho s}(f^{\lambda\mu}(\nu, r)v_{\nu r})^2+\delta_{\nu r}\delta_{\varrho s}f^{\lambda\mu}(\varrho, \varrho)\,f^{\lambda\mu}(\nu, \nu)v_{\varrho\varrho}\,v_{\nu\nu}}{[\varepsilon(s)+\varepsilon(\nu)+\omega_i^{\lambda\mu}-\zeta_n]^2}. \quad (9.91')$$

The structure of odd–odd deformed nuclei is affected by the quasiparticle–phonon interaction more than the structure of odd-A nuclei. The interaction decreases the resulting energies with respect to both the $\varepsilon(\varrho)+\varepsilon(\nu)$ values and the poles $\varepsilon(s)+\varepsilon(\nu)+\omega_i^{\lambda\mu}$, $\varepsilon(\varrho)+\varepsilon(r)+\omega_i^{\lambda\mu}$. The predominantly two-quasiparticle states are shifted with respect to the energies $\varepsilon(\varrho)+\varepsilon(\nu)$ by an amount approximately equal to the sum of similar shifts in corresponding odd-A nuclei. The two-quasiparticle component is, in such a case, approximately

$$(C^n_{\varrho\nu})^2 \approx (C^n_\varrho C^n_\nu)^2, \quad (9.92)$$

i.e., the role of admixtures has increased.

The spectra of odd–odd nuclei also contain the collective nonrotational states and the states with a complex structure. The total number of these states is the sum of the analogous states in corresponding odd-N and odd-Z nuclei. The structure of these states is also close to the structure of their odd-A analogs. The energies are, however, lowered somewhat more than in the neighboring odd-A nuclei. Thus, the general characteristics of the states in odd–odd nuclei can be deduced from the corresponding data on states in odd-N and odd-Z nuclei.

The low-lying part (1.0–1.5 MeV) of the spectrum of an odd–odd nucleus is more complex than the similar spectrum of odd-A or even–even nuclei. The odd–odd nuclei have many collective states and many states with a complex structure in this part of the spectrum. This general conclusion is supported by the experimental evidence.[534–536] The Coriolis coupling and the forces breaking the spin degeneracy further complicate the quantitative analysis of such spectra.

5. In this chapter, we have studied the structure of deformed nuclei using the model of pairing and multipole–multipole interaction. The importance of quasiparticle–phonon interaction was stressed. The chapter also discussed the changes in structure of the states with increasing excitation energy. The calculations have shown that the model gives a satisfactory description of the excited nonrotational states in even–even deformed nuclei. The concepts of quasiparticle, one-phonon, and two-phonon states are valid up to the excitation energies 2.5–3.0 MeV.

The quasiparticle–phonon interaction is responsible for the changing structure of the states with an increasing excitation energy. The existing experimental data support such a conclusion. However, it is not quite obvious that our model gives an adequate description of odd-A and odd–odd deformed nuclei above 1.0–1.5 MeV excitation energy. Similarly, the structure of even–even nuclei above 2.5–3.0 MeV is not sufficiently understood. Other forces have a similar tendency to mask the simple excitation modes. For example, the particle–rotation coupling, the interaction breaking the spin degeneracy, etc., all such forces lead to complications of the involved nuclear states.

We would expect that some new fundamental features of the nuclear interaction will emerge at higher excitation energies. The experimental facts, not described in the framework of the present model, may therefore give an important clue to the nature of these unknown components of nuclear forces.

The most important conclusions coming from the semimicroscopic treatment of nuclear structure just described do not depend on the detailed form of the residual forces employed. Instead, they are related to the general symmetry properties of these forces and to the nuclear average field potential.

CHAPTER 10

VIBRATIONAL STATES IN SPHERICAL NUCLEI

§ 1. One-phonon States

1. In this chapter we shall study the vibrational states in spherical nuclei. The experimental data on one- and two-phonon states in even–even nuclei and on the vibrational states in odd-A nuclei will be systemized, discussed, and compared with the predictions of the pairing plus multipole–multipole force model. The main equations have already been obtained in § 4 of Chapter 8. Many formulae describing the β- and γ-transitions in deformed nuclei are easily modified for application in spherical nuclei. Therefore, we shall restrict ourselves to the very specific features of spherical nuclei.

As usual, the nuclear Hamiltonian contains the average field, the pairing superfluid forces, and the multipole–multipole forces. We shall also consider the spin–quadrupole forces, the surface δ-interaction, and the finite range forces.

The one-phonon states are calculated in the model with pairing plus multipole–multipole forces ($P+QQ$). In the approximation $\varkappa^{(\lambda)} \equiv \varkappa_n^{(\lambda)} = \varkappa_p^{(\lambda)} = \varkappa_{np}^{(\lambda)}$ the nuclear model Hamiltonian has the form

$$H_v = \sum_j \varepsilon(j) \sum_m \alpha_{jm}^{+} \alpha_{jm} - \frac{1}{2} \sum_{\lambda=2,\,3} \frac{\varkappa^{(\lambda)}}{2\lambda+1} \times \sum_{j,\,j',\,j_2,\,j_2'} \mathfrak{f}^{\lambda}(j, j')\, \mathfrak{f}^{\lambda}(j_2, j_2') u_{jj'} u_{j_2 j_2'} \times$$
$$\times \sum_{\mu} (-1)^{\lambda-\mu} g_{jj'}^{\lambda i} g_{j_2 j_2'}^{\lambda i'} \{Q_i^{+}(\lambda\mu)\, Q_{i'}^{+}(\lambda-\mu) + Q_{i'}(\lambda-\mu) Q_i(\lambda\mu)$$
$$+ (-1)^{\lambda-\mu}\, Q_i^{+}(\lambda\mu)\, Q_{i'}(\lambda\mu) + (-1)^{\lambda-\mu}\, Q_i^{+}(\lambda-\mu)\, Q_i(\lambda-\mu)\}. \tag{10.1}$$

The single-particle states are, for brevity, characterized by a single quantum number j; the quantum numbers Nl are, however, understood.

The matrix element $\mathfrak{f}^{\lambda}(j, j')$ equals

$$\mathfrak{f}^{\lambda}(j, j') \equiv \langle Nlj \,|\, i^{\lambda} Y_{\lambda} (\sqrt{m\overset{\circ}{\omega}_0}\, r)^{\lambda} \,|\, N'l'j' \rangle = \frac{i^{l'+\lambda-l}(-1)^{j'-1/2}}{\sqrt{4\pi}}\; \frac{1+(-1)^{l+l'-\lambda}}{2} \times$$
$$\times \sqrt{(2j+1)(2j'+1)}\, (j \tfrac{1}{2} j' - \tfrac{1}{2} \,|\, \lambda 0) \langle Nl \,|\, (\sqrt{m\overset{\circ}{\omega}_0}\, r)^{\lambda} \,|\, N'l' \rangle. \tag{10.2}$$

The radial integrals $\langle Nl \,|\, (\sqrt{m\overset{\circ}{\omega}_0}\, r)^{\lambda} \,|\, N'l' \rangle$, calculated using the harmonic oscillator wave functions, are collected in Table 10.1 (see refs. 151 and 421).

TABLE 10.1. *The radial matrix elements*

	N'	l'	$\langle Nl \| (\sqrt{m\mathring{\omega}_0} r)^\lambda \| N'l' \rangle$
	N	l	$N+\frac{3}{2}$
	$N\pm 2$	l	$-\frac{1}{2}\{(N+l+2\pm 1)(N-l+1\pm 1)\}^{1/2}$
$\lambda = 2$	$N\pm 2$	$l\pm 2$	$\frac{1}{2}\{(N+l+1\pm 2)(N+l+3\pm 2)\}^{1/2}$
	N	$l\pm 2$	$-\{(N+l+2\pm 1)(N-l+1\mp 1)\}^{1/2}$
	$N\mp 2$	$l\pm 2$	$\frac{1}{2}\{(N-l\mp 2)(N-l+2\mp 2)\}^{1/2}$
	$N\pm 1$	$l\pm 1$	$-\frac{1}{2}(3N-l+4\pm 1)\{\frac{1}{2}(N+l+2\pm 1)\}^{1/2}$
	$N\mp 1$	$l\pm 1$	$\frac{1}{2}(3N+l+5\mp 1)\{\frac{1}{2}(N-l+1\mp 1)\}^{1/2}$
	$N\mp 1$	$l\pm 3$	$-3\{\frac{1}{8}(N+l+2\pm 1)(N-l\mp 2)(N-l+2\mp 2)\}^{1/2}$
	$N\pm 1$	$l\pm 3$	$3\{\frac{1}{8}(N+l+1\pm 2)(N+l+3\pm 2)(N-l+1\mp 1)\}^{1/2}$
$\lambda = 3$	$N\pm 3$	$l\mp 1$	$-\{\frac{1}{8}(N+l+2\pm 1)(N-l\pm 2)(N-l+2\mp 2)\}^{1/2}$
	$N\pm 3$	$l\pm 1$	$\{\frac{1}{8}(N+l+1\pm 2)(N+l+3\pm 2)(N-l+1\pm 1)\}^{1/2}$
	$N\pm 3$	$l\pm 3$	$-\{\frac{1}{8}(N+l+1\pm 4)(N+l+3\pm 4)(N+l+2\pm 1)\}^{1/2}$
	$N\mp 3$	$l\pm 3$	$\{\frac{1}{8}(N-l\mp 4)(N-l+2\mp 4)(N-l+1\mp 1)\}^{1/2}$

Let us state here the basic formulae describing the one-phonon states. The ground state of an even–even nucleus is the phonon vacuum satisfying the condition

$$Q_i(\lambda\mu)\Psi = 0. \tag{10.3}$$

Formula (8.55) shows the explicit form of Ψ. The one-phonon states have the form

$$Q_i^+(\lambda\mu)\Psi, \tag{10.4}$$

where the phonon creation operator equals

$$Q_i^+(\lambda\mu) = \frac{1}{2}\sum_{j,j'}\{\psi_{jj'}^{\lambda i}A^+(j,j';\lambda\mu)-(-1)^{\lambda-\mu}\varphi_{jj'}^{\lambda i}A(j,j';\lambda-\mu)\}, \tag{10.5}$$

and the functions $\psi_{jj'}^{\lambda i}$ and $\varphi_{jj'}^{\lambda i}$ are calculated according to (8.161) and (8.161′). The phonon operators obey the commutation relations (8.128) and (8.128′). The one-phonon states are orthogonal to each other, and they are also orthogonal to the ground state.

The energies of one-phonon states are the solutions of the secular equations

$$\frac{1}{\varkappa_0^{(\lambda)}} = \frac{1}{2\lambda+1}\sum_{j,j'}\frac{(\mathfrak{f}^\lambda(j,j')u_{jj'})^2[\varepsilon(j)+\varepsilon(j')]}{[\varepsilon(j)+\varepsilon(j')]^2-(\omega_i^\lambda)^2} \equiv X^\lambda(\omega). \tag{10.6}$$

The summation in (10.6) contains all pairs (j, j') satisfying the selection rules of the matrix element $\mathfrak{f}^\lambda(j, j')$. For the quadrupole states, the important contribution in (10.6) comes from the diagonal terms corresponding to the incompletely filled neutron and proton subshells.

Figure 10.1 shows an example of the function $X^2(\omega)$ for the $\lambda = 2$ state in ^{110}Cd. The solutions of (10.6) correspond to the intersection of $X(\omega)$ with the horizontal line $1/\varkappa_0^{(2)}$. The first $I^\pi = 2^+$ state in ^{110}Cd is obviously collective; its energy is 2.2 MeV below the

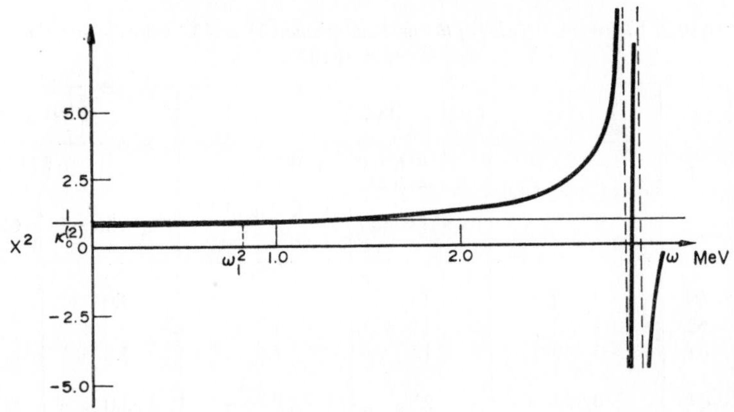

FIG. 10.1. The righthand side of the secular equation (10.6) for $I^\pi = 2^+$ states in ^{110}Cd. The solutions of (10.6) are the intersections of $X^2(\omega)$ and horizontal line $1/\varkappa_0^{(2)}$. The dashed vertical lines denote the diagonal poles of (10.6), the first pole corresponds to $(ppg_{9/2}g_{9/2})$, the second to $(nng_{7/2}g_{7/2})$.

lowest pole of (10.6). The $\varkappa_{\text{exp}}^{(2)}$ value corresponding to the experimental energy is very close to the critical value $\varkappa_{\text{cr}}^{(2)}$ [see (8.69′)]. The energy of the first state $\omega_1^{(2)}$ depends sensitively on $\varkappa^{(2)}$; a small change in the $\varkappa^{(2)}$ value causes large shifts of $\omega_1^{(2)}$. The accuracy of the quasiboson approximation is obviously insufficient in such a case. The example shown in Fig. 10.1 is rather typical; a similar situation is encountered in many spherical nuclei. The development of new methods, more general than the discussed quasiboson approximation, is therefore particularly important in the case of spherical nuclei.

We have seen in the preceding chapter that the spin–quadrupole forces affect the structure of one-phonon states in deformed nuclei. The quadrupole and spin–quadrupole vibrational states are coupled by the function $W^{2\mu i}$ determined in (8.141″). The function $W^{2\mu i}$ in deformed nuclei is 10–100 times smaller than the functions $X^{2\mu i}$ and $S^{2\mu i}$.

The function W is equal to the sum of the terms containing the products of quadrupole and spin–quadrupole matrix elements. The function W does not contain contributions of diagonal terms. However, the diagonal terms give an important contribution in the secular equation (10.6). Thus, in spherical nuclei, the function W^{2i} is quite small and the quadrupole and spin–quadrupole states become practically independent. The detailed calculations have shown that the spin–quadrupole forces are not responsible for the formation of the lowest 2^+ states in spherical nuclei. They can have, nevertheless, some effect on the structure of the 2^+ states.[130]

2. To calculate the energies and the wave functions of the vibrational states in spherical nuclei, we must use a set of single-particle levels and the corresponding single-particle wave functions; the values of interaction constants $\varkappa^{(\lambda)}$ must be also chosen.

The harmonic oscillator potential with spin–orbit coupling or the Woods–Saxon potential (1.31) and (1.32) are often used as the models of the nuclear average field. We have stressed previously that the empirical average field potential in spherical nuclei reflects only the interaction of nucleons in completely filled shells. The effect of nucleons in incompletely filled shells is not included. The average field potential is, therefore, not quite selfconsistent. To counteract this deficiency, it is sometimes necessary to change the potential parameters when nucleons are added into the last shell.

TABLE 10.2. *The characteristics of the first one-phonon 2^+ and 3^- states in nuclei with $60 \leqslant A \leqslant 102$*

Element	A	First 2^+ states			First 3^- states	
		ω_1^2 (experimental) (MeV)	$B(E2; 0^+ \to 2^+)_{s.p.u.}$		ω_1^3 (MeV)	
			Experimental	Calculated(540) $e_{eff}^{(2)} = 0.3$	Experimental	Calculated(539)
$_{28}$Ni	60	1.332	13	16	4.0	4.0
	62	1.172	11	15	3.8	3.9
	64	1.340	12	12	3.8	3.8
$_{30}$Zn	64	0.992	22	27	3.0	–
	66	1.037	19	26	2.8	–
	68	1.072	15	22	2.8	–
$_{32}$Ge	70	1.040	20	28	–	–
	72	0.835	26	38	2.52	2.5
	74	0.596	35	45	2.54	2.8
	76	0.563	29	41	2.69	3.0
$_{34}$Se	74	0.635	23	55	–	–
	76	0.559	50	50	2.60	2.8
	78	0.614	40	40	2.56	3.0
	80	0.666	28	33	–	–
	82	0.655	21	24	–	–
$_{36}$Kr	78	0.450	51	61	–	–
	80	0.618	34	36	–	–
	82	0.777	18	26	–	–
	84	0.880	15	22	–	–
$_{38}$Sr	86	1.078	–	–	–	–
	88	1.836	12	7	2.76	2.6
	90	0.830	–	–	–	–
$_{40}$Zr	90	2.180	3.5	12	2.75	2.6
	92	0.934	65	19	2.33	2.3
	94	0.920	65	22	2.06	2.0
	96	1.730	–	–	1.87	1.6
$_{42}$Mo	92	1.540	–	–	–	–
	94	0.871	21	21	–	–
	96	0.778	23	25	–	–
	98	0.787	21	30	–	–
	100	0.536	44	47	–	–
$_{44}$Ru	96	0.833	19	20	–	–
	98	0.660	35	28	–	–
	100	0.540	41	40	–	–
	102	0.475	51	50	2.22	2.2

TABLE 10.3. *The characteristics of the first one-phonon* 2^+ *and* 3^- *states in nuclei with* $100 < A < 150$

Element	A	First 2^+ states			First 3^- states			
		ω_1^2 (experimental) (MeV)	$B(E2; 0^+ \rightarrow 2^+)_{s.p.u.}$ Experimental	$B(E2; 0^+ \rightarrow 2^+)_{s.p.u.}$ Calculated(540) $e_{eff}^{(2)} = 0.3$	ω_1^3 (MeV) Experimental	ω_1^3 (MeV) Calculated(539)	$B(E3; 0^+ \rightarrow 3^-)_{s.p.u.}$ Experimental	$B(E3; 0^+ \rightarrow 3^-)_{s.p.u.}$ Calculated(539)
$_{46}$Pd	104	0.556	37	36	—	—	—	—
	106	0.512	43	41	2.09	2.3	—	—
	108	0.439	48	48	1.99	2.2	—	—
	110	0.374	54	56	—	—	—	—
$_{48}$Cd	108	0.633	35	27	—	—	—	—
	110	0.658	31	26	2.05	2.3	14	19
	112	0.617	33	28	1.97	2.2	14	18
	114	0.558	35	30	1.96	2.3	15	19
	116	0.513	36	31	1.90	2.4	15	20
$_{50}$Sn	114	1.299	12	10	2.28	2.4	29	11
	116	1.293	12	10	2.22	2.2	40	12
	118	1.230	14	10	2.29	2.2	30	13
	120	1.171	13	9	2.25	2.2	23	14
	122	1.140	15	8	2.49	2.4	36	16
	124	1.131	12	8	2.57	2.4	34	17
$_{52}$Te	122	0.564	37	34	—	—	—	—
	124	0.603	22	29	2.30	2.5	—	—
	126	0.667	28	24	2.40	2.6	—	—
	128	0.743	22	20	2.45	2.7	—	—
	130	0.840	17	17	—	—	—	—
$_{54}$Xe	128	0.441	—	—	—	—	—	—
	130	0.538	24	37	—	—	—	—
	132	0.668	14	30	—	—	—	—
	134	0.850	—	—	—	—	—	—
	136	1.320	—	—	—	—	—	—
$_{56}$Ba	132	0.464	36	51	—	—	—	—
	134	0.604	—	—	—	—	—	—
	136	0.818	—	—	2.44	2.3	—	—
	138	1.426	14	15	—	—	—	—
$_{58}$Ce	138	0.790	—	—	—	—	—	—
	140	1.596	17	13	2.47	2.1	—	—
	142	0.650	19	—	1.66	2.0	—	—
$_{60}$Nd	142	1.570	15	18	2.09	1.9	—	—
	144	0.697	20	46	1.51	1.8	—	—
	146	0.453	37	67	1.21	1.5	—	—
$_{62}$Sm	144	1.631	11	—	1.82	1.5	—	—
	146	0.747	—	—	—	1.3	—	—
	148	0.551	30	56	1.16	1.2	35	32

Both the harmonic oscillator(149, 420–423, 434, 537–539) and the Woods–Saxon potential(540, 541) have been used in calculations of the quadrupole and octupole one-phonon states. Different single-particle level schemes were used in the calculations.

The interaction constants $\varkappa^{(\lambda)}$ of the multipole–multipole forces are fitted by comparing the calculated and experimental energies of the lowest one-phonon states. The nuclei close to magic nuclei are often chosen for determination of the constants $\varkappa^{(\lambda)}$.

The actual values of the constants $\varkappa^{(\lambda)}$ depend on the number of subshells included in (10.6). We have to renormalize the constants $\varkappa^{(\lambda)}$ [see (9.12)] when this number is changed. Only the subshells of incompletely filled neutron and proton shells are usually taken into account in calculations of the quadrupole states. When additional shells are included in (10.6), the agreement with the experimental data is usually improved.

It is worth mentioning that the $\varkappa^{(\lambda)}A^{-(2\lambda+2)/3}$ values [see (9.13)] are identical in spherical and deformed nuclei when the same number of single-particle states (including degeneracy) is used in both systems.

TABLE 10.4. *The characteristics of the first* 2^+ *states in nuclei with* $190 \leq A \leq 220$

Element	A	ω_1^2 (experimental) (MeV)	$B(E2; 0^+ \rightarrow 2^+)_{s.p.u.}$	
			Experimental	Calculated(540)
$_{78}$Pt	190	0.30	–	–
	192	0.32	–	–
	194	0.33	59	37
	196	0.36	37	31
	198	0.41	39	24
$_{80}$Hg	196	0.43	–	–
	198	0.41	32	15
	200	0.37	24	16
	202	0.44	17	11
	204	0.43	–	–
$_{82}$Pb	202	0.96	–	–
	204	0.90	5	2
	206	0.80	4	1
	208	–	–	–
	210	0.80	–	–
	212	0.81	–	–
	214	0.84	–	–
$_{84}$Po	208	0.66	–	–
	210	1.18	–	–
	212	0.73	–	–
	214	0.61	–	–
	216	0.55	–	–
$_{86}$Rn	216	0.46	–	–
	218	0.32	–	–
	220	0.24	–	–

3. Let us discuss the one-phonon $\lambda = 2$, 2^+ states and the related $B(E2)$ probabilities. The lowest 2^+ states in many even–even spherical nuclei have been observed; the $B(E2)$ values for their excitation are also often known. Some of the data are collected in Tables 10.2–10.4 (they are taken from refs. 172, 434, and 542; the results of calculations appear in ref. 540). The most remarkable feature of this data is the strong increase of the energy ω_1^2 in magic or half-magic nuclei and the gradual decrease of ω_1^2 in transitional region.

Let us see how well the secular equation (10.6) with the quadrupole strength constant

$$\varkappa_0^{(2)} = \frac{k^{(2)}}{A^{7/3}}\,\text{MeV} \tag{10.7}$$

($k^{(2)}$ is a constant for all nuclei) describes the energies of the first 2^+ states. The calculations using the $P+QQ$ model, harmonic oscillator wave functions, and including only the subshells

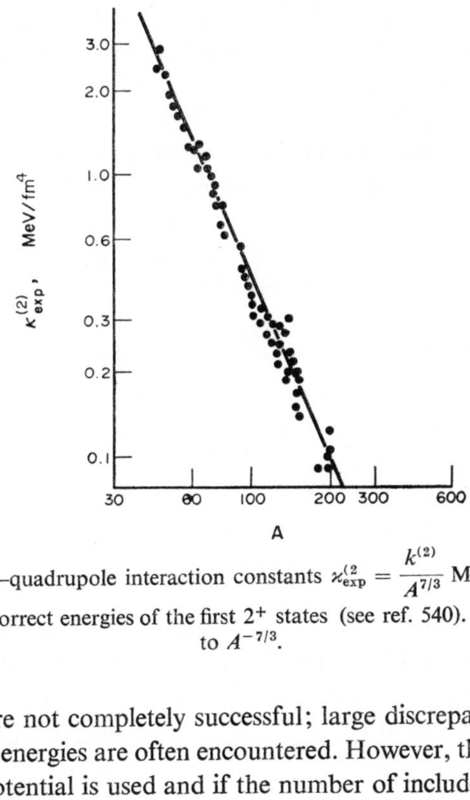

FIG. 10.2. The quadrupole–quadrupole interaction constants $\varkappa_{\text{exp}}^{(2)} = \dfrac{k^{(2)}}{A^{7/3}}$ MeV/fm^4. The black dots show the $\varkappa_{\text{exp}}^{(2)}$ values giving the correct energies of the first 2^+ states (see ref. 540). The full line is proportional to $A^{-7/3}$.

of incomplete shells, are not completely successful; large discrepancies between the calculated and experimental energies are often encountered. However, the deviations are reduced if the Woods–Saxon potential is used and if the number of included subshells is increased. Figure 10.2 shows the $\varkappa_{\text{exp}}^{(2)}$ values (at $\varkappa^{(2)} = \varkappa_{\text{exp}}^{(2)}$ the calculated and experimental $\omega_1^{(2)}$ energies coincide). The good agreement between $\varkappa_{\text{exp}}^{(2)}$ and the expected behavior (10.7) is apparent. However, the secular equation (10.6) has only imaginary solutions below the first pole in a number of cases. This feature clearly shows the inadequacy of the quasiboson approximation.

We have stressed repeatedly that the enhancement of reduced electromagnetic transition probabilities points to the coherent character of the interaction responsible for the considered excited state. Let us discuss the $E2$ transitions between the ground states and quadrupole states, having this point in mind. The reduced $E2$ transition probability for excitation of the one-phonon $(2i)$ state equals

$$B(E2; 0^+ \to 2^+) = 5\mathcal{M}^2; \tag{10.8}$$

with the matrix element

$$\mathcal{M} = \frac{e}{\sqrt{2}}\left\{(1+e_p^{(2)}) \sum_{j_1^p, j_1^p} p^2(j_1^p, j_2^p) g^{21}_{j_1^p j_2^p} u_{j_1^p j_2^p} + e_n^{(2)} \sum_{j_1^n, j_2^n} p^2(j_1^n, j_2^n) g^{21}_{j_1^n j_2^n} u_{j_1^n j_2^n}\right\}, \tag{10.9}$$

the function $g^{2i}_{j_1 j_2}$ is calculated according to (8.160). The single-particle $E2$ transition matrix element $p^2(j_1, j_2)$ and the matrix element $\mathfrak{f}^2(j_1, j_2)$ are identical. The summation in (10.9) therefore contains the squares of the single-particle matrix elements, and all terms contribute coherently in the $i = 1$ case.

The renormalization is included through the effective charge parameter, the usual approximation is $e^{(2)}_{\text{eff}} = e^{(2)}_n = e^{(2)}_p$. The $e^{(2)}_{\text{eff}}$ is chosen $e^{(2)}_{\text{eff}} = 1$ when only the subshells of the incomplete shell are included in the summations (10.6) and (10.9). If the number of included shells increases, the $e^{(2)}_{\text{eff}}$ value decreases.

The experimental and calculated $B(E2)_{\text{s.p.u.}}$ values are collected in Tables 10.2–10.4. The calculations[540] used the Woods–Saxon single-particle wave functions. The large $B(E2)_{\text{s.p.u.}}$ values, 10–50, show the collective character of the first 2^+ states. The $B(E2)_{\text{s.p.u.}}$ values are smaller, 5–15, in nuclei with a magic number of protons or neutrons. The $B(E2)$ values calculated using the $P+QQ$ model agree satisfactorily with the experimental data. The agreement is better if the Woods–Saxon potential is used instead of the harmonic oscillator potential.

The energies of the first 2^+ state were also calculated using the surface δ-interaction in combination with the Tamm–Dancoff method[133] and Gaussian forces together with both the Tamm–Dancoff method and the random phase approximation method.[226, 422] In all these cases, the calculated and experimental energies agree fairly well. Thus it is proven again that the radial dependence of the residual interaction is unimportant for successful description of the first 2^+ states.

4. Let us turn our attention to the one-phonon octupole $I^\pi = 3^-$ states in even–even spherical nuclei.

Part of the experimental material, i.e., the energies and the $B(E3)$ values, is shown in Tables 10.2 and 10.3. The first 3^- states are in majority of cases, collective states. However, the collectivity of the 3^- states is not as large as the collectivity of the 2^+ states. Consequently, the approximate second quantization method gives a more adequate description of the octupole states.

The ω_1^3 and $B(E3)$ values in spherical nuclei with $60 < A < 150$ and $190 < A < 210$ were calculated in ref. 539. The nuclei were divided into nine groups, each of the groups had a slightly different single-particle level scheme. The octupole interaction strength $\varkappa^{(3)}$ is close to the quantity used in deformed nuclei[129] (the A^{-3} dependence is being considered).

The multipole–multipole interaction constants can be divided into the isoscalar and isovector parts $\varkappa_0^{(\lambda)}$ and $\varkappa_1^{(\lambda)}$ [see (3.69)]:

$$\varkappa_n^{(\lambda)} = \varkappa_p^{(\lambda)} = \varkappa_0^{(\lambda)} + \varkappa_1^{(\lambda)}, \quad \varkappa_{np}^{(\lambda)} = \varkappa_0^{(\lambda)} - \varkappa_1^{(\lambda)}.$$

The calculations were performed with $\varkappa_1^{(3)} = 0$ and $\varkappa_1^{(3)} = -0.5\ \varkappa^{(3)}$.[539] The comparison of the experimental and theoretical energies and $B(E3)$ values did not bring any definite conclusion about the $\varkappa_1^{(3)}$ value.

The degree of collectivity of various 3^- states changes. The first 3^- states in strontium, zirconium, molybdenum, ruthenium, palladium, cadmium, and tin isotopes are strongly collective, their energies are 1–2 MeV lower than the first poles of the secular equation, the $B(E3)_{\text{s.p.u.}}$ values are in the interval 14–40. On the other hand, the first 3^- states in tellurium, platinum, and mercury isotopes are close to the lowest poles, the $B(E3)_{\text{s.p.u.}}$ have values 1–10. Such octupole states are apparently less collective.

Calculations of the octupole states in tin and zirconium include the octupole–octupole forces, surface δ-interaction, and standard δ-forces.[496] In all three cases the values of $\omega_1^{(3)}$ and $B(E3)$ are practically identical. This fact shows that the collective states are insensitive to the radial dependence of the interaction. The effective interaction constants and the single-particle energies and wave functions corresponding to lower poles of (10.6) are the determining factors for the $\omega_1^{(3)}$ and $B(E3)$ values.

5. The one-phonon states are superpositions of the two-quasiparticle states. The first 2^+ and 3^- states are collective. This means that many two-quasiparticle components contribute to the wave functions.

We have discussed in § 3 of Chapter 9 the experimental methods giving the information about the quasiparticle components of collective states. The β-decay to one-phonon states, γ-decay from higher lying states, and one-nucleon transfer reactions—all these processes offer a valuable information about the structure of collective states. The formulae in § 3 of Chapter 9 can be easily modified to describe the specific features of spherical nuclei.

The existing experimental $\log ft$ values qualitatively agree with our description of one-phonon states. For example, approximately 30 β-transitions from the 1^+ states in odd–odd nuclei populate the 0^+ ground state and the first excited 2^+ states in nuclei with $62 \leqslant A \leqslant 136$. The average $\log ft$ value is 4.9 for the ground-state transitions and 5.5 for the 2^+ one-phonon states transitions. The single-particle matrix elements diagonal in l contribute most to the β-decays populating 2^+ states; the l-forbidden matrix elements are quite small. Consequently, only one component of the one-phonon state determines the $\log ft$ value. The difference in $\log ft$ values for 0^+ and 2^+ states shows that the corresponding weight factors are considerably smaller than unity.

A large number of the experimental $\log ft$ values for the β-transitions from 1^+ to 0^+ and 2^+ states is analyzed in ref. 543. The calculations using the $P+QQ$ model with the approximate second quantization method correctly describe the reduction of the $\log ft$ values for the 2^+ states.

The direct one-nucleon transfer reactions offer many possibilities for determination of the quasiparticle components of the one-phonon states.

§ 2. Two-phonon States and the Anharmonicity of Vibrations

1. The collectivity of a state can be recognized not only by the enhanced $B(E\lambda)$ values but also by a certain prescribed sequence of the states. For example, the rotational states have energies proportional to $I(I+1)$; each rotational band can encompass a relatively large

interval of excitation energy. The collective vibrational states are characterized by equidistant levels corresponding to one-, two-, three-, and more phonon states. Only the lowest states of such sets are experimentally known. We do not know yet how such a sequence of vibrational states behaves at higher excitation energies.

In this section we shall study the two- and three-phonon states. We shall discuss the characteristics of such states, and particularly the features which are at variance with the simple harmonic theory. The methods describing the anharmonicity of vibrations will be developed.

Section 3 of Chapter 2 discussed the general properties of harmonic vibrations. Figure 2.9 shows schematically the spectra of quadrupole and octupole vibrational states. For the quadrupole states, the multipolarity equals 2 and the quantum number n_2 denotes the number of quadrupole phonons. The one-phonon $n_2 = 1$ state has $I^\pi = 2^+$ and the energy ω_1^2. At $n_2 = 2$ the degenerate triplet $I^\pi = 0^+, 2^+, 4^+$ has the energy $2\omega_1^2$, etc. The spectrum of octupole states has an analogous structure. Besides, we expect to encounter many-phonon states based on the different phonons; for example, the states with one quadrupole and one octupole phonon, etc.

The selection rules and the ratios of $E\lambda$ transition probabilities are identical in the semi-empirical theory of harmonic vibrations and in the phenomenological theory. According to (2.46) and (2.49), the selection and intensity rules have the following form:

$$\Delta n = \pm 1,$$
$$\sum_f B(E\lambda; n_\lambda I_i \to n_\lambda - 1, I_f) = n_\lambda B(E\lambda; n_\lambda = 1 \to n_\lambda = 0).$$

We shall denote the energies of two-phonon and three-phonon states by $E_{\text{vib}}(\lambda i, \lambda' i'; I^\pi)$ and $E_{\text{vib}}(\lambda_1 i_1, \lambda_2 i_2, \lambda_3 i_3; I^\pi)$. The numbers $\lambda_1 i_1, \lambda_2 i_2, \ldots,$ denote the multipolarity and the solution of the secular equation for the corresponding phonons. An extensive experimental material is available on the energies of two-phonon quadrupole states and on the relative intensities of $E2$ transitions. Table 10.5 contains some of this material; the experimental data are taken from refs. 172, 542, 544, and 545. In many cases, all three states of the two-phonon triplet were observed.

Table 10.5 contains the ratios $E_{\text{vib}}(21, 21; 0^+)/\omega_1^2$, $E_{\text{vib}}(21, 21; 2^+)/\omega_1^2$ and $E_{\text{vib}}(21, 21; 4^+)/\omega_1^2$, which should equal 2 according to the harmonic description. The table shows that such ratios are indeed close to 2, but large deviations from this value are also present.

The maximal deviations from the value 2 are found in $E_{\text{vib}}(21, 21; 0^+)/\omega_1^2$. For example, the first excited 0^+ state in ^{70}Ge is only slightly above ω_1^2. In ^{72}Ge, this state has actually the energy lower than ω_1^2. The 0^+ states contain, however, not only two-phonon states but also pairing vibrational states. This fact could explain the large deviations from the harmonicity observed in this case. The problem of 0^+ states is discussed in § 5 of Chapter 8. The nuclear surface vibrations and the pairing vibrations are without doubt coupled. Consequently, in a quantitative theory these two collective modes must be treated simultaneously.

The two-phonon triplet is split in all known cases. The sequence of levels $0^+, 2^+, 4^+$ is different in different nuclei, and the amount of splitting strongly fluctuates.

The existing experimental data on reduced $E2$ transition probabilities are best understood

TABLE 10.5. *Energies of quadrupole two-phonon states and their ratios to the energy of the first* 2^+ *state*

Element	A	$I^\pi = 0^+$		$I^\pi = 2^+$		$I^\pi = 4^+$	
		$E_{vib}(21, 21; 0^+)$ Experimental (MeV)	$\frac{E_{vib}(21, 21; 0^+)}{\omega_1^2}$	$E_{vib}(21, 21; 2^+)$ Experimental (MeV)	$\frac{E_{vib}(21, 21; 2^+)}{\omega_1^2}$	$E_{vib}(21, 21; 4^+)$ Experimental (MeV)	$\frac{E_{vib}(21, 21; 4^+)}{\omega_1^2}$
$_{28}$Ni	60	2.286	1.72	2.159	1.62	2.506	1.88
	62	2.047	1.75	2.304	1.97	2.340	1.99
$_{30}$Zn	64	1.904	1.92	1.800	1.82	2.321	2.34
	66	2.371	2.18	1.867	1.72	–	–
$_{32}$Ge	70	1.215	1.16	1.710	1.64	–	–
	72	0.690	0.83	1.465	1.76	1.729	2.06
$_{34}$Se	76	1.110	1.96	1.217	2.16	1.340	2.43
	78	1.510	2.44	1.310	2.13	1.510	2.44
$_{38}$Sr	86	–	–	1.857	1.73	2.232	2.07
$_{40}$Zr	92	1.382	1.47	1.830	1.95	1.494	1.59
$_{42}$Mo	94	–	–	1.580	1.81	1.573	1.80
	96	–	–	1.499	1.93	1.629	2.09
$_{44}$Ru	98	–	–	1.400	2.16	1.420	2.17
	100	1.130	2.09	1.362	2.52	1.227	2.27
	102	0.944	1.98	1.103	2.32	1.106	2.33
$_{46}$Pd	104	1.334	2.40	1.342	2.42	1.324	2.38
	106	1.134	2.21	1.128	2.20	1.229	2.40
	108	1.049	2.42	0.940	2.17	1.048	2.42
$_{48}$Cd	110	1.411	2.15	1.476	2.25	1.549	2.35
	112	1.224	1.98	1.312	2.13	1.414	2.29
	114	1.133	2.03	1.208	2.17	1.283	2.30
$_{52}$Te	122	1.350	2.40	1.250	2.22	1.170	2.08
$_{54}$Xe	132	–	–	1.298	1.94	1.441	2.15
$_{56}$Ba	134	1.360	2.26	1.167	1.93	1.400	2.31
$_{58}$Ce	140	1.902	1.19	–	–	2.083	1.31

when they are expressed as the ratios

$$\frac{B(E2;\,0_2^+\to 2_1^+)}{B(E2;\,2_1^+\to 0^+)},\quad \frac{B(E2;\,2_2^+\to 2_1^+)}{B(E2;\,2_1^+\to 0^+)}\quad\text{and}\quad \frac{B(E2;\,4_2^+\to 2_1^+)}{B(E2;\,2_1^+\to 0^+)}.$$

Figure 10.3 collects the experimental information about the second of the above-written ratios. This ratio equals 2 in the harmonic approximation. Large fluctuations of the quantity $\frac{B(E2;\,2_2^+\to 2_1^+)}{B(E2;\,2_1^+\to 0^+)}$ are apparent; the experimental ratios are typically smaller than 2. The other ratios $\frac{B(E2;\,0_2^+\to 2_1^+)}{B(E2;\,2_1^+\to 0^+)}$ and $\frac{B(E2;\,4_2^+\to 2_1^+)}{B(E2;\,2_1^+\to 0^+)}$ also strongly fluctuate.

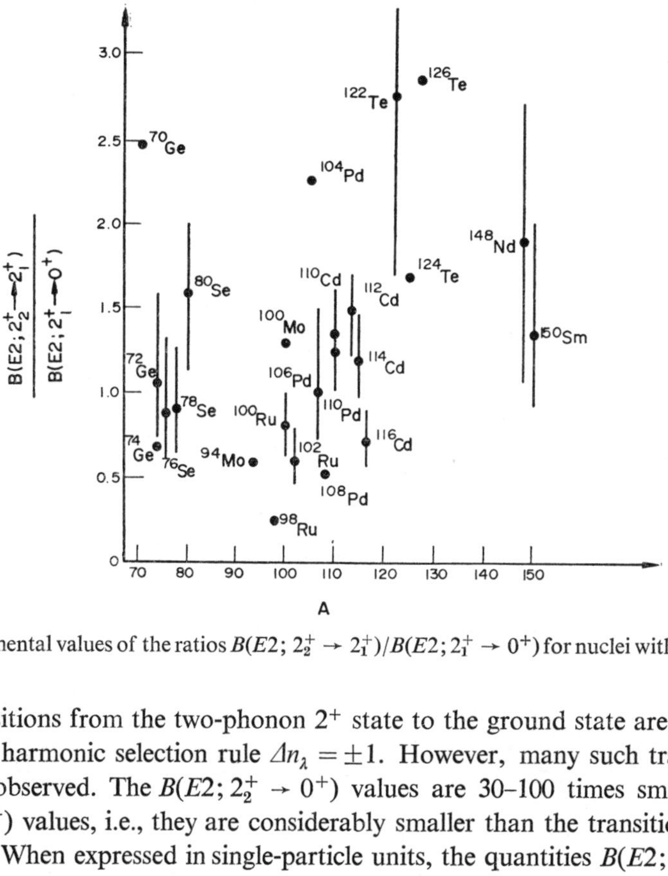

FIG. 10.3. Experimental values of the ratios $B(E2;\,2_2^+\to 2_1^+)/B(E2;\,2_1^+\to 0^+)$ for nuclei with $70 < A < 150$.

The $E2$ transitions from the two-phonon 2^+ state to the ground state are forbidden according to the harmonic selection rule $\Delta n_\lambda = \pm 1$. However, many such transitions have been actually observed. The $B(E2;\,2_2^+\to 0^+)$ values are 30–100 times smaller than the $B(E2;\,2_1^+\to 0^+)$ values, i.e., they are considerably smaller than the transitions from one-phonon states. When expressed in single-particle units, the quantities $B(E2;\,2_2^+\to 0^+)_{\text{s.p.u.}}$ equal 0.2–3.

The splitting of the quadrupole two-phonon triplets and the violation of the harmonic selection and intensity rules show that an improvement of the theoretical description is necessary.

2. The description of the collective states may be improved if the anharmonicity of the vibrational motion is taken into account. This is often achieved by including the quasiparticle–phonon interaction.[546–553]

Let us briefly describe the effect of the quasiparticle–phonon interaction in even–even spherical nuclei. We shall use formulae (8.154) and (8.156) and the secular equation (10.6). The nuclear model Hamiltonian including the quasiparticle–phonon interaction has the form

$$H_{vq} = \sum_j \varepsilon(j) \sum_m \alpha^+_{jm}\alpha_{jm} - \tfrac{1}{4} \sum_{\lambda=2,3} \frac{2\lambda+1}{\varkappa^{(\lambda)}} \sum_{i,i'} \sum_{\mu} \frac{1}{\sqrt{Y^i(\lambda)\,Y^{i'}(\lambda)}} \times$$
$$\times \{Q^+_i(\lambda\mu)\,Q_{i'}(\lambda\mu) + Q^+_{i'}(\lambda\mu)\,Q_i(\lambda\mu) + (-1)^{\lambda-\mu}\,Q^+_i(\lambda\mu)\,Q^+_{i'}(\lambda-\mu)$$
$$+(-1)^{\lambda-\mu}\,Q_{i'}(\lambda-\mu)\,Q_i(\lambda\mu)\} - \frac{1}{2\sqrt{2}} \sum_{\lambda=2,3} \sum_{\mu,i} \frac{1}{\sqrt{Y^i(\lambda)}} \times$$
$$\times \sum_{j,j'} f^\lambda(j,j')\, v_{jj'} \{[Q_i(\lambda-\mu) + (-1)^{\lambda-\mu}\,Q^+_i(\lambda\mu)]\, B(j,j';\lambda-\mu)$$
$$+ B(j,j';\lambda\mu)\,[Q_i(\lambda\mu) + (-1)^{\lambda-\mu}\,Q^+_i(\lambda-\mu)]\}. \tag{10.10}$$

The trial wave function is chosen as the series

$$\Psi_{i_0 n}(I^\pi) = C_{i_0 n}(\lambda_0) \Big\{ Q^+_{i_0}(\lambda_0\mu_0) + \sum_{\substack{\lambda_1,\lambda_2,i_1,i_2\\ \mu_1,\mu_2}} (\lambda_1\mu_1\lambda_2\mu_2 \,|\, \lambda_0\mu_0)\, \varDelta^{\lambda_1 i_1}_{\lambda_2 i_2}(\lambda_0 i_0, n)\, Q^+_{i_1}(\lambda_1\mu_1)\, Q^+_{i_2}(\lambda_2\mu_2)$$
$$+ \sum_{\substack{\lambda_1,\lambda_2,\lambda_3,\lambda'\\ \mu_1,\mu_2,\mu_3,\mu'}} \sum_{i_1,i_2,i_3} (\lambda_1\mu_1\lambda_2\mu_2 \,|\, \lambda'\mu')\,(\lambda'\mu'\lambda_3\mu_3 \,|\, \lambda_0\mu_0) \times$$
$$\times \varDelta^{\lambda_1 i_1}_{\lambda_2 i_2}(\lambda_3 i_3, \lambda', \lambda_0 i_0, n)\, Q^+_{i_1}(\lambda_1\mu_1)\, Q^+_{i_2}(\lambda_2\mu_2)\, Q^+_{i_3}(\lambda_3\mu_3) + \ldots \Big\} \Psi. \tag{10.11}$$

In order to explain the role of two-phonon admixtures as simply as possible, we shall restrict the summation over i to the $i = 1$ case (i.e., only the most collective phonon will be taken into account). We shall also, for simplicity, restrict the series (10.11) to only two terms. The wave function of a 2^+ state then has the simple form

$$\Psi^{1n}(2^+) = C^{1n} \Big\{ Q^+_1(2\mu_0) + \sum_{\mu_1,\mu_2} (2\mu_1 2\mu_2 \,|\, 2\mu_0)\, \varDelta_n Q^+_1(2\mu_1)\, Q^+_1(2\mu_2) \Big\} \Psi. \tag{10.12}$$

The normalization condition can be expressed as

$$1 = C^2_{1n}\{1 + 2\varDelta^2_n\}. \tag{10.12'}$$

We shall again use the variational method for determining the unknown energies and coefficients C_{1n} and $\varDelta_n$. By finding the expectation value of H_{vq} in the state (10.12) and by demanding that such a quantity is minimal, we obtain the equation

$$\delta\{(\Psi^*_{1n}(2^+)\, H_{vq} \Psi_{1n}(2^+)) - \zeta_n [C^2_{1n}(1 + 2\varDelta^2_n) - 1]\} = 0. \tag{10.13}$$

The Lagrange multiplier ζ_n determines the energies of the first and second 2^+ states ($n = 1, 2$).

After simple calculations, we can write

$$\zeta_1 = 1.5\omega_1 - \sqrt{0.25\omega_1^2 + L_{21}^2}, \tag{10.14}$$

$$\zeta_2 = 1.5\omega_1 + \sqrt{0.25\omega_1^2 + L_{21}^2}, \tag{10.14'}$$

$$C_{in}^{-2} = 1 + \frac{L_{21}^2}{(2\omega_1 - \zeta_n)^2}, \tag{10.15}$$

$$\Delta_n = \frac{1}{\sqrt{2}} \frac{L_{21}}{2\omega_1 - \zeta_n}, \tag{10.16}$$

where

$$L_{21} = 5 \sum_{j, j', j_2} \frac{v_{jj'}}{\sqrt{2Y^i(2)}} \mathfrak{f}(j, j') \left[\psi_{jj_2}^{21}\psi_{j_2j'}^{21} + \varphi_{jj_2}^{21}\varphi_{j_2j'}^{21} + \tfrac{1}{2}\left(\psi_{jj_2}^{21}\varphi_{j_2j'}^{21} + \psi_{j_2j'}^{21}\varphi_{jj_2}^{21}\right)\right] (-1)^{j+j_2} \begin{Bmatrix} 2 & 2 & 2 \\ j & j' & j_2 \end{Bmatrix}. \tag{10.17}$$

The $6j$ symbol $\begin{Bmatrix} 2 & 2 & 2 \\ j & j' & j_2 \end{Bmatrix}$ is defined in ref. 27; $\mathfrak{f}(j, j')$ denotes the single-particle quadrupole matrix element and ω_1 is the energy of the one-phonon state.

The harmonic approximation is obviously obtained when $L_{21} = 0$. In such a case, $\zeta_1 = \omega_1$, $\zeta_2 = 2\omega_2$; $C_{11}^2 = 1$, $2C_{11}^2\Delta_1^2 = 0$; $C_{21}^2 = 0$, $2C_{12}^2\Delta_2^2 = 1$.

From (10.14) it follows that, if $L_{21} \neq 0$, the first 2^+ state is somewhat lowered, i.e., $\zeta_1 < \omega_1$. At the same time, the second 2^+ state is shifted upwards, i.e., $\zeta_2 > 2\omega_1$. The two-phonon 0^+ and 4^+ states are not affected in this approximation, they remain degenerate. The degeneracy is broken when the three-phonon admixtures are also included. Therefore, the wave function (10.12) cannot be used for calculating the two-phonon triplet splitting. More realistic calculations taking into account the three-phonon states are very similar to the just-explained methods. However, the procedure of calculation is quite cumbersome. The effect was calculated in a number of papers, 546, 548, and 549. Unfortunately, the splitting of the triplet has not been correctly described anywhere.

The quasiparticle and phonon excitations are coupled to each other. Therefore, when calculating the properties of the two-, three-, and more phonon states we cannot restrict ourselves to the wave function (10.11). The pure two-quasiparticle, four-quasiparticle, etc., components must be also explicitly included. For example, to describe the 4^+ states we have to include the two-phonon quadrupole states, the two-quasiparticle states, and, probably, the one-phonon, $\lambda = 4$, states. The importance of two-quasiparticle components is different in different nuclei. The two-quasiparticle components are less important if the subshells $j = \frac{1}{2}$ or $j = \frac{3}{2}$ are at the Fermi level. The two-quasiparticle 4^+ states have a relatively high energy in such a case and therefore become unimportant. On the other hand, when the subshell with $j \geqslant \frac{5}{2}$ is at the Fermi level, the 4^+ and 2^+ poles have identical energies and the role of the two-quasiparticle states increases.

3. We have stated previously that the reduced probabilities $B(E2)$ for the transitions connecting the ground, one-phonon and two-phonon states do not obey the rules of harmonic approximation. Let us see how the wave function (10.12) behaves in this respect.

The $E2$ transition matrix elements and the corresponding $B(E2)$ values are easily calculated when the wave functions (10.12) are used. We shall denote the reduced transition probability

$B(E2; 2_1^+ \to 0^+)$ determined in the harmonic approximation by $B_0(E2)$. (The statistical factor $2I+1$ is not included). The reduced probability of the $E2$ transition between the first 2^+ state (wave function $\psi_{11}(2^+)$) and the nuclear ground state equals

$$B(E2; 2_1^+ \to 0^+) = B_0(E2)(1-2C_{11}^2\Delta_1^2) = B_0(E2)C_{11}^2, \tag{10.18}$$

and the corresponding energy ζ_1 is given by (10.14). The two-phonon admixtures thus reduce the $E2$ transition probability.

The $E2$ transition from the second 2^+ state to the ground state is no longer forbidden. Its reduced probability equals

$$B(E2; 2_2^+ \to 0^+) = B_0(E2)C_{12}^2. \tag{10.19}$$

The energy ζ_2 of the second 2^+ state is given by (10.14′). The $B(E2)$ values (10.19) are, however, rather small. They have approximately the same magnitude as the typical $B(E2)$ values for the single-particle transitions.

The reduced probability of the $E2$ transition between the second and first 2^+ state equals

$$B(E2; 2_2^+ \to 2_1^+) = 2B_0(E2)\,C_{11}^2(2C_{12}^2\Delta_2^2), \tag{10.20}$$

i.e., it is also somewhat reduced compared to the harmonic approximation.

From (10.18) and (10.20) it follows that

$$\frac{B(E2; 2_2^+ \to 2_1^+)}{B(E2; 2_1^+ \to 0^+)} = 2(2C_{12}^2\Delta_2^2) < 2. \tag{10.21}$$

Thus, the anharmonicity reduces this ratio. In most cases, the experimental ratios shown in Fig. 10.3 are also smaller than 2.

As in the case of energies, we have to include the three-phonon components in order to obtain more realistic values of the reduced transition probabilities $B(E2; 4_2^+ \to 2_1^+)$, $B(E2; 0_2^+ \to 2_1^+)$, $B(E_2; 2_2^+ \to 2_1^+)$, and $B(E_2; 2_1^+ \to 0^+)$. Such a procedure is rather complex and does not give a fully satisfactory description of the $E2$ transitions between the ground, one-, and two-phonon states.

4. The successful measurement of quadrupole moments of the first 2^+ states in even–even spherical nuclei gave a surprising result. By utilizing the reorientation effect it was possible to determine the quadrupole moments in a number of nuclei from ruthenium to samarium—all these nuclei belong to the traditionally "vibrational" region. These quadrupole moments are relatively large in absolute value, 0.2–1.8 barns,[554] and in most cases they have a negative sign.

These measurements gave another impetus to the development of the improved theory of nuclear vibrational motion. The quadrupole moments of 2^+ states were calculated in a number of papers, refs. 550, 551, 555–557. Both the phenomenological and the semi-microscopic methods have been used.

The harmonic vibrational model predicts vanishing quadrupole moments of the first 2^+ one-phonon states. The nonvanishing values of quadrupole moments are most naturally explained by inclusion of the anharmonic effects. Let us demonstrate the role of anharmonic corrections using as an example the simple combination (10.12) of the one- and two-phonon states.

The expectation value of the quadrupole moment operator in the state $\psi_{11}(2^+)$ is equal to

$$Q_1(2_1^+) = 16\sqrt{\frac{\pi}{35}}\,Q_0 C_{11}\sqrt{2}\Delta_1\sqrt{1-2C_{11}^2\Delta_1^2} = 16\sqrt{\frac{\pi}{35}}\,Q_0\sqrt{2}C_{11}^2\Delta_1, \tag{10.22}$$

where

$$Q_0 = \frac{e}{m\overset{\circ}{\omega}_0}\frac{1}{\sqrt{10}}\frac{1}{\sqrt{Y^1(2)}}\left\{\left(1+e_{\text{eff}}^{(2)}\right)\sum_{j_1^p, j_2^p}\frac{\left(\mathfrak{f}(j_1^p, j_2^p)u_{j_1^p j_2^p}\right)^2\left[\varepsilon(j_1^p)+\varepsilon(j_2^p)\right]}{[\varepsilon(j_1^p)+\varepsilon(j_2^p)]^2-(\omega_1)^2}\right.$$
$$\left.+e_{\text{eff}}^{(2)}\sum_{j_1^n, j_2^n}\frac{\left(\mathfrak{f}(j_1^n, j_2^n)u_{j_1^n j_2^n}\right)^2\left[\varepsilon(j_1^n)+\varepsilon(j_2^n)\right]}{[\varepsilon(j_1^n)+\varepsilon(j_2^n)]^2-(\omega_1)^2}\right\}. \tag{10.23}$$

The mentioned calculations explain the large quadrupole moments of the first 2^+ states in a number of even–even spherical nuclei. However, it is impossible to explain simultaneously the entire set of the experimental material, i.e., the quadrupole moments of the first 2^+ states, the splitting of the two-phonon triplet, and the intensities of the $E2$ transition with a single wave function containing one-, two-, and three-phonon components.

5. We have discussed the one- and two-phonon states. Let us now turn our attention to the available experimental evidence about the three- and many-phonon states in even–even spherical nuclei.

Some experimental facts suggest an existence of the three-phonon quadrupole states in spherical nuclei. Such states form a quintuplet $I^\pi = 0^+, 2^+, 3^+, 4^+$, and 6^+. The states are characterized by the strong $E2$ transitions connecting them with two-phonon states, while the transitions to the one-phonon state and to the ground state are weak.

The $I^\pi = 3^+$ states with an energy approximately equal to $3\omega_1^2$ have been observed[545] in ^{100}Ru, ^{102}Ru, ^{104}Pd, ^{110}Cd, and several other nuclei. The experimental ratios $B(E3; 3_3^+ \to 2_2^+)/B(E2; 3_3^+ \to 2_1^+)$ have values 24–40 and the ratios $B(E2; 3_3^+ \to 4_2^+)/B(E3; 2_3^+ \to 2_1^+)$ have values 6–20. Such experimental facts clearly show that these 3^+ states contain large components of the three-phonon states. The transitions from the 3^+ three-phonon states to the 2^+ one-phonon states are strictly forbidden in the harmonic approximation. The discovery of such transitions means that the 3^+ states have a complex structure and contain besides the 3^+ three-phonon states some other components.

The density of states is already large in the energy interval containing the three-, four-, and many-phonon states. The identification of such states is, therefore, quite difficult. Nevertheless, the problem of three-phonon states is very interesting. It would be important to find how pure the three-phonon states are and which states admix to different numbers of the three-phonon quintuplet. It seems that the $I^\pi = 6^+$ state should have the largest three-phonon component. The purity of four-, five-, and more phonon states is a completely open problem. Perhaps the states are so mixed with other states that the very concept of such states has no meaning.

§ 3. Quasiparticle–Phonon Intraction in Odd-*A* Spherical Nuclei

1. The quasiparticle–phonon interaction strongly affects the spectra of odd-A spherical nuclei. The related effects are often described by the phenomenological methods,[555, 558] treating the interaction of an odd nucleon with the core vibrations. In this section, we shall briefly describe the semimicroscopic treatment of the quasiparticle–phonon interaction. We shall discuss the characteristic features of odd-A spherical nuclei and some properties of the transitional nuclei.

Section 3 of Chapter 4 discussed the quasiparticle excited states in odd-A spherical nuclei. If the concept of odd particle plus phonon is used, then the quasiparticle excitations correspond to the situation where the core is kept in its ground state. In such a description, we expect at higher energies another group of states, the quasiparticle excitation plus one-phonon state of the core, and, later, the quasiparticle plus two-phonon state of the core, etc.

In the simplest model of a quasiparticle (angular momentum j_0) plus one-phonon state of the core (angular momentum λ) we expect a degenerate multiplet with angular momentum $I = j_0+\lambda, j_0+\lambda-1, \ldots, |j_0-\lambda|$. The energy of such a multiplet is equal to the phonon energy plus the quasiparticle excitation energy $\mathcal{E}_0(j_0)-\mathcal{E}_0(j^F)$. The quantity $\mathcal{E}_0(j^F)$ denotes the ground state energy (4.39′) of the odd-A nucleus. The quasiparticle–phonon interaction alters these simple features, the multiplet degeneracy is broken and the structure of the involved states becomes more complex.

The wave functions of low-lying states in nuclei with doubly magic core have the simplest structure. The average field potential is determined in such a way that the density matrix in magic nuclei is almost diagonal. The effect of the residual interaction is therefore minimal in nuclei with one nucleon (or one hole) above the doubly magic core. The low-lying states in such nuclei are one-particle (one-hole) states, one-particle (hole) plus phonon, etc. The one-nucleon transfer reactions[559] have proven that the states up to 1.5–2.0 MeV excitation energy in ^{207}Tl, ^{207}Pb, ^{209}Pb, and ^{209}Bi are practically pure one-particle (hole) states. At higher excitation energies, some states remain rather pure one-particle states, other become mixed and contain admixtures of the higher lying configurations. For example, the $p_{1/2}^{-1}$, $f_{5/2}^{-1}$, $p_{3/2}^{-1}$, and $h_{9/2}^{-1}$ states in ^{207}Pb are pure one-hole states, while the $f_{7/2}^{-1}$ and $h_{9/2}^{-1}$ states contain 20% and 35% of admixtures. The spectra of nuclei adjacent to ^{208}Pb are shown in Fig. 10.4.

The octupole one-phonon 3^- state at 2.615 MeV is the lowest excited state in ^{208}Pb. Consequently, we expect the multiplets particle (hole) plus octupole phonon to appear in the neighboring odd-A nuclei. The multiplets containing the particle (hole) with the ground-state quantum numbers must be, naturally, the lowest.

Such multiplets have been actually observed. In ^{207}Tl there is a doublet $I^\pi = \frac{5}{2}^-$ and $\frac{7}{2}^-$ with the energy around 2.6 MeV and the structure $Q_1(3)$ phonon plus $s_{1/2}^{-1}$ hole. The $I^\pi = \frac{5}{2}^+$ and $\frac{7}{2}^+$ doublet in ^{207}Pb has energies 2.625 and 2.664 MeV. This doublet contains $p_{1/2}^{-1}$ hole plus $Q_1(3)$ phonon. All levels of the septuplet $h_{9/2}+Q_1(3)$ in ^{209}Bi have been observed.[559] The center of mass of the septuplet is close to the 3^- phonon energy in ^{208}Pb; the levels are split by 250 keV. In ^{209}Pb we expect a septuplet $g_{9/2}+Q_1(3)$ phonon; the spins and parities are $I^\pi = \frac{3}{2}^- - \frac{15}{2}^-$.

The splitting of all observed particle (hole)+$Q_1(3)$ phonon multiplets is quite small. This means that the quasiparticle–phonon interaction is relatively unimportant in these nuclei.

The phonon plus excited particle (hole) states are expected at higher excitation energies in nuclei adjacent to ^{208}Pb. We also expect the three-, five-, and many-particle states, and the states containing a particle plus two phonons, etc. It would be very interesting to observe some of these higher lying states and study the gradual increase of admixtures when the excitation energy increases.

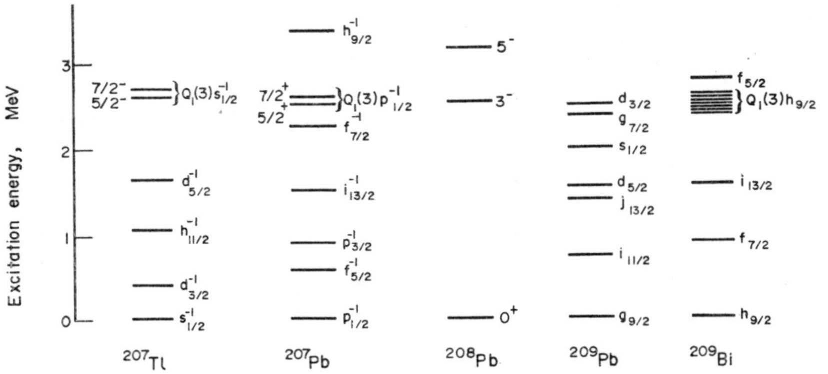

FIG. 10.4. Experimental spectra of ^{208}Pb and the adjacent nuclei. $g_{9/2}$, $i_{11/2}$, $h_{9/2}$, etc. denote the particle states; $s^{-1}_{1/2}$, $d^{-1}_{3/2}$, and $p^{-1}_{1/2}$, etc., denote the hole states. The values $Q_1(3)s^{-1}_{1/2}$, $Q_1(3)p^{-1}_{1/2}$, $Q_1(3)h_{9/2}$ denote the multiplets octupole phonons plus particle (hole).

The one-phonon octupole 3^- state in ^{208}Pb and the related particle (hole) plus $Q_1(3)$ phonon multiplets in neighboring nuclei are excited in inelastic processes. The reduced $E3$ transition probability for the excitation equals $B(E3)_{\text{s.p.u.}} = 30$–$40$. The excitation cross-sections of the multiplets are related by

$$\frac{d\sigma(j_i \to J_f)}{d\Omega} = \frac{2J_f+1}{(2\lambda+1)(2j_i+1)} \frac{d\sigma(0^+ \to 3^-)}{d\Omega} \tag{10.24}$$

to the quantity $\dfrac{d\sigma(0^+ \to 3^-)}{d\Omega}$, i.e., to the excitation cross-section of the 3^- state in ^{208}Pb. The large $B(E3)$ values support the particle–phonon interpretation of the multiplets.

Calculations[560] satisfactorily describe the energies and other properties of the particle (hole) plus phonon states in the nuclei adjacent to magic nuclei.

2. As the neutron and proton numbers of the studied nuclei move further away from the magic numbers, the deviations from the simple situation described above increase. The admixtures of higher configurations become more important, the splitting of the multiplets increases, and the spectra become more complex. Such a process, for example, can be exemplified by comparing the spectra of ^{211}Bi and ^{209}Bi.[561]

On the other hand, the structure of the states in odd-A nuclei becomes more complex when the excitation energy increases.

Let us see how far from the above described situation are the nuclei not close to magic. Figure 10.5 shows three examples of such nuclei. The energies and the $B(E2; 0^+ \rightarrow 2^+)_{\text{s.p.u.}}$ values in even–even nuclei are shown for comparison. In odd-A nuclei we can find the levels containing large components of the quasiparticle plus quadrupole phonon states. The corresponding $B(E2)$ values in odd-A nuclei are also shown. The $B(E2)$ values contain different statistical factors $(2I+1)$; this partially explains their fluctuation in odd-A nuclei.

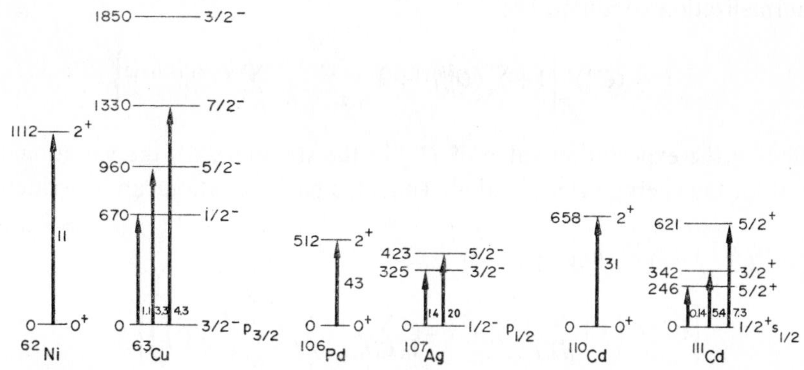

FIG. 10.5. Excited states in odd-A nuclei having large quasiparticle plus quadrupole phonon components. The quadrupole one-phonon states in adjacent even–even nuclei are also shown. The numbers at the arrows show the $B(E2)_{\text{s.p.u.}}$ values for excitation of the corresponding state. Experimental data of refs. 563 and 564 are used.

As we can see, the multiplets are strongly split; the center of gravity does not have the energy of one-phonon states in the neighboring even–even nuclei. The members of multiplets have a complex structure. For example, in ^{63}Cu the complex quadruplet $p_{3/2}+Q_1(2)$ has been observed. The levels of the quadruplet are split by 1 MeV. The lowest state, $I^\pi = \frac{1}{2}^-$ contains approximately 70% of the one-quasiparticle $p_{1/2}$ component. Similarly, the $I^\pi = \frac{5}{2}^-$ state contains approximately 45% of the one-quasiparticle $f_{5/2}$ component.

The one-nucleon transfer reaction data(562) show that the one-particle strength is distributed in odd-A spherical nuclei among several levels. Such a fragmentation is caused by the quasiparticle–phonon interaction.

The quasiparticle plus phonon states are also fragmented. For example, the $s_{1/2}+Q_1(2)$ state in ^{111}Cd is distributed between two $I^\pi = \frac{5}{2}^+$ states at 246 keV and 621 keV. The $B(E2)$ values show that the 621 keV state contains the major part of the $s_{1/2}+Q_1(2)$ state. To describe the multiplet splitting and the fragmentation of both the quasiparticle and quasiparticle plus phonon states correctly, the quasiparticle–phonon interaction must be taken into consideration.

3. The quasiparticle–phonon interaction in spherical nuclei is treated similarly to the same interaction in deformed nuclei. Usually, the interaction includes only one type of phonons ($\lambda = 2, i = 1$ or $\lambda = 3, i = 1$) in spherical nuclei. On the other hand, the quasiparticle plus two-phonon states are also often taken into account. Let us briefly describe the mathematical apparatus used in the studies of quasiparticle–phonon interaction in spherical nuclei.

The nuclear Hamiltonian (10.10) is used. The trial wave function of the odd-A spherical nuclei has the form

$$\Psi_n(I^\pi; j_0 m_0) = C^n_{j_0}\Big\{\alpha^+_{j_0 m_0} + \sum_{\lambda,\mu,i}\sum_{j,m} D^{\lambda i n}_{j_0 j}(jm\lambda\mu\,|\,j_0 m_0)\,\alpha^+_{jm}Q^+_i(\lambda\mu)$$
$$+\sum_{\lambda_1,\mu_1,\lambda_2,\mu_2,i_1,i_2}\sum_{j,m,I',M'}(jmI'M'\,|\,j_0m_0)\,\Delta^{\lambda_1 i_1\lambda_2 i_2 n}_{j_1 j I'}\times$$
$$\times(\lambda_1\mu_1\lambda_2\mu_2\,|\,I'M')\,\alpha^+_{jm}Q^+_{i_1}(\lambda_1\mu_1)\,Q^+_{i_2}(\lambda_2\mu_2)\Big\}\Psi \tag{10.25}$$

and the normalization condition is

$$1 = (C^n_{j_0})^2\Big\{1+\sum_{\lambda ij}(D^{\lambda in}_{j_0 j})^2+2\sum_{\lambda_1,\lambda_2,i_1,i_2}\sum_{j,I'}(\Delta^{\lambda_1 i_1\lambda_2 i_2 n}_{j_0 jI})^2\Big\}. \tag{10.25'}$$

After finding the expectation value of H_{vq} in the state (10.25), the variational method can be used for the energy determination. Thus, the basic equation can be written as

$$\delta\Big\{(\Psi^*_n(I^\pi; j_0m_0)\,H_{vq}\Psi_n(I^\pi; j_0m_0))-\eta_n\times$$
$$\times\Big[(C^n_{j_0})^2\Big(1+\sum_{j,\lambda,i}(D^{\lambda in}_{j_0j})^2+2\sum_{\lambda_1,\lambda_2,i_1,i_2}\sum_{j,I'}(\Delta^{\lambda_1 i_1\lambda_2 i_2 n}_{j_0 jI'})^2\Big)\Big]\Big\} = 0, \tag{10.26}$$

where the Lagrange multiplier η_n determines the resulting energies in odd-A spherical nuclei.

The secular equation

$$\varepsilon(j_0)-\eta_n = \frac{1}{2}\,\frac{1}{2j_0+1}\sum_{\lambda,i,j}\frac{2\lambda+1}{Y^i(\lambda)}\times$$
$$\times\frac{(\mathfrak{f}^\lambda(j_0, j)v_{j_0j})^2}{\varepsilon(j)+\omega^\lambda_i-\dfrac{1}{2j+1}\displaystyle\sum_{\lambda',i',j'}\frac{2\lambda'+1}{Y^{i'}(\lambda')}\,\frac{(\mathfrak{f}^{\lambda'}(j, j')v_{jj'})^2}{\varepsilon(j')+\omega^\lambda_i+\omega^{\lambda'}_{i'}-\eta_n}-\eta_n}. \tag{10.27}$$

is approximately obtained from (10.26). This secular equation can be written in the simple form

$$\varepsilon(j_0)-\eta_n = \frac{1}{2}\,\frac{1}{2j_0+1}\sum_{\lambda,i,j}\frac{2\lambda+1}{Y^i(\lambda)}\,\frac{(\mathfrak{f}^\lambda(j_0, j)v_{j_0j})^2}{\varepsilon(j)+\omega^\lambda_i-\eta_n} \tag{10.28}$$

if the trial function (10.25) does not contain the quasiparticle plus two-phonon components [i.e., if in (10.25) the quantity Δ is identically equal to zero]. Equation (10.28) is identical with (9.44) used in odd-A deformed nuclei.

As we have stated earlier, only one type of phonon is usually included in (10.27), either $\lambda = 2$, $i = 1$ or $\lambda = 3$, $i = 1$. Equation (10.27) can be, in such a case, rewritten as

$$\varepsilon(j_0)-\eta_n = \frac{1}{2}\,\frac{2\lambda+1}{2j_0+1}\sum_j\frac{1}{Y^1(\lambda)}\times$$
$$\times\frac{(\mathfrak{f}^\lambda(j_0, j)v_{j_0j})^2}{\varepsilon(j)+\omega^\lambda_1-\dfrac{2\lambda+1}{2j+1}\displaystyle\sum_{j'}\frac{1}{Y^1(\lambda)}\,\frac{(\mathfrak{f}^\lambda(j, j')v_{jj'})^2}{\varepsilon(j')+2\omega^\lambda_1-\eta_n}-\eta_n}. \tag{10.29}$$

The inclusion of two-phonon components in (10.29) causes a shift of the poles of (10.29) compared to the poles of (10.28).

The functions $C^n_{j_0}$, $D^{\lambda 1 n}_{j_0 j}$, $\Delta^{\lambda 1 \lambda 1 n}_{j_0 j I'}$ are easily calculated by using the normalization condition (10.25′) and secular equation (10.29). The final expressions will have the forms

$$(C^n_{j_0})^{-2} = 1+\sum_j \frac{2\lambda+1}{2j_0+1}\frac{(\mathfrak{f}^\lambda(j_0, j)v_{j_0 j})^2}{2Y^1(\lambda)} \frac{1+2\dfrac{2\lambda+1}{2j+1}\dfrac{1}{Y^1(\lambda)}\sum\limits_{j'}\dfrac{(\mathfrak{f}^\lambda(j, j')v_{jj'})^2}{[\varepsilon(j')+2\omega_1^\lambda-\eta_n]^2}}{\left[\varepsilon(j)+\omega_1^\lambda-\dfrac{1}{2j+1}\dfrac{2\lambda+1}{Y^1(\lambda)}\sum\limits_{j'}\dfrac{(\mathfrak{f}^\lambda(j, j')v_{jj'})^2}{\varepsilon(j')+2\omega_1^\lambda-\eta_n}-\eta_n\right]^2}, \tag{10.30}$$

$$D^{\lambda 1 n}_{j_0 j I'} = \sqrt{\frac{2\lambda+1}{2j_0+1}}\frac{(-1)}{\sqrt{2Y^1(\lambda)}} \frac{\mathfrak{f}^\lambda(j_0, j)v_{j_0 j}}{\varepsilon(j)+\omega_1^\lambda-\dfrac{1}{2j+1}\dfrac{2\lambda+1}{Y^1(\lambda)}\sum\limits_{j'}\dfrac{(\mathfrak{f}^\lambda(j, j')v_{jj'})^2}{\varepsilon(j')+2\omega_1^\lambda-\eta_n}-\eta_n}, \tag{10.31}$$

$$\Delta^{\lambda 1 \lambda 1 n}_{j_0 j I'} = \frac{2\lambda+1}{2Y^1(\lambda)}\sqrt{\frac{2I'+1}{2j_0+1}}\frac{1}{\varepsilon(j)+2\omega_1^\lambda-\eta_n}\sum_{j'}(-1)^{j+j_0}\begin{Bmatrix}\lambda & \lambda & I' \\ j & j_0 & j'\end{Bmatrix}\times$$
$$\times\frac{\mathfrak{f}^\lambda(j_0, j')\mathfrak{f}^\lambda(j', j)v_{j_0 j'}v_{j'j}}{\varepsilon(j')+\omega_1^\lambda-\dfrac{2\lambda+1}{2\lambda'+1}\dfrac{1}{Y^1(\lambda)}\sum\limits_{j_2}\dfrac{(\mathfrak{f}^\lambda(j', j_2)v_{j'j_2})^2}{\varepsilon(j_2)+2\omega_1^\lambda-\eta_n}-\eta_n}. \tag{10.32}$$

Similarly, the $E\lambda$ transition matrix elements, quadrupole and magnetic moments, β-decay matrix elements, etc., are easily calculated with the wave function (10.25) (see ref. 434).

The energies of low-lying states (below 1 MeV), the $B(E2)$ values, and other characteristics of many odd-A spherical nuclei were calculated in refs. 434 and 538. The parameter $\varkappa^{(\lambda)}$ is often chosen in such a way that the energy ω_1^λ in the adjacent even–even nucleus has the correct experimental value. The results of such calculations are in qualitative agreement with the experimental data. For example, it is possible to explain the enhancement of many $E2$ transitions in odd-A spherical nuclei. On the other hand, the $E2$ transitions in other nuclei are inhibited; such an inhibition is explained by the small value of the pairing factor $u_j v_{j'} - u_{j'} v_j$ in such cases.

Other forms of the residual interaction are also often employed in studies of odd-A spherical nuclei. For example, the surface δ-interaction or other methods explained in this monograph were used in modified or generalized form (see refs. 421, 553, and 565–572).

4. The reliable description of the transitional nuclei is more difficult than the description of the spherical or strongly deformed nuclei. The shape of such nuclei is not uniquely determined; the transitional nuclei are "soft" with respect to the β- or γ-vibrations. The rotational and vibrational motions are strongly coupled with each other and with the quasiparticle degrees of freedom. The vibrations themselves are anharmonic. Therefore, we do not have a consistent description of the low-lying states in transitional nuclei.

Rather popular is an empirical scheme describing the gradual development of rotational states out of the vibrational excitations. The process is illustrated in Fig. 10.6. The lowest 2^+ vibrational states ultimately give the first 2^+ rotational states. The 0^+ and 2^+ states of the two-phonon multiplet evolve in the β- and γ-vibrational states, and the 4^+ two-phonon state forms the $I^\pi K = 4^+0$ state of the ground-state rotational band.[71] Such a simple scheme is supported by a number of experimental clues.[573]

Some tendencies in transitional nuclei are quite apparent. For example, the energies of the first 2^+ states monotonically increase for the transitions from the deformed to the spherical nuclei. At the same time, the energies of the second 2^+ state somewhat decrease and move close to the twice vibrational frequency. The moments of inertia and the $B(E2; 2_1^+ \to 0^+)$ values decrease in a correlated way; the ratios $\frac{B(E2; 2_2^+ \to 2_1^+)}{B(E2; 2_0^+ \to 0^+)}$ sharply increase. Such tendencies are clearly recognizable in osmium isotopes between ^{182}Os and ^{192}Os.

The opposite trends are observed at the other end of the deformed region, i.e., when the deformation energy increases with the mass number.

The structure of the excited states in transitional nuclei is more complex than the structure of the states in deformed and spherical nuclei. Thus the transitional nuclei may have the

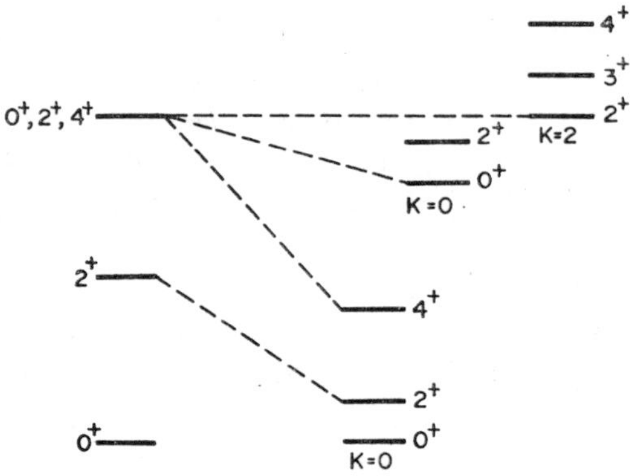

FIG. 10.6. Schematic illustration of the correspondence between the levels of harmonic vibrator and the lowest rotational bands in deformed nuclei, based on the ground state and β- and γ-vibrational states.

excited states with one equilibrium deformation while the ground state has another equilibrium deformation. We have discussed this problem in § 2 of Chapter 7.

It is obvious that we need a unifying method containing a description of both spherical and deformed nuclei for a quantitative description of transitional nuclei.

Several phenomenological methods can describe certain features of both nuclear types. The nuclear shape is, in such a case, characterized by the parameters β and γ; the moments of inertia, kinetic and potential energies, etc., are calculated for each value of the parameters. Afterwards, the Schrödinger equation in variables β and γ is solved.(553, 574, 575)

The development of a unified semiempirical theory of rotational and vibrational motions in even–even transitional nuclei made the first serious steps forward.(576–578)

The number of β-stable transitional nuclei is much smaller than the number of spherical or deformed nuclei. However, if we move further away from the line of β-stability, the number of transitional nuclei increases. This fact should give an additional impetus to the development of quantitative theory describing the transitional nuclei.

5. The phenomenological description of nuclear spectra is based on the division of nuclear motion into three modes: rotational, vibrational, and quasiparticle. The semimicroscopic method forms a basis for a unified description of the quasiparticle and vibrational modes.

This method also states prerequisites for inclusion of the rotational motion into the same scheme.

The description of excited states in terms of the quasiparticles and phonons is very general. Such a description is fully satisfactory in case of low-lying states in doubly magic nuclei and their immediate neighbors. Considering the nuclei further away from the doubly magic core, the description in terms of phonons and quasiparticles becomes more and more cumbersome. In reality, there is no consistent mathematical apparatus suitable for the transitional nuclei. The description of excited states in terms of phonons and quasiparticles becomes again adequate in strongly deformed nuclei. The quantitative description of the spectra is, however, complicated by the effects of Coriolis interaction.

The structure of the states becomes more complex at higher excitation energies. It is rather difficult to predict this tendency and estimate at what energy the description in terms of phonons and quasiparticles completely breaks down.

The whole set of existing experimental data shows that even the low-lying nuclear states are more complex than we thought few years ago. Consequently, the only way to a better understanding of such states and states with higher excitation energies is to study the nuclei in detail. This means that the nucleus is studied by the methods of traditional α-, β-, and γ-spectroscopy [including (n, γ) reactions], and by direct nuclear reactions using monochromatic beams of different projectiles including the heavy ions.

The studies of nuclear structure are inseparably linked with the understanding of the mechanism of nuclear reactions. The important progress in treatment of the direct nuclear reactions made it possible to use the reactions more effectively in nuclear structure studies. The theory of nuclear reactions must be, however, further developed. A unification of the nuclear structure with reaction theories is a necessary step in the development of nuclear physics.

In this monograph we have discussed the ground and low-lying nuclear states. Let us briefly discuss the level of knowledge at higher excitation energies. The region between 2–3 MeV and the nucleon binding energy is rather poorly known.

The fragmentation of one-, three-, or more-quasiparticle states in odd-A nuclei is probably very strong in this region of energies. The density of states is a rapidly-growing function of the excitation energy. The density of states with a certain value of spin and parity is at the neutron binding energy B_n (5–8 MeV) about 10^5–10^6 times larger than near the Fermi level. Such a large density is related to the fragmentation of the single-particle states. The development of the on-line mass separators will hopefully make it possible to study the levels at 2–5 MeV excitation in medium and heavy nuclei.

Neutron spectroscopy gives an information about the averaged properties of compound nuclei near the neutron binding energy. The utilization of the high resolution time of flight methods allows us to study different highly excited resonance states. Experimental data including the energies, spins, neutron widths, and partial γ- and α-widths are available for a number of resonances.

The attempt to include the highly excited states into the semiempirical description, i.e., to interpret these states in terms of quasiparticles and phonons, is very interesting. Such a treatment was suggested in ref. 579; the wave function of such a state contains the components with different numbers of quasiparticles and phonons. The effectiveness of such a method has been proven; the A-dependence of the strength function of the s-neutrons has been qualitatively explained without the optical model, and the correlations between various processes have been shown.

Studies of the analog states and fine structure of different giant resonances represent a possible source of the information about the quasicontinuous states. At the same time, the experiments with high-energy particles give more information about nuclear structure. Thus the scattering of fast electrons may show the deviation of nuclear charge distribution from the simple Fermi-type function.[580, 581] The K^- capture experiments suggest the existence of a neutron skin.[582]

The studies of the interaction of μ-mesons with nuclei give a very valuable information about the nuclear shape and structure of dipole and quadrupole states. The interaction of π-mesons with nuclei gives an information about the short-range correlations of the nucleons in nuclei. The direct knockout reactions caused by fast protons contribute information about nuclear clusters.[583–585]

The interactions of high-energy protons with nuclei are intensively studied. A number of interesting results has been obtained.[586] However, the multiplicity of secondary particle production increases with increasing proton energy (above 3–5 GeV). This effect makes the extraction of nuclear structure information rather difficult.

The nuclear mass distribution, the structure of phonon states, and other nuclear characteristics can be studied when beams of protons, deuterons, electrons, and other particles in a broad energy interval (up to several GeV) with sufficient energy resolution will be available. Intense K-meson beams are also necessary.

Studies of hypernuclei are also very promising. The Pauli principle does not affect the hyperon–nucleon interactions and, perhaps, we can study new features of such interactions in the future. The information obtained from heavier hypernuclei would be even more interesting. Up till now, the main emphasis was on studies of the nucleon–hyperon interaction and effects of three-particle forces, etc. The information about the bound excited states of hypernuclei and about the heavier hypernuclei is still missing.

APPENDIX: TRANSLATOR'S ADDENDUM

The monograph was written three years ago. During such a period, nuclear physics, being a viable scientific discipline, made a considerable progress. In the case of the topics described here, the progress was more of a quantitative nature. Additional nuclear states were observed and new theoretical calculations—many along the lines described here—were made.

While working on the translation I thought that the new development should be, in some way, included in the English edition. The reader who has worked his (or her) way through the whole book is certainly capable of reading the original scientific papers. Thus, I decided that the updating of the monograph can be achieved by simply adding a list of "recommended supplementary reading material." Such material, with brief comments, is collected below. It contains mostly review articles; the original references can be found in them. I have concentrated on the articles showing some new aspects of the problems discussed in the monograph. Thus not all topics are mentioned. The list of references is by no means complete, and I would like to apologize to those whose work is not included.

Quasiparticle Excited States

Professor Soloviev kindly agreed to revise for the English translation Tables 4.7–4.12, 4.16, and 4.17 in the main text. New experimental data on the quasiparticle excited states have been added; the corresponding new references are printed below each table.

In the meantime, paper 24 appeared in print.[A1] It contains a detailed analysis of the experimental situation for non-rotational states of deformed odd-A nuclei with $150 < A < 190$.

The same empirical material is analyzed in ref. A2 from a different point of view. The experimental data on quasiparticle excitations in odd-A deformed nuclei are used for constructing the empirical single-particle level scheme. The relation of such a scheme to the potential model calculation is also discussed.

Both papers again show that the techniques explained in this book describe quite well the low-lying non-rotational states in odd-A deformed nuclei.

Single-nucleon Transfer Reactions in Deformed Nuclei

The nuclear structure information obtained in one-nucleon transfer reactions was discussed in § 4 of Chapter 6. The same problem is treated in detail in the review article A3. The paper appeared almost simultaneously with the Russian edition of the monograph, but the subject is so important that I decided to include also the new review article.

Nuclear Deformation and Fission

The equilibrium deformation calculations and the related problems were discussed in Chapter 7. The problem was intensively studied in recent years, with particular emphasis on the properties of superheavy elements, fissioning isomers, and on the description of the fission process in general. Several review articles appeared in print, each usually reflecting the work of one group. The basic approach is, however, very similar in all of them; they are based on the "Strutinsky renormalization procedure."[(371)] The most recent reviews in this series are listed in ref. A4.

The experimental studies of the fissioning isomers also made a substantial progress. The rotational bands built on the isomeric states and the γ-transitions connecting the isomeric and normal states have been observed. The present situation was reviewed by Specht at the Munich meeting.[(A5)]

High Angular Momentum States in Even–Even Nuclei

The book dealt with the properties of rotational bands briefly in Chapter 2; the calculations of moments of inertia are described in Chapter 7. However, the new experimental and theoretical development in this field is intimately related to the problems discussed in the monograph.

The utilization of the heavy-ion compound nucleus reactions $(HI, x\,n\,\gamma)$ provides a systematic method of populating the very high spin nuclear states. Particularly, the ground-state rotational bands up to 20^+ were observed in a number of even–even rare-earth nuclei. At lower angular momentum, the moment of inertia smoothly increases (see Fig. 2.12). More sudden changes occur at higher angular momentum in several nuclei. The effective moment of inertia dramatically increases, sometimes nearly doubling over a spin change interval 4–6 units. Such an effect is usually called "backbending."

The behavior of rotational bands at lower angular momentum is adequately, though phenomenologically, described by the variable moment of inertia (VMI) model.[(A6)]

The sudden changes at high angular velocities may be caused by the effect of rotation on pairing correlations (Mottelson–Valatin effect[(408)]), discussed in Chapter 7. However, other explanations, usually based on various band crossings, have also been suggested. Additional experimental and theoretical work is needed to solve this problem. The present situation is described in review articles A7 and A8.

Rotation-aligned Coupling Schemes

Throughout the book we have used two single-particle schemes to describe the nuclear states. In spherical nuclei, each nucleon was characterized by its angular momentum; in deformed nuclei, each nucleon was characterized by the projection of angular momentum on the nuclear symmetry axis.

In studies of odd-A nuclei by the $(H\,I, x\,n)$ reactions a new pattern emerged. The resulting structures related to the particles with relatively large intrinsic angular momentum have been described in terms of different coupling schemes.[(A9), (A10)] In the new scheme, the particle has a tendency to align its angular momentum with the nuclear rotational axis. In the lowest state, the particle is maximally aligned and therefore it is maximally decoupled from the core.

A number of such states was observed, particularly in the transitional, or almost spherical, nuclei.[(A11)]

The 0^+ States and Two-nucleon Transfer Reactions

This problem was discussed in § 5 of Chapter 8. The intimate relation between the 0^+ states, the two-nucleon transfer reactions, and the treatment of pairing correlations has been emphasized. A similar theme is discussed in a recent specialized review article.[(A12)] Several low-lying 0^+ states were excited in (p, t) reactions in a number of deformed even–even nuclei in both rare-earth and actinide regions (a number of references are given in ref. A12, we include only the most recent paper[(A13)]. Various mechanisms have been proposed for the description of such states. Besides the collective modes, discussed in the monograph, other effects can also be important. For example, the existence of gaps in the single-particle spectra or the (p, t) strength of individual Nilsson orbitals were discussed.

It is obvious that a satisfactory description of all low-lying 0^+ states in deformed nuclei is not yet realized.

Octupole States in Deformed Nuclei

Further theoretical development[(A14)] of the methods described in Chapter 9 lead to the improved description of the octupole one-phonon states. First, the octupole–octupole interaction (3.62) has been modified. The new interaction has formally the same form as the original one, but acts in the "stretched" coordinate system, similar to (1.82). This modification made it possible to describe all octupole one-phonon states with essentially a single value of the constant $\varkappa^{(3)}$, independent of the projection μ. The $K^\pi = 3^-$ states became also collective, particularly in tungsten and osmium isotopes.

Another aspect is the inclusion of the Coriolis coupling. The theoretically calculated Coriolis matrix elements are large when the corresponding octupole states are collective. The Coriolis coupling changes the pattern of rotational bands; it also has the tendency to concentrate the $B(E3; 0^+ \rightarrow 3^-)$ transition probability in the first excited $I^\pi = 3^-$ state.

Finally, the inclusion of anharmonic effects made the description of octupole one-phonon states in the radium and light thorium isotopes possible.

Many octupole states have been observed since. Their properties generally very well agree with the theoretical predictions.

Particle-phonon Coupling in the Vicinity of ^{208}Pb.

As it was stated in Chapter 10, the odd-A nuclei adjacent to ^{208}Pb offer an excellent testing ground for the treatment of particle–phonon interaction. Numerous theoretical and experimental studies were devoted to this subject. Review A15 describes the present knowledge. This may serve as an example of the complex experimental and theoretical studies of one particular problem discussed in this book.

Highly Excited Nuclear States

The techniques developed in this book have been used for the description of nuclear states at energies near the neutron binding energy[(A16)] (see also ref. 579) as already mentioned in Chapter 10. The theory gives a qualitative explanation of the basic properties of highly excited nuclear states. The experimental data and the components of the model wave functions are related. The model is based on the quasiparticle–phonon interaction (9.38) and on a generalization of the wave function (9.41).

APPENDIX REFERENCES

A1. M. E. Bunker and C. W. Reich, *Rev. Mod. Phys.* **43,** 348 (1971).
A2. W. Ogle, S. Wahlborn, P. Piepenbring, and S. Frederiksson, *Rev. Mod. Phys.* **43,** 424 (1971).
A3. B. Elbek and P. O. Tjom, *Adv. Nucl. Phys.* **3,** 259 (1969).
A4. T. Johansson, S. G. Nilsson, and Z. Szymanski, *Annls. Phys. Paris* **5,** 388 (1970). J. R. Nix, *A. Rev. Nucl. Sci.* **22,** 65 (1972). M. Brack, J. Damgaard, H. C. Pauli, A. S. Jensen, V. Strutinsky, and C. Y. Wong, *Rev. Mod. Phys.* **44,** 320 (1972). H. C. Pauli, *Phys. Lett.* **7C,** 36 (1973).
A5. H. J. Specht, *International Conference on Nuclear Physics, Munich, 1973*, vol. 2, North-Holland, Amsterdam.
A6. S. M. Harris, *Phys. Rev.* **138B,** 509 (1965). M. Mariscotti, G. Scharff-Goldhaber, and B. Buck, *Phys. Rev.* **178,** 1869 (1969).
A7. A. Johnson and Z. Szymanski, *Phys. Lett.* **7C,** 182 (1973).
A8. R. A. Sorensen, *Rev. Mod. Phys.* **45,** 353 (1973).
A9. P. Vogel, *Phys. Lett.* **33B,** 400 (1970).
A10. F. S. Stephens, R. N. Diamond, and S. G. Nilsson, *Phys. Lett.* **44B,** 929 (1973).
A11. F. S. Stephens, *International Conference on Nuclear Physics, Munich 1973*, vol. 2, North-Holland, Amsterdam.
A12. R. A. Broglia, O. Hansen, and C. Riedel, *Adv. Nucl. Phys.* **6,** 287 (1973).
A13. M. A. Oothoudt and N. M. Hintz, *Nucl. Phys.* A, **213,** 221 (1973).
A14. K. Neergård and P. Vogel, *Nucl. Phys.* A, **145,** 33 (1970); A, **149,** 209, 217 (1970).
A15. A. Arima and I. Hamamoto, *A. Rev. Nucl. Sci.* **21,** 55 (1971).
A16. V. G. Soloviev, *Phys. Elem. Part. Atom. Nucleus* **3,** 770 (1972); V. G. Soloviev and L. A. Malov, *Nucl. Phys.* A, **196,** 433 (1972).

REFERENCES

1. M. MAYER, *Phys. Rev.* **78,** 16 (1950). O. HAXEL, J. JENSEN, and H. E. SUESS, *Z. Phys.* **128,** 295 (1950).
2. A. BOHR, *Mat.-fys. Medd. Dan. Vid. Selsk.* **26,** (14) (1952), A. BOHR and B. MOTTELSON, *Mat.-fys. Medd. Dan. Vid. Selsk.* **27** (16) (1953).
3. A. S. DAVYDOV and G. F. FILIPPOV, *Zh. éksp. teor. Fiz.* **35,** 440 (1958) (translation, *Soviet Phys. JETP* **8,** 303 (1959)); *Zh. éksp. teor. Fiz.* **36,** 1467 (1959), (translation, *Soviet Phys. JETP* **9,** 1061 (1959)). A. S. DAVYDOV, *Izv. AN USSR*, phys. ser. **23,** 792 (1959).
4. N. N. BOGOLYUBOV, *Lectures on Quantum Statistics*, Sovetskaya Shkola, Kiev, 1949 (translation, Gordon & Breach, New York, 1967).
5. J. BARDEEN, L. COOPER, and J. SCHRIEFFER, *Phys. Rev.* **108,** 1175 (1957).
6. N. N. BOGOLYUBOV, *Zh. éksp. teor. Fiz.* **34,** 58, 73 (1958) (translation, *Soviet Phys. JETP* **7,** 41, **51** (1958)). N. N. BOGOLYUBOV, V. V. TOLMACHEV, and D. V. SHIRKOV, *New Method in Superconductivity Theory*, Acad. Sci., USSR, Moscow, 1958 (translation, Consultants Bureau, New York, 1959).
7. L. D. LANDAU, *Zh. éksp. teor. Fiz.* **35,** 97 (1958) (translation, *Soviet Phys. JETP* **8,** 70 (1959)).
8. N. N. BOGOLYUBOV, *DAN USSR* **119,** 52 (1958); (translation, *Soviet Phys. Dokl.* **3,** 279 (1958)).
9. A. BOHR, B. MOTTELSON, and D. PINES, *Phys. Rev.* **110,** 936 (1958).
10. S. T. BELYAEV, *Mat.-fys. Medd. Dan. Vid. Selsk.* **31** (11) (1959).
11. V. G. SOLOVIEV, *Zh. éksp. teor. Fiz.* **35,** 823 (1958) **36,** 1869 (1959) (translation, *Soviet Phys. JETP* **8,** 572 (1959); **9,** 1331 (1959); *DAN USSR* **123,** 652 (1958) (translation, *Soviet Phys. Dokl.* **3,** 1197 (1958)). *Nucl. Phys.* **9,** 655 (1958/9).
12. V. G. SOLOVIEV, *Effect of Pairing Superfluid Correlations on the Properties of Atomic Nuclei*, Atomizdat, Moscow, 1963; *Selected Topics in Nuclear Theory*, p. 233; IAEA, Vienna, 1963.
13. A. B. MIGDAL, *Theory of Finite Fermi Systems and Properties of Atomic Nuclei*, Nauka, Moscow, 1965 (translation, Interscience, New York, 1967).
14. A. B. MIGDAL, *The Quasiparticle Method in Nuclear Theory*, Nauka, Moscow, 1967 (translation, Benjamin, New York, 1968).
15. J. BLATT and V. WEISSKOPF, *Theoretical Nuclear Physics*, Wiley, New York, 1952.
16. N. A. VOINOVA and B. S. DZHELEPOV, *Nuclear Isobars with Mass Number A = 182*, Nauka, Moscow, 1968.
17. A. BOHR and B. MOTTELSON, *Nuclear Structure*, vol. 1, Benjamin, New York and Amsterdam, 1969.
18. A. C. DAVYDOV, *Theory of Atomic Nucleus*, Fizmatgiz, Moscow, 1958.
19. M. PRESTON, *Physics of the Nucleus*, Addison-Wesley, Reading, Mass., 1962.
20. I. E. MCCARTHY, *Introduction to Nuclear Theory*, Wiley, New York, 1968.
21. D. THOULESS, *Quantum Mechanics of Many-body Systems*, Academic Press, New York, 1961.
22. G. E. BROWN, *Unified Theory of Nuclear Models and Forces*, North-Holland, Amsterdam, 1967.
23. K. BRUECKNER, Theory of nuclear matter, in *Theory of Nuclear Structure*, Dunod, Paris, 1959.
24. C. W. REICH and M. E. BUNKER, A survey of non-rotational states of deformed odd-A nuclei ($150 < A > 190$), *Rev. Mod. Phys.* **43,** 348 (1971).
25. E. P. GRIGORIEV, *Nuclear Structure*, p. 185, FAN Uz. SSR, Tashkent, 1969.
26. E. U. CONDON and G. H. SHORTLEY, *The Theory of Atomic Spectra*, Cambridge, 1935.
27. D. EDMONDS, *Angular Momentum in Quantum Mechanics*, Princeton Univ. Press, Princeton, 1960.
28. N. N. BOGOLYUBOV and D. V. SHIRKOV, *Introduction to the Theory of Quantized Fields*, Gostekhizdat, Moscow, 1957 (translation, Interscience, New York, 1959).
29. V. A. FOCK, *Z. Phys.* **61,** 126 (1930).
30. L. GOMES, J. WALECKA, and V. WEISSKOPF, *Ann. Phys.* **3,** 241 (1958).
31. D. I. BLOKHINTSEV, *Quantum Mechanics*, Vysshaya Shkola, Moscow, 1963 (translation, Allyn & Bacon, Boston, 1964).
32. A. S. DAVYDOV, *Quantum Mechanics*, Fizmatgiz, Moscow, 1963 (translation, Pergamon, Oxford, 1965).

33. L. D. LANDAU and E. M. LIFSHITZ, *Quantum Mechanics*, Fizmatgiz, Moscow, 1963 (translation, Pergamon, Oxford, 1965).
34. P. E. NEMIROVSKIJ, *Contemporary Models of Atomic Nucleus*, Atomizdat, Moscow, 1960.
35. J. ELLIOTT and A. LANE, *Encyclopedia of Physics* **39**, 240 (1957).
36. J. ELLIOTT, *Selected Topics in Nuclear Theory*, p. 157, IAEA, Vienna, 1963.
37. L. A. SLIV and YU. I. KHARITONOV, *Yad. Fiz.* **1**, 1129 (1965) (translation, *Soviet J. Nucl. Phys.* **1**, 804 (1965).
38. G. M. TEMMER, *Fundamentals in Nuclear Theory*, p. 163, IAEA, Vienna, 1967.
39. L. A. SLIV (ed.), *Gamma-Rays*, AN USSR, Moscow, 1961.
40. S. MOSZKOWSKI, *Phys. Rev.* **89**, 474 (1953).
41. S. G. NILSSON, *Mat.-fys. Medd. Dan. Vid. Selsk.* **29** (16) (1955).
42. C. GUSTAFSON, I. L. LAMM, B. NILSSON, and S. G. NILSSON, *Ark. Fys.* **36**, 613 (1967).
43. B. MOTTELSON and S. G. NILSSON, *Mat.-fys. Skr. Dan. Vid. Selsk.* **1** (8) (1959).
44. LIU YUAN, N. I. PYATOV, V. G. SOLOVIEV, I. N. SILIN, and V. I. FURMAN, *Zh. éksp. teor. Fiz.* **40**, 1503 (1961) (translation, *Soviet Phys. JETP* **13**, 1052 (1961)). V. P. GARISTOV and I. D. KHRISTOV, preprint P4–3584, JINR, Dubna 1967.
45. S. G. NILSSON and O. PRIOR, *Mat.-fys. Medd. Dan. Vid. Selsk.* **32** (16) (1960).
46. T. VERESH, V. G. SOLOVIEV, and T. SIKLOS, *Izv. AN USSR*, phys. ser. **26**, 1045 (1962).
47. L. A. MALOV, V. G. SOLOVIEV, and I. D. KHRISTOV, *Yad. Fiz.* **6**, 1186 (1967) (translation, *Soviet J. Nucl. Phys.* **6**, 863 (1968)).
48. T. D. NEWTON, *Can. J. Phys.* **38**, 700 (1960), Preprint CRT-886, Chalk River, Ontario, Canada, 1960.
49. D. A. ARSENIEV, L. A. MALOV, V. V. PASHKEVICH, and V. G. SOLOVIEV, *Izv. AN USSR*, phys. ser. **32**, 866 (1968).
50. R. H. LEMMER and A. E. GREEN, *Phys. Rev.* **119**, 1043 (1960).
51. P. E. NEMIROVSKIJ and V. A. CHEPURNOV, *Yad. Fiz.* **3**, 998 (1966) (translation, *Soviet J. Nucl. Phys.* **3**, 730 (1966)).
52. A. FAESSLER and R. SHELINE, *Phys. Rev.* **148**, 1003 (1966). P. RÖPER, *Z. Phys.* **195**, 316 (1966).
53. E. ROST, *Phys. Rev.* **154**, 994 (1967).
54. B. N. KALINKIN, JA. GRABOWSKI, and F. A. GAREEV, *Acta phys. pol.* **30**, 999 (1966). F. A. GAREEV, S. P. IVANOVA, and B. N. KALINKIN, *Izv. AN USSR*, phys. ser. **32**, 1960 (1968). F. A. GAREEV, L. I. VINOKUROV, and B. N. KALINKIN, preprint E4–3453, JINR, Dubna, 1967.
55. F. A. GAREEV, S. P. IVANOVA, B. N. KALINKIN, S. K. SLEPNEV, and M. G. GINZBURG, preprint P4–3607, JINR, Dubna, 1967.
56. F. A. GAREEV and S. P. IVANOVA, preprint P4–5221, JINR, Dubna, 1970.
57. V. V. PASHKEVICH and V. M. STRUTINSKY, *Yad. Fiz.* **9**, 56 (1969) (translation, *Soviet J. Nucl. Phys.* **9**, 35 (1969)).
58. A. S. DAVYDOV, *Excited States of Atomic Nuclei*, Atomizdat, Moscow, 1967.
59. J. ROGERS, *A. Rev. Nucl. Sci.* **15**, 241 (1965).
60. O. NATHAN and S. G. NILSSON, in *Alpha-, Beta- and Gamma-ray Spectroscopy* (ed. K. Siegbahn), vol. 1, p. 608, North-Holland, Amsterdam, 1965.
61. F. S. STEPHENS, N. LARK, and R. M. DIAMOND, *Phys. Rev. Lett.* **12**, 225 (1964); *Nucl. Phys.* **63**, 82 (1965).
62. A. S. DAVYDOV and A. A. CHABAN, *Nucl. Phys.* **20**, 499 (1960).
63. A. BOHR and B. R. MOTTELSON, *Atomnaya Energiya* **14**, 41 (1963).
64. G. ALAGA, K. ALDER, A. BOHR, and B. R. MOTTELSON, *Mat.-fys. Medd. Dan. Vid. Selsk.* **29** (9) (1955).
65. C. J. GALLAGHER, *Selected Topics in Nuclear Spectroscopy* (ed. B. J. Verhaar), p. 133, North-Holland, Amsterdam, 1964.
66. J. DE BOER and J. D. ROGERS, *Phys. Lett.* **3**, 304 (1963).
67. D. B. FOSSAN and B. HERSKING, *Nucl. Phys.* **40**, 24 (1963).
68. F. BOEHM, G. GOLDRING, G. B. HAGEMANN, G. D. SYMONS, and A. TVETER, *Phys. Lett.* **22**, 627 (1966).
69. B. S. DZHELEPOV, *Structure of Complex Nuclei*, p. 189, Atomizdat, Moscow, 1966 (translation, Consultants Bureau, New York, 1969).
70. A. HRYNKIEWICZ and S. OGAZA, *Structure of Complex Nuclei*, p. 272, Atomizdat, Moscow, 1966 (translation, Consultants Bureau, New York, 1969).
71. R. K. SHELINE, *Rev. Mod. Phys.* **32**, 1 (1960).
72. J. P. DAVIDSON, *Rev. Mod. Phys.* **37**, 105 (1065).
73. C. A. HERAS, S. M. ABECASIS, and H. E. BOSCH, *Phenomenological and Microscopic Analysis of Properties of Even–Even Nuclei*, Aeronautica Argentina DINFIA, Buenos Aires, 1965.

74. D. J. Rowe, *Fundamentals in Nuclear Theory*, pp. 93, 531, IAEA, Vienna, 1967.
75. K. T. Hecht, *Selected Topics in Nuclear Spectroscopy* (ed. B. J. Verhaar), p. 51, North-Holland, Amsterdam, 1964.
76. D. P. Grechukhin, *Yad. Fiz.* **4,** 691 (1966) (translation, *Soviet J. Nucl. Phys.* **4,** 490 (1967).
77. G. Scharff-Goldhaber and J. Weneser, *Phys. Rev.* **98,** 212 (1955).
78. A. K. Kerman and C. M. Shakin, *Phys. Lett.* **1,** 151 (1962).
79. A. S. Davydov, *Collective Excitation of Even Atomic Nuclei*, Ukr. NIINTI, Kiev, 1967.
80. V. M. Mikhailov, *Izv. AN USSR*, phys. ser. **28,** 308 (1964); **30,** 1334 (1966).
81. V. M. Mikhailov and M. A. Mikhailova, *Izv. AN USSR*, phys. ser. **31,** 1708 (1967).
82. O. B. Nielsen, *Proceedings of the Rutherford Jubilee International Conference*, p. 317, Heywood, 1962.
83. A. Faessler, W. Greiner, and R. Sheline, *Nucl. Phys.* **70,** 33 (1965).
84. E. R. Marshalek, *Phys. Rev.* **139,** B770 (1965); *Phys. Rev. Lett.* **16,** 190 (1966).
85. A. K. Kerman, *Mat.-fys. Medd. Dan. Vid. Selsk.* **30** (15) (1956).
86. B. R. Mottelson, *Proceedings of the International Conference of Nuclear Structure* (ed. J. Sanada), p. 87, Tokyo, 1967.
87. I. Hamamoto and T. Udagawa, *Nucl. Phys.* A, **126,** 241 (1969).
88. D. J. Rowe, *Nucl. Phys.* **61,** 1 (1965). R. T. Brookneier, S. Wahlborn, E. J. Seppi, and F. Boehm, *Nucl. Phys.* **63,** 102 (1965).
89. P. P. Day, C. A. Mallman, and E. D. Klema, *Nucl. Phys.* **25,** 226 (1961).
90. R. M. Diamond, B. Elbek, and F. S. Stephens, *Nucl. Phys.* **43,** 560 (1963).
91. Yu. T. Grin and I. M. Pavlichenkow, *Phys. Lett.* **9,** 249 (1964).
92. P. G. Hansen, O. B. Nielsen, and R. K. Sheline, *Nucl. Phys.* **12,** 389 (1959).
93. K. E. Lassila, M. H. Hull, H. M. Ruppal, F. A. McDonald, and G. Breit, *Phys. Rev.* **126,** 881 (1962).
94. M. Hamada and I. D. Johnston, *Nucl. Phys.* **34,** 383 (1962).
95. F. Tabakin, *Ann. Phys.* **30,** 51 (1964).
96. R. A. Bryan and B. L. Scott, *Phys. Rev.* **135,** B434 (1964). D. Y. Wong, *Nucl. Phys.* **55,** 212 (1964). A. E. S. Green and R. D. Sharma, *Phys. Rev. Lett.* **14,** 380 (1965).
97. V. V. Babikov, *Yad. Fiz.* **2,** 326 (1966) (translation, *Soviet J. Nucl. Phys.* **2,** 231 (1966)); *Nucl. Phys.* **76,** 665 (1966); *Izv. AN USSR*, phys. ser. **32,** 299 (1968).
98. J. Goldstone, *Proc. R. Soc.* A, **239,** 267 (1957).
99. S. A. Moszkowski and B. L. Scott, *Ann. Phys.* **11,** 65 (1960). G. M. Vagradov and D. A. Kirzhnitz, *Zh. éksp. teor. Fiz.* **38,** 1499 (1960) (translation, *Soviet Phys. JETP* **11,** 1082 (1960)).
100. H. A. Bethe, B. H. Brandow, and A. G. Petschek, *Phys. Rev.* **129,** 225 (1963).
101. R. Tamagaki, *Proceedings of the International Conference on Nuclear Structure* (ed. J. Sanada), p. 27, Tokyo, 1967.
102. H. A. Bethe, *Proceedings of the International Conference on Nuclear Structure* (ed. J. Sanada), p. 56, Tokyo, 1967.
103. D. M. Brink, *Structure of Complex Nuclei*, p. 111, Atomizdat, Moscow, 1966 (translation, Consultants Bureau, New York, 1969).
104. F. Villars, *Proceedings of the International School of Physics "Enrico Fermi", Course XXIII*, p. 1, 1963; *Course XXXVI*, p. 1, 1966, Academic Press, New York.
105. V. Gillet, *Proceedings of the International School of Physics "Enrico Fermi", Course XXXVI*, p. 43, 1966, Academic Press, New York.
106. D. M. Brink and E. Boeker, *Nucl. Phys.* A, **91,** 1, 27 (1967).
107. C. W. Nestor, D. Davies, S. Krieger, and M. Baranger, *Nucl. Phys.* A, **113,** 14 (1968).
108. K. A. Brueckner, A. M. Lockett, and M. Rotenberg, *Phys. Rev.* **121,** 255 (1961). R. J. Eden and V. J. Emery, *Proc. R. Soc.* A, **248,** 266 (1958). R. J. Eden, V. J. Emery, and S. Sampanthar, *Proc. R. Soc.* A, **253,** 177, 186 (1959).
109. I. Ulehla, *Structure of Complex Nuclei*, p. 5, Atomizdat, Moscow, 1966 (translation, Consultants Bureau, New York, 1969). J. Blank and I. Ulehla, *Czech. J. Phys.* **15,** 797 (1965).
110. A. K. Kerman, J. P. Svenne, and F. M. Villars, *Phys. Rev.* **147,** 710 (1966). C. M. Shakin, J. P. Svenne, and Y. R. Waghmare, *Phys. Lett.* **21,** 209 (1966).
111. M. Baranger, *International Nuclear Physics Conference*, p. 659, Academic Press, New York and London, 1967. R. Muthukrishnan and M. Baranger, *Phys. Lett.* **18,** 160 (1965). K. T. R. Davies, S. J. Krieger, and M. Baranger, *Nucl. Phys.* **84,** 545 (1966).
112. I. Vashakidze and V. Mamasakhlisov, *Yad. Fiz.* **6,** 732 (1967); **7,** 303 (1967) (translation, *Soviet J. Nucl. Phys.* **6,** 542 (1968); **7,** 205 (1968)).

113. A. Lande and J. P. Svenne, *Phys. Lett.* **25B**, 91 (1967).
114. A. Faessler, P. V. Sauer, and M. M. Stingl, *Z. Phys.* **212**, 1 (1968). A. Faessler and A. Plastino, *Z. Phys.* **220**, 88 (1969).
115. Yu. A. Simonov, *Yad. Fiz.* **3**, 630 (1966) (translation, *Soviet J. Nucl. Phys.* **3**, 461 (1966)). A. I. Baz and M. V. Zhukov, *Yad. Fiz.* **11**, 779 (1970) (translation, *Soviet J. Nucl. Phys.* **11**, 580 (1970)).
116. G. E. Brown and T. T. Kuo, *Nucl. Phys.* **85**, 40 (1966); A, **92**, 481 (1967). G. E. Brown, *Nuclear Structure, Dubna Symposium, 1968*, p. 563, IAEA, Vienna, 1968.
117. N. N. Bogolyubov, *DAN USSR* **119**, 224 (1958) (translation, *Soviet Phys. Dokl.* **3**, 292 (1958)).
118. A. Pawlikowski and W. Rybarska, *Acta phys. pol.* **27**, 537 (1964).
119. N. N. Bogolyubov and V. G. Soloviev, *DAN USSR* **124**, 1011 (1959) (translation, *Soviet Phys. Dokl.* **4**, 143 (1959)).
120. N. N. Bogolyubov, *Usp. fiz. Nauk* **67**, 549 (1959) (translation, *Soviet Phys. Usp.* **2**, 236 (1960)).
121. S. V. Tyablikov, *DAN USSR* **121**, 250 (1959) (translation, *Soviet Phys. Dokl.* **3**, 772 (1959)).
122. C. Bloch and A. Messiah, *Nucl. Phys.* **39**, 95 (1962).
123. V. G. Soloviev, preprint P–801, JINR, Dubna, 1961.
124. N. N. Bogolyubov, *Particles and Nucleus* **1**, 301 (1971).
125. A. Kerman and A. Klein, *Phys. Lett.* **1**, 185 (1962); *Phys. Rev.* **132**, 1326 (1963). A. Klein and A. Kerman, *Phys. Rev.* **138**, B 1323 (1965). A. Klein, L. Celenza, and A. Kerman, *Phys. Rev.* **140**, B245 (1965).
126. I. N. Mikhailov and E. Nadjakov, preprint P4–4293, JINR, Dubna, 1969. E. Nadjakov and I. N. Mikhailov, *Nucl. Phys.* A, **107**, 92 (1968).
127. S. T. Belyaev and V. G. Zelevinskij, preprint 298, Inst. Nucl. Phys. SO AN USSR, 1969.
128. A. M. Lane, *Nuclear Theory*, Benjamin, New York and Amsterdam, 1964.
129. V. G. Soloviev, *Atomic Energy Rev.* **3** (2) 117 (1965).
130. L. S. Kisslinger, *Nucl. Phys.* **35**, 114 (1962).
131. K. M. Zheleznova, N. I. Pyatov, and M. I. Cherney, *Izv. AN USSR*, phys. ser. **31**, 550 (1967). N. I. Pyatov, *Ark. Fys.* **36**, 667 (1967). N. I. Pyatov and M. I. Cherney, *Izv. AN USSR*, phys. ser. **31**, 1689 (1967).
132. I. M. Green and S. A. Moszkowski, *Phys. Rev.* **139**, B790 (1965).
133. A. Plastino, R. Arvieu, and S. A. Moszkowski, *Phys. Rev.* **145**, 837 (1966).
134. P. W. M. Glaudemans, B. H. Wildenthal, and J. B. McGrory, *Phys Lett.* **21**, 437 (1966).
135. A. Faessler, A. Plastino, and S. A. Moszkowski, *Phys. Rev.* **156**, 1064 (1967).
136. A. Faessler and A. Plastino, *Nuovo Cim.* **48B**, 429 (1967). *Z. Physik* **203**, 333 (1967); *Phys. Rev.* **156**, 1072 (1967); *Nucl. Phys.* A, **116**, 129 (1968).
137. A. B. Migdal, *Structure of Complex Nuclei*, p. 87, Atomizdat, Moscow 1966 (translation, Consultants Burzeau, New York, 1969); *Nuclear Structure, Dubna Symposium, 1968*, p. 541, IAEA, Vienna, 1968.
138. Z. Bochnacki, I. M. Holban, and I. N. Mikhailov, *Nucl. Phys.* A, **97**, 33 (1967). I. N. Mikhailov, *Nuclear Structure, Dubna Symposium, 1968*, p. 549, IAEA, Vienna, 1968.
139. Z. Bochnacki and S. Ogaza, *Nucl. Phys.* **69**, 186 (1965); **83**, 619 (1966).
140. V. G. Soloviev, *Structure of Complex Nuclei*, p. 38, Atomizdat, Moscow, 1966 (translation, Consultants Bureau, New York, 1969).
141. V. B. Belyaev, B. N. Zakhariev, and V. G. Soloviev, *Zh. éksp. teor. Fiz.* **38**, 952 (1960) (translation, *Soviet Phys. JETP* **11**, 686 (1960)).
142. B. H. Flowers, *Proceedings of the Rutherford Jubilee International Conference* (ed. B. Birks), p. 207, Heywood, London, 1961.
143. B. Bremond and J. G. Valatin, *Nucl. Phys.* **41**, 640 (1963). B. H. Flowers and M. Vujicic, *Nucl. Phys.* **49**, 586 (1963). P. Vogel, *Phys. Lett.* **10**, 314 (1964). J. M. Irvine, *Phys. Lett.* **13**, 144 (1964). M. K. Pal and M. K. Banerjee, *Phys. Lett.* **13**, 155 (1964).
144. H. K. Kissender and L. Münchow, *Phys. Lett.* **19**, 665 (1966); **25B**, 493 (1967).
145. G. N. Afanasiev, preprint P4–3431, JINR, Dubna, 1967. P. Camiz, A. Covello, and M. Jean, *Nuovo Cim.* **36**, 663 (1965); **42**, 199 (1966).
146. V. G. Soloviev, *DAN USSR* **133**, 325 (1960); **139**, 847 (1961) (translation, *Soviet Phys. Dokl.* **5**, 778 (1960); **6**, 707 (1962)).
147. V. G. Soloviev, *Izv. AN USSR* **25**, 1198 (1961).
148. S. Wahlborn, *Nucl. Phys.* **37**, 554 (1962).
149. L. S. Kisslinger and R. A. Sorensen, *Mat.-fys. Medd. Dan. Vid. Selsk.* **32** (9) (1960).
150. M. Baranger, *Phys. Rev.* **120**, 957 (1960).
151. S. Yoshida, *Phys. Rev.* **123**, 2122 (1961).

152. V. G. Soloviev, *Zh. éksp. teor. Fiz.* **40,** 654 (1961); (translation, *Soviet Phys. JETP* **13,** 456 (1961)).
153. V. G. Soloviev, *Mat.-fys. Skr. Dan. Vid. Selsk.* **1** (11) (1961).
154. A. B. Migdal, *Zh. éksp. teor. Fiz.* **37,** 249 (1959) (translation, *Soviet Phys. JETP* **10,** 176 (1960)).
155. Yu. T. Grin, S. I. Drozdov, and D. F. Zaretskij, *Zh. éksp. teor. Fiz.* **38,** 222, 1297 (1960) (translation, *Soviet Phys. JETP* **11,** 162, 936 (1960)).
156. S. I. Drozdov and D. F. Zaretskij, *Zh. éksp. teor. Fiz.* **40,** 286 (1961) (translation, *Soviet Phys. JETP* **13,** 194 (1961)).
157. B. R. Mottelson, *The Many-body Problem*, Grenoble Université, École d'Éte de Physique Theorique Les Houches, Wiley, New York, 1959.
158. J. C. Valatin, *Lectures in Theoretical Physics*, vol. IV, p. 1, Univ. of Colorado, Boulder, 1962.
159. Z. Szymanski, *Nuclear Structure*, vol. 1, p. 59, Federal Nucl. Energy Comm., Yugoslavia, 1966.
160. H. J. Mang and H. A. Weidenmüller, *A. Rev. Nucl. Sci.* **18,** 1 (1968).
161. V. A. Kravtzov, *Atomic Masses and Nuclear Binding Energies*, Atomizdat, Moscow, 1965.
162. J. H. Mattauch, W. Thiele, and A. H. Wapstra, *Nucl. Phys.* **67,** 1 (1965).
163. R. A. Demirkhanov, V. V. Dorokhov, and V. G. Soloviev, *Yad. Fiz.* **2,** 10 (1965) (translation, *Soviet J. Nucl. Phys.* **2,** 7 (1966)).
164. A. I. Vdovin, A. L. Komov, and L. A. Malov, preprint P4–5125, JINR, Dubna, 1970.
165. W. Stepien and Z. Szymanski, *Phys. Lett.* **26B,** 181 (1968).
166. H. Britt, R. Stokes, W. Gibbs, and J. Griffin, *Phys. Rev. Lett.* **11,** 343 (1963); *Phys. Rev.* **139,** B354 (1965).
167. V. G. Soloviev, *Zh. éksp. teor. Fiz.* **43,** 246 (1962) (translation, *Soviet Phys. JETP* **16,** 176 (1963)).
168. F. S. Stephens, F. Asaro, and I. Perlman, *Phys. Rev.* **113,** 212 (1959).
169. C. W. Reich and M. E. Bunker, *Izv. AN USSR*, phys. ser. **31,** 42 (1967).
170. C. W. Reich and M. E. Bunker, *Nuclear Structure, Dubna Symposium, 1968*, p. 119, IAEA, Vienna, 1968.
171. B. S. Dzhelepov, L. K. Peker, and V. O. Sergeev, *Decay Schemes of Radioactive Nuclei*, AN USSR, Moscow, 1963.
172. C. M. Lederer, J. M. Hollander, and I. Perlman, *Table of Isotopes*, 6th edn., Wiley, New York, 1967.
173. Nuclear Data Sheets. Academic Press, New York, 1959.
174. B. S. Dzhelepov, G. F. Dranitsina, and J. Zvolsky, *Nuclear Isobars with Mass Number A = 157*, Nauka, Moscow, 1966.
175. B. Harmatz, T. H. Handley, and J. W. Mihelich, *Phys. Rev.* **119,** 1345 (1960); **128,** 1186 (1962). B. Harmatz and T. H. Handley, *Nucl. Phys.* A, **121,** 481 (1968).
176. R. Kenefick and R. Sheline, *Phys. Rev.* **139,** B1479 (1965).
177. J. Borggreen, H. J. Frahm, N. J. Sigurd Hansen, and S. Bjørnholm, *Nucl. Phys.* **72,** 509 (1965).
178. A. A. Abdurazakov, Zh. T. Zhelev, V. G. Kalinnikov, J. Liptak, F. Molnar, U. K. Nazarov, and J. Urbanec, *Izv. AN USSR* **32,** 764 (1968). A. A. Abdurazakov, V. G. Kalinnikov, Zh. T. Zhelev, V. K. Nazarov, and J. Urbanec, preprint P6–3660 JINR, Dubna, 1968.
179. L. Funke, H. Graber, K. H. Kaun, H. Sodan, and L. Werner, *Nucl. Phys.* **70,** 353 (1965); **55,** 401 (1964).
180. H. Ryde, L. Persson, and K. Oelsner-Ryde, *Ark. Fys.* **23,** 195 (1962).
181. L. Funke, H. Graber, K. H. Kaun, H. Sodan, G. Geske, and J. Frána, *Nucl. Phys.* **84,** 424 (1966).
182. O. W. B. Schult, M. E. Bunker, D. W. Hafemeister, E. B. Shera, E. T. Jurney, J. W. Starner, A. Bäcklen, B. Fogelberg, U. Guber, B. Maier, H. Koch, W. Shelton, M. Minor, and R. Sheline, *Phys. Rev.* **154,** 1146 (1967).
183. V. A. Bondarenko, P. T. Prokofiev, and L. I. Simonova, *Izv. AN USSR*, phys. ser. **29,** 2168 (1965).
184. G. Markus, W. Michaelis, H. Schmidt, and C. Weilkamp, *Z. Physik* **206,** 84 (1967). W. Michaelis, F. Weller, H. Schmidt, G. Markus, and U. Fanger, *Nucl. Phys.* A, **119,** 609 (1968).
185. W. Kurcewicz, Z. Moroz, Z. Preibisz, and B. Schmidt-Nielsen, *Nucl. Phys.* A, **108,** 434 (1968).
186. T. Kutzarova, V. Zvolska, and M. Weiss, *Izv. AN USSR*, phys. ser. **32,** 126 (1968).
187. D. G. Burke, B. Zeidman, B. Elbek, B. Herskind, and M. Olesen, *Mat.-fys. Medd. Dan. Vid. Selsk.* **35** (2) (1966).
188. S. A. Baranov, V. M. Kulakov, and V. M. Shatinsky, *Nucl. Phys.* **56,** 252 (1964). S. A. Baranov, V. M. Kulakov, and S. N. Belenkij, *Zh. éksp. teor. Fiz.* **43,** 1135 (1962) (translation, *Soviet Phys. JETP* **16,** 851 (1963)).
189. C. M. Lederer, J. K. Poggenburg, F. Asaro, J. O. Rasmussen, and I. Perlman, *Nucl. Phys.* **84,** 481 (1966). I. Ahmand, A. M. Friedman, R. F. Barnes, R. H. Sjøblom, J. Milsted, and P. R. Fields,

Phys. Rev. **164,** 1537 (1967). A. Chetham-Strode, R. J. Silva, J. R. Tarrant, and I. R. Williams, *Nucl. Phys.* A, **107,** 645 (1968).
190. H. T. Motz and F. A. Rickey (personal communication).
191. T. H. Braid, R. R. Chasman, J. R. Erskine, and A. M. Friedman, *Phys. Lett.* **18,** 149 (1965); *Phys. Rev. C,* **1,** 275 (1970).
192. J. Kooi and A. H. Wapstra, *J. Inorg. Nucl. Chem.* **29,** 293 (1967). A. M. Friedman, I. Ahmad, J. Milsted, D. W. Engelkemeir, *Nucl. Phys.* A, **127,** 33 (1969).
193. F. N. Bakke, *Nucl. Phys.* **9,** 670 (1958/9).
194. L. Kristensen, M. Jørgensen, O. B. Nielson, and G. Sidenius, *Phys. Lett.* **8,** 57 (1964).
195. M. J. Emmott, J. R. Leigh, J. O. Newton, and D. Ward, *Phys. Lett.* **20,** 56 (1966). J. O. Newton, *Nucl. Phys.* A, **108,** 353 (1968).
196. K. Ya. Gromov, *Structure of Complex Nuclei,* p. 299, Atomizdat, Moscow, 1966 (translation, Consultants Bureau, New York, 1969).
197. H. S. Johansen, M. Jørgensen, O. B. Nielsen, and G. Sidenius, *Phys. Lett.* **8,** 61 (1964).
198. Z. Preibisz, W. Kurcewicz, K. Stryczniewicz, and J. Zylicz, *Phys. Lett.* **14,** 206 (1965).
199. V. Hnatowitz, K. Ya. Gromov, and M. Finger, *Izv. AN USSR* **31,** 587 (1967).
200. L. Funke, K. H. Kaun, P. Kemnitz, H. Sodan, G. Winter, M. Bonetz, and F. Stary, *Nucl. Phys.* A, **130,** 333 (1969).
201. N. D. Newby, *Phys. Rev.* **125,** 2063 (1962).
202. N. I. Pyatov, *Izv. AN USSR,* phys. ser. **27,** 1436 (1963). N. I. Pyatov and A. S. Chernyshov, *Izv. AN USSR,* phys. ser. **28,** 1173 (1964).
203. C. J. Gallagher and V. G. Soloviev, *Mat.-fys. Skr. Dan. Vid. Selsk.* **2** (2) (1962).
204. C. J. Gallagher, *Nucl. Phys.* **16,** 215 (1960); *Phys. Rev.* **126,** 1525 (1962).
205. S. B. Bjørnholm, *Radiochemical and Spectroscopic Studies of Nuclear Excitations in Even Isotopes of the Heaviest Elements,* Munksgaard, Copenhagen, 1965.
206. R. Sheline, *Nuclear Structure, Dubna Symposium, 1968,* p. 71, IAEA, Vienna, 1968.
207. N. I. Pyatov and V. G. Soloviev, *Izv. AN USSR,* phys. ser. **28,** 11, 1617 (1964).
208. V. G. Soloviev and T. Siklos, *Nucl. Phys.* **59,** 145 (1964).
209. R. G. Helmer and C. W. Reich, *Nucl. Phys.* A, **114,** 649 (1968).
210. Z. Preibisz and J. Zylicz, *Phys. Lett.* **9,** 258 (1964).
211. C. J. Gallagher and S. A. Moszkowski, *Phys. Rev.* **111,** 1282 (1958).
212. B. L. Cohen, J. B. Moorhead, and R. A. Mayer, *Phys. Rev.* **161,** 1257 (1967).
213. D. Göller, R. Schönebeerg, and F. Flammersfeld, *Z. Physik* **208,** 299 (1968).
214. B. L. Cohen, *Nuclear Structure, Dubna Symposium, 1968,* p. 3, IAEA, Vienna, 1968.
215. I. Khristov, preprint P4–4896, JINR, Dubna, 1970.
216. V. Gillet, B. Giraud, and M. Rho, *Nucl. Phys.* A, **103,** 257 (1967).
217. K. F. Alexander, *Nuclear Structure, Dubna Symposium, 1968,* p. 15, IAEA, Vienna, 1968.
218. L. K. Peker, *Structure of Complex Nuclei,* p. 325, Atomizdat, Moscow, 1966 (translation, Consultants Bureau, New York, 1969).
219. B. F. Bayman, *Application of Group Theory in Nuclear Spectroscopy,* Fizmatgiz, Moscow, 1961. Lecture notes, NORDITA, Copenhagen, 1957.
220. D. Kobe, *Ann. Phys.* **19,** 448 (1962); **20,** 279 (1962); **25,** 121 (1963); **28,** 400 (1964).
221. N. N. Bogolyubov, D. N. Zubarev, and Yu. A. Tzerkovnikov, *DAN USSR* **117,** 788 (1957) (translation, *Soviet Phys. Dokl.* **2,** 535 (1957)).
222. R. R. Chasman, *Phys. Rev.* **132,** 343 (1963); **156,** 1197 (1967); *Nucl. Phys.* **73,** 588 (1965).
223. R. W. Richardson, *Phys. Lett.* **3,** 277 (1963).
224. R. W. Richardson, *Phys. Lett.* **5,** 82 (1963). R. W. Richardson and N. Sherman, *Nucl. Phys.* **52,** 221, 253 (1964).
225. H. J. Mang, J. K. Poggenburg, and J. O. Rasmussen, *Nucl. Phys.* **64,** 353 (1965).
226. R. Arvieu, E. Baranger, M. Veneroni, M. Baranger, and V. Gillet, *Phys. Lett.* **4,** 119 (1963).
227. H. J. Lipkin, *Ann. Phys.* **9,** 272 (1960).
228. N. N. Bogolyubov, preprint D–781, JINR, Dubna, 1961.
229. J. P. Elliot and T. M. L. Skyrme, *Proc. R. Soc.* A, **232,** 561 (1955).
230. R. E. Peierls and J. Yoccoz, *Proc. Phys. Soc.* A, **70,** 381 (1957).
231. J. J. Griffin and I. A. Wheeler, *Phys. Rev.* **108,** 311 (1957).
232. L. Kisslinger, *Nucl. Phys.* **35,** 114 (1962).
233. I. N. Mikhailov, *Acta phys. pol.* **22,** 475 (1963).

234. M. K. Volkov, A. Pawlikowski, V. Rybarska, and V. G. Soloviev, *Izv. AN USSR*, phys. ser. **27**, 878 (1963).
235. I. N. Mikhailov, *Acta phys. pol.* **24**, 419 (1963).
236. B. F. Bayman, *Nucl. Phys.* **15**, 33 (1960).
237. I. N. Mikhailov, *Zh. éksp. teor. Fiz.* **45**, 1102 (1963) (translation, *Soviet Phys. JETP* **18**, 761 (1964)).
238. E. M. Bang and I. N. Mikhailov, *Izv. AN USSR*, phys. ser. **29**, 113 (1965).
239. K. Dietrich, H. J. Mang, and J. H. Pradal, *Phys. Rev.* **135**, B22 (1964).
240. M. Rho and J. O. Rasmussen, *Phys. Rev.* **135B**, 1295 (1964).
241. Giu Do Dang and A. Klein, *Phys. Rev.* **143**, 735 (1966).
242. Y. Nogami, *Phys. Rev.* **134**, B313 (1964).
243. M. Jean, *Nuovo Cim.* **40A**, 1224 (1965).
244. A. Covello and E. Salusti, *Phys. Rev.* **162**, 859 (1967).
245. F. Iwamoto and H. Onishi, *Prog. Theor. Phys.* **37**, 682 (1967).
246. T. D. Lee, *Phys. Rev.* **95**, 1329 (1954).
247. G. Källen and W. Pauli, *Mat.-fys. Medd. Dan. Vid. Selsk.* **30** (7) (1955).
248. H. Umezawa and A. Viskonti, *Nucl. Phys.* **1**, 20 (1956).
249. W. Thirring, *Helv. phys. acta* **28**, 344 (1955).
250. V. G. Soloviev, *Zh. éksp. teor. Fiz.* **32**, 1050 (1957) (translation, *Soviet Phys. JETP* **5**, 859 (1957)).
251. M. Ichimura, *Progr. Nucl. Phys.* **10**, 309 (1968).
252. V. G. Soloviev, *Nucl. Phys.* **18**, 161 (1960).
253. M. Ichimura, *Prog. Theor. Phys.* **31**, 575 (1964).
254. A. Kerman, R. Lawson, and M. Macfarlane, *Phys. Rev.* **124**, 162 (1961).
255. A. Pawlikowski and V. Rybarska, *Zh. éksp. teor. Fiz.* **43**, 543 (1962) (translation, *Soviet Phys. JETP* **16**, 376 (1962)).
256. A. Lande, *Ann. Phys.* **31**, 525 (1965).
257. I. Perlman and J. Rasmussen, *Encyclopedia of Physics* **42**, 109 (1957).
258. E. K. Hyde, I. Perlman, and G. T. Seaborg, *The Nuclear Properties of the Heavy Elements*, Prentice-Hall, Englewood Cliffs, New Jersey, 1964.
259. J. Rasmussen, *Alpha-, Beta- and Gamma-ray Spectroscopy* (ed. K. Siegbahn), vol. 1, p. 701, North-Holland, Amsterdam, 1965.
260. R. G. Thomas, *Progr. Theor. Phys.* **12**, 253 (1954).
261. H. J. Mang, *Z. Phys.* **148**, 572 (1957).
262. P. J. Brussaard and H. A. Tolhoek, *Physica* **24**, 263 (1958).
263. J. Rasmussen, *Phys. Rev.* **113**, 1593; **115**, 1675 (1959).
264. H. J. Mang, *Phys. Rev.* **119**, 1069 (1960). H. D. Zeh and H. J. Mang, *Nucl. Phys.* **29**, 529 (1962).
265. J. Rasmussen, *Nucl. Phys.* **44**, 93 (1963).
266. J. Rasmussen and B. Segall, *Phys. Rev.* **103**, 1298 (1956).
267. A. Bohr, P. O. Fröman, and B. R. Mottelson, *Mat.-fys. Medd. Dan. Vid. Selsk.* **29** (10) (1955).
268. V. G. Nosov, *DAN USSR* **112**, 414 (1957) (translation, *Soviet Phys. Dokl.* **2**, 48 (1957)).
269. V. M. Strutinsky, *Zh. éksp. teor. Fiz.* **32**, 1412 (1957) (translation, *Soviet Phys. JETP* **5**, 1150 (1957)).
270. P. O. Fröman, *Mat.-fys. Skr. Dan. Vid. Selsk.* **1** (3) (1957).
271. G. M. Andelson-Velskij, A. P. Birzgal, A. D. Pilia, A. L. Goldin, and K. A. Ter-Martyrosian, *Zh. éksp. teor. Fiz.* **35**, 184 (1958) (translation, *Soviet Phys. JETP* **8**, 127 (1959)).
272. R. Chasman and J. Rasmussen, *Phys. Rev.* **112**, 512 (1958).
273. V. G. Soloviev, *DAN USSR* **144**, 1281 (1962) (translation, *Soviet Phys. Dokl.* **6**, 707 (1962)). V. G. Soloviev, *Phys. Lett.* **1**, 202 (1962).
274. K. Harada, *Prog. Theor. Phys.* **26**, 667 (1961).
275. H. J. Mang and J Rasmussen, *Mat.-fys. Skr. Dan. Vid. Selsk.* **2** (3) (1962).
276. A. Sandulescu, *Nucl. Phys.* **48**, 345 (1963).
277. T. Veres, *Izv. AN USSR*, phys. ser. **27**, 891 (1963).
278. A. Sandulescu and F. I. Stancu, *Acta phys. pol.* **27**, 655 (1965).
279. S. A. Baranov, M. G. Gadzhiev, V. M. Kulakov, and V. M. Shatinskij, *Yad. Fiz.* **1**, 557 (1965) (translation, *Soviet J. Nucl. Phys.* **1**, 397 (1965). S. A. Baranov, Yu. F. Rodionov, V. M. Kulakov, and V. M. Shatinskij, *Yad. Fiz.* **4**, 1108 (1966) (translation, *Soviet J. Nucl. Phys.* **4**, 798 (1967)). I. Ahmad, A. M. Friedman, and J. P. Unik, *Nucl. Phys.* A, **119**, 27 (1968).
280. C. Gallagher and J. Rasmussen, *J. Inorg. Nucl. Chem.* **3**, 333 (1957). O. Prior, *Ark. Fys.* **6**, 15 (1959).
281. W. McHarris, F. S. Stephens, F. Asaro, and J. Perlman, *Phys. Rev.* **144**, 1031 (1966). F. Asaro,

M. C. Michel, S. G. Thompson, and I. Perlman, *Proceedings of the Rutherford Jubilee International Conference* (ed. B. Birks), p. 311, Heywood, London, 1961.
282. H. F. Schopper, *Weak Interactions and Nuclear Beta Decay*, North-Holland, Amsterdam, 1966.
283. A. Winther, *On the Theory of Nuclear β-Decay*, Munksgaard, Copenhagen. 1962.
284. *Alpha-, Beta- and Gamma-ray Spectroscopy* (ed. K. Siegbahn), vol. 2, North-Holland, Amsterdam, 1965.
285. H. Weidenmüller, *Rev. Mod. Phys.* **33**, 574 (1961).
286. R. J. Blin-Stoyle and S. C. K. Nair, *Adv. Phys.* **15**, 493 (1966).
287. *Beta-decay and Weak Interactions* (ed. B. Eman), D. Tadic, Zagreb, 1967.
288. Yu. G. Abov, P. A. Krupchitskij, and Yu. F. Oratovskij, *Yad. Fiz.* **1**, 479 (1965) (translation, *Soviet J. Nucl. Phys.* **1**, 341 (1965)). F. Boehm and E. Kankeleit, *Phys. Rev. Lett.* **14**, 312 (1965). P. Bock and F. Schopper, *Phys. Lett.* **16**, 284 (1965). E. Warning, F. Stecher-Rasmussen, W. Ratynski, and J. Kopecky, *Phys. Lett.* **25B**, 200 (1967). V. M. Lobashov, V. A. Nazarenko, L. F. Sayenko, L. M. Smotrizky, and G. I. Kharkovitch, *Proceedings of the International Conference on Nuclear Structure* (ed. J. Sanada), p. 443, Tokyo, 1967. E. Bodenstedt, L. Ley, H. O. Schlenz, and U. Wehmen, *Phys. Lett.* **29B**, 165 (1969).
289. B. S. Dzhelepov and L. N. Zyryanova, *The Effect of Atomic Electric Field on the Beta-decay*, AN USSR, Moscow, 1966.
290. A. H. Wapstra, C. J. Nijgh, and R. van Lieshout, *Nuclear Spectroscopy Tables*, North-Holland, Amsterdam, 1959.
291. L. N. Zyryanova, *The Unique Beta-decay*, AN USSR, Moscow, 1969.
292. P. G. Hansen, H. L. Nielsen, and K. Wilsky, *Izv. AN USSR*, phys. ser. **31**, 68 (1967). P. G. Hansen, H. L. Nielsen, K. Wilsky, and J. Treherne, *Phys. Lett.* **19**, 34 (1965). P. G. Hansen, H. L. Nielsen, K. Wilsky, and J. G. Guninghame, *Phys. Lett.* **24B**, 95 (1967).
293. V. G. Soloviev, *DAN USSR* **137**, 1350 (1961) (translation, *Soviet Phys. Dokl.* **6**, 346 (1962)).
294. P. G. Hansen, H. L. Nielsen, and K. Wilsky, *Nucl. Phys.* **54**, 657 (1964). H. Daniel, J. Huefner, Th. Lorentz, O. W. Schult, and U. Gruber, *Nucl. Phys.* **56**, 147 (1964).
295. K. Ya. Gromov, V. Hnatowitz, A. S. Danagulyan, A. T. Strigachev, and V. S. Sphinel, *Yad. Fiz.* **1**, 201 (1965) (translation, *Soviet J. Nucl. Phys.* **1**, 140 (1965)).
296. A. F. de Miranda and M. A. Preston, *Nucl. Phys.* **44**, 529 (1963). Gy. Bencze and A. Sandulescu, *Annalen Acad. Scient. Fennicae*, ser. A, VI; *Physica* **209**, 13 (1966).
297. J. Zylicz, P. G. Hansen, H. L. Nielsen, and K. Wilsky, *Ark. Fys.* **36**, 643 (1966).
298. B. N. Zakhariev, N. I. Pyatov, and V. I. Furman, *Zh. éksp. teor. Fiz.* **41**, 1669 (1961) (translation, *Soviet Phys. JETP* **14**, 1186 (1962)).
299. F. A. Gareev, S. P. Ivanova, and M. I. Cherney, *Yad. Fiz.* **9**, 308 (1969) (translation, *Soviet J. Nucl. Phys.* **9**, 182 (1969)).
300. Z. Bochnacki and S. Ogaza, *Nucl. Phys.* A, **102**, 529 (1967).
301. F. A. Gareev, B. N. Kalinkin, N. I. Pyatov, and M. I. Cherney, *Yad. Fiz.* **8**, 305 (1968) (translation, *Soviet J. Nucl. Phys.* **8**, 176 (1969)).
302. D. Bogdan, *Nucl. Phys.* **32**, 553 (1962); **48**, 273 (1963); **61**, 241 (1965). D. Bogdan, *Z. Phys.* **206**, 49 (1967). D. Bogdan and T. Badica, *Phys. Lett.* **12**, 40 (1964).
303. M. Sakai, *Nucl. Phys.* **33**, 96 (1962). M. Sakai and S. Yoshida, *Nucl. Phys.* **50**, 497 (1964).
304. L. Silverberg and A. Winther, *Phys. Lett.* **3**, 158 (1963).
305. Jun-Ichi Fujita, *Nucl. Phys.* **67**, 145 (1965).
306. Jun-Ichi Fujita, Y. Futami, and K. Ikeda, *Prog. Theor. Phys.* **38**, 107 (1967).
307. N. I. Pyatov, *Acta phys. pol.* **25**, 21 (1964). O. G. Gadetskij and N. I. Pyatov, *Izv. AN USSR*, phys. ser. **29**, 830 (1965).
308. G. Monsonego and R. Piepenbring, *Nucl. Phys.* **58**, 593 (1964). M. Vergnes and J. Rasmussen, *Nucl. Phys.* **62**, 233 (1965).
309. R. B. Begzhanov, V. M. Belenkij, and R. Kh. Safarov, *Numerical Values of the Reduced Electromagnetic Transition Probabilities in the Nilsson Model*, FAN Uz. SSR, Tashkent, 1967.
310. Ch. Perdrisat, *Rev. Mod. Phys.* **38**, 41 (1966).
311. H. Schulz and H. J. Wiebicke, preprint, JINR E4–4210 (1968), Dubna.
312. K. Löbner and S. Malmskog, *Nucl. Phys.* **80**, 505 (1966). S. Malmskog, *Ark. Fys.* **35**, 237 (1967).
313. N. I. Pyatov and M. I. Cherney, *Yad. Fiz.* **4**, 486 (1966) (translation, *Soviet J. Nucl. Phys.* **4**, 346 (1967)).
314. H. Ikegami and T. Udagawa, *Phys. Rev.* **133**, B1388 (1964).

315. E. E. BERLOVICH, *Izv. AN USSR*, phys. ser. **29,** 838 (1965); *Yad. Fiz.* **5,** 560 (1967) (translation, *Soviet J. Nucl. Phys.* **5,** 399 (1967)).
316. V. A. CHEPURNOV, *Yad. Fiz.* **7,** 1199 (1968) (translation, *Soviet J. Nucl. Phys.* **7,** 715 (1968)).
317. S. G. MALMSKOG, A. MARELIUS, and S WAHLBORN, *Nucl. Phys.* A, **103,** 481 (1967).
318. K. LÖBNER, P. GOUDSMIT, J. KONIJN, E. LINGEMAN, and P. POLAK, *Nucl. Phys.* A, **122,** 214 (1968).
319. K. E. LÖBNER, *Phys. Lett.* **26B,** 369 (1968).
320. A. ANDREEFF, R. KÄSTNER, P. MANFRASS, M. BONITZ, J. BONITZ, J. BORGGREEN, and N. J. SIGURD HANSEN, *Nucl. Phys.* A, **102,** 241 (1967).
321. G. BREIT, *Theory of Resonance Nuclear Reactions,* IL, Moscow, 1961.
322. W. TOBOCMAN, *The Theory of Direct Nuclear Reactions,* Oxford, 1961.
323. N. AUSTERN, *Selected Topics in Nuclear Theory,* p. 17, IAEA, Vienna, 1963.
324. I. S. SHAPIRO, *Theory of Direct Nuclear Reactions,* Gosatomizdat, Moscow, 1963. I. S. SHAPIRO, *Selected Topics in Nuclear Theory,* p. 85, IAEA, Vienna, 1963.
325. M. H. MACFARLANE and J. B. FRENCH, *Rev. Mod. Phys.* **32,** 567 (1960).
326. *Proceedings of the Conference on Direct Interactions and Nuclear Reaction Mechanisms* (ed. E. Clementel and C. Villi), Gordon & Breach, New York and London, 1962.
327. G. R. SATCHLER, *Ann. Phys.* **3,** 275 (1958); *Nucl. Phys.* **55,** 1 (1964).
328. S. I. DROZDOV, *Yad. Fiz.* **1,** 407 (1965) (translation, *Soviet J. Nucl. Phys.* **1,** 290 (1965)).
329. T. TAMURA, *Rev. Mod. Phys.* **37,** 679 (1965). *Proceedings of the International Conference on Nuclear Structure* (ed. J. Sanada), p. 288, Tokyo, 1967. *Nuclear Structure, Dubna Symposium, 1968,* p. 213, IAEA, Vienna, 1968.
330. K. A. GRIDNEV, L. V. KRASNOV, I. N. KUHTINA, V. K. LUKYANOV, V. I. NIKITINA, and V. I. FURMAN, preprint 2458, JINR, Dubna, 1965.
331. K. ALDER and A. WINTHER, *Mat.-fys. Medd. Dan. Vid. Selsk.* **32** (8) (1960). H. LÜTKEN and A. WINTHER, *Mat.-fys. Skr. Dan. Vid. Selsk.* **2** (6) (1964).
332. F. S. STEPHENS, R. M. DIAMOND, and I. PERLMAN, *Phys. Rev. Lett.* **3,** 435 (1959). B. ELBEK, M. KREGAR, and P. VEDELSBY, *Nucl. Phys.* **86,** 385 (1966). P. H. STELSON, *Nucl. Phys.* **32,** 652 (1962).
333. B. ELBEK, T. GROTDAL, K. NYBO, P. O. TJOM, and E. VEJE, *Proceedings of the International Conference on Nuclear Structure* (ed. J. Sanada), p. 180, Tokyo, 1967. P. O. TJOM and B. ELBEK, *Nucl. Phys.* A, **107,** 385 (1968).
334. V. B. BELYAEV and B. N. ZAKHARIEV, *Izv. AN USSR,* phys. ser. **25,** 1152 (1961).
335. V. G. SOLOVIEV, *Prog. Nuclear Physics* **10,** 239 (1968).
336. B. L. COHEN, *Phys. Rev.* **130,** 227 (1963).
337. V. K. LUKYANOV, *Izv. AN USSR,* phys. ser. **26,** 1096 (1962). V. K. LUKYANOV and I. PETKOV, *Yad. Fiz.* **6,** 988 (1967) (translation, *Soviet J. Nucl. Phys.* **6,** 720 (1968)).
338. P. J. IANO and N. AUSTERN, *Phys. Rev.* **151,** 853 (1966).
339. V. K. LUKYANOV and I. PETKOV, *Phys. Lett.* **28B,** 368 (1969).
340. H. SCHULZ and H. J. WIEBICKE, *Phys. Lett.* **29B,** 18 (1969).
341. H. WIEBICKE, O. DIMITRESCU, V. K. LUKYANOV, I. PETKOV, and G. SCHULTZ, Preprint P2–4687, JINR, Dubna, 1969.
342. P. O. TJOM and B. ELBEK, *Mat.-fys. Medd. Dan. Vid. Selsk.* **36** (8) (1967).
343. R. H. SIEMSSEN and J. R. ERSKINE, *Phys. Rev. Lett.* **19,** 90 (1967).
344. R. A. HARLAN and R. K. SHELINE, *Phys. Rev.* **168,** 1373 (1968).
345. F. A. RICKEY and R. K. SHELINE, *Phys. Rev. Lett.* **17,** 1057 (1966).
346. P. O. TJOM and B. ELBEK, *Mat.-fys. Medd. Dan. Vid. Selsk.* **37** (7) (1969).
347. W. N. SHELTON and R. SHELINE, *Phys. Rev.* **133,** B624 (1964).
348. D. G. BURKE and B. ELBEK, *Mat.-fys. Medd. Dan. Vid. Selsk.* **36** (6) (1967).
349. D. G. BURKE, D. E. NELSON, and C. W. REICH, *Nucl. Phys.* A, **124,** 683 (1969).
350. S. YOSHIDA, *Nucl. Phys.* **33,** 685 (1962).
351. M. EL NADI and H. SHERIF, *Nucl. Phys.* **55,** 489 (1964).
352. R. A. BROGLIA, C. RIEDEL, and B. SØRENSEN, *Nucl. Phys.* A, **107,** 1 (1968). R. A. BROGLIA, C. RIEDEL, B. SØRENSEN, and T. UDAGAWA, *Nucl. Phys.* A, **115,** 273 (1968).
353. O. NATHAN, *Structure of Complex Nuclei,* p. 158, Atomizdat, Moscow, 1966 (translation, Consultant Bureau, New York, 1964).
354. O. NATHAN, *Nuclear Structure, Dubna Symposium, 1968,* p. 191, IAEA, Vienna, 1968.
355. N. GLENDENNING, *Nucl. Phys.* **29,** 109 (1962); *Phys. Rev.* **137,** B102 (1965).
356. D. R. BÉS and Z. SZYMANSKI, *Nucl. Phys.* **28,** 42 (1961). Z. SZYMANSKI, *Nucl. Phys.* **28,** 63 (1961).
357. A. SOBICZEWSKI, *Nucl. Phys.* A, **93,** 501 (1967); A, **96,** 258 (1967).

358. E. MARSHALEK, L. W. PERSON, and R. K. SHELINE, *Rev. Mod. Phys.* **35,** 108 (1963).
359. S. DAS GUPTA and M. A. PRESTON, *Nucl. Phys.* **49,** 401 (1963). M. R. GUNYE, S. DAS GUPTA, and M. A. PRESTON, *Phys. Lett.* **13,** 246 (1964).
360. K. KUMAR and M. BARANGER, *Phys. Rev. Lett.* **12,** 73 (1964).
361. D. A. ARSENIEV, L. A. MALOV, V. V. PASHKEVICH, A. SOBICZEWSKI, and V. G. SOLOVIEV, *Yad. Fiz.* **8,** 883 (1968) (translation, *Soviet J. Nucl. Phys.* **8,** 514 (1969)).
362. D. A. ARSENIEV, A. SOBICZEWSKI, and V. G. SOLOVIEV, *Nucl. Phys.* A, **126,** 15 (1969).
363. D. A. ARSENIEV, A. SOBICZEWSKI, and V. G. SOLOVIEV, *Nucl. Phys.* A, **139,** 269 (1969).
364. K. DIETRICH, H. J. MANG, and J. PRADAL, *Z. Phys.* **190,** 357 (1966).
365. M. BARANGER and K. KUMAR, *Nucl. Phys.* **62,** 113 (1965); A, **110,** 529 (1968).
366. K. KUMAR and M. BARANGER, *Nucl. Phys.* A, **122,** 273 (1968).
367. Z. SZYMANSKI, *Nuclear Structure, Dubna Symposium, 1968,* p. 405, IAEA, Vienna, 1968.
368. K. KUMAR, *Nuclear Structure, Dubna Symposium, 1968,* p. 419, IAEA, Vienna, 1968.
369. U. MOSEL and W. GREINER, *Z. Phys.* **217,** 256 (1968); **222,** 261 (1969).
370. E. BALBUTZEV and I. N. MIKHAILOV, *Izv. AN USSR,* phys. ser. **30,** 1118 (1966).
371. V. M. STRUTINSKY, *Yad. Fiz.* **3,** 614 (1966) (translation, *Soviet J. Nucl. Phys.* **3,** 425 (1966)); *Nucl. Phys.* A, **95,** 420 (1967).
372. A. SOBICZEWSKI, F. A. GAREEV, and B. N. KALINKIN, *Phys. Lett.* **22,** 500 (1966). H. MELDNER, *Phys. Rev.* **178,** 1818 (1969); *Ark. Fys.* **36,** 593 (1967).
373. S. G. NILSSON, J. R. NIX, A. SOBICZEWSKI, Z. SZYMANSKI, S. WYCECH, C. GUSTAFSON, and P. MÖLLER, *Nucl. Phys.* A, **115,** 545 (1968). S. G. NILSSON, S. G. THOMPSON, and C. F. TSANG, *Phys. Lett.* **28 B,** 458 (1969).
374. S. G. NILSSON, *Proceedings of the International Conference on the Properties of Nuclear States,* p. 149, Montreal, Les Presses de L'Université de Montréal, 1969.
375. YU. A. MUZYCHKA, V. V. PASHKEVICH, and V. M. STRUTINSKY, *Yad. Fiz.* **8,** 716 (1968) (translation, *Soviet J. Nucl. Phys.* **8,** 417 (1969)).
376. S. G. NILSSON, CHIN FU TSANG, A. SOBICZEWSKI, Z. SZYMANSKI, S. WYCECH, C. GUSTAFSON, I. L. LAMM, P. MÖLLER, and B. NILSSON, *Nucl. Phys.* A, **131,** 1 (1969).
377. G. N. FLEROV, *Proc. Panel, Future of Nuclear Structure Studies,* p. 11, IAEA, Vienna, 1969.
378. G. N. FLEROV, *Proceedings of the International Conference on the Properties of Nuclear States,* p. 175, Montreal, Les Presses de L'Université de Montréal, 1969.
379. G. J. J. WESOLOWSKI, W. JOHNS, R. JEWELL, and F. GUY, *Phys. Lett.* **28B,** 544 (1969). O. OTGONSUREN, V. P. PERELYGIN, and G. N. FLEROV, *DAN USSR* **189,** 1200 (1969).
380. P. VOGEL, *Phys. Lett.* **25B,** 65 (1967); *Nucl. Phys.* A, **112,** 583 (1968).
381. P. O. FRÖMAN, *Mat. Fys. Skr. Dan. Vid. Selsk* **1,** (3) (1957). L. L. GOLDIN, G. I. NOVIKOVA, and K. A. TER-MARTYROSIAN, *Zh. éksp. teor. Fiz.* **36,** 502 (1959) (translation, *Soviet Phys. JETP* **36,** 356 (1959)).
382. D. L. HENDRIE, N. K. GLENDENNING, B. G. HARVEY, O. N. JARVIS, H. H. DUHM, J. SAUDINOS, and I. MAHONEY, *Phys, Lett.* **26B,** 127 (1968).
383. P. MÖLLER, B. NILSSON, S. G. NILSSON, A. SOBICZEWSKI, Z. SZYMANSKI, and S. WYCECH, *Phys. Lett.* **26B,** 418 (1968). B. NILSSON, *Nucl. Phys.* A, **129,** 445 (1969).
384. F. A. GAREEV, C. P. IVANOVA, and N. YU. SHIRIKOVA, preprint P4–4259, JINR, Dubna (1969). F. A. GAREEV, S. P. IVANOVA, and V. V. PASHKEVICH, *Yad. Fiz.* **11,** 1200 (1970) (translation, *Soviet J. Nucl. Phys.* **11,** 667 (1970).
385. M. HASSAN, Z. SKLADANOWSKI, and Z. SZYMANSKI, *Nucl. Phys.* **78,** 593 (1966).
386. V. G. SOLOVIEV, *Phys. Lett.* **21,** 311 (1966).
387. L. A. MALOV, C. M. POLIKANOV, and V. G. SOLOVIEV, *Yad. Fiz.* **4,** 528 (1966) (translation, *Soviet J. Nucl. Phys.* **4,** 376 (1967).
388. B. T. GEILIKMAN, *Proceedings of the International Conference on Nuclear Structure,* p. 874, North-Holland, Amsterdam, 1960.
389. V. M. STRUTINSKY, *Nucl. Phys.* A, **122,** 1 (1968).
390. V. V. PASHKEVICH, *Nucl. Phys.* A, **133,** 400 (1969).
391. J. E. LYNN, *Nuclear Structure, Dubna Symposium, 1968,* p. 463, IAEA, Vienna, 1968.
392. V. M. STRUTINSKY and S. BJØRNHOLM, *Nuclear Structure, Dubna Symposium 1968,* p. 431, IAEA, Vienna, 1968.
393. S. P. KAPITZA, N. S. RABOTNOV, G. N. SMIENKIN, A. S. SOLDATOV, L. M. USATCHEV, and YU. M. TZIPENYUK, *Zh. éksp. teor. Fiz. Lett.* **9,** 128 (1969) (translation *JETP Letters.* **9,** 73 (1969)). D. L. SHPAK, D. N. STEPANOV, G. N. SMIENKIN, *Yad. Fiz.* **9,** 970 (1969) (translation, *Soviet J. Nucl. Phys.* **9,** 551 (1969). C. D. JAMES and E. R. RAE, *Nucl. Phys.* A, **118,** 313 (1968).

394. S. M. Polikanov, V. A. Druin, V. A. Karnaukhov, V. L. Mikheev, A. A. Pleve, N. K. Skobelev, V. G. Subbotin, G. M. Ter-Akopian, and V. A. Fomichev, *Zh. éksp. teor. Fiz.* **42,** 1462 (1962) (translation, *Soviet Phys. JETP* **15,** 1016 (1962)). G. N. Flerov, S. M. Polikanov, K. A. Gavrilov, V. L. Mikheev, V. P. Perelygin, and A. A. Pleve, *Zh. éksp. teor. Fiz.* **45,** 1396 (1963) (translation, *Soviet Phys. JETP* **18,** 964 (1964)).
395. S. M. Polikanov, *Usp. fiz. Nauk* **94,** 44 (1968) (translation, *Soviet Phys. Usp.* **11,** 22 (1968)).
396. S. M. Polikanov, *Nuclear Structure, Dubna Symposium, 1968*, p. 449, IAEA, Vienna, 1968.
397. A. M. Baldin, S. F. Semenko, and B. A. Tulupov, *Yad. Fiz.* **8,** 326 (1968) (translation, *Soviet J. Nucl. Phys.* **8,** 187 (1969)).
398. J. R. Nix and G. E. Walker, *Nucl. Phys.* A, **132,** 60 (1969).
399. K. Alexander, W. Neubert, H. Rotter, S. Chojnacki, Ch. Droste, and T. Morek, *Nucl. Phys.* A, **133,** 77 (1969).
400. Ch. Droste, W. Neubert, T. Morek, S. Chojnacki, Z. Wilhelmi, and K. Alexander, preprint P6–4539, JINR, Dubna (1969).
401. W. Neubert, Ch. Droste, T. Morek, S. Chojnacki, K. Alexander, and Z. Wilhelmi, preprint P6–4591, JINR, Dubna (1969). Ch. Droste, W. Neubert, T. Morek, Z. Wilhelmi, S. Chojnacki, and K. Alexander, preprint P6–4592, JINR, Dubna (1969).
402. A. Bohr and B. Mottelson, *Mat.-fys. Medd. Dan. Vid. Selsk.* **30** (1) (1955).
403. J. Griffin and M. Rich, *Phys Rev.* **118,** 850 (1960).
404. Liu Yuan, *Izv. AN USSR*, phys. ser. **28,** 18 (1964).
405. R. Kh. Safarov, *Izv. AN Uz. SSR*, math. phys. ser. **1,** 86 (1965).
406. Yu. K. Khokhlov, *Zh. éksp. teor. Fiz.* **37,** 1136 (1959) (translation, *Soviet Phys. JETP* **10,** 808 (1960)).
407. I. N. Mikhailov, *DAN USSR* **154,** 68 (1964) (translation, *Soviet Phys. Dokl.* **9,** 42 (1964)).
408. B. R. Mottelson and J. G. Valatin, *Phys. Rev. Lett.* **5,** 511 (1960).
409. Yu. T. Grin, *Zh. éksp. teor. Fiz.* **41,** 445 (1961) (translation, *Soviet Phys. JETP* **14,** 320 (1962)). Yu. T. Grin and I. M. Pavlitchenkov, *Zh. éksp. teor. Fiz.* **43,** 465 (1962) (translation, *Soviet Phys. JETP* **16,** 333 (1963)). Yu. T. Grin and A. T. Larkin, *Yad. Fiz.* **2,** 40 (1965) (translation, *Soviet J. Nucl. Phys.* **2,** 27 (1966)).
410. K. Y. Chan and J. G. Valatin, *Nucl. Phys.* **82,** 222 (1966). K. Y. Chan, *Nucl. Phys.* **85,** 261 (1966).
411. K. Sugawara, *Prog. Theor. Phys.* **35,** 44 (1966).
412. A. Faessler, W. Greiner, and R. K. Sheline, *Nucl. Phys.* **62,** 241 (1965). T. Udagawa and R. K. Sheline, *Phys. Rev.* **147,** 671 (1966).
413. Z. Galasiewicz, *Nucl. Phys.* **32,** 140 (1962).
414. J. Krumlinde, *Nucl. Phys.* A, **121,** 306 (1968).
415. E. Marshalek, *Phys. Rev.* **158,** 993 (1967).
416. N. I. Pyatov and M. I. Cherney, preprint E4–4523, P4–4533, JINR, Dubna, 1969. B. L. Birbrair, *Izv. AN USSR*, phys. ser. **33,** 1343 (1969).
417. K. Ikeda, M. Kobayasi, T. Marumori, T. Shiozaki, and S. Takagi, *Prog. Theor. Phys.* **22,** 663 (1959). M. Kobayasi and T. Marumori, *Prog. Theor. Phys.* **23,** 387 (1960). T. Marumori, *Prog. Theor. Phys.* **24,** 331 (1960).
418. S. T. Belyaev, *Zh. éksp. teor. Fiz.* **39,** 1387 (1960) (translation, *Soviet Phys. JETP* **12,** 968 (1961)). S. T. Belyaev, *Selected Topics in Nuclear Theory*, p. 291, IAEA, Vienna, 1963.
419. D. J. Thouless, *Nucl. Phys.* **22,** 78 (1961).
420. T. Tamura and T. Udagawa, *Prog. Theor. Phys.* **25,** 1051 (1961); **26,** 947 (1961).
421. S. Yoshida, *Nucl. Phys.* **38,** 380 (1962).
422. A. Arvieu and M. Veneroni, *Compt. rend.* **250,** 992, 2155 (1960) A. Arvieu, E. Salusti, and M. Veneroni, *Phys. Lett.* **8,** 334 (1964).
423. B. L. Birbrair, K. I. Erokhina, and I. Kh. Lemberg, *Izv. AN USSR*, phys. ser. **27,** 150 (1963).
424. D. Bés and Z. Szymanski, *Nuovo Cim.* **26,** 787 (1962).
425. D. F. Zaretskij and M. G. Urin, *Zh. éksp. teor. Fiz.* **41,** 898 (1961); **42,** 304 (1962) (translation, *Soviet Phys. JETP* **14,** 641 (1962); **15,** 211 (1962).
426. V. G. Soloviev and P. Vogel, *Phys. Lett.* **6,** 126 (1963).
427. E. R. Marshalek and J. O. Rasmussen, *Nucl. Phys.* **43,** 438 (1963).
428. V. G. Soloviev, P. Vogel, and A. A. Korneichuk, *DAN USSR* **154,** 72 (1963) (translation, *Soviet Phys. Dokl.* **9,** 45 (1964)); *Izv. AN USSR*, phys. ser. **28,** 1599 (1964).
429. Liu Yuan, V. G. Soloviev, and A. A. Korneichuk, *Zh. éksp. teor. Fiz.* **47,** 252 (1964) (translation, *Soviet Phys. JETP* **20,** 169 (1965)). V. G. Soloviev, *DAN USSR* **159,** 310 (1964) (translation, *Soviet Phys. Dokl.* **9,** 993 (1965)).

430. D. R. Bés, *Nucl. Phys.* **49,** 544 (1963).
431. V. G. Soloviev, *Nucl. Phys.* **69,** 1 (1965).
432. D. R. Bés, P. Federman, E. Maqueda, and A. Zuker, *Nucl. Phys.* **65,** 1 (1965).
433. L. A. Malov, V. G. Soloviev, and P. Vogel, *Phys. Lett.* **22,** 441 (1966).
434. L. S. Kisslinger and R. A. Sørensen, *Rev. Mod. Phys.* **35,** 853 (1963).
435. D. R. Bés and R. A. Sørensen, *Adv. Nucl. Phys.* **2,** 129 (1969).
436. A. A. Vlasov, *Theory of Many-particle Systems,* GITTL, Moscow, 1950.
437. R. V. Jolos and V. G. Soloviev, *Phys. Elem. Part. Atom Nucleus* **1,** 365 (1971).
438. J. Högaasen-Feldman, *Nucl. Phys.* **28,** 258 (1961).
439. A. Bohr, *Nuclear Structure, Dubna Symposium, 1968,* p. 179, IAEA, Vienna, 1968.
440. V. P. Krainov and V. V. Malov, *Yad. Fiz.* **6,** 252 (1967) (translation, *Soviet J. Nucl. Phys.* **6,** 183 (1968)).
441. S. P. Kamerdzhiev, *Yad. Fiz.* **9,** 324 (1969) (translation, *Soviet J. Nucl. Phys.* **9,** 190 (1969)).
442. P. W. Anderson, *Phys. Rev.* **112,** 1900 (1958).
443. D. Bohm and D. Pines, *Phys. Rev.* **92,** 609 (1953). K. Sawada, *Phys. Rev.* **106,** 372 (1957).
444. I. E. Tamm, *J. Phys. (USSR)* **9,** 449 (1945).
445. S. M. Dancoff, *Phys. Rev.* **78,** 382 (1950).
446. V. A. Fock, *Sov. Phys.* **6,** 425 (1934).
447. J. P. Elliott and D. H. Flowers, *Proc. R. Soc. London.* A, **24,** (2) 57 (1957).
448. G. Brown and M. Bolsterli, *Nucl. Phys.* **29,** 89 (1962).
449. K. Hara, *Prog. Theor. Phys.* **32,** 88 (1964).
450. K. Ikeda, T. Udagawa, and H. Yamaura, *Prog. Theor. Phys.* **33,** 22 (1965).
451. P. Vogel, *Izv. AN USSR,* phys. ser. **30,** 1095 (1966).
452. J. Schwinger, *Proc. Natn. Acad. Sci. USA* **37,** 452, 455 (1951). F. Dyson, *Phys. Rev.* **75,** 486 (1949).
453. N. N. Bogolyubov, *DAN USSR,* **99,** 226 (1964).
454. M. K. Polivanov, *DAN USSR* **100,** 1061 (1955); V. G. Soloviev, *DAN USSR* **111,** 578 (1956).
455. B. L. Bonch-Bruevich, *Zh. eksp. teor. Fiz.* **38,** 21 (1955); **20,** 342 (1956) (translation, *Soviet Phys. JETP* **1,** 169 (1955); **3,** 278 (1956)).
456. N. N. Bogolyubov and S. V. Tyablikov, *DAN USSR* **126,** 53 (1959) (translation, *Soviet Phys. Dokl.* **4,** 589 (1960).
457. V. M. Galitskij and A. B. Migdal, *Zh. eksp. teor. Fiz.* **34,** 139 (1958) (translation, *Soviet Phys. JETP* **7,** 96 (1958).
458. D. N. Zubarev, *Usp. fiz. Nauk* **71,** 71 (1960) (translation, *Soviet Phys. Usp.* **3,** 320 (1961)).
459. V. L. Bonch-Bruevich and S. V. Tyablikov, *Green's Function Method in Statistical Physics,* Fizmatgiz, Moscow, 1961 (translation, North-Holland, Amsterdam, 1962).
460. A. A. Abrikosov, L. P. Gorkov, and I. E. Dyzaloshinskij, *Methods of Quantum Field Theory in Statistical Physics,* Fizmatgiz, Moscow, 1962 (translation, Prentice-Hall, New Jersey, 1963).
461. D. Pines, *Many-body Problems,* Benjamin, New York, 1961.
462. S. V. Tyablikov, *Methods of Quantum Theory of Magnetism,* Nauka, Moscow, 1965.
463. L. P. Gorkov, *Zh. éksp. teor. Fiz.* **34,** 735 (1958) (translation, *Soviet Phys. JETP* **4,** 505 (1958)).
464. J. O. Rasmussen, *Nuclear Structure, Dubna Symposium, 1968,* p. 169, IAEA, Vienna 1958.
465. D. R. Bés and R. A. Broglia, *Nucl. Phys.* **80,** 289 (1966).
466. A. A. Kuliev and N. I. Pyatov, *Izv. AN USSR,* phys. ser. **35,** 831 (1968).
467. S. T. Belyaev, *Yad. Fiz.* **4,** 936 (1966) (translation, *Soviet J. Nucl. Phys.* **4,** 671 (1967). S. T. Belyaev and B. A. Rumyantsev, *Proceedings of the International Conference on Nuclear Structure* (ed. J. Sanada), p. 125, Tokyo, 1967.
468. S. T. Belyaev, *Nuclear Structure, Dubna Symposium, 1968,* p. 155, IAEA Vienna, 1968.
469. O. Mikoshiba, R. K. Sheline, T. Udagava, and S. Yoshida, *Nucl. Phys.* A, **101,** 202 (1967).
470. G. Ripka and R. Padjen, *Nucl. Phys.* A, **132,** 489 (1969).
471. B. Sørensen, *Nucl. Phys.* **A, 93,** 1 (1967).
472. R. V. Jolos, *Phys. Lett.* **30B,** 390 (1969).
473. K. M. Zheleznova, A. A. Korneichuk, V. G. Soloviev, P. Vogel, and G. Jungklaussen, preprint D-2157, JINR, Dubna, 1965.
474. A. A. Korneichuk, L. A. Malov, S. I. Fedotov, and G. Schultz, *Yad. fiz.* **9,** 750 (1969) (translation, *Soviet J. Nucl. Phys.* **9,** 436 (1969)).
475. L. A. Malov, V. G. Soloviev, and U. M. Finer, *DAN USSR* **186,** 299 (1969) (translation, *Soviet Phys. Dokl.* **14,** 461 (1969)); *Izv. AN USSR* phys. ser. **33,** 1244 (1969).
476. A. L. Komov, L. A. Malov, and V. G. Soloviev, preprint P4–5126, JINR, Dubna, 1970.

477. L. A. Malov, V. G. Soloviev, and S. I. Fedotov, *DAN USSR* **189**, 987 (1969) (translation, *Soviet Phys. Dokl.* **14**, 1186 (1970)).
478. J. H. Bjerregaard, O. Hansen, O. Nathan, and S. Hinds, *Nucl Phys.* **86**, 145 (1966).
479. R. Graetzer, G. Hagemann, K. Hagemann, and B. Elbek, *Nucl. Phys.* **76**, 1 (1966).
480. A. A. Abdurazakov, J. Vrzal, K. Ya. Gromov, Zh. Zhelev, V. G. Kalinnikov, J. Liptak, Li-San-Gun, F. N. Mukhtasimov, M. K. Nazarov, and J. Urbanec, preprint P6–3464, JINR, Dubna, 1967.
481. N. A. Bonch-Osmolovskaya, J. Vrzal, E. P. Grigoriev, J. Liptak, G. Pfrepper, J. Urbanec, and D. Khristov, *Izv. AN USSR*, phys. ser. **32,** 98 (1968).
482. J. Vrzal, K. Ya. Gromov, J. Liptak, F. Molnar, V. A. Morozov, J. Urbanec, and V. G. Chumin, *Izv. AN USSR*, phys. ser. **31,** 604 (1967).
483. N. Kaffrell and G. Herrmann, *Nucl. Phys.* A, **118,** 78 (1968).
484. J. Zylicz, M. Jørgensen, O. B. Nielsen, and O. Skilbreid, *Nucl. Phys.* **81,** 88 (1966).
485. C. Gunther and D. Parsignaul, *Phys. Rev.* **153,** 1297 (1967).
486. E. Veje, B. Elbek, B. Herskind, and M. Olesen, *Nucl. Phys.* A, **109**, **489** (1968).
487. L. V. Groshev, A. M. Demidov, V. A. Ivanov, V. N. Lucenko, V. I. Pelekhov, and N. Shadiev, *Izv. AN USSR*, phys. ser. **29**, 772 (1965).
488. K. Kemp and G. Hagemann, *Nucl. Phys.* A, **97,** 666 (1967).
489. O. Ottenson and R. Helmer, *Phys. Rev.* **164,** 1485 (1967).
490. L. V. Groshev, A. M. Demidov, V. A. Ivanov, V. N. Lucenko, and V. I. Pelekhov, *Izv. AN USSR*, phys. ser. **28,** 1244 (1964).
491. T. Grotdal, K. Nybø, T. Thorsteinsen, and B. Elbek, *Nucl. Phys.* A. **110,** 385 (1968). R. I. Keddy, Y. Yoshisawa, B. Elbek, B. Herskind, and M. Olesen, *Nucl. Phys.* A, **113**, 676 (1968).
492. S. Bjørnholm, J. Dubois, and B. Elbek, *Nucl. Phys.* A, **118,** 421 (1968). S. Bjørnholm, J. Borggreen, D. Davies, N. I. Hansen, I. Pedersen, and H. L. Nielsen, *Nucl. Phys.* A, **118**, 261 (1968).
493. F. S. Stephens, F. Asaro, S. Fried, and I. Perlman, *Phys. Rev. Lett.* **15**, 420 (1965). C. M. Lederer, I. M. Jaklevic and S. G. Prussin, *Nucl. Phys.* A, **135,** 36 (1969). C. J. Orth, *Phys. Rev.* **148,** 1226 (1966).
494. P. T. Prokofiev and G. L. Rezvaya, *Izv. AN USSR*, phys. ser. **33,** 1655 (1969).
495. B. S. Dzhelepov, V. E. Ter-Kersesyantz, and S. A. Shestopalova, *Izv. AN USSR*, phys. ser. **31,** 1633 (1967); **33,** 2 (1969). N. A. Bonch-Osmolovskaya, Kh. Ballynd, A. Zylinski, A. Plokhocki, and Z. Preibisz, preprint P6–4773, JINR, Dubna, 1969.
496. P. Vogel, *Nuclear Structure, Dubna Symposium, 1968*, p. 59, IAEA, Vienna, 1968. K. Neegård and P. Vogel, *Phys. Lett.* **30B,** 75 (1969).
497. P. Vogel, *Yad. Fiz.* **1,** 752 (1965) (translation, *Soviet J. Nucl. Phys.* **1,** 538 (1965)).
498. L. A. Borisoglebskij, *Usp. fiz. Nauk* **81,** 271 (1963) (translation, *Soviet Phys. Usp.* **6,** 715 (1964)).
499. E. L. Church and J. Weneser, *Phys. Rev.* **103,** 1035 (1956).
500. J. O. Rasmussen, *Nucl. Phys.* **19,** 85 (1960).
501. C. Günther, H. Ryde, and K. Krien, *Nucl. Phys.* **A, 122,** 401 (1968). E. P. Grigoriev, A. V. Zolotavin, V. O. Sergeev, and M. I. Sovtzov, *Meeting on Nuclear Spectroscopy and Nuclear Theory*, p. 88, JINR, Dubna, 1969.
502. A. Sandulescu and O. Dumitrescu, *Phys. Lett.* **19,** 404 (1964). A. Sandulescu, *Nucl. Phys.* **48,** 435 (1963). M. I. Cristu, O. Dumitrescu, N. I. Pyatov, and A. Sandulescu, *Nucl. Phys.* A, **130,** 31 (1969).
503. V. G. Soloviev, D. Bogdan, S. Salageanu, and N. Decui, *Nuovo Cim.* **51B,** 66 (1967).
504. A. A. Kuliev and N. I. Pyatov, *Nucl. Phys.* A, **106,** 689 (1968).
505. J. Kern, O. Mikoshiba, R. K. Sheline, and S. Yoshida, *Nucl. Phys.* A, **104,** 642 (1967).
506. V. G. Soloviev, *Phys. Lett.* **16,** 308 (1956).
507. V. G. Soloviev and P. Vogel, *Nucl. Phys.* A, **92,** 449 (1967). V. G. Soloviev and P. Vogel, *DAN USSR* **171,** 69 (1966) (translation, *Soviet Phys. Dokl.* **11,** 940 (1967)).
508. L. A. Malov and V. G. Soloviev, *Yad. Fiz.* **5,** 566 (1967) (translation, *Soviet J. Nucl. Phys.* **5,** 503 (1967)).
509. V. G. Soloviev, P. Vogel, and G. Jungklaussen, *Izv. AN USSR*, phys. ser. **31,** 518 (1967).
510. L. A. Malov, V. G. Soloviev, and S. I. Fedotov, *Izv. AN USSR*, phys. ser. **35**, 747 (1971).
511. D. R. Bés and Cho Yi-chung, *Nucl. Phys.* **86,** 581 (1966).
512. V. G. Soloviev, *Nuclear Structure, Dubna Symposium, 1968,* p. 101, IAEA, Vienna 1968.
513. K. Ya. Gromov, B. S. Dzhelepov, V. Zvolska, J. Zvolsky, N. A. Lebedev, and J. Urbanec, *Izv. AN USSR*, phys. ser. **26,** 1023 (1962). Yu. Gangrskij and I. Kh. Lemberg, *Izv. AN USSR*, phys. ser. **26,** 1027 (1962).
514. O. W. Schult, B. P. Maier, and U. Gruber, *Z. Phys.* **182,** 171 (1964).

515. R. M. DIAMOND, B. ELBEK, and F. S. STEPHENS, *Nucl. Phys.* **43,** 560 (1963).
516. K. M. BISGARD, L. J. KIELSEN, E. STABELL, and P. OSTERGARD, *Nucl. Phys.* **71,** 192 (1965). K. M. BISGARD and E. VEJE, *Nucl. Phys.* A, **103,** 545 (1967). J. O. NEWTON, *Nucl. Phys.* A, **108,** 353 (1968).
517. B. HARMATZ and T. H. HANDLEY, *Nucl. Phys.* A, **121,** 481 (1968).
518. G. MARKUS, W. MICHAELIS, H. SCHMIDT, and C. WETTKAMP, *Z. Phys.* **206,** 84 (1967). B. C. DUTTA, T. VON EGIDY, TH. W. ELZE, and W. KAISER, *Z. Phys.* **207,** 153 (1967). W. MICHAELIS, F. WELLER, U. FANGER, R. GAETA, G. MARKUS, H. OTTMAR, and H. SCHMIDT, *Nucl. Phys.* A, **143,** 225 (1970).
519. G. MARGUIER and R. CHERY, *Rapport Annuel Inst. de Physique Nucleare*, p. 20, 1969.
520. L. D. MCISAAC, R. G. HELMER, and C. W. REICH, *Nucl. Phys.* A, **132,** 28 (1969).
521. A. TVETER and B. HERSKIND, *Nucl. Phys.* A, **134,** 599 (1669).
522. K. YA. GROMOV, ZH. T. ZHELEV, V. ZVOLSKA, and V. G. KALINNIKOV, *Yad. Fiz.* **2,** 783 (1965) (translation, *Soviet J. Nucl. Phys.* **2,** 559 (1966)).
523. S. A. BARANOV, V. M. SHATINSKY, and V. M. KULAKOV, *Yad. Fiz.* **10,** 889 (1969) (translation, *Sov. J. Nucl. Phys.* **10,** 513 (1970)).
524. V. HNATOWITZ and K. YA. GROMOV, *Yad. Fiz.* **3,** 8 (1966) (translation, *Sov. J. Nucl. Phys.* **3,** 5 (1966)).
525. D. D. BOGDANOV, SH. DAROCI, V. A. KARNAUKHOV, L. A. PETROV, and G. M. TER-AKOPYAN, *Yad. Fiz.* **6,** 893 (1967) (translation, *Sov. J. Nucl. Phys.* **6,** 650 (1968)). V. A. KARNAUKHOV, *Yad. Fiz.* **10,** 450 (1969) (translation, *Sov. J. Nucl. Phys.* **10,** 257 (1970)).
526. P. G. HANSEN, *Proceedings of the International Conference on Properties of Nuclear States*, p. 189, Montreal, Les Presses de L'Université de Montréal, 1969.
527. R. V. JOLOS, V. G. SOLOVIEV, and K. M. ZHELEZNOVA, *Phys. Lett.* **25B,** 393 (1967). R. V. JOLOS, U. M. FINER, V. G. SOLOVIEV, and K. M. ZHELEZNOVA, *Phys. Lett.* **27B,** 614 (1968).
528. R. A. MEYER, *Phys. Rev.* **170,** 1089 (1968).
529. L. L. RIEDINGER, N. R. JOHNSON, and J. H. HAMILTON, *Phys. Rev.* **179,** 1214 (1969).
530. V. G. SOLOVIEV, *Yad. Fiz.* **10,** 296 (1969) (translation, *Sov. J. Nucl. Phys.* **10,** 171 (1969)).
531. N. KAFFREL and G. HERRMANN, *Phys. Lett.* **34B,** 46 (1971).
532. B. S. DZHELEPOV and S. A. SHESTOPALOVA, *Nuclear Structure, Dubna Symposium, 1958*, p. 39, IAEA, Vienna, 1968.
533. V. G. SOLOVIEV, *Phys. Lett.* **21,** 320 (1966).
534. P. T. PROKOFIEV and L. I. SIMONOVA, *Yad. Fiz.* **5,** 697 (1967) (translation, *Soviet. J. Nucl. Phys.* **5,** 493 (1967)).
535. R. K. SHELINE, C. WATSON, B. P. MAIER, U. GRUBER, R. H. KOCH, O. SCHULT, H. A. MOTZ, E. T. JURNEY, G. L. STRUBLE, T. EGIDY, TH. ELZE, and E. BIEBER, *Phys. Rev.* **143,** 857 (1966). K. FRANSSON, H. RYDE, T. HERSKIND, G. D. SYMONS, and A. TVETER, *Nucl. Phys.* A, **106,** 369 (1968).
536. J. TREHERNE, J. VANHORENBEECK, and J. VALENTIN, *Nucl. Phys.* A, **131,** 193 (1969).
537. T. TAMURA and T. UDAGAWA, *Nucl. Phys.* **53,** 33 (1964).
538. R. A. SØRENSEN, *Phys. Rev.* **133,** B281 (1964).
539. C. J. VEJE, *Mat. Fys. Medd. Dan. Vid. Selsk.* **35,** (1) (1966).
540. R. A. UHER and R. A. SØRENSEN, *Nucl. Phys.* **86,** 1 (1966).
541. S. P. KAMERDZHIEV, *Yad. Fiz.* **7,** 706 (1968) (translation, *Sov. J. Nucl. Phys.* **7,** 430 (1968)).
542. N. SHADIEV, preprint IAE-1167, 196.
543. Y. FUTAMI and M. SAKAI, *Nucl. Phys.* A, **92,** 91 (1967).
544. N. M. ANTONOVA, E. P. GRIGORIEV, G. S. KATYKHIN, L. F. PROTASOVA, J. VRZAL, J. LIPTAK, and J. URBANEC, *Izv. AN URSS*, phys. ser. **33,** 27 (1969).
545. J. LIPTAK, J. VRZAL, E. P. GRIGORIEV, G. S. KATYKHIN, and J. URBANEC, *Czech. J. Phys.* B, **19,** 1127 (1969).
546. S. T. BELYAEV and V. G. ZELEVINSKY, *Nucl. Phys.* **39,** 582 (1962). S. T. BELYAEV and V. G. ZELEVINSKY, *Izv. AN USSR*, phys. ser. **28,** 127 (1964).
547. T. MARUMORI, M. YAMAMURA, and A. TOKUNAGA, *Prog. Theor. Phys.* **31,** 1009 (1964).
548. A. K. KERMAN and A. KLEIN, *Phys. Rev.* **132,** 1326 (1963); **138,** B1323 (1965). G. DO DANG and A. KLEIN, *Phys. Rev.* **133,** B257 (1964).
549. A. TOKUNAGA, *Prog. Theor. Phys.* **37,** 315 (1967). M. YAMAMURA, A. TOKUNAGA, and T. MARUMORI, *Prog. Theor. Phys.* **37,** 336 (1967).
550. T. TAMURA and T. UDAGAWA, *Phys. Rev.* **150,** 783 (1966).
551. E. B. BALBUCEV and R. V. JOLOS, *Yad. Fiz.* **7,** 788 (1968) (translation, *Soviet J. Nucl. Phys.* **7,** 481 (1968)).
552. T. KISHIMOTO and K. IKEDA, *Prog. Theor. Phys.* **40,** 499 (1968).
553. D. R. BÉS and G. G. DUSSEL, *Nucl. Phys.* A, **135,** 1, 25 (1969).

554. J. DE BOER, *Proceedings of the International Conference on Nuclear Structure* (ed. J. Sanada), p. 199, Tokyo, 1967.
555. G. ALAGA, *Proceedings of the International School of Physics "Enrico Fermi", Course XL*, p. 28, 1968, Academic Press, New York, 1969.
556. G. DO DANG, R. DREIZLER, A. KLEIN, and C. S. WU, *Phys. Rev. Lett.* **17,** 709 (1966).
557. R. A. BROGLIA, B. SØRENSEN, and T. UDAGAWA, *Phys. Lett.* **26B,** 250 (1968).
558. A. DE SHALIT, *Selected Topics in Nuclear Theory*, p. 209, IAEA, Vienna, 1963.
559. N. STEIN, *Proceedings of the International Conference on Properties of Nuclear States*, p. 337, Montreal, Les Presses de l'Université de Montréal, 1969.
560. I. HAMAMOTO, *Nucl. Phys.* A, **126,** 545 (1969).
561. S. GORODETZKY, F. A. BECK, T. BYRSKI, and A. KNIPPER, *Nucl. Phys.* A, **117,** 208 (1968).
562. M. H. MACFARLANE, *Proceedings of the International Conference on Properties of Nuclear States*, p. 385, Montreal, Les Presses de l'Université de Montréal, 1969.
563. D. G. ALKHAZOV, K. I. EROKHINA, and I. KH. LEMBERG, *Izv. AN USSR*, phys. ser. **28,** 1667 (1964). J. MCDONALD and D. PORTER, *Nucl. Phys.* A, **109,** 529 (1968). J. L. BLACK, W. GRUHLE, and P. W. HEIKKINEN, *Phys. Lett.* **22,** 598 (1967). G. BERZINS, M. E. BUNKER, and J. W. STARNER, *Nucl. Phys.* A, **126,** 273 (1969).
564. A. G. BLAIR, *Phys. Lett.* **9,** 37 (1964). M. A. ESWARAN, H. E. GOVE, A. E. LITHERLAND, and C. BROUDE, *Nucl. Phys.* **66,** 401 (1965).
565. S. T. BELYAEV and V. G. ZELEVINSKY, *Yad. Fiz.* **1,** 13 (1965) (translation, *Soviet J. Nucl. Phys.* **1,** 16 (1965)).
566. G. DO DANG, *Nucl. Phys.* **62,** 153 (1965). M. YAMAMURA, *Prog. Theor. Phys.* **33,** 199 (1965).
567. W. BERES, *Nucl. Phys.* **68,** 49 (1965); **75,** 255 (1966). P. DOLESCHALL and I. LOVAS, *Phys. Lett.* **21,** 546 (1966).
568. A. GOSWAMI and A. I. SHERWOOD, *Nucl. Phys.* **89,** 465 (1966). *Phys. Rev.* **161,** 1232 (1967). A. COVELI O and G. SATORIS, *Nucl. Phys.* A, **93,** 481 (1967).
569. V. K. THANKAPPAN and W. W. TRUE, *Phys. Rev.* **137,** B793 (1965).
570. L. KISSLINGER and K. KUMAR, *Phys. Rev. Lett.* **19,** 1239 (1967).
571. E. E. BERLOVICH, *Nuclear Structure*, p. 15, AN UzSSR, Tashkent, 1969.
572. R. ARVIEU, O. BOHIGAS, and C. QUESNE, *Nucl. Phys.* A, **143,** 577 (1970).
573. L. K. PEKER, *Nuclear Structure*, p. 133, AN UzSSR, Tashkent, 1969. G. SCHARF-GOLDHABER, *Proceedings of the International Conference on Nuclear Structure* (ed. J. Sanada), p. 150, Tokyo, 1967. M. SAKAI, *ibid.*, p. 576.
574. K. KUMAR and M. BARANGER, *Phys. Rev. Lett.* **17,** 1146 (1966).
575. N. S. RABOTNOV and N. S. SEREGIN, *Phys. Lett.* **29B,** 162 (1969).
576. T. MARUMORI, M. YAMAMURA, Y. SHONO, A. TOKUNAGA, and Y. MIYANISHI, *Proceedings of the International Conference on Nuclear Structure* (ed. J. Sanada), p. 581, Tokyo, 1967. T. MARUMORI, M. YAMAMURA, Y. MIYANISHI, and S. NISIYAMA, *Yad. Fiz.* **9,** 501 (1969) (translation, *Soviet J. Nucl. Phys.* **9,** 287 (1969)).
577. R. A. SØRENSEN, *Proceedings of the International Conference on Properties of Nuclear States*, p. 85, Montreal, Les Presses de l'Université de Montréal, 1969.
578. R. M. DREIZLER, A. KLEIN, and T. K. DAS, *Phys. Lett.* **31B,** 333 (1970).
579. V. G. SOLOVIEV, *Yad. Fiz.* **13,** 48 (1971) (translation, *Soviet J. Nucl. Phys.* **13,** 32 (1971)). *Izv. AN USSR*, phys. ser. **35,** 666 (1971)).
580. R. F. FROSCH, R. HOFSTADTER, J. S. MCCARTHY, G. K. NÖLDEKE, K. J. VAN OOSTRUM, M. R. YEARIAN, B. C. CLARK, R. HERMAN, and D. G. RAVENHALL, *Phys. Rev.* **174,** 1380 (1968). V. M. KHVASTUNOV, N. G. AFANASIEV, V. D. AFANASIEV, I. S. GULKAROV, A. S. OMELAENKO, G. A. SAVICKIJ, A. A. KHOMICH, N. G. SHEVCHENKO, and V. S. ROMANOV, *Yad. Fiz.* **10,** 217 (1969) (translation, *Soviet J. Nucl. Phys.* **10,** 122 (1966)).
581. T. DE FOREST and J. D. WALECKA, *Adv. Phys.* **15,** 1 (1966). V. K. LUKYANOV, I. ZH. PETKOV, and YU. S. POL, *Yad. Fiz.* **9,** 349 (1969) (translation, *Soviet J. Nucl. Phys.* **9,** 204 (1969)). V. K. LUKYANOV and YU. S. POL, *Yad. Fiz.* **11,** 556 (1970) (translation, *Soviet J. Nucl. Phys.* **11,** 312 (1970)).
582. D. H. WILKINSON, *Proc. Panel*, "*Future of Nuclear Structure Studies*", p. 25, IAEA, Vienna, 1969.
583. V. V. BALASHOV and R. A. ERAMZHIAN, *Atomic Energy Rev.* **5** (3) 3 (1967). H. UBERALL, *Acta phys. austriaca* **30,** 89 (1969).
584. T. I. KAPALEISHVILI, *Nucl. Phys.* B, **1,** 335 (1967).
585. L. S. AZHGIREI, I. K. VZOROV, V. P. ZRELOV, M. G. MESHCHERYAKOV, B. S. NEGANOV, and A. F. SHABUDIN, *Zh. éksp. teor. Fiz.* **33,** 1185 (1957); **36,** 1631 (1959) (translation, *Soviet Phys. JETP* **6,** 911

(1958); **9**, 1163 (1959)). V. I. Komarov, G. E. Kasarev, and O. V. Savchenko, *Yad. Fiz.* **11**, 711 (1970).

586. V. S. Barashenkov and V. D. Toneev, preprint P2–4292, JINR, Dubna, 1969. V. S. Barashenkov, K. K. Gudina, S. M. Eliseev, A. S. Ilinov, and V. D. Toneev, *Acta Phys. Ac. Sci. Hungaricae* **29**, suppl. 3, 285 (1970).

INDEX

OTHER TITLES IN THE SERIES IN NATURAL PHILOSOPHY

OTHER TITLES IN THE SERIES

14105930

35—